Studies in Theoretical Physics, Volume 1

Fundamental mathematical methods

Studies in Theoretical Physics, Volume 1

Fundamental mathematical methods

Daniel Erenso

Department of Physics and Astronomy Middle Tennessee State Univeristy, Murfreesboro, Tennessee 37132

Victor Montemayor

Germantown Academy, Fort Washington, PA 19034

IOP Publishing, Bristol, UK

ISBN   978-0-7503-3135-7 (ebook)
ISBN   978-0-7503-3133-3 (print)
ISBN   978-0-7503-3136-4 (myPrint)
ISBN   978-0-7503-3134-0 (mobi)

DOI   10.1088/978-0-7503-3135-7

Version: 20220701

IOP ebooks

British Library Cataloguing-in-Publication Data: A catalogue record for this book is available from the British Library.

Published by IOP Publishing, wholly owned by The Institute of Physics, London

IOP Publishing, Temple Circus, Temple Way, Bristol, BS1 6HG, UK

US Office: IOP Publishing, Inc., 190 North Independence Mall West, Suite 601, Philadelphia, PA 19106, USA

*DBE dedicates this book to his family and friends, particularly his father
(Bekele Erenso Seta) and his friend (Birhanu Weldemichael).
VJM dedicates this book to his family.*

Contents

Preface

As with many textbooks, this book began as a series of lecture notes. In 1991, one of us (Victor Montemayor) designed and started teaching a new two-semester sequence of courses at Middle Tennessee State University (MTSU) intended for sophomore physics majors. The second of us (Daniel Erenso) started sharing the teaching responsibilities of this sequence about 13 years later. This sequence quickly became a foundation on which the physics major at MTSU was built.

The purpose of this sequence was to give our physics majors the mathematical background they needed to be successful in the upper-division physics courses, which means that an emphasis was placed on applications whenever possible. The prerequisites to these courses were one year of introductory physics (preferably but not necessarily calculus-based) and one year of calculus. The classes met twice per week for 1.5 hours per class.

This book is divided into two parts, corresponding to the material covered in the two semesters of the course at MTSU. The book starts off where the second semester of calculus tends to end up—with infinite series (chapter 1). It then moves on to discuss complex numbers and function of a complex variable, including complex infinite series (chapter 2). Chapters 3 and 4 cover essential topics in linear algebra: applications of vectors and matrices. It also sets the stage for future work in introducing the topics of linear independence, spanning, and basis sets. This completes what may be considered the foundation of the book—basic mathematics that many of the students would already have seen parts of in their mathematics and physics courses in high school and in their freshman year at college.

Chapter 5 introduces the students to partial differentiation and to the all-important total differential, and chapter 6 starts the discussion of solving differential equations—an important topic that occupies much of this book. Chapter 6 discusses the solution of linear ordinary differential equations of various orders.

Chapter 7 serves to reinforce the idea of integration and how to 'build' an integral in applications from mechanics and electricity and magnetism. It also introduces the students to 3D curvilinear coordinates systems—cylindrical coordinates and spherical coordinates—and to the idea of differential area and volume elements within these systems.

Chapter 8 discusses vector analysis and the powerful Green's theorem, the divergence theorem, and Stokes' theorem. These eight chapters are usually covered in the first semester of the academic year.

The topics we cover in the second semester start with discussing the calculus of variations in chapter 9, which is an essential topic for upper-division courses in mechanics. In this chapter, the fundamental Brachistochrone problem is solved, and the students are introduced to Laplace's equations in mechanics.

Chapter 10 returns to linear algebra and a discussion of the eigenvalue problem with a non-trivial application in mechanics of a system of coupled oscillators. Chapter 11 introduces some useful integral functions encountered in various areas of physics.

Chapters 12 and 13 again pick up the important topic of the solutions of differential equations. First, chapter 12 introduces the powerful technique of power-series solutions to differential equations and, in the process, introduces the students to Legendre polynomials, spherical harmonics and Bessel functions. Picking up with the ideas introduced in chapter 4, the fact that these infinite sets of functions form basis sets in their respective spaces is introduced and then exploited in chapter 13, in which solutions to Laplace's and Poisson's equations are discussed in the context of systems displaying rectangular, cylindrical and spherical symmetry.

Chapter 14 is a brief introduction to complex analysis and to the idea of contour integration in the complex plane. It then goes on to introduce Cauchy's theorem and discusses consequences of that theorem.

Finally, chapters 15 and 16 introduce the idea of integral transforms—the Laplace and Fourier transforms—and their applications, culminating with a discussion of Fourier transforms and the Heisenberg Uncertainty Principle.

As mentioned at the beginning of this preface, this book represents lectures in a two-semester, 3-hours per week course for sophomore physics majors who have already completed a year-long course in calculus and in physics. All of the material in this book has been routinely covered during the two-semester sequence for the past 30 years. However, many institutions do not have the luxury of being able to dedicate two semesters to such a course, and a one-semester course may not be taken until the junior or senior year. The coverage in a one-semester course could easily be selected from this book.

For example, if the course were to emphasize differential equations, the course could start with topics from chapters 1 and 2, since power series and complex numbers are essential to the solution of differential equations, and then jump to chapter 6 to cover the solution to ordinary differential equations. Going back to chapter 5 on partial differentiation and then picking up the topics of cylindrical and spherical coordinates from chapter 7 would then allow the students to move on to power-series solutions in chapter 12 and topics from chapter 13 on partial differential equations.

One of us (VM) is currently teaching the material in this book to advanced and very motivated high-school seniors who have already completed a year's worth of calculus and physics. Of course, the pace of coverage is not the same as that in university courses, so all of the topics cannot be covered in a year-long course. The high-school course tends to cover most—but not all—of the topics in chapters 1 through 10, Fourier series (spatial and temporal arguments) from chapter 16, Legendre polynomials and spherical harmonics in chapter 12, and finally the solution to Laplace's equation in rectangular and spherical coordinates in chapter 13. In a good year we can also cover Fourier transforms and convolution in chapter 16 at the end of the year.

Besides the topics included in this book, two features make this book unique from other similar standard mathematical methods textbooks. The first notable feature is that the book presents many example problems illustrating the mathematical methods applied to real physical problems presented in almost all sections of each chapter. Unlike other standard textbooks, each chapter in this book also introduces Mathematica to resolve the example problems. The last section in each chapter is dedicated to serving this purpose. It introduces the basic commands one can use in Mathematica and it shows how to resolve the example problems and get the same result. The second unique feature is, as this book, *Studies in Theoretical Physics: Volume 1: Fundamental mathematical methods*, serves as a foundation for the subsequent five volumes that one of us (DE) is currently working on. These volumes include *Studies in Theoretical Physics, Volume 2: Advanced mathematical methods; Studies in Theoretical Physics, Volume 3: Classical mechanics; Studies in Theoretical Physics, Volume 4: Electricity and magnetism, Studies in Theoretical Physics, Volume 5: Quantum mechanics; Studies in Theoretical Physics, Volume 6: General relativity*. The mathematical methods covered in Volume 1 are linked, applied, and cross-referenced in these five volumes making this volume the foundation and glue that integrates all books.

The material in this book constitutes a very challenging coverage of the application of mathematics for students interested in the mathematical sciences. Our experience in teaching this material over the past 30 years has been that, given that the material is covered carefully and efficiently by a passionate instructor, the students really value the material covered and the learning process that they go through as they work their way through the course. Our hope is that other students working their way through the material in this book find the experience equally rewarding.

Acknowledgement

VJM wishes to acknowledge the support and encouragement of the Department of Physics & Astronomy at MTSU as the original notes for this manuscript were first being developed, as well as the tolerance of his friends and family as he spent too many hours for over a decade with this work and away from them. DBE also wishes to acknowledge the support of the Department of Physics & Astronomy at MTSU. We both acknowledge Mr Tesfu Kassaye Woldeyes, who has edited the entire book and shared his professional expertise in the topics covered and the example problems worked out. Mr Woldeyes has more than twenty-five years of experience teaching physics at the introductory and upper-division levels.

Author biographies

Daniel Erenso

Dr Daniel Erenso is a professor of physics at Middle Tennessee State University (MTSU), Murfreesboro, Tennessee, USA. He joined MTSU in 2003 after he received his PhD in theoretical physics from the University of Arkansas, Fayetteville, Arkansas. Before he came to the USA, Dr Erenso received a BSc (1990) and MSc (1997) in physics from Addis Ababa University in his native country Ethiopia. He also received an Advanced Diploma in Condensed Matter Physics from Abdul Salam International Center for Theoretical Physics (ASICTP), Trieste, Italy, in 1999. Dr Erenso has taught, researched, and mentored at different universities in his native and adopted counties for more than two decades. Since he began his service at MTSU, Dr Erenso has taught several introductory and upper-level physics courses. For the excellence and dedication that Dr Erenso demonstrated, he has received the MTSU, College of Basic & Applied Sciences Excellence in Teaching Award in 2011. More recently, Dr Erenso has shown his dedication and hard work by going above and beyond to serve students during the COVID-19 pandemic (https://mtsunews.com/tag/daniel-erenso/). Inspired by this pandemic, Dr Erenso has published a book, Virtual and Real Labs for Introductory Physics II Optics, modern physics, and electromagnetism. Dr Erenso also has two more books: Studies in Theoretical Physics, Volume 1: Fundamental Mathematical Methods, which is this book, and the other is Studies in Theoretical Physics, Volume 2: Advanced Mathematical Methods, which is under production for publication by the Institute of Physics (IOP) in Bristol, UK.

At MTSU, Dr Erenso has also maintained an active research program with undergraduate students. He has been a research advisor for several undergraduate students. His research interest includes theoretical and experimental physics. By training, Dr Erenso is a theoretical physicist in quantum optics. However, since he came to MTSU, he has extended his research to experimental biophysics and quantum optics/quantum information. Dr Erenso has published more than 36 and presented over 60 research works at national and international venues. For his outstanding research accomplishment, Dr Erenso received Sigma Xi the Scientific Research Society Aubrey E Harvey Outstanding Graduate Research Award from the University of Arkansas in 2003, MTSU Foundations Special Project Award in 2005, MTSU, College of Basic & Applied Sciences Distinguished Research Award in 2016 and a nomination for American Physical Society (APS) Prize for a Faculty Member for Research in an Undergraduate Institution in 2020.

Dr Erenso is a member of several professional societies such as the American Physical Society (APS) and the Optical Society of America (OSA) and serves as an invited reviewer to several international journals. His excellence in teaching and research earned him the Fulbright Scholar Award in 2016 for teaching and research at the AAU physics department.

Victor Montemayor

 Dr Victor Montemayor teaches Physics and Advanced Mathematics at Germantown Academy (GA) in Fort Washington, PA. He retired from Middle Tennessee State University (MTSU) in 2015 after serving for 25 years as Professor of Physics. Dr Montemayor has received numerous awards for Teaching, Learning, and Innovative Educational Technology, from both GA and MTSU. He was also the recipient of the 2013 Ernest L Boyer International Award for Excellence in Teaching, Learning, and Technology.

Dr Montemayor received BS degrees in Physics and Mathematics from Bucknell University and the PhD degree in Theoretical Physics from the University of Toledo. He was a visiting scientist at the Hahn-Meitner Institute for Nuclear Research in Berlin, Germany and at the Institute for Atomic Research of the Hungarian Academy of Sciences in Debrecen, Hungary prior to accepting the position at MTSU. Dr Montemayor has enjoyed teaching many different courses in physics, from introductory to advanced undergraduate-level courses. He was chair of the Committee on Medical Physicists as Educators and a member of the Education Council of the American Association of Physicists in Medicine for 11 years. Dr Montemayor is currently finishing another book introducing undergraduate science majors to the field of Medical Physics.

IOP Publishing

Studies in Theoretical Physics, Volume 1
Fundamental mathematical methods
Daniel Erenso and Victor Montemayor

Chapter 1

Series and convergence

This chapter assumes you have some basic knowledge of sequence, series, and tests for convergence of series, and it summarizes the basics with more emphasis on examples. We begin with a summary of sequence and series, focusing on geometric series and the methods of test of convergence of series. Following this, we introduce the Taylor series and Maclaurin's series representation for some real essential functions. The last section presents the basic commands in Mathematica that we can use to test the convergence of a series and series expansion of real functions. The example problems considered in this chapter will also be resolved in this section using Mathematica.

1.1 Sequence and series

A sequence: A sequence is an ordered list of objects (or events). Like a set, it contains members (also called elements, or terms), and the number of ordered elements (possibly infinite) is called the length of the sequence. Unlike a set, order matters, and elements can appear multiple times at different positions in the sequence. A sequence is a discrete function.

A series: The sum of terms of a sequence is a series. More precisely, if $\{x_1, x_2, x_3, ...\}$ is a sequence, one may consider the sequence of partial sums $(S_1, S_2, S_3, ...S_n.)$, with

$$S_n = x_1 + x_2 + x_3...+x_n = \sum_{k=1}^{n} x_k. \tag{1.1}$$

Geometric series: Let's consider two simple idealistic biological and physical processes that can be described by geometric progressions.

i. **Bacteria growth**: Bacterial growth is proliferation of bacterium into two daughter cells, in a process called binary fission. Depending on the environment, in particular the temperature, the growth of bacteria could be slow or fast. Figure 1.1 shows a goofy bacteria growth. At $t = 0$ there was only one bacteria, after an hour this bacteria becomes two, after two hours the two bacteria becomes four, the four bacteria becomes eight, and so on. We may describe growth of this bacteria by a sequence

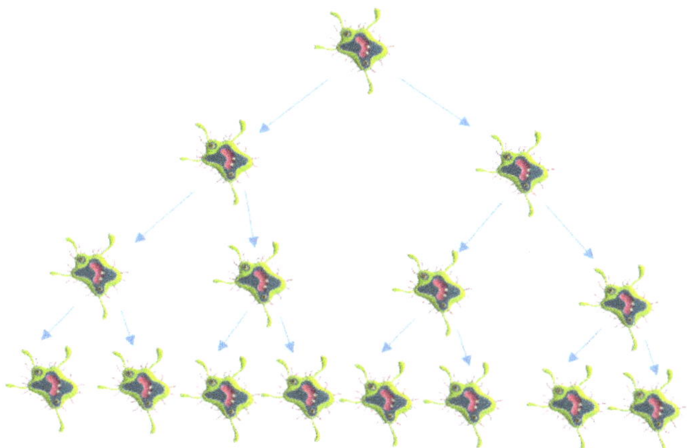

Figure 1.1. Bacteria growth.

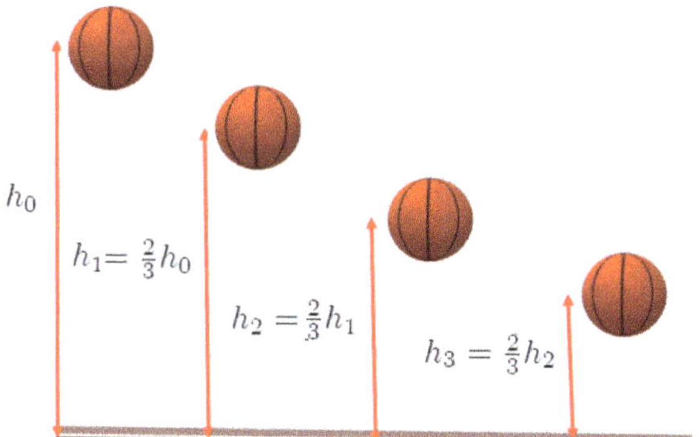

Figure 1.2. A bouncing basketball.

$$\{1, 2, 4, 8, 16, \ldots\} \tag{1.2}$$

ii. **A bouncing ball**: Suppose we bounce a basketball really hard on a basketball court. The ball bounces a maximum height h_0 for the first bounce and then afterwards each time about 2/3 of the height of the previous bounce as shown in figure 1.2. This means

$$\begin{cases}
t = t_0, & h_0, \\[2mm]
t = t_1, & h_1 = \dfrac{2}{3}h_0, \\[2mm]
t = t_2, & h_2 = \dfrac{2}{3}h_1 = \dfrac{4}{9}h_0, \\[2mm]
t = t_3, & h_3 = \dfrac{2}{3}h_2 = \dfrac{8}{27}h_0, \\[2mm]
t = t_4, & h_4 = \dfrac{2}{3}h_3 = \dfrac{16}{81}h_0,
\end{cases} \tag{1.3}$$

and it can be described by the sequence

$$\left\{ h_0, \frac{2}{3}h_0, \frac{4}{9}h_0, \frac{8}{27}h_0, \frac{16}{81}h_0, \ldots , \right\} \tag{1.4}$$

To determine the total number of bacteria or the total height the ball bounces in a given period of time, we add the terms in the above expressions. Suppose we denote the sum of these terms by S_n, we may write the total number of bacteria as

$$S_n = 2[1 + 2 + 2^2 + 2^3 + \ldots 2^n]$$
$$\Rightarrow S_n = 2 \sum_{k=0}^{n} 2^k = 2 \sum_{k=0}^{n} r^k, \tag{1.5}$$

where $r = 2$. Similarly for the total height that the ball bounces as

$$S_n = h_0\left[1 + \frac{2}{3} + \left(\frac{2}{3}\right)^2 + \left(\frac{2}{3}\right)^3 + \ldots\left(\frac{2}{3}\right)^n \right]$$
$$\Rightarrow S_n = h_0 \sum_{k=0}^{n} \left(\frac{2}{3}\right)^k = h_0 \sum_{k=0}^{n} r^k, \tag{1.6}$$

where $r = 2/3$.

In general as $n \to \infty$, the infinite series

$$S = \lim_{n\to\infty} S_n = a \sum_{k=0}^{\infty} r^k \tag{1.7}$$

is known as *geometric series*. For the case where $|r| < 1$, the geometric series becomes a convergent series and we can derive a simple expression. To this end, we consider the two series

$$\sum_{k=0}^{n} r^k = 1 + r + r^2 + r^3 + r^4 + \ldots r^n \tag{1.8}$$

and

$$\sum_{k=0}^{n} r^{k+1} = r + r^2 + r^3 + r^4 + \ldots r^n + r^{n+1}, \tag{1.9}$$

so that upon subtracting these equations, we find

$$\sum_{k=0}^{n} r^k - \sum_{k=0}^{n} r^{k+1} = (1 - r) \sum_{k=0}^{n} r^k$$

$$= 1 + (r + r^2 + r^3 + r^4 + \ldots r^n)$$
$$- (r + r^2 + r^3 + r^4 + \ldots r^n) - r^{n+1} \tag{1.10}$$

$$\Rightarrow (1 - r) \sum_{k=0}^{n} r^k = 1 - r^{n+1}.$$

Then for $|r| < 1$, one can easily write

$$\sum_{k=0}^{n} r^k = \frac{1 - r^{n+1}}{1 - r}, \tag{1.11}$$

so that in the limit as $n \to \infty$, one finds

$$\sum_{k=0}^{\infty} r^k = \frac{1}{1 - r}. \tag{1.12}$$

Then the geometric series becomes

$$S = \lim_{n \to \infty} S_n = a \sum_{k=0}^{\infty} r^k = \frac{a}{1 - r}. \tag{1.13}$$

Example 1.1. Two trains move toward one another with a speed v_o. A bird (an 'ideal physics bird') flies back and forth between the two trains with a speed $v_b = \alpha v_o$, where $\alpha > 1$. Find the total distance traveled by the bird before the trains crash if they start a distance D_o apart.

Solution: Consider the first four consecutive back and forth movement of the bird as shown in figure 1.3. The bird starts from Train 1 (T_1) and moves toward Train 2 (T_2). Let the time it takes to reach T_2 be t_1 and during this time it has traveled a distance x_1 as measured from the initial position of T_1. Then we can express the distance x_1

$$x_1 = D_0 - v_0 t_1. \tag{1.14}$$

Since the bird is moving with a constant speed αv_o, the distance the bird traveled x_1, over the time t_1, is related to its speed by

$$\alpha v_o = \frac{x_1}{t_1} \Rightarrow t_1 = \frac{x_1}{\alpha v_o} \Rightarrow v_0 t_1 = \frac{x_1}{\alpha}. \tag{1.15}$$

Substituting equation (1.15) into equation (1.14), we find

$$x_1 = D_0 - \frac{x_1}{\alpha} \Rightarrow x_1 = \frac{\alpha}{1 + \alpha} D_0. \tag{1.16}$$

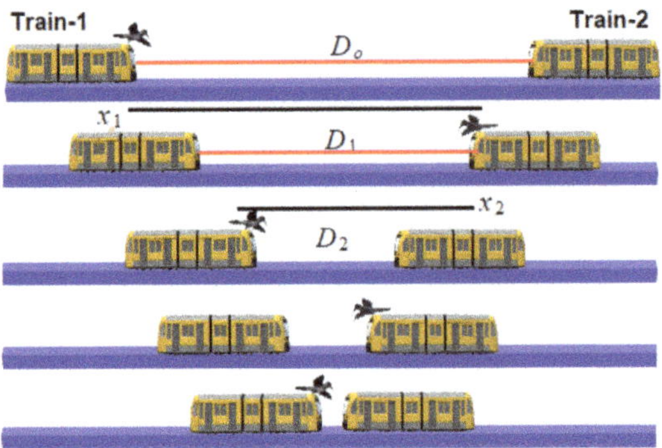

Figure 1.3. The motion of the bird and the two trains.

Let the distance traveled by each train during the time t_1 be y_1. Since each train is moved with a constant speed v_o, we have

$$v_o = \frac{y_1}{t_1} \Rightarrow y_1 = v_o t_1 = \frac{x_1}{\alpha} = \frac{1}{1+\alpha} D_0. \tag{1.17}$$

This indicates during the time, the two trains are closer to one another by a distance of $2y_1$. Then from figure 1.3, the new distance between the two trains after time t_1 (when the bird is just about to make her flight back to T_1), D_1, can be expressed as

$$D_1 = D_0 - 2y_1 \Rightarrow D_1 = D_0 - 2\left(\frac{1}{1+\alpha}\right)D_0$$
$$\Rightarrow D_1 = \frac{\alpha - 1}{\alpha + 1} D_0. \tag{1.18}$$

Following a similar procedure and referring to figure 1.3, the distance the bird traveled when it flies back to the first train x_2, can be expressed as

$$x_2 = D_1 - v_0 t_2, \tag{1.19}$$

so that using

$$\alpha v_o = \frac{x_2}{t_2} \Rightarrow t_2 = \frac{x_2}{\alpha v_o}, \tag{1.20}$$

we find

$$x_2 = D_1 - v_0 t_2 = D_1 - \frac{x_2}{\alpha} \Rightarrow x_2 = \frac{\alpha}{1+\alpha} D_1. \tag{1.21}$$

Substituting equation (1.18) into equation (1.21), one finds

$$x_2 = \left(\frac{\alpha}{1+\alpha}\right)\left(\frac{\alpha - 1}{\alpha + 1}\right)D_0. \tag{1.22}$$

In a similar way one can show that the new distance between the two trains is given by

$$D_2 = D_1 - 2y_2 \Rightarrow D_2 = D_1 - 2\left(\frac{\alpha}{1+\alpha}\right)D_1$$

$$\Rightarrow D_2 = \frac{\alpha-1}{\alpha+1}D_1 = \left(\frac{\alpha-1}{\alpha+1}\right)^2 D_0, \tag{1.23}$$

so that for x_3, one can write

$$x_3 = \frac{\alpha}{1+\alpha}D_2 = \left(\frac{\alpha}{1+\alpha}\right)\left(\frac{\alpha-1}{\alpha+1}\right)^2 D_0. \tag{1.24}$$

Then the total distance X the bird flies before the two trains crash can be expressed as an infinite series,

$$X = x_1 + x_2 + x_3 + \ldots = \frac{\alpha}{1+\alpha}D_0 + \left(\frac{\alpha}{1+\alpha}\right)\left(\frac{\alpha-1}{\alpha+1}\right)D_0$$

$$+ \left(\frac{\alpha}{1+\alpha}\right)\left(\frac{\alpha-1}{\alpha+1}\right)^2 D_0 \ldots \tag{1.25}$$

$$\Rightarrow X = \frac{\alpha}{1+\alpha}D_0\left[\left(\frac{\alpha-1}{\alpha+1}\right)^0 + \left(\frac{\alpha-1}{\alpha+1}\right)^1 + \left(\frac{\alpha-1}{\alpha+1}\right)^2 + \ldots\right],$$

which can be put in the form

$$X = a\sum_{n=0}^{\infty} r^n, \tag{1.26}$$

where

$$a = \frac{\alpha}{1+\alpha}D_0, \ r = \frac{\alpha-1}{\alpha+1} < 1. \tag{1.27}$$

Equation (1.26) is a geometric series and its sum can easily be determined using the relation in equation (1.13),

$$X = a\sum_{k=0}^{\infty} r^k = \frac{a}{1-r} = \frac{\alpha}{1+\alpha}D_0\left(\frac{1}{1-\dfrac{\alpha-1}{\alpha+1}}\right) = \frac{1}{2}\alpha D_0. \tag{1.28}$$

The result in equation (1.28) is the total distance that the bird had traveled before the two trains crashed.

1.2 Testing series for convergence

If an infinite series, like the geometric series we studied in the previous section, has a finite sum, it is convergent; otherwise, it is divergent. There are different ways to test the convergence of an infinite series. Here we discuss some of the standard methods used to test the convergence or divergence of a series. To this end, let's consider a series given by

$$a_0 + a_1 + a_2 + \ldots\ldots = \sum_{n=0}^{\infty} a_n. \tag{1.29}$$

(1) *Preliminary test*: If a_n does not tend to zero as $n \to \infty$,

$$\lim_{n\to\infty} a_n \neq 0, \tag{1.30}$$

the series is a divergent series.

(2) *The comparison test*: Consider another series

$$\sum_{n=0}^{\infty} b_n = b_0 + b_1 + b_2 + \ldots. \tag{1.31}$$

with all positive terms (i.e., $b_n > 0$) that is convergent. Then when you compare the nth term in the series in equation (1.31) (b_n) with the corresponding nth term with the series in equation (1.29) (a_n), if $0 \le a_n \le b_n$ for all n, we have

$$\sum_{n=0}^{\infty} a_n \le \sum_{n=0}^{\infty} b_n, \tag{1.32}$$

and the series

$$\sum_{n=0}^{\infty} a_n \tag{1.33}$$

is a convergent series. On the other hand if there exists a series

$$\sum_{n=0}^{\infty} c_n = c_0 + c_1 + c_2 + \ldots, \tag{1.34}$$

that is a divergent series and for all the terms, $a_n \ge c_n$, the series

$$\sum_{n=0}^{\infty} a_n \tag{1.35}$$

is a divergent series. In applying the comparison test to test the convergence of a given series, we can use *the geometric series* we saw in the previous section as a convergent series

$$\sum_{n=0}^{\infty} b_n = \sum_{n=0}^{\infty} r^n \text{ (for } |r| < 1) \tag{1.36}$$

and the series

$$1 + \frac{1}{2} + \frac{1}{3} + \frac{1}{4} + \ldots = \sum_{n=0}^{\infty} \frac{1}{n+1}, \tag{1.37}$$

known as *the harmonic series,* which is a divergent series

$$\sum_{n=0}^{\infty} c_n = \sum_{n=0}^{\infty} \frac{1}{n+1}. \tag{1.38}$$

(3) *The ratio test*: In the ratio test, for the infinite series in equation (1.29), we first evaluate the ratio defined by

$$\rho = \lim_{n \to \infty} \left| \frac{a_{n+1}}{a_n} \right|. \tag{1.39}$$

The convergence of the series will then be determined by the value of the ratio ρ, according to

$$\rho = \begin{cases} < 1, & \text{the series is convergent,} \\ > 1, & \text{the series is divergent,} \\ = 1, & \text{use a different test.} \end{cases} \tag{1.40}$$

(4) *The integral test*: In the series in equation (1.29), if $0 < a_{n+1} \leq a_n$ for $n > N$, where N is a sufficiently large number, then one can use the integral test,

$$I = \int^{\infty} a_n dn. \tag{1.41}$$

The series in equation (1.29) is convergent if this integral is finite; otherwise, it is a divergent series. In equation (1.41), only the upper limit of the integration is included. This means the integral needs to be evaluated only at the upper limit to test the convergence.

(5) *Absolute convergence test*: Suppose the terms a_n in the series in equation (1.29) may vary in sign, some are positive and others are negative. If the sum of the absolute values of each term in the series in equation (1.29) converges, then the infinite series is said to be an *absolutely convergent series*. In other words, if

$$|a_0| + |a_1| + |a_2| + \ldots = \sum_{n=0}^{\infty} |a_n| \tag{1.42}$$

is convergent, the series

$$a_0 + a_1 + a_2 + \ldots = \sum_{n=0}^{\infty} a_n \tag{1.43}$$

is a convergent series. On the other hand when the series

$$a_0 + a_1 + a_2 + \ldots = \sum_{n=0}^{\infty} a_n, \tag{1.44}$$

is convergent, the convergence of the series

$$|a_0| + |a_1| + |a_2| + \ldots = \sum_{n=0}^{\infty} |a_n| \tag{1.45}$$

is conditional and is not always guaranteed. The alternating harmonic series

$$\sum_{n=0}^{\infty} a_n = \sum_{n=0}^{\infty} (-1)^n \frac{1}{n+1} = 1 - \frac{1}{2} + \frac{1}{3} - \frac{1}{4} + \ldots, \tag{1.46}$$

is a good example. This series is a convergent series since $|a_n|$ is a monotonically decreasing function for $n > N$, where N is a sufficiently large number,

$$0 < |a_{n+1}| < |a_n| \text{ for } n > N. \tag{1.47}$$

On the other hand the harmonic series

$$\sum_{n=0}^{\infty} c_n = \sum_{n=0}^{\infty} |a_n| = \sum_{n=0}^{\infty} \frac{1}{n+1} = 1 + \frac{1}{2} + \frac{1}{3} + \frac{1}{4} + \ldots, \tag{1.48}$$

is a divergent series.

Example 1.2. Consider the infinite series given by

$$1 + \frac{1}{6} + \frac{2}{120} + \frac{6}{5040} + \frac{24}{362\,880}, \ldots, \tag{1.49}$$

(a) Rewrite this series using standard summation notation.
(b) Use the ratio test to check the convergence of this series.

Solution:
   (a) We first rewrite the series using some general expression for a_n that generates the nth term in the series. In most cases we look for some kind of relation that can be expressed using factorial and/or exponential function. To this end, we note that

$$1 + \frac{1}{6} + \frac{2}{120} + \frac{6}{5040} + \ldots = \frac{1}{0!} + \frac{1}{3!} + \frac{2 \times 1}{5!} + \frac{3 \times 2}{7!}$$
$$+ \frac{4 \times 3 \times 2}{9!} \ldots$$
$$= \frac{0!}{1!} + \frac{1!}{3!} + \frac{2!}{5!} + \frac{3!}{7!} + \frac{4!}{9!}$$
$$\ldots \frac{n!}{(2n + 1)!} \ldots, \tag{1.50}$$

and the series can be expressed as

$$1 + \frac{1}{6} + \frac{2}{120} + \frac{6}{5040} + \frac{24}{362\,880}, \ldots = \sum_{n=0}^{\infty} a_n, \tag{1.51}$$

where

$$a_n = \frac{n!}{(2n + 1)!}. \tag{1.52}$$

Note that by definition $0! = 1$.

(b) We use the ratio test to determine if the series is convergent or divergent. Using the result we found in part (a), the ratio, ρ,

$$\rho = \lim_{n \to \infty} \left| \frac{a_{n+1}}{a_n} \right| = \lim_{n \to \infty} \left| \frac{\dfrac{(n + 1)!}{(2(n + 1) + 1)!}}{\dfrac{n!}{(2n + 1)!}} \right|$$

$$= \lim_{n \to \infty} \left| \frac{\dfrac{(n + 1)(n!)}{(2n + 3)(2n + 2)(2n + 1)!}}{\dfrac{n!}{(2n + 1)!}} \right| \tag{1.53}$$

$$= \lim_{n \to \infty} \left| \frac{n + 1}{(2n + 3)(2n + 2)} \right| \Rightarrow \rho \simeq \lim_{n \to \infty} \left| \frac{n}{4n^2} \right|$$

$$= \lim_{n \to \infty} \left| \frac{1}{4n} \right| = 0.$$

This result shows, $\rho < 1$, and the series is a convergent series according to equation (1.40).

1.3 Series representations of real functions

This section introduces how we express a given well-defined real function in a series form. We will study three forms of series representation of a function $P(x)$: *Power series, Taylor series, and Maclaurin series.* The series representation of this function could be convergent or divergent. The convergence or divergence is determined by the domain of the variable, x, that the function depends on. We will see how we can determine the range of the domain for which the series is convergent (*i.e., the interval of convergence*).

i. *Power series:* A series in which the nth term is a constant b_n times a function $(x - a)^n$,

$$P(x) = \sum_{n=0}^{\infty} b_n(x - a)^n = b_0 + b_1(x - a) + b_2(x - a)^2 + ..., \tag{1.54}$$

where a is real constant and x is real variable (i.e., $x \in \mathbb{R}$). Note that a power series can be convergent or divergent.

ii. *Taylor series:* Taylor series is a power series in which the coefficient in the nth term of the series b_n is given by the nth derivative of the function $P(x)$ evaluated at $x = a$ and divided by n factorial,

$$P(x) = \sum_{n=0}^{\infty} b_n(x - a)^n, \quad \text{where } b_n = \frac{1}{n!} \frac{d^n P(x)}{dx^n} \bigg|_{x=a}. \tag{1.55}$$

iii. *Maclaurin's series:* When $x = a = 0$, a Taylor series becomes a Maclaurin series,

$$P(x) = \sum_{n=0}^{\infty} b_n x^n, \quad \text{where } b_n = \frac{1}{n!} \frac{d^n P(x)}{dx^n} \bigg|_{x=0}. \tag{1.56}$$

Any function $f(x)$ that is *differentiable for all values of x in the specified domain,* can be expressed in *Taylor* or *Maclaurin series.* That means if

$$\frac{d^n f(x)}{dx^n} \text{ exists for all } n \geq 0 \text{ and } x \in \mathbb{R}, \tag{1.57}$$

we can write

$$f(x) = \sum_{n=0}^{\infty} b_n(x - a)^n, \quad \text{where, } b_n = \frac{1}{n!} \frac{d^n f(x)}{dx^n} \bigg|_{x=a}. \tag{1.58}$$

Taylor Series expansion for some basic functions for $a = 0$:

$$\sin(x) = \sum_{n=0}^{\infty} \frac{(-1)^n}{(2n + 1)!} x^{2n+1}$$

$$= x - \frac{x^3}{3!} + \frac{x^5}{5!} + ... \text{ convergent for all } x \in \mathbb{R} \tag{1.59}$$

$$\cos(x) = \sum_{n=0}^{\infty} \frac{(-1)^n}{(2n)!} x^{2n}$$

$$= 1 - \frac{x^2}{2!} + \frac{x^4}{4!} - \ldots \text{convergent for all } x \in \mathbb{R}$$

$$e^x = \sum_{n=0}^{\infty} \frac{1}{n!} x^n$$

$$= 1 + x + \frac{x^2}{2!} + \ldots \text{convergent for all } x \in \mathbb{R},$$

$$\ln(1 + x) = \sum_{n=1}^{\infty} \frac{(-1)^{n+1}}{n} x^n$$

$$= x - \frac{x^2}{2} + \frac{x^3}{3} - \ldots \text{convergent for } -1 < x \leq 1,$$

$$(1 + x)^p = \sum_{n=0}^{\infty} \binom{p}{n} x^n$$

$$= 1 + px + \frac{p(p-1)}{2!} x^2 + \ldots \text{convergent for all } |x| < 1,$$

where

$$\binom{p}{n} = \frac{p(p-1)(p-2)(p-3)\ldots(p-n+1)}{n!}, \tag{1.60}$$

and it is called the *binomial coefficient*.

Example 1.3.

(a) Show that Maclaurin's series expansion for the function

$$f(x) = \ln(1 + x) \tag{1.61}$$

is expressible as

$$\ln(1 + x) = \sum_{n=1}^{\infty} \frac{(-1)^{n+1}}{n} x^n. \tag{1.62}$$

(b) Find the interval of convergence for this series.

Solution:

(a) Using equation (1.58)

$$f(x) = \sum_{n=0}^{\infty} b_n x^n, \text{ where } b_n = \frac{1}{n!} \frac{d^n f(x)}{dx^n} \bigg|_{x=0} \tag{1.63}$$

one can write

$$b_0 = \frac{1}{0!} \ln(1 + x)|_{x=0} \tag{1.64}$$
$$\Rightarrow b_0 = 0,$$

$$b_1 = \frac{1}{1!} \frac{d}{dx} \ln(1 + x)\bigg|_{x=0} = \frac{1}{1 + x}\bigg|_{x=0} \tag{1.65}$$
$$\Rightarrow b_1 = 1,$$

$$b_2 = \frac{1}{2!} \frac{d^2}{dx^2} \ln(1 + x)\bigg|_{x=0} = \frac{1}{2!} \frac{d}{dx}\left(\frac{1}{1 + x}\right)\bigg|_{x=0} = \frac{1}{2!} \frac{-1}{(1 + x)^2}\bigg|_{x=0} \tag{1.66}$$
$$\Rightarrow b_2 = \frac{(-1)^1}{2},$$

$$b_3 = \frac{1}{3!} \frac{d^3}{dx^3} \ln(1 + x)\bigg|_{x=0} = \frac{1}{3!} \frac{d}{dx}\left(\frac{-1}{(1 + x)^2}\right)\bigg|_{x=0}$$
$$= \frac{1}{3!} \frac{2}{(1 + x)^3}\bigg|_{x=0} \tag{1.67}$$
$$\Rightarrow b_3 = \frac{(-1)^2}{3},$$

and

$$b_4 = \frac{1}{4!} \frac{d4}{dx^4} \ln(1 + x)\bigg|_{x=0} = \frac{1}{4!} \frac{d}{dx}\left(\frac{2}{(1 + x)^3}\right)\bigg|_{x=0}$$
$$= \frac{1}{4!} \frac{-6}{(1 + x)^4}\bigg|_{x=0} = \frac{1}{4 \times 3!} \frac{(-1)^3 3!}{(1 + x)^4}\bigg|_{x=0} \tag{1.68}$$
$$\Rightarrow b_4 = \frac{(-1)^3}{4}.$$

One can easily see from the above results that

$$b_n = \begin{cases} \dfrac{(-1)^{n-1}}{n}, & \text{for } n = 1, 2, 3..., \\ 0, & n = 0, \end{cases} \tag{1.69}$$

which leads to

$$\ln(1 + x) = \sum_{n=0}^{\infty} b_n x^n = \sum_{n=1}^{\infty} \frac{(-1)^{n-1}}{n} x^n. \tag{1.70}$$

This is Maclaurin's series for the function $f(x) = \ln(1 + x)$.

(b) To find the interval of convergence we use the ratio test. The series converges when

$$\rho = \lim_{n \to \infty} \left| \frac{a_{n+1}(x)}{a_n(x)} \right| < 1. \tag{1.71}$$

Noting that

$$a_n(x) = \frac{(-1)^{n-1}}{n} x^n, \Rightarrow a_{n+1}(x) = \frac{(-1)^n}{n+1} x^{n+1}, \tag{1.72}$$

we find

$$\rho = \lim_{n \to \infty} \left| \frac{(-1)^n}{n+1} x^{n+1} \frac{n}{(-1)^{n-1} x^n} \right| = \lim_{n \to \infty} \left| -\frac{nx}{n+1} \right| = |x| < 1 \tag{1.73}$$

$$\Rightarrow -1 < x < 1.$$

We note that at $x = \pm 1$, the series becomes

$$\sum_{n=1}^{\infty} \frac{(-1)^{n-1}}{n} x^n = \begin{cases} -\displaystyle\sum_{n=1}^{\infty} \frac{(-1)^n}{n}, & \text{for } x = 1, \\ -\displaystyle\sum_{n=1}^{\infty} \frac{1}{n}, & \text{for } x = -1, \end{cases} \tag{1.74}$$

and the ratio test cannot tell us whether the series is a convergent or divergent series. Using a different test, it can be shown that the series converges for $x = 1$ and diverges for $x = -1$ (See Problem 12). Therefore, the interval of convergence for Maclaurin's series is $-1 < x \leq 1$.

Example 1.4. In 1905, Albert Einstein published his Special Theory of Relativity which explains the physics dealing with objects traveling very close to the speed of light (that is, objects traveling with speeds in excess of about $0.1c = 3 \times 10^7 \text{ms}^{-1}$), where $c = 3.0 \times 10^8 \text{ms}^{-1}$ is the speed of light in free space. Suppose you (standing on planet earth) spotted an Alien spaceship traveling with a speed v east (x-direction) as shown in figure 1.4. In this theory the mass of an object (the alien) is found not to be a constant, but rather to increase as the object's (the spaceship) speed increases. If an object is at rest next to you, then you would measure its mass to be the so-called *rest mass* m_0; if it's moving past you at some speed, then you would measure its mass to be the relativistic mass m. (The rest mass m_0 is simply the mass that we're used to thinking of—it's the relativistic mass m that's new....). For example, suppose the spaceship stopped at your place and the alien stepped out to greet you. If you ask the alien to stand on a scale and you measure his mass, that is the rest mass m_0. The relativistic mass m, for the alien is the mass that you would measure while he is sitting in his spaceship and traveling with speed v. Einstein found that, if an object of rest mass m_0 travels with a speed v past an observer, then that observer would measure the object's (relativistic) mass m to be

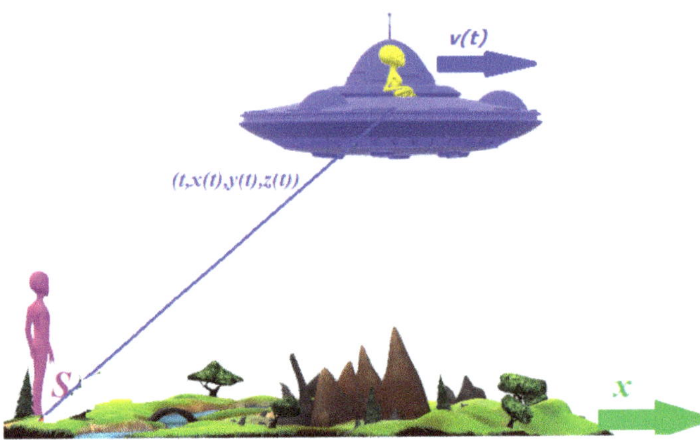

Figure 1.4. An alien spaceship traveling with a speed v along east (x-direction) relative to a human observer on the Earth.

$$m = \frac{m_0}{\sqrt{1 - \dfrac{v^2}{c^2}}}.$$

(1.75)

In Special Theory of Relativity the total energy of an object with rest mass m_0 moving with a speed v is given by

$$E = \frac{m_0 c^2}{\sqrt{1 - \dfrac{v^2}{c^2}}}.$$

This total energy is the sum of the kinetic energy KE and the rest-mass energy $E_o = m_0 c^2$ (this is the object's energy when the object is not moving, so $v = 0$):

$$E = KE + E_o = KE + m_0 c^2 = \frac{m_0 c^2}{\sqrt{1 - \dfrac{v^2}{c^2}}}.$$

(1.76)

Show that the relativistic kinetic energy as deduced from these equations reduces to the expected (classical) form in the non-relativistic limit—that is, for $v < < c$. That means we want to show that the kinetic energy of the object is given by

$$T \simeq \frac{1}{2} m_0 v^2,$$

(1.77)

for $v < < c$ which is what we know from Introductory Physics.

Solution: The kinetic energy KE in terms of the total relativistic kinetic energy and rest-mass energy can be expressed as

$$KE = \frac{m_0 c^2}{\sqrt{1 - \dfrac{v^2}{c^2}}} - m_0 c^2 = m_0 c^2 [(1 - \beta^2)^{-1/2} - 1]$$

(1.78)

where $\beta = v/c < 1$. Applying the series expansion

$$(1 + x)^p = \sum_{n=0}^{\infty} \binom{p}{n} x^n = 1 + px + \frac{p(p-1)}{2!} x^2 \ldots, \qquad (1.79)$$

for $x = -\beta^2$ and $p = -1/2$, we may write

$$(1 - \beta^2)^{-1/2} = 1 + \frac{1}{2}\beta^2 + \frac{-\frac{1}{2}(-\frac{1}{2} - 1)}{2!}\beta^4 + \ldots \qquad (1.80)$$

In the classical limit $v < < c \Rightarrow \beta = v/c < < 1$, we can drop all the terms with β^n for $n \geq 4$,

$$(1 - \beta^2)^{1/2} \simeq 1 + \frac{1}{2}\beta^2 = 1 + \frac{1}{2}\frac{v^2}{c^2}. \qquad (1.81)$$

Therefore, the relativistic kinetic energy in the classical limit becomes

$$KE \simeq m_0 c^2 \left[1 + \frac{1}{2}\frac{v^2}{c^2} - 1 \right] \Rightarrow KE \simeq \frac{1}{2}m_0 v^2. \qquad (1.82)$$

Example 1.5. Two identical small balls of charge Q and mass m are hung from two silk threads of length L. The threads are attached at a common point on the ceiling. The balls hang in equilibrium separated by a distance x as shown in figure 1.5. Show that for small charges Q and large lengths L the equilibrium position is given by

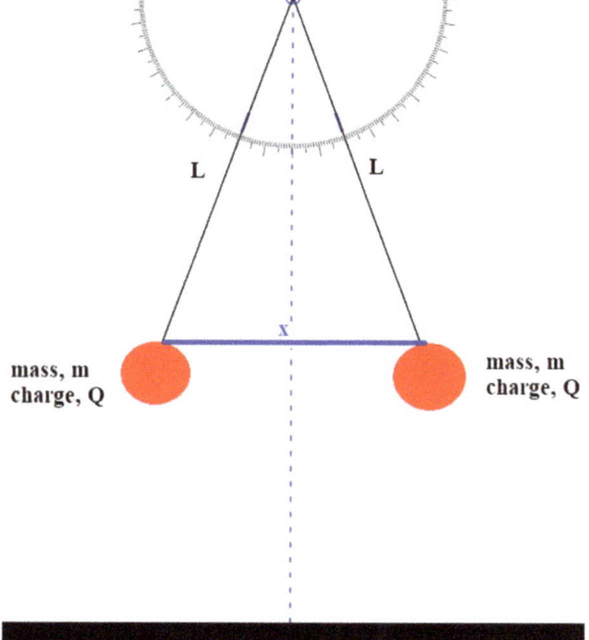

Figure 1.5. Two balls with identical mass m and charge Q at equilibrium.

$$x = \left(\frac{Q^2 L}{2\pi\epsilon_0 mg} \right)^{1/3}. \tag{1.83}$$

Solution: If the balls are at equilibrium the net force acting on each ball must be zero. There are three forces acting on each of the balls. These are the repulsive electrostatic force (F_e), the tension force (T), and Earth's gravitational force (W) as shown in figure 1.6. The magnitude of the electrical (F_e) and the gravitational (W) forces are given by

$$F_e = \frac{1}{4\pi\epsilon_0} \frac{Q^2}{x^2}, \quad W = mg, \tag{1.84}$$

respectively. Since the balls are at equilibrium, the net force on each ball must add up to zero. Considering the ball on the right side in figure 1.6, for the net force in the y-direction, we have

$$\sum F_y = 0 \Rightarrow T_y - W = 0 \Rightarrow T \cos\theta = mg, \tag{1.85}$$

and in the x-direction

$$\sum F_x = 0 \Rightarrow F_e - T_x = 0 \Rightarrow T \sin(\theta) = \frac{1}{4\pi\epsilon_0} \frac{Q^2}{x^2}. \tag{1.86}$$

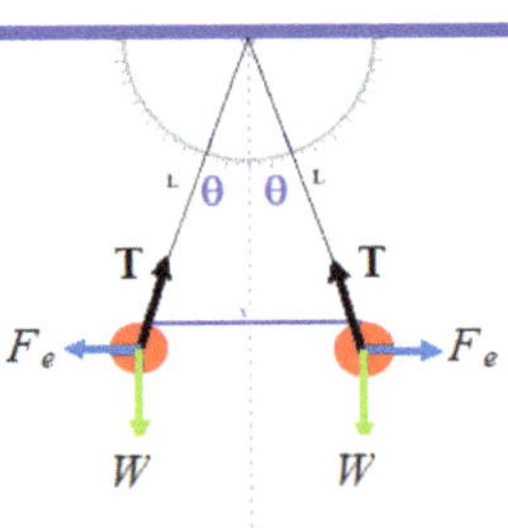

Figure 1.6. Two charged balls suspended from a massless string.

Upon dividing equation (1.86) by (1.85), we find

$$\tan(\theta) = \frac{\sin(\theta)}{\cos(\theta)} = \frac{Q^2}{4\pi\epsilon_0 mg x^2}.$$ (1.87)

Apply the Maclaurin series expansion

$$\sin(\theta) = \theta - \frac{\theta^3}{3!} + \frac{\theta^5}{5!} - \frac{\theta^7}{7!}\ldots$$ (1.88)

$$\cos(\theta) = 1 - \frac{\theta^2}{2!} + \frac{\theta^4}{4!} - \frac{\theta^6}{6!}\ldots$$ (1.89)

for small angle θ (which is the case for $x < <L$ since for small charge the electrostatic force is weak compared with the gravitational force and the separation distance x is very small), we have

$$\cos(\theta) \simeq 1 \Rightarrow \tan(\theta) = \frac{\sin(\theta)}{\cos(\theta)} \simeq \sin(\theta)$$

so that equation (1.86) becomes

$$\sin(\theta) \simeq \frac{Q^2}{4\pi\epsilon_0 mg x^2} \Rightarrow \frac{x}{2L} \simeq \frac{Q^2}{4\pi\epsilon_0 mg x^2} \Rightarrow x \simeq \left(\frac{Q^2 L}{2\pi\epsilon_0 mg}\right)^{1/3}.$$ (1.90)

1.4 Sequence, series and Mathematica

Convergence: In Mathematica we can test the convergence of a series

$$\sum_{n=0}^{\infty} a_n$$

by executing the command

$$\text{SumConvergence}[a, n].$$

We use this command to illustrate that the harmonic series diverges while the alternating harmonic series converges $\text{SumConvergence}\left[\frac{1}{n+1}, n\right]$ False $\text{SumConvergence}\left[\frac{(-1)^n}{n+1}, n\right]$ True.

Series expansion: In Mathematica the command

$$\text{Series}[f, \{x, x_0, n\}]$$

generates a power series expansion for $f(x)$ about the point $x = x_0$ to order $(x - x_0)^n$. The series expansion for the problem we considered in Example 1.3 and those listed in equations (1.59)–(1.3), using Mathematica, are given by

$$\text{Series[Log[1 + }x], \{x, 0, 6\}]\; x - \frac{x^2}{2} + \frac{x^3}{3} - \frac{x^4}{4} + \frac{x^5}{5} + O[x]^6 \tag{1.91}$$

$$\text{Series[Cos[}x], \{x, 0, 6\}]\; 1 - \frac{x^2}{2} + \frac{x^4}{24} + O[x]^6 \tag{1.92}$$

$$\text{Series[Sin[}x], \{x, 0, 6\}]\; x - \frac{x^3}{6} + \frac{x^5}{120} + O[x]^6 \tag{1.93}$$

$$\text{Series[Tan[}x], \{x, 0, 6\}]\; x + \frac{x^3}{3} + \frac{2x^5}{15} + O[x]^7 \tag{1.94}$$

$$\text{Series[Exp[}x], \{x, 0, 6\}]\; 1 + x + \frac{x^2}{2} + \frac{x^3}{6} + \frac{x^4}{24} + \frac{x^5}{120} + \frac{x^6}{720} + O[x]^7 \tag{1.95}$$

$$\text{Series[}(1 + x)^p, \{x, 0, 3\}]\; 1 + px + \frac{1}{2}(-1 + p)px^2 + \frac{1}{6}(-2 + p)(-1 + p) \tag{1.96}$$
$$px^3 + O[x]^4.$$

1.5 Homework assignment

Problem 1. Land for sell: Imagine you are given a specific procedure to sell a big chunk of land to investors. This land is triangularly shaped plateau. The triangle can be approximated as an equilateral triangle. The size of this land, approximately, is about α square kilometers. The procedure for selling the land reads as follows:

    (a) Connect the midpoints of the sides of the triangular chunk of land to form a total of four smaller triangularly shaped chunks of land with equal sides (i.e., four equilateral triangles, see figure 1.7).

Figure 1.7. The four smaller equilateral triangles formed by connecting the middle of the three sides of an equilateral triangle with area α square kilometers.

(b) Sell the middle triangularly shaped chunk of land.

(c) For each of the other three remaining unsold chunks of land, draw lines connecting the midpoints. We then have four small chunks of land from each. Again we sell each middle triangular chunk of land for another investor.

Find the infinite series for the total size (i.e., area) of land sold to the investors if the process described above continued indefinitely.

Problem 2. Find the limit of the sequence a_n given by

$$a_n = \frac{n^2 + 5n^3}{2n^3 + 3\sqrt{4 + n^6}}. \tag{1.97}$$

Problem 3. Use the ratio test to find whether the following series converge or diverge

(a)

$$\sum_{n=0}^{\infty} \frac{5^n (n!)^2}{(2n)!}. \tag{1.98}$$

(b)

$$\sum_{n=0}^{\infty} \frac{\sqrt{(2n)!}}{n!}. \tag{1.99}$$

Problem 4. Find the interval of convergence for the series

$$\sum_{n=0}^{\infty} \frac{(x - 1)^n}{2^n}. \tag{1.100}$$

Problem 5. Prove that

$$\sum_{i=1}^{n} i = \frac{n(n + 1)}{2} \tag{1.101}$$

is true for any integer $n > 0$ (even or odd).

For the functions listed in Problems 6 and 7: (a) Find the first four non-zero terms of Maclaurin's series; (b) Find the general term and write the series in summation form; (c) Check your result in (a) by computer (see last section) ; (d) Use a computer to plot the functions and several approximating partial sum of the series.

Problem 6.

$$f(x) = \frac{x}{\sqrt{1 - x^2}}.$$
(1.102)

Problem 7.

$$g(x) = \frac{1 - x}{1 + x}$$
(1.103)

Hint: Express the function as sum or product of two functions.

Problem 8. The monkey on a tire swing: A monkey on a tire swing is displaced by an applied force of magnitude F directed horizontally. The swing has displaced by an angle θ from the vertical. The center of gravity of the monkey and the tire have moved a distance of x from the vertical. At this position, the monkey is at equilibrium. Let us assume that the weight of the monkey and the tire be W and the length of the rope that the tire is hanging from is l with negligible mass. (See figure 1.8).

Note: We can approximate the distance of the center of mass to the point where the rope is tied to the tree by l.

(a) Find the ratio of the applied force to the weight (F/W) in terms of the angle θ. Express the result in Maclaurin's series.

(b) It is not easy to estimate or measure the angle θ as the rope is tied to a very tall tree. Nevertheless, we can estimate x and l. Find F/W in terms of x and l. Express the result as a series of power of x/l.

Problem 9. Evaluate the following indeterminate functions using L'Hospital's rule
(a)

$$f(x) = \lim_{x \to \infty} \frac{\ln (x - a)}{\sqrt{x - a}},$$
(1.104)

where $a > 0$.

Figure 1.8. A monkey on tire swing.

(b)

$$g(x) = \lim_{x \to 1} (x - 1) \ln [4(x - 1)].$$ (1.105)

Problem 10. Use the series you know to show that

$$\frac{\pi^2}{3!} - \frac{\pi^4}{5!} + \frac{\pi^6}{7!}...= 1.$$ (1.106)

Problem 11. Prove that an alternating series of the form

$$\sum_{n=1}^{\infty}(-1)^{n+1}a_n,$$ (1.107)

with $a_n > 0$ is a convergent series if a_n is monotonically decreasing (i.e., $a_{n+1} < a_n$).

Problem 12. In Example 1.3 we have shown that the series expansion for the function $\ln(1 + x)$ at $x = 0$

$$\ln(1 + x) = \sum_{n=0}^{\infty} b_n x^n = \sum_{n=1}^{\infty} \frac{(-1)^{n-1}}{n} x^n$$ (1.108)

converges when $-1 < x < 1$ using the ratio test. But the ratio test fails when $x = \pm 1$. Using any other test, show that the series

$$\begin{cases} -\sum_{n=1}^{\infty} \frac{(-1)^n}{n}, & \text{is convergent for } x = 1, \\ -\sum_{n=1}^{\infty} \frac{(1)^n}{n}, & \text{is divergent for } x = -1. \end{cases}$$ (1.109)

Problem 13. One of the convergence tests that we did not discuss is the Cauchy root test. It states: In the series

$$\sum_{n=0}^{\infty} a_n = a_0 + a_1 + a_3 ...,$$ (1.110)

if $(a_n)^{1/n} < r < 1$ for all sufficiently large n with r independent of n, then the series is a convergent series. Conversely, if $(a_n)^{1/n} > 1$, then $a_n > 1$ and the series is a divergent series. Verify the Cauchy root test using the Comparison test.

Problem 14. Show that the harmonic series

$$\sum_{n=0}^{\infty} c_n = \sum_{n=0}^{\infty} \frac{1}{n + 1} = 1 + \frac{1}{2} + \frac{1}{3} + \frac{1}{4} + ...,$$ (1.111)

is a divergent series.

IOP Publishing

Studies in Theoretical Physics, Volume 1
Fundamental mathematical methods
Daniel Erenso and Victor Montemayor

Chapter 2

Complex numbers, functions, and series

When we punch the real number $x = -1$ in a calculator and find its square roots, the calculator displays an error message. That is because, in reality, the square root for a negative number does not exist. For Mathematicians and Physicists, the square root of a negative number does exist. It is just not in a real set of numbers but rather in a universal set of numbers known as complex numbers. This chapter focuses on the algebra of complex numbers, such as how we determine roots of complex numbers and properties of functions of a complex variable and its series expansion.

2.1 Complex numbers

A complex number z is defined by

$$z = a + ib, \tag{2.1}$$

with a and b as real numbers and i is known as the *imaginary* unit, which is defined as,

$$i^2 = -1 \Rightarrow \pm i = (-1)^{1/2}. \tag{2.2}$$

For any complex number $z = a + ib$, a is known as the *real part* and b is the *imaginary part*,

$$\mathrm{Re}\, z = a, \ \mathrm{Im}\, z = b. \tag{2.3}$$

Cartesian and polar representations:
Any complex number z can be represented using Cartesian coordinates (x, y) or polar coordinates (r, θ). In Cartesian coordinate representation the real part a is the x coordinate and the imaginary part b is the y coordinate,

$$a = x, \ b = y. \tag{2.4}$$

In figure 2.1 a complex number z is represented by Cartesian and Polar coordinates. Using simple trigonometric relations, we can easily see that these coordinates are related by

doi:10.1088/978-0-7503-3135-7ch2

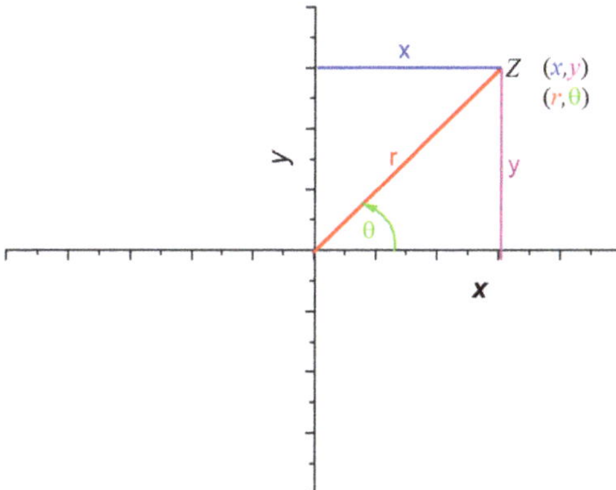

Figure 2.1. Cartesian (rectangular) and polar representation of an arbitrary complex number $z = x + iy$.

$$a = x = r \cos(\theta), \; b = y = r \sin(\theta), \tag{2.5}$$

where r is called the modulus (or magnitude) and θ is called the phase of the complex number z. Thus one can express the complex number $z = a + ib$ as

$$z = x + iy = r \cos(\theta) + ir \sin(\theta). \tag{2.6}$$

Euler's equation and exponential representation:
Euler's equation, that relates an exponential function and trigonometric functions, is given by

$$\exp(i\theta) = \cos(\theta) + i \sin(\theta). \tag{2.7}$$

We will derive Euler's equation when we introduce the series expansion of complex functions. For now, we see its application in carrying out complex numbers algebra. Using Euler's equation, one can also represent the complex number in equation (2.6), in an exponential form as

$$z = x + iy = r(\cos(\theta) + i \sin(\theta)) = r \exp(i\theta). \tag{2.8}$$

The magnitude r (which is also represented by $|z|$) and the phase angle θ for the complex number z are related to the Cartesian coordinates (x, y) by

$$|z| = \sqrt{x^2 + y^2} = \sqrt{r^2 \cos^2(\theta) + r^2 \sin^2(\theta)} = r,$$
$$\theta = \tan^{-1}\left(\frac{y}{x}\right). \tag{2.9}$$

Note that the phase angle is always measured from the positive x-axis in a counterclockwise direction (see figure 2.1).

The Complex conjugate
The complex conjugate to a complex number $z = x + iy$ is denoted by z^* and defined as

$$z^* = x - iy. \tag{2.10}$$

Upon multiplying z with its conjugate z^*, we find

$$\sqrt{zz^*} = \sqrt{(x + iy)(x - iy)} = \sqrt{x^2 + y^2} = |z|, \tag{2.11}$$

which is the magnitude of the complex number z and it is real.

Example 2.1. Find the real and imaginary parts of the complex number

$$z_1 = \frac{3 + 4i}{3 - 4i}. \tag{2.12}$$

Solution: To find the real and imaginary parts, first we need to write it in the form $z = x + iy$. This can be done by making the denominator real. In order to make the denominator real we multiply it (and also the numerator) by its complex conjugate,

$$z_1 = \frac{3 + 4i}{3 - 4i} = \frac{(3 + 4i)(3 + 4i)}{(3 - 4i)(3 + 4i)} = \frac{-7 + 24i}{25} = \frac{-7}{25} + \frac{24}{25}i$$

$$\Rightarrow \operatorname{Re} z = -\frac{7}{25}, \ \operatorname{Im} z = \frac{24}{25}. \tag{2.13}$$

Example 2.2. Find the real and imaginary parts of the complex conjugate for

$$z_2 = \frac{\sqrt{3} + 2i \sin(\theta)}{4 + 2e^{4i\theta}}, \tag{2.14}$$

where θ is real.

Solution: The complex conjugate z^* is determined by replacing i with $-i$ everywhere in the expression,

$$z_2 = \frac{\sqrt{3} + 2i \sin(\theta)}{4 + 2e^{4i\theta}} \Rightarrow z_2^* = \frac{\sqrt{3} - 2i \sin(\theta)}{4 + 2e^{-4i\theta}}. \tag{2.15}$$

In order to make the denominator real we multiply it with its complex conjugate and we should do the same for the numerator

$$z_2^* = \frac{\left(\sqrt{3} - 2i \sin(\theta)\right)(4 + 2e^{-4i\theta})^*}{(4 + 2e^{-4i\theta})(4 + 2e^{-4i\theta})^*}. \tag{2.16}$$

Noting that for the denominator

$$4 + 2e^{-4i\theta} \Rightarrow (4 + 2e^{-4i\theta})^* = (4)^* + (2e^{-4i\theta})^* = 4 + 2e^{4i\theta}$$

$$\Rightarrow (4 + 2e^{-4i\theta})(4 + 2e^{-4i\theta})^* = 16 + 8(e^{-4i\theta} + e^{4i\theta}) + 4 \tag{2.17}$$

$$\Rightarrow (4 + 2e^{-4i\theta})(4 + 2e^{-4i\theta})^* = 20 + 8(e^{-4i\theta} + e^{4i\theta}).$$

Applying Euler's formula one can write,

$$e^{4i\theta} = \cos(4\theta) + i\sin(4\theta),\ e^{-4i\theta} = \cos(4\theta) - i\sin(4\theta)$$
$$e^{-4i\theta} + e^{4i\theta} = 2\cos(4\theta),$$

(2.18)

so that equation (2.17) becomes

$$(4 + 2e^{-4i\theta})(4 + 2e^{-4i\theta})^* = 20 + 16\cos(4\theta).$$

(2.19)

Substituting this result into equation (2.16), one can then write

$$z_2^* = \frac{\left(\sqrt{3} - 2i\sin(\theta)\right)(4 + 2e^{4i\theta})}{20 + 16\cos(4\theta)}.$$

(2.20)

Applying Euler's relation, we have

$$2e^{4i\theta} = 2\cos(4\theta) + 2i\sin(4\theta),$$

(2.21)

so that

$$z_2^* = \frac{\left(\sqrt{3} - 2i\sin(\theta)\right)(4 + 2\cos(4\theta) + 2i\sin(4\theta))}{20 + 16\cos(4\theta)}$$
$$= \left\{\sqrt{3}(4 + 2\cos(4\theta)) + 4\sin(4\theta)\sin(\theta) + 2i\sqrt{3}\sin(4\theta)\right.$$
$$\left. - 2i\sin(\theta)(4 + 2\cos(4\theta))\right\}/\{20 + 16\cos(4\theta)\}$$
$$\Rightarrow z_2^* = \frac{\sqrt{3}(4 + 2\cos(4\theta)) + 4\sin(4\theta)\sin(\theta)}{20 + 16\cos(4\theta)}$$
$$+ i\frac{2\sqrt{3}\sin(4\theta) - 2\sin(\theta)(4 + 2\cos(4\theta))}{20 + 16\cos(4\theta)}.$$

(2.22)

Therefore, the real and imaginary parts of z_2^* are

$$\mathrm{Re}\,(z_2^*) = \frac{\sqrt{3}(4 + 2\cos 4\theta) + 4\sin 4\theta \sin(\theta)}{20 + 16\cos(4\theta)},$$

(2.23)

$$\mathrm{Im}\,(z_2^*) = \frac{2\sqrt{3}\sin 4\theta - 2\sin(\theta)(4 + 2\cos 4\theta)}{20 + 16\cos(4\theta)}.$$

(2.24)

Example 2.3. Consider the complex function

$$Z(x) = 2f(x) - \frac{3}{2}ig(x),$$

(2.25)

where $f(x)$ and $g(x)$ are complex functions of the real variable x.
    (a) Find an expression for the complex conjugate of the function $Z(x)$.
    (b) Evaluate $Z^*(x)$ for $x = 1$ if

$$f(x) = 3ix^2 \text{ and } g(x) = 2x + 4\sqrt{2}\, i. \tag{2.26}$$

Solution:
    (a) Since $f(x)$ and $g(x)$ are complex functions, the complex conjugate of $Z(x)$
    should be written as

$$Z^*(x) = 2f^*(x) + \frac{3}{2}ig^*(x) \tag{2.27}$$

    where $f^*(x)$ and $g^*(x)$ are the complex conjugates of $f(x)$ and $g(x)$,
    respectively.
    (b) Noting that for the given functions

$$f(x) = 3ix^2, \quad g(x) = 2x + 4\sqrt{2}\, i, \tag{2.28}$$

    the complex conjugates are

$$f^*(x) = -3ix^2, \quad g^*(x) = 2x - 4\sqrt{2}\, i \tag{2.29}$$

    one can write

$$
\begin{aligned}
Z^*(x) &= 2f^*(x) + \frac{3}{2}ig^*(x) \\
&= 2(-3ix^2) + \frac{3}{2}i\left(2x - 4\sqrt{2}\, i\right) = -6ix^2 + 3xi + 6\sqrt{2}. \\
&\Rightarrow Z^*(1) = 6\sqrt{2} - 3i,
\end{aligned}
\tag{2.30}
$$

    where we substituted $x = 1$.

Example 2.4. Evaluate $f(z)$

$$f(z) = \frac{z - 1}{z}, \tag{2.31}$$

at $z_0 = 2 - i$ and find the real and imaginary parts.

Solution: We can write $f(z_0)$ as

$$f(z_0) = \frac{2 - i - 1}{2 - i} = \frac{1 - i}{2 - i} \tag{2.32}$$

which can be rewritten as

$$f(z_0) = \frac{(1 - i)(2 + i)}{(2 - i)(2 + i)} = \frac{3 - i}{5} \tag{2.33}$$

and the real and imaginary parts become

$$\mathrm{Re}\,(f(z_0)) = 3/5, \ \mathrm{Im}\,(f(z_0)) = -1/5. \tag{2.34}$$

Example 2.5. Find the values of the real quantities x and y for

$$(5 + i)x - 5yi = 4 + 3i. \tag{2.35}$$

Solution: To determine the values of the real quantities x and y, we first need to write the left-hand side in the form $z = a + ib$,

$$(5 + i)x - 5yi = 5x + i(x - 5y) = 4 + 3i. \tag{2.36}$$

Two complex number z_1 and z_2 are equal if and only if their corresponding Re and Im parts are equal. Thus

$$5x = 4, \ x - 5y = 3 \Rightarrow x = 4/5, \ y = -11/25. \tag{2.37}$$

2.2 Complex infinite series

In chapter 1, we saw the series for real function $f(x)$, which depends on a single variable x, and introduced how to determine the interval of convergence for the series. This section will introduce the series for complex function $f(z)$, which depends on two real variables, x and y. Since such a series is a function of x and y on the complex plane, we have a circle of convergence instead of an interval of convergence.

Definition and convergence

An infinite series

$$\sum_{n=0}^{\infty} z_n = z_0 + z_1 + z_2 + z_3 \ldots z_n + \ldots \tag{2.38}$$

is said to be a complex infinite series if

$$z_n = a_n + ib_n, \tag{2.39}$$

where a_n and b_n are real numbers. An infinite complex series is convergent if and only if both the real and the imaginary parts of the complex series are convergent,

$$\sum_{n=0}^{\infty} a_n = a_0 + a_1 + a_2 + a_3 \ldots a_n + \ldots = A \tag{2.40}$$

and

$$\sum_{n=0}^{\infty} b_n = b_0 + b_1 + b_2 + b_3...b_n + ... = B, \tag{2.41}$$

where A and B are real constants.

Test of convergence

For a complex infinite series
    (a) You can use any convergence test that involves the absolute value sign, which in this case will mean the magnitude of a complex quantity. This is applied to the full complex infinite series.
    (b) You can also use any convergence test to separately test the convergence of the real and imaginary parts of the complex series. The full series converges only if both the real and imaginary parts of the series converge separately.

In the previous chapter we have seen a real power series

$$\sum_{n=0}^{\infty} a_n x^n = a_0 + a_1 x + a_2 x^2 + a_3 x^3...a_n x^n + ..., \tag{2.42}$$

where a_n and x are real. A power series

$$\sum_{n=0}^{\infty} c_n(z - z_0)^n = c_0 + c_1(z - z_0) + c_2(z - z_0)^2 + c_3(z - z_0)^3...c_n(z - z_0)^n + ... \tag{2.43}$$

is a series of complex function when z is a complex variable and z_0 or c_n are generally complex constants. For complex series, we describe the region of convergence by a circle, and the radius of this circle gives the radius of convergence for the series. Applying the ratio test, the region of convergence for the series in equation (2.43) can be determined by solving the equation,

$$\rho = \lim_{n \to \infty} \left| \frac{c_{n+1}(z - z_0)^{n+1}}{c_n(z - z_0)^n} \right| < 1 \Rightarrow \lim_{n \to \infty} \left| \frac{c_{n+1}}{c_n} \right| |z - z_0| < 1, \tag{2.44}$$

for z. For example, for

$$\frac{c_{n+1}}{c_n} = \frac{n}{n+1}, \tag{2.45}$$

we have

$$\lim_{n \to \infty} \left| \frac{c_{n+1}}{c_n} \right| = 1, \tag{2.46}$$

which leads to

$$\rho = |z - z_0| < 1. \tag{2.47}$$

Substituting $z = x + iy$ and $z_0 = a + ib$, one can then write

$$z - z_0 = x + iy - (a + ib) = (x - a) + i(y - b)$$
$$\Rightarrow |z - z_0| = \sqrt{(x - a)^2 + (y - b)^2} < 1 \tag{2.48}$$
$$\Rightarrow (x - a)^2 + (y - b)^2 < 1^2.$$

We recall that the equation

$$(x - a)^2 + (y - b)^2 = R^2, \tag{2.49}$$

defines a circle with radius, R, centered at (a, b). Thus the series converges for all complex numbers z inside the circle (disk) of radius $R = 1$ on the complex plane. This circle (disk) is known as the *circle (disk) of convergence* for the infinite complex series in equation (2.43). The radius of the circle is called *the radius of convergence*.

Example 2.6. Determine the convergence of the following complex infinite series

$$\sum_{n=0}^{\infty} z_n = 1 - \frac{i}{2} - \frac{1}{4!} + \frac{i}{6!} - \frac{1}{8!} \tag{2.50}$$

Solution: This series can be expressed as

$$\sum_{n=0}^{\infty} z_n = \sum_{n=0}^{\infty} \frac{(-i)^n}{(2n)!}. \tag{2.51}$$

Applying the ratio test we have

$$\lim_{n \to \infty} \left| \frac{z_{n+1}}{z_n} \right| = \lim_{n \to \infty} \left| \frac{(-i)^{n+1}}{(2(n+1))!} \Big/ \frac{(-i)^n}{(2n)!} \right| = \lim_{n \to \infty} \left| \frac{(-i)^{n+1}}{(2(n+1))!} \frac{(2n)!}{(-i)^n} \right|$$
$$= \lim_{n \to \infty} \left| \frac{-i}{(2n+2)(2n+1)} \right| = \lim_{n \to \infty} \left| \frac{1}{(2n+2)(2n+1)} \right| \tag{2.52}$$
$$\Rightarrow \lim_{n \to \infty} \left| \frac{z_{n+1}}{z_n} \right| = 0 < 1,$$

which means the series is convergent.

Example 2.7. Find and sketch the circle of convergence in the complex plane for the following complex infinite series:

$$\sum_{n=0}^{\infty} c_n z^n = \sum_{n=0}^{\infty} \frac{(z - 2 + i)^n}{2^n}. \tag{2.53}$$

Solution: We recall, a complex series

$$\sum_{n=0}^{\infty} a_n(z - z_0)^n, \tag{2.54}$$

is convergent when

$$\rho = \lim_{n \to \infty} \left| \frac{a_{n+1}(z - z_0)^{n+1}}{a_n(z - z_0)^n} \right| < 1. \tag{2.55}$$

Noting that for the given series

$$a_n(z - z_0)^n = \frac{(z - 2 + i)^n}{2^n},$$

$$a_{n+1}(z - z_0)^{n+1} = \frac{(z - 2 + i)^{n+1}}{2^{n+1}}, \tag{2.56}$$

we can write

$$\rho = \lim_{n \to \infty} \left| \frac{(z - 2 + i)^{n+1}}{2^{n+1}} \middle/ \frac{(z - 2 + i)^n}{2^n} \right| < 1,$$

$$\Rightarrow \rho = \lim_{n \to \infty} \left| \frac{(z - 2 + i)^{n+1}}{2^{n+1}} \frac{2^n}{(z - 2 + i)^n} \right| < 1 \Rightarrow \left| \frac{z - 2 + i}{2} \right| < 1, \tag{2.57}$$

$$\Rightarrow |z - 2 + i| < 2 \Rightarrow |z - (2 - i)| < 2.$$

Substituting $z = x + iy$, one can write

$$|(x - 2) + i(y - (-1))| < 2 \Rightarrow \sqrt{(x - 2)^2 + (y - (-1))^2} < 2$$

$$\Rightarrow (x - 2)^2 + (y - (-1))^2 = 2^2.$$

Comparing this result with the general equation for a circle in equation (2.49), the series is convergent inside a disk of radius $R = 2$ centered at $z_0 = 2 - i$ on the complex plane. The region bounded by the circle is shown in figure 2.2.

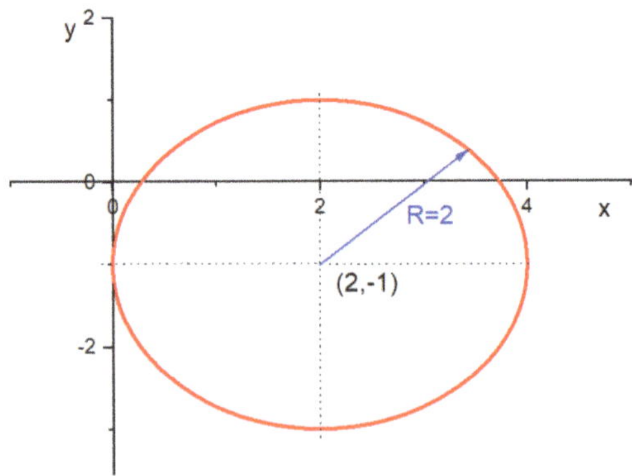

Figure 2.2. The circle of convergence.

Example 2.8. Derive
(a) Euler's formula

$$e^{i\theta} = \cos(\theta) + i\sin(\theta), \tag{2.58}$$

using series expansion of complex variables,
(b) the double-angle formulae for $\sin(2\theta)$ and $\cos(2\theta)$

$$\sin(2\theta) = 2\sin(\theta)\cos(\theta),$$
$$\cos(2\theta) = \cos^2(\theta) - \sin^2(\theta). \tag{2.59}$$

Solution:
(a) We recall from chapter 1 for a real differentiable function $f(x)$, *Maclaurin's series* is given by

$$f(x) = \sum_{n=0}^{\infty} b_n x^n, \quad \text{where } b_n = \frac{1}{n!}\frac{d^n f(x)}{dx^n}\bigg|_{x=0}. \tag{2.60}$$

This can be generalized to a complex function $f(z)$ as

$$f(z) = \sum_{n=0}^{\infty} c_n z^n, \quad \text{where } c_n = \frac{1}{n!}\frac{d^n f(z)}{dz^n}\bigg|_{z=0}. \tag{2.61}$$

Thus for the complex exponential function

$$f(z) = \exp(z), \tag{2.62}$$

one can write

$$\exp(z) = \sum_{0}^{\infty} \frac{z^n}{n!} = 1 + z + \frac{z^2}{2!} + \frac{z^3}{3!} + \dots. \tag{2.63}$$

For $z = i\theta$, where θ is real, we have

$$\exp(i\theta) = \sum_{0}^{\infty} \frac{(i\theta)^n}{n!} = 1 + i\theta - \frac{\theta^2}{2!} - i\frac{\theta^3}{3!} + \dots \tag{2.64}$$

so that separating the odd and even terms in series one can write

$$\exp(i\theta) = \left[1 - \frac{\theta^2}{2!} + \frac{\theta^4}{4!} + \dots\right] + \left[i\theta - i\frac{\theta^3}{3!} + i\frac{\theta^3}{5!}\dots\right]$$
$$= \left[1 - \frac{\theta^2}{2!} + \frac{\theta^4}{4!} + \dots\right] + i\left[\theta - \frac{\theta^3}{3!} + \frac{\theta^3}{5!}\dots\right]. \tag{2.65}$$

Applying the series expansion for sine and cosine real functions from chapter 1,

$$\sin(\theta) = \theta - \frac{\theta^3}{3!} + \frac{\theta^5}{5!} - \frac{\theta^7}{7!} \cdots = \sum_{n=0}^{\infty} \frac{(-1)^n}{(2n-1)!}\theta^{2n+1}, \tag{2.66}$$

$$\cos(\theta) = 1 - \frac{\theta^2}{2!} + \frac{\theta^4}{4!} - \frac{\theta^6}{6!} \cdots = \sum_{n=0}^{\infty} \frac{(-1)^n}{(2n)!}\theta^{2n}, \tag{2.67}$$

one can easily see that

$$\exp(i\theta) = \cos(\theta) + i\sin(\theta), \tag{2.68}$$

which is Euler's formula.

(b) Using Euler's formula, one can write

$$[\exp(i\theta)]^2 = [\exp(i2\theta)] = \cos(2\theta) - i\sin(2\theta), \tag{2.69}$$

and

$$[\exp(i\theta)]^2 = [\cos(\theta) - i\sin(\theta)]^2$$
$$\Rightarrow [\exp(i\theta)]^2 = \cos^2(\theta) - \sin^2(\theta) - 2i\sin(\theta)\cos(\theta). \tag{2.70}$$

Combining equations (2.69) and (2.70), we have

$$\cos(2\theta) - i\sin(2\theta) = \cos^2(\theta) - \sin^2(\theta) - 2i\sin(\theta)\cos(\theta), \tag{2.71}$$

which leads to

$$\cos(2\theta) = \cos^2(\theta) - \sin^2(\theta), \quad \sin(2\theta) = 2\sin(\theta)\cos(\theta). \tag{2.72}$$

2.3 Powers and roots of complex numbers

The powers and roots of a complex number are easily determined by applying Euler's formula. Let us consider a complex number,

$$z = x + iy = re^{i\theta} = re^{i(\theta + 2\pi k)}, \tag{2.73}$$

where

$$r = \sqrt{x^2 + y^2}, \quad \theta = \tan^{-1}\left(\frac{y}{x}\right), \tag{2.74}$$

and $k = 0, 1, 2, 3....$ This complex number to the power n (z^n) can then be written as

$$z^n = (x + iy)^n = (re^{i\theta})^n = r^n e^{in\theta}$$
$$\Rightarrow z^n = (x + iy)^n = r^n[\cos(n\theta) + i\sin(n\theta)]. \tag{2.75}$$

Similarly for z to the power of $1/n$ ($z^{1/n}$), which is the nth roots to the complex number z, one finds

$$z^{1/n} = (x + iy)^{1/n} = r^{1/n}e^{i\theta_k} = r^{1/n}[\cos(\theta_k) + i\sin(\theta_k)], \tag{2.76}$$

where

$$\theta_k = \frac{\theta + 2\pi k}{n}, \quad \text{for } k = 0, 1, 2, \ldots, n-1. \tag{2.77}$$

Here we would like to emphasize that since

$$\begin{aligned} \exp[in(\theta + 2\pi k)] &= \cos[n(\theta + 2\pi k)] + i\cos[n(\theta + 2\pi k)] \\ &= \cos[n\theta] + i\cos[n\theta] = \exp[in\theta], \end{aligned} \tag{2.78}$$

for n equal to zero or positive integer, it does not make any difference whether we included the $2\pi k$ or not. However, for $1/n$ that is non-integer, we must always keep the $2\pi k$ in representing a complex number using polar coordinates or Euler's formula.

Example 2.10. Solve the equation

$$z^5 = 32 \tag{2.79}$$

and show the solutions on a complex plane.

Solution: We can rewrite the given equation as

$$\begin{aligned} z^5 &= 32\exp[i(0 + 2\pi k)] = 2^5\exp[i(0 + 2\pi k)] \\ &\Rightarrow z = r^{1/5}e^{i\theta_k}, \end{aligned} \tag{2.80}$$

where

$$r^{1/5} = (2^5)^{1/5} = 2, \ \theta_k = \frac{2\pi k}{5}, \quad \text{for } k = 0, 1, 2, 3, 4, \tag{2.81}$$

according to the relations in equations (2.76) and (2.77). Then by substituting the allowed values of k, we have

$$\begin{aligned} k &= 0 \Rightarrow \theta_0 = 0, \\ k &= 1 \Rightarrow \theta_1 = \frac{2\pi}{5} = 72°, \\ k &= 2 \Rightarrow \theta_2 = \frac{4\pi}{5} = 144°, \\ k &= 3 \Rightarrow \theta_3 = \frac{6\pi}{5} = 216°, \\ k &= 4 \Rightarrow \theta_4 = \frac{8\pi}{5} = 288° \end{aligned} \tag{2.82}$$

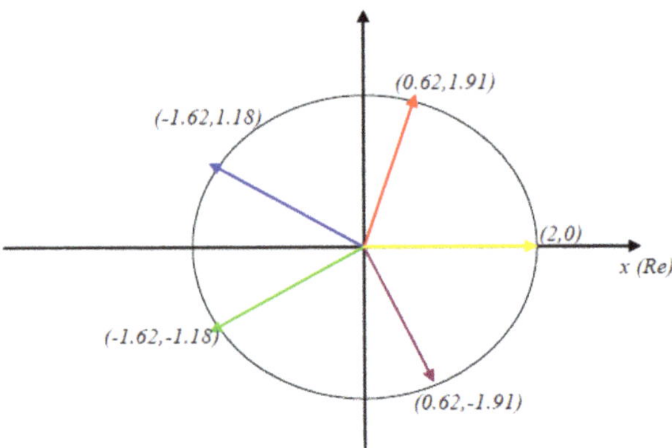

Figure 2.3. The root on the complex plane. The circle has a radius $r = 2$

so that the solutions to the equation becomes

$$z_0 = 2[\cos(0) + i \sin(0)] = 2 + i0$$

$$z_1 = 2\left[\cos\left(\frac{2\pi}{5}\right) + i \sin\left(\frac{2\pi}{5}\right)\right] = 0.62 + i1.90$$

$$z_2 = 2\left[\cos\left(\frac{4\pi}{5}\right) + i \sin\left(\frac{4\pi}{5}\right)\right] = -1.62 + i1.18$$

$$z_3 = 2\left[\cos\left(\frac{6\pi}{5}\right) + i \sin\left(\frac{6\pi}{5}\right)\right] = -1.62 - i1.18$$

$$z_4 = 2\left[\cos\left(\frac{8\pi}{5}\right) + i \sin\left(\frac{8\pi}{5}\right)\right] = 0.62 - i1.90.$$

$$(2.83)$$

These roots are shown on the complex plane in figure 2.3. Note that if we calculate θ_k for $k = 5$, we find $\theta_5 = 2\pi$ which leads to

$$z_5 = 2[\cos(2\pi) + i \sin(2\pi)] = 2 + i0 = z_0.$$

This is the reason why we set the maximum value for $k = n - 1$ for the nth roots for the complex number z.

2.4 Algebraic versus transcendental functions

The logarithmic and the exponential function are examples of transcendental functions. The term transcendental is often used to describe trigonometric functions, such as sine, cosine, tangent, cotangent, secant and cosecant. A function that is not transcendental is said to be algebraic. Examples of algebraic functions are rational and square root functions. A composition of transcendental functions can give an algebraic function. For example, let us consider the function,

$$f(x) = \cos[\arcsin(x)], \tag{2.84}$$

which is a composition of cosine and arcsine functions. Introducing the variable

$$\theta = \arcsin(x) \Rightarrow \sin(\theta) = x,$$

one finds

$$f(x) = \cos(\theta) = \sqrt{1 - \sin^2(\theta)} = \sqrt{1 - x^2}, \tag{2.85}$$

which is a square root function.

Euler's formula, trigonometric, and hyperbolic functions

We recall Euler's formula in terms of a real variable x is given by

$$\exp(ix) = \cos(x) + i\sin(x) \tag{2.86}$$

and taking the complex conjugate, we have

$$\exp(-ix) = \cos(x) - i\sin(x). \tag{2.87}$$

Adding and subtracting these two equations leads to

$$\cos(x) = \frac{e^{ix} + e^{-ix}}{2} \tag{2.88}$$

and

$$\sin(x) = \frac{e^{ix} - e^{-ix}}{2i}, \tag{2.89}$$

respectively. Replacing x by a complex number z, one can then write

$$\cos(z) = \frac{e^{iz} + e^{-iz}}{2}, \quad \sin(z) = \frac{e^{iz} - e^{-iz}}{2i}. \tag{2.90}$$

Consider the case where z is a complex number just only with imaginary part (i.e., $z = iy$) also called a pure complex number. Substituting $z = iy$ into equation (2.90), the sine and cosine functions become

$$\cos(iy) = \frac{e^{-y} + e^{y}}{2} = \cosh(y),$$

$$\sin(iy) = \frac{e^{-y} - e^{y}}{2i} = i\frac{e^{y} - e^{-y}}{2} = i\sinh(y), \tag{2.91}$$

where the functions

$$\cosh(y) = \frac{e^{y} + e^{-y}}{2}, \quad \sinh(y) = \frac{e^{y} - e^{-y}}{2}, \tag{2.92}$$

which depend on the real variable y, are called the cosine hyperbolic (cosh) and sine hyperbolic (sinh) functions, respectively. These functions can also be generalized for a complex number z as

$$\sinh(z) = \frac{e^z - e^{-z}}{2}, \quad \cosh(z) = \frac{e^z + e^{-z}}{2}. \tag{2.93}$$

Other hyperbolic functions that are named and defined in a way similar to the trigonometric functions include

$$\tanh(z) = \frac{\sinh(z)}{\cosh(z)}, \quad \coth(z) = \frac{1}{\tanh(z)},$$
$$\operatorname{csch}(z) = \frac{1}{\sinh(z)}, \quad \operatorname{sech}(z) = \frac{1}{\cosh(z)}. \tag{2.94}$$

Logarithmic functions

Next, we want to introduce how we do the algebra of a Logarithmic function of complex functions, like the nth roots of a complex function. To this end, let us consider the complex number

$$z = x + iy = re^{i(\theta + 2\pi k)}, \quad \text{where } k = 0, 1, 2 \dots. \tag{2.95}$$

The logarithm of z to the base e, which is referred to as the natural logarithm and denoted by $\ln z$, becomes

$$\ln z = \ln\left[re^{i(\theta + 2\pi k)}\right] = \ln r + \ln\left[e^{i(\theta + 2\pi k)}\right] = \ln r + i(\theta + 2\pi k)\ln e$$
$$\Rightarrow \ln z = \ln r + i(\theta + 2\pi k). \tag{2.96}$$

This result gives infinite values for the function $\ln z$ since the imaginary part differs for different values of k and is represented by different points in the complex plane. Usually, we are interested in *the principal value* for the logarithm of a complex function, which is when $k = 0$,

$$\ln z = \ln r + i\theta. \tag{2.97}$$

Example 2.11. Evaluate the complex expression

$$\cos[i \ln 2]. \tag{2.98}$$

Solution: Using Euler's formula

$$\cos(z) = \frac{e^{iz} + e^{-iz}}{2}, \tag{2.99}$$

for $z = \cos[i \ln 2]$, one can write

$$\cos(z) = \cos[i \ln 2] = \frac{e^{i(i \ln 2)} + e^{-i(i \ln 2)}}{2} = \frac{e^{-\ln 2} + e^{\ln 2}}{2}. \tag{2.100}$$

Applying the relations

$$\ln a^b = b \ln a, \quad x = e^{\ln x} \tag{2.101}$$

we have

$$e^{\ln 2} = 2 \text{ and } e^{-\ln 2} = e^{\ln 2^{-1}} = 2^{-1} = \frac{1}{2}, \tag{2.102}$$

so that upon substituting equation (2.102) into equation (2.100), we find

$$\cos\left[i \ln 2\right] = \frac{\frac{1}{2} + 2}{2} = \frac{5}{4}. \tag{2.103}$$

Example 2.12. Find the principal value of the complex logarithm

$$w = \ln(1 - i). \tag{2.104}$$

Solution: To find the principal value first we need to express $z = 1 - i$ using polar coordinates. To this end, we note that

$$r = |1 - i| = \sqrt{1^2 + (-1)^2} = \sqrt{2},$$
$$\theta = \tan^{-1}\left(\frac{-1}{1}\right) + 2\pi k = \frac{7}{4}\pi + 2\pi k, \tag{2.105}$$

where $k = 0, 1, 2, \ldots$ But for the principal value for a logarithmic function, we use only the result when $k = 0$, which is $\theta = \frac{7}{4}\pi$. Thus using

$$z = 1 - i = \sqrt{2}\, e^{i\frac{7}{4}\pi}, \tag{2.106}$$

one can write

$$\ln z = \ln\left(\sqrt{2}\, e^{i\frac{7}{4}\pi}\right) = \ln 2^{1/2} + i\frac{7\pi}{4} = \frac{\ln 2}{2} + i\frac{7\pi}{4}. \tag{2.107}$$

Therefore the principal value is found to be

$$w = \ln z = \ln(1 - i) = \frac{\ln 2}{2} + i\frac{7\pi}{4}. \tag{2.108}$$

Noting that the angle $\theta = \frac{7\pi}{4}$ is measured from the positive x-axis in a counterclockwise direction. But we may express this same angle as

$$\alpha = \theta - 2\pi = -\frac{\pi}{4},$$

measured in a clockwise direction from the positive x-axis. If we use this angle, the complex number becomes

$$z = 1 - i = \sqrt{2}\,e^{-i\frac{\pi}{4}} \tag{2.109}$$

and the result becomes

$$w = \ln z = \ln\,(1 - i) = \frac{\ln 2}{2} - i\frac{\pi}{4}. \tag{2.110}$$

But on the complex plane the results in equations (2.108) and (2.110) do not represent the same complex number. Therefore, it is advised to use the standard measurement of the phase angle, which is from the positive x-axis in the counter-clockwise direction when we deal with logarithmic functions of complex variables.

Example 2.13. Find the principal value of the complex quantity

$$z = (1 - i)^i. \tag{2.111}$$

Solution: Applying the relation

$$z = e^{\ln z}$$

one can write

$$z = e^{\ln\,(1-i)^i} = e^{i\,\ln\,(1-i)}. \tag{2.112}$$

In view of the result in Example 2.13 (equation (2.108)), equation (2.112) becomes

$$z = e^{i\left(\frac{\ln 2}{2} + i\frac{7\pi}{4}\right)} = e^{i\frac{\ln 2}{2}}e^{-\frac{7\pi}{4}} = e^{-\frac{7\pi}{4}}\left[\cos\left(\frac{\ln 2}{2}\right) + i\,\sin\left(\frac{\ln 2}{2}\right)\right]. \tag{2.113}$$

Example 2.14. Find the value for the following complex expression

$$w = \sin^{-1}\left[\left(\frac{\sqrt{3} + i}{\sqrt{3} - i}\right)^{12}\right]. \tag{2.114}$$

Solution: Since w is sign inverse of

$$\left(\frac{\sqrt{3} + i}{\sqrt{3} - i}\right)^{12}, \tag{2.115}$$

we can write

$$\sin\,(w) = \left(\frac{\sqrt{3} + i}{\sqrt{3} - i}\right)^{12}. \tag{2.116}$$

For the complex numbers

$$z_1 = \sqrt{3} + i \text{ and } z_2 = \sqrt{3} - i \tag{2.117}$$

one can write

$$z_1 = r_1 e^{i\theta_1}, \; z_2 = r_2 e^{i\theta_2}, \tag{2.118}$$

where

$$r_1 = |\sqrt{3} + i| = 2, \; \theta_1 = \tan^{-1}\left[\frac{1}{\sqrt{3}}\right] = \frac{\pi}{6},$$

$$r_2 = |\sqrt{3} - i| = 2, \; \theta_2 = \tan^{-1}\left[\frac{-1}{\sqrt{3}}\right] = \frac{11\pi}{6}. \tag{2.119}$$

Then using the results in equations (2.117)–(2.119), equation (2.116) can be rewritten as

$$\sin(w) = \left(\frac{z_1}{z_2}\right)^{12} = \left(\frac{2e^{i\frac{\pi}{6}}}{2e^{i\frac{11\pi}{6}}}\right)^{12} = \left[e^{-i\frac{10\pi}{6}}\right]^{12}$$

$$\Rightarrow \sin(w) = e^{-20i\pi} = e^{-10i(2\pi)} = 1 \Rightarrow w = \frac{\pi}{2} \pm 2k\pi, \tag{2.120}$$

where $k = 0, 1, 2, \ldots$.

Example 2.15. Applying Euler's formula, evaluate the integral,

$$I = \int_{-\pi}^{\pi} \sin(nx)\cos(mx)dx, \tag{2.121}$$

where n and m are positive integers.

Solution: Applying Euler's formula, we can write

$$\sin(nx) = \frac{e^{inx} - e^{-inx}}{2i}, \; \cos(mx) = \frac{e^{imx} + e^{-imx}}{2} \tag{2.122}$$

so that

$$\sin(nx)\cos(mx) = \left(\frac{e^{inx} - e^{-inx}}{2i}\right)\left(\frac{e^{imx} + e^{-imx}}{2}\right)$$

$$= \frac{e^{i(n+m)x} - e^{-i(n+m)x} + e^{i(n-m)x} - e^{-i(n-m)x}}{4i} \tag{2.123}$$

$$= \frac{1}{2}\left[\frac{e^{i(n+m)x} - e^{-i(n+m)x}}{2i} + \frac{e^{i(n-m)x} - e^{-i(n-m)x}}{2i}\right].$$

In view of the relations

$$\frac{e^{i(n + m)x} - e^{-i(n + m)x}}{2i} = \sin\left[(n + m)x\right],$$

$$\frac{e^{i(n-m)x} - e^{-i(n-m)x}}{2i} = \sin\left[(n - m)x\right], \tag{2.124}$$

we can rewrite

$$\sin(nx)\cos(mx) = \frac{1}{2}[\sin\left[(n + m)x\right] + \sin\left[(n - m)x\right]]. \tag{2.125}$$

Then substituting this expression into equation (2.121), we have

$$I = \frac{1}{2}\int_{-\pi}^{\pi}[\sin\left[(n + m)x\right] + \sin\left[(n - m)x\right]]dx, \tag{2.126}$$

which leads to

$$I = -\frac{1}{2}\left[\frac{\cos\left[(n + m)x\right]}{n + m} + \frac{\cos\left[(n - m)x\right]}{n - m}\right]\Bigg|_{-\pi}^{\pi}. \tag{2.127}$$

Since cosine is an even function,

$$\cos(u) = \cos(-u), \tag{2.128}$$

equation (2.127) reduces to

$$I = \frac{\cos\left[(n + m)\pi\right] - \cos\left[(n + m)\pi\right]}{n + m} + \frac{\cos\left[(n - m)\pi\right] - \cos\left[(n - m)\pi\right]}{n - m} = 0. \tag{2.129}$$

Therefore

$$I = \int_{-\pi}^{\pi}\sin(nx)\cos(mx)dx = 0. \tag{2.130}$$

Orthogonal and orthonormal set of functions

The set of functions $\{f_1(x), f_2(x), ... f_j(x), ...\}$ satisfying the condition

$$\int_{x_1}^{x_2} f_i^*(x)f_j(x)dx = C\delta_{ij} = \begin{cases} C & i = j \\ 0 & i \neq j \end{cases} \tag{2.131}$$

are known as the *orthogonal set of functions* in the domain $x_1 \leqslant x \leqslant x_2$. Note that δ_{ij} is the *Kronecker delta* defined by

$$\delta_{ij} = \begin{cases} 1, & i = j, \\ 0, & i \neq j, \end{cases} \tag{2.132}$$

and C is a constant. When $C = 1$, the set of functions form an *orthonormal set of functions*. We often encounter such kinds of functions in quantum mechanics.

Example 2.16. Evaluate the integral,

$$I = \int_{-\pi}^{\pi} \sin(nx) \sin(mx)dx, \tag{2.133}$$

where n and m are positive integers. Show that the set of functions

$$\{\sin(x), \sin(2x), \sin(3x)...\}$$

form an *orthogonal set of functions*. Find the multiplying factor that makes this set of functions an *orthonormal set of functions*.

Solution:
(a) Recalling that

$$\sin(nx) = \frac{e^{inx} - e^{-inx}}{2i}, \quad \sin(mx) = \frac{e^{imx} - e^{-imx}}{2i} \tag{2.134}$$

we can write

$$\sin(nx)\sin(mx) = \left(\frac{e^{inx} - e^{-inx}}{2i}\right)\left(\frac{e^{imx} - e^{-imx}}{2i}\right)$$

$$= \frac{-e^{-i(n-m)x} - e^{i(n-m)x} + e^{i(n+m)x} + e^{-i(n+m)x}}{-4}$$

$$= \frac{1}{2}\left[\frac{e^{i(n-m)x} + e^{-i(n-m)x}}{2} - \frac{e^{i(n+m)x} + e^{-i(n+m)x}}{2}\right] \tag{2.135}$$

$$\Rightarrow \sin(nx)\sin(mx) = \frac{1}{2}\{\cos[(n-m)x] - \cos[(n+m)x]\}.$$

Then substituting this equation into equation (2.133), we have

$$I = \frac{1}{2}\int_{-\pi}^{\pi} \{\cos[(n-m)x] - \cos[(n+m)x]\}dx$$

$$\Rightarrow I = \frac{1}{2}\left\{\frac{\sin[(n-m)x]}{n-m} - \frac{\sin[(n+m)x]}{n+m}\right\}_{-\pi}^{\pi}. \tag{2.136}$$

The sine function is an odd function

$$\sin(-u) = -\sin(u). \tag{2.137}$$

This implies that

$$\sin[-(n-m)\pi] = -\sin[(n-m)\pi], \ \sin[-(n+m)\pi] = -\sin[(n+m)\pi] \tag{2.138}$$

and equation (2.136) becomes

$$I = \frac{\sin[(n-m)\pi]}{n-m} - \frac{\sin[(n+m)\pi]}{n+m}. \tag{2.139}$$

Since it is always true that

$$\frac{\sin(k\pi)}{k} = 0, \ \text{for } k = 1, 2, 3, \ldots, \tag{2.140}$$

for n and m positive integers and $n \neq m$, one can easily see that

$$\frac{\sin[(n-m)x]}{n-m} = \frac{\sin[(n+m)x]}{n+m} = 0. \tag{2.141}$$

Thus one can rewrite equation (2.139) as

$$I = \begin{cases} \dfrac{\sin[(n-n)\pi]}{n-n}, & n = m \\ 0, & n \neq m \end{cases}. \tag{2.142}$$

For $n = m$, we find 0/0 and one can apply L'Hospital's rule. Introducing the variable $u = n - m$, for $n = m$, we find

$$\frac{\sin[(n-m)\pi]}{n-m} = \lim_{u \to 0} \frac{\dfrac{d}{du}\sin(u\pi)}{\dfrac{d}{du}u} = \lim_{u \to 0} \frac{\pi \cos(u\pi)}{1} = \pi. \tag{2.143}$$

Therefore, the result of the integral is

$$I = \int_{-\pi}^{\pi} \sin(nx)\sin(mx)dx = \pi\delta_{n,m} = \begin{cases} \pi & n = m \\ 0 & n \neq m \end{cases} \tag{2.144}$$

which shows the sine functions form an orthogonal set of functions. To form an orthonormal set of functions one must multiply each sine function by a factor of $1/\sqrt{\pi}$. Thus it is important to note that

$$\{\sin(x), \ \sin(2x), \ \sin(3x), \ \sin(4x), \ \ldots\} \tag{2.145}$$

form an orthogonal set of functions while

$$\left\{ \frac{1}{\sqrt{\pi}} \sin(x), \frac{1}{\sqrt{\pi}} \sin(2x), \frac{1}{\sqrt{\pi}} \sin(3x), \frac{1}{\sqrt{\pi}} \sin(4x), \ldots \right\} \qquad (2.146)$$

form an orthonormal set of functions in the domain $[-\pi, \pi]$.

2.5 Complex numbers, functions and Mathematica

The following commands are the basics that can be used with complex numbers and functions.

- Re $[z]$ gives the real part of the complex number z.
- Im $[z]$ gives the imaginary part of the complex number z.
- Conjugate $[z]$ gives the complex conjugate of the complex number z.
- Abs $[z]$ gives the absolute value of the real or complex number z.
- ComplexExpand $[expr]$ expands a complex mathematical expression assuming that all variables are real. This expansion gives you the simplified form with the real and imaginary parts of the complex expression.
- Solve $[expr, vars, dom]$ solves the expression for the variables over the specified domain. Common choices of dom are Reals, Integers and Complexes.

In the following we do some of the examples from the previous sections using Mathematica

- Example 2.1

  $$z_1 = \frac{3 + 4i}{3 - 4i}$$

  $$-\frac{7}{25} + \frac{24i}{25}$$

  Conjugate[z_1]

  $$-\frac{7}{25} - \frac{24i}{25}$$

  Re[z_1]

  $$-\frac{7}{25}$$

  Im[z_1]

  $$\frac{24}{25}$$

  Abs[z_1]

  $$1$$

- Example 2.11
  ComplexExpand[Cos[iLog[2]]]
  $$\frac{5}{4}$$

- Example 2.12
  ComplexExpand[Log[1 − i]]

$$-\frac{i\pi}{4} + \frac{\text{Log}[2]}{2}$$

- Example 2.13
  ComplexExpand[(1 − i)^i]

$$e^{\pi/4}\text{Cos}\left[\frac{\text{Log}[2]}{2}\right] + ie^{\pi/4}\text{Sin}\left[\frac{\text{Log}[2]}{2}\right]$$

- Example 2.14 and Mathematica solution

$$\text{ArcSin}\left[\text{ComplexExpand}\left[\left(\frac{\sqrt{3}+I}{\sqrt{3}-I}\right)^{12}\right]\right]$$
$$\frac{\pi}{2}$$

- Example 2.10
  Solve[z^5 == 32.0, z, Complexes]

$$\{\{z \to -1.618\,03 - 1.175\,57i\}, \{z \to -1.618\,03 + 1.175\,57i\},$$
$$\{z \to 0.618\,034 - 1.902\,11i\}, \{z \to 0.618\,034 + 1.902\,11i\}, \{z \to 2.\,\}\}$$

  Solve[z^5 == 32.0, z, Reals]
  $\{\{z \to 2.\,\}\}$

2.6 Homework assignment

Problem 1. For the complex number

$$z = \sqrt{2}\,e^{-\frac{i\pi}{4}} \tag{2.147}$$

(a) find x, y, r, θ and write it in the form

$$z = (x, y),\ z = x + iy,\ z = (r, \theta),\ \text{and}\ z = r(\cos\theta + i\sin(\theta)), \tag{2.148}$$

(b) find z^* and also write it in the form
$$z = (x, y),\ z = x + iy,\ z = (r, \theta),\ \text{and}\ z = r(\cos\theta + i\sin(\theta)), \tag{2.149}$$

(c) show both z and z^* on the complex plane.

Problem 2. Write the complex expression

$$z = \frac{1}{(2 - 3i)^2} \tag{2.150}$$

in the form $z = a + bi$. Check your result using Mathematica.

Problem 3. Prove that

$$\left(\frac{z_1}{z_2}\right)^* = \frac{z_1^*}{z_2^*},$$

$$(z_1 z_2)^* = z_1^* z_2^*,$$

$$(z_1 \pm z_2)^* = z_1^* \pm z_2^*.$$

(2.151)

Hint: it is easier to prove the statements about the quotient and the product using polar coordinates representation of the complex numbers (i.e., $z = re^{i\theta}$); for the sum or difference it is easier to use the rectangular form $x + iy$.

Problem 4.
(a) Find the absolute value (the magnitude) of the complex number

$$z = (1 + 2i)^3.$$

(2.152)

(b) Solve for all possible values of the real numbers x and y in the equation

$$2ix + 3 = y - i.$$

(c) Check your result using Mathematica.

Problem 5. *Physical application*: A particle moves in the $(x - y)$ plane so that its position (x, y) as a function of time is given by

$$z = \cos(\omega t) + i\sin(\omega t),$$

where ω is known as the angular frequency and t is time and both are real.
(a) Find the velocity (v) and acceleration (a) of the particle

$$v = \frac{dz}{dt}, \quad a = \frac{d^2z}{dt^2}.$$

(b) Find the magnitude (absolute value) of the velocity and acceleration
(c) Describe the motion of the particle.

Problem 6. Test each of the following series for convergence.
(a)

$$\sum_{n=0}^{\infty} \frac{1}{(1+i)^n} = 1 + \frac{1}{(1+i)} + \frac{1}{(1+i)^2} + \cdots \frac{1}{(1+i)^n} + \cdots$$

(b)

$$\sum_{n=0}^{\infty}\left(\frac{1-i}{1+i}\right)^{n} = 1 + \frac{1-i}{1+i} + \left(\frac{1-i}{1+i}\right)^{2} + \ldots\left(\frac{1-i}{1+i}\right)^{n} + \ldots$$

Problem 7. Find the disk of convergence for each of the following power series of complex variable

(a)

$$\sum_{n=0}^{\infty}\frac{(n!)^{3}z^{n}}{(3n)!} = 1 + \frac{z}{3!} + \frac{(2!)^{3}z^{2}}{6!} + \ldots\frac{(n!)^{3}z^{n}}{(3n)!}\ldots$$

(b)

$$\sum_{n=1}^{\infty}\frac{(z-i)^{n}}{n} = \frac{z-i}{1} + \frac{(z-i)^{2}}{2} + \frac{(z-i)^{3}}{3} + \ldots\frac{(z-i)^{n}}{n}\ldots$$

Problem 8.

(a) Show that for any real y, $|e^{iy}| = 1$. Hence show that $|e^{z}| = e^{x}$ for every complex z

$$|e^{z}| = e^{x}.$$

(b) For the complex numbers z_1 and z_2, prove that

$$|z_1 z_2| = |z_1||z_2|$$

$$\left|\frac{z_1}{z_2}\right| = \frac{|z_1|}{|z_2|}.$$

Hint: write the numbers in $re^{i\theta}$ form.

(c) Using the relations in part (a) and (b) find the value for

$$W = \left|\frac{1-i}{1+i}\right|.$$

Problem 9. For the complex expression

$$z^5 = 32i,$$

(a) Find the values for z.
(b) Check your result using Mathematica.
(c) Show your results for z on the complex plane.

Problem 10. Find the formula for $\sin(3\theta)$ and $\cos(3\theta)$.
 Hint: refer to Example 2.9.

Problem 11. Find the values in rectangular form $x + iy$ for the complex expressions
 (a)

$$z = \sin(\pi - i \ln 3),$$

 (b)

$$z = \tan i,$$

 (c)

$$z = \cos(2i \ln i),$$

 (d)

$$z = \tanh^{-1}(i\sqrt{3}).$$

 (e) Check the results you found for (a)–(d) using Mathematica.

Problem 12. *Physical application*: Consider the RLC-ac circuit shown in figure 2.4. The circuit has a resistor and an inductor connected in series and then a capacitor in parallel with them (R and L in series, and then C in parallel with them).
 For an ac-source voltage given by

$$V(t) = V_{\max} \sin(\omega t)$$

from introductory physics, the impedance for a resistor (Z_R), capacitor (Z_C), and inductor (Z_L), are given by

$$Z_R = R, \ Z_C = 1/(i\omega C), \ \text{and} \ Z_L = i\omega L,$$

respectively. Note that R is the resistance of the resistor, C is the capacitance of the capacitor, L is the inductance of the inductor, and ω is the angular frequency of the

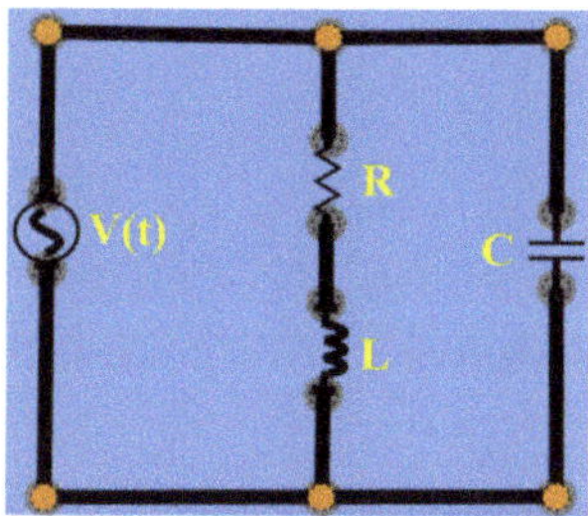

Figure 2.4. Ac RLC circuit. The resistor and the inductor are in series.

ac-source voltage. In ac circuit, for series connection of two impedances, Z_1 and Z_2 the equivalence impedance Z_{12} is given by

$$Z_{12} = Z_1 + Z_2$$

whereas for parallel connection

$$\frac{1}{Z_{12}} = \frac{1}{Z_1} + \frac{1}{Z_2}.$$

(a) Find the equivalent impedance of the circuit.
(b) Find the magnitude of the equivalent impedance of the circuit.
(c) A circuit is said to be at resonance when the total impedance of the circuit is real. Find ω in terms of R, L, and C at resonance.

IOP Publishing

Studies in Theoretical Physics, Volume 1
Fundamental mathematical methods
Daniel Erenso and Victor Montemayor

Chapter 3

Vectors

Some quantities used to describe the physical properties of an object requires a number and unit. For example, we can describe the mass, density, charge, and energy of an object as 100 kg, 1000 kg m^{-3}, 10 C, and 200 J, respectively. Such quantities are known as scalar quantities. However, other properties such as velocity, acceleration, force, electric field, magnetic field require direction, number, and units. Quantities that can only be described using magnitude (number and unit) and direction are vector quantities. This chapter introduces how to carry out vector algebra. We focus on how to add or multiply vectors. In vector multiplications, we shall consider multiplications resulting in scalar (inner product) or another vector (vector or cross product). We also see how we use vectors to define equations of a line or a plane.

3.1 Vector fundamentals

A vector is a quantity described by magnitude and direction and commonly represented using either a boldface letter (for example, **A**) or letters with an arrow on top of it ($\vec{A}$). Here we will use the second way of representation. A vector generally can have N components depending on the existing number of dimensions. For example, in the case of three-dimensional space, a vector may have a maximum three non-zero components. The components of a vector depend on the type of coordinate system used to represent the vector. In Cartesian coordinates, a vector $\vec{A}$ has three components along the x-direction A_x, the y-direction A_y, and the z-direction A_z. The complete description of a vector, besides the components, requires a set of three unit vectors that are perpendicular to each other. Unit vector is a vector with a magnitude of one that is pointing along a given direction. We represent unit vectors using letters with a hat. For example, in the Cartesian coordinates system, the unit vectors directed in the positive-x, positive-y, and positive-z directions are represented by $\hat{x}$, $\hat{y}$, and $\hat{z}$, respectively. In some textbooks, these unit vectors are represented by i, j, and k. We can express a vector $\vec{A}$ in terms of these unit vectors as

$$\vec{A} = A_x\hat{x} + A_y\hat{y} + A_z\hat{z} = A_x i + A_y j + A_z k \qquad (3.1)$$

doi:10.1088/978-0-7503-3135-7ch3

or simply

$$\vec{A} = \left(A_x, A_y, A_z\right). \tag{3.2}$$

Unit vectors are not limited to only unit vectors along x-, y-, and z-directions. We can define a unit vector along any vector direction. The unit vector along the direction of the vector $\vec{A}$ denoted by $\hat{A}$ is given by

$$\hat{A} = \frac{\vec{A}}{|\vec{A}|}, \tag{3.3}$$

where $|\vec{A}|$ is the magnitude of the vector $\vec{A}$ shown in figure 3.1 is given by

$$A = \sqrt{A_x^2 + A_y^2 + A_z^2} \qquad \text{(3–Dimensions).} \tag{3.4}$$

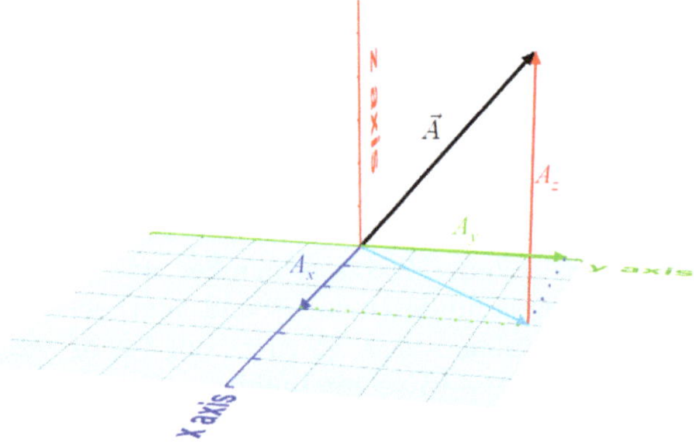

Figure 3.1. Components of a 3-dimensional vector in Cartesian coordinates.

For a 2-dimensional vector shown in figure 3.2 one can write the magnitude as

$$A = \sqrt{A_x^2 + A_y^2} \qquad \text{(2–Dimensions).}$$

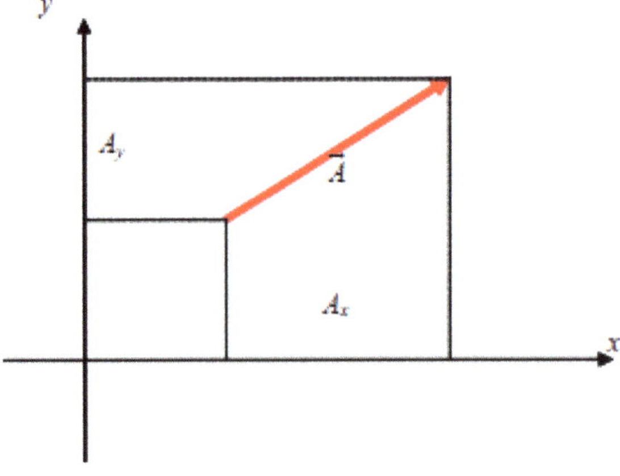

Figure 3.2. A two-dimensional vector and its components on the x–y plane.

3.2 Vector addition

Consider the 3-dimensional vectors $\vec{A}$ and $\vec{B}$ with components in Cartesian coordinates

$$\vec{A} = A_x\hat{x} + A_y\hat{y} + A_z\hat{z}, \; \vec{B} = B_x\hat{x} + B_y\hat{y} + B_z\hat{z}. \tag{3.5}$$

The addition of these two vectors gives a vector $\vec{C}$

$$\begin{aligned} \vec{C} &= \vec{A} + \vec{B} \\ &= A_x\hat{x} + A_y\hat{y} + A_z\hat{z} + B_x\hat{x} + B_y\hat{y} + B_z\hat{z} \\ &= (A_x + B_x)\hat{x} + (A_y + B_y)\hat{y} + (A_z + B_z)\hat{z}, \end{aligned} \tag{3.6}$$

with components given by

$$C_x = A_x + B_x, \; C_y = A_y + B_y, \; C_z = A_z + B_z. \tag{3.7}$$

Geometrically a vector is represented by an arrow with a length proportional to its magnitude. In such representation, the vector addition

$$\vec{C} = \vec{A} + \vec{B}$$

can be made by first placing the tail of vector $\vec{B}$ at the head of vector $\vec{A}$ then drawing an arrow pointing from the tail of vector $\vec{A}$ to the head of vector $\vec{B}$ as shown in figure 3.3. Vector addition is commutative. This means the vector $\vec{C}$ obtained by placing the tail of vector $\vec{A}$ at the head of vector $\vec{B}$ and then drawing an arrow pointing from the tail of vector $\vec{B}$ to the head of vector $\vec{A}$ is the same as the vector $\vec{C}$ shown in figure 3.3.

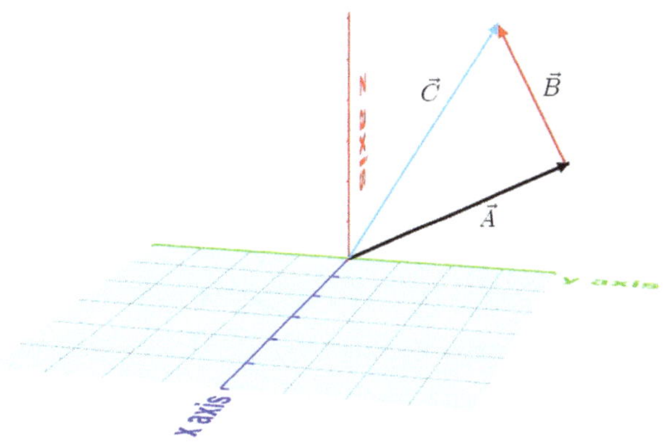

Figure 3.3. Addition of two vectors.

3.3 Vector multiplication

We can multiply two vectors in two different ways. One way of multiplying two vectors results in a scalar. Such multiplication of vectors is known as a scalar product (or dot product, or inner product). The other way of multiplying two vectors leads to another vector and is known as the cross product (vector product).

A. *The scalar product*: Let us consider the two vectors

$$\vec{A} = A_x\hat{x} + A_y\hat{y} + A_z\hat{z}, \ \vec{B} = B_x\hat{x} + B_y\hat{y} + B_z\hat{z},$$

shown in figure 3.4. The scalar product of these two vectors, represented by $\vec{A} \cdot \vec{B}$, is given by

$$\vec{A} \cdot \vec{B} = A_x B_x + A_y B_y + A_z B_z. \tag{3.8}$$

In terms of the magnitude of the two vectors and the angle θ see (figure 3.4), the scalar product of these two vectors can also be expressed as

$$\vec{A} \cdot \vec{B} = |\vec{A}||\vec{B}| \ \cos\theta. \tag{3.9}$$

In view of equation (3.9), for the Cartesian unit vectors $\hat{x}$, $\hat{y}$, and $\hat{z}$, one can easily find that

$$\hat{x} \cdot \hat{y} = \hat{x} \cdot \hat{z} = \hat{z} \cdot \hat{y} = 0,$$
$$\hat{x} \cdot \hat{x} = \hat{y} \cdot \hat{y} = \hat{z} \cdot \hat{z} = 1, \tag{3.10}$$

where we used the fact that the angle between the Cartesian unit vectors is 90°. A scalar product of two vectors in terms of their components and the magnitudes of the two vectors are often used to determine the direction (the angle) of one vector as measured relative to the other,

$$\theta = \cos^{-1}\left(\frac{\vec{A} \cdot \vec{B}}{|\vec{A}||\vec{B}|}\right) = \cos^{-1}\left(\frac{A_x B_x + A_y B_y + A_z B_z}{|\vec{A}||\vec{B}|}\right). \tag{3.11}$$

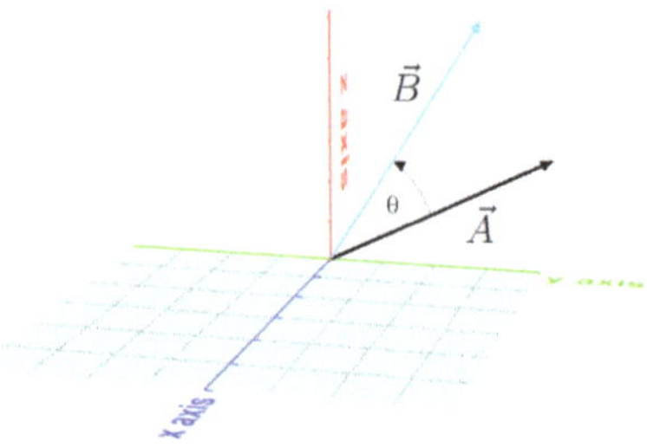

Figure 3.4. Vector $\vec{B}$. directed at angle θ measured in a counterclockwise direction from vector $\vec{A}$.

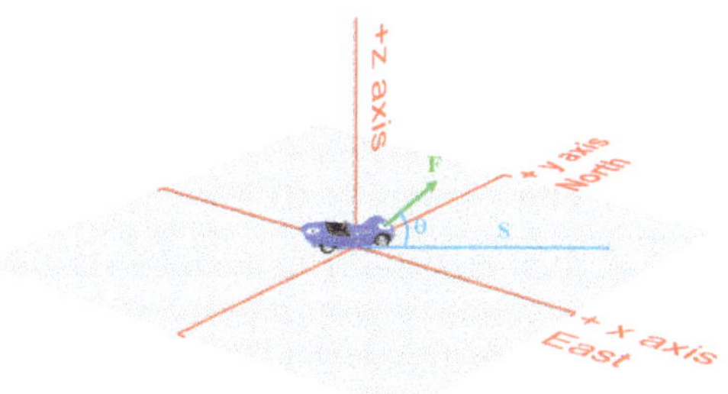

Figure 3.5. A force $\vec{F}$ pulling a car.

Application: Suppose we pull the car shown in figure 3.5 with a rope by applying a force $\vec{F}$ and displaced it by $\vec{S}$. The force is directed at an angle θ measured from the direction of the displacement. The work done W, which is a scalar quantity, is given by the scalar product of the force and the displacement vector

$$W = |\vec{F}||\vec{S}| \, \cos\,(\theta) = \vec{F} \cdot \vec{S}. \tag{3.12}$$

B. *The vector product (the cross product)*: The vector product of the two vectors $\vec{A}$ and $\vec{B}$ shown in figure 3.4 is represented by $\vec{A} \times \vec{B}$. It results in a third vector $\vec{C}$ perpendicular to the plane formed by the two vectors. If the angle between the two vectors is θ, the *magnitude* of $\vec{C}$ is given by

$$C = |\vec{A} \times \vec{B}| = |\vec{A}||\vec{B}| \, \sin\,(\theta) \tag{3.13}$$

and the direction can be determined using the *right-hand rule.*
C. *The right-hand rule*: To find the direction of $\vec{C} = \vec{A} \times \vec{B}$, using our right hand we point our index finger along the direction of the first vector ($\vec{A}$), the middle finger along the direction of the second vector ($\vec{B}$), then our extended thumb points in the direction of vector $\vec{C}$.
D. *Right-handed coordinate systems*: Generally all the coordinate systems we use in Physics are all right-hand coordinate system in which the right-hand rule is valid. For example, in a Cartesian coordinate system we have

$$\hat{x} \times \hat{y} = \hat{z}, \, \hat{y} \times \hat{z} = \hat{x}, \, \hat{z} \times \hat{x} = \hat{y}. \tag{3.14}$$

In view of equation (3.13), one can also see that

$$\hat{x} \times \hat{x} = \hat{y} \times \hat{y} = \hat{z} \times \hat{z} = 0. \tag{3.15}$$

Using the relations in equations (3.14) and (3.15) we can find the components of the vector $\vec{C} = \vec{A} \times \vec{B}$. To this end, we note that

$$\vec{C} = \vec{A} \times \vec{B} = \left(A_x\hat{x} + A_y\hat{y} + A_z\hat{z}\right) \times \left(B_x\hat{x} + B_y\hat{y} + B_z\hat{z}\right)$$
$$= A_xB_x(\hat{x} \times \hat{x}) + A_xB_y(\hat{x} \times \hat{y}) + A_xB_z(\hat{x} \times \hat{z})$$
$$+ A_yB_x(\hat{y} \times \hat{x}) + A_yB_y(\hat{y} \times \hat{y}) + A_yB_z(\hat{y} \times \hat{z}) \tag{3.16}$$
$$+ A_zB_x(\hat{z} \times \hat{x}) + A_zB_y(\hat{z} \times \hat{y}) + A_zB_z(\hat{z} \times \hat{z}),$$

and using the relations in equations (3.14) and (3.15), we find

$$\vec{C} = \vec{A} \times \vec{B} = A_xB_y\hat{z} + A_xB_z(-\hat{y})$$
$$+ A_yB_x(-\hat{z}) + A_yB_z(\hat{x}) + A_zB_x(\hat{y}) + A_zB_y(-\hat{x}) \tag{3.17}$$
$$\Rightarrow \vec{C} = C_x\hat{x} + C_y\hat{y} + C_z\hat{z},$$

where

$$C_x = A_yB_z - A_zB_y,$$
$$C_y = A_zB_x - A_xB_z, \tag{3.18}$$
$$C_z = A_xB_y - A_yB_z.$$

Application: Consider a mass m attached to a rigid rod with negligible mass. This mass is acted by a force $\vec{F}$ as shown in figure 3.6. The mass's position is described by a vector $\vec{r}$ measured from the z-axis (axis of rotation). The magnitude of the torque τ due to this force, from introductory physics, is given by

$$\tau = r_\perp F = rF\sin(\theta) = |\vec{r}||\vec{F}|\sin(\theta) = |\vec{r} \times \vec{F}|,$$

which generally expressed, in a vector form, as

$$\vec{\tau} = \vec{r} \times \vec{F}.$$

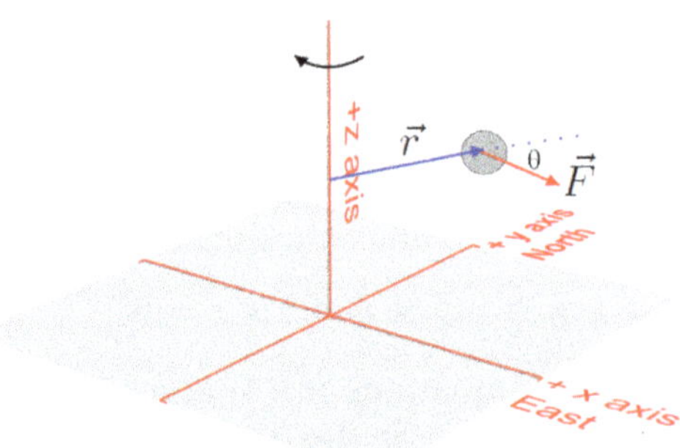

Figure 3.6. A mass m attached to a massless ridged rod rotating about the z-axis due to a force $\vec{F}$.

Example 3.1. A force of magnitude $F = 4.0N$ pointing in the *positive-x direction* is applied to a large solid object at the position $\vec{r} = (0, 2, 0)m$. Find the resulting torque exerted on the object about the origin using the
    (a) magnitude and direction form of the cross product,
    (b) component form of the cross product.

Solution:
    (a) The torque is given by

$$\vec{\tau} = \vec{r} \times \vec{F}, \tag{3.19}$$

where the force vector, pointing along the positive x-direction, is given by

$$\vec{F} = (4, 0, 0)N = 4N\hat{x} \tag{3.20}$$

and the position vector

$$\vec{r} = (0, 2, 0)m = 2m\hat{y}. \tag{3.21}$$

We use the relation for the magnitude of the cross product of two vectors given by

$$|\vec{\tau}| = |\vec{r}||\vec{F}| \sin(\theta), \tag{3.22}$$

where θ is the angle between the position and the force vector. The magnitude of the position ($|\vec{r}|$) and force ($|\vec{F}|$) vectors are

$$|\vec{r}| = \sqrt{r_x^2 + r_y^2 + r_z^2} = 2m \tag{3.23}$$

and

$$|\vec{F}| = \sqrt{F_x^2 + F_y^2 + F_z^2} = 4N. \tag{3.24}$$

Noting that the angle between $\vec{r}$ and $\vec{F}$ is

$$\theta = \frac{\pi}{2}, \tag{3.25}$$

we find for the magnitude of the torque,

$$|\vec{\tau}| = |\vec{F}||\vec{r}| \sin(\theta) = 8Nm. \tag{3.26}$$

Using the right-hand rule, we find the direction of the torque to be along the negative z-direction. Then one can write torque as

$$\vec{\tau} = -8Nm\hat{z}. \tag{3.27}$$

(b) The force vector, pointing along the positive x-direction is

$$\vec{F} = (4,\, 0,\, 0)N = 4N\hat{x} \tag{3.28}$$

and the position vector is

$$\vec{r} = (0,\, 2,\, 0)m = 2m\hat{y}. \tag{3.29}$$

Then using the vector form of the torque,

$$\vec{\tau} = \vec{r} \times \vec{F}. \tag{3.30}$$

In view of the relations in equation (3.17) and (3.18), one can write

$$\vec{\tau} = \vec{r} \times \vec{F} = \tau_x\hat{x} + \tau_y\hat{y} + \tau_z\hat{z}, \tag{3.31}$$

where

$$\begin{aligned} \tau_x &= r_yF_z - r_zF_y,\ \tau_y = r_zF_x - r_xF_z, \\ \tau_z &= r_xF_y - r_yF_z. \end{aligned} \tag{3.32}$$

Noting that $r_x = r_z = 0$, $r_y = 2\,m$, and $F_x = 4N$, $F_z = F_y = 0$, one finds,

$$\tau_x = 0,\ \tau_y = 0,\ \tau_z = -8Nm\hat{z}. \tag{3.33}$$

As we will see in chapter 4, we can use the determinant to find the vector product of two vectors,

$$\vec{A} \times \vec{B} = \begin{vmatrix} \hat{x} & \hat{y} & \hat{z} \\ A_x & A_y & A_z \\ B_x & B_y & B_z \end{vmatrix} \tag{3.34}$$

which gives

$$\vec{\tau} = \vec{r} \times \vec{F} = \begin{vmatrix} \hat{x} & \hat{y} & \hat{z} \\ 0 & 2 & 0 \\ 4 & 0 & 0 \end{vmatrix} = \hat{x}\begin{vmatrix} 2 & 0 \\ 0 & 0 \end{vmatrix} - \hat{y}\begin{vmatrix} 0 & 0 \\ 4 & 0 \end{vmatrix} + \hat{z}\begin{vmatrix} 0 & 2 \\ 4 & 0 \end{vmatrix} \tag{3.35}$$

$$\vec{\tau} = \vec{r} \times \vec{F} = -8Nm\hat{z}.$$

3.4 Vectors and equations of a line and a plane

Equation of a line

Let us consider two points on the x–y plane with coordinates (x_0, y_0) and (x_1, y_1). From introductory algebra, a point with coordinates (x, y) on a line connecting these points (see figure 3.7) is defined by the equation

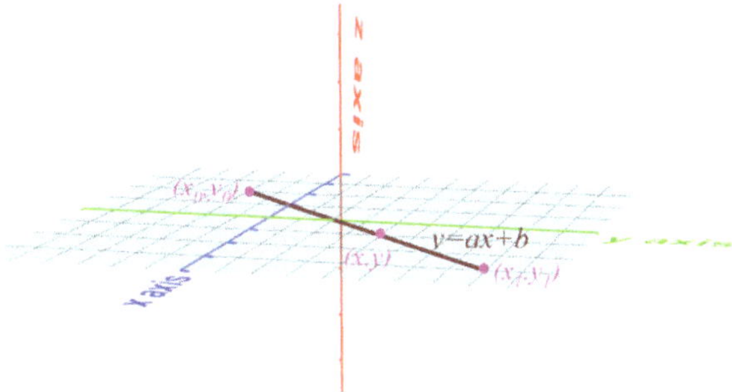

Figure 3.7. A line on the x–y plane.

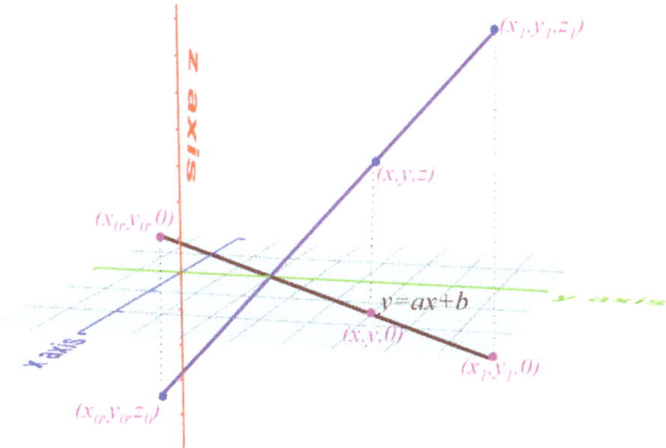

Figure 3.8. A line in 3-dimensional space.

$$\frac{y - y_0}{x - x_0} = \frac{y_1 - y_0}{x_1 - x_0} = m_1 \Rightarrow \frac{y - y_0}{m_1(x - x_0)} = 1, \tag{3.36}$$

which is commonly expressed as

$$\frac{y - y_0}{x - x_0} = m_1 \Rightarrow y - y_0 = m_1(x - x_0) \Rightarrow y = m_1 x + a_1, \tag{3.37}$$

where

$$m_1 = \frac{y_1 - y_0}{x_1 - x_0}, \tag{3.38}$$

is the slope and

$$a_1 = y_0 - m_1 x_0, \tag{3.39}$$

is the y-intercept of the line.

For the 3-dimensional case shown in figure 3.8, clearly the projection of the blue line on the x–y plane is defined by equation (3.36). Following a similar approach one can easily see that for the projection of the same blue line on the x–z plane,

$$\frac{z - z_0}{x - x_0} = \frac{z_1 - z_0}{x_1 - x_0} = m_2 \Rightarrow \frac{z - z_0}{m_2(x - x_0)} = 1, \tag{3.40}$$

and y–z plane

$$\frac{z - z_0}{y - y_0} = \frac{z_1 - z_0}{y_1 - y_0} = m_3 \Rightarrow \frac{z - z_0}{m_3(y - y_0)} = 1. \tag{3.41}$$

Now combining equations (3.40) and (3.41), one can write

$$\frac{z - z_0}{m_2(x - x_0)} = \frac{z - z_0}{m_3(y - y_0)} \Rightarrow \frac{y - y_0}{m_2} = \frac{x - x_0}{m_3}, \tag{3.42}$$

and combining equations (3.36) and (3.40)

$$\frac{y - y_0}{m_1(x - x_0)} = \frac{z - z_0}{m_2(x - x_0)} \Rightarrow \frac{y - y_0}{m_1} = \frac{z - z_0}{m_2} \tag{3.43}$$

$$\Rightarrow y - y_0 = \frac{\dfrac{z - z_0}{m_2}}{m_1}. \tag{3.44}$$

Then from equations (3.42) and (3.43) follows

$$\frac{x - x_0}{m_3} = \frac{y - y_0}{m_2} = \frac{z - z_0}{\dfrac{m_2}{m_1}} \tag{3.45}$$

or

$$\frac{x - x_0}{a} = \frac{y - y_0}{b} = \frac{z - z_0}{c} \tag{3.46}$$

where we introduced the constants $a = m_3$, $b = m_2$, and $c = m_2/m_1$. This equation for a line in a 3-dimensional space is known as *symmetric equations. In a 2-dimensional case where*, for example $c = 0$, we should write the symmetric equation for a line as

$$\frac{x - x_0}{a} = \frac{y - y_0}{b}, \; z = z_0. \tag{3.47}$$

One can use vectors to define an equation of a line. The point P_0 with the coordinates (x_0, y_0, z_0) can be described by a vector

$$\vec{r}_0 = x_0\hat{x} + y_0\hat{y} + z_0\hat{z}. \tag{3.48}$$

Similarly, any other point P with coordinates (x, y, z) can also be described by a vector

$$\vec{r} = x\hat{x} + y\hat{y} + z\hat{z}. \tag{3.49}$$

Then using the vector $\vec{R}$ directed from point P_0 to point P, one can write (figure 3.9)

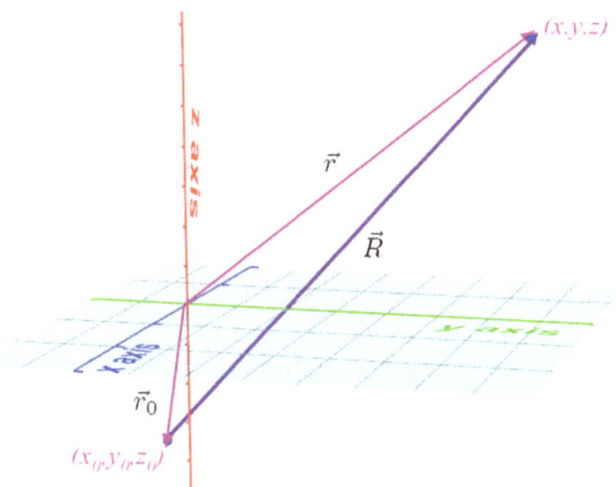

Figure 3.9. Vector representation of 3-D line.

$$\vec{r} = \vec{r}_0 + \vec{R}. \tag{3.50}$$

In terms of the unit vector $\hat{R}$, we note that

$$\vec{R} = R\hat{R}. \tag{3.51}$$

For a line the unit vector $\hat{R}$ remains constant. By changing the magnitude and keeping the direction the same for the vector $\vec{R}$, the point defined by the vector $\vec{r}$ would be confined on a line. Then one can express the vector $\vec{R}$ as

$$\vec{R} = t\vec{A}, \tag{3.52}$$

where t is some parameter and

$$\vec{A} = a\hat{x} + b\hat{y} + c\hat{z}. \tag{3.53}$$

We can easily check that the unit vector $\hat{R}$ is independent of the parameter t and remains a constant. Noting that

$$R = |t\vec{A}| = t\,|\vec{A}| = t\sqrt{a^2 + b^2 + c^3}, \tag{3.54}$$

we find

$$\hat{R} = \frac{\vec{R}}{R} = a\hat{x} + b\hat{y} + c\hat{z}, \tag{3.55}$$

which is a constant and independent of the parameter t. Then equation of a line for a 3-dimensional case can be expressed as

$$\begin{aligned}
\vec{r} &= \vec{r}_0 + \vec{R}(t) = \vec{r}_0 + t\vec{A} \\
&= x_0\hat{x} + y_0\hat{y} + z_0\hat{z} + at\hat{x} + bt\hat{y} + ct\hat{z} \\
&= (x_0 + at)\hat{x} + \left(y_0 + bt\right)\hat{y} + (z_0 + ct)\hat{z},
\end{aligned} \tag{3.56}$$

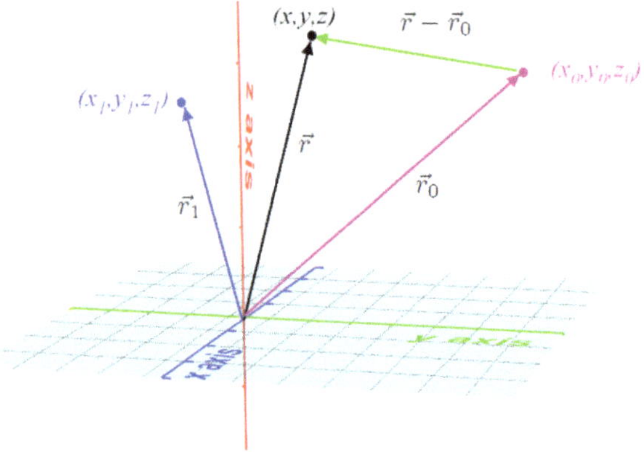

Figure 3.10. A plane formed by three points in space.

which we can rewrite in the form

$$\vec{r} = \vec{r}_0 + \vec{A}t \Rightarrow \begin{cases} x = x_0 + at, \\ y = y_0 + bt, \\ z = z_0 + ct. \end{cases} \tag{3.57}$$

These equations are known as *parametric equations*. One can easily show from these equations that

$$t = \frac{x - x_0}{a} = \frac{y - y_0}{b} = \frac{z - z_0}{c}, \tag{3.58}$$

which is the symmetric equations.

Equation of a plane

To form a plane, we need three points in space. Let us consider the three points shown in figure 3.10. The first point described by the coordinates (x_0, y_0, z_0) can be represented by a vector

$$\vec{r}_0 = x_0\hat{x} + y_0\hat{y} + z_0\hat{z}, \tag{3.59}$$

the second point with coordinates (x_1, y_1, z_1) is represented by a vector

$$\vec{r}_1 = x_1\hat{x} + y_1\hat{y} + z_1\hat{z}, \tag{3.60}$$

and the third point with coordinates (x, y, z) is also represented by a vector

$$\vec{r} = x\hat{x} + y\hat{y} + z\hat{z}. \tag{3.61}$$

The vector product of the two vectors $\vec{r}_0$ and $\vec{r}_1$ is given by

$$\vec{r}_0 \times \vec{r}_1 = (y_0 z_1 - y_1 z_0)\hat{x} + (z_0 x_1 - x_1 z_0)\hat{y} + (x_0 y_1 - y_0 x_1)\hat{z},$$

which we rewrite as

$$\vec{N} = \vec{r_0} \times \vec{r_1} = a\hat{x} + b\hat{y} + c\hat{z}, \tag{3.62}$$

where

$$a = y_0 z_1 - y_1 z_0, \ b = z_0 x_1 - x_1 z_0, \ c = x_0 y_1 - y_0 x_1. \tag{3.63}$$

We recall the vector product of two vectors gives a third vector normal to the plane formed by the two vectors. This means the vector $\vec{N}$ is normal to the plane formed by the two vectors $\vec{r_1}$ and $\vec{r_0}$. Then a vector $\vec{R}$ defined by

$$\vec{R} = \vec{r} - \vec{r_0} = (x - x_0)\hat{x} + (y - y_0)\hat{y} + (z - z_0)\hat{z}, \tag{3.64}$$

that belongs to the plane formed by the vectors $\vec{r_1}$ and $\vec{r_0}$ must be perpendicular to the vector $\vec{N} = \vec{r_0} \times \vec{r_1}$. For two vectors perpendicular to one another, the scalar product is zero. Therefore

$$\vec{R} \cdot \vec{N} = a(x - x_0) + b(y - y_0) + c(z - z_0) = 0, \tag{3.65}$$

which is the equation of the plane formed by the three points shown in figure 3.10. We can also put this equation in the form

$$ax + by + cz = d, \tag{3.66}$$

where

$$d = ax_0 + by_0 + cz_0. \tag{3.67}$$

Example 3.2. The position of a particle at time $t = 0$ is given by $(1, 2, 3)m$. The particle moves with a constant velocity $(0, 2, 1)\text{ms}^{-1}$. What is the equation of the particle's trajectory?

Solution: The initial position of the particle can be expressed, using Cartesian unit vectors, as

$$\vec{r_0} = \hat{x} + 2\hat{y} + 3\hat{z}, \tag{3.68}$$

at a later time t, let the position of the particle be described by the vector

$$\vec{r} = x(t)\hat{x} + y(t)\hat{y} + z(t)\hat{z}. \tag{3.69}$$

Since the particle is traveling with a constant velocity, the vector

$$\vec{r} - \vec{r_0} = [x(t) - 1]\hat{x} + [y(t) - 2]\hat{y} + [z(t) - 3]\hat{z}, \tag{3.70}$$

must be proportional to the velocity vector

$$\vec{v} = 0\hat{x} + (2\text{ms}^{-1})\hat{y} + (1\text{m s}^{-1})\hat{z}. \tag{3.71}$$

Which means, we can write

$$
\begin{aligned}
\vec{r} - \vec{r}_0 &= t\vec{v} \Rightarrow [x(t) - 1]\hat{x} + [y(t) - 2]\hat{y} + [z(t) - 3]\hat{z} \\
&= \left(2\text{ms}^{-1}\right)t\hat{y} + \left(1\text{ms}^{-1}\right)t\hat{z} \\
\Rightarrow \vec{r} &= 1\hat{x} + 2\hat{y} + 3\hat{z} + \left(2\text{ms}^{-1}\right)t\hat{y} + \left(1\text{ms}^{-1}\right)t\hat{z} \\
\Rightarrow \vec{r}(t) &= \vec{r}_0 + \vec{v}t,
\end{aligned}
\tag{3.72}
$$

where

$$
\vec{r}_0 = \hat{x} + 2\hat{y} + 3\hat{z}, \quad \vec{v} = (2\text{ms}^{-1})\hat{y} + (1\text{ms}^{-1})\hat{z}.
\tag{3.73}
$$

Example 3.3. Consider the particle in the previous example. What is the distance of closest approach (the 'impact parameter') of the particle to the origin?

Solution: Let the point along the trajectory of the particle closest to the origin be (x', y', z') (see figure 3.11) be

$$
\vec{r}' = x'\hat{x} + y'\hat{y} + z'\hat{z}.
\tag{3.74}
$$

For the closest distance, which is given by $|\vec{r}'|$, the vector $\vec{r}'$ must be normal to the vector that defines the trajectory of the particle. Noting that the vector $\vec{r}' - \vec{r}_0$ defines the trajectory of the particle, for the closest approach to the origin, one can write

$$
\vec{r}' \cdot (\vec{r}' - \vec{r}_0) = 0.
\tag{3.75}
$$

Since $\vec{r}'$ describes a point on the trajectory of the particle defined by

$$
\vec{r}(t) = \vec{r}_0 + \vec{v}t,
\tag{3.76}
$$

where the particle becomes closest to the origin, we may write

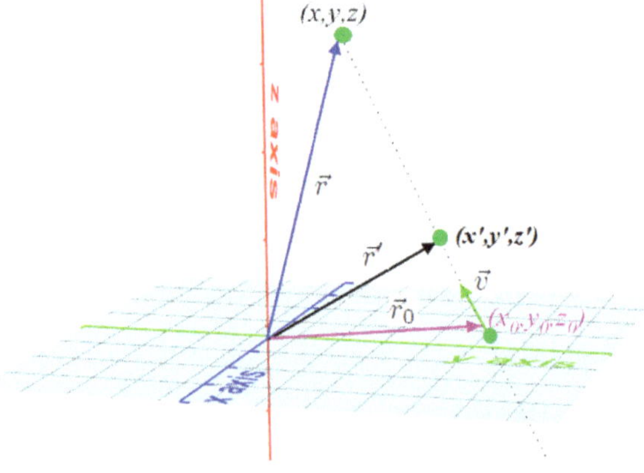

Figure 3.11. A particle moving with a constant velocity, $\vec{v}$.

$$\vec{r}' = \vec{r}_0 + \vec{v}t' \Rightarrow \vec{r}' - \vec{r}_0 = \vec{v}t', \tag{3.77}$$

so that

$$(\vec{r}_0 + \vec{v}t') \cdot \vec{v}t' = (\vec{r}_0 \cdot \vec{v})t' + (\vec{v} \cdot \vec{v})t'^2 = 0. \tag{3.78}$$

Noting that

$$\begin{aligned}
\vec{v} \cdot \vec{v} &= (2\text{ms}^{-1})^2 + (1\text{ms}^{-1})^2 = 5(\text{ms}^{-1})^2, \\
\vec{r}_0 \cdot \vec{v} &= r_{0x}v_x + r_{0y}v_y + r_{0z}v_z = 2 \times 2 + 3 \times 1 = 7\text{m}^2\text{s}^{-1},
\end{aligned} \tag{3.79}$$

for

$$(\vec{r}_0 \cdot \vec{v})t' + (\vec{v} \cdot \vec{v})t'^2 = 0, \tag{3.80}$$

one finds

$$7t' + 5t'^2 = 0 \Rightarrow t' = 0, \ t' = -\frac{7}{5}. \tag{3.81}$$

The acceptable value is $t' = 0$. Thus the shortest distance from the origin is given by the magnitude of the vector

$$\vec{r}' = \vec{r}_0 + \vec{v}t' = \vec{r}_0 \Rightarrow r' = \sqrt{1^2 + 2^2 + 3^2} = \sqrt{14}, \tag{3.82}$$

which means the particle is already at the closest distance to begin with.

Example 3.4. Find the symmetric and parametric equations of the line through $(0, -2, 4)$ and $(3, -2, 1)$.

Solution: From equation (3.57), the symmetric equations of a line are given by

$$\frac{x - x_0}{a} = \frac{y - y_0}{b} = \frac{z - z_0}{c}, \tag{3.83}$$

where

$$\vec{A} = a\hat{x} + b\hat{y} + c\hat{z}, \tag{3.84}$$

is a vector parallel to the line passing through $(0, -2, 4)$ and $(3, -2, 1)$. Noting that this vector can be

$$\begin{aligned}
\vec{A} &= (3 - 0)\hat{x} + (-2 + 2)\hat{y} + (1 - 4)\hat{z} \\
&\Rightarrow \vec{A} = 3\hat{x} + 0\hat{y} - 3\hat{z},
\end{aligned} \tag{3.85}$$

we find

$$a = 3, \, b = 0, \, c = -3. \tag{3.86}$$

Thus in view of equation (3.57), the symmetric equations for the line becomes

$$\frac{x - 0}{3} = -\frac{z - 4}{3}, \, y = -2. \tag{3.87}$$

The parametric equations using equation (3.58) can be written as

$$\vec{r} = \vec{r_0} + \vec{A}t \text{ or } \begin{cases} x = x_0 + at \\ y = y_0 + bt \\ z = z_0 + ct \end{cases} \Rightarrow \vec{r} = \begin{cases} x = 3t, \\ y = -2, \\ z = 4 - 3t. \end{cases} \tag{3.88}$$

Example 3.5. Find the shortest distance from the origin to the plane

$$3x - 2y - 6z = 7. \tag{3.89}$$

Solution: Comparing the equation,

$$3x - 2y - 6z = 7, \tag{3.90}$$

to the generic equation for a plane

$$ax + by + cz = d, \tag{3.91}$$

and taking into account the results in equations (3.62)–(3.67), a vector that is normal to the plane can be written as

$$\vec{N} = 3\hat{x} - 2\hat{y} - 6\hat{z}. \tag{3.92}$$

The unit vector directed along this normal vector is given by

$$\hat{N} = \frac{\vec{N}}{|\vec{N}|} = \frac{3\hat{x} - 2\hat{y} - 6\hat{z}}{\sqrt{3^2 + (-2)^2 + (-6)^2}}$$

$$\Rightarrow \hat{N} = \frac{3}{7}\hat{x} - \frac{2}{7}\hat{y} - \frac{6}{7}\hat{z}. \tag{3.93}$$

We now pick any point on the plane with coordinates (x_0, y_0, z_0) as shown in figure 3.12. This point must satisfy the equation

$$3x - 2y - 6z = 7. \tag{3.94}$$

Let this point be $(3, 1, 0)$, which we write as

$$\vec{r_0} = 3\hat{x} + 1\hat{y}. \tag{3.95}$$

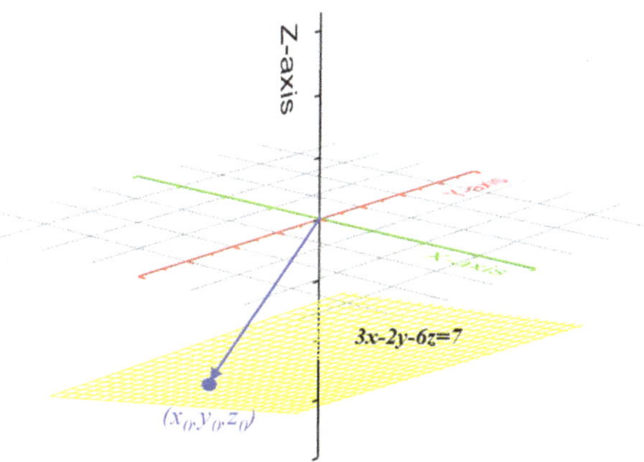

Figure 3.12. A plane defined by the equation $3x - 2y - 6z = 7$.

Then the shortest distance from the origin to the plane is the component of the vector $\vec{r}_0$ along the direction of $\hat{N}$. We can then write this distance as

$$d = \vec{r}_0 \cdot \hat{N} = \frac{3 \times 3}{7} - \frac{2 \times 1}{7} \Rightarrow d = 1 \text{units}. \tag{3.96}$$

Example 3.6. Find the equation of the plane containing the three points $(0, 1, 1)$, $(2, 1, 3)$, and $(4, 4, 1)$.

Solution: We first construct two vectors $\vec{A}$ and $\vec{B}$ that lie on the plane formed by these three points and let these vectors be the vectors pointing from $(0, 1, 1)$ to $(2, 1, 3)$ and from $(0, 1, 1)$ to $(4, 4, 1)$, respectively. Then we can write

$$\begin{aligned} \vec{A} &= (2 - 0)\hat{x} + (1 - 1)\hat{y} + (3 - 1)\hat{z} = 2\hat{x} + 2\hat{z}, \\ \vec{B} &= (4 - 0)\hat{x} + (4 - 1)\hat{y} + (1 - 1)\hat{z} = 4\hat{x} + 3\hat{y}. \end{aligned} \tag{3.97}$$

The vector $\vec{N}$ that is normal to the plane formed by these two vectors (i.e., the three points) can be determined using the cross product of the two vectors

$$\vec{N} = \vec{A} \times \vec{B} = \begin{vmatrix} \hat{x} & \hat{y} & \hat{z} \\ 2 & 0 & 2 \\ 4 & 3 & 0 \end{vmatrix}$$

$$\Rightarrow \vec{N} = \hat{x} \begin{vmatrix} 0 & 2 \\ 3 & 0 \end{vmatrix} - \hat{y} \begin{vmatrix} 2 & 2 \\ 4 & 0 \end{vmatrix} + \hat{z} \begin{vmatrix} 2 & 0 \\ 4 & 3 \end{vmatrix} \tag{3.98}$$

$$\Rightarrow \vec{N} = -6\hat{x} + 8\hat{y} + 6\hat{z}.$$

We recall that the equation of a plane is given by

$$a(x - x_0) + b(y - y_0) + c(z - z_0) = 0, \tag{3.99}$$

where

$$\vec{N} = a\hat{x} + b\hat{y} + c\hat{z} \tag{3.100}$$

and

$$\vec{r}_0 = x_0\hat{x} + y_0\hat{y} - z_0\hat{z} \tag{3.101}$$

is a point on the plane as measured from the origin. Then we can write the equation of the plane formed by the three points as

$$-6(x - x_0) + 8(y - y_0) + 6(z - z_0) = 0, \tag{3.102}$$

where

$$\vec{r}_0 = x_0\hat{x} + y_0\hat{y} - z_0\hat{z} \tag{3.103}$$

which can take the values $\vec{r}_0 = (0,\ 1,\ 1)$, $(2,\ 1,\ 3)$, or $(4,\ 4,\ 1)$.

3.5 Vectors and Mathematica

In Mathematica we can carry out the following basic vector operations.
   A vector in three dimensions.

$$\begin{aligned} In:\ & v = \{1,\ 3,\ 2\} \\ Out:\ & \{1,\ 3,\ 2\} \end{aligned} \tag{3.104}$$

A vector u in the direction opposite to v with twice the magnitude.

$$\begin{aligned} In:\ & u = -2v \\ Out:\ & \{-2,\ -6,\ -4\} \end{aligned} \tag{3.105}$$

Reassigning the first component of u to negative.

$$\begin{aligned} In:\ & u[[1]] = -u[[1]];\ u \\ Out:\ & \{2,\ -6,\ -4\} \end{aligned} \tag{3.106}$$

The scalar (dot) product of u and v.

$$\begin{aligned} In:\ & u\ .\ v \\ Out:\ & -24 \end{aligned} \tag{3.107}$$

The norm (magnitude) of v.

$$In: \quad \mathrm{Norm}[v]$$
$$Out: \quad \sqrt{14}$$
(3.108)

The unit vector in the same direction as v.

$$In: \quad \mathrm{Normalize}[v]$$
$$Out: \quad \left\{\frac{1}{\sqrt{14}}, \frac{3}{\sqrt{14}}, \sqrt{\frac{2}{7}}\right\}$$
(3.109)

Verifying that the norm is 1.

$$In: \quad \mathrm{Norm}[\%]$$
$$Out: \quad 1$$
(3.110)

Orthogonal vector operations

The projection of u onto v.

$$In: \quad p = \mathrm{Projection}[u, v]$$
$$Out: \quad \left\{-\frac{12}{7}, -\frac{36}{7}, -\frac{24}{7}\right\}$$
(3.111)

p is a scalar multiple of v.

$$In: \quad p/v$$
$$Out: \quad \left\{-\frac{12}{7}, -\frac{12}{7}, -\frac{12}{7}\right\}$$
(3.112)

$u - p$ is orthogonal to v.

$$In: \quad (u - p) . v$$
$$Out: \quad 0$$
(3.113)

Starting from the set of vectors $\{u, v\}$, this finds an orthonormal set of two vectors.

$$In: \quad \mathrm{Orthogonalize}[\{u, v\}]$$
$$Out: \quad \left\{\left\{\frac{1}{\sqrt{14}}, -\frac{3}{\sqrt{14}}, -\sqrt{\frac{2}{7}}\right\}, \left\{\sqrt{\frac{13}{14}}, \frac{3}{\sqrt{182}}, \sqrt{\frac{2}{91}}\right\}\right\}$$
(3.114)

When one of the vectors is linearly dependent on the vectors preceding it, the corresponding position in the result will be a zero vector.

In: Orthogonalize[$\{v, p, u\}$]

$$Out: \left\{\left\{\frac{1}{\sqrt{14}}, \frac{3}{\sqrt{14}}, \sqrt{\frac{2}{7}}\right\}, \{0, 0, 0\}, \left\{\sqrt{\frac{13}{14}}, -\frac{3}{\sqrt{182}}, -\sqrt{\frac{2}{91}}\right\}\right\} \qquad (3.115)$$

Scalar product of vectors:

$$
\begin{aligned}
&In: \quad \{a, b, c\} \cdot \{x, y, z\} \\
&Out: \quad ax + by + cz
\end{aligned}
\qquad (3.116)
$$

The cross product of two vectors:

$$
\begin{aligned}
&In: \quad \mathrm{Cross}[\{a, b, c\}, \{x, y, z\}] \\
&Out: \quad \{-cy + bz, \, cx - az, \, -bx + ay\}
\end{aligned}
\qquad (3.117)
$$

3.6 Homework assignment

Problem 1. Find the angles between $\vec{A} = -2i + j - 2k$ and $\vec{B} = 2i - 2j$.

Problem 2. Show that the vectors $\vec{A} = 2i - j + 4k$ and $\vec{B} = 5i + 2j - 2k$ are orthogonal (perpendicular) and find a third vector perpendicular to both.

Problem 3. Show that the lines that join $(0, 0, 0)$ to $(1, 2, -1)$ and $(1, 1, 1)$ to $(2, 3, 4)$ intersect and find the acute angle between them.

Problem 4. A particle is traveling along the line $(x - 3)/2 = (y + 1)/(-2) = z - 1$. Write the equation of this path in the form $\vec{r} = \vec{r_0} + \vec{A}t$. Find the distance of closest approach of the particle to the origin (that is, the distance from the origin to the line). If t represents time, show that the time of the closest approach is

$$t = -\frac{\vec{A} \cdot \vec{r_0}}{A^2}.$$

Use this value to check your answer for the distance of closest approach. If P is the point of closest approach, what is $\vec{A} \cdot \vec{r}$?

Problem 5. The vectors $\vec{A} = ai + bj$ and $\vec{B} = ui + vj$ form two sides of a parallelogram. Show that the area of the parallelogram is given by the absolute value of the following determinant

$$\begin{vmatrix} a & b \\ u & v \end{vmatrix}. \qquad (3.118)$$

IOP Publishing

Studies in Theoretical Physics, Volume 1
Fundamental mathematical methods
Daniel Erenso and Victor Montemayor

Chapter 4

Matrices and determinants

Matrices and determinants are some of the essential mathematical methods that we need in most branches of physics. In particular in quantum mechanics, classical mechanics, and electrodynamics. Most quantum mechanics problems often require solving the energy, the angular momentum (orbital or spin angular momentum) eigenvalue equations that involve matrices and determinants, eigenvectors, and eigenvalues. Maxwell stress tensor and the electromagnetic field tensor are a few that we encounter in electrodynamics involving matrices. Even in classical mechanics, we have what is known as the moment of inertial tensor, which also involves matrices. This chapter will introduce matrices, matrices operation, finding the determinant of a square matrix, and applications of matrices.

4.1 Important terminologies

A matrix

A rectangular array of quantities usually enclosed in large parentheses with n rows and m columns.

$$A = \begin{bmatrix} a_{11} & a_{12} \cdot\cdot & a_{1m-1} & a_{1m} \\ a_{21} & a_{22} \cdot\cdot & a_{2m-1} & a_{2m} \\ \cdot & \cdot & \cdot & \cdot \\ \cdot & \cdot & \cdot & \cdot \\ \cdot & \cdot & \cdot & \cdot \\ a_{n-11} & a_{n-12} \cdot\cdot\cdot & a_{n-1m-1} & a_{n-1m} \\ a_{n1} & a_{n2} \cdot\cdot\cdot & a_{nm-1} & a_{nm} \end{bmatrix}. \tag{4.1}$$

When $n = m$, the matrix is called a square matrix. In the previous chapter, we were introduced to vectors. It is important to note that vectors can be represented using matrices. For example, the position vector $\vec{r}$ in Cartesian coordinates

$$\vec{r} = x\hat{x} + y\hat{y} + z\hat{z}, \tag{4.2}$$

can be represented using a column matrix (i.e., $n = 3$ and $m = 1$),

$$r = \begin{bmatrix} x \\ y \\ z \end{bmatrix}. \tag{4.3}$$

A column matrix is a matrix with n rows and one column

$$A = \begin{bmatrix} a_{11} \\ a_{21} \\ \cdot \\ \cdot \\ \cdot \\ a_{n-11} \\ a_{n1} \end{bmatrix}. \tag{4.4}$$

Transpose of a matrix: The Transpose of a matrix A in equation (4.1), which we denote by A^T is obtained by writing the rows of matrix A as columns. In other words, if the element of the matrix A is represented by A_{ij}, (i.e., ith row and jth column), then the corresponding element of the transposed matrix $(A^T)_{ij}$ is given by

$$(A^T)_{ij} = A_{ji}. \tag{4.5}$$

Thus for the matrix in equation (4.1), the transpose matrix can be written as

$$A^T = \begin{bmatrix} a_{11} & a_{21} \cdot \cdot & a_{n-11} & a_{n1} \\ a_{12} & a_{22} \cdot \cdot & a_{n-12} & a_{n2} \\ \cdot & \cdot & \cdot & \cdot \\ \cdot & \cdot & \cdot & \cdot \\ a_{1m-1} & a_{2m-1} \cdot \cdot & a_{n-1m-1} & a_{nm-1} \\ a_{1m} & a_{2m} \cdot \cdot & a_{n-1m} & a_{nm} \end{bmatrix}. \tag{4.6}$$

For example, for the matrices

$$A = \begin{pmatrix} 2 & 3 & 1 \\ 2 & 1 & 0 \end{pmatrix}, \ B = \begin{pmatrix} 2 & 4 \\ 1 & -1 \\ 3 & -1 \end{pmatrix} \tag{4.7}$$

the transpose matrices are given by

$$A^T = \begin{pmatrix} 2 & 2 \\ 3 & 1 \\ 1 & 0 \end{pmatrix}, \ B^T = \begin{pmatrix} 2 & 1 & 3 \\ 4 & -1 & -1 \end{pmatrix}. \tag{4.8}$$

In elementary algebra, we know that 1 is an identity of multiplication since multiplying a number or a variable by 1 produces no change to the number or the variable,

$$1 \times A = A. \tag{4.9}$$

Similarly, when the matrix is square, there exists what we call the identity matrix I that we can multiply the square matrix with and brings no change

$$IA = AI = A, \tag{4.10}$$

where A is a square matrix.

4.2 Matrix arithmetic and manipulation

Given two or more matrices, the answer to when we can and cannot do matrix arithmetic is determined by the matrices' dimensions.

Addition and subtraction

Any two matrices A and B can be added or subtracted if they have the same number of rows and columns. In other words, two matrices having different numbers of rows or columns cannot be added or subtracted. When this condition met, then the elements of the matrix $C = A + B$ are given by

$$C_{ij} = A_{ij} + B_{ij}. \tag{4.11}$$

As an example, let us consider the matrices A, B, C, and D

$$A = \begin{pmatrix} 2 & 3 & 1 \\ 2 & 1 & 0 \end{pmatrix}, \quad B = \begin{pmatrix} 2 & 4 \\ 1 & -1 \\ 3 & -1 \end{pmatrix},$$

$$C = \begin{pmatrix} 2 & 1 & 3 \\ 4 & -1 & -2 \\ -1 & 0 & 1 \end{pmatrix}, \quad D = \begin{pmatrix} -2 & 0 & 1 \\ 1 & -1 & 2 \\ 3 & 1 & 0 \end{pmatrix}. \tag{4.12}$$

From these matrices, we can add/subtract only matrices C and D

$$\begin{aligned} C + D &= \begin{pmatrix} 2 & 1 & 3 \\ 4 & -1 & -2 \\ -1 & 0 & 1 \end{pmatrix} + \begin{pmatrix} -2 & 0 & 1 \\ 1 & -1 & 2 \\ 3 & 1 & 0 \end{pmatrix} \\ &= \begin{pmatrix} 2-2 & 1+0 & 3+1 \\ 4+1 & -1-1 & -2+2 \\ -1+3 & 0+1 & 1+0 \end{pmatrix} = \begin{pmatrix} 0 & 1 & 4 \\ 5 & -2 & 0 \\ 2 & 1 & 1 \end{pmatrix}. \end{aligned} \tag{4.13}$$

Multiplication by a scalar

Suppose we have a matrix A of any dimension with elements A_{ij} and a scalar α. Then the elements for the matrix $B = \alpha A$ is given by

$$B_{ij} = \alpha A_{ij}. \tag{4.14}$$

For example, for $\alpha = 2$ and

$$A = \begin{pmatrix} 2 & 3 & 1 \\ 2 & 1 & 0 \end{pmatrix}, \tag{4.15}$$

one can write

$$B = \alpha A = 2A = \begin{pmatrix} 2 \times 2 & 3 \times 2 & 1 \times 2 \\ 2 \times 2 & 1 \times 2 & 0 \times 2 \end{pmatrix}$$

$$\Rightarrow B = \begin{pmatrix} 4 & 6 & 2 \\ 4 & 2 & 0 \end{pmatrix}. \tag{4.16}$$

Multiplication of matrices

Multiplication of two matrices, A and B, is a bit tricky. The order of multiplication (which comes first and which comes second) is critical, and we even decide whether we can do the multiplication or not in the first place. In general multiplication of matrices is not commutative

$$AB \neq BA.$$

The multiplication of the two matrices A and B in the order

$$C = AB,$$

exists if and only if the number of columns of the first matrix (A) is equal to the number of rows of the second matrix (B). The element C_{ij} (in row i and column j) of the product matrix $C = AB$ is given by

$$(ab)_{ij} = a_{i1}b_{k1} + a_{i2}b_{2j} + \ldots a_{in}b_{nj} = \sum_{k=1}^{n} a_{ik}b_{kj}, \tag{4.17}$$

where n is the number of columns in the first matrix A, which is equal to the number of rows in the second matrix B. As an example, let us consider the matrices in equation (4.12). From these matrices, we can make the multiplications

$$AB = \begin{pmatrix} 2 & 3 & 1 \\ 2 & 1 & 0 \end{pmatrix} \begin{pmatrix} 2 & 4 \\ 1 & -1 \\ 3 & -1 \end{pmatrix}$$

$$= \begin{pmatrix} (2 \times 2) + (3 \times 1) + (1 \times 3) & (2 \times 4) + (3 \times (-1)) + (1 \times (-1)) \\ (2 \times 2) + (1 \times 1) + (0 \times 3) & (2 \times 4) + (1 \times (-1)) + (0 \times (-1)) \end{pmatrix} \tag{4.18}$$

$$= \begin{pmatrix} 10 & 4 \\ 5 & 7 \end{pmatrix}$$

and

$$CD = \begin{pmatrix} 2 & 1 & 3 \\ 4 & -1 & -2 \\ -1 & 0 & 1 \end{pmatrix}\begin{pmatrix} -2 & 0 & 1 \\ 1 & -1 & 2 \\ 3 & 1 & 0 \end{pmatrix} = \begin{pmatrix} (cd)_{11} & (cd)_{12} & (cd)_{13} \\ (cd)_{21} & (cd)_{22} & (cd)_{23} \\ (cd)_{31} & (cd)_{32} & (cd)_{33} \end{pmatrix}. \qquad (4.19)$$

However, we cannot make the matrix multiplications BC or BD.

The commutator

Earlier, we have seen that the multiplication of two matrices is not commutative,

$$CD \neq DC. \qquad (4.20)$$

If the matrices C and D are square matrices with the same dimension, the difference

$$CD - DC = [C, D] \qquad (4.21)$$

is known as *the commutator*. We often see the commutator in quantum mechanics. In quantum mechanics, physically observable quantities, such as position, momentum, and energy, are represented with operators, and the matrix representations of these operators, generally, do not commute.

Finally, it is essential to note that matrix multiplication is associative and is also distributive with respect to matrix addition; that is,

$$F(GH) = (FG)H \qquad (4.22)$$

and

$$F(G + H) = FG + FH \qquad (4.23)$$

provided that the indicated multiplications and additions are possible.

4.3 Matrix representation of a set of linear equations

Matrices and determinants can be used to solve a set of linear equations. Next, we will see how to represent a set of linear equations using matrices. To this end, let us consider a row matrix

$$A = \begin{bmatrix} a & b & c \end{bmatrix} \qquad (4.24)$$

and a column matrix

$$\begin{bmatrix} x \\ y \\ z \end{bmatrix}. \qquad (4.25)$$

Note that the multiplication

$$AB = \begin{bmatrix} a & b & c \end{bmatrix}\begin{bmatrix} x \\ y \\ z \end{bmatrix} = ax + by + cz = k \qquad (4.26)$$

leads to a linear equation. Now consider the set of linear equations

$$
\begin{aligned}
a_1 x + b_1 y + c_1 z &= k_1, \\
a_2 x + b_2 y + c_2 z &= k_2, \\
a_3 x + b_3 y + c_3 z &= k_4,
\end{aligned}
\tag{4.27}
$$

which we can rewrite, using matrices, as

$$
\begin{bmatrix} a_1 x + b_1 y + c_1 z \\ a_2 x + b_2 y + c_2 z \\ a_3 x + b_3 y + c_3 z \end{bmatrix} = \begin{bmatrix} k_1 \\ k_2 \\ k_3 \end{bmatrix}.
\tag{4.28}
$$

Applying the properties of matrices in addition, one can rewrite this equation as

$$
\begin{bmatrix} a_1 x + b_1 y + c_1 z + d_1 w \\ 0 \\ 0 \end{bmatrix} + \begin{bmatrix} 0 \\ a_2 x + b_2 y + c_2 z \\ 0 \end{bmatrix} + \begin{bmatrix} 0 \\ 0 \\ a_3 x + b_3 y + c_3 z \end{bmatrix} = \begin{bmatrix} k_1 \\ k_2 \\ k_3 \end{bmatrix}.
\tag{4.29}
$$

The properties of matrices in multiplication allow us to write,

$$
\begin{aligned}
\begin{bmatrix} a_1 x + b_1 y + c_1 z \\ 0 \\ 0 \end{bmatrix} &= \begin{bmatrix} a_1 & b_1 & c_1 \\ 0 & 0 & 0 \\ 0 & 0 & 0 \end{bmatrix} \begin{bmatrix} x \\ y \\ z \end{bmatrix}, \\[2mm]
\begin{bmatrix} 0 \\ a_2 x + b_2 y + c_2 z \\ 0 \end{bmatrix} &= \begin{bmatrix} 0 & 0 & 0 \\ a_2 & b_2 & c_2 \\ 0 & 0 & 0 \end{bmatrix} \begin{bmatrix} x \\ y \\ z \end{bmatrix}, \\[2mm]
\begin{bmatrix} 0 \\ 0 \\ a_3 x + b_3 y + c_3 z \end{bmatrix} &= \begin{bmatrix} 0 & 0 & 0 \\ 0 & 0 & 0 \\ a_3 & b_3 & c_3 \end{bmatrix} \begin{bmatrix} x \\ y \\ z \end{bmatrix},
\end{aligned}
\tag{4.30}
$$

so that equation (4.29) can be put in the form

$$
\begin{bmatrix} a_1 & b_1 & c_1 \\ 0 & 0 & 0 \\ 0 & 0 & 0 \end{bmatrix} \begin{bmatrix} x \\ y \\ z \end{bmatrix} + \begin{bmatrix} 0 & 0 & 0 \\ a_2 & b_2 & c_2 \\ 0 & 0 & 0 \end{bmatrix} \begin{bmatrix} x \\ y \\ z \end{bmatrix} + \begin{bmatrix} 0 & 0 & 0 \\ 0 & 0 & 0 \\ a_3 & b_3 & c_3 \end{bmatrix} \begin{bmatrix} x \\ y \\ z \end{bmatrix} = \begin{bmatrix} k_1 \\ k_2 \\ k_3 \end{bmatrix}.
\tag{4.31}
$$

Upon factoring out the column matrix, we have

$$
\begin{aligned}
&\left\{ \begin{bmatrix} a_1 & b_1 & c_1 \\ 0 & 0 & 0 \\ 0 & 0 & 0 \end{bmatrix} + \begin{bmatrix} 0 & 0 & 0 \\ a_2 & b_2 & c_2 \\ 0 & 0 & 0 \end{bmatrix} = \begin{bmatrix} c_1 \\ c_2 \\ c_3 \end{bmatrix} \Rightarrow \begin{bmatrix} a_1 & b_1 & c_1 \\ a_2 & b_2 & c_2 \\ a_3 & b_3 & c_3 \end{bmatrix} \begin{bmatrix} x \\ y \\ z \end{bmatrix} = \begin{bmatrix} k_1 \\ k_2 \\ k_3 \end{bmatrix} \right. \\[2mm]
&\left. + \begin{bmatrix} 0 & 0 & 0 \\ 0 & 0 & 0 \\ a_3 & b_3 & c_3 \end{bmatrix} \right\} \begin{bmatrix} x \\ y \\ z \end{bmatrix}
\end{aligned}
\tag{4.32}
$$

$$
\Rightarrow MR = K,
$$

where

$$R = \begin{bmatrix} x \\ y \\ z \end{bmatrix}, \quad k = \begin{bmatrix} k_1 \\ k_2 \\ k_3 \end{bmatrix}, \tag{4.33}$$

and

$$M = \begin{bmatrix} a_1 & b_1 & c_1 \\ a_2 & b_2 & c_2 \\ a_3 & b_3 & c_3 \end{bmatrix}, \tag{4.34}$$

which is known as the *coefficient matrix*. We can augment the matrices M and k and write a matrix

$$A = \begin{bmatrix} a_1 & b_1 & c_1 & \vdots & k_1 \\ a_2 & b_2 & c_2 & \vdots & k_2 \\ a_3 & b_3 & c_3 & \vdots & k_3 \end{bmatrix}, \tag{4.35}$$

which is referred as the *augmented matrix*. The augmented matrix is the matrix representation to the set of linear equations in equation (4.27).

4.4 Solving a set of linear equations using matrices

In the previous section, we saw how we represent a set of linear equations using matrices. Next, we will see the methods to determine the solutions to a set of linear equations we represented using matrices. There are two different methods. The first involves *the Gaussian elimination method* and *row echelon form*, whereas the second method involves *Cramer's rule*. We first discuss the Gaussian method, and then we come to the second method after introducing the determinant of a square matrix.

Gaussian Elimination method and Row Echelon Form

This method was named after the German mathematician and scientist Carl Friedrich Gauss. Before we outline the steps we must follow in this method; we need to explain what it means when we say that an augmented coefficient matrix is in a row echelon or reduced row echelon form. An augmented coefficient matrix is in row echelon form if all of the following conditions are satisfied:
- All non-zero rows (rows with at least one non-zero element) are above any rows of all zeroes.
  Note: All zero rows, if any, belong at the bottom of the matrix.
- The leading coefficient (the first non-zero number from the left, also called the pivot) of a non-zero row is always strictly to the right of the leading coefficient of the row above it.
- All entries in a column below a leading entry are zeroes (implied by the first two criteria).

A matrix in row echelon form and every leading coefficient is one is in *reduced row echelon form* (*row canonical form*). This method uses the following two steps to solve a set of linear equations expressed as augmented matrix.

Apply any of the following elementary row operations to the augmented matrix.
   (a) Multiply any row of the matrix by a non-zero constant,
   (b) Interchange any two rows,
   (c) Replace any row with a linear combination of the matrix rows (as long as the linear combination includes the row being replaced). In other words, we can add to or subtract from a multiple of one row to another,
   (d) You may repeat any of the above operations until the augmented matrix is in *reduced row echelon form*.

Example 4.1. Which of the following matrices

$$B = \begin{bmatrix} 0 & 1 & 4 & 0 \\ 0 & 0 & 1 & 0 \\ 0 & 0 & 0 & 1 \\ 0 & 0 & 0 & 0 \end{bmatrix}, \; C = \begin{bmatrix} 1 & 1 & 1 & 1 \\ 0 & 9 & 0 & 2 \\ 0 & 0 & 0 & 3 \end{bmatrix}, \; D = \begin{bmatrix} 1 & 2 & 3 & 4 \\ 0 & 3 & 7 & 2 \\ 0 & 2 & 0 & 0 \end{bmatrix} \tag{4.36}$$

are in *reduced row echelon form, row echelon form, or neither?*

Solution: The matrix

$$B = \begin{bmatrix} 0 & 1 & 4 & 0 \\ 0 & 0 & 1 & 0 \\ 0 & 0 & 0 & 1 \\ 0 & 0 & 0 & 0 \end{bmatrix} \tag{4.37}$$

is in *reduced row echelon form*. The justifications are

1. All zero rows belong at the bottom of the matrix.
2. The leading coefficient (i.e., the pivot) in the non-zero rows are all to the right of the leading coefficient of the row above it.
3. All entries in a column below a leading entry are zeroes.
4. Every leading coefficient in a row is 1.
The matrix

$$C = \begin{bmatrix} 1 & 1 & 1 & 1 \\ 0 & 9 & 0 & 2 \\ 0 & 0 & 0 & 3 \end{bmatrix} \tag{4.38}$$

is in row echelon form since it satisfies the first three conditions above. However, it is not in reduced row echelon form as its leading coefficients in rows 2 and 3 are not 1. The matrix

$$D = \begin{bmatrix} 1 & 2 & 3 & 4 \\ 0 & 3 & 7 & 2 \\ 0 & 2 & 0 & 0 \end{bmatrix} \tag{4.39}$$

is not in *row echelon form* as the leading coefficient of row 3 (which is 2) is not to the right of the leading coefficient of the row above it (row 2, which is 3).

Rank of a Matrix

The rank of a matrix is the number of non-zero rows after the matrix has been row-reduced and put in reduced row echelon form. The rank of the coefficient matrix M in equation (4.34), the augmented matrix A in equation (4.35), and the number of unknowns n determine the solutions to the set of linear equations.

(a) If (rank M) < (rank A), the equations are inconsistent, and there is no solution.

(b) If (rank M) = (rank A) = n (number of unknowns), there is one solution for each unknown.

(c) If (rank M) = (rank A) = $R < n$, then R unknowns can be found in terms of the remaining $n - R$ unknowns.

Example 4.2. Consider the circuit shown in figure 4.1.

(a) Applying Kirchhoff's voltage and current rules, find a set of linear equations.

(b) Find the coefficient and augmented matrices from the set of linear equations.

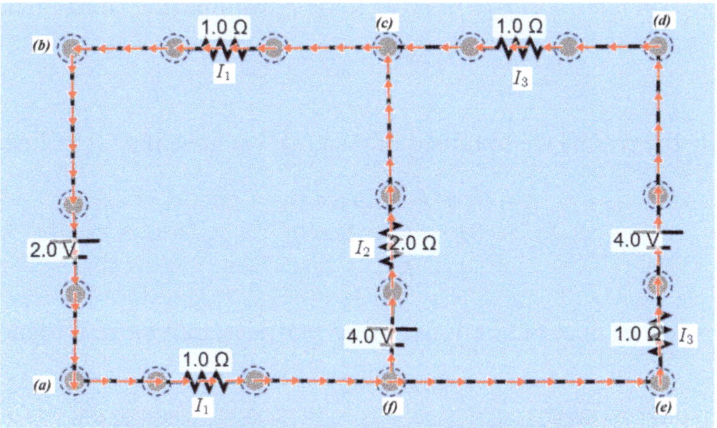

Figure 4.1. A dc electrical circuit (built using PhTH simulation).

(c) Find the values of the three branch currents I_1, I_2, and I_3 using Gaussian elimination method.

(d) Find the ranks for the coefficient and augmented matrices. Are the ranks consistent with what we discussed earlier?

Kirchhoff's voltage law: Start at one point in a circuit and go around any closed loop in the circuit. The voltage at the starting point of the loop must be the same as the voltage at the ending point since they are the same point in the circuit. Another way of stating this is that the net change in voltage as we go around any closed loop in the circuit must be zero (so the starting potential must be the same as the ending potential). Kirchhoff's voltage law is just a restatement of the conservation of energy.

Kirchhoff's current law: The net current flowing into any circuit node must equal the net current flowing out of that same node. This law states that there is no place from which charge magically appears in the circuit and there is no place that the charge disappears. Kirchhoff's current law is a statement of conservation of charge in an electric circuit.

Solution:

(a) Applying Ohm's law

$$V_R = I_R R,$$

and Kirchhoff's voltage law for the closed-loop *abcfa* in figure 4.1, one can write

$$2 + I_1 + 2I_2 - 4 + I_1 = 0 \Rightarrow I_1 + I_2 = 1. \tag{4.40}$$

Similarly, for closed-loop *cdefc*, we find

$$I_3 - 4 + I_3 + 4 - 2I_2 = 0 \Rightarrow -I_2 + I_3 = 0. \tag{4.41}$$

Using *Kirchhoff's current law* at node c, we find the third linear equation,

$$I_1 = I_2 + I_3 \Rightarrow I_1 - I_2 - I_3 = 0. \tag{4.42}$$

(b) Using the results we obtained in part (a), we have the set of linear equations

$$\begin{aligned} I_1 + I_2 + 0I_3 &= 1, \\ 0I_1 - I_2 + I_3 &= 0, \\ I_1 - I_2 - I_3 &= 0. \end{aligned} \tag{4.43}$$

These set of linear equations, using matrices, can be rewritten as

$$MI = K, \tag{4.44}$$

where coefficient matrix M is given by

$$M = \begin{bmatrix} 1 & 1 & 0 \\ 0 & -1 & 1 \\ 1 & -1 & -1 \end{bmatrix} \tag{4.45}$$

and

$$I = \begin{bmatrix} I_1 \\ I_2 \\ I_3 \end{bmatrix}, \; K = \begin{bmatrix} 1 \\ 0 \\ 0 \end{bmatrix}. \tag{4.46}$$

Then the augmented matrix becomes

$$A = \begin{bmatrix} 1 & 1 & 0 & 1 \\ 0 & -1 & 1 & 0 \\ 1 & -1 & -1 & 0 \end{bmatrix}. \tag{4.47}$$

(c) We apply elementary row operations to write the augmented matrix

$$A = \begin{bmatrix} 1 & 1 & 0 & 1 \\ 0 & -1 & 1 & 0 \\ 1 & -1 & -1 & 0 \end{bmatrix} \tag{4.48}$$

in *reduced row echelon form.* To this end:
Replacing row 3 by (row 3 - row 1),

$$A = \begin{bmatrix} 1 & 1 & 0 & 1 \\ 0 & -1 & 1 & 0 \\ 0 & -2 & -1 & -1 \end{bmatrix}. \tag{4.49}$$

Replacing row 3 by (row 3 -2 row 2),

$$A = \begin{bmatrix} 1 & 1 & 0 & 1 \\ 0 & -1 & 1 & 0 \\ 0 & 0 & -3 & -1 \end{bmatrix}. \tag{4.50}$$

Divide row 3 by −3,

$$A = \begin{bmatrix} 1 & 1 & 0 & 1 \\ 0 & -1 & 1 & 0 \\ 0 & 0 & 1 & \dfrac{1}{3} \end{bmatrix}. \tag{4.51}$$

Replacing row 2 by (row 2 - row 3),

$$A = \begin{bmatrix} 1 & 1 & 0 & 1 \\ 0 & -1 & 0 & -\dfrac{1}{3} \\ 0 & 0 & 1 & \dfrac{1}{3} \end{bmatrix}. \tag{4.52}$$

Divide row 2 by −1,

$$A = \begin{bmatrix} 1 & 1 & 0 & 1 \\ 0 & 1 & 0 & \dfrac{1}{3} \\ 0 & 0 & 1 & \dfrac{1}{3} \end{bmatrix}. \tag{4.53}$$

Replacing row 1 by (row 2 - row 1),

$$A = \begin{bmatrix} 1 & 0 & 0 & \dfrac{2}{3} \\ 0 & 1 & 0 & \dfrac{1}{3} \\ 0 & 0 & 1 & \dfrac{1}{3} \end{bmatrix}. \tag{4.54}$$

The augmented matrix described by equation (4.48) and the corresponding augmented matrix in reduced row echelon form described by equation (4.54) are row-equivalent but not the same. Recalling that

$$MI = K, \tag{4.55}$$

and the augmented matrix A constructed from the matrices M and K, one can write

$$\begin{bmatrix} 1 & 0 & 0 \\ 0 & 1 & 0 \\ 0 & 0 & 1 \end{bmatrix} \begin{bmatrix} I_1 \\ I_2 \\ I_3 \end{bmatrix} = \begin{bmatrix} \dfrac{2}{3} \\ \dfrac{1}{3} \\ \dfrac{1}{3} \end{bmatrix} \Rightarrow \begin{bmatrix} I_1 \\ I_2 \\ I_3 \end{bmatrix} = \begin{bmatrix} \dfrac{2}{3} \\ \dfrac{1}{3} \\ \dfrac{1}{3} \end{bmatrix} \tag{4.56}$$

$$\Rightarrow I_1 = \frac{2}{3}A, \; I_2 = \frac{1}{3}A, \; I_3 = \frac{1}{3}A.$$

(d) The number of non-zero rows for the row-reduced augmented matrix

$$A = \begin{bmatrix} 1 & 0 & 0 & \dfrac{2}{3} \\ 0 & 1 & 0 & \dfrac{1}{3} \\ 0 & 0 & 1 & \dfrac{1}{3} \end{bmatrix} \tag{4.57}$$

is 3 and the rank is 3. For the reduced coefficient matrix

$$M = \begin{bmatrix} 1 & 0 & 0 \\ 0 & 1 & 0 \\ 0 & 0 & 1 \end{bmatrix}, \tag{4.58}$$

we also have three non-zero rows, and the rank is 3. There are three unknowns, and this means that (rank M) = (rank A) = 3 (number of unknowns). Therefore we find one solution for each unknown.

4.5 Determinant of a square matrix

The determinant of any square matrix with $n \times n$ dimension,

$$A = \begin{bmatrix} a_{11} & a_{12} & \cdots & a_{1j} & \cdots & a_{1n-1} & a_{1n} \\ a_{21} & a_{22} & \cdots & a_{2j} & \cdots & a_{2n-1} & a_{2n} \\ \vdots & & & & & & \\ a_{i1} & a_{i2} & \cdots & a_{ij} & \cdots & a_{in-1} & a_{in} \\ \vdots & \vdots & & \vdots & & \vdots & \vdots \\ a_{n-11} & a_{n-12} & a_{n-1j} & a_{n-1n-1} & a_{n-1n} \\ a_{n1} & a_{n2} & a_{nj} & a_{nn-1} & a_{nn} \end{bmatrix}, \tag{4.59}$$

is given by

$$\det|A| = \sum_{j=1}^{n} (-1)^{i+j} a_{ij} M_{ij}, \tag{4.60}$$

where $i = 1, 2, 3 \ldots n$, M_{ij} is the *minor* of a_{ij}, and it is the determinant of the matrix A with the ith row and the jth column is removed. Let us begin with a 2×2 square matrix

$$A = \begin{bmatrix} a_{11} & a_{12} \\ a_{21} & a_{22} \end{bmatrix}, \tag{4.61}$$

the determinant is given by

$$\det|A| = \sum_{j=1}^{2} a_{ij}(-1)^{i+j} M_{ij}, \tag{4.62}$$

where $i = 1$ or 2. For $i = 1$, the determinant can be expressed as

$$|A| = \sum_{j=1}^{2} (-1)^{1+j} a_{1j} M_{1j} = (-1)^{1+1} a_{11} M_{11} + (-1)^{1+2} a_{12} M_{12}. \tag{4.63}$$

Noting that the minors

$$M_{11} = a_{22}, \; M_{12} = a_{21},$$

(4.64)

we find

$$\det|A| = a_{11}a_{22} - a_{12}a_{21}.$$

(4.65)

If one chose $i = 2$,

$$\det|A| = \sum_{j=1}^{2}(-1)^{2+j}a_{2j}M_{2j} = (-1)^{2+1}a_{21}M_{21} + (-1)^{2+2}a_{22}M_{22},$$

(4.66)

and using the minors

$$M_{21} = a_{12}, \; M_{22} = a_{11},$$

(4.67)

we also find

$$\det|A| = -a_{21}a_{12} + a_{22}a_{11} = a_{11}a_{22} - a_{12}a_{21},$$

(4.68)

which is the same as the result in equation (4.65). Next we shall consider a 3×3 matrix

$$A = \begin{bmatrix} a_{11} & a_{12} & a_{13} \\ a_{21} & a_{22} & a_{23} \\ a_{31} & a_{32} & a_{33} \end{bmatrix}.$$

(4.69)

The determinant is given by

$$\det|A| = \sum_{j=1}^{3} a_{ij}(-1)^{i+j}M_{ij},$$

(4.70)

where $i = 1, 2,$ or 3. For $i = 1$, equation (4.70) becomes

$$\begin{aligned} \det|A| &= (-1)^{1+1}a_{11}M_{11} + (-1)^{1+2}a_{12}M_{12} + (-1)^{1+3}a_{13}M_{13} \\ &\quad + (-1)^{1+4}a_{14}M_{14} \\ &\Rightarrow \det|A| = a_{11}M_{11} - a_{12}M_{12} + a_{13}M_{13}, \end{aligned}$$

(4.71)

where the minors are given by the determinants of the 2×2 matrices

$$\begin{aligned} M_{11} &= \begin{vmatrix} a_{22} & a_{23} \\ a_{32} & a_{33} \end{vmatrix} = a_{22}a_{33} - a_{23}a_{32}, \\[6pt] M_{12} &= \begin{vmatrix} a_{21} & a_{23} \\ a_{31} & a_{33} \end{vmatrix} = a_{21}a_{33} - a_{31}a_{23}, \\[6pt] M_{13} &= \begin{vmatrix} a_{21} & a_{22} \\ a_{31} & a_{23} \end{vmatrix} = a_{31}a_{23} - a_{22}a_{31}. \end{aligned}$$

(4.72)

Example 4.3. Find the determinant of the matrix

$$C = \begin{bmatrix} \hat{x} & \hat{y} & \hat{z} \\ A_x & A_y & A_z \\ B_x & B_y & B_z \end{bmatrix}, \tag{4.73}$$

where $\hat{x}$, $\hat{y}$, and $\hat{z}$ are the unit vectors, (A_x, A_y, A_z) and (B_x, B_y, B_z) are components of the vectors $\vec{A}$ and $\vec{B}$ in Cartesian coordinate system.

Solution: Using equation (4.60), we write the determinant of the given matrix as

$$\det|C| = \sum_{j=1}^{3}(-1)^{1+j}c_{ij}M_{ij} \tag{4.74}$$
$$= (-1)^{1+1}c_{11}M_{11} + (-1)^{1+2}c_{12}M_{12} + (-1)^{1+3}c_{13}M_{13},$$

where we chose $i = 1$. Note that for a 3-dimensional square matrix $i = 1, 2,$ or 3. Using

$$c_{11} = \hat{x}, \ c_{12} = \hat{y}, \ \text{and } c_{13} = \hat{z}, \tag{4.75}$$

and the corresponding minors M_{11}, M_{12}, and M_{13}

$$M_{11} = \begin{vmatrix} A_y & A_z \\ B_y & B_z \end{vmatrix} = A_y B_z - A_z B_y, \tag{4.76}$$

$$M_{12} = \begin{vmatrix} A_x & A_z \\ B_x & B_z \end{vmatrix} = A_x B_z - A_z B_x, \tag{4.77}$$

$$M_{13} = \begin{vmatrix} A_x & A_y \\ B_x & B_y \end{vmatrix} = A_x B_y - A_y B_x. \tag{4.78}$$

Equation (4.74) can be rewritten as

$$\det|C| = (-1)^{1+1}\hat{x}(A_y B_z - A_z B_y) + (-1)^{1+2}\hat{y}(A_x B_z - A_z B_x)$$
$$+ (-1)^{1+3}\hat{z}(A_x B_y - A_y B_x)$$
$$\Rightarrow \det|C| = (A_y B_z - A_z B_y)\hat{x} + (A_z B_x - A_x B_z)\hat{y}$$
$$+ (A_x B_y - A_y B_x)\hat{z}. \tag{4.79}$$

The result in equation (4.79) shows that the determinant of the matrix C is the vector product of $\vec{A}$ and $\vec{B}$,

$$\vec{A} \times \vec{B} = \left(A_y B_z - A_z B_y\right)\hat{x} + (A_z B_x - A_x B_z)\hat{y} + \left(A_x B_y - A_y B_x\right)\hat{z}$$

$$= \begin{vmatrix} \hat{x} & \hat{y} & \hat{z} \\ A_x & A_y & A_z \\ B_x & B_y & B_z \end{vmatrix} . \tag{4.80}$$

Properties of determinants

(a) **Multiplying by a constant**: Suppose a row or a column of a square matrix A is multiplied by a constant k, the determinant of the new matrix is k times the determinant of the matrix A. For example, for the 2-dimensional square matrix,

$$A = \begin{bmatrix} a_{11} & a_{12} \\ a_{21} & a_{22} \end{bmatrix}, \tag{4.81}$$

we may write a new matrix B given by

$$B = \begin{bmatrix} ka_{11} & a_{12} \\ ka_{21} & a_{22} \end{bmatrix} \text{ or } B = \begin{bmatrix} ka_{11} & ka_{12} \\ a_{21} & a_{22} \end{bmatrix}, \tag{4.82}$$

then

$$\det |B| = k \det |A|. \tag{4.83}$$

This implies, for a square matrix, when all elements of a row (or a column) are zero (i.e., $k = 0$), the determinant of the matrix is zero. For example, for a 2×2 matrix,

$$A = \begin{bmatrix} 0 & a_{12} \\ 0 & a_{22} \end{bmatrix} \text{ or } A = \begin{bmatrix} a_{11} & a_{12} \\ 0 & 0 \end{bmatrix} \Rightarrow \det |A| = 0. \tag{4.84}$$

(b) When two rows (or columns) of a square matrix are proportional, the determinant of the matrix is zero. Again for a 2×2 matrix,

$$B = \begin{bmatrix} a_{11} & a_{12} = ka_{11} \\ a_{21} & a_{22} = ka_{21} \end{bmatrix} \text{ or } B = \begin{bmatrix} a_{11} & a_{12} \\ a_{21} = ka_{11} & a_{22} = ka_{12} \end{bmatrix} \tag{4.85}$$

$$\Rightarrow \det |B| = 0,$$

where k is the proportionality constant.

(c) If two rows (or two columns) of a square matrix are interchanged, the determinant of the matrix changes sign. For example, for a 2×2 matrix

$$A = \begin{bmatrix} a_{11} & a_{12} \\ a_{21} & a_{22} \end{bmatrix}, \quad B = \begin{bmatrix} a_{12} & a_{11} \\ a_{22} & a_{21} \end{bmatrix}$$
$$\Rightarrow \det|A| = a_{11}a_{22} - a_{21}a_{12}, \quad \det|B| = a_{12}a_{21} - a_{11}a_{22}$$
$$\Rightarrow \det|B| = -(a_{11}a_{22} - a_{21}a_{12}) = -\det|A|. \tag{4.86}$$

Example 4.4. Evaluate the determinant of the matrix,

$$A = \begin{pmatrix} 2 & 2 & 0 \\ 0 & -2 & 2 \\ 1 & -1 & -1 \end{pmatrix}. \tag{4.87}$$

Solution: Applying the relation in equations (4.71) and (4.72), we have

$$\det|A| = 2 \begin{vmatrix} -2 & 2 \\ -1 & -1 \end{vmatrix} - 2 \begin{vmatrix} 0 & 2 \\ 1 & -1 \end{vmatrix} + 0 \begin{pmatrix} 0 & -2 \\ 1 & -1 \end{pmatrix} \tag{4.88}$$

and using equation (4.65), we find

$$\det|A| = 2[(-2 \times -1) - (2 \times -1)] - 2[(0 \times -1) - (2 \times 1)] = 12. \tag{4.89}$$

4.6 Cramer's rule

Before we stated Cramer's rule for solving a set of linear equations, let us find the solutions to the set of linear equations,

$$a_{11}x_1 + a_{12}x_2 = k_1, \tag{4.90}$$

$$a_{21}x_1 + a_{22}x_2 = k_2. \tag{4.91}$$

To this end, multiplying equation (4.90) by a_{22} and equation (4.91) by a_{12}, we have

$$a_{11}a_{22}x_1 + a_{12}a_{22}x_2 = k_1 a_{22},$$
$$a_{12}a_{21}x_1 + a_{12}a_{22}x_2 = k_2 a_{12}, \tag{4.92}$$

so that upon subtracting the two equations, we find

$$x_1 = \frac{k_1 a_{22} - k_2 a_{12}}{a_{11}a_{22} - a_{12}a_{21}}. \tag{4.93}$$

Similarly, multiplying equation (4.90) by a_{21}, equation (4.91) by a_{11}, and subtracting the resulting equations, one can easily show that

$$x_2 = \frac{k_2 a_{11} - k_1 a_{21}}{a_{11} a_{22} - a_{12} a_{21}}. \tag{4.94}$$

Using matrices, we can rewrite equations (4.90) and (4.91) in a matrix form

$$MR = K, \tag{4.95}$$

where the matrices M, R and K are given by

$$M = \begin{bmatrix} a_{11} & a_{12} \\ a_{21} & a_{22} \end{bmatrix}, \ R = \begin{bmatrix} x_1 \\ x_2 \end{bmatrix}, \ K = \begin{bmatrix} k_1 \\ k_2 \end{bmatrix}. \tag{4.96}$$

Consider the coefficient matrix M, and two matrices M_1 and M_2 constructed by replacing the first and second columns of the matrix M by the matrix K, respectively,

$$M = \begin{bmatrix} a_{11} & a_{12} \\ a_{21} & a_{22} \end{bmatrix}, \ M_1 = \begin{bmatrix} k_1 & a_{12} \\ k_2 & a_{22} \end{bmatrix}, \ M_2 = \begin{bmatrix} a_{11} & k_1 \\ a_{21} & k_2 \end{bmatrix}. \tag{4.97}$$

The determinants of these matrices are given by

$$\det | M | = \begin{vmatrix} a_{11} & a_{12} \\ a_{21} & a_{22} \end{vmatrix} = a_{11} a_{22} - a_{12} a_{21}, \tag{4.98}$$

$$\det M_1 = \begin{vmatrix} k_1 & a_{12} \\ k_2 & a_{22} \end{vmatrix} = k_1 a_{22} - k_2 a_{12}, \tag{4.99}$$

$$\det M_2 = \begin{vmatrix} a_{11} & k_1 \\ a_{21} & k_2 \end{vmatrix} = k_2 a_{11} - k_1 a_{21}. \tag{4.100}$$

Upon dividing equations (4.99) and (4.99) by equation (4.100), we find

$$\frac{\det | M_1 |}{\det | M |} = \frac{k_1 a_{22} - k_2 a_{12}}{a_{11} a_{22} - a_{12} a_{21}}, \ \frac{\det | M_2 |}{\det | M |} = \frac{k_2 a_{11} - k_1 a_{21}}{a_{11} a_{22} - a_{12} a_{21}}. \tag{4.101}$$

We can easily see that the results in equation (4.101) are the solutions to the two linear equations in equations (4.93) and (4.94). Therefore, one can express these solutions in terms of the determinant of the three matrices as

$$x_1 = \frac{\det | M_1 |}{\det | M |}, \ x_2 = \frac{\det | M_2 |}{\det | M |}.$$

Cramer's rule states that this can be generalized to n set of linear equations. For a set of n linear equations with n unknowns, which we can express using matrices as,

$$MR = K,$$

where

$$M = \begin{bmatrix} a_{11} & a_{12} & \cdots & a_{1j} & \cdots & a_{1n-1} & a_{1n} \\ a_{21} & a_{22} & \cdots & a_{2j} & \cdots & a_{2n-1} & a_{2n} \\ \vdots & \vdots & & & & \vdots & \vdots \\ \vdots & \vdots & & & & \vdots & \vdots \\ a_{n1} & a_{n2} & & a_{nj} & & a_{nn-1} & a_{nn} \end{bmatrix}, \quad R = \begin{bmatrix} x_1 \\ x_2 \\ \vdots \\ \vdots \\ x_n \end{bmatrix}, \quad K = \begin{bmatrix} k_1 \\ k_2 \\ \vdots \\ \vdots \\ k_n \end{bmatrix} \tag{4.102}$$

the solutions are given by

$$x_j = \frac{\det \left| M_j \right|}{\det \left| M \right|}, \tag{4.103}$$

where $\det \left| M \right|$ is the determinant of the matrix $M \neq 0$ and $\det \left| M_j \right|$ is the determinant of the matrix M with the jth column is replaced by the elements of the column matrix K.

Example 4.5. Find the values of the three branch currents I_1, I_2, and I_3 in Example 4.2 using Cramer's rule.

Solution: We recall that the coefficient matrix M is given by

$$M = \begin{bmatrix} 1 & 1 & 0 \\ 0 & -1 & 1 \\ 1 & -1 & -1 \end{bmatrix} \tag{4.104}$$

and the column matrix K by

$$K = \begin{bmatrix} 1 \\ 0 \\ 0 \end{bmatrix}. \tag{4.105}$$

Replacing the first, the second, and the third columns of M by K, we find the three matrices

$$M_1 = \begin{bmatrix} 1 & 1 & 0 \\ 0 & -1 & 1 \\ 0 & -1 & -1 \end{bmatrix} \quad M_2 = \begin{bmatrix} 1 & 1 & 0 \\ 0 & 0 & 1 \\ 1 & 0 & -1 \end{bmatrix}, \text{ and } M_3 = \begin{bmatrix} 1 & 1 & 1 \\ 0 & -1 & 0 \\ 1 & -1 & 0 \end{bmatrix}. \tag{4.106}$$

For the determinants of these matrices, one can easily find

$$\det \left| M \right| = \begin{vmatrix} -1 & 1 \\ -1 & -1 \end{vmatrix} - \begin{vmatrix} 0 & 1 \\ 1 & -1 \end{vmatrix} = 3, \tag{4.107}$$

$$\det \left| M_1 \right| = \begin{vmatrix} -1 & 1 \\ -1 & -1 \end{vmatrix} - \begin{vmatrix} 0 & 1 \\ 0 & -1 \end{vmatrix} = 2, \tag{4.108}$$

$$\det |M_2| = \begin{vmatrix} 0 & 1 \\ 0 & -1 \end{vmatrix} - \begin{vmatrix} 0 & 1 \\ 1 & -1 \end{vmatrix} = 1, \tag{4.109}$$

$$\det |M_3| = \begin{vmatrix} -1 & 0 \\ -1 & 0 \end{vmatrix} - \begin{vmatrix} 0 & 0 \\ 1 & 0 \end{vmatrix} + \begin{vmatrix} 0 & -1 \\ 1 & -1 \end{vmatrix} = 1. \tag{4.110}$$

Therefore, the three currents are given by

$$I_1 = \frac{\det |M_1|}{\det |M|} = \frac{2}{3}, \; I_2 = \frac{\det |M_2|}{\det |M|} = \frac{1}{3}, \; I_3 = \frac{\det |M_2|}{\det |M|} = \frac{1}{3}. \tag{4.111}$$

These are the same as the results we found in Example 4.2 using the Gaussian elimination method.

4.7 The adjoint and inverse of a matrix

The cofactor and adjoint matrices

In the previous section, we were introduced to how we find the determinant of a square matrix A which is given by

$$\det|A| = \sum_{j=1}^{n}(-1)^{i+j}a_{ij}M_{ij}, \tag{4.112}$$

where M_{ij} is the determinant of the matrix constructed from the matrix A with ith row and the jth column removed. The matrix constructed with the elements defined by

$$[cof(A)]_{ij} = (-1)^{i+j}M_{ij} \tag{4.113}$$

is known as *the cofactor matrix $cof(A)$*. The transpose of the cofactor matrix gives the adjoint matrix $adj(A)$

$$adj(A) = [cof(A)]^{T}. \tag{4.114}$$

As an example let's consider a 2×2 matrix

$$A = \begin{bmatrix} a_{11} & a_{12} \\ a_{21} & a_{22} \end{bmatrix}. \tag{4.115}$$

Noting that

$$\begin{aligned}
[cof(A)]_{12} &= (-1)^{1+2}M_{12} = -a_{21}, \\
[cof(A)]_{21} &= (-1)^{2+1}M_{21} = -a_{12}, \\
[cof(A)]_{11} &= (-1)^{1+1}M_{11} = a_{22}, \\
[cof(A)]_{22} &= (-1)^{2+2}M_{22} = a_{11},
\end{aligned} \tag{4.116}$$

one can write for the cofactor matrix,

$$cof(A) = \begin{bmatrix} [cof(A)]_{11} & [cof(A)]_{12} \\ [cof(A)]_{21} & [cof(A)]_{22} \end{bmatrix} = \begin{bmatrix} a_{22} & -a_{21} \\ -a_{12} & a_{11} \end{bmatrix}, \tag{4.117}$$

and for the adjoint matrix

$$adj(A) = [cof(A)]^T = \begin{bmatrix} a_{22} & -a_{12} \\ -a_{21} & a_{11} \end{bmatrix}. \tag{4.118}$$

Inverse of a square matrix

For a square matrix A with a non-zero determinant (i.e., $\det|A| \neq 0$), the inverse matrix A^{-1} exists such that

$$A^{-1}A = AA^{-1} = I, \tag{4.119}$$

where I is the identity matrix which depends on the dimension of the matrix A,

$$I = \begin{bmatrix} 1 & 0 \\ 0 & 1 \end{bmatrix}, \text{ for } 2 \times 2, \; I = \begin{bmatrix} 1 & 0 & 0 \\ 0 & 1 & 0 \\ 0 & 0 & 1 \end{bmatrix}, \text{ for } 3 \times 3$$

$$I = \begin{bmatrix} 1 & 0 & 0 & 0 \\ 0 & 1 & 0 & 0 \\ 0 & 0 & 1 & 0 \\ 0 & 0 & 0 & 1 \end{bmatrix}, \text{ for } 4 \times 4. \tag{4.120}$$

For an invertible square matrix, we can determine the inverse using two methods. The first method is row reduction, which requires using the Gaussian elimination method and writing the matrix in reduced row echelon form. The second method involves using the adjoint matrix.

    a. **Row reduction:** Consider the 2 by 2 square matrix,

$$A = \begin{bmatrix} a_{11} & a_{12} \\ a_{21} & a_{22} \end{bmatrix}. \tag{4.121}$$

In this method we start from the matrix

$$\begin{bmatrix} a_{11} & a_{12} & | & 1 & 0 \\ a_{21} & a_{22} & | & 0 & 1 \end{bmatrix} \tag{4.122}$$

and do elementary row operation until we end up with

$$\begin{bmatrix} 1 & 0 & | & b_{11} & b_{12} \\ 0 & 1 & | & b_{21} & b_{22} \end{bmatrix}, \tag{4.123}$$

so that the inverse of the matrix A is given by

$$A^{-1} = \begin{bmatrix} b_{11} & b_{12} \\ b_{21} & b_{22} \end{bmatrix}. \tag{4.124}$$

b. **Using the adjoint matrix:** In this method, using the adjoint matrix, the inverse of an invertible matrix A ($\det|A| \neq 0$) can be expressed as

$$A^{-1} = \frac{adj(A)}{\det|A|} = \frac{[cof(A)]^T}{\det|A|}. \tag{4.125}$$

We can easily derive this relation using a two-by-two matrix. To this end, consider an invertible matrix A

$$A = \begin{bmatrix} a_{11} & a_{12} \\ a_{21} & a_{22} \end{bmatrix}. \tag{4.126}$$

Let the inverse matrix A^{-1} be

$$A^{-1} = \begin{pmatrix} b_{11} & b_{12} \\ b_{21} & b_{22} \end{pmatrix} \tag{4.127}$$

so that

$$\begin{aligned}
A^{-1}A &= AA^{-1} = I \\
&\Rightarrow \begin{bmatrix} a_{11} & a_{12} \\ a_{21} & a_{22} \end{bmatrix}\begin{bmatrix} b_{11} & b_{12} \\ b_{21} & b_{22} \end{bmatrix} = \begin{bmatrix} 1 & 0 \\ 0 & 1 \end{bmatrix} \\
&\Rightarrow \begin{bmatrix} a_{11}b_{11} + a_{12}b_{21} & a_{11}b_{12} + a_{12}b_{22} \\ a_{21}b_{11} + a_{22}b_{21} & a_{21}b_{12} + a_{22}b_{22} \end{bmatrix} = \begin{bmatrix} 1 & 0 \\ 0 & 1 \end{bmatrix}.
\end{aligned} \tag{4.128}$$

From equation (4.128) follows that

$$a_{11}b_{12} + a_{12}b_{22} = 0, \; a_{21}b_{11} + a_{22}b_{21} = 0 \tag{4.129}$$

and

$$a_{11}b_{11} + a_{12}b_{21} = 1, \; a_{21}b_{12} + a_{22}b_{22} = 1. \tag{4.130}$$

Solving for b_{12} and b_{21} from equation (4.129), we find

$$b_{12} = -\frac{a_{12}}{a_{11}}b_{22}, \; b_{21} = -\frac{a_{21}}{a_{22}}b_{11}. \tag{4.131}$$

Upon substituting equation (4.131) into equation (4.130), one can easily show that b_{11} and b_{22},

$$\begin{aligned}
b_{11} &= \frac{a_{22}}{a_{22}a_{11} - a_{12}a_{21}}, \\
b_{22} &= \frac{a_{11}}{a_{22}a_{11} - a_{12}a_{21}}.
\end{aligned} \tag{4.132}$$

Using equation (4.131) and equation (4.132), we find,

$$b_{12} = - \frac{a_{12}}{a_{22}a_{11} - a_{12}a_{21}},$$
$$b_{21} = - \frac{a_{21}}{a_{22}a_{11} - a_{12}a_{21}}.$$

(4.133)

Replacing the results in equations (4.132) and (4.133) for the elements of the inverse matrix, one can write

$$A^{-1} = \begin{pmatrix} b_{11} & b_{12} \\ b_{21} & b_{22} \end{pmatrix} = \begin{pmatrix} \dfrac{a_{22}}{a_{22}a_{11} - a_{12}a_{21}} & -\dfrac{a_{12}}{a_{22}a_{11} - a_{12}a_{21}} \\ -\dfrac{a_{21}}{a_{22}a_{11} - a_{12}a_{21}} & \dfrac{a_{11}}{a_{22}a_{11} - a_{12}a_{21}} \end{pmatrix}$$

$$\Rightarrow A^{-1} = \frac{\begin{pmatrix} a_{22} & -a_{12} \\ -a_{21} & a_{11} \end{pmatrix}}{a_{22}a_{11} - a_{12}a_{21}}.$$

(4.134)

We recall from equation (4.118), the adjoint of the matrix A is

$$adj(A) = [cof(A)]^T = \begin{pmatrix} a_{22} & -a_{12} \\ -a_{21} & a_{11} \end{pmatrix},$$

(4.135)

and the determinant is

$$\det|A| = a_{22}a_{11} - a_{12}a_{21}.$$

(4.136)

Therefore, using these equations, one can express the inverse of the matrix A in equation (4.134) as

$$A^{-1} = \frac{adj(A)}{\det|A|}.$$

(4.137)

The result in equation (4.137) can be generalized to any $n \times n$ invertible square matrix.

Example 4.6. Find the inverse of the matrix

$$A = \begin{pmatrix} -1 & 2 & 3 \\ 2 & 0 & -4 \\ 1 & -1 & 1 \end{pmatrix},$$

(4.138)

using
    a. the row reduction method,
    b. the adjoint matrix method.

Solution:

a. In the row reduction method we start from

$$\left[\begin{array}{ccc|ccc} -1 & 2 & 3 & 1 & 0 & 0 \\ 2 & 0 & -4 & 0 & 1 & 0 \\ 1 & -1 & 1 & 0 & 0 & 1 \end{array}\right], \tag{4.139}$$

and find

$$\left[\begin{array}{ccc|ccc} 1 & 0 & 0 & a_{11} & a_{12} & a_{13} \\ 0 & 1 & 0 & a_{21} & a_{22} & a_{23} \\ 0 & 0 & 1 & a_{31} & a_{32} & a_{33} \end{array}\right], \tag{4.140}$$

by performing elementary row operations. Then the inverse matrix is given by

$$A^{-1} = \begin{pmatrix} a_{11} & a_{12} & a_{13} \\ a_{21} & a_{22} & a_{23} \\ a_{31} & a_{32} & a_{33} \end{pmatrix}. \tag{4.141}$$

We perform the following elementary row operations.

Add row 1 to row 3:

$$\left[\begin{array}{ccc|ccc} -1 & 2 & 3 & 1 & 0 & 0 \\ 2 & 0 & -4 & 0 & 1 & 0 \\ 0 & 1 & 4 & 1 & 0 & 1 \end{array}\right]. \tag{4.142}$$

Add row 3 to row 2:

$$\left[\begin{array}{ccc|ccc} -1 & 2 & 3 & 1 & 0 & 0 \\ 2 & 1 & 0 & 1 & 1 & 1 \\ 0 & 1 & 4 & 1 & 0 & 1 \end{array}\right]. \tag{4.143}$$

Multiply row 2 by 2 and subtract the result from row 1:

$$\left[\begin{array}{ccc|ccc} -5 & 0 & 3 & -1 & -2 & -2 \\ 2 & 1 & 0 & 1 & 1 & 1 \\ 0 & 1 & 4 & 1 & 0 & 1 \end{array}\right]. \tag{4.144}$$

Divide row 1 by -5 and row 2 by 2:

$$\left[\begin{array}{ccc|ccc} 1 & 0 & -3/5 & 1/5 & 2/5 & 2/5 \\ 1 & 1/2 & 0 & 1/2 & 1/2 & 1/2 \\ 0 & 1 & 4 & 1 & 0 & 1 \end{array}\right]. \tag{4.145}$$

Subtract row 1 from row 2:

$$\left[\begin{array}{ccc|ccc} 1 & 0 & -3/5 & 1/5 & 2/5 & 2/5 \\ 0 & 1/2 & 3/5 & 3/10 & 1/10 & 1/10 \\ 0 & 1 & 4 & 1 & 0 & 1 \end{array}\right]. \tag{4.146}$$

Multiply row 2 by 2:

$$\left[\begin{array}{ccc|ccc} 1 & 0 & -3/5 & 1/5 & 2/5 & 2/5 \\ 0 & 1 & 6/5 & 3/5 & 1/5 & 1/5 \\ 0 & 1 & 4 & 1 & 0 & 1 \end{array}\right]. \tag{4.147}$$

Subtract row 2 from row 3:

$$\left[\begin{array}{ccc|ccc} 1 & 0 & -3/5 & 1/5 & 2/5 & 2/5 \\ 0 & 1 & 6/5 & 3/5 & 1/5 & 1/5 \\ 0 & 0 & 14/5 & 2/5 & -1/5 & 4/5 \end{array}\right]. \tag{4.148}$$

Divide row 1 by $-3/5$, row 2 by $6/5$, and row 3 by $14/5$:

$$\left[\begin{array}{ccc|ccc} -5/3 & 0 & 1 & -1/3 & -2/3 & -2/3 \\ 0 & 5/6 & 1 & 1/2 & 1/6 & 1/6 \\ 0 & 0 & 1 & 1/7 & -1/14 & 2/7 \end{array}\right]. \tag{4.149}$$

Subtract row 3 from row 2:

$$\left[\begin{array}{ccc|ccc} -5/3 & 0 & 1 & -1/3 & -2/3 & -2/3 \\ 0 & 5/6 & 0 & 5/14 & 5/21 & -5/42 \\ 0 & 0 & 1 & 1/7 & -1/14 & 2/7 \end{array}\right]. \tag{4.150}$$

Subtract row 3 from row 1:

$$\left[\begin{array}{ccc|ccc} -5/3 & 0 & 0 & -10/21 & -25/42 & -20/21 \\ 0 & 5/6 & 0 & 5/14 & 5/21 & -5/42 \\ 0 & 0 & 1 & 1/7 & -1/14 & 2/7 \end{array}\right]. \tag{4.151}$$

Divide row 1 by $-5/3$ and row 2 by $5/6$:

$$\left[\begin{array}{ccc|ccc} 1 & 0 & 0 & 2/7 & 5/14 & 4/7 \\ 0 & 1 & 0 & 3/7 & 2/7 & -1/7 \\ 0 & 0 & 1 & 1/7 & -1/14 & 2/7 \end{array}\right]. \tag{4.152}$$

Then the inverse matrix is given by

$$A^{-1} = \begin{pmatrix} 2/7 & 5/14 & 4/7 \\ 3/7 & 2/7 & -1/7 \\ 1/7 & -1/14 & 2/7 \end{pmatrix}. \tag{4.153}$$

b. Here we first need to find the adjoint matrix of A. To find the adjoint matrix, we need to find the cofactor matrix. Let this cofactor matrix be

$$cof(A) = \begin{pmatrix} c_{11} & c_{12} & c_{13} \\ c_{21} & c_{22} & c_{23} \\ c_{31} & c_{32} & c_{33} \end{pmatrix} \tag{4.154}$$

where the elements

$$c_{ij} = (-1)^{i+j} M_{ij} \tag{4.155}$$

in which M_{ij} is the minor for the matrix A. Using

$$A = \begin{pmatrix} -1 & 2 & 3 \\ 2 & 0 & -4 \\ 1 & -1 & 1 \end{pmatrix} \tag{4.156}$$

we can write

$$M_{11} = \begin{vmatrix} 0 & -4 \\ -1 & 1 \end{vmatrix} = -4 \Rightarrow c_{11} = (-1)^2 M_{11} = -4, \tag{4.157}$$

$$M_{12} = \begin{vmatrix} 2 & -4 \\ 1 & 1 \end{vmatrix} = 6 \Rightarrow c_{12} = (-1)^3 M_{12} = -6, \tag{4.158}$$

$$M_{13} = \begin{vmatrix} 2 & 0 \\ 1 & -1 \end{vmatrix} = -2 \Rightarrow c_{13} = (-1)^4 M_{11} = -2, \tag{4.159}$$

$$M_{21} = \begin{vmatrix} 2 & 3 \\ -1 & 1 \end{vmatrix} = 5 \Rightarrow c_{21} = (-1)^3 M_{21} = -5 \tag{4.160}$$

$$M_{22} = \begin{vmatrix} -1 & 3 \\ 1 & 1 \end{vmatrix} = -4 \Rightarrow c_{22} = (-1)^4 M_{22} = -4 \tag{4.161}$$

$$M_{23} = \begin{vmatrix} -1 & 2 \\ 1 & -1 \end{vmatrix} = -1 \Rightarrow c_{23} = (-1)^5 M_{23} = 1 \tag{4.162}$$

$$M_{31} = \begin{vmatrix} 2 & 3 \\ 0 & -4 \end{vmatrix} = -8 \Rightarrow c_{31} = (-1)^4 M_{31} = -8, \tag{4.163}$$

$$M_{32} = \begin{vmatrix} -1 & 3 \\ 2 & -4 \end{vmatrix} = -2 \Rightarrow c_{32} = (-1)^5 M_{32} = 2, \tag{4.164}$$

$$M_{33} = \begin{vmatrix} -1 & 2 \\ 2 & 0 \end{vmatrix} = -4 \Rightarrow c_{33} = (-1)^6 M_{33} = -4. \tag{4.165}$$

There follows that

$$cof(A) = \begin{bmatrix} -4 & -6 & -2 \\ -5 & -4 & 1 \\ -8 & 2 & -4 \end{bmatrix} \tag{4.166}$$

and

$$[cof(A)]^T = \begin{bmatrix} -4 & -5 & -8 \\ -6 & -4 & 2 \\ -2 & 1 & -4 \end{bmatrix}. \tag{4.167}$$

The determinant of A is given by

$$\begin{aligned} \det|A| &= \begin{vmatrix} -1 & 2 & 3 \\ 2 & 0 & -4 \\ 1 & -1 & 1 \end{vmatrix} \\ &= -\begin{vmatrix} 0 & -4 \\ -1 & 1 \end{vmatrix} - 2\begin{vmatrix} 2 & -4 \\ 1 & 1 \end{vmatrix} + 3\begin{vmatrix} 2 & 0 \\ 1 & -1 \end{vmatrix} \\ &\Rightarrow \det|A| = -14. \end{aligned} \tag{4.168}$$

Then the inverse matrix

$$A^{-1} = \frac{[cof(A)]^T}{\det|A|}, \tag{4.169}$$

becomes

$$A^{-1} = -\frac{1}{14}\begin{bmatrix} -4 & -5 & -8 \\ -6 & -4 & 2 \\ -2 & 1 & -4 \end{bmatrix} = \begin{pmatrix} 2/7 & 5/14 & 4/7 \\ 3/7 & 2/7 & -1/7 \\ 1/7 & -1/14 & 2/7 \end{pmatrix}, \tag{4.170}$$

which is the same as the result we found in part (a).

4.8 Orthogonal matrices and the rotation matrix

Orthogonal matrices

For a square matrix M, if the inverse matrix is equal to its transpose matrix, matrix M is called an orthogonal matrix,

$$M^{-1} = M^T. \tag{4.171}$$

The transformation of vectors by orthogonal matrices keeps the magnitude of the vectors unchanged. A well-known orthogonal matrix is the rotation matrix. Consider a vector in the x–y–z Cartesian coordinate system,

$$\vec{r} = x\hat{x} + y\hat{y} + z\hat{z}, \tag{4.172}$$

shown in figure 4.2. Suppose we rotate this coordinate system by an angle θ about the z-axis in the counterclockwise direction. Let this coordinate system be x'–y'–z' (see figure 4.2). We are interested in finding the components of the vector

$$\vec{r}' = x'\hat{x}' + y'\hat{y}' + z'\hat{z}' \tag{4.173}$$

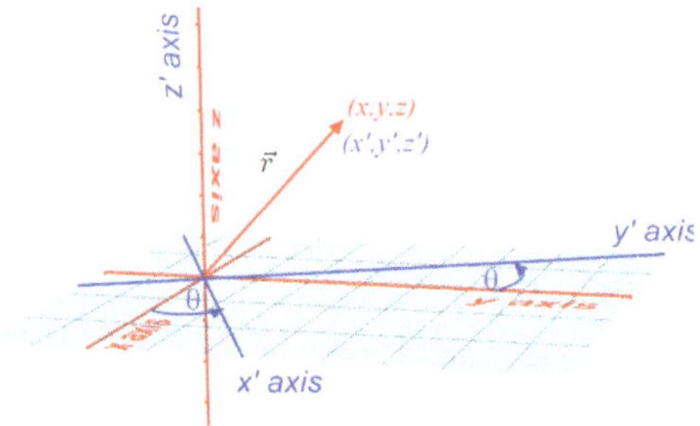

Figure 4.2. Cartesian coordinates (x, y, z) rotated about the z-axis by an angle θ in a counterclockwise direction to form (x', y', z') Cartesian coordinates.

in the x'–y'–z' coordinate system. Since the rotation is about the z-axis, there is no change in the z-coordinate, which means

$$z' = z. \tag{4.174}$$

Let the component of the vector $\vec{r}$ (the projection) on the x–y plane be $\vec{R}$

$$\vec{R} = x\hat{x} + y\hat{y} \tag{4.175}$$

and similarly for $\vec{r}'$ be $\vec{R}'$

$$\vec{R}' = x'\hat{x}' + y'\hat{y}'. \tag{4.176}$$

The vector $\vec{R}$ is shown in figure 4.3. In terms of the components (x, y) of the vector $\vec{R}$, one can express the components (x', y') for the vector $\vec{R}'$ as

$$\begin{aligned} x' &= x\cos(\theta) + y\sin(\theta), \\ y' &= -x\sin(\theta) + y\cos(\theta). \end{aligned} \tag{4.177}$$

We already know that for a rotation about the z-axis

$$z' = z. \tag{4.178}$$

Using matrix representation, we can put equations (4.177) and (4.178) in the form

$$\begin{pmatrix} x' \\ y' \\ z' \end{pmatrix} = \begin{pmatrix} \cos(\theta) & \sin(\theta) & 0 \\ -\sin(\theta) & \cos(\theta) & 0 \\ 0 & 0 & 1 \end{pmatrix} \begin{pmatrix} x \\ y \\ z \end{pmatrix} \Rightarrow r' = Rr, \tag{4.179}$$

where the matrix $R(\theta) = R$ is the rotation matrix given by

$$R = \begin{pmatrix} \cos(\theta) & \sin(\theta) & 0 \\ -\sin(\theta) & \cos(\theta) & 0 \\ 0 & 0 & 1 \end{pmatrix}. \tag{4.180}$$

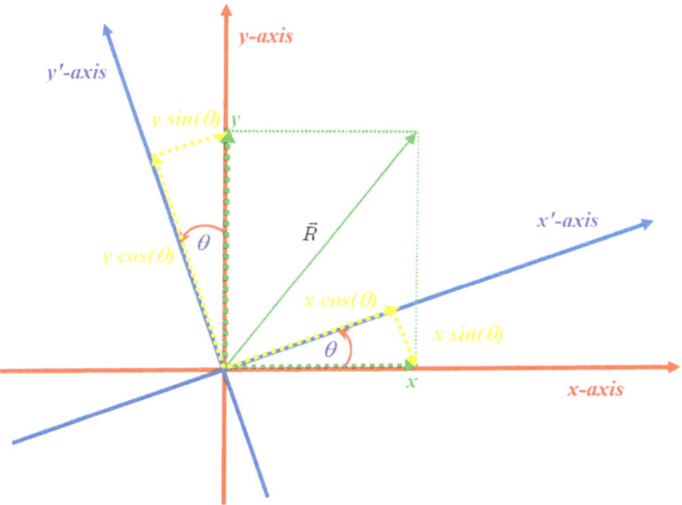

Figure 4.3. The projection of the vector $\vec{r}$ on the x–y plane ($\vec{R}$) and the projection of the vector $\vec{r}'$ on the x'–y' plane ($\vec{R}'$).

The transpose of this matrix is

$$R^T = \begin{pmatrix} \cos(\theta) & -\sin(\theta) & 0 \\ \sin(\theta) & \cos(\theta) & 0 \\ 0 & 0 & 1 \end{pmatrix}, \tag{4.181}$$

and it can be shown that the inverse is given by

$$R^{-1} = \begin{pmatrix} \cos(\theta) & -\sin(\theta) & 0 \\ \sin(\theta) & \cos(\theta) & 0 \\ 0 & 0 & 1 \end{pmatrix}, \tag{4.182}$$

which shows that

$$R^{-1} = R^T \tag{4.183}$$

and the *rotation matrix is an orthogonal matrix*.

Linear operators

An operator $\hat{F}(x)$ is a linear operator when the following two conditions are met:

(a)

$$\hat{F}(x)[g(ax)] = a\hat{F}(x)[g(x)] \tag{4.184}$$

(b)

$$\hat{F}(x)[g_1(x) + g_2(x)] = \hat{F}(x)g_1(x) + \hat{F}(x)g_2(x) \tag{4.185}$$

where a is a constant and $g_1(x)$, $g_2(x)$ are some scalar functions. If any one or these conditions are violated, then the operator is not a linear operator!

Example 4.7. Consider the operator $\hat{F}(x)$ when it acts on the function $g(x) = x$, it squares the function and adds a constant β, which means

$$\hat{F}(x)[g(x)] = \hat{F}(x)[x] = x^2 + \beta. \qquad (4.186)$$

Is this operator a linear operator?

Solution: We need to check the two conditions stated above. Upon checking the first condition, we find

$$\begin{aligned}
\hat{F}(x)[g(ax)] &= \hat{F}(x)[ax] = a^2 x^2 + \beta, \\
&\Rightarrow \hat{F}(x)[g(ax)] = a^2 x^2 + \beta \neq a\hat{F}(x)[g(x)] = a(x^2 + \beta).
\end{aligned} \qquad (4.187)$$

Without checking the second condition, we can conclude that the function is not linear. But let us also check the second condition,

$$\begin{aligned}
\hat{F}(x)[g(x_1) + g(x_2)] &= \hat{F}(x)[x_1 + x_2] = (x_1 + x_2)^2 + \beta \\
&\Rightarrow \hat{F}(x)[g(x_1) + g(x_2)] = x_1^2 + 2x_1 x_2 + x_2^2 + \beta
\end{aligned} \qquad (4.188)$$

on the other hand

$$\begin{aligned}
\hat{F}(x)(x_1) + \hat{F}(x)g(x_2) &= x_1^2 + \beta + x_2^2 + \beta = x_1^2 + x_2^2 + 2\beta \\
&\Rightarrow \hat{F}(x)[g(x_1) + g(x_2)] \neq \hat{F}(x) \\
&\qquad g(x_1) + \hat{F}(x)g(x_2).
\end{aligned} \qquad (4.189)$$

Note: Generally, an operation that involves squaring $\hat{F} = (\ )^2$ is not a linear operation.

4.9 Linear dependence and independence

A set of linear equations or vectors could be linearly dependent or independent. In order to obtain a complete solution to a set of linear equations using matrices, the equations must be linearly independent. Sometimes in some physical problems, it may be necessary to express a given vector as a set of linearly independent vectors. This section will see how linear independence in a set of linear equations, vectors, and functions can be verified and how we can obtain a set of linearly independent orthogonal vectors.

Set of linear equations and vectors

The solutions to a set of linear equations with n unknowns depend on the linear independence of the equations determined by the rank of the augmented (A) and coefficient matrices (M). We recall that the rank of a matrix is the number of non-zero rows remaining when a matrix has been row-reduced.

    a. If (rank M) < (rank A), the equations are inconsistent and there is no solution.

    b. If (rank M) = (rank A) = n (number of unknowns), there is one solution.

    c. If (rank M) = (rank A) = $m < n$, then m unknowns can be determined in terms of the remaining $n - m$ unknowns.

Next, we shall see if these conditions can be related to linear dependency in a set of linear equations. We shall examine this using a problem involving Kirchhoff's laws.

Example 4.8. Let us reconsider the electrical circuit in Example 4.2 (see figure 4.4). Find the values of the three branch currents I_1, I_2, and I_3 by solving the system of three simultaneous equations obtained by applying *only* Kirchhoff's voltage law to three different loops. In this example, the current rule is excluded intensionally to illustrate the linear independence of a set of linear equations.

Solution: We already obtained two equations in Example 4.2 using the loops *abcfa*

$$2 + I_1 + 2I_2 - 4 + I_1 = 0 \Rightarrow I_1 + I_2 = 1, \tag{4.190}$$

and *cdefc*

$$I_3 - 4 + I_3 + 4 - 2I_2 = 0 \Rightarrow -I_2 + I_3 = 0. \tag{4.191}$$

Applying Kirchhoff's voltage law to a third loop *abcdefa*, we find a third equation

$$2 + I_1 + I_3 - 4 + I_3 + I_1 = 0 \Rightarrow I_1 + I_3 = 1. \tag{4.192}$$

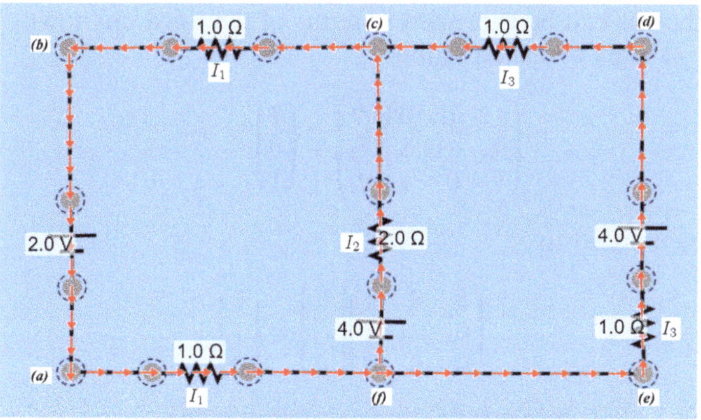

Figure 4.4. A dc electrical circuit (built using PhTH simulation).

Equations (4.190)–(4.192), using matrices, can be put in the form,

$$\begin{pmatrix} 1 & 1 & 0 \\ 0 & -1 & 1 \\ 1 & 0 & 1 \end{pmatrix}\begin{pmatrix} I_1 \\ I_2 \\ I_3 \end{pmatrix} = \begin{pmatrix} 1 \\ 0 \\ 1 \end{pmatrix}. \tag{4.193}$$

There follows that for the augmented matrix, we have

$$A = \begin{pmatrix} 1 & 1 & 0 & | & 1 \\ 0 & -1 & 1 & | & 0 \\ 1 & 0 & 1 & | & 1 \end{pmatrix} \tag{4.194}$$

and for the coefficient matrix

$$M = \begin{pmatrix} 1 & 1 & 0 \\ 0 & -1 & 1 \\ 1 & 0 & 1 \end{pmatrix}. \tag{4.195}$$

To find the rank of the augmented and the coefficient matrix, we carry out elementary row operation. To this end, for the augmented matrix, add row two to row one,

$$\begin{pmatrix} 1 & 0 & 1 & | & 1 \\ 0 & -1 & 1 & | & 0 \\ 1 & 0 & 1 & | & 1 \end{pmatrix}, \tag{4.196}$$

and subtract row one from row three,

$$\begin{pmatrix} 1 & 0 & 1 & | & 1 \\ 0 & -1 & 1 & | & 1 \\ 0 & 0 & 0 & | & 0 \end{pmatrix}. \tag{4.197}$$

We note that the augmented matrix cannot be row reduced any further. Then the rank of the coefficient matrix is 2, and also, the rank of the augmented matrix is 2. That means (rank M)=(rank A), but the number of unknown variables is 3. Thus only two unknowns can be expressed in terms of the third unknown. We can easily see this if we go back to the equation

$$\begin{pmatrix} 1 & 1 & 0 \\ 0 & -1 & 1 \\ 1 & 0 & 1 \end{pmatrix}\begin{pmatrix} I_1 \\ I_2 \\ I_3 \end{pmatrix} = \begin{pmatrix} 1 \\ 0 \\ 1 \end{pmatrix}, \tag{4.198}$$

which has row reduced to

$$\begin{pmatrix} 1 & 0 & 1 \\ 0 & -1 & 1 \\ 0 & 0 & 0 \end{pmatrix}\begin{pmatrix} I_1 \\ I_2 \\ I_3 \end{pmatrix} = \begin{pmatrix} 1 \\ 1 \\ 0 \end{pmatrix},$$

which indicates that instead of three equations, we have only two linearly independent equations,

$$I_1 + I_3 = 1, \; I_2 - I_3 = 1 \Rightarrow I_1 = 1 - I_3, \; I_2 = 1 + I_3. \tag{4.199}$$

Let us consider the results in equations (4.190)–(4.192). We determined these equations by intensionally omitting the current rule and using only the voltage rule for three closed loops. Let us introduce three vectors from these three equations defined by

$$\vec{A} = (1,\, 1,\, 0),\; \vec{B} = (0,\, -1,\, 0),\; \vec{C} = (1,\, 0,\, 1). \tag{4.200}$$

For these vectors, we can see that

$$\vec{A} + \vec{B} = \vec{C} \Rightarrow \vec{A} + \vec{B} - \vec{C} = 0 \tag{4.201}$$

and the three vectors are linearly dependent vectors. This shows that the third equation, we determine using the third loop, is not an independent equation, so we must always use both the voltage and current rules.

In general a set of N-dimensional vectors $\{\vec{A}_1,\, \vec{A}_2,\, \vec{A}_3,\, ...\vec{A}_N\}$ are linearly independent, when

$$k_1\vec{A}_1 + k_2\vec{A}_2 + k_3\vec{A}_3 + ... + k_n\vec{A}_n \neq 0, \tag{4.202}$$

for $k_i \neq 0$. Let us consider the three set of vectors in Cartesian coordinates,

$$\begin{aligned}
\vec{A}_1 &= a_{11}\hat{x} + a_{12}\hat{y} + a_{13}\hat{z}, \\
\vec{A}_2 &= a_{21}\hat{x} + a_{22}\hat{y} + a_{23}\hat{z}, \\
\vec{A}_3 &= a_{31}\hat{x} + a_{32}\hat{y} + a_{33}\hat{z}.
\end{aligned} \tag{4.203}$$

which are *not linearly independent*. We can write these vectors, using matrices, as

$$\begin{aligned}
&k_1\vec{A}_1 + k_2\vec{A}_2 + k_3\vec{A}_3 = 0 \\
&\Rightarrow k_1(a_{11}\hat{x} + a_{12}\hat{y} + a_{13}\hat{z}) + k_2(a_{21}\hat{x} + a_{22}\hat{y} + a_{23}\hat{z}) \\
&\quad + k_3(a_{31}\hat{x} + a_{32}\hat{y} + a_{33}\hat{z}) = 0 \\
&\Rightarrow \begin{pmatrix} k_1a_{11} & k_1a_{12} & k_1a_{13} \\ k_2a_{21} & k_2a_{21} & k_2a_{23} \\ k_3a_{31} & k_3a_{32} & k_3a_{33} \end{pmatrix} \begin{pmatrix} \hat{x} \\ \hat{y} \\ \hat{z} \end{pmatrix} = 0.
\end{aligned} \tag{4.204}$$

This equality holds when

$$\begin{vmatrix} k_1a_{11} & k_1a_{12} & k_1a_{13} \\ k_2a_{21} & k_2a_{21} & k_2a_{23} \\ k_3a_{31} & k_3a_{32} & k_3a_{33} \end{vmatrix} = 0. \tag{4.205}$$

(*See Problem 5 for the proof*). Using the properties of determinants, one can show that

$$\begin{vmatrix} k_1 a_{11} & k_1 a_{12} & k_1 a_{13} \\ k_2 a_{21} & k_2 a_{21} & k_2 a_{23} \\ k_3 a_{31} & k_3 a_{32} & k_3 a_{33} \end{vmatrix} = k_1 \begin{vmatrix} a_{11} & a_{12} & a_{13} \\ k_2 a_{21} & k_2 a_{21} & k_2 a_{23} \\ k_3 a_{31} & k_3 a_{32} & k_3 a_{33} \end{vmatrix}$$

$$= (k_1 k_2) \begin{vmatrix} a_{11} & a_{12} & a_{13} \\ a_{21} & a_{21} & a_{23} \\ a_{31} & a_{32} & a_{33} \end{vmatrix} = (k_1 k_2 k_3) \begin{vmatrix} a_{11} & a_{12} & a_{13} \\ a_{21} & a_{21} & a_{23} \\ a_{31} & a_{32} & a_{33} \end{vmatrix} = 0 \tag{4.206}$$

$$\Rightarrow (k_1 k_2 k_3) = 0 \text{ or } \begin{vmatrix} a_{11} & a_{12} & a_{13} \\ a_{21} & a_{21} & a_{23} \\ a_{31} & a_{32} & a_{33} \end{vmatrix} = 0.$$

Therefore, for the three vectors to be *linearly independent*, $k_i \neq 0$, for $i = 1, 2, 3$ and also

$$\begin{vmatrix} a_{11} & a_{12} & a_{13} \\ a_{21} & a_{21} & a_{23} \\ a_{31} & a_{32} & a_{33} \end{vmatrix} \neq 0. \tag{4.207}$$

This condition is applicable when the dimension of the vector is the same as the number of vectors. In other words, when the coefficient matrix is a square.

Example 4.9. Consider the set of three mutually orthogonal Cartesian unit vectors. Prove that these three unit vectors form a linearly independent set of vectors.

Solution: For the three unit vectors, we can write

$$\begin{aligned} \vec{A}_1 &= \hat{x}, \Rightarrow a_{11} = 1, \, a_{12} = a_{13} = 0, \\ \vec{A}_2 &= \hat{y}, \, a_{211} = 0, \, a_{22} = a_{23} = 0, \\ \vec{A}_3 &= \hat{z} \Rightarrow a_{31} = 1, \, a_{32} = a_{33} = 0, \end{aligned} \tag{4.208}$$

so that

$$\begin{vmatrix} k_1 a_1 & k_1 a_2 & k_1 a_2 \\ k_2 b_1 & k_2 b_2 & k_2 b_2 \\ k_3 c_1 & k_3 c_2 & k_3 c_2 \end{vmatrix} = \begin{vmatrix} k_1 & 0 & 0 \\ 0 & k_2 & 0 \\ 0 & 0 & k_3 \end{vmatrix} = k_1 k_2 k_3. \tag{4.209}$$

Therefore

$$\begin{vmatrix} k_1 a_1 & k_1 a_2 & k_1 a_2 \\ k_2 b_1 & k_2 b_2 & k_2 b_2 \\ k_3 c_1 & k_3 c_2 & k_3 c_2 \end{vmatrix} = k_1 k_2 k_3, \tag{4.210}$$

can be zero if and only if $k_1 = 0$, $k_2 = 0$, or $k_3 = 0$. This means that the Cartesian unit vectors are *linearly independent*.

Set of linear functions

Consider the set of function, $\{f_1(x), f_2(x), f_3(x), \ldots f_n(x)\}$ where the derivatives each of these function up to the order of $n - 1$ exist (i.e., $\{f'_m(x), f''_m(x), f'''_m(x), \ldots f_m^{n-1}(x)\}$ exist). When the determinant

$$W = \begin{vmatrix} f_1(x) & f_2(x)\ldots & f_n(x) \\ f'_1(x) & f'_2(x)\ldots & f'_n(x) \\ \vdots & \vdots & \vdots \\ f_1^{n-1}(x) & f_2^{n-1}(x)\ldots & f_n^{n-1}(x) \end{vmatrix} \neq 0, \tag{4.211}$$

the functions are *linearly independent*. The determinant W is called the *Wronskian* of the functions.

Example 4.10. Assuming that ω is a constant, verify if the set of functions $\{\sin(\omega t), \sin^2(\omega t)\}$ is a linearly dependent or independent set.

Solution: We can write the Wronskian for the two functions as

$$W = \begin{vmatrix} f_1(x) & f_2(x) \\ f'_1(x) & f'_2(x) \end{vmatrix} = \begin{vmatrix} \sin(\omega t) & \sin^2(\omega t) \\ \dfrac{d}{dt}[\sin(\omega t)] & \dfrac{d}{dt}[\sin^2(\omega t)] \end{vmatrix}$$

$$= \begin{vmatrix} \sin(\omega t) & \sin^2(\omega t) \\ \omega \cos(\omega t) & 2\omega \sin(\omega t)\cos(\omega t) \end{vmatrix} \tag{4.212}$$

$$= 2\omega \sin^2(\omega t)\cos(\omega t) - \omega \sin^2(\omega t)\cos(\omega t)$$

$$\Rightarrow W = \omega \sin^2(\omega t)\cos(\omega t) \neq 0$$

and the functions are independent.

Basis vectors

In general, any set of linearly independent vectors are mutually orthogonal set of vectors. Otherwise, the vectors in the set can be combined to form a set of mutually orthogonal vectors using a method called Gram–Schmidt orthogonalization, which we will see in the next section.

A Basis set of vectors: A set of vectors

$$\{\vec{r}_1, \vec{r}_2, \vec{r}_3 \ldots \vec{r}_N\}$$

is said to be basis set of vectors in a vector space when the vectors in the set are linearly independent. The unit vectors in the Cartesian coordinate system form a basis set of vectors in 3-D vector space.

A spanning set of vectors: a set of vectors spans a vector space if all the vectors in the space can be written as a linear combination of the basis set of vectors for that vector space.

Example 4.11. A two-dimensional vector space is defined by the plane formed by three points in space described by three vectors

$$\vec{r}_1 = (9, 0, 7), \ \vec{r}_2 = (0, -9, 13), \ \vec{r}_3 = (0, 0, 0). \tag{4.213}$$

For the set of vectors

$$\vec{A}_1 = (1, 4, -5), \ \vec{A}_2 = (5, 2, 1), \ \vec{A}_3 = (2, -1, 3), \ \vec{A}_4 = (3, -6, 11), \tag{4.214}$$

show that,

(a) the vectors $\vec{r}_1$ and $\vec{r}_2$ are a basis set of vectors for the two-dimensional vector space,

(b) the four vectors are a spanning set of vectors in the two-dimensional space formed by the two vectors $\vec{r}_1$ and $\vec{r}_2$.

Solution:

(a) First let's find the equation of the plane forming this two-dimensional vector space. To this end, we note that the vector normal to the plane formed by $\vec{r}_1 = (9, 0, 7)$ and $\vec{r}_2 = (0, -9, 13)$ is given by

$$\vec{N} = \vec{r}_1 \times \vec{r}_2 = \begin{vmatrix} \hat{x} & \hat{y} & \hat{z} \\ 9 & 0 & 7 \\ 0 & -9 & 13 \end{vmatrix} = 63\hat{x} - 117\hat{y} + 81\hat{z}. \tag{4.215}$$

Suppose we peak a point on this plane described by

$$\vec{r} = (x, y, z), \tag{4.216}$$

and chose our second point as $\vec{r}_3 = (0, 0, 0)$, which is on the plane, then one can write

$$\vec{r} - \vec{r}_3 = x\hat{x} + y\hat{y} + z\hat{z}, \tag{4.217}$$

and the equation of the plane defined by

$$(\vec{r} - \vec{r}_3) \cdot \vec{N} = 0$$
$$\Rightarrow [x\hat{x} + y\hat{y} + z\hat{z}] \cdot (63\hat{x} - 117\hat{y} + 81\hat{z}) = 0 \tag{4.218}$$
$$\Rightarrow 63x - 117y + 81z = 0.$$

This 2-D vector space, defined by this equation, is shown in figure 4.5.

(b) If the two vectors are basis vectors, these vectors must be linearly independent in this 2-D vector space. That means for

$$k_1\vec{r}_1 + k_2\vec{r}_2 \neq 0 \tag{4.219}$$

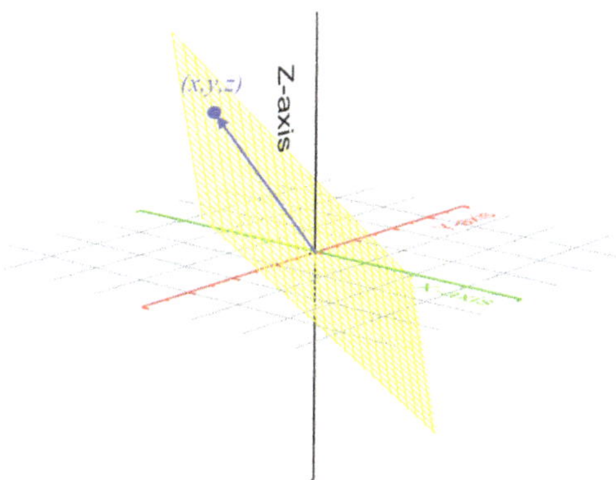

Figure 4.5. The 2-D vector space (the plane with cyan grid) defined by three points in 3-D space described by the vectors $\vec{r}_1 = (9, 0, 7)$, $\vec{r}_2 = (0, -9, 13)$, and $\vec{r}_3 = (0, 0, 0)$.

for non-zero values for k_1 and k_2. But what we find is

$$k_1(9, 0, 7) + k_2(0, -9, 13) = 0 \Rightarrow (9k_1, -9k_2, 7k_1 - 13k_2) = (0, 0, 0)$$
$$\Rightarrow k_1 = 0, \; k_2 = 0,$$

(4.220)

which verifies that equation (4.219) is true (figure 4.5).

(c) When we perform row reduction for the matrix formed by the four vectors,

$$M = \begin{bmatrix} 1 & 4 & -5 \\ 5 & 2 & 1 \\ 2 & -1 & 3 \\ 3 & -6 & 11 \end{bmatrix},$$

(4.221)

we find the basis vectors

$$M' = \begin{bmatrix} 9 & 0 & 7 \\ 0 & -9 & 13 \\ 0 & 0 & 0 \\ 0 & 0 & 0 \end{bmatrix}.$$

(4.222)

All the row operations are reversible and we can write the vectors

$$\vec{A} = (1, 4, -5), \; \vec{B} = (5, 2, 1),$$
$$\vec{C} = (2, -1, 3), \; \vec{D} = (3, -6, 11),$$

(4.223)

in terms of the basis set of vectors

$$\vec{r_1} = (9, 0, 7), \vec{r_2} = (0, -9, 13). \tag{4.224}$$

This verifies that the vectors $\vec{A}$, $\vec{B}$, $\vec{C}$ and $\vec{D}$ are a spanning set of vectors in the two-dimensional space formed by the two basis vectors $\vec{r_1}$ and $\vec{r_2}$.

4.10 Gram–Schmidt orthogonalization

We call a set of vectors orthonormal when they are mutually orthogonal (perpendicular) and normalized (i.e., each vector has a unit magnitude). The Cartesian coordinate unit vectors $\hat{x}$, $\hat{y}$, $\hat{z}$ or example, form an orthonormal set of vectors. The Gram–Schmidt orthogonalization is a method we can use to find orthonormal basis vectors from a set of basis vectors. Let us consider a set of three basis vectors $\vec{A}$, $\vec{B}$, and $\vec{C}$, shown in figure 4.6. To find an orthonormal set of basis vectors, using the Gram–Schmidt orthogonalization method, we follow the steps:

1. Normalize $\vec{A}$ - we get the first element in the orthonormal set of basis vectors,

$$\hat{A_1} = \frac{\vec{A}}{|\vec{A}|}. \tag{4.225}$$

We can equally choose vector $\vec{B}$ or $\vec{C}$.

2. Decompose the second vector into components parallel and perpendicular to vector $\vec{A}$. The second vector $\vec{B}$ can be expressed as a sum of these two components as shown in figure 4.7,

$$\vec{B} = \vec{B}_{\hat{A_1}}^{\parallel} + \vec{B}_{\hat{A_1}}^{\perp}, \tag{4.226}$$

where $\vec{B}_{\hat{A}}^{\parallel}$ is the parallel and $\vec{B}_{\hat{A}}^{\perp}$ is the perpendicular component to $\hat{A_1}$. Applying scalar product of two vectors, one can express the component of the second vector $\vec{B}$ along the direction of $\hat{A_1}$ as

$$\vec{B}_{\hat{A_1}}^{\parallel} = (\vec{B} \cdot \hat{A_1})\hat{A_1}. \tag{4.227}$$

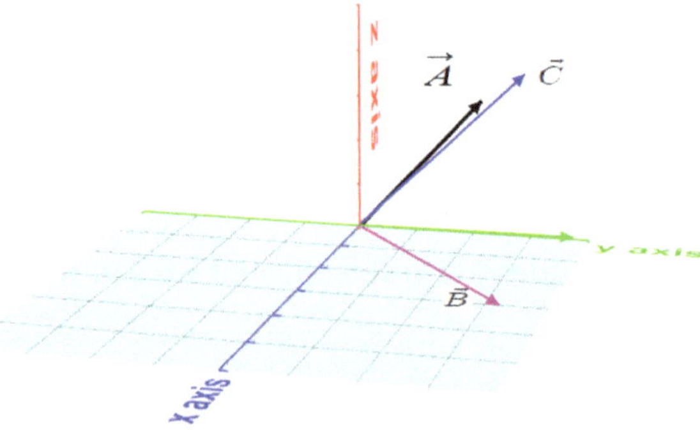

Figure 4.6. A set of three basis vectors $\vec{A}$, $\vec{B}$, and $\vec{C}$.

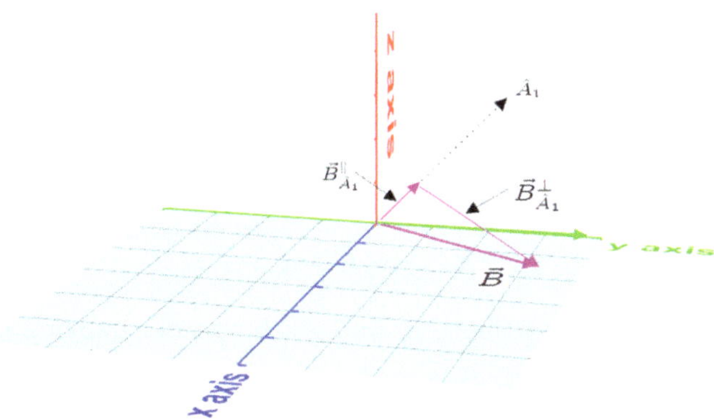

Figure 4.7. The components of vector $\vec{B}$ along the directions parallel and perpendicular to the vector $\vec{A}$.

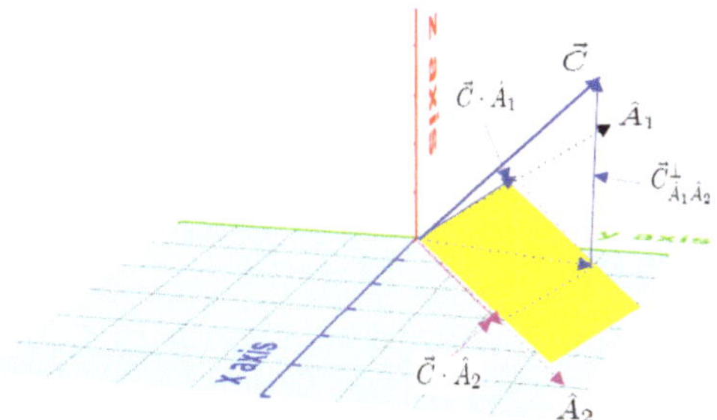

Figure 4.8. Decomposing vector $\vec{C}$ along the directions $\hat{A}_1$, $\hat{A}_2$ (which are perpendicular to one another), and normal to the plane formed by these two vectors.

Then we subtract this component from $\vec{B}$

$$\vec{A}_2 = \vec{B}_{\hat{A}_1}^{\perp} = \vec{B} - \vec{B}_{\hat{A}_1}^{\parallel} = \vec{B} - (\vec{B} \cdot \hat{A}_1)\hat{A}_1, \qquad (4.228)$$

and normalize the resulting vector,

$$\hat{A}_2 = \frac{\vec{A}_2}{|\vec{A}_2|} = \frac{\vec{B} - (\vec{B} \cdot \hat{A}_1)\hat{A}_1}{|\vec{B} - (\vec{B} \cdot \hat{A}_1)\hat{A}_1|}. \qquad (4.229)$$

This is the second orthonormal vector in the set. It is orthogonal to $\hat{A}_1$ and it has a unit magnitude.

3. We now have two orthonormal vectors $\hat{A}_1$ and $\hat{A}_2$. We can decompose the third vector $\vec{C}$ into three components one is parallel to $\hat{A}_1$, the second is parallel to $\hat{A}_2$, and the third is perpendicular to both $\hat{A}_1$ and $\hat{A}_2$ (or the plane formed by these two vectors) as shown in figure 4.8,

$$\vec{C} = \vec{C}_{\hat{A}_1}^{\parallel} + \vec{C}_{\hat{A}_2}^{\parallel} + \vec{C}_{\hat{A}_1\hat{A}_2}^{\perp}. \qquad (4.230)$$

Noting that the components of the third vector $\vec{C}$ along the directions of $\hat{A}_1$ and $\hat{A}_2$ are given by

$$\vec{C}_{\hat{A}_1} = (\vec{C} \cdot \hat{A}_1)\hat{A}_1, \quad \vec{C}_{\hat{A}_2} = (\vec{C} \cdot \hat{A}_2)\hat{A}_2, \tag{4.231}$$

one can write

$$\vec{A}_3 = \vec{C}_{\hat{A}_1\hat{A}_2}^{\perp} = \vec{C} - \vec{C}_{\hat{A}_1}^{\parallel} - \vec{C}_{\hat{A}_2}^{\parallel},$$

which is orthogonal to both $\hat{A}_1$ and $\hat{A}_2$. Upon normalizing this vector

$$\hat{A}_3 = \frac{\vec{C} - (\vec{C} \cdot \hat{A}_1)\hat{A}_1 - (\vec{C} \cdot \hat{A}_2)\hat{A}_2}{|\, \vec{C} - (\vec{C} \cdot \hat{A}_1)\hat{A}_1 - (\vec{C} \cdot \hat{A}_2)\hat{A}_2 \,|}, \tag{4.232}$$

we find the third orthonormal vector in the set.

For more than three sets of basis vectors by following these steps one can form a set of orthonormal basis vectors.

Example 4.12. For the set of basis vectors,

$$\vec{A} = (0, 2, 0, 0), \ \vec{B} = (3, -4, 0, 0), \ \vec{C} = (1, 2, 3, 4), \tag{4.233}$$

use the Gram–Schmidt orthogonalization method to find an orthonormal set of vectors.

Solution: Following step 1, we normalize vector $\vec{A}$

$$\hat{A}_1 = \frac{\vec{A}}{|\,\vec{A}\,|} = \frac{(0, 2, 0, 0)}{2} = (0, 1, 0, 0). \tag{4.234}$$

Then according to step 2, we find the component of the vector $\vec{B}$ along $\hat{A}_1$,

$$\vec{B}_{\hat{A}_1} = (\vec{B} \cdot \hat{A}_1)\hat{A}_1 = [(3, -4, 0, 0) \cdot (0, 1, 0, 0)](0, 1, 0, 0) = (0, -4, 0, 0) \tag{4.235}$$

and determine the second orthonormal vector in the set,

$$\hat{A}_2 = \frac{\vec{B} - (\vec{B} \cdot \hat{A}_1)\hat{A}_1}{|\, \vec{B} - (\vec{B} \cdot \hat{A}_1)\hat{A}_1 \,|} = \frac{(3, -4, 0, 0) - (0, -4, 0, 0)}{|\,(3, -4, 0, 0) - (0, -4, 0, 0)\,|} = \frac{(3, 0, 0, 0)}{|\,(3, 0, 0, 0)\,|} \tag{4.236}$$

$$\Rightarrow \hat{A}_2 = (1, 0, 0, 0).$$

To find the third orthonormal vector we use step-3. We first find the components of vector $\vec{C}$ along the direction of $\hat{A}_1$ and $\hat{A}_2$,

$$\vec{C} \cdot \hat{A}_1 = (1, 2, 3, 4) \cdot (0, 1, 0, 0) = 2,$$
$$\vec{C} \cdot \hat{A}_2 = (1, 2, 3, 4) \cdot (1, 0, 0, 0) = 1. \tag{4.237}$$

Then noting that the component of $\vec{C}$ that is normal to these two vectors is

$$\begin{aligned}
\vec{A}_3 &= \vec{C} - (\vec{C} \cdot \hat{A}_1)\hat{A}_1 - (\vec{C} \cdot \hat{A}_2)\hat{A}_2 \\
&= (1, 2, 3, 4) - 2(0, 1, 0, 0) - (1, 0, 0, 0) \\
&\Rightarrow \vec{A}_3 = (0, 0, 3, 4),
\end{aligned} \tag{4.238}$$

one finds for the third orthonormal basis vector

$$\hat{A}_3 = \frac{\vec{A}_3}{\mid \vec{A}_3 \mid} = \frac{(0, 0, 3, 4)}{\mid (0, 0, 3, 4) \mid} = \left(0, 0, \frac{3}{5}, \frac{4}{5}\right). \tag{4.239}$$

4.11 Matrices and Mathematica

The following commands are some of the basic commands that we can use to carry out various operations associated with matrices.

A 2×2 matrix:

$$\begin{aligned}
Input &: m = \{\{a_{11}, a_{12}\}, \{a_{21}, a_{22}\}\} \\
Out &: \{\{a_{11}, a_{12}\}, \{a_{21}, a_{22}\}\}
\end{aligned} \tag{4.240}$$

Writing in a matrix form:

$$\begin{aligned}
Input &: \mathrm{MatrixForm}[\{\{a_{11}, a_{12}\}, \{a_{21}, a_{22}\}\}] \\
Out &: \begin{pmatrix} a_{11} & a_{12} \\ a_{21} & a_{22} \end{pmatrix}
\end{aligned} \tag{4.241}$$

Here is the first row:

$$\begin{aligned}
Input &: m[[1]] \\
Out &: \{a_{11}, a_{12}\}
\end{aligned} \tag{4.242}$$

Here is the element m_{12}:

$$\begin{aligned}
Input &: m[[1, 2]] \\
Out &: a_{12}
\end{aligned} \tag{4.243}$$

The rank of the matrix:

$$\begin{aligned}
Input &: \mathrm{MatrixRank}[m] \\
Out &: 2
\end{aligned} \tag{4.244}$$

A 3×3 matrix A:

$$\begin{aligned}
Input &: \mathrm{MatrixForm}[A = \{\{1, 2, 3\}, \{4, 5, 6\}, \{7, 8, 9\}\}] \\
Out &: \begin{pmatrix} 1 & 2 & 3 \\ 4 & 5 & 6 \\ 7 & 8 & 9 \end{pmatrix}
\end{aligned} \tag{4.245}$$

The rank of matrix A:

$$Input: \text{MatrixRank}[A]$$
$$Out: 2$$

(4.246)

Row reduction of a square matrix A:

$$Input: \text{RowReduce}[A]//\text{MatrixForm}$$
$$Out: \begin{pmatrix} 1 & 0 & -1 \\ 0 & 1 & 2 \\ 0 & 0 & 0 \end{pmatrix}$$

(4.247)

Row reduction on a rectangular matrix:

$$Input: \text{RowReduce}[\{\{1, 2, 3, 1, 0, 0\}, \{4, 5, 6, 0, 1, 0\}, \{7, 8, 9, 0, 0, 1\}\}]$$
$$Out: \left\{ \left\{ 1, 0, -1, 0, -\frac{8}{3}, \frac{5}{3} \right\}, \left\{ 0, 1, 2, 0, \frac{7}{3}, -\frac{4}{3} \right\}, \{0, 0, 0, 1, -2, 1\} \right\}$$

(4.248)

$$Input: \text{MatrixForm}[\%]$$
$$Out: \begin{pmatrix} 1 & 0 & -1 & 0 & -\frac{8}{3} & \frac{5}{3} \\ 0 & 1 & 2 & 0 & \frac{7}{3} & -\frac{4}{3} \\ 0 & 0 & 0 & 1 & -2 & 1 \end{pmatrix}$$

(4.249)

Transpose of matrix m:

$$Input: \text{MatrixForm}[\text{Transpose}[m]]$$
$$Out: \begin{pmatrix} a_{11} & a_{21} \\ a_{12} & a_{22} \end{pmatrix}$$

(4.250)

Conjugate and transpose of matrix m:

$$Input: \text{MatrixForm}[\text{ConjugateTranspose}[m]]$$
$$Out: \begin{pmatrix} \text{Conjugate}\,a_{11} & \text{Conjugate}\,a_{21} \\ \text{Conjugate}\,a_{12} & \text{Conjugate}\,a_{22} \end{pmatrix}$$

(4.251)

The minors for matrix m:

$$Input: \text{MatrixForm}[\text{Minors}[m]]$$
$$Out: \begin{pmatrix} a_{11} & a_{12} \\ a_{21} & a_{22} \end{pmatrix}$$

(4.252)

A 3×3 matrix and its minors:

$$Input: (\text{mat} = \text{Array}[a_{\#\#}\&, \{3, 3\}])//\text{MatrixForm}$$

$$Out: \begin{pmatrix} a_{1,1} & a_{1,2} & a_{1,3} \\ a_{2,1} & a_{2,2} & a_{2,3} \\ a_{3,1} & a_{3,2} & a_{3,3} \end{pmatrix} \tag{4.253}$$

$$In: \text{Minors}[\text{mat}]//\text{MatrixForm}$$

$$Out: \begin{pmatrix} -a_{1,2}a_{2,1} + a_{1,1}a_{2,2} & -a_{1,3}a_{2,1} + a_{1,1}a_{2,3} & -a_{1,3}a_{2,2} + a_{1,2}a_{2,3} \\ -a_{1,2}a_{3,1} + a_{1,1}a_{3,2} & -a_{1,3}a_{3,1} + a_{1,1}a_{3,3} & -a_{1,3}a_{3,2} + a_{1,2}a_{3,3} \\ -a_{2,2}a_{3,1} + a_{2,1}a_{3,2} & -a_{2,3}a_{3,1} + a_{2,1}a_{3,3} & -a_{2,3}a_{3,2} + a_{2,2}a_{3,3} \end{pmatrix} \tag{4.254}$$

The inverse matrix m:

$$In: \text{MatrixForm}[\text{Inverse}[m]]$$

$$Out: \begin{pmatrix} \dfrac{a_{22}}{-a_{12}a_{21} + a_{11}a_{22}} & -\dfrac{a_{12}}{-a_{12}a_{21} + a_{11}a_{22}} \\ -\dfrac{a_{21}}{-a_{12}a_{21} + a_{11}a_{22}} & \dfrac{a_{11}}{-a_{12}a_{21} + a_{11}a_{22}} \end{pmatrix} \tag{4.255}$$

Two component vector (a column matrix):

$$In: \text{MatrixForm}[v = \{x, y\}]$$

$$Out: \begin{pmatrix} x \\ y \end{pmatrix} \tag{4.256}$$

Multiplying with a scalar and adding vectors component by component (p and q are treated as scalars):

$$In: \text{MatrixForm}[v + \{xp, yp\} + \{xpp, ypp\}]$$

$$Out: \begin{pmatrix} x + xp + xpp \\ y + yp + ypp \end{pmatrix} \tag{4.257}$$

Multiply a matrix (m) by a vector (v):

$$In: \text{MatrixForm}[m. v]$$

$$Out: \begin{pmatrix} xa_{11} + ya_{12} \\ xa_{21} + ya_{22} \end{pmatrix} \tag{4.258}$$

Multiply a matrix (m) by a matrix (m):

$$In: \text{MatrixForm}[m. m]$$

$$Out: \begin{pmatrix} a_{11}^2 + a_{12}a_{21} & a_{11}a_{12} + a_{12}a_{22} \\ a_{11}a_{21} + a_{21}a_{22} & a_{12}a_{21} + a_{22}^2 \end{pmatrix} \tag{4.259}$$

Multiply a vector (v) by a matrix (m):

$$In: \text{MatrixForm}[v.\,m]$$

$$Out: \begin{pmatrix} xa_{11} + ya_{21} \\ xa_{12} + ya_{22} \end{pmatrix} \tag{4.260}$$

Matrix and vectors multiplication resulting in a scalar:

$$In: \text{MatrixForm}[v.\,m.\,v]$$
$$Out: x(xa_{11} + ya_{21}) + y(xa_{12} + ya_{22}) \tag{4.261}$$

Note: Because of the way the Wolfram Language uses lists to represent vectors and matrices, you never have to distinguish between 'row' and 'column' vectors.

Finding the orthonormal set of vectors: Starting from the set of vectors $\{u, v\}$, we find the orthonormal set of two vectors using the following commands:

$$In: \text{Orthogonalize}[\{u, v\}]$$

$$Out: \left\{ \frac{u}{\sqrt{u.\,u}}, \frac{v - \dfrac{u\dfrac{u}{\sqrt{u.\,u}}.\,v}{\sqrt{u.\,u}}}{\sqrt{\left(v - \dfrac{u\dfrac{u}{\sqrt{u.\,u}}.\,v}{\sqrt{u.\,u}}\right).\,\left(v - \dfrac{u\dfrac{u}{\sqrt{u.\,u}}.\,v}{\sqrt{u.\,u}}\right)}} \right\} \tag{4.262}$$

$$In: v = \{1, 3, 2\}$$
$$Out: \{1, 3, 2\}$$

$$In: u = \{2, -6, -4\}$$
$$Out: \{2, -6, -4\} \tag{4.263}$$

$$In: \text{Orthogonalize}[\{u, v\}]//\text{MatrixForm}$$

$$Out: \begin{pmatrix} \dfrac{1}{\sqrt{14}} & -\dfrac{3}{\sqrt{14}} & -\sqrt{\dfrac{2}{7}} \\ \sqrt{\dfrac{13}{14}} & \dfrac{3}{\sqrt{182}} & \sqrt{\dfrac{2}{91}} \end{pmatrix} \tag{4.264}$$

When one of the vectors is linearly dependent on the vectors preceding it, the corresponding position in the result will be a zero vector:

$$In: p = \text{Projection}[u, v]$$

$$Out: \left\{ -\frac{12}{7}, -\frac{36}{7}, -\frac{24}{7} \right\} \tag{4.265}$$

$$In: \text{Orthogonalize}[\{v, p, u\}]//\text{MatrixForm}$$

$$Out: \begin{pmatrix} \dfrac{1}{\sqrt{14}} & \dfrac{3}{\sqrt{14}} & \sqrt{\dfrac{2}{7}} \\ 0 & 0 & 0 \\ \sqrt{\dfrac{13}{14}} & -\dfrac{3}{\sqrt{182}} & -\sqrt{\dfrac{2}{91}} \end{pmatrix} \tag{4.266}$$

2-D rotation matrix that rotates 2-D vectors counterclockwise by θ radians about the z-axis:

$$In: \text{RotationMatrix}[\theta]//\text{MatrixForm}$$

$$Out: \begin{pmatrix} \text{Cos}[\theta] & -\text{Sin}[\theta] \\ \text{Sin}[\theta] & \text{Cos}[\theta] \end{pmatrix} \tag{4.267}$$

When $\theta = 30°$:

$$In: \text{RotationMatrix}[30\text{Degree}]// \text{MatrixForm}$$

$$Out: \begin{pmatrix} \dfrac{\sqrt{3}}{2} & -\dfrac{1}{2} \\ \dfrac{1}{2} & \dfrac{\sqrt{3}}{2} \end{pmatrix} \tag{4.268}$$

Apply rotation by θ to a unit vector in the x direction:

$$In: \text{RotationMatrix}[\theta].\ \{1, 0\}$$
$$Out: \{\text{Cos}[\theta], \text{Sin}[\theta]\} \tag{4.269}$$

3-D rotation around the z-axis:

$$In: \text{RotationMatrix}[\theta, \{0, 0, 1\}]// \text{MatrixForm}$$

$$Out: \begin{pmatrix} \text{Cos}[\theta] & -\text{Sin}[\theta] & 0 \\ \text{Sin}[\theta] & \text{Cos}[\theta] & 0 \\ 0 & 0 & 1 \end{pmatrix} \tag{4.270}$$

3-D rotation around the y-axis:

$$In: \text{RotationMatrix}[\theta, \{0, 1, 0\}]// \text{MatrixForm}$$

$$Out: \begin{pmatrix} \text{Cos}[\theta] & 0 & \text{Sin}[\theta] \\ 0 & 1 & 0 \\ -\text{Sin}[\theta] & 0 & \text{Cos}[\theta] \end{pmatrix} \tag{4.271}$$

3-D rotation around the x−axis:

$$In: \text{RotationMatrix}[\theta, \{1, 0, 0\}]// \text{MatrixForm}$$

$$Out: \begin{pmatrix} 1 & 0 & 0 \\ 0 & \cos[\theta] & -\sin[\theta] \\ 0 & \sin[\theta] & \cos[\theta] \end{pmatrix} \tag{4.272}$$

4.12 Homework assignment

Problem 1. Solve the following set of equations for the four unknowns by reducing the matrix using elementary row operations...

$$x - 2y - 4 = 0,\ 2z + w = 5,\ w + y = 2,\ 3x - 1 = 5z. \tag{4.273}$$

Problem 2. Solve the following dc circuit problems shown in figures 4.9 and 4.10 using row reduction of matrices.

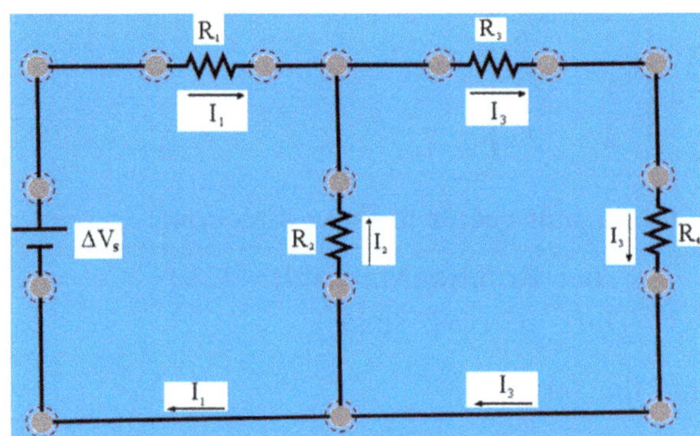

Figure 4.9. An electrical circuit: $\Delta V_s = 100V$, $R_1 = 5.2k\Omega$, $R_2 = 2.2k\Omega$, $R_3 = 1.2k\Omega$, and $R_4 = 3.2k\Omega$.

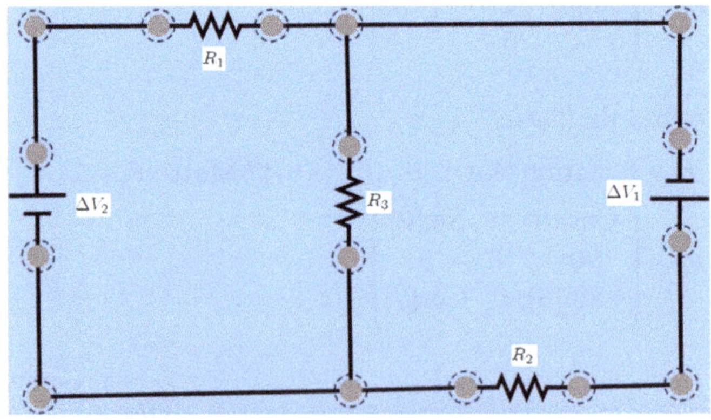

Figure 4.10. An electrical circuit: $\Delta V_1 = 20V$, $\Delta V_2 = 15V$ $R_1 = 1.0k\Omega$, $R_2 = 2.0k\Omega$, and $R_3 = 3.0k\Omega$.

Problem 3.

(a) Provide a clear statement of what is meant by the specific gravity (*sg*) of a substance?

(b) Provide a clear statement of what is Archimedes' Principle?

(c) An object is composed of x grams of lead ($sg = 11$) and y grams of tin ($sg = 7$). The object has a mass of 82 *grams* in air and an apparent mass of 77 *grams* when suspended in oil with $sg = 0.5$. Find the values of x and y.

Problem 4. The half-life of a radioactive substance is the time it takes for half of it to decay into some other substances called decay products. Suppose that a radioactive sample consists of components A and B having half-lives $2hr$ and $3hr$, respectively. Assume that the decay products are gases that escape at once. The sample is found to have 56 *grams* mass after $12hr$ and 12 *grams* after $18hr$. Find the masses of A and B that were initially present in the sample.

Problem 5. For the linear equations

$$a_{11}x_1 + a_{12}x_2 = 0,$$
$$a_{21}x_1 + a_{22}x_2 = 0, \tag{4.274}$$

which can be put in the form

$$Ar = k, \tag{4.275}$$

where

$$A = \begin{bmatrix} a_{11} & a_{12} \\ a_{21} & a_{22} \end{bmatrix}, \; r = \begin{bmatrix} x_1 \\ x_2 \end{bmatrix}, \; k = \begin{bmatrix} 0 \\ 0 \end{bmatrix}. \tag{4.276}$$

Show that the determinant of the matrix A is zero.

Problem 6.

(a) Given the following sets of equations

$$x' = \frac{1}{2}(x + y\sqrt{3}), \; y' = \frac{1}{2}(-x\sqrt{3} + y), \; \text{and } z' = z \tag{4.277}$$

and

$$x'' = \frac{1}{2}(-x' + y'\sqrt{3}), \; y'' = -\frac{1}{2}(x'\sqrt{3} + y'), \; \text{and } z'' = z' \tag{4.278}$$

write each set of equations using matrices and then solve for x'' and y''' in terms of x and y. These equations represent rotations about the z-axis. Find the rotation angles by comparing the matrices with the rotation matrix derived in section 4.8.

(b) Prove that the matrices corresponding to the two sets of equations given in (a) can correspond to rotation matrices by demonstrating that they are orthogonal matrices.

Problem 7. Show that

$$f(\vec{r}) = \vec{A} \cdot \vec{r} + 3 \tag{4.279}$$

is a linear or non-linear function.

Problem 8. Show that

$$f(\vec{r}) = \vec{A} \times \vec{r} \tag{4.280}$$

is a linear or non-linear vector function.

Problem 9. Define the integral with respect to x from 0 to 1; the object being operated on is a function of x. Show that this integral operator is a linear or non-linear operator. (This means we want to show that the operator

$$\int_0^1 (\ )dx \tag{4.281}$$

is a linear or non-linear operator).

Problem 10.

    (a) For the matrix

$$A = \begin{bmatrix} 1 & 0 & 5i \\ -2i & 2 & 0 \\ 1 & 1+i & 0 \end{bmatrix} \tag{4.282}$$

    find the transpose, the inverse, the complex conjugate, and the transpose conjugate.

    (b) Show that the product AA^T is a symmetric matrix.

Problem 11. For each of the following problems, write and row reduce the augmented matrix to determine whether the given set of equations have exactly one solution, no solution, or an infinite set of solutions.

    (a)

$$\begin{aligned} 2x + y - z &= 2 \\ 4x + y - 2z &= 3 \end{aligned} \tag{4.283}$$

    (b)

$$\begin{aligned} 2x + 5y + z &= 2 \\ x + y + 2z &= 1 \\ x + 5z &= 3. \end{aligned} \tag{4.284}$$

Problem 12. Find the rank of the matrix

$$M = \begin{pmatrix} 1 & 1 & 4 & 3 \\ 3 & 1 & 10 & 7 \\ 4 & 2 & 14 & 10 \\ 2 & 0 & 6 & 4 \end{pmatrix}. \tag{4.285}$$

Problem 13. Write the vectors

$$\vec{A} = (1, 4, -5), \ \vec{B} = (5, 2, 1), \ \vec{C} = (2, -1, 3), \ \vec{D} = (3, -6, 11) \tag{4.286}$$

as a linear combination of the vectors

$$\vec{a} = (9, 0, 7), \ \vec{b} = (0, -9, 13). \tag{4.287}$$

Problem 14. For each of the following problems, show that the given functions are linearly independent functions.

(a)

$$f_1(x) = \sin(x), f_2(x) = \cos(x). \tag{4.288}$$

(b)

$$f_1(x) = x, f_2(x) = e^x, f_3(x) = xe^x. \tag{4.289}$$

Problem 15. Show that the matrix

$$M = \begin{pmatrix} \dfrac{1 + i\sqrt{3}}{4} & \dfrac{\sqrt{3}}{2\sqrt{2}}(1 + i) \\[3ex] -\dfrac{\sqrt{3}}{2\sqrt{2}}(1 + i) & \dfrac{\sqrt{3} + i}{4} \end{pmatrix} \tag{4.290}$$

is a unitary matrix. That means we need to show that

$$M^{-1} = M^T. \tag{4.291}$$

Problem 16. In algebra based or calculus based physics you were introduced to Kirchhoff's voltage and current rules. The following problem is based on the application of these rules to the circuit shown in figure 4.11. You must use the currents (I_1, I_2, and I_3) as labeled in the figure!

(a) Applying Kirchhoff's voltage rule to two different loops and current rule to one node, find three independent sets of linear equations.

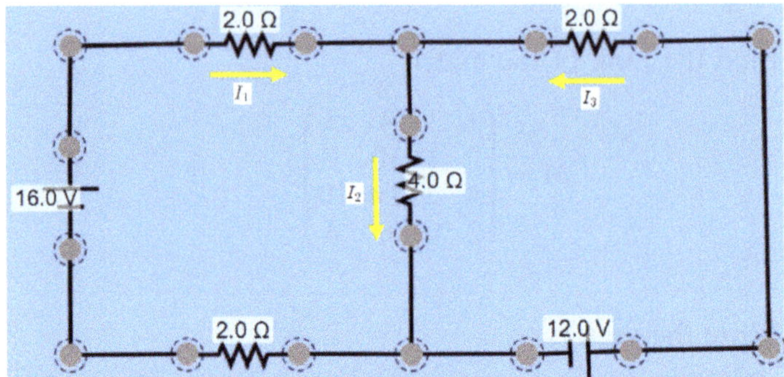

Figure 4.11. Mixed dc circuit.

NB: If you do not do this part correctly, you will not get the right answers to the questions (b) to (g).

(b) Express the three equations using matrix representation.

(c) Find the rank of the augmented and the coefficient matrices *by performing row operations*. Does this set of equations give exactly one solution, no solution, or an infinite set of solutions?

(d) Find the determinant of the coefficient matrix.

(e) Find the cofactor matrix for the coefficient matrix.

(f) Find the inverse matrix for the coefficient matrix.

(g) *Applying the inverse matrix* find the current I_1, I_2, and I_3.

Problem 17. Suppose you made two successive counterclockwise rotations of the $x-y-z$ coordinate system. The first is a rotation about the x-axis by angle θ_1 that leads to the $x'-y'-z'$ coordinate system as shown in figure 4.12. The second rotation is a rotation about the x'-axis by an angle θ_2 that leads to $x''-y''-z''$ as shown in figure 4.13

    (a) Find the two sets of equations for these two rotations and write these equations using matrices.

    (b) Find the matrix equation that relates (x'', y'', z'') with (x, y, z), for $\theta_1 = \pi/3$ and $\theta_2 = 2\pi/3$.

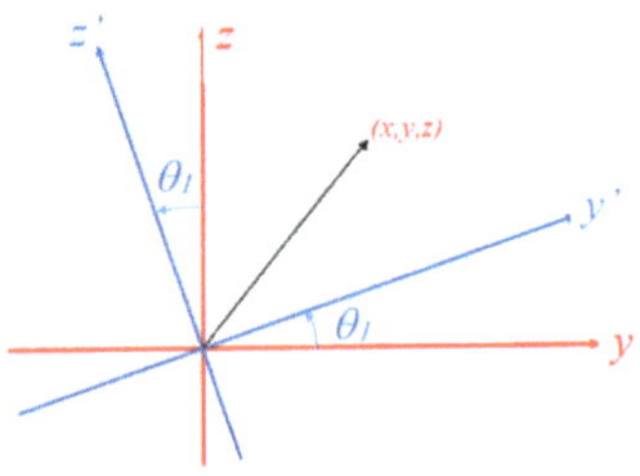

Figure 4.12. A rotation about the x-axis by an angle θ_1 in a counterclockwise direction.

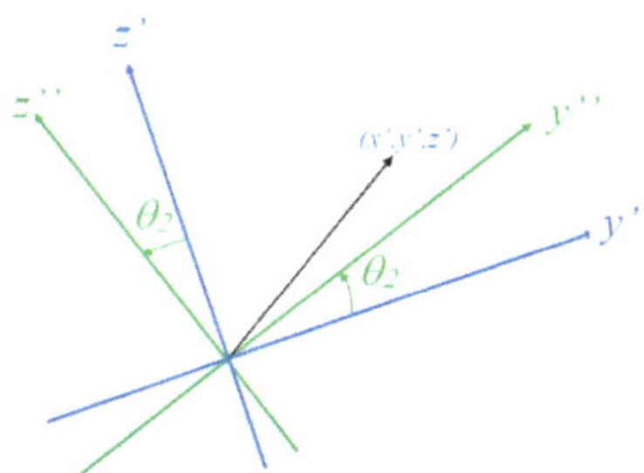

Figure 4.13. A rotation about the x'-axis by an angle θ_2 in a counterclockwise direction.

Problem 18. For the set of basis vectors

$$\vec{A} = (0, 0, 0, 7),\ \vec{B} = (2, 0, 0, 5),\ \vec{C} = (3, 1, 1, 4)$$

find an orthonormal set of vectors *using the Gram–Schmidt method.*

Problem 19. In quantum mechanics the Pauli spin matrices for spin-1/2 particles (like the electron), are defined by

$$\sigma_x = \frac{\hbar}{2}\begin{bmatrix} 0 & 1 \\ 1 & 0 \end{bmatrix},\ \sigma_y = \frac{\hbar}{2}\begin{bmatrix} 0 & -i \\ i & 0 \end{bmatrix}, \tag{4.292}$$

where $\hbar = h/2\pi$ and h is Planck's constant.

(a) Find the determinant, the cofactor, adjoint, and inverse for these matrices.
(b) Find the matrices defined by

$$\sigma_+ = \sigma_x + i\sigma_y,\ \sigma_- = \sigma_x - i\sigma_y. \tag{4.293}$$

(c) Identify the Hermitian matrices among σ_x, σ_y, σ_+, and σ_-. Explain the reason why these matrices are Hermitian.

IOP Publishing

Studies in Theoretical Physics, Volume 1
Fundamental mathematical methods
Daniel Erenso and Victor Montemayor

Chapter 5

Introduction to differential calculus I

In advanced undergraduate or graduate physics courses, we often deal with differential calculus. In this chapter, we will get introduced to the basics of differential calculus. These include partial differentiation and total differential of multivariable functions and their applications to real physical problems. In particular, we see how we determine extremum (maximum and minimum) points in single and multivariable functions and optimizations. We use real physical problems to illustrate these applications. At the end of the chapter, we will also see how we do partial differentiation and optimizations using Mathematica.

5.1 Partial differentiation

Before we get into partial differentiation, let us see how we differentiate a function of a single variable $(z(x))$,

$$\frac{dz(x)}{dx},$$

what it means, and what it represents or tells us about this function. To this end, consider a curved path on the x–z plane defined by the function $z(x, b)$ as shown in figure 5.1. The y-coordinate b is a constant and the function $z(x, b) = z(x)$ depends only on the variable x. In fact, as shown in figure 5.1, it defines the points in 3-D space with coordinates $(x, b, z(x, b))$ on the curved line. Suppose we draw a tangent line to this curved path at a point with coordinates $(x, b, z(x, b))$ (see figure 5.1). From introductory algebra, we know that the slope m of this tangent line is determined by

$$m = \frac{\Delta z}{\Delta x},$$

since this line is on the x–z plane. In the limit as $\Delta x \to 0$,

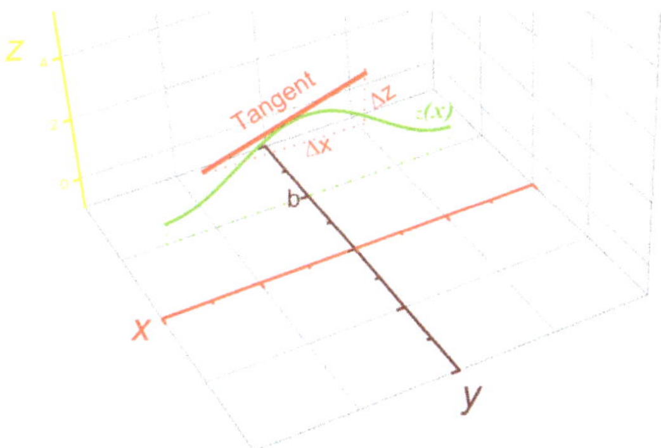

Figure 5.1. A curved path on the x–z plane defined by the function $z(x)$.

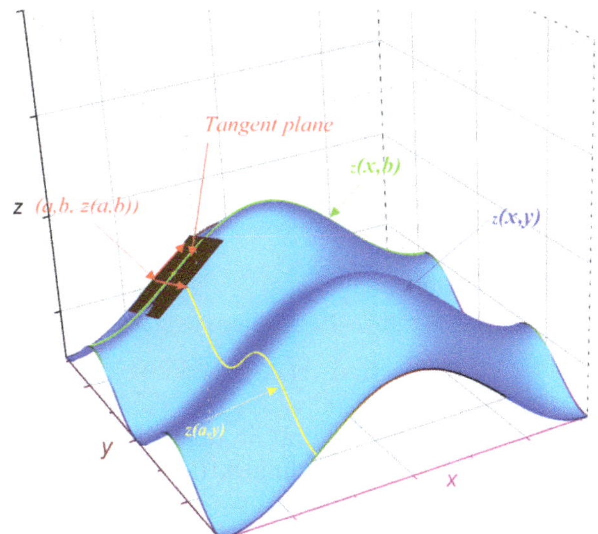

Figure 5.2. A curved surface in 3-D space.

$$m = \lim_{\Delta x \to 0} \frac{\Delta z(x)}{\Delta x} = \lim_{\Delta x \to 0} \left[\frac{z(x + \Delta x, b) - z(x, b)}{\Delta x} \right] = \frac{dz(xb)}{dx} \qquad (5.1)$$

gives the derivative of the function $z(x)$ at point $(x, b, z(x, b))$ on the curve defined by the function $z(x, b)$. Therefore, the differentiation of the function $z(x, b)$

$$\frac{dz(x, b)}{dx} = \lim_{\Delta x \to 0} \left[\frac{z(x + \Delta x, b) - z(x, b)}{\Delta x} \right],$$

tells us by how much the curved line is sloppy at the point $(x, b, z(x, b))$.

Now instead of a curved line let's consider a curved surface defined by the function of two variables $z(x, y)$ shown in figure 5.2. The differentiation of the function $z(x, y)$, when y is kept constant at the point $(x, y, z(x, y))$ on the curved surface shown in figure 5.2 is the partial differentiation with respect to x and is given by

$$\frac{\partial z(x, y)}{\partial x} = \lim_{\Delta x \to 0} \frac{\Delta z(x, y)}{\Delta x} = \lim_{\Delta x \to 0} \left[\frac{z(x + \Delta x, y) - z(x, y)}{\Delta x} \right]. \tag{5.2}$$

Similarly, the partial differentiation of the function $z(x, y)$ with respect to y is obtained by keeping x constant and is given by

$$\frac{\partial z(x, y)}{\partial y} = \lim_{\Delta y \to 0} \frac{\Delta z(x, y)}{\Delta y} = \lim_{\Delta y \to 0} \left[\frac{z(x, y + \Delta y) - z(x, y)}{\Delta y} \right]. \tag{5.3}$$

The two partial differentiation in equation (5.2) and (5.3) tells the slope along the x-direction and y-direction of a plane tangent to the curved surface at the point $(x, y, z(x, y))$, respectively. Generally, for a function $f(x_1, x_2, ...x_n)$ of n independent variables $\{x_1, x_2, ...x_n\}$, the partial differentiation with respect to one of these variables x_i is given by

$$\frac{\partial}{\partial x_i} f(x_1, x_2, ...x_n)$$

$$= \lim_{\Delta x_i \to 0} \left[\frac{f(x_1 + \Delta x_1, x_2 + \Delta x_2, ...x_i + \Delta x_i. \, .x_n + \Delta x_n) - f(x_1, x_2, ...x_i. \, ..x_n)}{\Delta x_i} \right], \tag{5.4}$$

where all $x \neq x_i$ are kept constant.

Example 5.1. Consider the function $z(x, y) = x^2 + y^2$, where x and y are independent.

    (a) Sketch this function to show that it defines a surface in a 3-D space described by the Cartesian coordinates (x, y, z), where the z coordinate depends on the x and y coordinates.

    (b) Evaluate the partial differentiations

$$\frac{\partial z}{\partial x}, \frac{\partial z}{\partial y}, \frac{\partial^2 z}{\partial x \partial y} = \frac{\partial}{\partial x}\left[\frac{\partial z}{\partial y}\right], \frac{\partial^2 z}{\partial x^2} = \frac{\partial}{\partial x}\left[\frac{\partial z}{\partial x}\right],$$

$$\frac{\partial^2 z}{\partial y \partial x} = \frac{\partial}{\partial y}\left[\frac{\partial z}{\partial x}\right], \tag{5.5}$$

at the point $(x, y) = (1, 0)$.

Solution:

    (a) The 3-D plot of the function $z(x, y) = x^2 + y^2$ is shown in figure 5.3.

    (b) Using the function $z(x, y) = x^2 + y^2$, we can evaluate the partial differentiations at point $(x, y) = (1, 0)$ as:

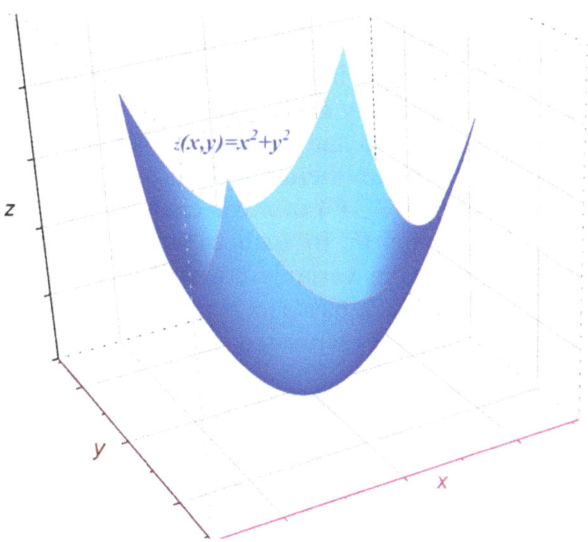

Figure 5.3. A surface defined by the function $z(x, y) = x^2 + y^2$.

$$\frac{\partial z}{\partial x} = \frac{\partial}{\partial x}(x^2 + y^2) = 2x \Rightarrow \left[\frac{\partial z}{\partial x}\right]_{x=1,y=0} = 2,$$

$$\frac{\partial z}{\partial y} = \frac{\partial}{\partial y}(x^2 + y^2) = 2y \Rightarrow \left[\frac{\partial z}{\partial y}\right]_{x=1,y=0} = 0. \tag{5.6}$$

These two results tells us the slope of the tangent plane at a point with coordinates $(1, 0, z(1, 0) = 1)$ on the surface. For the higher derivatives, we find

$$\frac{\partial^2 z}{\partial x \partial y} = \frac{\partial}{\partial x}\left\{\frac{\partial}{\partial y}(x^2 + y^2)\right\} = \frac{\partial}{\partial x}(2y) = 0$$

$$\Rightarrow \left[\frac{\partial^2 z}{\partial x \partial y}\right]_{x=1,y=0} = 0,$$

$$\frac{\partial^2 z}{\partial x^2} = \frac{\partial}{\partial x}\left\{\frac{\partial}{\partial x}(x^2 + y^2)\right\} = \frac{\partial}{\partial x}(2x) = 2$$

$$\Rightarrow \left[\frac{\partial^2 z}{\partial x^2}\right]_{x=1,y=0} = 2, \tag{5.7}$$

$$\frac{\partial^2 z}{\partial y \partial x} = \frac{\partial}{\partial y}\left\{\frac{\partial}{\partial x}(x^2 + y^2)\right\} = \frac{\partial}{\partial y}(2x) = 0$$

$$\Rightarrow \left[\frac{\partial^2 z}{\partial y \partial x}\right]_{x=1,y=0} = 0.$$

In some practical application, we may be given a function, $f(x_1, x_2, \ldots x_n)$, where the set of variables $x_1, x_2, \ldots x_n$ are not all independent variables. For such a function, we may be interested in how the property of the system defined by the function changes when we change one of the variables and keep the remaining variables constant. We often encounter such functions in thermodynamics. For example, we may be interested in how the pressure $P(T, V)$ of a gas changes with a change in temperature T and a constant volume V. In such cases, the partial differentiation is expressed using subscripts for the variables that need to be kept constant. For example, for a function of two variables $f(x_1, x_2)$ the partial derivative of $f(x_1, x_2)$ with respect to x_1 for x_2 constant is expressed as,

$$\left(\frac{\partial f(x_1, x_2)}{\partial x_1} \right)_{x_2}. \tag{5.8}$$

Note that the function could be expressed in terms of other variables, for example $f(x_1, x_3)$, and we first must express $f(x_1, x_3)$ as $f(x_1, x_2)$.

Example 5.2. For $z(x, y) = x^2 + 2y^2$, $x = r \cos(\theta)$, and $y = r \sin(\theta)$, find the following partial derivative

$$\left(\frac{\partial z}{\partial r} \right)_x. \tag{5.9}$$

Solution: First, we express the function $z(x, y)$ as $z(x, r)$. To this end, we note that

$$x^2 + y^2 = r^2 \Rightarrow y = \sqrt{r^2 - x^2}, \tag{5.10}$$

and

$$z(x, r) = x^2 + 2y^2 = x^2 + 2(r^2 - x^2) = 2r^2 - x^2. \tag{5.11}$$

Then one can easily show that

$$\left(\frac{\partial z(x, r)}{\partial r} \right)_x = \left(\frac{\partial}{\partial r}(2r^2 - x^2) \right)_x = 4r. \tag{5.12}$$

Example 5.3. An ideal gas is defined to be a gas that obeys the ideal gas equation of state,

$$PV = nRT \tag{5.13}$$

and

$$U = \frac{3}{2}nRT, \tag{5.14}$$

where P is the pressure, V is the volume, n mass of the ideal gas measured in moles, T is the temperature, U is the internal energy, and R the universal gas constant. Note that P, V, T, and U are the thermodynamic variables. Obviously, as we can see from these two equations these variables are not independent sets of variables. Find expressions for the following partial differentiations:

$$\left(\frac{\partial P}{\partial V}\right)_T, \left(\frac{\partial P}{\partial T}\right)_V, \left(\frac{\partial^2 P}{\partial V^2}\right)_T, \left(\frac{\partial^2 P}{\partial T^2}\right)_V, \left(\frac{\partial P}{\partial U}\right)_V, \left(\frac{\partial P}{\partial V}\right)_U. \tag{5.15}$$

Solution: Using the ideal gas equations

$$PV = nRT, \quad U = \frac{3}{2}nRT, \tag{5.16}$$

we write

$$P(T, V) = \frac{nRT}{V}, \tag{5.17}$$

so that

$$\left(\frac{\partial P}{\partial V}\right)_T = \left(\frac{\partial P(V, T)}{\partial V}\right)_T = \frac{\partial}{\partial V}\left[\frac{nRT}{V}\right] = -\frac{nRT}{V^2},$$

$$\left(\frac{\partial^2 P}{\partial V^2}\right)_T = \left(\frac{\partial^2 P(V, T)}{\partial V^2}\right)_T = \frac{\partial}{\partial V}\left[\frac{\partial P(V, T)}{\partial V}\right] = \frac{\partial}{\partial V}\left[-\frac{nRT}{V^2}\right] = \frac{2nRT}{V^3},$$

$$\left(\frac{\partial P}{\partial T}\right)_V = \left(\frac{\partial P(V, T)}{\partial T}\right)_V = \frac{\partial}{\partial T}\left[\frac{nRT}{V}\right] = \frac{nR}{V}, \tag{5.18}$$

$$\left(\frac{\partial^2 P}{\partial T^2}\right)_V = \left(\frac{\partial^2 P(V, T)}{\partial T^2}\right)_V = \frac{\partial}{\partial T}\left[\frac{\partial P}{\partial T}\right] = \frac{\partial}{\partial T}\left[\frac{nR}{V}\right] = 0.$$

For the last two differentiations, we need to express the pressure P in terms of the temperature T and the internal energy U,

$$U = \frac{3}{2}nRT, \quad T = \frac{2U}{3nR} \Rightarrow P = \frac{nRT}{V} = \frac{nR}{V} \times \frac{2U}{3nR}$$

$$\Rightarrow P(U, V) = \frac{2U}{3V}. \tag{5.19}$$

Using equation (5.19), one can write

$$\left(\frac{\partial P}{\partial V}\right)_U = \left(\frac{\partial P(V, U)}{\partial V}\right)_U = \frac{\partial}{\partial V}\left[\frac{2U}{3V}\right] = -\frac{2U}{3V^2},$$

$$\left(\frac{\partial P}{\partial U}\right)_V = \left(\frac{\partial P(U, V)}{\partial U}\right)_V = \frac{\partial}{\partial U}\left[\frac{2U}{3V}\right] = \frac{2}{3V}. \tag{5.20}$$

You will often encounter such types of derivatives in thermodynamics.

5.2 Total differential

Let us reconsider the plane defined by the function $z(x, y)$ shown in figure 5.4. Suppose a guy begins walking from the point described by the coordinates $(x_1, y_1, z_1(x_1, y_1))$ to the point with coordinates $(x_2, y_2, z_2(x_2, y_2))$ along the path shown in the figure. First, he walked along the x-direction, and as a result, his coordinate changed to $(x_2, y_1, z_1(x_2, y_1))$ and then along the y-direction ending up at $(x_2, y_2, z_2(x_2, y_2))$. We are interested in finding the total altitude he climbed (i.e., the displacement in z, Δz_{total}) (figure 5.4). Recalling that the slope of the line along the x-direction is given by

$$m_x = \frac{z(x_2, y_1) - z_1(x_1, y_1)}{x_2 - x_1} = \frac{\Delta z}{\Delta x}, \tag{5.21}$$

along this path, he has climbed

$$\Delta z_1 = m_x(x_2 - x_1) = m_x \Delta x. \tag{5.22}$$

Similarly using the slope for the path along the y-direction

$$m_y = \frac{z_2(x_2, y_2) - z(x_2, y_1)}{y_2 - y_1} = \frac{\Delta z}{\Delta y}, \tag{5.23}$$

along this path, he has climbed

$$\Delta z_2 = m_y(y_2 - y_1) = m_y \Delta y. \tag{5.24}$$

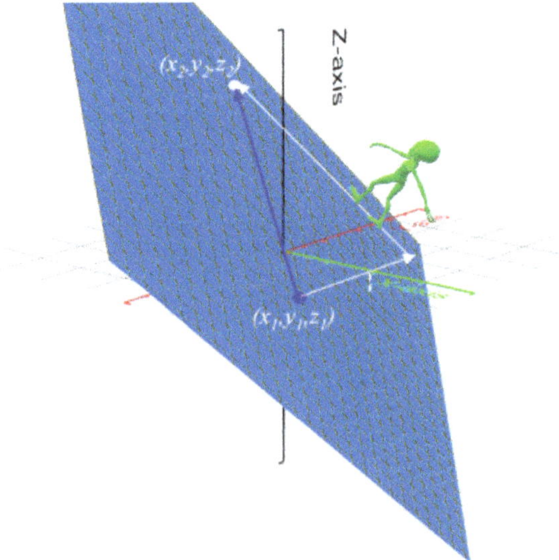

Figure 5.4. A guy climbing from a point (x_1, y_1, z_1) to a point (x_2, y_2, z_2) along the line on a plane defined by the function $z(x, y)$.

His elevation from where he was initially becomes

$$\Delta z_{\text{total}} = m_x \Delta x + m_y \Delta y, \tag{5.25}$$

or more generally

$$\Delta z(x, y) = \left(\frac{\Delta z(x, y)}{\Delta x}\right)\Delta x + \left(\frac{\Delta z(x, y)}{\Delta y}\right)\Delta y. \tag{5.26}$$

Now, instead of on a plane, let us consider a guy walking on a curved surface, like on a mountain with hills and valleys, as shown in figure 5.5. Since this is a curved surface, one must know how the elevation changes at the infinitessimal scale. That means

$$\lim_{\Delta z \to 0} \Delta z(x, y) = \left(\lim_{\Delta x \to 0} \frac{\Delta z(x, y)}{\Delta x}\right)\lim_{\Delta x \to 0} \Delta x$$

$$+ \left(\lim_{\Delta y \to 0} \frac{\Delta z(x, y)}{\Delta y}\right)\lim_{\Delta y \to 0} \Delta y, \tag{5.27}$$

which leads to

$$dz = \frac{\partial z(x, y)}{\partial x}dx + \frac{\partial z(x, y)}{\partial y}dy. \tag{5.28}$$

The elevation z depends on the guy's location on the surface (mountain) (x, y) because of the hills and valleys. The guy's total distance depends on what he traveled along the x-direction, y-direction, and the slope in these directions at the point $(x, y, z(x, y))$.

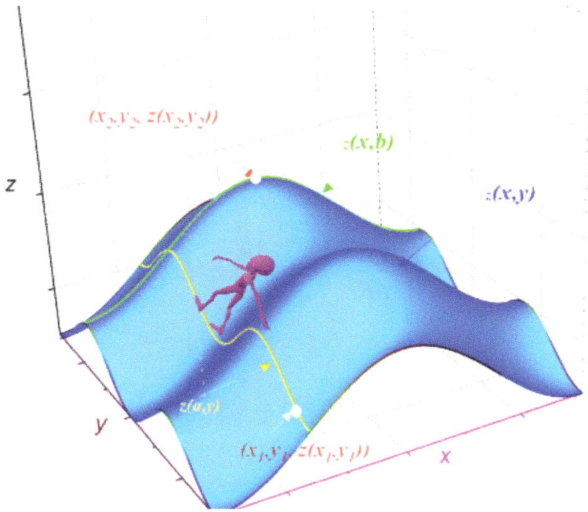

Figure 5.5. Walking on a curved surface.

As a real physical example let's consider a plane surface where you deposited a charge. Unfortunately this charge is not distributed uniformly over the surface. That means the charge depends on where you are on this surface, in other words the charge $Q(x, y)$ depends on both the x and y coordinates. Thus the infinitesimal charge $\Delta Q(x, y)$ on the surface over an area $(\Delta x) \times (\Delta y)$ depends on the length (Δx) and how the charge changes over Δx $\left(\frac{\Delta Q(x)}{\Delta x}\right)$ and on the width (Δy) and how the charge changes over Δy $\left(\frac{\Delta Q(x)}{\Delta x}\right)$,

$$\Delta Q(x, y) = \frac{\Delta Q(x, y)}{\Delta x}\Delta x + \frac{\Delta Q(x, y)}{\Delta y}\Delta y. \tag{5.29}$$

In the limit as the interval length Δx and width Δy of the area goes to zero, we find

$$dQ(x, y) = \frac{\partial Q(x, y)}{\partial x}dx + \frac{\partial Q(x, y)}{\partial y}dy, \tag{5.30}$$

which is the differential of the function $Q(x, y)$.

Example 5.3. *Application of the total differential*: The height and radius of a right-circular cylinder are measured to be

$$H \pm \Delta H = 7.6 \pm 0.2 \text{ cm}, \quad R \pm \Delta R = 3.7 \pm 0.2 \text{ cm}. \tag{5.31}$$

Find an approximate value for the uncertainty in the cylinder's volume ΔV.

Solution: The volume of a cylinder is given by

$$V(R, H) = \pi R^2 H. \tag{5.32}$$

The total differential for $V(R, H)$ can be expressed as

$$dV = \frac{\partial V}{\partial R}dR + \frac{\partial V}{\partial H}dH, \tag{5.33}$$

replacing dV, dR, and dH by ΔV, ΔR, and ΔH, respectively, one can express the uncertainty in volume as

$$\Delta V = \frac{\partial V}{\partial R}\Delta R + \frac{\partial V}{\partial H}\Delta H. \tag{5.34}$$

Upon evaluating

$$\frac{\partial V}{\partial R} = \frac{\partial}{\partial R}[\pi R^2 H] = 2\pi R H,$$
$$\frac{\partial V}{\partial H} = \frac{\partial}{\partial H}[\pi R^2 H] = \pi R^2, \tag{5.35}$$

for $H = 7.6$ and $R = 3.7$, we have

$$\left[\frac{\partial V}{\partial R}\right]_{H=7.6, R=3.7} = 176.7,$$

$$\left[\frac{\partial V}{\partial H}\right]_{H=7.6, R=3.7} = 43.01, \tag{5.36}$$

so that using these results and the values $\Delta R = \pm 0.2$ and $\Delta H = \pm 0.2$, the uncertainty in the volume becomes

$$\Delta V = \frac{\partial V}{\partial R}\Delta R + \frac{\partial V}{\partial H}\Delta H = 176.6 \times (\pm 0.2) + 42.9 \times (\pm 0.2) \tag{5.37}$$

$$\Rightarrow \Delta V \simeq \pm 43.9. \tag{5.38}$$

Since the measured uncertainties in H and R are in one significant figures or digits, the calculated uncertainty in the volume should be written in one significant figure,

$$\Delta V = \pm 40 \text{ cm}^3. \tag{5.39}$$

Using

$$V(R, H) = \pi R^2 H = 326.9 \text{ cm}^3, \tag{5.40}$$

one can express the volume with the uncertainty as

$$V \pm \Delta V = 330 \pm 40 \text{ cm}^3. \tag{5.41}$$

5.3 The multivariable form of the chain rule

In a function of one variable $y(x)$, the variable x could be a function of another variable such as time t, $(x(t))$. In such cases, to find the function $y(x)$ time dependence, we need to use the chain rule for a function of one variable. We recall from the previous section, the total differential for $y(x)$ can be written as

$$dy = \frac{dy}{dx}dx, \tag{5.42}$$

so that diving this by dt, we find the chain rule for a function of single variable

$$\frac{dy}{dt} = \frac{dy}{dx}\frac{dx}{dt}. \tag{5.43}$$

We often need to apply the chain rule for the function of one variable and function of more variables. For a function of two variable, $z(x, y)$, we recall that the total differential is given by

$$dz = \frac{\partial z}{\partial x}dx + \frac{\partial z}{\partial y}dy. \tag{5.44}$$

When both x and y depend on another variable, like time, t, dividing this equation by dt, we find the chain rule for a function of two variables

$$\frac{dz}{dt} = \frac{\partial z}{\partial x}\frac{dx}{dt} + \frac{\partial z}{\partial y}\frac{dy}{dt}. \tag{5.45}$$

One can generalize this and write

$$\frac{df}{dt} = \frac{\partial f}{\partial x_1}\frac{dx_1}{dt} + \frac{\partial f}{\partial x_2}\frac{dx_2}{dt} + \dots \frac{\partial f}{\partial x_n}\frac{dx_n}{dt} = \sum_{i=1}^{n} \frac{\partial f}{\partial x_i}\frac{dx_i}{dt}, \tag{5.46}$$

for a function of multivariable $f(x_1, x_2, \dots x_n)$.

Example 5.4. Consider the function

$$z = \exp(-x^2 - y^2), \tag{5.47}$$

where

$$x(t) = e^t \text{ and } y(t) = e^{-t}. \tag{5.48}$$

Find

$$\frac{dz}{dt}. \tag{5.49}$$

Solution: Since x and y are a function of the variable t, one can write

$$\frac{dz}{dt} = \frac{\partial z}{\partial x}\frac{dx}{dt} + \frac{\partial z}{\partial y}\frac{dy}{dt}. \tag{5.50}$$

Using the results

$$\frac{\partial z}{\partial x} = -2x \exp(-x^2 - y^2), \quad \frac{\partial z}{\partial y} = -2y \exp(-x^2 - y^2),$$

$$\frac{dx}{dt} = e^t = x, \quad \frac{dy}{dt} = -e^{-t} = -y, \tag{5.51}$$

one easily finds

$$\frac{dz}{dt} = -2x \exp(-x^2 - y^2)x - 2y \exp(-x^2 - y^2)(-y)$$

$$\Rightarrow \frac{dz}{dt} = 2(y^2 - x^2)\exp[-x^2 - y^2]. \tag{5.52}$$

Example 5.5. Let us consider the projectile motion of a ball shown in figure 5.6. The ball has a mass m and is thrown with a speed v_i from the point $(z_i, x_i = 0)$ at an angle

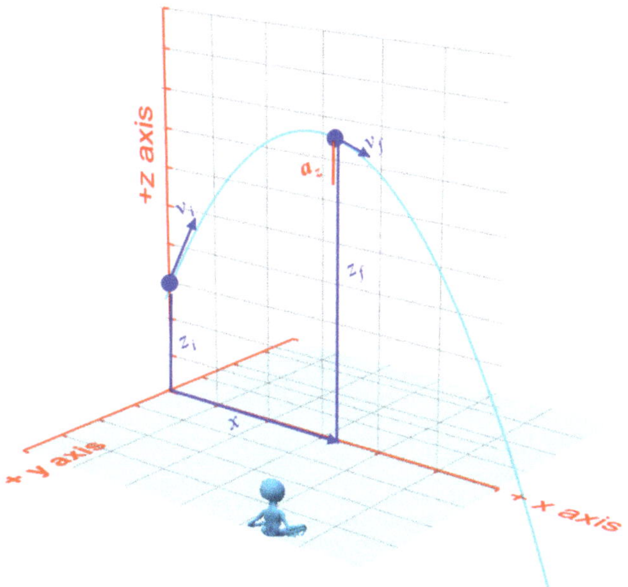

Figure 5.6. The projectile motion of a ball.

θ measured from the horizontal ($+x$-axis). Show that the trajectory of this projectile motion is given by the equation,

$$z(x) = ax^2 + bx + c, \qquad (5.53)$$

where

$$a = -\frac{g}{2v_i \cos^2(\theta)}, \; b = v_i \tan(\theta), \; c = z_i. \qquad (5.54)$$

Note that x is the horizontal distance of the ball measured on the ground and $z(x)$ is the height of the ball from the ground.

Solution: From Newton's second law

$$F_x = ma_x \text{ and } F_z = ma_z. \qquad (5.55)$$

Since for a projectile motion there is no force on the x-direction, $F_x = 0$, we find

$$a_x = \frac{dv_x}{dt} = 0 \; \Rightarrow \; v_x(t) = v_{ix} \; \Rightarrow \; \frac{dx}{dt} = v_i \cos(\theta)$$

$$\Rightarrow x(t) = x_i + v_i \cos(\theta)t = v_i \cos(\theta)t, \qquad (5.56)$$

where we set the value $x_i = 0$. This shows that the x coordinate depends on time t. In the z-direction, there is net gravitational force, $F_z = -mg$ which leads to

$$F_z = ma_z \Rightarrow a_z = \frac{dv_z}{dt} = -g \Rightarrow v_z(t) = v_{iz} - gt$$

$$\Rightarrow \frac{dz}{dt} = v_i \sin(\theta) - gt. \tag{5.57}$$

Using the chain rule for a function of one variable, one can write

$$\frac{dz}{dt} = \frac{dz}{dx}\frac{dx}{dt}. \tag{5.58}$$

From the result in equation (5.56), we have

$$t = \frac{x(t)}{v_i \cos(\theta)}, \tag{5.59}$$

so that equation (5.57) becomes

$$\frac{dz}{dt} = v_i \sin(\theta) - \frac{x(t)g}{v_i \cos(\theta)}. \tag{5.60}$$

From equation (5.56), we have

$$\frac{dx}{dt} = v_i \cos(\theta). \tag{5.61}$$

Substituting equations (5.60) and (5.61) into equation (5.58), we find

$$v_i \sin(\theta) - \frac{g}{v_i \cos(\theta)}x = \frac{dz}{dx}v_i \cos(\theta), \tag{5.62}$$

from which follows

$$\frac{dz}{dx} = v_i \tan(\theta) - \frac{g}{v_i \cos^2(\theta)}x. \tag{5.63}$$

Upon integrating this equation with respect to x,

$$\int_{z_i}^{z} dz = \int_{0}^{x}\left[v_i \tan(\theta) - \frac{gx}{v_i \cos^2(\theta)}\right]dx, \tag{5.64}$$

one finds

$$z(x) - z_i = v_i \tan(\theta)x - \frac{g}{v_i \cos^2(\theta)}\frac{x^2}{2}. \tag{5.65}$$

This can be put in the form

$$z(x) = ax^2 + bx + c, \tag{5.66}$$

where

$$a = -\frac{g}{2v_0 \cos^2(\theta)}, \quad b = v_0 \tan(\theta), \quad c = z_i. \tag{5.67}$$

Example 5.6. *Implicit differentiation and application of the total differential:* Consider the equation

$$y^3 - x^2 y = 8, \tag{5.68}$$

where y is a function of x. Evaluate,

$$\frac{dy}{dx} \text{ and } \frac{d^2 y}{dx^2}, \tag{5.69}$$

at the point $(x, y) = (3, -1)$.

Solution: The easy way to find both

$$\frac{dy}{dx} \text{ and } \frac{d^2 y}{dx^2} \tag{5.70}$$

is to apply implicit differentiation. To this end, let

$$z(x, y) = y^3 - x^2 y = 8, \tag{5.71}$$

so that the total differential

$$dz = \frac{\partial z}{\partial x} dx + \frac{\partial z}{\partial y} dy = 0 \Rightarrow \frac{dz}{dx} = \frac{\partial z}{\partial x} + \frac{\partial z}{\partial y} \frac{dy}{dx} = 0, \tag{5.72}$$

becomes

$$-2xy + (3y^2 - x^2)\frac{dy}{dx} = 0 \Rightarrow \frac{dy}{dx} = \frac{2xy}{3y^2 - x^2}. \tag{5.73}$$

Then at $(x, y) = (3, -1)$, we find

$$\frac{dy}{dx} = 1. \tag{5.74}$$

To find the second derivative of the function, we use the result in equation (5.73),

$$\frac{dy}{dx} = \frac{2xy}{3y^2 - x^2} = f(x, y). \tag{5.75}$$

To this end, we note that

$$\frac{d^2 y}{dx^2} = \frac{\partial f(x, y)}{\partial x} + \left[\frac{\partial f(x, y)}{\partial y} \right] \frac{dy}{dx}. \tag{5.76}$$

Upon carrying out the partial differentiations, we have

$$\frac{\partial f(x, y)}{\partial x} = \frac{2y}{3y^2 - x^2} + \frac{4x^2 y}{(3y^2 - x^2)^2} \tag{5.77}$$

and

$$\frac{\partial f(x, y)}{\partial y} = \frac{2x}{3y^2 - x^2} - \frac{12xy^2}{(3y^2 - x^2)^2},$$ (5.78)

so that equation (5.76) becomes

$$\frac{d^2y}{dx^2} = \frac{2y}{3y^2 - x^2} + \frac{4x^2y}{(3y^2 - x^2)^2} + \left(\frac{2x}{3y^2 - x^2} - \frac{12xy^2}{(3y^2 - x^2)^2}\right)\frac{dy}{dx}.$$ (5.79)

This can be rewritten as

$$\frac{d^2y}{dx^2} = \frac{1}{x}\left(\frac{2yx}{3y^2 - x^2}\right) + \frac{1}{y}\left(\frac{2xy}{3y^2 - x^2}\right)^2$$
$$+ \left[\frac{1}{y}\left(\frac{2xy}{3y^2 - x^2}\right) - \frac{3}{x}\left(\frac{2xy}{3y^2 - x^2}\right)^2\right]\frac{dy}{dx},$$ (5.80)

and using the result in equation (5.73), we find

$$\frac{d^2y}{dx^2} = \frac{1}{x}\frac{dy}{dx} + \frac{1}{y}\left(\frac{dy}{dx}\right)^2 + \left[\frac{1}{y}\left(\frac{dy}{dx}\right) - \frac{3}{x}\left(\frac{dy}{dx}\right)^2\right]\frac{dy}{dx}.$$ (5.81)

Recalling the result we obtained for equation (5.73) at $(x, y) = (3, -1)$,

$$\frac{dy}{dx} = 1,$$ (5.82)

we find from equation (5.81)

$$\left[\frac{d^2y}{dx^2}\right]_{(x,y)=(3,-1)} = \frac{1}{3} - 1 - 1 - 1 = -\frac{8}{3}.$$ (5.83)

5.4 Extremum (max/min) problems

For a function of single variable $y(x)$, the first derivative

$$f(x) = \frac{dy}{dx},$$ (5.84)

tells us the slope of the tangent line at the point on the curve with the coordinate $(x, y(x))$. The tangent line to extremum point (see figure 5.7) on a curve is a horizontal line, the slope of which is zero. Therefore, if the x coordinates of the extremum point on the curve defined by $y(x)$ is x_i, since the slope of the tangent line at $(x_i, y(x_i))$ is zero,

$$f(x_i) = \frac{dy}{dx}\bigg|_{x=x_i} = 0.$$ (5.85)

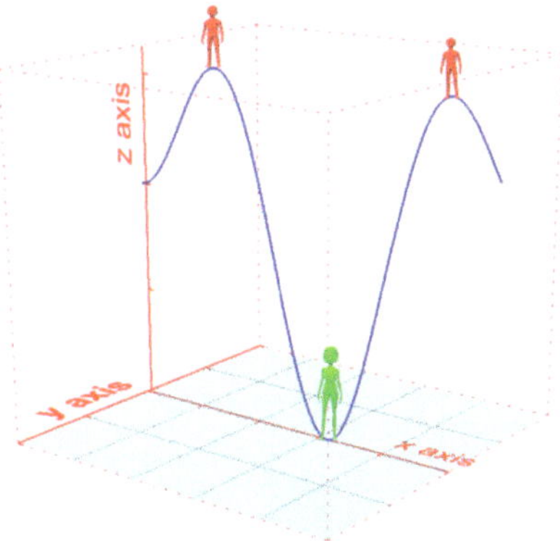

Figure 5.7. Extremum points: (a) the point where the two guys standing are maximum points. (b) The point where the girl is standing is a minimum point.

Note that

$$f(x_i) = \left.\frac{dy}{dx}\right|_{x=x_i} = 0, \tag{5.86}$$

does not always represent a maximum or minimum point. The condition described by equation (5.86) is a necessary condition for an extremum at x_i, but is not a sufficient condition. In order guarantee the existence an extremum, one should calculate the second derivative of the function at x_i and note the sign of the result. If the result is positive, x_i represents local or global minimum. If it is negative, x_i represents local or global maximum. If the result is zero, x_i represents neither minimum nor maximum point. For example, for the function,

$$y(x) = x^3 \Rightarrow \frac{dy}{dx} = 0 \Rightarrow 3x^2 = 0$$
$$\Rightarrow x = 0, \tag{5.87}$$

the point $(0, 0)$ is neither maximum nor minimum, as shown in figure 5.8, and such points are known as inflection points.

For a function of two variables $z(x, y)$, the extremum points with coordinates $(x_i, y_i, z(x_i, y_i))$ represent the coordinates for maxima on the hills or the minima in the valleys on the surface defined by the function $z(x, y)$. The tangent plane to these extremum points are horizontal planes (parallel to the x–y plane) (figure 5.9). For this plane the slopes along the x and the y must be zero. Mathematically, these slopes are defined by the partial derivative of the function $z(x, y)$ evaluated at the maxima or minima points with coordinates (x_i, y_i). Thus at the extremum (maxima or minima) points

$$\left.\frac{\partial z(x, y)}{\partial x}\right|_{x=x_i} = 0 \text{ and } \left.\frac{\partial z(x, y)}{\partial y}\right|_{y=y_i} = 0. \tag{5.88}$$

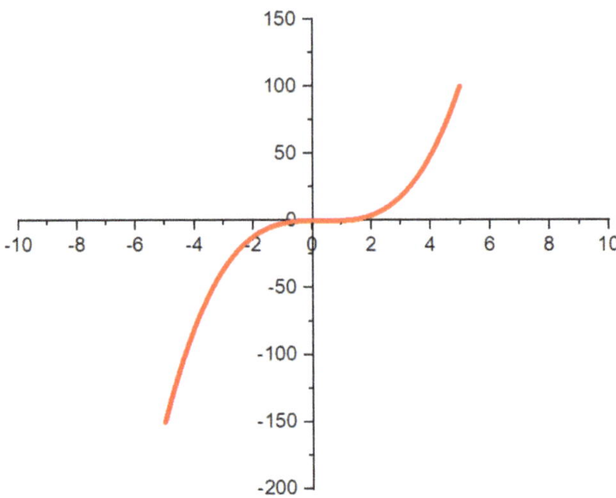

Figure 5.8. The point with coordinates (0,0) is neither maximum nor minimum. It is an inflection point.

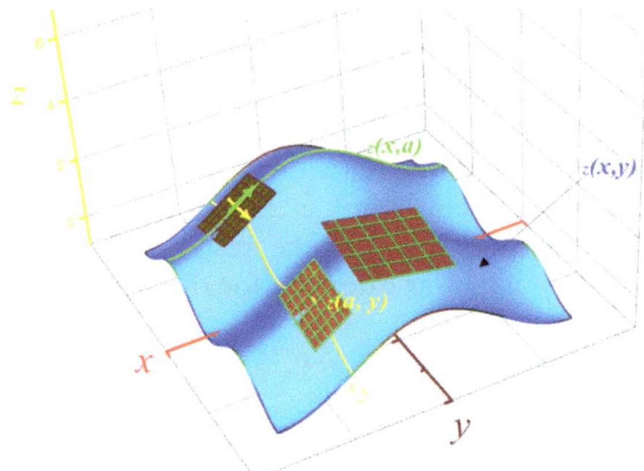

Figure 5.9. The slope for the tangent plane at the maxima or minima is zero both in the x and y directions.

These two equations define the horizontal tangent plane at the extremum points on the surface defined by $z(x, y)$. But it does not mean that whenever equation (5.88) is true, the point represents a maximum or minimum point. For example, for the function

$$z(x, y) = x^2 + y^3, \tag{5.89}$$

we find

$$\left. \frac{\partial z(x, y)}{\partial x} \right|_{x=x_i} = 0 \Rightarrow 2x = 0 \Rightarrow x = 0,$$

$$\left. \frac{\partial z(x, y)}{\partial y} \right|_{y=y_i} = 0 \Rightarrow 3y^2 \Rightarrow y = 0. \tag{5.90}$$

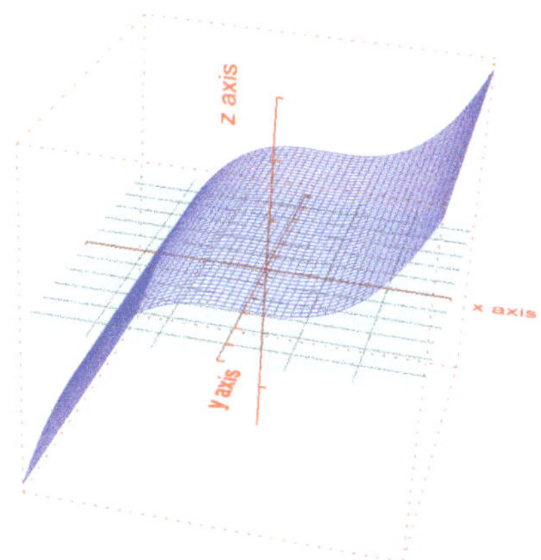

Figure 5.10. A surface defined by a function $z(x, y) = x^2 + y^3$.

The point (0,0) does represent neither a maximum (top of a hill) nor a minimum (bottom of a valley) point, as shown in figure 5.10.

Example 5.7. An aquarium with rectangular sides and bottom has a fixed volume V_0. Find its proportions so that it will require the least amount of glass for construction.

Solution: As shown in figure 5.11, the aquarium's length, width, and height are l, w, and h, respectively. The volume, which is fixed, is given by

$$V(l, w, h) = lwh = V_0. \tag{5.91}$$

The aquarium has rectangular sides made of glass. The surface area, $A(l, w, h)$ of the five sides (excluding the open top side), is given by

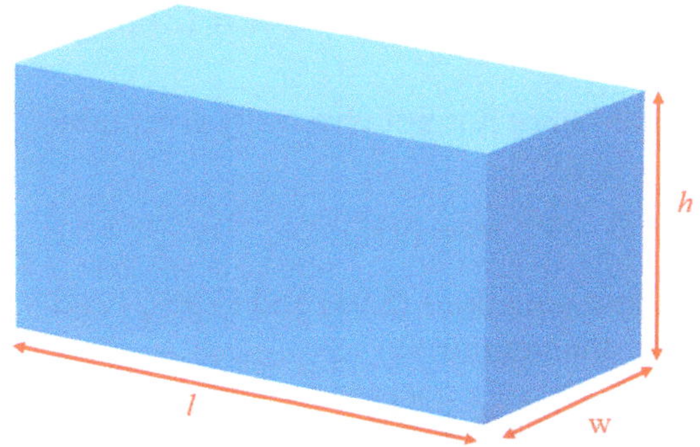

Figure 5.11. The aquarium has length l, width w and height h.

$$A(l, w, h) = lw + 2(lh + wh). \tag{5.92}$$

We are interested in constructing an aquarium with the least amount of glass. It means the glass sides' total area of the aquarium must be minimized while the volume remains constant. As we can see from equation (5.92) l, w, and h are not independent variables. We can reduce the area from a function of three variables to two variables by expressing h in terms of l and w using equation (5.91) as

$$h = \frac{V_0}{lw}. \tag{5.93}$$

Substituting equation (5.93) into equation (5.92), we can then express the area in terms of two independent variables,

$$A(l, w) = lw + 2V_0\left(\frac{1}{l} + \frac{1}{w}\right). \tag{5.94}$$

For the minimum area, we must have

$$\frac{\partial A(l, w)}{\partial l} = 0, \tag{5.95}$$

and

$$\frac{\partial A(l, w)}{\partial w} = 0. \tag{5.96}$$

Using equation (5.94), one finds from equations (5.95) and (5.96),

$$w - \frac{2V_0}{l^2} = 0, \tag{5.97}$$

and

$$l - \frac{2V_0}{w^2} = 0, \tag{5.98}$$

respectively. Combining these two equations, we find

$$w\left(1 - \frac{w^3}{2V_0}\right) = 0 \Rightarrow w = 0, w = (2V_0)^{1/3}. \tag{5.99}$$

The result $w = 0$ is rejected since it is not physically acceptable. The acceptable result is

$$w = (2V_0)^{1/3}. \tag{5.100}$$

Substituting this result into equations (5.93) and (5.98), l and h of the aquarium can be expressed in terms of the volume as

$$l = \frac{2V_0}{w^2} = \frac{2V_0}{(2V_0)^{2/3}} = (2V_0)^{1/3}, \tag{5.101}$$

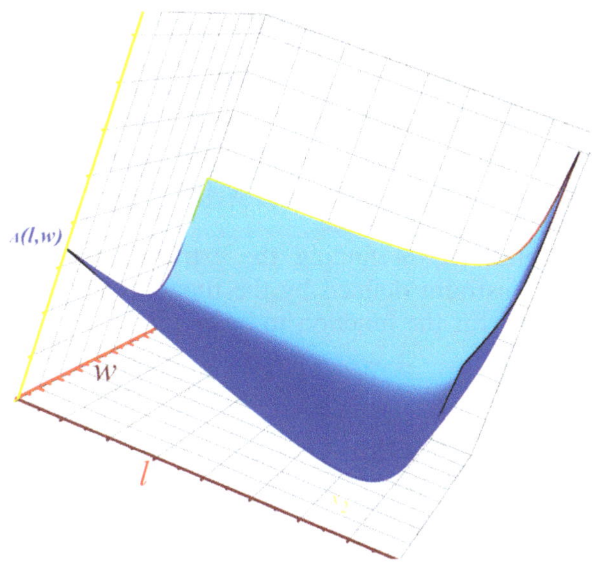

Figure 5.12. The 3-D plot for the function $A(l, w) = lw + 2V_0\left(\frac{1}{l} + \frac{1}{w}\right)$..

$$h = \frac{V_0}{(2V_0)^{1/3}(2V_0)^{1/3}} = \frac{1}{2}(2V_0)^{1/3}. \tag{5.102}$$

Thus the proportions can be expressed as

$$\frac{l:\, w:\, h}{(2V_0)^{1/3}} = 1:\, 1:\, \frac{1}{2}.$$

These are the proportions of the aquarium dimension for the least amount of glass for construction and a volume V_0. The plot for the area as a function of the length and width of the aquarium is shown in figure 5.12.

5.5 The method of Lagrangian multipliers

In Example 5.7 the volume of the rectangular aquarium, which is a function of three variables, is specified,

$$V(l, w, h) = lwh = V_0. \tag{5.103}$$

That means the volume of the rectangular aquarium given cannot be less or greater than V_0. We were asked to determine the minimum total surface area $A(l, w, h)$

$$A(l, w, h) = lw + 2(lh + wh). \tag{5.104}$$

This area appears to be a function of three variables. However, because of the condition set for the volume, we have seen that it is a function of two variables. Such types of conditions are constraints. The function that sets the volume to V_0 in equation (5.103) is the constraint for the area defined by the function in equation (5.104).

Suppose the function $f(x, y, z, \ldots)$ describes some physical observable for some system. In real physical problems the variables $x, y, z, \ldots$ are subject to constraints so that they are no longer independent. At least in principle, it is possible using the constraint to eliminate one variable and proceed with a new and smaller set of independent variables. *Lagrangian multipliers* are an alternate technique that may be applied to determine extremum points for the function when this elimination of variables is inconvenient or undesirable.

Suppose we are interested in finding the extremum points for the function $f(x, y, z)$ under the constraint defined by the function $\phi(x, y, z) = k$, where k is a constant. We note that for the function to be extremum

$$df = 0 \Rightarrow \frac{\partial f(x, y, z)}{\partial x}dx + \frac{\partial f(x, y, z)}{\partial y}dy + \frac{\partial f(x, y, z)}{\partial z}dz = 0, \qquad (5.105)$$

and the necessary condition is

$$\frac{\partial f(x, y, z)}{\partial x} = \frac{\partial f(x, y, z)}{\partial y} = \frac{\partial f(x, y, z)}{\partial z} = 0. \qquad (5.106)$$

Noting that for the constraint

$$d\phi(x, y, z) = dk = 0 \Rightarrow \frac{\partial \phi}{\partial x}dx + \frac{\partial \phi}{\partial y}dy + \frac{\partial \phi}{\partial z}dz = 0, \qquad (5.107)$$

and for a function

$$F(x, y, z, \ldots) = f(x, y, z) + \lambda\phi(x, y, z), \qquad (5.108)$$

for λ a constant, at the extremum points, we may write

$$F(x, y, z) = f(x, y, z) + \lambda\phi(x, y, z) = f(x, y, z) + \lambda k \Rightarrow dF = df = 0$$

$$\Rightarrow \left(\frac{\partial f}{\partial x} + \lambda\frac{\partial \phi}{\partial x}\right)dx + \left(\frac{\partial f}{\partial y} + \lambda\frac{\partial \phi}{\partial y}\right)dy + \left(\frac{\partial f}{\partial z} + \lambda\frac{\partial \phi}{\partial z}\right)dz = 0. \qquad (5.109)$$

There follows that

$$\frac{\partial f}{\partial x} + \lambda\frac{\partial \phi}{\partial x} = 0, \quad \frac{\partial f}{\partial y} + \lambda\frac{\partial \phi}{\partial y} = 0, \quad \text{and} \quad \frac{\partial f}{\partial z} + \lambda\frac{\partial \phi}{\partial z} = 0. \qquad (5.110)$$

Here we want only x, y, and z, so λ need not be determined. This method will fail if the coefficient of λ vanish at the extremum,

$$\frac{\partial \phi}{\partial x} = \frac{\partial \phi}{\partial y} = \frac{\partial \phi}{\partial z} = 0. \qquad (5.111)$$

The following steps can be used to find extremum points for a function $f(x, y, z, \ldots)$ with a constraint $\phi(x, y, z, \ldots) = k$

Step-1 Construct the function $F(x, y, z, \ldots) = f(x, y, z, \ldots) + \lambda\phi(x, y, z, \ldots)$, where λ is the (unknown) Lagrangian multiplier.

Step-2 Compute the partial derivatives of F with respect to each of the variables x, y, z, … and set them equal to zero.

Step-3 Solve the resulting system of simultaneous equations generated in step 2 along with the constraint equation $\phi(x, y, z, \ldots) = k$.

Example 5.8. Consider a wire bent to fit a curve defined by $y = 1 - x^2$ as shown in figure 5.13.

Suppose a string is stretched from the origin to a point (x, y) on the curve. Find the minimum length of the string.

Solution: We need to find the minimum value for the distance

$$d(x, y) = \sqrt{x^2 + y^2}, \tag{5.112}$$

with the constraint

$$y(x) = 1 - x^2 \Rightarrow \phi(x, y) = y + x^2 = 1. \tag{5.113}$$

Following the procedure for the method of Lagrangian multipliers:

Step-1 Write the function

$$F(x, y) = f(x, y) + \lambda\phi(x, y) = \sqrt{x^2 + y^2} + \lambda(y + x^2). \tag{5.114}$$

Step-2 Find the partial derivatives of F with respect to each of the variables x, y and set them equal to zero,

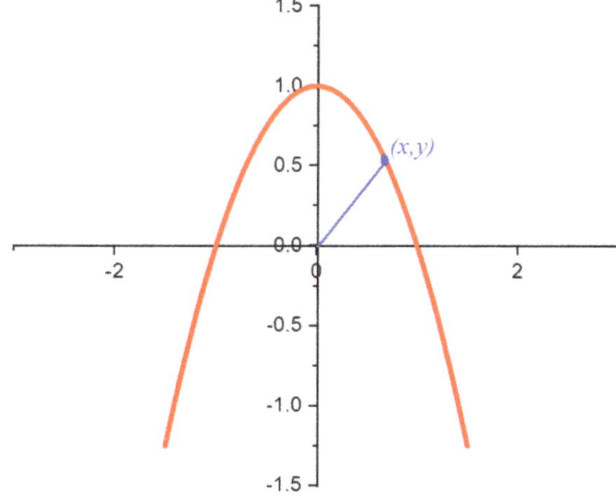

Figure 5.13. A wire bent to fit the function defined by $y(x) = 1 - x^2$.

$$\frac{\partial}{\partial x}F(x,\,y)=0 \Rightarrow \frac{x}{\sqrt{x^2+y^2}}+2\lambda x=0 \Rightarrow \frac{1}{\sqrt{x^2+y^2}}=-2\lambda,$$

$$\frac{\partial}{\partial y}F(x,\,y)=0 \Rightarrow \frac{y}{\sqrt{x^2+y^2}}+\lambda=0 \Rightarrow \frac{y}{\sqrt{x^2+y^2}}=-\lambda.$$

$$(5.115)$$

Step-3 Solve the resulting system of simultaneous equations generated in step 2 and the constraint $\phi(x,\,y)=k$ to determine the coordinates $(x,\,y)$ at which the function $d(x,\,y)$ takes a maxima or minima value,

$$\frac{1}{\sqrt{x^2+y^2}}=-2\lambda, \ \frac{y}{\sqrt{x^2+y^2}}=-\lambda, \ y+x^2=1. \tag{5.116}$$

These three equations can be combined to find the values for x and y

$$\frac{1}{\sqrt{x^2+y^2}}=\frac{2y}{\sqrt{x^2+y^2}}, \ y=1-x^2 \Rightarrow 1=2(1-x^2) \tag{5.117}$$

$$\Rightarrow x=\pm\frac{1}{\sqrt{2}}, \ y=\frac{1}{2}. \tag{5.118}$$

Then the minimum distance is found to be

$$d_{\min}=\sqrt{x^2+y^2}=\sqrt{3}\big/2. \tag{5.119}$$

Example 5.9. *The aquarium problem*: An aquarium with rectangular sides and bottom has a fixed volume V_0. Find its proportions so that it will require the least amount of glass for construction using Lagrangian multipliers (see figure 5.11).

Solution: In order to determine the minimum area of the glasses used using the method of Lagrangian multiplies, we need to write the constraint. The constraint is the volume

$$V_0=lwh \Rightarrow \phi(l,\,w,\,h)=lwh=V_0=\text{constant.} \tag{5.120}$$

The surface area of the aquarium is

$$A(l,\,w,\,h)=lw+2(wh+lh) \Rightarrow f(l,\,w,\,h)=lw+2(wh+lh). \tag{5.121}$$

Constructing the function $F(l,\,w,\,h)$,

$$F(l,\,w,\,h)=f(l,\,w,\,h)+\lambda\phi(l,\,w,\,h)$$
$$\Rightarrow F(l,\,w,\,h)=lw+2(wh+lh)+\lambda lwh. \tag{5.122}$$

We then minimize this function setting the partial differentiation to zero

$$\frac{\partial F(l,\, w,\, h)}{\partial l} = 0 \Rightarrow w + 2h + \lambda wh = 0, \tag{5.123}$$

$$\frac{\partial F(l,\, w,\, h)}{\partial w} = 0 \Rightarrow l + 2h + \lambda lh = 0, \tag{5.124}$$

and

$$\frac{\partial F(l,\, w,\, h)}{\partial h} = 0 \Rightarrow 2w + 2l + \lambda lw = 0. \tag{5.125}$$

We solve equations (5.123)–(5.125) and the constraint,

$$lwh = V_0, \tag{5.126}$$

to determine l, w, and h. Since we do not need the solution for λ, from equation (5.123), we can express λ as

$$\lambda = -\frac{w + 2h}{wh}, \tag{5.127}$$

and substitute it into equations (5.124) and (5.125), so that

$$l + 2h - \left(\frac{w + 2h}{w}\right)l = 0 \Rightarrow l\left[\frac{w + 2h}{w} - 1\right] = 2h$$
$$\Rightarrow w = l, \tag{5.128}$$

and

$$2w + 2l - \left(\frac{w + 2h}{wh}\right)lw = 0 \Rightarrow l\left[\frac{w + 2h}{h} - 2\right] = 2w$$
$$\Rightarrow l = 2h. \tag{5.129}$$

Now substituting these results in equation (5.126), we have

$$4h^3 = V_0 \Rightarrow h = \left(\frac{V_0}{4}\right)^{1/3} \tag{5.130}$$

so that

$$w = l = 2h = (2V_0)^{1/3}, \tag{5.131}$$

which is the same as the result we found in Example 5.7.

Example 5.10. *Particle in a box-Quantum mechanics*: Consider the quantum mechanical problem of a particle with mass m in a box. The box is a rectangular box with length l, width w, and height h. The ground state energy is given by

$$E(l, w, h) = \frac{\pi^2 \hbar}{2m}\left(\frac{1}{l^2} + \frac{1}{w^2} + \frac{1}{h^2}\right). \tag{5.132}$$

Find the proportions to the sides of the box that minimize the ground state energy subject to the constraint that the volume is a constant,

$$V(l, w, h) = lwh = V_0. \tag{5.133}$$

Solution: Constructing the function $F(l, w, h)$,

$$F(l, w, h) = f(l, w, h) + \lambda\phi(l, w, h)$$

$$\Rightarrow F(l, w, h) = \frac{\pi^2 \hbar}{2m}\left(\frac{1}{l^2} + \frac{1}{w^2} + \frac{1}{h^2}\right) + \lambda lwh. \tag{5.134}$$

We minimize this function by setting the partial differentiation with respect to l, w, and h to zero,

$$\frac{\partial F(l, w, h)}{\partial l} = 0 \Rightarrow -\frac{\pi^2 \hbar}{ml^3} + \lambda wh = 0, \tag{5.135}$$

$$\frac{\partial F(l, w, h)}{\partial w} = 0 \Rightarrow -\frac{\pi^2 \hbar}{mw^3} + \lambda lh = 0, \tag{5.136}$$

and

$$\frac{\partial F(l, w, h)}{\partial h} = 0 \Rightarrow -\frac{\pi^2 \hbar}{mh^3} + \lambda lw = 0. \tag{5.137}$$

We solve the above three equations and the constraint,

$$lwh = V_0, \tag{5.138}$$

to determine l, w, and h. Since we do not need the solution for λ, using the first equation, we have

$$\lambda = \frac{\pi^2 \hbar}{ml^3 wh} \tag{5.139}$$

so that using this expression for λ, we find

$$-\frac{\pi^2 \hbar}{mw^3} + \lambda lh = 0 \Rightarrow \frac{\pi^2 \hbar}{mw^3} = \frac{\pi^2 \hbar}{ml^2 w} \Rightarrow w = l, \tag{5.140}$$

and

$$-\frac{\pi^2 \hbar}{mh^3} + \lambda lw = 0 \Rightarrow \frac{\pi^2 \hbar}{mh^3} = \frac{\pi^2 \hbar}{ml^2 h} \Rightarrow h = l. \tag{5.141}$$

Now substituting these results into the constraint for the volume, we have

$$l^3 = V_0 \Rightarrow l = (V_0)^{1/3}, \tag{5.142}$$

which leads to

$$w = l = h = (V_0)^{1/3}.$$

Therefore, the box must be a cube, $w:l:h = 1$ in order to minimize the energy of the particle.

5.6 Change of variables

Problems in most branches of physics require solving differential equations. These differential equations could be a function of single or multi-variable. To find the solutions to these differential equations, a change of variables sometimes is necessary for different reasons. Making variables change when the differential equation is a multi-variable function requires partial differentiation of the function. We illustrate partial differentiation in a change of variables in a second-order differential equation for the single and multi-variable function in the following examples.

Function of single variable differential equation

Example 5.11. The *Bessel differential equation* is given by

$$x^2\frac{d^2y}{dx^2} + x\frac{dy}{dx} - (1 - x)y = 0. \tag{5.143}$$

In this equation $y = y(x)$ which is function of single variable. Find the corresponding differential equation for the change of variable defined by

$$u = 2\sqrt{x}, \; y = y(u). \tag{5.144}$$

Solution: Using partial differentiation, we can express

$$\frac{dy}{dx} = \frac{\partial y}{\partial u}\frac{du}{dx}. \tag{5.145}$$

Since $y = y(u)$ is a function of one variable $u = 2\sqrt{x}$, this can be rewritten as

$$\frac{dy}{dx} = \frac{dy}{du}\frac{d}{dx}(2\sqrt{x}) = \frac{dy}{du}\frac{1}{\sqrt{x}}. \tag{5.146}$$

Using the result in equation (5.146), the second derivative in the Bessel equation can be expressed as

$$\frac{d^2y}{dx^2} = \frac{d}{dx}\left[\frac{dy}{du}\frac{1}{\sqrt{x}}\right] = \frac{1}{\sqrt{x}}\frac{d}{dx}\left[\frac{dy}{du}\right] + \frac{dy}{du}\frac{d}{dx}\left[\frac{1}{\sqrt{x}}\right]. \tag{5.147}$$

Noting that

$$\frac{d}{dx}\left[\frac{dy}{du}\right] = \frac{d}{du}\left[\frac{dy}{du}\right]\frac{du}{dx} = \frac{d^2y}{du^2}\frac{1}{\sqrt{x}} \tag{5.148}$$

and

$$\frac{d}{dx}\left[\frac{1}{\sqrt{x}}\right] = -\frac{1}{2x\sqrt{x}} \tag{5.149}$$

equation (5.147) becomes

$$\frac{d^2y}{dx^2} = \frac{1}{\sqrt{x}}\frac{d}{dx}\left[\frac{dy}{du}\right] + \frac{dy}{du}\frac{d}{dx}\left[\frac{1}{\sqrt{x}}\right] = \frac{1}{x}\frac{d^2y}{du^2} - \frac{1}{2x\sqrt{x}}\frac{dy}{du}. \tag{5.150}$$

Substituting equations (5.146) and (5.150) into equation (5.143), the Bessel equation takes the form

$$x^2\left(\frac{1}{x}\frac{d^2y}{du^2} - \frac{1}{2x\sqrt{x}}\frac{dy}{du}\right) + x\left(\frac{dy}{du}\frac{1}{\sqrt{x}}\right) - (1 - x)y = 0 \tag{5.151}$$

which can be simplified into

$$x\frac{d^2y}{du^2} + \frac{\sqrt{x}}{2}\frac{dy}{du} - (1 - x)y = 0. \tag{5.152}$$

Replacing $x = u^2/4$ in equation (5.152), we find

$$\frac{u^2}{4}\frac{d^2y}{du^2} + \frac{u}{4}\frac{dy}{du} - \left(1 - \frac{u^2}{4}\right)y = 0$$

$$\Rightarrow u^2\frac{d^2y}{du^2} + u\frac{dy}{du} + (u^2 - 4)y = 0. \tag{5.153}$$

Function of multivariable differential equation

In classical or quantum mechanical studies of the electrical or magnetic properties of particles or systems in physics often involve solving a second-order partial differential equation for some form of multivariable scalar or vector function. One of such differential equations that is commonly used is the Laplace equation. The Laplace equation in Cartesian coordinates for some scalar function $V(x, y, z)$ is given by

$$\frac{\partial^2 V(x, y, z)}{\partial x^2} + \frac{\partial^2 V(x, y, z)}{\partial y^2} + \frac{\partial^2 V(x, y, z)}{\partial z^2} = 0.$$

In order to determine the solution to such differential equations, it may be necessary to change the variables from Cartesian to cylindrical (r, φ, z) or spherical (r, θ, φ) coordinates. Such a change of variables is determined by the equations that relate a point in space defined by the Cartesian coordinates to cylindrical or spherical coordinates. In cylindrical and Cartesian coordinates a point in space is shown in figure 5.14. These coordinates are related by

$$x = s \cos(\theta),\ y = s \sin(\theta),\ z = z. \tag{5.154}$$

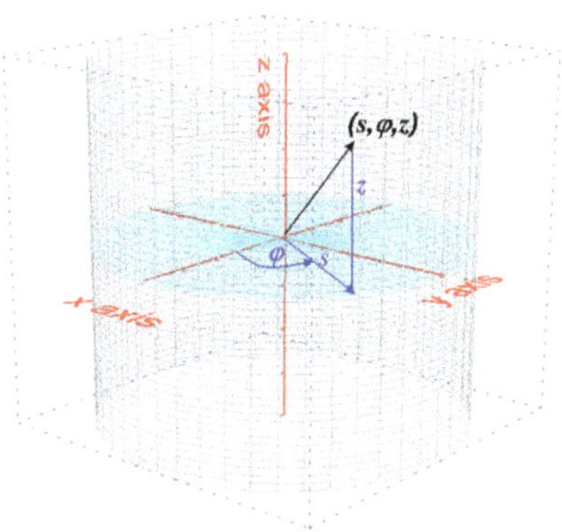

Figure 5.14. A point in space described by spherical coordinates, s, φ, and z.

Figure 5.15 shows a point in space described by Cartesian and spherical coordinates, which are related by

$$x = r \cos(\varphi) \sin(\theta),\ y = r \sin(\varphi) \sin(\theta),\ z = r \cos(\theta). \tag{5.155}$$

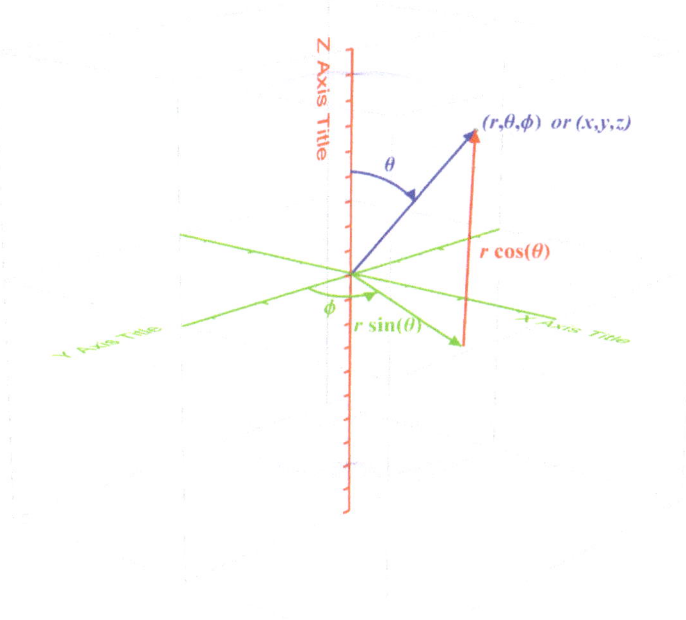

Figure 5.15. A point in space in spherical and Cartesian coordinates.

We will see how we change variables for the two-dimensional Laplace equation in Cartesian to cylindrical coordinates in the following example.

Example 5.12. In classical electrodynamics, in a two-dimensional space where there are no free charges, the electrical potential, $V(x, y)$, satisfies the two-dimensional Laplace equation given by

$$\frac{\partial^2 V(x, y)}{\partial x^2} + \frac{\partial^2 V(x, y)}{\partial y^2} = 0. \tag{5.156}$$

Suppose the two-dimensional space is a dielectric disk of radius R, and the potential at the edge of the disk is zero ($V(x, y) = 0$, for $x^2 + y^2 = R^2$). It is much easier for such a kind of problem if we solve the Laplace equation in polar coordinates. Transform the Laplace equation into polar coordinates using,

$$x = r \cos(\theta), \; y = r \sin(\theta), \; V = V(r, \theta). \tag{5.157}$$

Solution: There are two different ways of changing this equation from Cartesian to polar coordinates. We will use the quick way. To this end, we note that

$$\frac{\partial V(r, \theta)}{\partial r} = \frac{\partial V(x, y)}{\partial x}\frac{\partial x}{\partial r} + \frac{\partial V(x, y)}{\partial y}\frac{\partial y}{\partial r}, \tag{5.158}$$

so that using equation (5.157) for x and y, one finds

$$\frac{\partial V(r, \theta)}{\partial r} = \cos(\theta)\frac{\partial V(x, y)}{\partial x} + \sin(\theta)\frac{\partial V(x, y)}{\partial y}. \tag{5.159}$$

Similarly, for partial differentiation of $V(r, \theta)$ with respect to θ, we have

$$\begin{aligned}
\frac{\partial V(r, \theta)}{\partial \theta} &= \frac{\partial V(x, y)}{\partial x}\frac{\partial x}{\partial \theta} + \frac{\partial V(x, y)}{\partial y}\frac{\partial y}{\partial \theta} \\
&= -r\sin(\theta)\frac{\partial V(x, y)}{\partial x} + r\cos(\theta)\frac{\partial V(x, y)}{\partial y}
\end{aligned} \tag{5.160}$$

dividing this equation by r, one finds

$$\frac{1}{r}\frac{\partial V(r, \theta)}{\partial \theta} = -\sin(\theta)\frac{\partial V(x, y)}{\partial x} + \cos(\theta)\frac{\partial V(x, y)}{\partial y}. \tag{5.161}$$

Now multiplying equation (5.159) by $\cos(\theta)$ and equation (5.161) by $\sin(\theta)$, we find

$$\cos(\theta)\frac{\partial V(r, \theta)}{\partial r} = \cos^2(\theta)\frac{\partial V(x, y)}{\partial x} + \sin(\theta)\cos(\theta)\frac{\partial V(x, y)}{\partial y} \tag{5.162}$$

and

$$\frac{\sin(\theta)}{r}\frac{\partial V(r,\theta)}{\partial\theta} = -\sin^2(\theta)\frac{\partial V(x,y)}{\partial x} + \sin(\theta)\cos(\theta)\frac{\partial V(x,y)}{\partial y}. \qquad (5.163)$$

Subtracting equation (5.163) from equation (5.162) leads to

$$\frac{\partial V(x,y)}{\partial x} = \cos(\theta)\frac{\partial V(r,\theta)}{\partial r} - \frac{\sin(\theta)}{r}\frac{\partial V(r,\theta)}{\partial\theta}. \qquad (5.164)$$

Similarly, multiplying equation (5.159) by $\sin(\theta)$ and equation (5.161) by $\cos(\theta)$, we have

$$\sin(\theta)\frac{\partial V(r,\theta)}{\partial r} = \sin(\theta)\cos(\theta)\frac{\partial V(x,y)}{\partial x} + \sin^2(\theta)\frac{\partial V(x,y)}{\partial y} \qquad (5.165)$$

and

$$\frac{\cos(\theta)}{r}\frac{\partial V(r,\theta)}{\partial\theta} = -\sin(\theta)\cos(\theta)\frac{\partial V(x,y)}{\partial x} + \cos^2(\theta)\frac{\partial V(x,y)}{\partial y}, \qquad (5.166)$$

and adding these two equations results in

$$\frac{\partial V(x,y)}{\partial y} = \sin(\theta)\frac{\partial V(r,\theta)}{\partial r} + \frac{\cos(\theta)}{r}\frac{\partial V(r,\theta)}{\partial\theta}. \qquad (5.167)$$

From equation (5.164) and equation (5.167) follows that

$$\frac{\partial}{\partial y} = \sin(\theta)\frac{\partial}{\partial r} + \frac{\cos(\theta)}{r}\frac{\partial}{\partial\theta}, \qquad (5.168)$$

$$\frac{\partial}{\partial x} = \cos(\theta)\frac{\partial}{\partial r} - \frac{\sin(\theta)}{r}\frac{\partial}{\partial\theta} \qquad (5.169)$$

so that one can write

$$\frac{\partial^2}{\partial y^2} = \left[\sin(\theta)\frac{\partial}{\partial r} + \frac{\cos(\theta)}{r}\frac{\partial}{\partial\theta}\right]\left[\sin(\theta)\frac{\partial}{\partial r} + \frac{\cos(\theta)}{r}\frac{\partial}{\partial\theta}\right] \qquad (5.170)$$

and

$$\frac{\partial^2}{\partial x^2} = \left[\cos(\theta)\frac{\partial}{\partial r} - \frac{\sin(\theta)}{r}\frac{\partial}{\partial\theta}\right]\left[\cos(\theta)\frac{\partial}{\partial r} - \frac{\sin(\theta)}{r}\frac{\partial}{\partial\theta}\right]. \qquad (5.171)$$

On carrying out the partial differentiation in the first square bracket on the second square bracket in equation (5.170), we have

$$\frac{\partial^2}{\partial y^2} = \sin^2(\theta)\frac{\partial^2}{\partial r^2} + \sin(\theta)\cos(\theta)\frac{\partial}{\partial r}\left[\frac{1}{r}\frac{\partial}{\partial\theta}\right] + \frac{\cos(\theta)}{r}\frac{\partial}{\partial\theta}\left[\sin(\theta)\frac{\partial}{\partial r}\right]$$
$$+ \frac{\cos(\theta)}{r^2}\frac{\partial}{\partial\theta}\left[\cos(\theta)\frac{\partial}{\partial\theta}\right], \qquad (5.172)$$

and noting that

$$\frac{\partial}{\partial r}\left[\frac{1}{r}\frac{\partial}{\partial \theta}\right] = \frac{1}{r}\frac{\partial^2}{\partial r \partial \theta} - \frac{1}{r^2}\frac{\partial}{\partial r},$$

$$\frac{\partial}{\partial \theta}\left[\sin(\theta)\frac{\partial}{\partial r}\right] = \cos(\theta)\frac{\partial}{\partial r} + \sin(\theta)\frac{\partial^2}{\partial \theta \partial r},$$

$$\frac{\partial}{\partial \theta}\left[\cos(\theta)\frac{\partial}{\partial \theta}\right] = -\sin(\theta)\frac{\partial}{\partial \theta} + \cos(\theta)\frac{\partial^2}{\partial \theta^2},$$

(5.173)

one finds

$$\frac{\partial^2}{\partial y^2} = \sin^2(\theta)\frac{\partial^2}{\partial r^2} + \frac{\sin(\theta)\cos(\theta)}{r}\frac{\partial^2}{\partial r \partial \theta} - \frac{\sin(\theta)\cos(\theta)}{r^2}\frac{\partial}{\partial \theta}$$
$$+ \frac{\cos^2(\theta)}{r}\frac{\partial}{\partial r} + \frac{\sin(\theta)\cos(\theta)}{r}\frac{\partial^2}{\partial \theta \partial r} - \frac{\sin(\theta)\cos(\theta)}{r^2}\frac{\partial}{\partial \theta} + \frac{\cos^2(\theta)}{r^2}\frac{\partial^2}{\partial \theta^2}$$

(5.174)

that simplifies into

$$\frac{\partial^2}{\partial y^2} = \sin^2(\theta)\frac{\partial^2}{\partial r^2} + \frac{2\sin(\theta)\cos(\theta)}{r}\frac{\partial^2}{\partial r \partial \theta} - \frac{2\sin(\theta)\cos(\theta)}{r^2}\frac{\partial}{\partial \theta}$$
$$+ \frac{\cos^2(\theta)}{r}\frac{\partial}{\partial r} + \frac{\cos^2(\theta)}{r^2}\frac{\partial^2}{\partial \theta^2}.$$

(5.175)

Following a similar procedure in equation (5.171), we have

$$\frac{\partial^2}{\partial x^2} = \cos^2(\theta)\frac{\partial^2}{\partial r^2} - \sin(\theta)\cos(\theta)\frac{\partial}{\partial r}\left[\frac{1}{r}\frac{\partial}{\partial \theta}\right]$$
$$- \frac{\sin(\theta)}{r}\frac{\partial}{\partial \theta}\left[\cos(\theta)\frac{\partial}{\partial r}\right] + \frac{\sin(\theta)}{r^2}\frac{\partial}{\partial \theta}\left[\sin(\theta)\frac{\partial}{\partial \theta}\right]$$

(5.176)

that leads to

$$\frac{\partial^2}{\partial x^2} = \cos^2(\theta)\frac{\partial^2}{\partial r^2} - \frac{2\sin(\theta)\cos(\theta)}{r^2}\frac{\partial}{\partial \theta} + \frac{2\sin(\theta)\cos(\theta)}{r}\frac{\partial^2}{\partial r \partial \theta}$$
$$+ \frac{\sin^2(\theta)}{r}\frac{\partial}{\partial r} + \frac{\sin^2(\theta)}{r^2}\frac{\partial^2}{\partial \theta^2}.$$

(5.177)

Now adding equations (5.175) and (5.177), we find

$$\frac{\partial^2}{\partial x^2} + \frac{\partial^2}{\partial y^2} = \frac{\partial^2}{\partial r^2} + \frac{1}{r}\frac{\partial}{\partial r} + \frac{1}{r^2}\frac{\partial^2}{\partial \theta^2},$$

(5.178)

which can be rewritten as

$$\frac{\partial^2}{\partial x^2} + \frac{\partial^2}{\partial y^2} = \frac{1}{r}\frac{\partial}{\partial r}\left(r\frac{\partial}{\partial r}\right) + \frac{1}{r^2}\frac{\partial^2}{\partial \theta^2}.$$

(5.179)

Therefore, the two-dimensional Laplace equation in polar coordinates can be expressed as

$$\frac{1}{r}\frac{\partial}{\partial r}\left(r\frac{\partial V(r,\theta)}{\partial r}\right) + \frac{1}{r^2}\frac{\partial^2 V(r,\theta)}{\partial \theta^2} = 0. \tag{5.180}$$

5.7 Partial differentiation and Mathematica

The following are the basic commands that we can use to do partial differentiation in Mathematica.

- $D[f, \{x, n\}]$ gives the multiple derivative $\partial^n f/\partial x^n$
- $D[f, x, y, ...]$ gives the partial derivative $(\partial/\partial x)(\partial/\partial y)f$
- $D[f, \{x, n\}, \{y, m\}, ...]$ gives the multiple partial derivative $(\partial^m/\partial x^m)(\partial^m/\partial y^m)f$
- $D[f, \{\{x_1, x_2, ...\}\}]$ for a scalar f gives the vector derivative $(\partial f/\partial x_1, \partial f/\partial x_2, ...)$
- $D[f, x]$ gives the partial derivative $\partial f/\partial x$.

Here are a few examples

- Derivative with respect to x:

    $D[x^\wedge n, x]$

    nx^{-1+n}

- Fourth derivative with respect to x:

    $D[Sin[x]^\wedge 10, \{x, 4\}]$

    $5040Cos[x]^4Sin[x]^6 - 4680Cos[x]^2\,Sin[x]^8 + 280Sin[x]^{10}$

- Derivative of order n with respect to x:

    $D[Cos[x], \{x, n\}]$

    $Cos\left[\frac{n\pi}{2} + x\right]$

- Derivative with respect to x and y:

    $D[Sin[xy]/(x^\wedge 2 + y^\wedge 2), x, y]$

    $-\frac{2x^2Cos[xy]}{\left(x^2+y^2\right)^2} - \frac{2y^2Cos[xy]}{\left(x^2+y^2\right)^2} + \frac{Cos[xy]}{x^2+y^2} + \frac{8xySin[xy]}{\left(x^2+y^2\right)^3} - \frac{xySin[xy]}{x^2+y^2}$

- Derivative involving a symbolic function f:

    $D[xf[x]f'[x], x]$

    $f[x]f'[x] + xf'^2 + xf[x]f''[x]$

- Derivative using the symbol ∂:

    $\partial_{x,x}ArcTan[x]$

    $-\frac{2x}{\left(1+x^2\right)^2}$

Finding maximum and minimum points

- FindMinimum$[f, x]$ searches for a local minimum in f, starting from an automatically selected point.
- FindMinimum$[f, x, \{x_0\}]$ searches for a local minimum in f, starting from the point $x = x_0$.

- FindMinimum$\big[f, \{\{x, x_0\}, \{y, y_0\}, \ldots\}\big]$ searches for a local minimum in a function of several variables.
- FindMinimum$\big[\{f, cons\}, \{\{x, x_0\}, \{y, y_0\}, \ldots\}\big]$ searches for a local minimum subject to the constraints *cons*.
- FindMinimum[$\{f, cons\}, \{x, y, \ldots\}$] starts from a point within the region defined by the constraints.
- FindMaximum[f, x] searches for a local maximum in f, starting from an automatically selected point.
- FindMaximum[$f, \{x, x_0\}$] searches for a local minimum in f, starting from the point $x = x_0$.
- FindMaximum$\big[f, \{\{x, x_0\}, \{y, y_0\}, \ldots\}\big]$ searches for a local minimum in a function of several variables.
- FindMaximum$\big[\{f, cons\}, \{\{x, x_0\}, \{y, y_0\}, \ldots\}\big]$ searches for a local minimum subject to the constraints *cons*.
- FindMaximum[$\{f, cons\}, \{x, y, \ldots\}$] starts from a point within the region defined by the constraints.

Here are some examples of minimum and maximum points
- Find a local minimum, starting the search at $x = 2$:
    FindMinimum[xCos[x], {x, 2}]
    {−3.288 37, {x → 3.425 62}}
    Plot[xCos[x], {x, 0, 20}] (figure 5.16)
- Extract the value of x at the local minimum:
    x/.Last[FindMinimum[xCos[x], {x, 2}]]
    3.425 62
- Find a local minimum, starting at $x = 7$, subject to constraints $1 \leqslant x \leqslant 15$:
    FindMinimum[{xCos[x], 1 ⩽ x ⩽ 15}, {x, 7}]
    {−9.477 29, {x → 9.529 33}}
    FindMinimum[3x⁴ − 28x³ + 84x² − 96x + 42, {x, 1.1}]
    {5 .,{x → 1. }}
    Minimize[3x⁴ − 28x³ + 84x² − 96x + 42, {x}]
    {−22, {x → 4}}
    Plot[3x⁴ − 28x³ + 84x² − 96x + 42, {x, 0, 5}] (figure 5.17)
    Maximize[−2x^2 − 3x + 5, x]

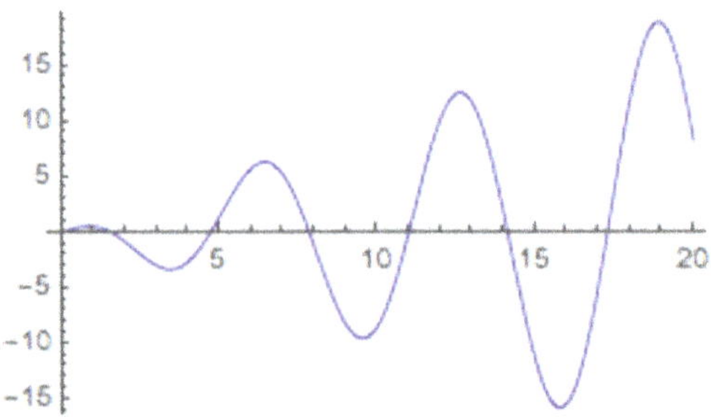

Figure 5.16. The graph for the function $f(x) = x\cos(x)$.

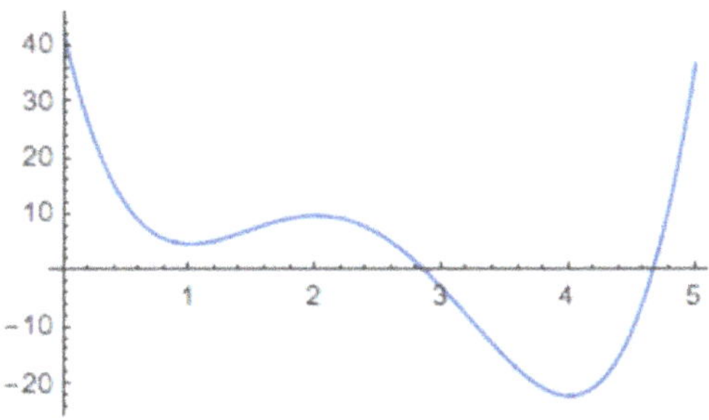

Figure 5.17. The graph for the function $f(x) = 3x^4 - 28x^3 + 84x^2 - 96x + 42$.

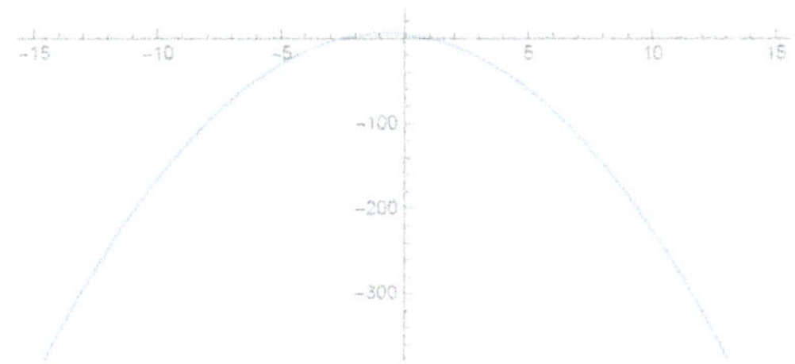

Figure 5.18. The graph for the function $f(x) = -2x^2 - 3x + 5$.

$$\left\{ \frac{49}{8}, \left\{ x \to -\frac{3}{4} \right\} \right\}$$

Plot$[-2x\wedge 2 - 3x + 5, \{x, -3, 3\}]$ (figure 5.18)

5.8 Homework assignments

Problem 1. For the function

$$z(u, v, w) = \ln \sqrt{u^2 + v^2 + w^2} \tag{5.181}$$

find the following partial derivatives:

$$\frac{\partial z}{\partial u}, \frac{\partial z}{\partial v}, \text{ and } \frac{\partial z}{\partial w}. \tag{5.182}$$

Verify your result using Mathematica.

Problem 2. For the function

$$u(x, y) = e^x \cos(y) \tag{5.183}$$

verify that
   (a)

$$\frac{\partial^2 u}{\partial x \partial y} = \frac{\partial^2 u}{\partial y \partial x} \tag{5.184}$$

(b)

$$\frac{\partial^2 u}{\partial x^2} + \frac{\partial^2 u}{\partial y^2} = 0. \tag{5.185}$$

Problem 3. Using the functions
$$z(x, y) = x^2 + 2y^2, \quad x = r\cos(\theta), \quad y = r\sin(\theta) \tag{5.186}$$
find

(a)

$$\left(\frac{\partial z}{\partial \theta}\right)_x, \tag{5.187}$$

(b)

$$\frac{\partial z}{\partial y \partial \theta}. \tag{5.188}$$

Problem 4. The electrostatic potential energy U of two point charges with charge q_1 and q_2, placed at a distance r apart is given by

$$U = k\frac{q_1 q_2}{r},$$

where $k = 1/4\pi\epsilon_0$ is a constant. Find the relative error in the electrostatic potential energy U in *the worst case if the relative error* in q_1 is 3%; in q_2 is 4%, and in r is 5%.

Problem 5. Use differential to show that for large n and small a

$$\sqrt{n + a} - \sqrt{n} \simeq \frac{a}{2\sqrt{n}}. \tag{5.189}$$

Problem 6. The acceleration due to gravity can be found from the length l and period T of a pendulum (see figure 5.19); the formula is $g = 4\pi^2 l/T^2$. Find the relative error in g in the worst case if the relative error in l is 5%, and the relative error in T is 2%.

Problem 7. For an ideal gas of N molecules, the number of molecules with speeds $\leqslant v$ is given by the formula

$$n(v) = \frac{4a^3 N}{\sqrt{\pi}} \int_0^v x^2 e^{-a^2 x^2} dx \tag{5.190}$$

where a is a constant and N is the total number of molecules. If $N = 10^{26}$, estimate the number of molecules with speeds between $v = 1/a$ and $1.01/a$.

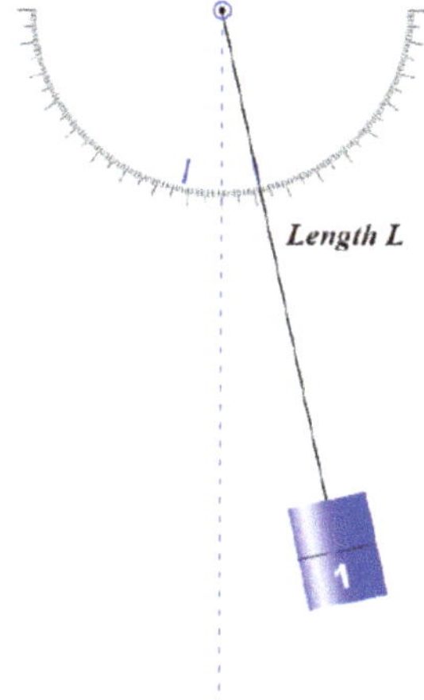

Figure 5.19. Simple pendulum of length L.

Problem 8. Given $z = xe^{-y}$, $x = \cosh(t)$, $y = \cos(t)$, find $\frac{dz}{dt}$.

Problem 9. If $xy^3 - yx^3 = 6$ is the equation of the curve, find the slope and the equation of the tangent line at the point $(1, 2)$. Computer plot the curve and the tangent line.

Problem 10. In Problem 9 find d^2y/dx^2 at $(1, 2)$.

Problem 11. Intercontinental ballistic missiles (ICBM) are very long-range (greater than 5500 km or 3500 miles) ballistic missiles typically designed for nuclear weapons delivery, that is delivering one or more nuclear warheads. Assume that these missiles can be described by a circular cylinder with one conical end and one flat end as shown in figure 5.20. The cylindrical part carries mostly of the fuel and the conical

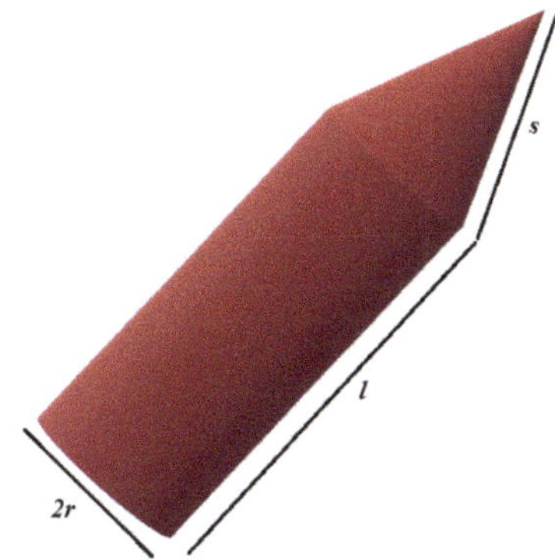

Figure 5.20. Model missile.

part carries the nuclear weapon. Therefore, maximized volume (both for the cylindrical part and the conical part) is required in order to form a long range high carrying capacity missiles. Suppose the surface area of the missile is specified (i.e., $A = A_0$). Using *Lagrangian multipliers* find the proportions (i.e., $r: l: s$) that will maximize the volume of the missile.

IOP Publishing

Studies in Theoretical Physics, Volume 1
Fundamental mathematical methods
Daniel Erenso and Victor Montemayor

Chapter 6

Introduction to differential calculus II

In the previous chapter, we introduced the basics of differential calculus and its applications. We introduced how to differentiate and also find the differential to functions of single and multivariable variables. This chapter will introduce a reverse process in differential calculus and its application in real physical problems. We will introduce the different methods used to determine single or multivariable variable functions satisfying a differential equation (DE). A *differential equation* is an equation involving derivatives of functions with respect to one or more independent variables. We often encounter linear or non-linear differential equations for scalar (e.g., an electrical potential) or vector (e.g., electric field) functions in real physical problems. We will focus on methods of solving linear differential equations. We also see how to solve such differential equations using Mathematica at the end of the chapter.

6.1 First-order ordinary DE

A differential equation is an equation involving derivatives of functions with respect to one or more independent variables. If there is only one independent variable, the equation is an ordinary differential equation (ODE). The order of an ODE or any DE is the order of the highest-order derivative in the DE. An nth-order ODE for the function $y(x)$ (with x as independent and y as dependent variable) has the generic form

$$a_0(x)y + a_1(x)\frac{dy}{dx} + a_2(x)\frac{d^2y}{dx^2} + a_3(x)\frac{d^3y}{dx^3} + a_4(x)\frac{d^4y}{dx^4} + \ldots a_n(x)\frac{d^ny}{dx^n} \quad (6.1)$$

$$= f(x),$$

or more concisely,

$$\sum_{i=0}^{n} a_i(x)\frac{d^i y}{dx^i} = f(x).$$ (6.2)

Note that in an ODE, $a_i(x)$ is generally a function of the independent variable x. When $a_i(x)$ = constant for all i, the ODE is known as a linear ordinary differential equation (LODE). It is the focus of this chapter to introduce the methods to determine the solutions to LODEs. The general solution to an nth-order ODE contains n independent, arbitrary constants. Generally, a particular solution to a DE is the general solution evaluated for particular values of its arbitrary constants. We determine the values for the constants by applying boundary conditions (BC) that pertain to the DE, usually obtained from an analysis of the physical system modeled by the DE.

The first-order LODE, generally, has the form

$$a_0 y(x) + a_1\frac{dy(x)}{dx} = f(x),$$ (6.3)

where a_0 and a_1 are constants. We can determine the solutions to first-order LODEs by direct integration. For $a_1 \neq 0$, one can put equation (6.3) in the form

$$\frac{a_0}{a_1}y(x) + \frac{dy(x)}{dx} = \frac{1}{a_1}f(x).$$ (6.4)

Noting that

$$e^{-\beta x}e^{\beta x} = 1 \Rightarrow \frac{d}{dx}[e^{-\beta x}e^{\beta x}] = 0$$

$$\Rightarrow e^{\beta x}\frac{de^{-\beta x}}{dx} + e^{-\beta x}\frac{de^{\beta x}}{dx} = 0 \Rightarrow -\beta e^{-\beta x}e^{\beta x} + e^{-\beta x}\frac{de^{\beta x}}{dx} = 0$$ (6.5)

$$\Rightarrow \beta = e^{-\beta x}\frac{de^{\beta x}}{dx},$$

for

$$\beta = \frac{a_0}{a_1},$$

one can then write

$$\frac{a_0}{a_1} = e^{-\frac{a_0}{a_1}x}\frac{de^{\frac{a_0}{a_1}x}}{dx}.$$ (6.6)

Substituting this relation into equation (6.4), we have

$$y(x)e^{-\frac{a_0}{a_1}x}\frac{de^{\frac{a_0}{a_1}x}}{dx} + \frac{dy(x)}{dx} = \frac{1}{a_1}f(x)$$ (6.7)

so that multiplying both sides of this equation by $e^{\frac{a_0}{a_1}x}$, we find

$$y(x)\frac{d\left(e^{\frac{a_0}{a_1}x}\right)}{dx} + e^{\frac{a_0}{a_1}x}\frac{dy(x)}{dx} = \frac{1}{a_1}f(x)e^{\frac{a_0}{a_1}t}. \tag{6.8}$$

In view of the product rule

$$\frac{d}{dx}\left[f_1(x)f_2(x)\right] = f_1(x)\frac{df_2(x)}{dx} + f_1(x)\frac{df_2(x)}{dx}, \tag{6.9}$$

one can then rewrite equation (6.8) as,

$$\frac{d}{dx}\left[y(x)e^{\frac{a_0}{a_1}x}\right] = \frac{1}{a_1}f(x)e^{\frac{a_0}{a_1}x}. \tag{6.10}$$

Integrating equation (6.10),

$$\int_{x_0}^{x}\frac{d}{dx'}\left[y(x')e^{\frac{a_0}{a_1}x'}\right]dx' = \frac{1}{a_1}\int_{x_0}^{x}f(x')e^{\frac{a_0}{a_1}x'}dx', \tag{6.11}$$

we find

$$y(x')e^{\frac{a_0}{a_1}x'}\Big|_{x_0}^{x} = \frac{1}{a_1}\int_{x_0}^{x}f(x')e^{\frac{a_0}{a_1}x'}dx'$$

$$y(x)e^{\frac{a_0}{a_1}x} - y(x_0)e^{\frac{a_0}{a_1}x_0} = \frac{1}{a_1}\int_{x_0}^{x}f(x)e^{\frac{a_0}{a_1}x}dx \tag{6.12}$$

$$\Rightarrow y(x)e^{\frac{a_0}{a_1}x} = y(x_0)e^{\frac{a_0}{a_1}x_0} + \frac{1}{a_1}\int_{x_0}^{x}f(x')e^{\frac{a_0}{a_1}x'}dx',$$

so that upon multiplying this equation by $e^{-\frac{a_0}{a_1}x}$, the solution to equation (6.3) can be expressed as

$$y(x) = e^{-\frac{a_0}{a_1}x}\left[y(x_0)e^{\frac{a_0}{a_1}x_0} + \frac{1}{a_1}\int_{x_0}^{x}f(x')e^{\frac{a_0}{a_1}x'}dx'\right]. \tag{6.13}$$

In equation (6.13), a_0 and a_1 are constants in the DE and x_0 and $y(x_0)$ are also constants defined by the boundary conditions of the function $y(x)$. It means what is the behavior of the function ($y(x_0)$) at the boundary x_0 and it must be given, or one should be able to find the value from the conditions set for the system modeled by the DE. Suppose we write

$$\int_{x_0}^{x}f(x')e^{\frac{a_0}{a_1}x'}dx' = \int^{x}f(x')e^{\frac{a_0}{a_1}x'}dx' - \int_{x_0}f(x')e^{\frac{a_0}{a_1}x'}dx'$$

$$= \int f(x)e^{\frac{a_0}{a_1}x}dx - C(x_0), \tag{6.14}$$

where we have replaced the result of the integral when it is evaluated at x_0 by a constant $C(x_0)$. Thus equation (6.13) can be expressed as

$$y(x) = e^{-\frac{a_0}{a_1}x}\left[y(x_0)e^{\frac{a_0}{a_1}x_0} - C(x_0) + \frac{1}{a_1}\int f(x)e^{\frac{a_0}{a_1}x}dx\right]. \tag{6.15}$$

Thus by introducing another constant C_1 for

$$C_1 = y(x_0)e^{\frac{a_0}{a_1}x_0} - C(x_0), \tag{6.16}$$

the general solution to the first-order LODE in equation (6.3) is given by

$$y(x) = C_1e^{-\frac{a_0}{a_1}x} + \frac{e^{-\frac{a_0}{a_1}x}}{a_1}\int f(x)e^{\frac{a_0}{a_1}x}dx. \tag{6.17}$$

Note that for $f(x) = 0$, the differential equation becomes

$$a_0y(x) + a_1\frac{dy(x)}{dx} = 0, \tag{6.18}$$

and the solution is given by

$$y(x) = C_1e^{-\frac{a_0}{a_1}x}. \tag{6.19}$$

In the following examples, we will see how we apply equation (6.17) to solve real physical problems described by a first-order LODE.

Example 6.1. A copper rod of length, h, mass, m, and electric resistance, R, slides with negligible friction on metal rails that have negligible electric resistance (see figure 6.1). A metallic rod with negligible electric resistance connects the rails. A uniform magnetic field, $\vec{B}$, points out of the page that fills the entire region. The magnetic force on the rod due to the induced current as a result of Faraday's law, can be shown to be,

$$F_m(t) = \frac{h^2v(t)B^2}{R}, \tag{6.20}$$

where $v(t)$ is the velocity of the rod at time t and directed to the left. Using Lenz's law, it can also be verified that the induced current direction in the rod is upward, which

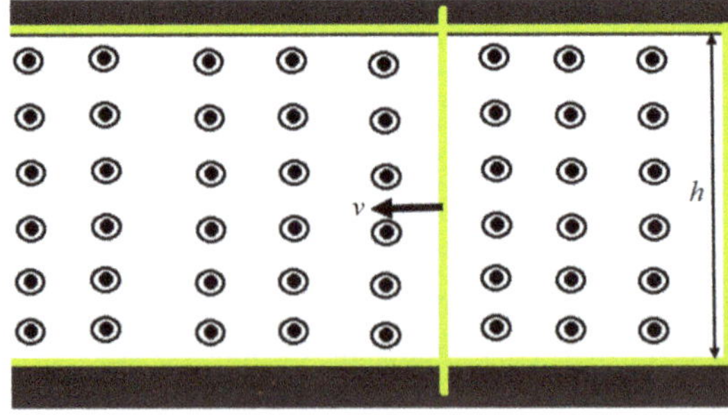

Figure 6.1. A rod sliding on u-shaped wire in a uniform magnetic field.

results in the direction of the magnetic force on the rod to be to the right. The rod starts with speed, v_0, at $t = 0$, and is left to slide. What is its speed at a later time t?

Solution: Using Newton's second law, one can write

$$m\frac{dv(t)}{dt} = -F_m = -\frac{l^2 v(t)B^2}{R} \Rightarrow \frac{l^2 B^2}{mR}v(t) + \frac{dv(t)}{dt} = 0. \tag{6.21}$$

that can be put in the form

$$a_0 v(t) + a_1\frac{dv(t)}{dt} = f(t), \tag{6.22}$$

where

$$a_0 = \frac{l^2 B^2}{mR}, \ a_1 = 1, f(t) = 0.$$

In view of equation (6.17), the solution to equation (6.22) can be expressed as

$$v(t) = C_1 e^{-\frac{a_0}{a_1}t} + \frac{e^{-\frac{a_0}{a_1}t}}{a_1}\int f(t)e^{\frac{a_0}{a_1}t}dt, \tag{6.23}$$

so that substituting the values of a_0, a_1, and $f(t)$ in equation (6.24), one finds for the speed

$$v(t) = C_1 e^{-\frac{a_0}{a_1}t} = C_1 e^{-\frac{l^2 B^2 t}{mR}}. \tag{6.24}$$

Noting that initial velocity $v(0) = v_0$, the velocity of the rod becomes

$$v(t) = v_0 e^{-\frac{l^2 B^2 t}{mR}}. \tag{6.25}$$

Example 6.2. An object dropped from a height y_0. It has a vertically downward acceleration given by $a = ge^{-kt}$, where k is a (real) positive constant and g is the acceleration due to gravity at the Earth's surface.
   (a) Find the general solution for the distance fallen.
   (b) Using the appropriate initial conditions find the distance fallen by the object described.
   (c) Comment on the limiting behavior of your answer, and physically describe the system for these limits.

Solution:
   (a) We recall that the acceleration a and the velocity v are related by the equation

$$a = \frac{dv}{dt},\tag{6.26}$$

and using the expression for the given acceleration, we find

$$\frac{dv}{dt} = ge^{-kt} \Rightarrow a_0 v(t) + a_1\frac{dv(t)}{dt} = f(t),\tag{6.27}$$

where

$$a_0 = 0,\, a_1 = 1,\, f(t) = ge^{-kt}.\tag{6.28}$$

In view of equation (6.17), the solution to equation (6.27) can be expressed as

$$v(t) = C_1 e^{-\frac{a_0}{a_1}t} + \frac{e^{-\frac{a_0}{a_1}t}}{a_1}\int f(t)e^{\frac{a_0}{a_1}t}dt\tag{6.29}$$

and substituting equation (6.28), we find

$$v(t) = C_1 + \int ge^{-kt}dt.\tag{6.30}$$

Upon carrying out the integration, the velocity becomes,

$$v(t) = C_1 - \frac{g}{k}e^{-kt}.\tag{6.31}$$

To find the position, we note that the velocity is related to the displacement by,

$$v(t) = \frac{dy}{dt} \Rightarrow \frac{dy}{dt} = C_1 - \frac{g}{k}e^{-kt}\tag{6.32}$$

which we may put in form,

$$a_0 y(t) + a_1\frac{dy(t)}{dt} = f(t),\tag{6.33}$$

where

$$a_0 = 0,\, a_1 = 1,\, f(t) = C_1 - \frac{g}{k}e^{-kt}.\tag{6.34}$$

Using the relation in equation (6.17), we can write the solution to equation (6.33) as,

$$y(t) = C_2 e^{-\frac{a_0}{a_1}t} + \frac{e^{-\frac{a_0}{a_1}t}}{a_1}\int f(t)e^{\frac{a_0}{a_1}t}dt,\tag{6.35}$$

so that using equation (6.34), we have

$$y(t) = C_2 + \int \left(C_1 - \frac{g}{k} e^{-kt} \right) dt, \tag{6.36}$$

and upon carrying out the integration, one finds

$$y(t) = C_2 + C_1 t + \frac{g}{k^2} e^{-kt}. \tag{6.37}$$

(b) We are given the initial conditions (boundary conditions) $(v(0) = 0)$ and assume that the object is dropped from an initial position $(y(0) = y_0)$. Then evaluating equations (6.31) and (6.37) at $t = 0$, we find

$$C_1 = \frac{g}{k}, \quad C_2 = y_0 - \frac{g}{k^2}, \tag{6.38}$$

so that equations (6.31) and (6.37) become,

$$v(t) = \frac{g}{k}(1 - e^{-kt}),$$

$$y(t) = y_0 + \frac{g}{k}t - \frac{g}{k^2}(1 - e^{-kt}), \tag{6.39}$$

and the distance the object has fallen, as a function of time, becomes

$$d(t) = y_0 - y(t) = -\frac{g}{k}t + \frac{g}{k^2}(1 - e^{-kt}). \tag{6.40}$$

(c) For small t (using Taylor series expansion), we can make the approximation

$$e^{-kt} \simeq 1 - kt + \frac{1}{2}(kt)^2 \Rightarrow 1 - e^{-kt} \simeq kt - \frac{1}{2}(kt)^2, \tag{6.41}$$

and find for the distance the object has fallen

$$d(t) \simeq -\frac{g}{k}t + \frac{g}{k^2}\left[kt - \frac{1}{2}(kt)^2 \right] \Rightarrow d(t) \simeq -\frac{1}{2}gt^2, \tag{6.42}$$

and for the speed

$$v(t) \simeq \frac{g}{k}\left[kt - \frac{1}{2}(kt)^2 \right] = g\left(t - \frac{1}{2}kt^2 \right) \Rightarrow v(t) \simeq gt, \tag{6.43}$$

where we have dropped the term kt^2 in equation (6.43) since it is negligibly small. The results in equations (6.42) and (6.43) showing that the object is under free-fall near the beginning. On the other hand for large time t_0 where

$$e^{-kt_0} \simeq 0 \tag{6.44}$$

the velocity and the distance fallen become

$$v(t) = \frac{g}{k}[1 - e^{-kt}] \simeq \frac{g}{k} \tag{6.45}$$

and

$$d(t) = -\frac{g}{k}t + \frac{g}{k^2}[1 - \exp(-kt)] \simeq -\frac{g}{k}t_0 + \frac{g}{k^2}. \tag{6.46}$$

The results in equations (6.45) and (6.46) show that the object has reached a constant speed

$$v_0 = \frac{g}{k}, \tag{6.47}$$

and at sufficiently large time $t = t_0$, since its acceleration

$$a = \frac{dv(t)}{dt} \simeq \frac{d}{dt}\left(\frac{g}{k}\right) = 0. \tag{6.48}$$

Equation (6.48) shows that the velocity is constant, and the object has reached a terminal speed.

6.2 The first-order ODE and exact total differential

In the previous section, we have seen how the solution to a first-order LODE

$$a_0(x)y(x) + a_1(x)\frac{dy(x)}{dx} = f(x), \tag{6.49}$$

can be determined when the coefficients a_0 and a_1 are constant. Generally, a first-order DE could be ODE where the coefficients are a function of one variable (only x) as described by equation (6.49). However, it is possible to rewrite equation (6.49) as

$$\frac{a_1(x)}{y}\frac{dy(x)}{dx} = -\left[a_0(x) - \frac{f(x)}{y}\right]$$

$$\Rightarrow \frac{a_1(x)}{y}dy = -\left[a_0(x) - \frac{f(x)}{y}\right]dx \tag{6.50}$$

$$\Rightarrow P(x, y) + Q(x, y)\frac{dy}{dx} = 0$$

$$\Rightarrow Q(x, y)\frac{dy}{dx} = -P(x, y)$$

where

$$Q(x, y) = \frac{a_1(x)}{y}, \ P(x, y) = a_0(x) - \frac{f(x)}{y}, \tag{6.51}$$

which are functions of two variable. Generally, it is possible to have a DE like in equation (6.50) in which the coefficients are a function of two variables (x and y). In this section, we will develop the method to solve such kind of first-order DEs. To this end, we note that equation (6.50) can be rewritten as

$$dF(x, y) = P(x, y)dx + Q(x, y)dy = 0, \tag{6.52}$$

which is the total differential of some constant function,

$$F(x, y) = \text{constant}, \tag{6.53}$$

that depends on both x and y and

$$P(x, y) = \frac{\partial F(x, y)}{\partial x}, \quad Q(x, y) = \frac{\partial F(x, y)}{\partial y}. \tag{6.54}$$

This indicates that if one can determine this constant function from the total differential derived from the first-order DE in equation (6.50), one can solve the DE by solving for y from the function. However, this depends on some conditions that the total differential for the function needs to satisfy. One may ask what these conditions are. Before we get into it, let us determine the total differential for a function that is not a constant in the following example.

Example 6.3. Find total differential dF for the function

$$F(x, y) = 3x^2 + 2x \sin(2y). \tag{6.55}$$

Solution: We recall from the last chapter, the total differential dF for a function of two variables, $F(x, y)$, is given by

$$dF = \frac{\partial F}{\partial x}dx + \frac{\partial F}{\partial y}dy = P(x, y)dx + Q(x, y)dy. \tag{6.56}$$

For the function in equation (6.55), we have

$$P(x, y) = \frac{\partial F}{\partial x} = \frac{\partial}{\partial x}[3x^2 + 2x \sin(2y)] = 6x + 2 \sin(2y), \tag{6.57}$$

$$Q(x, y) = \frac{\partial F}{\partial y} = \frac{\partial}{\partial y}[3x^2 + 2x \sin(2y)] = 4x \cos(2y), \tag{6.58}$$

so that total differential becomes

$$dF = [6x + 2 \sin(2y)]dx + [4x \cos(2y)]dy. \tag{6.59}$$

In Example 6.3 we found the total differential

$$dF = P(x, y)dx + Q(x, y)dy, \tag{6.60}$$

where

$$P(x, y) = \frac{\partial F}{\partial x} = 6x + 2\sin(2y),$$
$$Q(x, y) = \frac{\partial F}{\partial y} = 4x\cos(2y). \tag{6.61}$$

Taking the partial derivative of the functions $P(x, y)$ and $Q(x, y)$ with respect to y and x, respectively, we find

$$\frac{\partial P}{\partial y} = \frac{\partial F}{\partial y \partial x} = \frac{\partial}{\partial y}[6x + 2\sin(2y)] = 4\cos(2y),$$
$$\frac{\partial Q}{\partial x} = \frac{\partial F}{\partial x \partial y} = \frac{\partial}{\partial x}[4x\cos(2y)] = 4\cos(2y). \tag{6.62}$$

This shows that for the total differential

$$dF = P(x, y)dx + Q(x, y)dy, \tag{6.63}$$

we found

$$\frac{\partial P(x, y)}{\partial y} = \frac{\partial Q(x, y)}{\partial x}. \tag{6.64}$$

Such kind of differential is called an *exact differential*. Therefore, Example 6.3 indicates that, when the total differential

$$dF = P(x, y)dx + Q(x, y)dy, \tag{6.65}$$

is an *exact differential*, satisfying the condition

$$\frac{\partial P(x, y)}{\partial y} = \frac{\partial Q(x, y)}{\partial x}, \tag{6.66}$$

then a function $F(x, y)$ exists and it can be derived from

$$P(x, y) = \frac{\partial F}{\partial x}, \quad Q(x, y) = \frac{\partial F}{\partial y},$$

by direct integrations.

Example 6.4. Consider the first-order DE

$$(x^3 + 2y)\frac{dy}{dx} = 3x(2 - xy). \tag{6.67}$$

(a) Write this DE in the form

$$dF(x, y) = P(x, y)dx + Q(x, y)dy = 0, \qquad (6.68)$$

where $F(x, y)$= constant.

(b) Show that the total differential in (a) is an exact differential and find the function

$$F(x, y) = \text{constant.} \qquad (6.69)$$

(c) Using the result in part (b) find the solution to the given differential equation.

Solution:

(a) Multiplying equation (6.67) by dx, we have

$$(x^3 + 2y)dy = 3x(2 - xy)dx, \qquad (6.70)$$

which can be rewritten as

$$dF(x, y) = P(x, y)dx + Q(x, y)dy = 0, \qquad (6.71)$$

where

$$P(x, y) = 3x(2 - xy), \quad Q(x, y) = -(x^3 + 2y). \qquad (6.72)$$

(b) Now taking the partial derivative of the functions $P(x, y)$ and $Q(x, y)$ with respect to y and x, respectively, we find

$$\frac{\partial P}{\partial y} = \frac{\partial}{\partial y}[3x(2 - xy)] = -3x^2, \qquad (6.73)$$

$$\frac{\partial Q}{\partial x} = \frac{\partial}{\partial x}[-(x^3 + 2y)] = -3x^2. \qquad (6.74)$$

There follows that

$$\frac{\partial P}{\partial y} = \frac{\partial Q}{\partial x} \qquad (6.75)$$

which proves the total differential is an exact and there exists a function

$$F(x, y) = \text{constant,} \qquad (6.76)$$

such that

$$P = \frac{\partial F}{\partial x} \text{ and } Q = \frac{\partial F}{\partial y}. \tag{6.77}$$

Using the equation for the function $P(x, y)$, we have

$$P = \frac{\partial F}{\partial x} = 3x(2 - xy) \Rightarrow \int_{x_0}^{x} \frac{\partial F(x', y)}{\partial x'} dx' = \int_{x_0}^{x} 3x'(2 - x'y)dx' \tag{6.78}$$

so that upon carrying out the integration, one finds

$$F(x', y) \big|_{x_0}^{x} = 3x'^2 - x'^3 y \big|_{x_0}^{x}$$
$$F(x, y) - F(x_0, y) = 3x^2 - x^3 y - 3x_0^2 + x_0^3 y. \tag{6.79}$$

This equation can be rewritten as

$$F(x, y) = 3x^2 - x^3 y + [F(x_0, y) - 3x_0^2 + x_0^3 y] = 3x^2 - x^3 y + g(y), \tag{6.80}$$

where we introduced the function

$$g(y) = F(x_0, y) - 3x_0^2 + x_0^3 y, \tag{6.81}$$

which depends only on y, since x_0 is a constant. Now substituting equation (6.80) into the equation for $Q(x, y)$,

$$Q(x, y) = \frac{\partial F}{\partial y} = -(x^3 + 2y), \tag{6.82}$$

we find

$$\frac{\partial}{\partial y}[3x^2 - x^3 y + g(y)] = -(x^3 + 2y) \Rightarrow -x^3 + \frac{dg(y)}{dy} = -x^3 - 2y$$

$$\Rightarrow \frac{dg(y)}{dy} = -2y \Rightarrow \int_{y_0}^{y} \frac{dg(y')}{dy'} dy' \tag{6.83}$$

$$= -2 \int_{y_0}^{y} y' dy'.$$

Upon integration, one can easily show that

$$g(y') \big|_{y_0}^{y} = -y'^2 \big|_{y_0}^{y} \Rightarrow g(y) - g(y_0) = -y^2 + y_0^2$$
$$g(y) = -y^2 + y_0^2 + g(y_0) = -y^2 + C, \tag{6.84}$$

where

$$C = y_0^2 + g(y_0),$$

is a constant independent of x and y. Therefore, upon substituting equation (6.84) into (6.80), the constant is found to be

$$F(x, y) = 3x^2 - x^3y - y^2 + C = \text{constant}. \tag{6.85}$$

(c) Noting that

$$3x^2 - x^3y - y^2 + C = \text{constant} \tag{6.86}$$

can be rewritten as

$$y^2 + x^3y - 3x^2 + C_1 = 0 \Rightarrow ay^2 + by + c = 0, \tag{6.87}$$

where

$$a = 1, \; b = x^3, \; c = -3x^2 + C_1. \tag{6.88}$$

Equation (6.87) is a quadratic equation, the solution of which is given by

$$y = \frac{-b \pm \sqrt{b^2 - 4ac}}{2a}. \tag{6.89}$$

Upon substituting the results of equation (6.88) in equation (6.89), we find the solution to the DE in equation (6.67) to be

$$y(x) = \frac{1}{2}\left[-x^3 \pm \sqrt{x^6 + 4(3x^2 - C_1)} \right]. \tag{6.90}$$

Example 6.5. For the DE

$$(y^2 - 3x^2e^{3y})\frac{dy}{dx} = 2xe^{3y} + e^x, \tag{6.91}$$

Do part (a)–(c) in Example 6.4.

Solution:

(a) Multiplying the given first-order differential equation by dx, we find

$$(y^2 - 3x^2e^{3y})dy = (2xe^{3y} + e^x)dx \tag{6.92}$$

which can be put in the form

$$dF(x, y) = P(x, y)dx + Q(x, y)dy = 0, \tag{6.93}$$

where

$$P(x, y) = 2xe^{3y} + e^x, \; Q(x, y) = -(y^2 - 3x^2e^{3y}). \tag{6.94}$$

(b) For an exact differential, we must be able to show that

$$\frac{\partial P(x, y)}{\partial y} = \frac{\partial Q(x, y)}{\partial x}. \tag{6.95}$$

Using equation (6.94), we find

$$\frac{\partial P(x, y)}{\partial y} = \frac{\partial}{\partial y}(2xe^{3y} + e^x) = 6xe^{3y},$$

$$\frac{\partial Q(x, y)}{\partial x} = -\frac{\partial}{\partial x}(y^2 - 3x^2e^{3y}) = 6xe^{3y}, \tag{6.96}$$

that shows

$$\frac{\partial P(x, y)}{\partial y} = \frac{\partial Q(x, y)}{\partial x}. \tag{6.97}$$

Therefore, the differential is exact, and there exists a function

$$F(x, y) = \text{constant} \tag{6.98}$$

such that

$$P(x, y) = \frac{\partial F}{\partial x}, \ Q(x, y) = \frac{\partial F}{\partial y}. \tag{6.99}$$

This leads to

$$P(x, y) = \frac{\partial F}{\partial x} \Rightarrow \int_{x_0}^{x} \frac{\partial F}{\partial x'}dx' = \int_{x_0}^{x} (2x'e^{3y} + e^{x'})dx'$$
$$\Rightarrow F(x, y) - F(x_0, y) = x^2e^{3y} + e^x - (x_0^2e^{3y} + e^{x_0}) \tag{6.100}$$
$$\Rightarrow F(x, y) = x^2e^{3y} + e^x + F(x_0, y) - (x_0^2e^{3y} + e^{x_0})$$
$$\Rightarrow F(x, y) = x^2e^{3y} + e^x + g(y),$$

where we introduced the function defined by

$$g(y) = F(x_0, y) - (x_0^2e^{3y} + e^{x_0}). \tag{6.101}$$

Using the result in equation (6.100) and $Q(x, y) = -(y^2 - 3x^2e^{3y})$, one can write

$$\frac{\partial F}{\partial y} = Q(x, y) \Rightarrow \frac{\partial}{\partial y}[x^2e^{3y} + e^x + g(y)] = 3x^2e^{3y} - y^2$$
$$\Rightarrow \frac{dg(y)}{dy} = -y^2. \tag{6.102}$$

Integrating this equation yields

$$g(y) = -\frac{y^3}{3} + C, \tag{6.103}$$

where C is a constant independent of both x and y. Now substituting this result into equation (6.100), the solution to the exact differential in equation (6.93) is found to be

$$F(x, y) = x^2 e^{3y} + e^x - \frac{1}{3}y^3 + C = \text{constant}. \tag{6.104}$$

Note that we have defined C representing all the constants independent of both variables x and y.

(c) Noting that

$$x^2 e^{3y} + e^x - \frac{1}{3}y^3 + C = \text{constant} \tag{6.105}$$

can be rewritten as

$$e^x + x^2 e^{3y} - \frac{1}{3}y^3 = C_1, \tag{6.106}$$

which cannot be solved for y.

6.3 First-order ODE and non-exact total differential

We recall that the differential equation

$$dF = P(x, y)dx + Q(x, y)dy \tag{6.107}$$

is said to be an *exact differential*, when

$$\frac{\partial P(x, y)}{\partial y} = \frac{\partial Q(x, y)}{\partial x}. \tag{6.108}$$

For an *exact differential* there exists a function, $F(x, y)$, such that

$$P(x, y) = \frac{\partial F(x, y)}{\partial x}, \quad Q(x, y) = \frac{\partial F(x, y)}{\partial y}$$

$$\Rightarrow \frac{\partial^2 F(x, y)}{\partial y \partial x} = \frac{\partial^2 F(x, y)}{\partial x \partial y}. \tag{6.109}$$

Sometimes, we may find a first-order ODE with variable coefficients that may lead to a total differential that is not exact. Let us consider the first-order ODE

$$a_0(x)y(x) + a_1(x)\frac{dy(x)}{dx} = f(x) \tag{6.110}$$

which can be rewritten as

$$\frac{dy(x)}{dx} = -\left[\frac{a_0(x)}{a_1(x)}y(x) - \frac{f(x)}{a_1(x)}\right] \Rightarrow Q(x, y)\frac{dy(x)}{dx} = -P(x, y) \tag{6.111}$$

$$\Rightarrow dF(x, y) = P(x, y)dx + Q(x, y)dy = 0,$$

where

$$P(x, y) = \frac{a_0(x)}{a_1(x)}y - \frac{f(x)}{a_1(x)}, \quad Q(x, y) = 1,$$

$$\Rightarrow \frac{\partial P(x, y)}{\partial y} = \frac{\partial}{\partial y}\left[\frac{a_0(x)}{a_1(x)}y - \frac{f(x)}{a_1(x)}\right] = \frac{a_0(x)}{a_1(x)}, \tag{6.112}$$

$$\frac{\partial Q(x, y)}{\partial x} = 0,$$

and

$$\frac{\partial P(x, y)}{\partial y} \neq \frac{\partial Q(x, y)}{\partial x}, \tag{6.113}$$

for $a_0(x) \neq 0$. This indicates that the resulting differential form of the first-order ODE is not exact, and the method we introduced in the previous section does not work. However, we can still use this method if we can find a function that can make the ODE in equation (6.110) exact. Next, we will determine this function known as *the integration factor*.

The integration factor: The first-order ODE in equation (6.110) can put in the form

$$\frac{dy}{dx} + P(x)y = Q(x), \tag{6.114}$$

where we introduced the functions defined by

$$P(x) = \frac{a_0(x)}{a_1(x)}, \quad Q(x) = \frac{f(x)}{a_1(x)}. \tag{6.115}$$

Equation (6.114) is the generic form of a first-order ODE. We have seen that this ODE does not lead to an exact differential. However, let say that if we multiply the entire equation (6.114) by a function $V(x)$ gives us a total differential that is exact. Thus all we need to do is find this function. To this end, multiplying equation (6.114) by $V(x)$, we have

$$V(x)\frac{dy}{dx} + V(x)P(x)y = V(x)Q(x)$$

$$\Rightarrow V(x)\frac{dy}{dx} + V(x)P(x)y - V(x)Q(x) = 0 \tag{6.116}$$

which can put in the form

$$dF(x, y) = [V(x)P(x)y - V(x)Q(x)]dx + V(x)dy = 0$$
$$\Rightarrow dF(x, y) = P(x, y)dx + Q(x, y)dy = 0, \tag{6.117}$$

where

$$P(x, y) = V(x)P(x)y - V(x)Q(x), \quad Q(x, y) = V(x). \tag{6.118}$$

We recall that for a differential to be exact, we must have

$$\frac{\partial P(x, y)}{\partial y} = \frac{\partial Q(x, y)}{\partial x}. \tag{6.119}$$

Noting that

$$\frac{\partial P(x, y)}{\partial y} = \frac{\partial}{\partial y}[V(x)P(x)y - V(x)Q(x)] \Rightarrow \frac{\partial P(x, y)}{\partial y} = V(x)P(x),$$

$$\frac{\partial Q(x, y)}{\partial x} = \frac{dV(x)}{dx}. \tag{6.120}$$

If the function makes the differential exact, we must have

$$\frac{\partial P(x, y)}{\partial y} = \frac{\partial Q(x, y)}{\partial x} \Rightarrow V(x)P(x) = \frac{dV(x)}{dx} \Rightarrow P(x)dx = \frac{dV(x)}{V(x)}. \tag{6.121}$$

Upon integrating this equation, one finds

$$V(x) = e^{\int P(x)dx}. \tag{6.122}$$

The function $V(x)$ given by the integral expression equation (6.122), known as *the integration factor*, is the multiplying factor to the ODE in equation (6.114) that makes differential exact. Thus multiplying equation (6.114) by the integration factor, we find

$$e^{\int P(x)dx}\frac{dy}{dx} + yP(x)e^{\int P(x)dx} = e^{\int P(x)dx}Q(x). \tag{6.123}$$

Applying the product rule for differentiation, we have

$$e^{\int P(x)dx}\frac{dy}{dx} + yP(x)e^{\int P(x)dx} = \frac{d}{dx}\left[ye^{\int P(x)dx}\right],$$

so that equation (6.123) can be rewritten as

$$\frac{d}{dx}\left[ye^{\int P(x)dx}\right] = e^{\int P(x)dx}Q(x). \tag{6.124}$$

Upon integrating equation (6.124), one can write

$$ye^{\int P(x)dx} = C + \int \left[e^{\int P(x)dx}Q(x)\right]dx,$$

$$\Rightarrow y(x) = Ce^{-\int P(x)dx} + e^{-\int P(x)dx} \int \left[e^{\int P(x)dx}Q(x)\right]dx, \tag{6.125}$$

where C is the constant of integration. Equation (6.125) is the general solution to the first-order ODE in equation (6.114) and C is determined from the given boundary condition set to the specific problem that the DE is representing. In the following examples, we will see its application. But let's examine equation (6.125) if it gives the general solution we determined for a first-order ODE with constant coefficients. We recall, for an ODE with constant coefficients

$$a_0 y(x) + a_1 \frac{dy(x)}{dt} = f(x), \tag{6.126}$$

the solution is given by

$$y(x) = Ce^{-\frac{a_0}{a_1}x} + \frac{e^{-\frac{a_0}{a_1}x}}{a_1} \int f(x) e^{\frac{a_0}{a_1}x} dx. \tag{6.127}$$

Noting that equation (6.126) can be rewritten as equation (6.114)

$$\frac{dy}{dx} + P(x)y = Q(x), \tag{6.128}$$

where

$$P(x) = \frac{a_0}{a_1}, \quad Q(x) = \frac{f(x)}{a_1}. \tag{6.129}$$

Upon substituting equation (6.129) into equation (6.127), we find

$$y(x) = Ce^{-\frac{a_0}{a_1}\int dx} + e^{-\frac{a_0}{a_1}\int dx} \int \left[\frac{f(x)}{a_1} e^{\frac{a_0}{a_1}\int dx} \right] dx$$

$$\Rightarrow y(x) = Ce^{-\frac{a_0}{a_1}x} + \frac{e^{-\frac{a_0}{a_1}x}}{a_1} \int e^{\frac{a_0}{a_1}x} f(x) dx. \tag{6.130}$$

where we used

$$\int \frac{a_0}{a_1} dx = \frac{a_0}{a_1} \int dx = \frac{a_0}{a_1} x, \tag{6.131}$$

since a_0 and a_1 are constants. The result in equation (6.130) is the same as equation (6.127).

Example 6.6. Consider the first-order DE

$$2x \frac{dy}{dx} + 4x^2 = y. \tag{6.132}$$

(a) Express this DE in the form

$$\frac{dy}{dx} + P(x)y = Q(x). \tag{6.133}$$

(b) Using the general solution (equation (6.125)), find the function $y(x)$ subject to the boundary condition,

$$y(1) = \frac{26}{3}. \tag{6.134}$$

Solution:

(a) Dividing equation (6.132) by $2x$, one can write

$$\frac{dy}{dx} + 2x = \frac{1}{2x}y \Rightarrow \frac{dy}{dx} - \frac{1}{2x}y = -2x \Rightarrow \frac{dy}{dx} + P(x)y = Q(x), \tag{6.135}$$

we have

$$P(x) = -\frac{1}{2x}, \quad Q(x) = -2x. \tag{6.136}$$

(b) Using the function $P(x)$, we find

$$\int P(x)dx = -\int \frac{1}{2x}dx = -\frac{1}{2}\ln x. \tag{6.137}$$

Substituting this result and $Q(x) = -2x$ into equation (6.125), we have

$$\begin{aligned}
y(x) &= e^{\frac{1}{2}\ln x} \int \left[e^{-\frac{1}{2}\ln x}(-2x)\right]dx + Ce^{\frac{1}{2}\ln x} \\
&= -2e^{\ln(x^{1/2})} \int e^{\ln(x^{-1/2})}xdx + Ce^{\ln(x^{1/2})} \\
&= -2x^{1/2} \int x^{-1/2}xdx + Cx^{1/2}, \\
\Rightarrow y(x) &= -\frac{4}{3}x^2 + Cx^{1/2},
\end{aligned} \tag{6.138}$$

where we used the relations

$$c\ln(a) = \ln(a^c), \quad e^{\ln(a)} = a. \tag{6.139}$$

Using the given boundary condition,

$$y(1) = \frac{26}{3}, \tag{6.140}$$

the constant of integration C is found to be

$$-\frac{4}{3} + C = \frac{26}{3} \Rightarrow C = 10, \tag{6.141}$$

and equation (6.138) becomes

$$y(x) = \left(10 - \frac{4}{3}x^{3/2}\right)\sqrt{x} = 10\sqrt{x} - \frac{4}{3}x^2. \tag{6.142}$$

This is the solution to equation (6.132) under the given boundary condition.

6.4 Higher-order ODE

The previous section introduced various methods for solving the first-order ODE. Next, we will study how to solve higher-order ($n \geq 2$) ODEs. We recall that an nth-order ODE is given by

$$a_0(x)y + a_1(x)\frac{dy}{dx} + a_2(x)\frac{d^2y}{dx^2} + a_3(x)\frac{d^3y}{dx^3}...a_{n-1}(x)\frac{d^{n-1}y}{dx^{n-1}}$$
$$+ a_n(x)\frac{d^ny}{dx^n} = f(x). \tag{6.143}$$

We will develop the methods for solving this equation in two parts. In the first part, we will consider various techniques for homogenous ordinary differential equations (HODE), where $f(x) = 0$. In the second part, we will develop the techniques for non-homogeneous ordinary differential equations (NHODE) built upon the techniques for HODE.

A. HODE with constant coefficient
When all the coefficients

$$a_0(x) = a_0, \ a_1(x) = a_1, \ a_2(x) = a_2...a_n(x) = a_n, \tag{6.144}$$

are constant, the nth-order HODE in equation (6.143) becomes

$$a_0y + a_1\frac{dy}{dx} + a_2\frac{d^2y}{dx^2} + a_3\frac{d^3y}{dx^3}...a_{n-1}\frac{d^{n-1}y}{dx^{n-1}}$$
$$+ a_n\frac{d^ny}{dx^n} = 0. \tag{6.145}$$

We are interested in how the general solution to such a HODE can be determined by solving what is known as the *indicial equation* derived from the HODE.

Indicial equation: We guess the solution to equation (6.145) to be

$$y(x) = Ae^{kx}. \tag{6.146}$$

Substituting this into equation (6.145), one can easily find

$$(a_0 + a_1k + a_2k^2 + a_3k^3...a_{n-1}k^{n-1} + a_nk^n)Ae^{kx} = 0, \tag{6.147}$$

which leads to

$$(a_0 + a_1k + a_2k^2 + a_3k^3...a_{n-1}k^{n-1} + a_nk^n) = 0. \tag{6.148}$$

This equation is called the *indicial equation*. The solution to this equation gives n roots that could be real or complex. If these roots are *non-degenerate* (i.e., no identical roots)

$$k = k_1, k_2 ... k_n, \tag{6.149}$$

the general solution to equation (6.145) is given by

$$y(x) = C_1 e^{k_1 x} + C_2 e^{k_2 x} + ... C_n e^{k_n x}. \tag{6.150}$$

As we saw in the solutions to the first-order ODE, the constants C_1, C_2, ..., C_n, are determined from the boundary conditions pertinent to the particular problem.

Example 6.7. An inductor with inductance L and a capacitor with capacitance C are connected in a series circuit, as shown in figure 6.2. At $t = 0$, the charge on the capacitor plates is Q_o, and the electric current in the circuit is I_o. Find an expression for the charge on the capacitor, $Q(t)$. Note that the potential difference across a capacitor ΔV_C is given by

$$\Delta V_C = \frac{Q(t)}{C}, \tag{6.151}$$

and across an inductor ΔV_L

$$\Delta V_L = -L \frac{dI(t)}{dt} \tag{6.152}$$

Solution: Using Kirchhoff's voltage rule, we can write

$$\frac{Q(t)}{C} + L \frac{dI(t)}{dt} = 0. \tag{6.153}$$

Since electric current is related to charge by

$$I(t) = \frac{dQ}{dt},$$

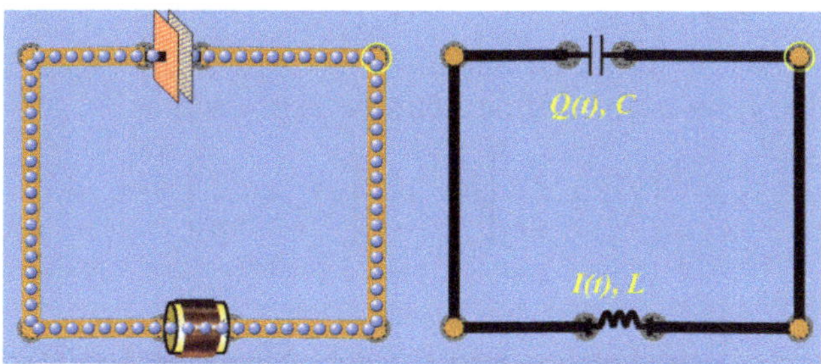

Figure 6.2. An LC circuit. A capacitor with capacitance C and charge Q_0 at the initial time is connected to an inductor with inductance L.

one can express equation (6.152), in terms Q as

$$\frac{d^2Q}{dt^2} + \frac{1}{LC}Q = 0 \Rightarrow \frac{d^2Q}{dt^2} + \omega^2 Q = 0 \tag{6.154}$$

where we introduced the constant,

$$\omega = \frac{1}{\sqrt{LC}}. \tag{6.155}$$

Equation (6.154) is a second-order HODE with constant coefficients. Substituting

$$Q(t) = Ae^{kt} \Rightarrow \frac{d^2}{dt^2}(Ae^{kt}) = k^2 Ae^{kt} \tag{6.156}$$

into equation (6.154), we find the indicial equation

$$k^2 + \omega^2 = 0 \Rightarrow k_1 = i\omega, \; k_2 = -i\omega. \tag{6.157}$$

The general solution can then be expressed as

$$Q(t) = Ae^{k_1 t} + Be^{-k_2 t} = Ae^{i\omega t} + Be^{-i\omega t}. \tag{6.158}$$

At $t = 0$ the capacitor is fully charged and the charge is Q_0, this means

$$Q(0) = Ae^{i\omega t} + Be^{-i\omega t} \, |_{t=0} = Q_0 \Rightarrow A + B = Q_0. \tag{6.159}$$

We are also given that the current at $t = 0$ is I_0. Using the definition of an electrical current, we find

$$I(0) = \frac{dQ}{dt}\bigg|_{t=0} = I_0$$

$$\Rightarrow \frac{d}{dt}[Ae^{i\omega t} + Be^{-i\omega t}]\bigg|_{t=0} = i\omega(Ae^{i\omega t} - Be^{-i\omega t}) \, |_{t=0} = I_0 \tag{6.160}$$

$$\Rightarrow A - B = \frac{I_0}{i\omega}.$$

Combining equations (6.159) and (6.160), one can easily show that

$$A = \frac{1}{2}\left[Q_0 + \frac{I_0}{i\omega}\right], \; B = \frac{1}{2}\left[Q_0 - \frac{I_0}{i\omega}\right]. \tag{6.161}$$

Therefore, the charge at a given time t can be expressed as

$$Q(t) = \frac{1}{2}\left[Q_0 + \frac{I_0}{i\omega}\right]e^{i\omega t} + \frac{1}{2}\left[Q_0 - \frac{I_0}{i\omega}\right]e^{-i\omega t}, \tag{6.162}$$

which we rewrite as

$$Q(t) = Q_0\left[\frac{e^{i\omega t} + e^{-i\omega t}}{2}\right] + \frac{I_0}{\omega}\left[\frac{e^{i\omega t} - e^{-i\omega t}}{2i}\right]. \tag{6.163}$$

Applying Euler's formula, equation (6.163) can be put in the form

$$Q(t) = Q_0 \cos(\omega t) + \frac{I_0}{\omega} \sin(\omega t). \tag{6.164}$$

We can also put equation (6.164) in a different form by replacing Q_0 and I_0/ω with

$$Q_0 = E \sin(\varphi), \quad \frac{I_0}{\omega} = E \cos(\varphi), \tag{6.165}$$

respectively, in equation (6.164),

$$Q(t) = E[\sin(\varphi)\cos(\omega t) + \cos(\varphi)\sin(\omega t)] = E \sin(\omega t + \varphi). \tag{6.166}$$

Note that E and φ are constants which we can express in terms of Q_0 and I_0/ω,

$$Q_0 = E \sin(\varphi), \quad \frac{I_0}{\omega} = E \cos(\varphi)$$
$$\Rightarrow \tan(\varphi) = \frac{Q_0}{I_0/\omega}, \quad E = \sqrt{Q_0^2 + \left(\frac{I_0}{\omega}\right)^2}. \tag{6.167}$$

Furthermore, if we introduce another constant defined in a slightly different way

$$Q_0 = G \cos(\varphi), \quad \frac{I_0}{\omega} = G \sin(\varphi), \tag{6.168}$$

one can rewrite equation (6.164) as

$$Q(t) = G[\cos(\varphi)\cos(\omega t) + \sin(\varphi)\sin(\omega t)]$$
$$\Rightarrow Q(t) = G \cos(\omega t - \varphi) = G \cos(\varphi - \omega t), \tag{6.169}$$

where

$$\tan(\varphi) = \frac{I_0/\omega}{Q_0}, \quad G = \sqrt{Q_0^2 + \left(\frac{I_0}{\omega}\right)^2}. \tag{6.170}$$

The current

$$I(t) = \frac{dQ}{dt} \tag{6.171}$$

can be determined using any one of the equations for the charge in equations (6.163), (6.164), (6.166), or (6.169). If we chose equation (6.164), we find

$$I(t) = \frac{d}{dt}\left[Q_0 \cos(\omega t) + \frac{I_0}{\omega} \sin(\omega t)\right]$$
$$\Rightarrow I(t) = -Q_0\omega \sin(\omega t) + I_0 \cos(\omega t), \tag{6.172}$$

for the current in the circuit.

In Example 6.7, we saw that the solution to the second-order DE in equation (6.154) can be expressed in four different forms given by equations (6.163), (6.164),

(6.166), and (6.169). In view of this result, we can make two generalizations to the solution of a second-order HODE that can be put in form

$$\frac{d^2y}{dx^2} \pm \omega^2 y = 0, \tag{6.173}$$

when ω is a real constant. For the plus case,

$$\frac{d^2y}{dx^2} + \omega^2 y = 0, \tag{6.174}$$

as we saw in Example 6.7, the roots to the indicial equation

$$k^2 + \omega^2 = 0, \tag{6.175}$$

are complex, $k_1 = i\omega$ and $k_2 = -i\omega$ and the general solution can be expressed in four different forms given by:

$$y(x) = \begin{cases} Ae^{i\omega x} + Be^{-i\omega x}, \\ C\cos(\omega x) + D\sin(\omega x), \\ E\sin(\omega x + \varphi), \\ G\cos(\omega x - \varphi). \end{cases} \tag{6.176}$$

On the other hand, for the minus case,

$$\frac{d^2y}{dx^2} - \omega^2 y = 0, \tag{6.177}$$

the roots to the corresponding indicial equation

$$k^2 - \omega^2 = 0, \tag{6.178}$$

are real, $k_1 = \omega$ and $k_2 = -\omega$. It can easily be shown from equation (6.176) that the solution can be expressed as

$$y(x) = \begin{cases} Ae^{\omega x} + Be^{-\omega x}, \\ C\cosh(\omega x) + D\sinh(\omega x), \\ E\sinh(\omega x + \varphi), \\ G\cosh(\omega x - \varphi). \end{cases} \tag{6.179}$$

We just need to recall the relations

$$\begin{aligned} \cos(i\omega x) &= \frac{e^{i(i\omega t)} + e^{-i(i\omega t)}}{2} = \frac{e^{\omega t} + e^{-\omega t}}{2} = \cosh(\omega x), \\ \sin(i\omega x) &= \frac{e^{i(i\omega t)} - e^{-i(i\omega t)}}{2i} = i\frac{e^{-\omega t} - e^{-\omega t}}{2} = i\sinh(\omega x), \end{aligned} \tag{6.180}$$

and note the factor i is included in the constant D.

Example 6.8. *The Schröedinger equation*: The one-dimensional time-independent Schröedinger equation for a particle traveling along the x-direction with a potential energy function $U(x)$ is given by

$$-\frac{\hbar^2}{2m}\frac{d^2\Psi(x)}{dx^2} + U(x)\Psi(x) = E\Psi(x), \tag{6.181}$$

where $\hbar = h/2\pi$ (h is Planck's constant), m is the mass of the particle, and E is the total energy of the particle, and $\Psi(x)$ is the wave function which is in general a complex function. Consider a particle of mass m and energy E traveling in the potential shown in figure (6.3). The x-axis is divided into two regions by the potential energy function. Region I has $U(x) = 0$, which means that there are no forces acting on the particle. In this region the particle is a free particle. Region II has $U(x) = U_o$, where U_o is a constant such that $U_o > E$. This is a forbidden region for the particle according to classical physics. The differential defined by

$$dP(x) = \mid \Psi(x)\mid^2 dx$$

gives the probability of finding the particle anywhere between x and $x + dx$. Solve the Schröedinger equation for the wave function of the particle in each of the two regions I ($x < 0$) and II ($x > 0$) shown in figure 6.3. Comment on your answers.

Solution: In region I, the potential energy $U(x) = 0$, and the Schröedinger equation becomes

$$-\frac{\hbar^2}{2m}\frac{d^2\Psi(x)}{dx^2} = E\Psi(x) \Rightarrow \frac{d^2\Psi(x)}{dx^2} + \frac{2mE}{\hbar^2}\Psi(x) = 0$$

$$\Rightarrow \frac{d^2\Psi(x)}{dx^2} + k^2\Psi(x) = 0, \tag{6.182}$$

where

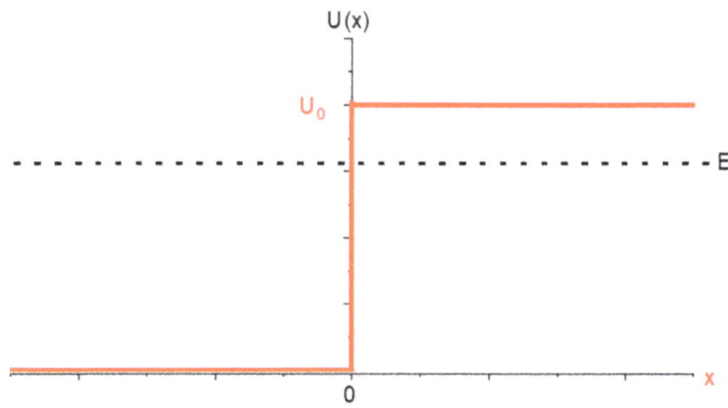

Figure 6.3. The potential energy function $U(x)$.

$$k^2 = \frac{2mE}{\hbar^2}. \tag{6.183}$$

In view of equation (6.174) and the corresponding solution in equation (6.176), one can write the solution to equation (6.182) as

$$\Psi(x) = A_1 e^{ikx} + A_2 e^{-ikx}. \tag{6.184}$$

In region *II*, the potential energy $U(x) = U_0$, $E < U_0$, and the Schröedinger equation becomes

$$-\frac{\hbar^2}{2m}\frac{d^2\Psi(x)}{dx^2} + U_0\Psi(x) = E\Psi(x) \Rightarrow \frac{d^2\Psi(x)}{dx^2} - \frac{2m(U_0 - E)}{\hbar^2}\Psi(x)$$

$$\Rightarrow \frac{d^2\Psi(x)}{dx^2} - q^2\Psi(x) = 0, \tag{6.185}$$

where

$$q^2 = \frac{2m(U_0 - E)}{\hbar^2}. \tag{6.186}$$

Similarly, in view of equation (6.177) and equation (6.179), one can write the solution to equation (6.185) as

$$\Psi(x) = A_1 e^{qx} + A_2 e^{-qx}. \tag{6.187}$$

Mathematically, the function we determined as a solution to the differential equation must be finite in the given domain, $(-\infty, \infty)$. Physically the probability of finding the particle which is proportional to $|\Psi(x)|^2$ cannot be infinity. Therefore, the acceptable solution to the differential equation is when $A_1 = 0$ as the function diverges in the region $x > 0$,

$$\Psi(x) = A_2 e^{-qx}. \tag{6.188}$$

This shows that the probability of the particle to exist in the classically forbidden region $(x > 0)$,

$$dP(x) = |\Psi(x)|^2\, dx = |A_2|^2\, e^{-2qx}dx, \tag{6.189}$$

is not zero.

Indicial equation with degenerate roots

As we stated earlier, the general solution in equation (6.150) is valid when there are no repeated roots (non-degenerate roots). However, sometimes the roots of the indicial equation can have degenerate roots. For example, for the second-order HODE,

$$\frac{d^2y}{dx^2} - 2a\frac{dy}{dx} + a^2y = 0, \tag{6.190}$$

the roots of the indicial equation

$$k^2 - 2ak + a^2 = 0, \tag{6.191}$$

are degenerate

$$k_1 = k_2 = a. \tag{6.192}$$

When the roots are degenerate, the general solution in equation (6.150) can be used with some changes that depend on the degeneracy in the roots of the indicial equation. For example, for the HODE in equation (6.190), the values for the roots are identical ($k_1 = k_2 = a$), where the degeneracy, $m = 2$, the general solution should be written as

$$y(x) = (A + Bx)e^{ax}. \tag{6.193}$$

Note that for $m = 2$, the coefficient to e^{ax} is a polynomial function, $P_1(x) = A + Bx$, of degree $l = m - 1 = 1$. Similarly, for a HODE that leads to an indicial equation, with degeneracy $m = 3$ ($k_1 = k_2 = k_3 = a$), the general solution is given by

$$y(x) = (A + Bx + Cx^2)e^{ax} = P_2(x)e^{ax}. \tag{6.194}$$

In this case the coefficient $P_2(x)$ is a polynomial function of degree $l = m - 1 = 2$.

Thus we can make the following changes to the general solution in equation (6.150) concerning the degenerate roots to the indicial equation:

(a) For an nth-order differential equation that leads to an indicial equation with degenerate roots $m = n$, the general solution is given by

$$y(x) = (A + Bx + Cx^2...+Dx^{n-1})e^{ax}. \tag{6.195}$$

(b) For an nth-order differential equation that leads to an indicial equation with, for example, three different sets of roots. Each set could be with degenerate or non-degenerate roots. Suppose the first set of roots has $m = m_1$ degenerate roots (i.e., $k_1 = k_2 = k_3 = ...,k_{m_1} = a$), the second set has $m = m_2$ degenerate roots (i.e., $k_1' = k_2' = k_3' = ...,k_{m_2}' = a'$), and the third set consists of m_3 non-degenerate roots ($k_1'' = a_1''$, $k_2'' = a_2''$, $k_3'' = a_3''...k_{m_3}'' = a_{m_3}''$). Note that

$$m_1 + m_2 + m_3 = n. \tag{6.196}$$

In such cases, the general solution for the HODE equation is given by

$$y(x) = P_{m_1-1}(x)e^{ax} + P_{m_2-1}(x)e^{a'x} + C_1''e^{a_1''x} + C_2''e^{a_2''x}$$
$$+ ...C_{m_3}''e^{a_{m_3}''x}, \tag{6.197}$$

where

$$P_{m_1-1}(x) = C_0 + C_1x + C_2x^2...+C_{m_1-1}x^{m_1-1},$$
$$P_{m_2-1}(x) = C_0' + C_1'x + C_2'x^2...+C_{m_2-1}'x^{m_2-1}, \tag{6.198}$$

are polynomials of degree $m_1 - 1$ and $m_2 - 1$, respectively.

Example 6.9. For each of the following DE, find the indicial equation, list the roots of the indicial equation $\{k_i\}$, and write out the general solution.

(a)

$$\frac{d^3y}{dx^3} + \frac{dy}{dx} = 0. \tag{6.199}$$

(b)

$$2\frac{d^4s}{dt^4} - 3\frac{d^3s}{dt^3} = 0. \tag{6.200}$$

Solution:

(a) Noting that

$$y(x) = Ae^{kx} \Rightarrow \frac{dy}{dx} = ky(x) \Rightarrow \frac{d^3y}{dx^3} = k^3 y(x)$$
$$\Rightarrow \frac{d^3y}{dx^3} + \frac{dy}{dx} = (k^3 + k)y(x) = 0, \tag{6.201}$$

the indicial equation becomes

$$k^3 + k = 0. \tag{6.202}$$

The is a cubic equation with three distinct roots

$$k^3 + k = 0 \Rightarrow k(k^2 + 1) = 0 \Rightarrow k_1 = 0,\ k_2 = i,\ k_3 = -i. \tag{6.203}$$

The roots do not show any degeneracy and the solution can be written as

$$y(x) = A + Be^{ix} + Ce^{-ix} = A + D\cos(x) + E\sin(x). \tag{6.204}$$

(b) Similarly, noting that

$$s(t) = Ae^{kt}, \Rightarrow \frac{d^3s}{dt^3} = k^3 s(t) \Rightarrow \frac{d^4s}{dt^4} = k^4 s(t)$$
$$\Rightarrow 2\frac{d^4s}{dt^4} - 3\frac{d^3s}{dt^3} = (2k^4 - 3k^3)s(t) = 0, \tag{6.205}$$

the roots to the indicial equation are found to be

$$2k^4 - 3k^3 = k^3(2k - 3) = 0 \Rightarrow k_1 = \frac{3}{2}, \; k_2 = k_3 = k_4 = 0. \tag{6.206}$$

The results in equation (6.206) show that the indicial equation has two sets of solutions. In the first set, there are three degenerate roots (k_2, k_3, k_4) with degeneracy $m = 3$, and in the second set, there is one non-degenerate root (k_1). Thus applying the relation in equation (6.197), the general solution can be written as

$$s(t) = (C_1 + C_2 t + C_3 t^2)e^{at} + C_4 e^{k_1 t}, \tag{6.207}$$

where $a = k_2 = k_3 = k_4 = 0$ is the value for the degenerate roots and k_1 is the value for the non-degenerate root. Using these values the general solution to the DE becomes

$$s(t) = C_1 + C_2 t + C_3 t^2 + C_4 e^{\frac{3}{2}t}. \tag{6.208}$$

Example 6.10. *Damped simple harmonic motion*: An ideal spring of spring constant k hangs vertically from a support rod and has a mass m attached to its lower end and suspended in a viscous fluid. The mass is pulled down some distance from its equilibrium position at $y = 0$ and released. Discuss the resulting motion of the mass, if the viscous fluid exerts a damping force given by

$$f_y(t) = -bv_y(t), \tag{6.209}$$

where b is a damping constant and v_y is the velocity of the mass at time t (figure 6.4).

Solution: The equation of motion for the mass is given by Newton's second law

$$F_{\text{net}} = ma_y = m\frac{d^2 y}{dt^2}, \tag{6.210}$$

where y is the displacement from this equilibrium position. Since the mass initially was at equilibrium in the fluid, we do not consider the buoyant and gravitational force on the mass. Thus the displacement from the equilibrium potion depends on the net force acting on the mass, which is the vector sum of the spring and the damping forces

$$F_{\text{net}} = -ky - bv_y = -ky - b\frac{dy}{dt}. \tag{6.211}$$

Then the equation of motion becomes,

$$m\frac{d^2 y}{dt^2} = -ky - b\frac{dy}{dt} \Rightarrow \frac{d^2 y}{dt^2} + 2\gamma\frac{dy}{dt} + \omega^2 y = 0, \tag{6.212}$$

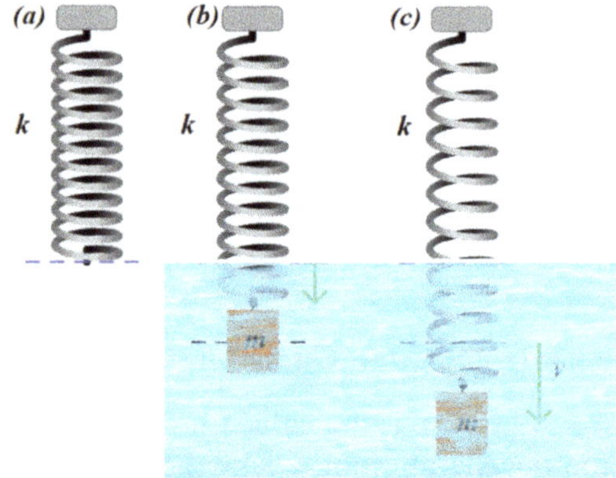

Figure 6.4. A mass m attached to a spring with a spring constant k. Initially it was at equilibrium when it was suspending inside water as shown in (b). Then it is displaced from its equilibrium position and released. It starts to oscillate up and down and at time t its displacement from the equilibrium position is $y(t)$ as shown in (c).

where

$$\omega^2 = \frac{k}{m}, \gamma = \frac{b}{2m}. \tag{6.213}$$

The indicial equation following from equation (6.212) becomes

$$\lambda^2 + 2\gamma\lambda + \omega^2 = 0, \tag{6.214}$$

and the solutions

$$\lambda_1 = -\gamma + D, \lambda_2 = -\gamma - D, \tag{6.215}$$

where

$$D = \sqrt{\gamma^2 - \omega^2}. \tag{6.216}$$

In the following we analyze the resulting values for D for three different cases which determine the solution to equation (6.212).

1. *Case 1* $(\gamma = \omega)$: For this case, equation (6.216) results in

$$D = \sqrt{\gamma^2 - \omega^2} = 0, \tag{6.217}$$

and the roots to the indicial equation becomes

$$\lambda_1 = \lambda_2 = -\gamma. \tag{6.218}$$

This leads to the general solution given by

$$y(t) = (At + B)e^{-\gamma t}, \tag{6.219}$$

which describes what is known as a *critically damped motion.*

2. *Case 2 $(\gamma > \omega)$* : We note from equation (6.216), D is real

$$D = \sqrt{\gamma^2 - \omega^2} < \gamma \tag{6.220}$$

which means the solution is given by

$$y(t) = Ae^{-(\gamma-D)t} + Be^{-(\gamma+D)t}. \tag{6.221}$$

This represents an *over-damped motion.*

3. *Case 3 $(\gamma < \omega)$*: In this case we can rewrite equation (6.216) as

$$D = \sqrt{\gamma^2 - \omega^2} = \sqrt{-(\omega^2 - \gamma^2)} = i\Omega, \tag{6.222}$$

where $\omega^2 - \gamma^2 > 0$ and

$$\Omega = \sqrt{\omega^2 - \gamma^2}, \tag{6.223}$$

is real. Then the roots for the indicial equation becomes

$$\lambda_1 = -\gamma + i\Omega, \; \lambda_2 = -\gamma - i\Omega \tag{6.224}$$

and the solution is given by

$$\begin{aligned} y(t) &= Ae^{-(\gamma-i\Omega)t} + Be^{-(\gamma+i\Omega)t} \\ &\Rightarrow y(t) = e^{-\gamma t}(Ae^{i\Omega t} + Be^{-i\Omega t}). \end{aligned} \tag{6.225}$$

In view of the differential equation (6.174) and the corresponding solution in equation (6.176), one can also rewrite equation (6.225) as

$$y(t) = \begin{cases} e^{-\gamma t}[C\cos(\Omega t) + D\sin(\Omega t)], \\ e^{-\gamma t}E\sin(\Omega t + - \gamma), \\ e^{-\gamma t}F\cos(\Omega t - \gamma). \end{cases} \tag{6.226}$$

A motion represented by such an equation is known as *under-damped oscillatory motion.*

B. NHODE with constant coefficients

We recall that an nth-order NHODE with constant coefficient has the form

$$a_0 y + a_1\frac{dy}{dx} + a_2\frac{d^2y}{dx^2} + a_3\frac{d^3y}{dx^3}....a_{n-1}\frac{d^{n-1}y}{dx^{n-1}} + a_n\frac{d^n y}{dx^n} = f(x). \tag{6.227}$$

Next, we will see the method for solving such differential equations. We built this method on what we developed for HODE and an educated guess to a particular solution to the non-homogenous differential equation. We follow the following successive steps to determine the solution to the NHODE.

(1) Find the general solution to the HODE (i.e., with $f(x) = 0$). This solution is called the *complementary function,* or *the homogeneous solution,* denoted by $y_h(x)$. Note that this solution contains n constants of integration for the nth-order HODE.

(2) Find a particular solution $y_p(x)$ that satisfies the full NHODE. Note that this function contains no arbitrary constants of integration.

(3) The general solution to the NHODE is then given by

$$y(x) = y_h(x) + y_p(x).$$

(6.228)

(4) Any boundary conditions given to the problem must then be applied to the general solution $y(x)$.

Example 6.11. Find the general solution to the differential equation,

$$\frac{d^2s}{dt^2} - 3\frac{ds}{dt} + 2s = 16.$$

(6.229)

Solution:

1. Find the homogeneous solution by solving

$$\frac{d^2s}{dt^2} - 3\frac{ds}{dt} + 2s = 0.$$

(6.230)

Noting that the indicial equation is

$$k^2 - 3k + 2 = 0 \Rightarrow k_1 = 2, k_2 = 1$$

(6.231)

the homogeneous solution can be written as

$$s_h = C_1 e^{2t} + C_2 e^t.$$

(6.232)

2. By inspecting the non-homogeneous ODE, we guess a particular solution of the form

$$s_p = b \Rightarrow \frac{ds_p}{dt} = \frac{d^2s_p}{dt^2} = 0,$$

(6.233)

so that when we substitute it back into the non-homogeneous ODE in equation (6.229), we find

$$2b = 16 \Rightarrow b = 8.$$

(6.234)

Thus the particular solution is found to be

$$s_p = 8.$$

(6.235)

3. Write the general solution

$$s(t) = s_h(t) + s_p(t) \Rightarrow s(t) = C_1 e^{2t} + C_2 e^t + 8. \tag{6.236}$$

4. Apply the boundary conditions if there are any. Here we are not given any.

Example 6.12. Find the general solution to the differential equation,

$$\frac{d^2 R}{dx^2} + 4R = \sin(3x). \tag{6.237}$$

Solution:

1. Find the homogeneous solution for

$$\frac{d^2 R}{dx^2} + 4R = 0. \tag{6.238}$$

Noting that the roots to the indicial equation are

$$k^2 + 4 = 0 \Rightarrow k_1 = 2i, \; k_2 = -2i, \tag{6.239}$$

one can write the homogeneous solution as

$$R_h(x) = C_1 \cos(2x) + C_2 \sin(2x). \tag{6.240}$$

2. By inspecting the given non-homogeneous ODE, we make a guess (an educated guess) to the particular solution that has the form

$$R_p = b \sin(3x). \tag{6.241}$$

Note that this guess will be valid because $\pm 3i$ is different from the roots of the indicial equation. See the next example.

Since in the particular solution, we are not allowed to have an arbitrary constant of integration, we must determine b. Upon substituting equation (6.241) into equation (6.237), this constant is found to be

$$-9b \sin(3x) + 4b \sin(3x) = \sin(3x) \Rightarrow -5b \sin(3x) = \sin(3x)$$

$$\Rightarrow b = -\frac{1}{5}. \tag{6.242}$$

Thus the particular solution becomes

$$R_p = -\frac{1}{5}\sin(3x).$$

(6.243)

3. Write the general solution

$$R(x) = R_h(x) + R_p(x)$$

$$\Rightarrow R(x) = C_1 \cos(2x) + C_2 \sin(2x) - \frac{1}{5}\sin(3x).$$

(6.244)

4. Apply the boundary conditions, if there are any, to find the constants C_1 and C_2. No boundary conditions are given here.

Example 6.13. Find the general solution to the differential equation

$$\frac{d^2y}{dx^2} + 4y = \sin(2x).$$

(6.245)

Solution:

1. Find the solution for the homogeneous OLDE

$$\frac{d^2y}{dx^2} + 4y = 0.$$

(6.246)

The roots to the indicial equation

$$k^2 + 4 = 0$$

(6.247)

are

$$k_1 = 2i, \ k_2 = -2i$$

(6.248)

and the homogeneous part of the solution given by

$$y_h(x) = C_1 \cos(2x) + C_2 \sin(2x),$$

(6.249)

or preferably

$$y_h(x) = C_1'e^{2ix} + C_2'e^{-2ix}.$$

(6.250)

2. When the non-homogenous part $f(x)$ is a trigonometric or hyperbolic function (i.e., $\sin(ax)$, $\cos(ax)$, $\sinh(ax)$, or $\cosh(ax)$), before we try to find the particular solution it is recommended to express the functions in terms of exponential functions applying Euler's formula. Thus for the function in equation (6.245), one should write

$$f(x) = \sin (2x) = \frac{1}{2i}(e^{2ix} - e^{-2ix}). \tag{6.251}$$

Now comparing equations (6.250) and (6.251), we note that values to the exponents to the homogeneous solution (the roots of the indicial equation) are the same as the exponents to the function, $f(x)$. Under these circumstances, one can make an educated guess to the particular solution based on the property of an exponential function and the order of the differential equation. For example, for the DE in equation (6.245), if one guesses a particular solution

$$y_p(x) = b_1 e^{2ix} + b_2 e^{-2ix},$$

which has the same form as the homogeneous solution in equation (6.250), we find

$$\frac{d^2 y_p}{dx^2} + 4y_p = -4y_p + 4y_p = 0.$$

Therefore, we must guess a function that must have a different form than the homogenous solution. In this case, our educated guess would be

$$y_p(x) = x(b_1 e^{2ix} + b_2 e^{-2ix}), \tag{6.252}$$

that leads to

$$\frac{dy_p}{dx} = b_1 e^{2ix} + b_2 e^{-2ix} + 2ix(b_1 e^{2ix} - b_2 e^{-2ix})$$

$$\Rightarrow \frac{d^2 y_p}{dx^2} = 2i(b_1 e^{2ix} - b_2 e^{-2ix}) + 2i(b_1 e^{2ix} - b_2 e^{-2ix})$$

$$- 4x(b_1 e^{2ix} + b_2 e^{-2ix}) \tag{6.253}$$

$$\Rightarrow \frac{d^2 y_p}{dx^2} + 4y_p = 4i(b_1 e^{2ix} - b_2 e^{-2ix}).$$

Substituting this result and equation (6.251) into equation (6.245), we find

$$4i(b_1 e^{2ix} - b_2 e^{-2ix}) = \frac{1}{2i}(e^{2ix} - e^{-2ix}).$$

There follows that

$$b_1 = b_2 = -\frac{1}{8}. \tag{6.254}$$

Therefore, the particular solution becomes

$$y_p(x) = -\frac{x}{8}[e^{2ix} + e^{-2ix}], \tag{6.255}$$

which we can put in the form

$$y_p(x) = -\frac{x}{4}\left[\frac{e^{2ix} + e^{-2ix}}{2}\right] = -\frac{x}{4}\cos(2x). \tag{6.256}$$

3. Write the general solution,

$$y(x) = y_h(x) + y_p(x) = C_1 \cos(2x) + C_2 \sin(2x) - \frac{x}{4}\cos(2x). \tag{6.257}$$

4. In this example, we are not given boundary conditions.

6.5 The particular solution and the method of superposition

In Example 6.13, the particular solution in equation (6.252) can be put as the sum of two particular solutions,

$$y_p(x) = b_1 x e^{2ix} + b_2 x e^{-2ix} = y_{p_1}(x) + y_{p_2}(x), \tag{6.258}$$

where

$$y_{p_1}(x) = -\frac{1}{8}x e^{2ix}, \; y_{p_2}(x) = -\frac{1}{8}x e^{-2ix}. \tag{6.259}$$

Equation (6.258), more generally, can be put in the form

$$y_{p_1}(x) = P_n(x)e^{2ix}, \; y_{p_2}(x) = Q_n(x)e^{-2ix}, \tag{6.260}$$

in which $P_n(x)$ and $Q_n(x)$ are polynomial functions

$$P_n(x) = -\frac{1}{8}x, \; Q_n(x) = -\frac{1}{8}x \tag{6.261}$$

of degree $n = 1$. In Example 6.13, we also wrote the function $f(x) = \sin(2x)$, as a sum of two functions

$$\frac{d^2 y_p(x)}{dx^2} + 4y_p(x) = f(x) = f_1(x) + f_2(x), \tag{6.262}$$

where

$$f_1(x) = \frac{1}{2i}e^{2ix}, \; f_2(x) = -\frac{1}{2i}e^{-2ix}, \tag{6.263}$$

or more generally

$$f_1(x) = P'_m(x)e^{2ix}, \; f_2(x) = Q'_m(x)e^{-2ix}, \tag{6.264}$$

where $P'_m(x)$ and $Q'_m(x)$ are polynomial functions of degree m. In the case of Example 6.13

$$P'_m(x) = \frac{1}{2i}, \quad Q'_m(x) = -\frac{1}{2i}, \tag{6.265}$$

which are polynomials with degree $m = 0$. In a similar way, if we were given the differential equations of the form

$$\frac{d^2y(x)}{dx^2} + 4y(x) = \frac{1}{2i}e^{2ix} = f_1(x) = P'_m(x)e^{2ix}, \tag{6.266}$$

and

$$\frac{d^2y(x)}{dx^2} + 4y(x) = -\frac{1}{2i}e^{-2ix} = f_2(x) = Q'_m(x)e^{-2ix}, \tag{6.267}$$

we would have guessed the particular solutions

$$y_{p_1}(x) = P_n(x)e^{2ix}, \tag{6.268}$$

and

$$y_{p_2}(x) = Q_n(x)e^{-2ix}, \tag{6.269}$$

respectively, such that

$$\frac{d^2y_{p_1}(x)}{dx^2} + 4y_{p_1}(x) = \frac{1}{2i}e^{2ix} = f_1(x) = P'_m(x)e^{2ix}, \tag{6.270}$$

and

$$\frac{d^2y_{p_2}(x)}{dx^2} + 4y_{p_2}(x) = -\frac{1}{2i}e^{-2ix} = f_2(x) = Q'_m(x)e^{-2ix}, \tag{6.271}$$

where $P_n(x) = b_1 x$ and $Q_n(x) = b_2 x$ are polynomial functions of degree $n = 1$. Upon adding equations (6.270) and (6.271), we find

$$\frac{d^2}{dx^2}\left[y_{p_1}(x) + y_{p_2}(x)\right] + 4\left[y_{p_1}(x) + y_{p_2}(x)\right]$$

$$= \frac{1}{2i}e^{2ix} - \frac{1}{2i}e^{-2ix} = f_1(x) + f_2(x) = P'_m(x)e^{2ix} + Q'_m(x)e^{-2ix}, \tag{6.272}$$

$$\Rightarrow \frac{d^2y_p(x)}{dx^2} + 4y_p(x) = f(x),$$

where

$$y_p(x) = y_{p_1}(x) + y_{p_2}(x) = b_1 x e^{2ix} + b_2 x e^{-2ix}$$

$$= P_n(x)e^{2ix} + Q_n(x)e^{-2ix}, \tag{6.273}$$

$$f(x) = f_1(x) + f_2(x) = \frac{1}{2i}e^{2ix} - \frac{1}{2i}e^{-2ix} = P'_m(x)e^{2ix} + Q'_m(x)e^{-2ix}.$$

Equation (6.272) is the same exact differential equation in equation (6.262) with the general solution

$$y(x) = y_h(x) + y_p(x) = y_h(x) + y_{p_1}(x) + y_{p_2}(x),$$ (6.274)

where $y_h(x)$ is the homogeneous part of the solution.

The result in equation (6.274) can be generalized for an nth-order NHODE,

$$a_0 y + a_1 \frac{dy}{dx} + a_2 \frac{d^2 y}{dx^2} + ... a_{n-1} \frac{d^{n-1} y}{dx^{n-1}} + a_n \frac{d^n y}{dx^n} = f(x) = f_1(x) + f_2(x) + ... + f_{n-1}(x) + f_n(x)$$ (6.275)

$$\Rightarrow \sum_{i=0}^{n} a_n \frac{d^n y}{dx^n} = \sum_{i=0}^{n} f_i(x),$$

the general solution is given by

$$y(x) = y_h(x) + \sum_{i=0}^{n} y_{p_i}(x),$$ (6.276)

where $y_{p_i}(x)$ is the particular solution to the differential equation

$$\sum_{i=0}^{n} a_n \frac{d^n y}{dx^n} = f_i(x).$$ (6.277)

This is known as *the principle of superposition.*

From what we introduced to finding the particular solution up to this point, we can make generalizations using a second-order NHODE. To this end, we note that

$$\sum_{i=0}^{2} a_n \frac{d^n y}{dx^n} = a_0 y + a_1 \frac{dy}{dx} + a_2 \frac{d^2 y}{dx^2} = f(x),$$ (6.278)

can be rewritten as

$$\left(\frac{d}{dx} - k_1 \right)\left(\frac{d}{dx} - k_2 \right) y(x) = f(x) = e^{\alpha x} P_m(x),$$ (6.279)

where

$$P_m(x) = c_0 + c_1 x + c_2 x^2 + ... c_m x^m = \sum_{i=0}^{m} c_i x^i.$$ (6.280)

k_1 and k_2 are the roots to the quadratic equation

$$a_2 k^2 + a_1 k + a_0 = 0,$$ (6.281)

which are given by

$$k_1 = \frac{-a_1 + \sqrt{a_1^2 - 4 a_0 a_2}}{2 a_2}, \quad k_2 = \frac{-a_1 - \sqrt{a_1^2 - 4 a_0 a_2}}{2 a_2}.$$ (6.282)

For example, for the differential equation

$$6y - 5\frac{dy}{dx} + \frac{d^2y}{dx^2} = f_i(x),$$ (6.283)

we have the indicial equation

$$k^2 - 5k + 6 = 0,$$ (6.284)

the roots of which are

$$k_1 = \frac{5 + \sqrt{25 - 24}}{2} = 3, \; k_2 = \frac{5 - \sqrt{25 - 24}}{2} = 2,$$ (6.285)

and we can rewrite the differential equation as

$$\begin{aligned}
\left(\frac{d}{dx} - 3\right)\left(\frac{d}{dx} - 2\right)y(x) &= \left(\frac{d}{dx} - 3\right)\left(\frac{dy}{dx} - 2y\right) \\
&= \frac{d}{dx}\left(\frac{dy}{dx} - 2y\right) - 3\left(\frac{dy}{dx} - 2y\right) \\
&= \frac{d^2y}{dx^2} - 2\frac{dy}{dx} - 3\frac{dy}{dx} + 6y \\
&\Rightarrow \frac{d^2y}{dx^2} - 5\frac{dy}{dx} + 6y = f_i(x).
\end{aligned}$$ (6.286)

For a function $f_i(x)$ that can be written in the form

$$f(x) = e^{\alpha x}[c_0 + c_1 x + c_2 x^2 + ...c_m x^m] = e^{\alpha x}\sum_{i=0}^{m} c_i x^i = e^{\alpha x}P_m(x),$$ (6.287)

the particular solution to such a differential equation is given by

$$y_p(x) = \begin{cases} e^{\alpha x}Q_m(x), & (\text{if } \alpha \neq k_1 \; \& \; \alpha \neq k_2 \Rightarrow k_1 \neq k_2), \\ xe^{\alpha x}Q_m(x), & (\text{if } \alpha = k_1 \text{ or } k_2 \; \& \; k_1 \neq k_2), \\ x^2 e^{\alpha x}Q_m(x), & (\text{if } \alpha = k_1 = k_2), \end{cases}$$ (6.288)

where

$$Q_m(x) = b_0 + b_1 x + b_2 x^2 + ...b_m x^m = \sum_{i=0}^{m} b_i x^i,$$ (6.289)

is a polynomial of the same degree as $P_m(x)$ with unknown coefficients b_i that must be determined by substituting the particular solution to the non-homogenous DE.

Example 6.14. Consider the differential equation

$$\frac{d^2y}{dx^2} - y = f(x),$$ (6.290)

where

$$f(x) = 2 \sin (x) + 4x \cos (x). \tag{6.291}$$

(a) Express the differential equation in the form

$$\left(\frac{d}{dx} - k_1\right)\left(\frac{d}{dx} - k_2\right) y(x) = f(x). \tag{6.292}$$

(b) Express the function

$$f(x) = 2 \sin (x) + 4x \cos (x) \tag{6.293}$$

in the form

$$f(x) = f_1(x) + f_1(x), \tag{6.294}$$

where

$$f_1(x) = e^{a_1 x} P_m^1(x), \, f_2(x) = e^{a_2 x} P_m^2(x) \tag{6.295}$$

where $P_m^1(x)$ and $P_m^2(x)$ are polynomials of degree m.
(c) Using the principle of superposition find the solution to the differential equation.

Solution:

(a) For the homogenous DE

$$\frac{d^2 y}{dx^2} - y = 0 \tag{6.296}$$

one finds the indicial equation

$$k^2 - 1 = 0. \tag{6.297}$$

The roots to this equation are

$$k_1 = 1, \, k_2 = -1, \tag{6.298}$$

and the differential equation can be rewritten as

$$\left(\frac{d}{dx} - 1\right)\left(\frac{d}{dx} + 1\right) y(x) = f(x), \tag{6.299}$$

where

$$f(x) = 2 \sin (x) + 4x \cos (x). \tag{6.300}$$

(b) Recalling that

$$\cos(x) = \frac{e^{ix} + e^{-ix}}{2}, \; \sin(x) = \frac{e^{ix} - e^{-ix}}{2i}, \tag{6.301}$$

one can write

$$f(x) = 2\left(\frac{e^{ix} - e^{-ix}}{2i}\right) + 4x\left(\frac{e^{ix} + e^{-ix}}{2}\right) \tag{6.302}$$

$$\Rightarrow f(x) = (2x - i)e^{ix} + (2x + i)e^{-ix} = f_1(x) + f_2(x),$$

where

$$f_1(x) = e^{ix}(2x - i), \; f_2(x) = e^{-ix}(2x + i). \tag{6.303}$$

We note that equation (6.303) can be put in the form

$$f_1(x) = e^{\alpha_1 x} P_m^1(x), \; f_2(x) = e^{\alpha_2 x} P_m^2(x), \tag{6.304}$$

where $P_m^1(x)$ and $P_m^2(x)$

$$P_m^1(x) = 2x - i, \; \alpha_1 = i, \; P_m^2(x) = 2x + i, \; \alpha_2 = -i \tag{6.305}$$

are polynomials of degree $m = 1$.

(c) Using the roots for the indicial equation

$$k_1 = 1, \; k_2 = -1, \tag{6.306}$$

the homogenous part of the solution can easily be written as

$$y_h(x) = C_1 e^x + C_2 e^{-x}. \tag{6.307}$$

For the particular solution, one can write two differential equations

$$\frac{d^2 y}{dx^2} - y = \left(\frac{d}{dx} + 1\right)\left(\frac{d}{dx} - 1\right) = f_1(x) = e^{\alpha_1 x} P_m(x) \tag{6.308}$$

and

$$\frac{d^2 y}{dx^2} - y = \left(\frac{d}{dx} + 1\right)\left(\frac{d}{dx} - 1\right) = f_2(x) = e^{\alpha_2 x} P_m(x). \tag{6.309}$$

Noting that

$$P_m^1(x) = 2x - i, \; P_m^2(x) = 2x + i,$$
$$\alpha_1 = i, \; k_1 = 1, \; k_2 = -1 \Rightarrow \alpha_1 \neq k_1 \neq k_2,$$
$$\alpha_2 = -i, \; k_1 = 1, \; k_2 = -1 \Rightarrow \alpha_2 \neq k_1 \neq k_2, \tag{6.310}$$

applying the relation in equation (6.288), one can write the particular solution to equations (6.308) and (6.309) as

$$y_{p_1}(x) = e^{ix}(b_0 + b_1 x), \tag{6.311}$$

and

$$y_{p_2}(x) = e^{-ix}(b_0' + b_1'x), \tag{6.312}$$

respectively. For $y_{p_1}(x)$, we have

$$\frac{dy_{p_1}}{dx} = b_1 e^{ix} + i(b_0 + b_1 x)e^{ix} = b_1 e^{ix} + iy_{p_1}(x),$$

$$\frac{d^2 y_{p_1}}{dx^2} = \frac{db_1 e^{ix}}{dx} + i\frac{dy_{p_1}}{dx} = ib_1 e^{ix} + i\left(b_1 e^{ix} + iy_{p_1}(x)\right) \tag{6.313}$$

$$= 2ib_1 e^{ix} - y_{p_1}(x).$$

So that upon substituting equations (6.313) into equations (6.308), one finds

$$2ib_1 e^{ix} - y_{p_1}(x) - y_{p_1}(x) = (2x - i)e^{ix}$$

$$\Rightarrow 2ib_1 e^{ix} - 2(b_0 + b_1 x)e^{ix} = (2x - i)e^{ix} \tag{6.314}$$

$$\Rightarrow (-2b_1 x + 2b_1 i - 2b_0)e^{ix} = (2x - i)e^{ix}.$$

There follows that

$$-2b_1 = 2 \Rightarrow b_1 = -1,$$

$$2b_1 i - 2b_0 = -i \Rightarrow -2i - 2b_0 = -i \Rightarrow b_0 = -\frac{i}{2}. \tag{6.315}$$

Then the first particular solution in equation (6.311) becomes

$$y_{p_1}(x) = (b_0 + b_1 x)e^{ix} = -\left(\frac{i}{2} + x\right)e^{ix}. \tag{6.316}$$

For $y_{p_2}(x)$, using equation (6.312), we have

$$\frac{dy_{p_2}}{dx} = b_1' e^{-ix} - iy_{p_2}(x) \Rightarrow \frac{d^2 y_{p_2}}{dx^2} = -2ib_1' e^{-ix} - y_{p_2}(x), \tag{6.317}$$

and substituting equation (6.311) into equation (6.309), we find

$$(-2b_1'x - 2b_1'i - 2b_0')e^{ix} = (2x + i)e^{-ix}, \tag{6.318}$$

which leads to

$$-2b_1' = 2, \Rightarrow b_1' = -1,$$

$$-2b_1'i - 2b_0' = i \Rightarrow 2i - 2b_0' = i \Rightarrow b_0' = \frac{i}{2}. \tag{6.319}$$

Using these results one can write the second particular solution as

$$y_{p_2}(x) = e^{-ix}(b_0' + b_1'x) = \left(\frac{i}{2} - x\right)e^{-ix}. \tag{6.320}$$

Applying the principle of superposition, the particular solutions in equations (6.316) and (6.320), the particular solution to the differential equation

$$\frac{d^2y}{dx^2} - y = f(x) = f_1(x) + f_2(x), \tag{6.321}$$

can be written as

$$y_p(x) = y_{p_1}(x) + y_{p_2}(x) = -\left(\frac{i}{2} + x\right)e^{ix} + \left(\frac{i}{2} - x\right)e^{-ix}. \tag{6.322}$$

Finally, using the results in equations (6.307) and (6.322), the general solution can be written as

$$y(x) = C_1 e^x + C_2 e^{-x} - \left(\frac{i}{2} + x\right)e^{ix} + \left(\frac{i}{2} - x\right)e^{-ix}, \tag{6.323}$$

which can also be put in the form

$$y(x) = C_1 e^x + C_2 e^{-x} + \sin(x) - 2x\cos(x). \tag{6.324}$$

The constants C_1 and C_2 are determined from the boundary conditions.

6.6 The method of successive integration

In this section, we will derive a general integral expression for the general solution to a second-order non-homogeneous ODE by a successive integration of the differential equation. We will also see how the general solutions we determined in the previous sections can also be obtained by applying the method of successive integration. To this end, we recall that the second-order ODE with constant coefficients from equation (6.278)

$$a_2\frac{d^2y}{dx^2} + a_1\frac{dy}{dx} + a_0 y = f(x). \tag{6.325}$$

We saw in the previous section, equation (6.325) can also be expressed as

$$\left(a_2\frac{d^2}{dx^2} + a_1\frac{d}{dx} + a_0\right)y = \left(\frac{d}{dx} - k_1\right)\left(\frac{d}{dx} - k_2\right)y = f(x), \tag{6.326}$$

where k_1 and k_2 are the roots to the quadratic equation (the indicial equation)

$$a_2 k^2 + a_1 k + a_0 = 0 \tag{6.327}$$

given by

$$k_1 = \frac{-a_1 + \sqrt{a_1^2 - 4a_0 a_2}}{2a_2}, \quad k_2 = \frac{-a_1 - \sqrt{a_1^2 - 4a_0 a_2}}{2a_2}. \tag{6.328}$$

Introducing a new function $u(x)$ defined by

$$u(x) = \left(\frac{d}{dx} - k_2\right)y(x), \tag{6.329}$$

equation (6.326) can be rewritten as

$$\left(\frac{d}{dx} - k_1\right)u(x) = \frac{du}{dx} - k_1 u = f(x).$$

(6.330)

Multiplying this equation by $e^{-k_1 x}$, we have

$$e^{-k_1 x}\frac{du}{dx} - k_1 e^{-k_1 x}u = e^{-k_1 x}f(x),$$

(6.331)

so that applying the product rule, one can write

$$\frac{d}{dx}[ue^{-k_1 x}] = e^{-k_1 x}f(x).$$

(6.332)

Integrating this equation with respect to x, we find

$$ue^{-k_1 x} = \int^x e^{-k_1 x'}f(x')dx' + C_1,$$

$$\Rightarrow u(x) = e^{k_1 x}\int^x e^{-k_1 x'}f(x')dx' + C_1 e^{k_1 x},$$

(6.333)

where C_1 is a constant of integration. Recalling that

$$u(x) = \left(\frac{d}{dx} - k_2\right)y(x) = \frac{dy(x)}{dx} - k_2 y(x),$$

(6.334)

in a similar way, multiplying this equation by $e^{-k_2 x}$, one can write

$$e^{-k_2 x}\frac{dy}{dx} - k_2 e^{-k_2 x}y = e^{-k_2 x}u(x) \Rightarrow \frac{d}{dx}[ye^{-k_2 x}] = e^{-k_2 x}u(x)$$

$$\Rightarrow ye^{-k_2 x} = \int^x e^{-k_2 x'}u(x')dx' + C_2,$$

(6.335)

$$\Rightarrow y(x) = e^{k_2 x}\int^x e^{-k_2 x'}u(x')dx' + C_2 e^{k_2 x}.$$

Now applying the result in equation (6.333), one can write for $u(x')$,

$$u(x') = e^{k_1 x'}\int^{x'} e^{-k_1 x''}f(x'')dx'' + C_1 e^{k_1 x'},$$

(6.336)

and substituting equation (6.336) into equation (6.335), we find

$$y(x) = C_1 e^{k_2 x}\int^x e^{(k_1 - k_2)x'}dx' + C_2 e^{k_2 x}$$

$$+ e^{k_2 x}\int^x \left[e^{(k_1 - k_2)x'}\int^{x'} e^{-k_1 x''}f(x'')dx''\right]dx',$$

(6.337)

where k_1 and k_2 are given by equation (6.328). Noting that the first and second terms are independent of the function $f(x)$ and the third term involving the function $f(x)$ is the result of the non-homogeneity of the DE, we can rewrite equation (6.337) as

$$y(x) = y_h(x) + y_p(x),$$ (6.338)

where the homogeneous part of the solution $y_h(x)$

$$y_h(x) = C_1 e^{k_2 x} \int^x e^{(k_1 - k_2)x'} dx' + C_2 e^{k_2 x},$$ (6.339)

and the particular solution $y_p(x)$

$$y_p(x) = e^{k_2 x} \int^x \left[e^{(k_1 - k_2)x'} \int^{x'} e^{-k_1 x''} f(x'') dx'' \right] dx'.$$ (6.340)

Next, we will use equations (6.339) and (6.340) to derive the solutions we determined in the previous sections.

A. **HODE with none degenerate roots:** We recall that for a second-order homogeneous ODE $(F(x) = 0)$

$$a_2 \frac{d^2 y}{dx^2} + a_1 \frac{dy}{dx} + a_0 y = 0,$$ (6.341)

and the indicial equation

$$a_2 k^2 + a_1 k + a_0 = 0.$$ (6.342)

The roots to this indicial equation are given by equation (6.328). When these roots are non-degenerate $(k_1 \neq k_2)$, we saw that the general solution to DE is given by

$$y(x) = C_1 e^{k_1 x} + C_2 e^{k_2 x}.$$ (6.343)

The result in equation (6.337), for $f(x) = 0$, becomes

$$y(x) = C_1 e^{k_2 x} \int^x e^{(k_1 - k_2)x''} dx'' + C_2 e^{k_2 x},$$ (6.344)

and upon carrying out the integration, one finds

$$y(x) = \frac{C_1 e^{k_2 x} e^{(k_1 - k_2)x}}{k_1 - k_2} + C_2 e^{k_2 x} = \frac{C_1 e^{k_1 x}}{k_1 - k_2} + C_2 e^{k_2 x}$$ (6.345)
$$= C_1 e^{k_1 x} + C_2 e^{k_2 x}.$$

where we have included the constant $1/(k_1 - k_2)$ in C_1. This is the same as equation (6.343).

B. **HODE with degenerate roots:** For a homogeneous ODE that leads to an indicial equation with degenerate roots (i.e., $k_1 = k_2$), equation (6.344) becomes

$$y(x) = C_1 e^{k_2 x} \int^x dx'' + C_2 e^{k_2 x} = (C_2 + C_1 x) e^{k_2 x},$$ (6.346)

which is in agreement with equation (6.193).

C. **Non-HODE:** We have seen that the particular solutions, $y_p(x)$, for the second-order non-HODE

$$a_2\frac{d^2y}{dx^2} + a_1\frac{dy}{dx} + a_0y = \left(\frac{d}{dx} - k_1\right)\left(\frac{d}{dx} - k_2\right) = e^{\alpha x}P_m(x), \qquad (6.347)$$

with a polynomial $P_m(x)$ of degree m, we saw that the particular solutions are given by

$$y_p(x) = \begin{cases} e^{\alpha x}Q_m(x), & (\text{when } \alpha \neq k_1 \ \& \ \alpha \neq k_2 \Rightarrow k_1 \neq k_2), \\ xe^{\alpha x}Q_m(x), & (\text{when } \alpha = k_1 \text{ or } k_2 \ \& \ k_1 \neq k_2), \\ x^2 e^{\alpha x}Q_m(x), & (\text{when } \alpha = k_1 = k_2), \end{cases} \qquad (6.348)$$

where $Q_m(x)$ is also a polynomial of degree m,

$$Q_m(x) = b_0 + b_1 x + b_2 x^2 + \dots b_m x^m = \sum_{i=0}^{m} b_i x^i. \qquad (6.349)$$

Next, we will derive these particular solutions from equation (6.340). To this end, we recall

$$f(x) = e^{\alpha x}P_m(x) = e^{\alpha x}(c_0 + c_1 x + c_2 x^2 + \dots c_m x^m) = e^{\alpha x}\sum_{i=0}^{m} c_i x^i. \qquad (6.350)$$

Then the particular solution in equation (6.340) can be expressed as,

$$y_p(x) = e^{k_2 x}\int^x \left[e^{(k_1-k_2)x'}\sum_{i=0}^{m} c_i \int^{x'} e^{(\alpha-k_1)x''}(x'')^i dx''\right]dx'$$
$$= \sum_{i=0}^{m} c_i e^{k_2 x}\int^x \left[e^{\beta x'}\int^{x'} e^{\gamma x''}(x'')^i dx''\right]dx', \qquad (6.351)$$

where

$$\gamma = \alpha - k_1 \text{ and } \beta = k_1 - k_2. \qquad (6.352)$$

(i) *When* $\alpha \neq k_1$ & $\alpha \neq k_2 \Rightarrow k_1 \neq k_2$: Under this condition, $\gamma = \alpha - k_1 \neq 0$ and one can use the relation

$$\frac{d}{d\gamma}(e^{\gamma x}) = xe^{\gamma x}, \ \frac{d^2}{d\gamma^2}(e^{\gamma x}) = x^2 e^{\gamma x}, \ \frac{d^3}{d\gamma^3}(e^{\gamma x}) = x^3 e^{\gamma x}\dots$$
$$\Rightarrow \frac{d^i}{d\gamma^i}(e^{\gamma x}) = x^i e^{\gamma x}, \qquad (6.353)$$

and show that

$$\int^{x'} e^{\gamma x''}(x'')^i dx'' = \frac{d^i}{d\gamma^i}\int^{x'} e^{\gamma x''}dx'' = \frac{d^i}{d\gamma^i}(e^{\gamma x'}). \tag{6.354}$$

Substituting this equation into equation (6.351), one can write the particular solution as

$$\begin{aligned} y_p(x) &= e^{k_2 x}\sum_{i=0}^{m}c_i\int^{x}\left[e^{\beta x'}\frac{d^i}{d\gamma^i}\left(\frac{e^{\gamma a x'}}{\gamma}\right)\right]dx' \\ &= e^{k_2 x}\sum_{i=0}^{m}c_i\frac{d^i}{d\gamma^i}\int^{x}\frac{e^{(\gamma+\beta)x'}}{\gamma}dx'. \end{aligned} \tag{6.355}$$

Upon carrying out the integration, we find

$$y_p(x) = e^{k_2 x}\sum_{i=0}^{m}c_i\frac{d^i}{d\gamma^i}\left[\frac{e^{(\gamma+\beta)x}}{\gamma(\gamma+\beta)}\right] = e^{k_2 x}\sum_{i=0}^{m}\frac{c_i}{\beta}\frac{d^i}{d\gamma^i}\left[\left(\frac{1}{\gamma}-\frac{1}{\gamma+\beta}\right)e^{(\gamma+\beta)x}\right]. \tag{6.356}$$

Now let us carry out the differentiation

$$c_i\frac{d^i}{d\gamma^i}\left[\frac{e^{(\gamma+\beta)x}}{\gamma(\gamma+\beta)}\right] = \frac{c_i}{\beta}\frac{d^i}{d\gamma^i}\left[\left(\frac{1}{\gamma}-\frac{1}{\gamma+\beta}\right)e^{(\gamma+\beta)x}\right], \tag{6.357}$$

for $i = 0$, 1, and 2, and see if we can find some kind of pattern. To this end, we note that

$$\begin{aligned} i = 0 &\Rightarrow \frac{c_0}{\beta}\frac{d^0}{d\gamma^0}\left[\left(\frac{1}{\gamma}-\frac{1}{\gamma+\beta}\right)e^{(\gamma+\beta)x}\right] = \frac{c_0}{\beta}\left(\frac{1}{\gamma}-\frac{1}{\gamma+\beta}\right)e^{(\gamma+\beta)x}, \\[2mm] i = 1 &\Rightarrow \frac{c_1}{\beta}\frac{d}{d\gamma}\left[\left(\frac{1}{\gamma}-\frac{1}{\gamma+\beta}\right)e^{(\gamma+\beta)x}\right] \\[2mm] &= \frac{c_1}{\beta}\left[-\frac{1}{\gamma^2}+\frac{1}{(\gamma+\beta)^2}+\left(\frac{1}{\gamma}-\frac{1}{\gamma+\beta}\right)x\right]e^{(\gamma+\beta)x} \\[2mm] &= \frac{c_1}{\beta}\left[-\frac{1}{\gamma^2}+\frac{1}{(\gamma+\beta)^2}+\frac{\beta}{\gamma(\gamma+\beta)}x\right]e^{(\gamma+\beta)x}, \\[2mm] i = 2 &\Rightarrow \frac{c_2}{\beta}\frac{d^2}{d\gamma^2}\left[\left(\frac{1}{\gamma}-\frac{1}{\gamma+\beta}\right)e^{(\gamma+\beta)x}\right] \\[2mm] &= \frac{c_2}{\beta}\frac{d}{d\gamma}\left[\left(-\frac{1}{\gamma^2}+\frac{1}{(\gamma+\beta)^2}+\left(\frac{1}{\gamma}-\frac{1}{\gamma+\beta}\right)x\right)e^{(\gamma+\beta)x}\right] \\[2mm] &= \frac{c_2}{\beta}\left[\frac{2}{\gamma^2}-\frac{2}{(\gamma+\beta)^3}+\left(-\frac{2}{\gamma^2}+\frac{2}{(\gamma+\beta)^2}\right)x+\frac{\beta}{\gamma(\gamma+\beta)}x^2\right]e^{(\gamma+\beta)x}. \end{aligned} \tag{6.358}$$

In view of the results in Eq. (6.358), one can write

$$\sum_{i=0}^{m} c_i \frac{d^i}{d\gamma^i}\left[\frac{e^{(\gamma+\beta)x}}{\gamma(\gamma+\beta)}\right] = \sum_{i=0}^{m} \frac{c_i}{\beta}\frac{d^i}{d\gamma^i}\left[\left(\frac{1}{\gamma}-\frac{1}{\gamma+\beta}\right)e^{(\gamma+\beta)x}\right] = \frac{c_0}{\beta}\left(\frac{1}{\gamma}-\frac{1}{\gamma+\beta}\right)e^{(\gamma+\beta)x}$$

$$+ \frac{c_1}{\beta}\left[-\frac{1}{\gamma^2}+\frac{1}{(\gamma+\beta)^2}+\frac{\beta}{\gamma(\gamma+\beta)}x\right]e^{(\gamma+\beta)x}$$

$$+ \frac{c_2}{\beta}\left[\frac{2}{\gamma^2}-\frac{2}{(\gamma+\beta)^3}+\left(-\frac{2}{\gamma^2}+\frac{2}{(\gamma+\beta)^2}\right)x\right.$$

$$\left.+\frac{\beta}{\gamma(\gamma+\beta)}x^2\right]e^{(\gamma+\beta)x} \ldots \tag{6.359}$$

$$\Rightarrow \sum_{i=0}^{m} c_i \frac{d^i}{d\gamma^i}\left[\frac{e^{(\gamma+\beta)x}}{\gamma(\gamma+\beta)}\right] = \sum_{i=0}^{m} \frac{c_i}{\beta}\frac{d^i}{d\gamma^i}\left[\left(\frac{1}{\gamma}-\frac{1}{\gamma+\beta}\right)e^{(\gamma+\beta)x}\right]$$

$$= \left\{\left[\frac{c_0}{\gamma(\gamma+\beta)}+\frac{c_1}{\beta}\left(-\frac{1}{\gamma^2}+\frac{1}{(\gamma+\beta)^2}\right)+\frac{c_2}{\beta}\left(\frac{2}{\gamma^2}-\frac{2}{(\gamma+\beta)^3}\right)\ldots\right]\right.$$

$$\left.+\left[\frac{c_1}{\gamma(\gamma+\beta)}+\frac{c_2}{\beta}\left(-\frac{2}{\gamma^2}+\frac{2}{(\gamma+\beta)^2}\right)\ldots\right]x+\left[\frac{c_2}{\gamma(\gamma+\beta)}\ldots\right]x^2\ldots\right\}e^{(\gamma+\beta)x}. \tag{6.360}$$

Introducing the constant defined by

$$b_0 = \frac{c_0}{\gamma(\gamma+\beta)}+\frac{c_1}{\beta}\left(-\frac{1}{\gamma^2}+\frac{1}{(\gamma+\beta)^2}\right)+\frac{c_2}{\beta}\left(\frac{2}{\gamma^2}-\frac{2}{(\gamma+\beta)^3}\right)\ldots,$$

$$b_1 = \frac{c_1}{\gamma(\gamma+\beta)}+\frac{c_2}{\beta}\left(-\frac{2}{\gamma^2}+\frac{2}{(\gamma+\beta)^2}\right)\ldots, \tag{6.361}$$

$$b_2 = \frac{c_2}{\gamma(\gamma+\beta)}\ldots,$$

one can easily see that

$$\sum_{i=0}^{m} c_i \frac{d^i}{d\gamma^i}\left[\frac{e^{(\gamma+\beta)x}}{\gamma(\gamma+\beta)}\right] = \sum_{i=0}^{m} \frac{c_i}{\beta}\frac{d^i}{d\gamma^i}\left[\left(\frac{1}{\gamma}-\frac{1}{\gamma+\beta}\right)e^{(\gamma+\beta)x}\right]$$

$$= (b_0 + b_1 x + b_2 x^2 + \ldots b_m x^m)e^{(\gamma+\beta)x} = Q_m(x)e^{(\gamma+\beta)x}, \tag{6.362}$$

where $Q_m(x)$ is a polynomial of degree m. Substituting Eq. (6.362), $\gamma = \alpha - k_1$, and $\beta = k_1 - k_2$ into Eq. (6.356), the particular solution becomes

(ii) *When $\alpha = k_1$ & $\alpha \neq k_2$:* For this case we have $\gamma = \alpha - k_1 = 0$ and equation (6.351) becomes

$$y_p(x) = e^{k_2 x}Q_m(x)e^{(\alpha - k_1 + k_1 - k_2)x} = Q_m(x)e^{\alpha x}. \tag{6.363}$$

Upon carrying out the integration with respect to x'', we have

$$y_p(x) = \sum_{i=0}^{m} \frac{c_i}{i+1}e^{k_2 x}\int^x [e^{\beta x'}(x')^{i+1}]dx', \tag{6.364}$$

which leads to

$$y_p(x) = e^{k_2 x}\sum_{j=1}^{m+1} \frac{c_{j-1}}{j}\int^x e^{\beta x'}(x')^i dx', \tag{6.365}$$

where we have introduced the summation index defined by $j = i + 1$. Applying the relation in equation (6.353), one can show that

$$y_p(x) = e^{k_2 x}\sum_{j=1}^{m+1} \frac{c_{j-1}}{j}\int^x e^{\beta x'}(x')^i dx' = e^{k_2 x}\sum_{j=1}^{m+1} \frac{c_{j-1}}{j}\frac{d^j}{d\beta^j}\int^x e^{\beta x''}dx''$$

$$\Rightarrow y_p(x) = e^{k_2 x}\sum_{j=1}^{m+1} \frac{c_{j-1}}{j}\frac{d^j}{d\beta^j}\left[\frac{e^{\beta x}}{\beta}\right]. \tag{6.366}$$

Following a similar procedure we used in (ii), one can show that

$$\sum_{j=1}^{m+1} \frac{c_{j-1}}{j}\frac{d^j}{d\beta^j}\left[\frac{e^{\beta x}}{\beta}\right] = (b_1 x + b_2 x^2 + ...b_{m+1}x^{m+1})e^{\beta x}$$

$$= x(b_0 + b_1 x + b_2 x^2 ...b_m x^m)e^{\beta x} = xe^{\beta x}Q_m(x). \tag{6.367}$$

Thus upon substituting $\beta = k_1 - k_2 = \alpha - k_2$, the particular solution in equation (6.366) becomes

$$y_p(x) = e^{k_2 x}xe^{\beta x}Q_m(x) = xe^{\alpha x}Q_m(x), \tag{6.368}$$

where $P_m(x)$ is a polynomial of degree m. This result is the same as the solution in equation (6.348) when $\alpha = k_1$ & $\alpha \neq k_2$ or $\alpha = k_2$ & $\alpha \neq k_1$.

(iii) *When* $\alpha = k_1 = k_2$: Ffor this case we have $\gamma = \alpha - k_1 = 0$ and $\beta = k_1 - k_2 = 0$ and equation (6.351) becomes

$$y_p(x) = \sum_{i=0}^{m} \frac{c_i}{i}e^{k_2 x}\int^x (x')^{i+1}dx' = e^{\alpha x}\sum_{j=1}^{m+1} \frac{c_{j-1}}{j}\int^x (x')^j dx', \tag{6.369}$$

where $j = i + 1$. Upon integrating this equation, we have

$$y_p(x) = e^{ax} \sum_{j=1}^{m+1} \frac{c_{j-1}}{j(j+1)}(x)^{j+1}, \tag{6.370}$$

which can be rewritten as

$$y_p(x) = (c_2 x^2 + c_2 x^3 + \ldots c_n x^{m+2})e^{ax} = x^2 e^{ax} Q_m(x), \tag{6.371}$$

where $Q_m(x)$ is a polynomial of degree m. This result is also the same as the solution in equation (6.348) when $\alpha = k_1 = k_2$.

Example 6.15. *A forced harmonic oscillator*: Consider an ideal spring with spring constant, $k = 8\,\mathrm{Nm^{-1}}$ with a mass $m = 2\,\mathrm{kg}$ attached to its one end with the other end fixed. The spring and mass are sitting on a frictionless surface. The mass is constantly acted by a periodic external force given by, $\vec{F_e} = 2\sin(3t)\hat{x}$. Find the displacement of the mass as a function of time. Initially, the mass is at rest, with the spring at its equilibrium position (i.e., the spring is neither stretched nor compressed) (figure 6.5).

Solution: In order to find the displacement, one needs to solve the equation of motion of the mass. Applying Newton's second law, the equation of motion for the mass, m, can be written as

$$m\frac{d^2x}{dt^2} = F_e - F_s = F_e - kx \Rightarrow \frac{d^2x}{dt^2} + \frac{k}{m}x = \frac{F_e}{m}. \tag{6.372}$$

Using the values provided for the external force $F_e = 2\sin(3t)$, the spring constant $k = 8\,\mathrm{Nm^{-1}}$, and mass $m = 2\,\mathrm{kg}$, the equation of motion becomes

$$\frac{d^2x}{dt^2} + 4x = \sin(3t)$$

$$\Rightarrow \left(\frac{d^2}{dt^2} + 4\right)x = \left(\frac{d}{dt} + 2i\right)\left(\frac{d}{dt} - 2i\right)x = \sin(3t). \tag{6.373}$$

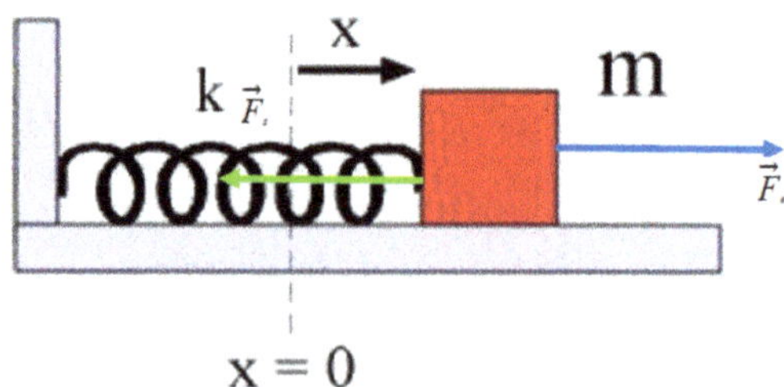

Figure 6.5. A harmonic oscillator acted on by an oscillating external force, $\vec{F_e} = 2\sin(3t)\hat{x}$.

In view of the general solutions in equations (6.339) and (6.340), the homogeneous and the particular solutions to equation (6.373) can be expressed as

$$x_h(t) = C_1 e^{k_2 t} \int^t e^{(k_1 - k_2)t'} dt' + C_2 e^{k_2 t} \tag{6.374}$$

and

$$x_p(t) = e^{k_2 t} \int^t \left[e^{(k_1 - k_2)t'} \int^{t'} e^{-k_1 t''} f(t'') dt'' \right] dt'. \tag{6.375}$$

From equation (6.373), we have

$$k_1 = 2i, \ k_2 = -2i, \ \text{and} \ f(t) = \sin(3t) = \frac{e^{3it} - e^{-3it}}{2i}, \tag{6.376}$$

so that equation (6.374) becomes

$$x_h(t) = C_1 e^{-2it} \int^t e^{4it'} dt' + C_2 e^{-2it} = \frac{C_1}{4i} e^{2it} + C_2 e^{-2it}. \tag{6.377}$$

Using Euler's formula this can be rewritten as

$$\begin{aligned} x_h(t) &= \frac{C_1}{4i}[\cos(2t) + i\sin(2t)] + C_2[\cos(2t) - i\sin(2t)] \\ &= \left(\frac{C_1}{4i} + C_2\right)\cos(2t) + \left(\frac{C_1}{4i} - C_2\right)i\sin(2t), \\ &\Rightarrow x_h(t) = C_1' \cos(2t) + C_2' \sin(2t), \end{aligned} \tag{6.378}$$

where we introduced the constants defined by

$$C_1' = \frac{C_1}{4i} + C_2, \ C_2' = \left(\frac{C_1}{4i} - C_2\right)i. \tag{6.379}$$

Substituting equation (6.376) into equation (6.375), the particular solution can be rewritten as

$$\begin{aligned} x_p(t) &= e^{-2it} \int^t \left[e^{4it'} \int^{t'} e^{-2it''} \left(\frac{e^{3it''} - e^{-3it''}}{2i} \right) dt'' \right] dt' \\ &= \frac{e^{-2it}}{2i} \int^t \left[e^{4it'} \int^{t'} \left(e^{it''} - e^{-5it''} \right) dt'' \right] dt'. \end{aligned} \tag{6.380}$$

Upon carrying out the inner integral, we have

$$\begin{aligned} x_p(t) &= \frac{e^{-2it}}{2i} \int^t \left[e^{4it'} \left(\frac{e^{it'}}{i} + \frac{e^{-5it'}}{5i} \right) \right] dt' \\ &= \frac{e^{-2it}}{2i} \int^t \left[\left(\frac{e^{5it'}}{i} + \frac{e^{-it'}}{5i} \right) \right] dt', \end{aligned} \tag{6.381}$$

so that one can also carry out similar integration to this equation and find

$$x_p(t) = \frac{e^{-2it}}{2i}\left(\frac{e^{5it}}{5i^2} - \frac{e^{-it}}{5i^2}\right) = -\frac{1}{5}\left(\frac{e^{3it} - e^{-3it}}{2i}\right) = -\frac{1}{5}\sin(3t). \tag{6.382}$$

Using the results in equations (6.378) and (6.382), the general solution becomes

$$x(t) = x_h(t) + x_p(t) = C_1' \cos(2t) + C_2' \sin(2t) - \frac{1}{5}\sin(3t). \tag{6.383}$$

Initially ($t = 0$), the spring is neither stretched nor compressed. This means for the position of the mass given by equation (6.383), we find

$$x(0) = 0 \Rightarrow C_1' = 0 \Rightarrow x(t) = C_2' \sin(2t) - \frac{1}{5}\sin(3t), \tag{6.384}$$

and using this expression for the displacement, one can write for the velocity

$$v(t) = \frac{dx(t)}{dt} = 2C_2' \cos(2t) - \frac{3}{5}\cos(3t). \tag{6.385}$$

At the initial time, the mass is at rest,

$$v(0) = 0 \Rightarrow 2C_2' - \frac{3}{5} = 0 \Rightarrow C_2' = \frac{3}{10}. \tag{6.386}$$

Therefore, with the given initial conditions (equation (6.384)) and (equation 6.385) become

$$\begin{aligned}
x(t) &= \frac{3}{10}\sin(2t) - \frac{1}{5}\sin(3t), \\
v(t) &= \frac{3}{5}[\cos(2t) - \cos(3t)].
\end{aligned} \tag{6.387}$$

6.7 Introduction to partial differential equations

In the previous sections, we saw how homogeneous or non-homogeneous ODEs of functions of one variable can be used in describing a simple physical system like the simple harmonic oscillator. We have also introduced different methods for solving such differential equations. However, most physical systems or processes are described by functions of two or more variables determined from a homogeneous or non-homogeneous *partial differential equation* (PDE). A PDE is a differential equation involving partial derivatives of an unknown function with respect to two or more independent variables. Various physical processes require the use of partial differential equations. Some are listed below:

1. **Poisson's equation:** Consider a system with a non-uniform mass or charge distribution described by a mass or charge density $\rho(x, y, z)$. The

gravitational potential energy due to the mass and the electric potential due to the charge of this system, which we may describe by a function $V(x, y, z)$, is determined by solving the PDE given by

$$\frac{\partial^2 V(x, y, z)}{\partial x^2} + \frac{\partial^2 V(x, y, z)}{\partial y^2} + \frac{\partial^2 V(x, y, z)}{\partial z^2} = \rho\left(x, y, z\right). \tag{6.388}$$

This PDE is known as *Poisson's equation*. If $\rho(x, y, z) = 0$, it is called *Laplace's equation*.

2. **The Schrödinger equation:** In classical mechanics, we solve Newton's equation of motion to determine quantities associated with the properties of a macroscopic system subjected to some external force. On the other hand, for a microscopic system, such as an electron in an atom, one must use the Schrödinger equation given by the PDE

$$-\frac{\hbar^2}{2m}\left[\frac{\partial^2 \Psi(x, y, z, t)}{\partial x^2} + \frac{\partial^2 \Psi(x, y, z, t)}{\partial y^2} + \frac{\partial^2 \Psi(x, y, z, t)}{\partial z^2}\right]$$
$$+ U(x, y, z)\Psi\left(x, y, z, t\right) = -i\hbar\frac{\partial}{\partial t}\Psi\left(x, y, z, t\right), \tag{6.389}$$

where $\Psi(x, y, z, t)$ is the wave function, $U(x, y, z)$ is the potential energy, $\hbar = h/2\pi$ (h is Planck's constant), and m is the mass of the object.

3. **The wave equation:** A wave is a disturbance in a continuous medium that propagates at a constant velocity, v. The displacement, caused by the disturbance, $u(x, y, z, t)$, is measured from the equilibrium state that obeys the wave equation given by the PDE

$$\frac{\partial^2 u(x, y, z, t)}{\partial x^2} + \frac{\partial^2 u(x, y, z, t)}{\partial y^2} + \frac{\partial^2 u(x, y, z, t)}{\partial z^2} = \frac{1}{v^2}\frac{\partial^2}{\partial t^2}u\left(x, y, z, t\right). \tag{6.390}$$

This section introduces the method for solving PDE and finding the unknown function. This method is known as *the separation of variables*. The following successive steps, building upon the methods for solving ODE introduced in the previous sections, are used in this method.

1. Assume a product form for the solution. That means if the PDE involves a function $f(x, y)$, we assume the function is expressible as a product of two functions $f(x, y) = g_1(x)g_2(y)$.
2. Substitute the product function into the PDE and simplify.
3. Divide both sides of the resulting DE by the product function.
4. Try to separate the variables, getting only terms involving one variable on one side of the equation.
5. The two sides of the equation must then be equal to the same constant, called the *separation constant*. (Think about what to name this constant—you can save yourself much unnecessary work by choosing a good name!)

6 Solve the resulting ordinary DEs using methods introduced in the previous sections.

7 The general solution is then a linear combination of all possible product-form solutions to the PDE, including all possible values of the separation constant!

8. Apply any BCs to the general solution.

Example 6.16. *A standing wave*: Figure 6.6 shows a string with linear mass density μ, a frictionless pulley, a vibrator, and an object with mass m. One end of the string connects to a vibrator that generates a wave with frequency f. The other end passes over the pulley and connects to the object. As a result of the interference between the traveling and reflected waves, we observe a standing wave on the string shown by the green lines in figure 6.6.

    (a) Show that the string supports *a wave motion by deriving the wave equation*.

$$\frac{\partial^2 u(y,\, t)}{\partial y^2} = \frac{1}{v^2}\frac{\partial^2 u(y,\, t)}{\partial t^2}, \tag{6.391}$$

where

$$v = \sqrt{\frac{T}{\mu}}, \tag{6.392}$$

T is the tension on the string.

    (b) Solve the 1-D traveling wave equation for a string of length L at the initial time $t = 0$ was undisturbed (at equilibrium state) and also has both ends fixed at all times. In other words, solve the wave equation in equation (6.391) subjected to the boundary conditions

$$u(y,\, t = 0) = 0,\ u(y = 0,\, t) = 0,\ u(y = L,\, t) = 0. \tag{6.393}$$

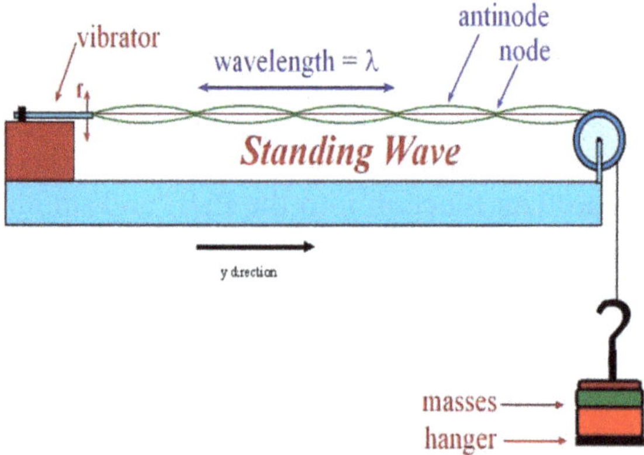

Figure 6.6. A standing wave on a string.

Solution:

(a) Because of the weight of the mass hanging on one end, the string experiences a tension force $T = mg$. When the string is continuously disturbed by the vibration on the other end, it will be displaced from the equilibrium creating a traveling wave on the string. The traveling and the reflected wave interfere to form a standing wave, as shown in figure 6.6. At a given instant of time t, a closer look at an infinitesimal length of the string Δy is shown in figure 6.7. If the string is displaced from the equilibrium and continuously vibrating up and down, the net force responsible for such motion on the segment between y and $y + \Delta y$ can be expressed as

$$\Delta F = T[\sin(\alpha) - \sin(\theta)]. \tag{6.394}$$

If we assume the displacement is small, we can replace sin by tan,

$$\Delta F \simeq T(\tan(\alpha) - \tan(\theta)). \tag{6.395}$$

Noting that

$$\tan(\alpha) = \left.\frac{\partial u}{\partial y}\right|_{y=y+\Delta y} , \quad \tan(\theta) = \left.\frac{\partial u}{\partial y}\right|_{y=y} , \tag{6.396}$$

we may write

$$\Delta F = T\left(\left.\frac{\partial u}{\partial y}\right|_{y=y+\Delta y} - \left.\frac{\partial u}{\partial y}\right|_{y=y}\right). \tag{6.397}$$

We recall that the first-order derivative of a function $u(y)$ at y,

$$\left.\frac{\partial u}{\partial y}\right|_{y=y} = \lim_{\Delta y \to 0}\left[\frac{u(y + \Delta y) - u(y)}{\Delta y}\right], \tag{6.398}$$

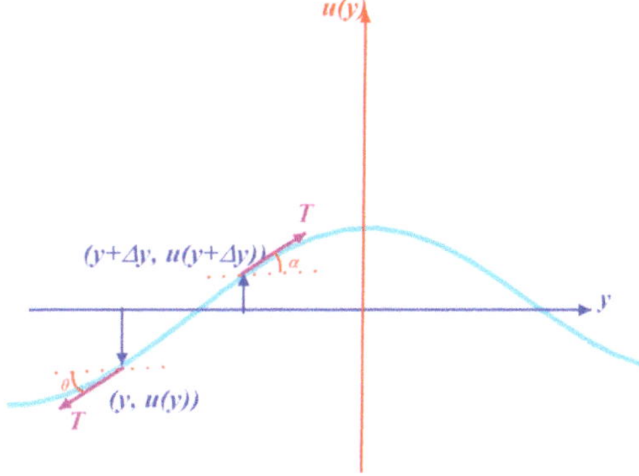

Figure 6.7. A snapshot of a standing wave on a string.

and the second-order derivative

$$\frac{\partial^2 u}{\partial y^2}\bigg|_{y=y} = \lim_{\Delta y \to 0}\left[\frac{\dfrac{\partial u}{\partial y}\bigg|_{y+\Delta y} - \dfrac{\partial u}{\partial y}\bigg|_{y}}{\Delta y} \right]$$

$$\Rightarrow \frac{\partial^2 u}{\partial y^2}\bigg|_{y=y}\Delta y = \lim_{\Delta y \to 0}\left[\frac{\partial u}{\partial y}\bigg|_{y+\Delta y} - \frac{\partial u}{\partial y}\bigg|_{y} \right], \tag{6.399}$$

so that for small Δy, we can write

$$\Delta F = T\left(\frac{\partial u}{\partial y}\bigg|_{y=y+\Delta y} - \frac{\partial u}{\partial y}\bigg|_{y=y} \right) \Rightarrow \Delta F = T\frac{\partial^2 u}{\partial y^2}\Delta y. \tag{6.400}$$

Noting that for the transverse motion $u(y, t)$ (the up-down motion as the wave propagates along the y-direction), Newton's second law can be written as

$$\Delta F = ma = m\frac{\partial^2 u}{\partial t^2}. \tag{6.401}$$

If the string has a linear mass density μ, we can write $m = \mu\Delta y$ so that

$$\Delta F = \mu\Delta y\frac{\partial^2 u}{\partial t^2}. \tag{6.402}$$

Combining the two expressions for the force ΔF, we find

$$T\frac{\partial^2 u}{\partial y^2}\Delta y = \mu\Delta y\frac{\partial^2 u}{\partial t^2}, \tag{6.403}$$

which leads to

$$\frac{\partial^2 u}{\partial y^2} = \frac{1}{v^2}\frac{\partial^2 u}{\partial t^2}, \tag{6.404}$$

where

$$v = \sqrt{\frac{T}{\mu}}. \tag{6.405}$$

Equation (6.404) is the one-dimensional wave equation. It describes a wave traveling in the y-direction with a speed v given by equation (6.405). The speed depends on the magnitude of the tension on the string T and the linear mass density of the string μ .

(b)

1. Assume the solution of the PDE can be expressed as

$$u(y, t) = Y(y)T(t). \tag{6.406}$$

2. Substitute equation (6.406) into the PDE in equation (6.404)

$$T(t)\frac{d^2 Y(y)}{dy^2} = \frac{Y(y)}{v^2}\frac{d^2 T(t)}{dt^2}. \tag{6.407}$$

3. Dividing equation (6.407) by equation (6.406)

$$\frac{1}{Y(y)}\frac{d^2 Y(y)}{dy^2} = \frac{1}{v^2 T(t)}\frac{d^2 T(t)}{dt^2}. \tag{6.408}$$

4. The two sides of equation (6.408) must then be equal to the same constant,

$$\frac{1}{Y(y)}\frac{d^2 Y(y)}{dy^2} = \frac{1}{v^2 T(t)}\frac{d^2 T(t)}{dt^2} = -k^2, \tag{6.409}$$

where k^2 is the separation constant. There follows that

$$\frac{d^2 Y(y)}{dy^2} + k^2 Y(y) = 0, \quad \frac{d^2 T(t)}{dt^2} + k^2 v^2 T(t) = 0. \tag{6.410}$$

5. Solve equation (6.410) using the methods introduced in the previous sections,

$$Y(y) = A \cos{(ky)} + B \sin{(ky)}, \; T(t) = C \cos{(kvt)} + D \sin{(kvt)}. \tag{6.411}$$

6. Write the general solution, which is given by

$$u(y, t) = \sum_k [A_k \cos{(ky)} + B_k \sin{(ky)}][C_k \cos{(kvt)} + D_k \sin{(kvt)}]. \tag{6.412}$$

7. Apply the boundary conditions. Using the first two boundary conditions, we find

$$u(y, t = 0) = 0 \Rightarrow \sum_k C_k[A_k \cos{(ky)} + B_k \sin{(ky)}] = 0 \Rightarrow C_k = 0$$

$$u(y = 0, t) = 0 \Rightarrow \sum_k A_k(C_k \cos{(kvt)} + D_k \sin{(kvt)}) = 0 \Rightarrow A_k = 0. \tag{6.413}$$

Upon setting $A_k = C_k = 0$, in equation (6.412), one can write the solution to the PDE as

$$u(y, t) = \sum_k B_k D_k \sin(ky) \sin(kvt) = \sum_k B_k' \sin(ky) \sin(kvt), \qquad (6.414)$$

where we replaced $B_k' = B_k D_k$. Then applying the last boundary condition $u(y = L, t) = 0$, one finds

$$u(L, t) = \sum_k B_k' \sin(kL) \sin(kvt) = 0 \Rightarrow \sin kL = 0$$

$$\Rightarrow kL = n\pi \Rightarrow k = \frac{n\pi}{L}, \qquad (6.415)$$

where $n = 1, 2, 3... $ .Therefore, the general solution becomes

$$u(y, t) = \sum_{n=1} B_n' \sin\left(\frac{n\pi}{L}y\right) \sin\left(\frac{n\pi v}{L}t\right). \qquad (6.416)$$

6.8 Linear differential equations and Mathematica

In Mathematica, we can solve differential equations using the following basic commands:

- Solve[*eqn,u,x*]: solves a differential equation for the function u, with independent variable x.
- DSolve [*eqn, u, {x, x*$_{min}$, x_{max}}]: solves a differential equation for x between x_{min} and x_{max} .
- DSolve [{*eqn*$_1$, *eqn*$_2$, ...}, {u_1, u_2, ...}, ...]: solves a list of differential equations.
- DSolve [*eqn, u, x*$_1$, x_2, ...]: solves a partial differential equation.
- DSolve [*eqn, u,* {x_1, x_2, ...}$\in\Omega$]: solves the partial differential equation *eqn* over the region Ω .

Next, using some of these commands, we will see how we solve some of the problems we considered the examples in the previous sections.

- Example 6.4
  Simplify [DSolve[$(x^3 + 2y[x])y'[x] == 3x(2 - xy[x])$, $y[x]$, x]]

$$\left\{\left\{y[x] \to \tfrac{1}{2}\left(-x^3 - \sqrt{12x^2 + x^6 + 4C[1]}\right)\right\},$$
$$\left\{y[x] \to \tfrac{1}{2}\left(-x^3 + \sqrt{12x^2 + x^6 + 4C[1]}\right)\right\}\right\}$$

- Example 6.5
  DSolve $\quad$ [$(y[x]^2 - 3x^2 \text{Exp}[3y[x]])y'[x] == 2x\,\text{Exp}[3y[x]] + \text{Exp}[x]$, $y[x]$, x]

$$\text{Solve}\left[e^x + e^{3y[x]}x^2 - \frac{y[x]^3}{3} == C[1], \ y[x]\right]$$

- Example 6.6
  Simplify [DSolve[$\{2xy'[x] + 4x^2 == y[x], \ y[1] == 26/3\}, \ y[x], \ x$]]

  $$\left\{\left\{y[x] \rightarrow 10\sqrt{x} - \frac{4x^2}{3}\right\}\right\}$$

- Example 6.7
  Simplify   [DSolve[$\{Q''[t] + \omega^2 Q[t] == 0, \ Q[0] == \text{Qo}, \ Q'[0] == \text{Io}\}, \ Q[t], \ t$]]

  $$\left\{\left\{Q[t] \rightarrow \text{QoCos}[t\omega] + \frac{\text{IoSin}[t\omega]}{\omega}\right\}\right\}$$

- Example 6.9(b)
  Simplify [DSolve[$2s'''[t] - 3s''[t] == 0, \ s[t], \ t$]]

  $$\left\{\left\{s[t] \rightarrow \frac{8}{27}e^{3t/2}C[1] + C[2] + t(C[3] + tC[4])\right\}\right\}$$

  Note that the factor 8/27 can be absorbed in the constant $C[1]$ and we find the same result as in Example 6.9(b).
- Example 6.11
  Simplify [DSolve[$s''[t] + 3s'[t] + 2s[t] == 16, \ s[t], \ t$]]
  $\{\{s[t] \rightarrow 8 + e^{-2t}C[1] + e^{-t}C[2]\}\}$
- Example 6.12
  Simplify [DSolve[$R''[x] + 4R[x] == \text{Sin}[3x], \ R[x], \ x$]]

  $$\left\{\left\{R[x] \rightarrow C[1]\text{Cos}[2x] + C[2]\text{Sin}[2x] - \frac{1}{5}\text{Sin}[3x]\right\}\right\}$$

- Example 6.14 Simplify [DSolve[$y''[x] - y[x] == 2\text{Sin}[x] + 4x\text{Cos}[x], \ y[x], \ x$]]
  $y[x] \rightarrow e^x C[1] + e^{-x}C[2] - 2x\text{Cos}[x] + \text{Sin}[x]$
- Example 6.9(b)
  Simplify [DSolve[$2s'''[t] - 3s''[t] == 0, \ s[t], \ t$]]
  $\left\{\left\{s[t] \rightarrow \frac{8}{27}e^{3t/2}C[1] + C[2] + t(C[3] + tC[4])\right\}\right\}$

6.9 Homework assignment

Problem 1.
Mothballs are small balls of chemical pesticide and deodorant, sometimes used when storing clothing and other articles susceptible to damage from mold or moth larvae (especially clothes moths like Tineola bisselliella). The chemical on a mothball evaporates at a rate proportional to the exposed surface. Suppose a spherical mothball of radius $r = 0.5$cm shrunk to a radius $r = 0.4$cm in 6 months due to the evaporation. How long will it take,
(a) for the mothball radius to shrunk to $r = 0.25$cm?
(b) for the mothball volume to shrunk to half of what it was initially?

Problem 2.
From the special theory of relativity, the momentum of an object with rest mass m_0 and traveling with a speed v near the speed of light c is given by

$$p = \frac{m_0 v}{\sqrt{1 - \dfrac{v^2}{c^2}}}. \tag{6.417}$$

If the object is subject to an external constant force F along the direction of the momentum, according to Newton's second law

$$\frac{dp}{dt} = \frac{d}{dt}\left(\frac{m_0 v}{\sqrt{1 - \dfrac{v^2}{c^2}}}\right) = F. \tag{6.418}$$

(a) Find the speed as a function of time $v(t)$ and show that

$$\lim_{t \to \infty} v(t) = c. \tag{6.419}$$

(b) Find the distance traveled by the object in time t if it starts from rest.

Problem 3.
For the differential equation

$$\frac{dy}{dx} = \frac{2xy^2 + x}{x^2 y - y}, \tag{6.420}$$

(a) find the solution,
(b) find the particular solution for the boundary condition $y = 0$ when $x = \sqrt{2}$,
(c) computer plot a slope field and the solution curve.

Problem 4.
For the curve defined by the function

$$y = kx^2, \tag{6.421}$$

(a) find the orthogonal trajectory
(b) computer plot the curves and the orthogonal trajectory.

Problem 5.
Consider the first-order differential

$$2x\frac{dy}{dx} + y = 2x^{5/2}, \tag{6.422}$$

(a) show that the differential equation is an exact differential equation or not,
(b) write the differential equation in the form

$$\frac{dy}{dx} + P(x)y = Q(x), \tag{6.423}$$

and determine the solution using the relation derived for such differential equation.
(c) verify your solution using computer programming (Mathematica is recommended).

Problem 6.

Do parts (a)–(c) in Problem 5 for the differential equation

$$\frac{dy}{dx} + y \tanh x = 2e^x. \tag{6.424}$$

Problem 7.

Consider a 10^9 gallons volume salty lake. Suppose a salty water with a salt density

$$\rho_{sw} = 5 \times 10^3 \text{lb/gal} \tag{6.425}$$

begins to flow at a rate of

$$\frac{dV_{in}}{dt} = 4 \times 10^5 \text{gal}\Big/\text{h} \tag{6.426}$$

into the salty lake. At the same time the salty water in the lake begins to flow out at a rate of

$$\frac{dV_{out}}{dt} = 10^5 \text{gal}\Big/\text{h}. \tag{6.427}$$

(a) Determine the DE describing the volume of the salty water, $V(t)$, in the lake. Solve the DE to find the volume of the salty water as a function of time.
(b) Find the DE that describes the mass, $m(t)$ of the salt in the salty lake. Solve the DE to find the mass of the salt in the lake as a function of time if the mass of the salt in the lake at the initial time, before the inflow and outflow begins, is $m(0) = 10^7 \text{lb}$.

Problem 8.

Find the general solution for the non-homogeneous DE

$$\frac{d^2y}{dx^2} + \frac{dy}{dx} - 2y = e^{2x}. \tag{6.428}$$

Problem 9.

Find the general solution for the non-homogeneous DE

$$5\frac{d^2y}{dx^2} + 6\frac{dy}{dx} + 2y = x^2 + 6x. \tag{6.429}$$

Problem 10.

Find the general solution for the non-homogeneous DE

$$\left(\frac{d^2}{dx^2} + 1\right)y = 8x \sin x. \tag{6.430}$$

Problem 11.

Using the method of superposition find the general solution to the DE

$$\frac{d^2y}{dx^2} - 5\frac{dy}{dx} + 6y = 2e^x + 6x - 5. \tag{6.431}$$

Problem 12.

In classical physics, the motion of a particle can be described for all time once the 'initial conditions' (position, velocity) of the particle's motion are known. The so-called 'equation of motion' in this case is Newton's second law. In order to apply Newton's second law, one must be able to specify all of the forces acting on the particle in question. In quantum mechanics, the equation of motion is called the Schrödinger equation. This equation is a partial differential equation involving a function called the wave function for the particle, denoted $\Psi(x,\, t)$ (for 1-D problems), which contains all observable information about the particle's motion (position, velocity, momentum, energy,...). Instead of precisely specifying all future motion of the particle, in quantum mechanics the quantity $|\Psi(x,\, t)|^2$ is used to predict the probability of a particular future motion. The 1-D Schrödinger equation is given by

$$-\frac{\hbar^2}{2m}\frac{\partial^2\Psi(x,\, t)}{\partial x^2} + U(x)\Psi(x,\, t) = i\hbar\frac{\partial}{\partial t}\Psi(x,\, t) \tag{6.432}$$

where m is the mass of the particle, $\hbar$ is Planck's constant ($1.05 \times 10 - 34J.\ s$), i is the pure imaginary number, and $U(x)$ is the net potential energy function of the particle at position x. (Note that, since $F_x = -dU(x)/dx$, specifying the potential energy function in quantum mechanics is equivalent to specifying all of the forces acting on the particle in classical mechanics.) The form of the Schrödinger equation given above is called the time-dependent (1-D) Schrödinger equation (TDSE) since it involves the full time-dependent wave function, $\Psi(x, t)$.

(a) Using the separation of variables technique, and letting the separation constant be denoted E (which turns out to be the total energy of the particle), show that the resulting differential equation that is independent of time—the so-called time-independent Schrödinger equation (TISE)—has the (1-D) form

$$-\frac{\hbar^2}{2m}\frac{d^2\psi(x, t)}{dx^2} + U(x)\psi(x, t) = E\psi(x, t) \tag{6.433}$$

where we have set $\Psi(x, t) = \psi(x)T(t)$. In applications, the potential energy function $U(x)$ must be specified, and the resulting TISE is solved for the spatial wave function $\psi(x)$.

(b) Also, show that the total wave function in this case has the form

$$\Psi(x, t) = \psi(x)e^{-iEt/\hbar}. \tag{6.434}$$

Problem 13.
Show that the solution to the DE

$$\frac{d^2y}{dx^2} - \omega^2 y = 0, \tag{6.435}$$

is given by

$$y(x) = \begin{cases} Ae^{\omega x} + Be^{-\omega x}, \\ C\cosh(\omega x) + D\sinh(\omega x), \\ E\sinh(\omega x + \varphi), \\ G\cosh(\omega x - \varphi), \end{cases} \tag{6.436}$$

gives

$$y(x) = \begin{cases} Ae^{i\omega x} + Be^{-i\omega x}, \\ C\cos(\omega x) + D\sin(\omega x), \\ E\sin(\omega x + \varphi), \\ G\cos(\omega x - \varphi), \end{cases} \tag{6.437}$$

when ω is replaced by $i\omega$ which is the solution to the DE

$$\frac{d^2y}{dx^2} + \omega^2 y = 0. \tag{6.438}$$

Problem 14.
Find the current and the charge in Example 6.7 if a resistor R is included in series with C and L.

Problem 15.
Derive the integral expression for the third-order and fourth-order non-homogeneous ODE with constant coefficients and, based on the results you obtained, write an expression for an nth-order non-homogeneous ODE with constant coefficients.

IOP Publishing

Studies in Theoretical Physics, Volume 1
Fundamental mathematical methods
Daniel Erenso and Victor Montemayor

Chapter 7

Integral calculus-scalar functions

In this chapter we will be introduced to single and multiple integrals involving scalar functions. We focus on the methods for carrying out such integrals. In almost all branches of physics, integrations of scalar functions are often used to determine various quantities describing the static or dynamic properties of an object. The object could be a microscopic particle like an electron in an atom or massive objects like a planet in the solar system. The simplicity of carrying out integrations of scalar functions depends on the geometry of the object and system that object is existing in. The methods of integrations in Cartesian coordinates for rectangular geometry and the methods of integrations in curvilinear coordinates for curved geometry can greatly simplify the integrations. We will introduce the methods of integrations of single and multiple integrals in Cartesian and curvilinear coordinates and its applications in real physical problems.

7.1 Integration in Cartesian coordinates

We are interested in the methods of integrations of functions of single, double, and triple variables in Cartesian coordinates.

A. Single integral—*Length*

Consider the two points P_1 and P_2 in the x–y plane shown in figure 7.1 described by the Cartesian coordinates $(x_1, y_1(x_1))$ and $(x_2, y_2(x_2))$, respectively. We took two paths to get from the first to the second point. The first is a straight line path shown in blue and the second is a curved path shown in cyan. Along the first path, the slope is constant for all x from the first to the second point but not along the second path. In either case, we can determine the slope using

$$m(x) = \lim_{\Delta x \to 0} \frac{\Delta y(x)}{\Delta x} = \frac{dy(x)}{dx}. \tag{7.1}$$

For these paths, the net distance traveled along the y-direction (Δy) determined from the line integral

doi:10.1088/978-0-7503-3135-7ch7

$$\Delta y = y_2(x_2) - y_1(x_1) = \int_{y_1(x_1)}^{y_2(x_2)} dy. \tag{7.2}$$

One can also use

$$dy(x) = m(x)dx, \tag{7.3}$$

and write

$$\Delta y = \int_{x_1}^{x_2} m(x)dx. \tag{7.4}$$

Similarly, along the x-direction, the net distance is given by

$$\Delta x = x_2 - x_1 = \int_{y_1}^{y_2} m'(y)dy, \tag{7.5}$$

where

$$m'(y) = \lim_{\Delta y \to 0} \frac{\Delta x(y)}{\Delta y} = \frac{dx(y)}{dy}. \tag{7.6}$$

The results for Δx and Δy are going to be the same for both paths. It is independent of the path we took from P_1 to P_2. However, for the total distance (*arc length*) l, for these two paths, is different, as one can imagine. For the first path this length can easily be determined using Pythagorean theorem

$$l = \sqrt{(y_2 - y_1)^2 + (x_2 - x_1)^2} = \sqrt{(\Delta y)^2 + (\Delta x)^2}. \tag{7.7}$$

That can also be put in the form

$$l = \Delta x \sqrt{\left(\frac{\Delta y}{\Delta x}\right)^2 + 1} = \Delta y \sqrt{\left(\frac{\Delta x}{\Delta y}\right)^2 + 1}. \tag{7.8}$$

For the second path, since the slope is not constant, one needs to divide the path into an infinitesimal interval, Δl_i, find the lengths using the Pythagorean theorem, and sum it up all to find the total distance,

$$\begin{aligned} l &= \sum_i \lim_{\Delta l_i \to 0} \Delta l_i = \sum_i \lim_{\Delta x_i \to 0} \Delta x_i \sqrt{\left(\frac{\Delta y_i}{\Delta x_i}\right)^2 + 1} \\ &= \sum_i \lim_{\Delta y_i \to 0} \Delta y_i \sqrt{\left(\frac{\Delta x_i}{\Delta y_i}\right)^2 + 1}. \end{aligned} \tag{7.9}$$

Noting that integration of a function $f(x)$ from p_1 to p_2 is defined as

$$\sum_i \lim_{\Delta f_i \to 0} \Delta f_i \rightarrow \int_{p_1}^{p_2} df, \tag{7.10}$$

and differentiation as

$$\frac{df(x)}{dx} = \lim_{\Delta x \to 0} \frac{\Delta f(x)}{\Delta x}, \tag{7.11}$$

one can write

$$l = \int_1^2 dl = \int_{x_1}^{x_2} \sqrt{\left[\left(\frac{dy}{dx}\right)^2 + 1\right]}\, dx = \int_{y_1}^{y_2} \sqrt{1 + \left(\frac{dx}{dy}\right)^2}\, dy. \tag{7.12}$$

This is the arc length (or the total distance) for any path one can follow to go from P_1 to P_2.

B. Double integral—*Area*

Let us add two points P_3 and P_4 (to figure 7.1) with coordinates (x_1, y_2) and (x_2, y_1), respectively, to form the rectangular region shown in figure 7.2. The area of this rectangle depends on the length L (the distance between P_1 and P_4) and the width W (the distance between P_1 and P_3)

$$A = LW. \tag{7.13}$$

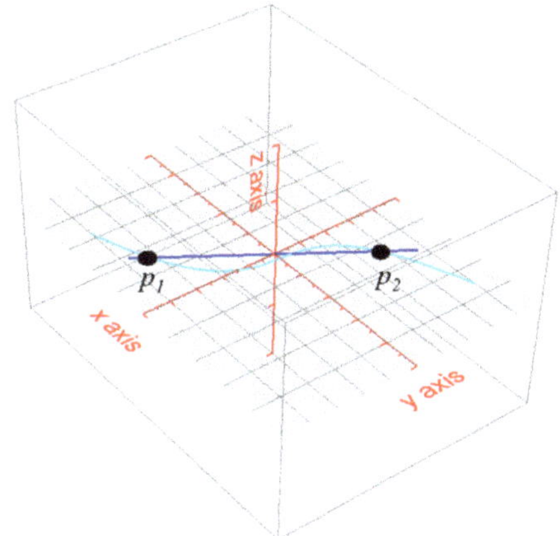

Figure 7.1. Two points on the x–y plane connect by two different paths.

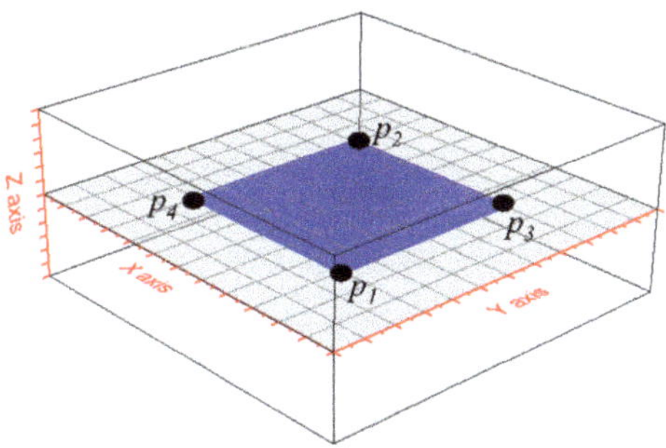

Figure 7.2. A rectangular region on the x–y plane.

The points along the length of the rectangle L are defined by the equation

$$x(y) = y_1 = \text{ constant or } x(y) = y_2 = \text{constant}, \qquad (7.14)$$

and those along the width W, by

$$y(x) = x_1 = \text{ constant or } y(x) = x_2 = \text{constant}. \qquad (7.15)$$

There follows that

$$\frac{dx}{dy} = 0, \qquad (7.16)$$

for points along the length of the rectangle, and

$$\frac{dy}{dx} = 0, \qquad (7.17)$$

along the width. Thus the length L and the width W of the rectangle, applying the relations in equation (7.12), can be expressed as

$$
\begin{aligned}
L &= \int_{x_1}^{x_2} \sqrt{\left[\left(\frac{dy}{dx}\right)^2 + 1\right]}\, dx = \int_{x_1}^{x_2} dx, \\
W &= \int_{y_1}^{y_2} \sqrt{1 + \left(\frac{dx}{dy}\right)^2}\, dy = \int_{y_1}^{y_2} dy.
\end{aligned}
\qquad (7.18)
$$

Using the line integrals in equation (7.18), the area in terms of a double (or surface) integral is expressible as,

$$
\begin{aligned}
A = lw &= \int_{x_1}^{x_2} dx \int_{y_1}^{y_2} dy = \int_{x_1}^{x_2} \int_{y_1}^{y_2} dxdy = \int_{y_1}^{y_2} dy \int_{x_1}^{x_2} dx \\
&= \int_{y_1}^{y_2} \int_{x_1}^{x_2} dydx.
\end{aligned}
\qquad (7.19)
$$

Note that the order of integration does not bring any difference in the result. This result is applicable for any arbitrary shape with boundaries defined by some functions of the coordinates x or y. Consider the region bounded by the two curves shown in figure 7.3. Let the points on the lower boundary be defined by the function

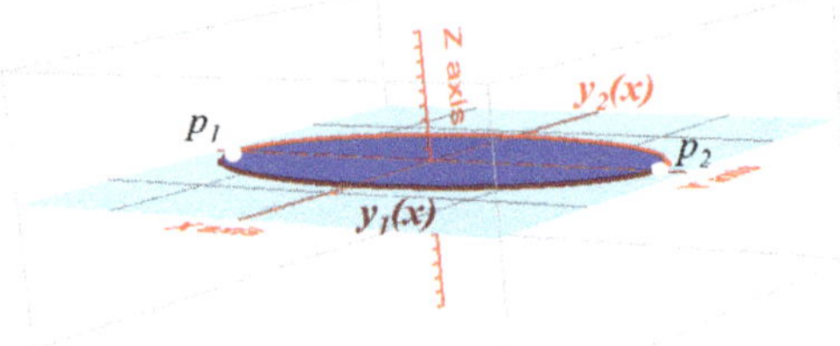

Figure 7.3. Area bounded by a curved region. The curved boundaries are defined by the functions $y_1(x)$ and $y_2(x)$.

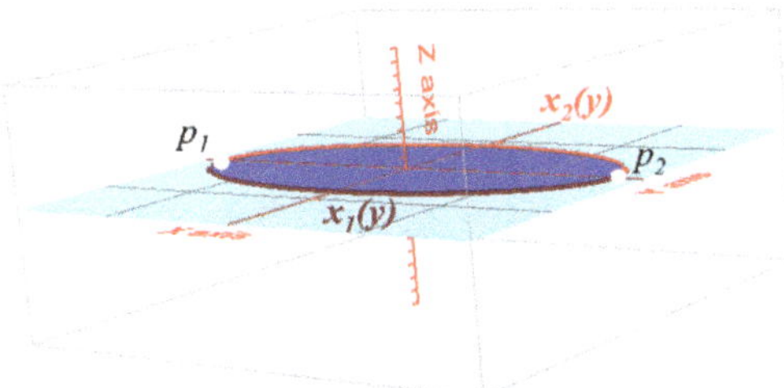

Figure 7.4. Area bounded by a curved region. The curved boundaries are defined by the functions $x_1(y)$ and $x_2(y)$.

$y_1(x)$ and on the upper boundary by the function $y_2(x)$. In this case, the double integral for the area in equation (7.19) should be expressed as

$$A = \int_{x_1}^{x_2} \left[\int_{y_1(x)}^{y_2(x)} dy \right] dx. \tag{7.20}$$

On the other hand, if the points in the lower boundary and the upper boundary are defined by the functions $x_1(y)$ and $x_2(y)$, respectively, instead (see figure 7.4), we need to switch the order of integration, and the area is given by

$$A = \int_{x_1}^{x_2} \left[\int_{x_1(y)}^{x_2(y)} dy \right] dx. \tag{7.21}$$

Note that even though we switched the order of integrations, the result should be the same.

C. Triple integral—Volume

Now we want to move the rectangle in figure 7.2 formed by the four points P_1, P_2, P_3, and P_4 vertically down below the x–y plane so that the coordinates for these points become (x_1, y_1, z_1), (x_2, y_2, z_1), (x_1, y_2, z_1) and (x_2, y_1, z_1), respectively. We also add other points P_5, P_6, P_7, and P_8 with coordinates (x_1, y_1, z_2), (x_2, y_2, z_2), (x_1, y_2, z_2) and (x_2, y_1, z_2), respectively (i.e., the same rectangle moved by z_2 above the x–y plane). These eight points form the corners of a rectangular box shown in figure 7.5. For a rectangular box the volume is given by

$$V = Ah, \tag{7.22}$$

where A is the base area given by equation (7.19) and h is the height of the box given by

$$h = z_2(x, y) - z_1(x, y) = \Delta z = \int_{z_2(x, y)}^{z_1(x, y)} dz. \tag{7.23}$$

Therefore, the volume of the rectangular box, using equations (7.19) and (7.23), can be expressed as

$$V = Ah = \int\int\int dxdydz = \int_{x_1}^{x_2} \left[\int_{x_1(y)}^{x_2(y)} \left[\int_{z_1(x, y)}^{z_2(x, y)} dz \right] dy \right] dx. \tag{7.24}$$

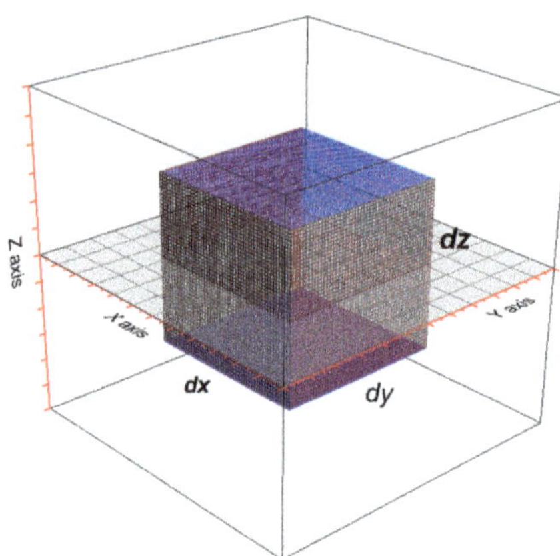

Figure 7.5. A rectangular box.

Example 7.1. *Line and surface integral*: Consider a wire bent into the arc of a circle of radius R, as shown in figure 7.6.

(a) Suppose we start walking from the point $(R, 0)$ on the arc and reach another point along this arc. By the time we reached this point, we have moved a distance of half of the radius along the x-direction. Determine your distance along the y-axis.

(b) Find the length of the wire.

(c) Find the area bounded by the curve, the x-axis, and the y-axis.

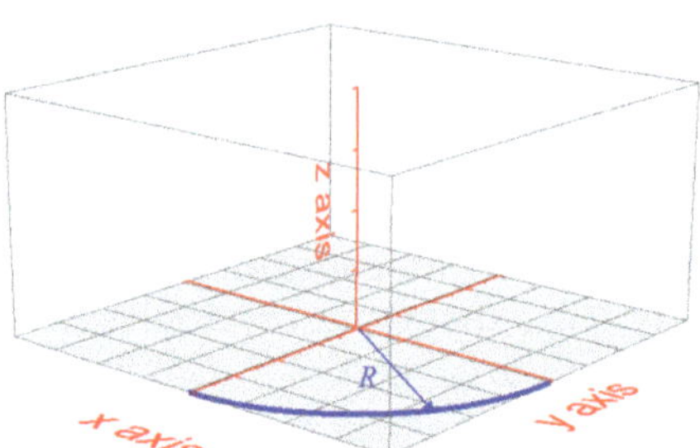

Figure 7.6. A wire bent into an arc of a circle with radius R on the x–y plane.

Solution:

(a) The distance along the y-axis (Δy),

$$\Delta y = \int_{x_1}^{x_2} m(x)dx, \tag{7.25}$$

where $m(x)$ is the slope of the wire given by

$$m(x) = \frac{dy(x)}{dx}.$$

The function defining the curve shown in figure 7.6 is

$$y(x) = \sqrt{R^2 - x^2}, \tag{7.26}$$

and the slope becomes

$$m(x) = \frac{dy(x)}{dx} = -\frac{x}{\sqrt{R^2 - x^2}}. \tag{7.27}$$

Since we started at $(R, 0)$ and reached to a point on the arc where the x coordinate is $R/2$, we have $x_1 = R$ and $x_2 = R/2$. The distance you traveled along the y-direction becomes

$$\Delta y = -\int_R^{R/2} \frac{x\,dx}{\sqrt{R^2 - x^2}} = (R^2 - x^2)^{1/2}\big|_R^{R/2} = \sqrt{\frac{3}{4}}\,R. \tag{7.28}$$

(b) The length of the wire is given by equation (7.12)

$$l = \int_{x_1}^{x_2} \sqrt{\left[\left(\frac{dy}{dx}\right)^2 + 1\right]}\,dx. \tag{7.29}$$

Using the result in equation (7.27), we may write

$$l = \int_{x_1}^{x_2} \sqrt{\left[\left(-\frac{x}{\sqrt{R^2 - x^2}}\right)^2 + 1\right]}\,dx = \int_{x_1}^{x_2} \frac{R}{\sqrt{R^2 - x^2}}\,dx. \tag{7.30}$$

Noting that, this time in order to keep the arc length positive, we should have $x_1 = 0$ and $x_2 = R$. Therefore, introducing the transformation defined by

$$x = R\sin(\theta) \Rightarrow dx = R\cos(\theta)d\theta \tag{7.31}$$

the length of the wire can be rewritten as

$$l = \int_0^{\pi/2} \frac{R^2\cos(\theta)}{\sqrt{R^2(1 - \sin^2(\theta))}}\,d\theta = R\int_0^{\pi/2} d\theta = \frac{R\pi}{2}. \tag{7.32}$$

Note that we have expressed the limits of integration in terms of the variable θ. That means for $x = 0$

$$x = R\sin(\theta) \Rightarrow \theta = 0 \tag{7.33}$$

and $x = R$

$$x = R\sin(\theta) \Rightarrow \theta = \sin^{-1}(1) = \pi/2. \tag{7.34}$$

(c) Let us find the area using the surface integral in equation (7.21)

$$A = \int_{x_1}^{x_2} \left[\int_{x_1(y)}^{x_2(y)} dy \right] dx. \tag{7.35}$$

The x-axis, the y-axis, and the arc are defined by the equations $y = 0$, $x = 0$, and $x(y) = \sqrt{R^2 - y^2}$, respectively and equation (7.35) becomes

$$A = \int_0^R \left(\int_0^{\sqrt{R^2-y^2}} dx \right) dy = \int_0^R \sqrt{R^2 - y^2}\, dy. \tag{7.36}$$

Introducing the transformation of variable defined by

$$y = R \sin(\theta) \Rightarrow dy = R \cos\theta\, d\theta, \tag{7.37}$$

and noting that

$$y = 0 \Rightarrow \theta = 0,\ y = R \Rightarrow \theta = \pi/2, \tag{7.38}$$

we find the area

$$A = \int_0^{\pi/2} \sqrt{R^2 - R^2 \sin(\theta)^2}\, R \cos\theta\, d\theta = R^2 \int_0^{\pi/2} \cos^2\theta\, d\theta = \frac{\pi R^2}{4}. \tag{7.39}$$

Example 7.2. *Volume integral*: A triangle in the x–y plane has vertices at the points $(x, y, z) = (0, 0, 0)$, $(1, 0, 0)$, and $(0, 2, 0)$. Find the volume of the space that lies above this triangle but below the plane $z = 2 + x + y$ (see figure 7.7). All distance units are in centimeters.

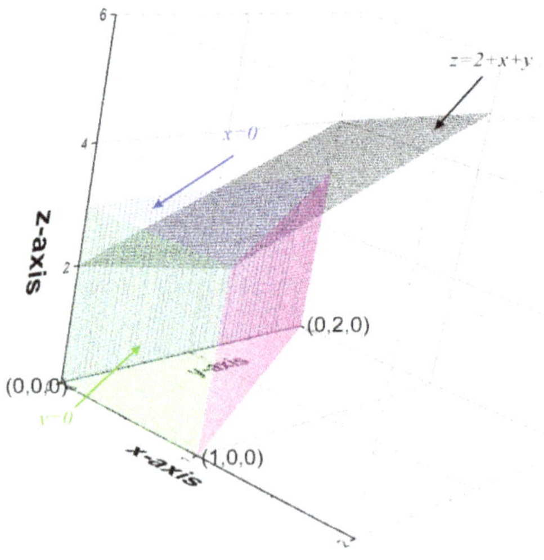

Figure 7.7. The volume of the region above a triangle with vertices $(0, 0, 0)$, $(1, 0, 0)$, and $(0, 2, 0)$; and below the plane defined by the equation $z = 2 + x + y$.

Solution: In view of equation (7.24), the volume is given by

$$V = \int_{x_1}^{x_2} \left[\int_{x_1(y)}^{x_2(y)} \left[\int_{z_1(x,\,y)}^{z_2(x,\,y)} dz \right] dy \right] dx. \tag{7.40}$$

The equation of the line joining the points $(1, 0, 0)$ and $(0, 2, 0)$ is given by

$$\frac{y-2}{x-0} = \frac{2-0}{0-1} \Rightarrow y = -2x + 2. \tag{7.41}$$

Then the volume in equation (7.40), for the space above the triangle (with vertices $(0, 0, 0)$, $(1, 0, 0)$, and $(0, 2, 0)$) on the x–y plane ($z = 0$) and below the plane defined by $z = 2 + x + y$ becomes,

$$V = \int_{x=0}^{x=1} \left\{ \int_{y=0}^{y=-2x+2} \left[\int_{z=0}^{z=2+x+y} dz \right] dy \right\} dx. \tag{7.42}$$

Integrating with respect to z,

$$V = \int_{x=0}^{x=1} \left\{ \int_{y=0}^{y=-2x+2} (2 + x + y)dy \right\} dx, \tag{7.43}$$

with respect to y,

$$V = \int_{x=0}^{x=1} \left(2y + xy + \frac{y^2}{2} \right) \Bigg|_0^{-2x+2} dx$$

$$= \int_{x=0}^{x=1} \left[2(-2x + 2) + x(-2x + 2) + \frac{(-2x + 2)^2}{2} \right] dx \tag{7.44}$$

$$= \int_{x=0}^{x=1} [6 - 6x]\, dx,$$

and then with respect to x, results in

$$V = \int_{x=0}^{x=1} [6 - 6x]dx = 6x - 3x^2 \, \big|_0^1 = 3 \text{ cm}^2. \tag{7.45}$$

7.2 Physical applications

Several physical quantities require carrying out single and multiple integrals. Here are some examples from electrostatics and mechanics.

    I. Electrostatics

      A. **Electrostatic potential**.

        Let us consider a point charge with charge, q, placed at the origin in free space with electrical permittivity ϵ_0. The electrostatic potential $V(\vec{r})$ due to this charge at a point in space described by the position vector, $\vec{r}$, measured from the charge is given by

$$V(\vec{r}) = \frac{q}{4\pi\epsilon_0 \, | \, \vec{r} \, |}. \tag{7.46}$$

When the charge is not a point charge, and the charge is a line, surface, or volume charge with no uniform distribution, the electrostatic potential, $V(r)$, is determined by carrying out single or multiple integrations that depend on the given charge distributions. In the region where we have the charge, let us consider an infinitely small charge dq' at a point described by the position vector, $\vec{r}'$, from the origin of the coordinate system. The corresponding infinitesimal electrostatic potential, $dV(\vec{r})$, at a point in space described by the position vector, $\vec{r}$ (also measured from the origin), depends on the distance of this point from the position of dq'. We determine this distance, $R(\vec{r}')$ from the magnitude of the position vector director directed from the position of dq' to the position of the point of interest for the electrostatic potential

$$R(r') = | \, \vec{r} - \vec{r}' \, |. \tag{7.47}$$

In view of equation (7.46), one can then express $dV(r)$ as

$$dV(r) = \frac{dq'}{4\pi\epsilon_0 \, | \, \vec{r} - \vec{r}' \, |}. \tag{7.48}$$

Line charge density: Suppose we have a charge distributed on a finite length of wire with length, L. Also, the charge distribution is not uniform, and a line charge density describes $\lambda(\vec{r}')$ as,

$$\lambda(\vec{r}') = \frac{dq'}{dr'} \Rightarrow dq' = \lambda(\vec{r}')dr'. \tag{7.49}$$

We can use this relation to find the total electrostatic potential due to the entire charge by integrating equation (7.48) over the entire length L

$$V(r) = \int \frac{dq'}{4\pi\epsilon_0 \, | \, \vec{r} - \vec{r}' \, |} = \frac{1}{4\pi\epsilon_0} \int_L \frac{\lambda(\vec{r}')}{| \, \vec{r} - \vec{r}' \, |}dr'. \tag{7.50}$$

Surface charge density: Any external charge that one may put on a metallic ball is distributed only on the sphere's surface, no matter the magnitude of the charge. This charge distributes itself uniformly or non-uniformly depending on the absence or presence of charges in surrounding regions of the sphere. Let us consider the general case of non-uniform distribution described by a surface charge density,

$$\sigma(\vec{r}') = \frac{dq'}{da'} \Rightarrow dq' = \sigma(\vec{r}')da', \tag{7.51}$$

where dq' is the infinitesimal charge on the infinitesimal area da'. In this case, the electrostatic potential due to the entire charge on the surface is given by

$$V(r) = \frac{1}{4\pi\epsilon_0} \int\!\!\!\int_{\text{Area}} \frac{\sigma(\vec{r}')}{|\vec{r} - \vec{r}'|} da'. \tag{7.52}$$

This is a surface integral that requires multiple integrations. Note that the infinitesimal area, da', on the surface of the sphere with radius r', in spherical coordinates can be expressed as

$$da' = r'^2 \sin(\theta')d\theta' d\varphi'. \tag{7.53}$$

Volume charge density: Similarly, for a charge described by a volume charge density,

$$\rho(\vec{r}') = \frac{dq'}{dv'} \Rightarrow dq' = \rho(\vec{r}')dv', \tag{7.54}$$

where dq' is the infinitesimal charge in the infinitesimal volume dv', the electrostatic potential at a point $\vec{r}$ for the entire volume of charge is given by a volume integral (triple integral)

$$V(r) = \frac{1}{4\pi\epsilon_0} \int\!\!\!\int\!\!\!\int_{\text{Volume}} \frac{\rho(\vec{r}')}{|\vec{r} - \vec{r}'|} dv'. \tag{7.55}$$

B. **The total charge:** The total charge, Q, distributed on a line is determined by a single integral

$$Q = \int_{\text{Line}} \lambda(\vec{r}')dr', \tag{7.56}$$

on a surface and volume by multiple integrations given by

$$Q = \int\!\!\!\int_{\text{Area}} \sigma(\vec{r}')da', \tag{7.57}$$

and

$$Q = \int\!\!\!\int\!\!\!\int_{\text{Volume}} \rho(\vec{r}')dv', \tag{7.58}$$

respectively.

II. Classical mechanics

A. *Moment of inertial, I:* In classical mechanics, we often see what is known as *a moment of inertia*. It is a property of a distribution of mass in space that measures its resistance to rotational acceleration about an axis. *Newton's first law,* which describes the inertia of a body in linear motion, can be extended to the inertia of a body rotating about an axis using the moment of inertia. That is, an object that is rotating at constant angular velocity will remain rotating unless acted upon by an external torque. In this way, the moment of inertia plays the same role in rotational dynamics as mass does in linear dynamics, describing the relationship between angular momentum and angular velocity, torque, and angular acceleration. The letter I and sometimes J is usually used to refer to the moment of inertia. For a discrete mass that consists of N entities, each with mass m_i', and with a corresponding perpendicular distance $r_{i\perp}'$ from the axis of rotation, the moment of inertial is given by

$$I = \sum_{i=1}^{N} m_i' r_{i\perp}'^2. \tag{7.59}$$

For continuum mass distribution in a volume, V, described by the mass density, $\rho(\vec{r})$, we can determine the moment of inertia using the volume integral

$$I = \int_M dm' r_\perp'^2 = \iiint_V \rho(\vec{r}')r_\perp'^2 dv' = \iiint_V \rho(x', y', z')r_\perp'^2 dx'dy'dz', \tag{7.60}$$

where $r_\perp'$ is the perpendicular distance from the axis of rotation and

$$\rho(\vec{r}') = \frac{dm}{dv'} = \frac{dm}{dx'dy'dz'} \Rightarrow dm = \rho(\vec{r}')dv' = \rho(x', y', z')dx'dy'dz' \tag{7.61}$$

is the volume mass density in Cartesian coordinates.

B. **The total mass:** The total mass, M, in a volume V with mass density, $\rho(\vec{r}')$ is given by

$$M = \iiint_V \rho(\vec{r}')dx'dy'dz'. \tag{7.62}$$

C. **Center of mass:** The center of mass (gravity) for a given mass with uniform or non-uniform distribution is a point where the mass can be at rotational equilibrium about an axis passing through this point. The center of mass, in Cartesian coordinates, $(\bar{x}, \bar{y}, \bar{z})$ can be determined using the integrals

$$\bar{x} \int dm = \int x dm, \ \bar{y} \int dm = \int y dm, \ \text{and} \ \bar{z} \int dm = \int z dm. \tag{7.63}$$

For a mass distributed on a volume with volume mass density

$$\rho(x, y, z) = \frac{dm}{dv} = \frac{dm}{dxdydz} \Rightarrow dm = \rho(x, y, z)dxdydz \tag{7.64}$$

the center of mass can be expressed

$$\bar{x} \;=\; \frac{1}{M} \iiint x\rho(x,\, y,\, z)dxdydz, \quad \bar{y} \;=\; \frac{1}{M} \iiint y\rho(x,\, y,\, z)dxdydz,$$

$$\bar{z} \;=\; \frac{1}{M} \iiint z\rho(x,\, y,\, z)dxdydz, \tag{7.65}$$

where the total mass M is given by

$$M = \iiint \rho(x,\, y,\, z)dxdydz.$$

If the mass is distributed along a line on the x-axis with linear mass density, we have

$$\lambda(x) = \frac{dm}{dx} \;\Rightarrow\; dm = \lambda(x)dx, \tag{7.66}$$

and the center of mass becomes

$$\bar{x} = \frac{1}{M} \int \lambda(x)xdx, \quad \bar{y} = 0, \;\; \text{and } \bar{z} = 0, \tag{7.67}$$

where

$$M = \int \lambda(x)dx.$$

Similarly, for surface mass distribution on the x-y plane, with surface mass density

$$\sigma(x,\, y) = \frac{dm}{da} = \frac{dm}{dxdy} \;\Rightarrow\; dm = \sigma(x,\, y)dxdy, \tag{7.68}$$

the center of mass is given by

$$\bar{x} = \frac{1}{M} \int\!\!\int x\sigma(x,\, y)dxdy, \quad \bar{y} = \frac{1}{M} \int\!\!\int y\sigma(x,\, y)dxdy, \;\; \text{and } \bar{z} = 0, \tag{7.69}$$

where

$$M = \int\!\!\int \sigma(x,\, y)dxdy$$

The following examples illustrate the application of single and multiple integrations to determine some of the quantities described above.

Example 7.3. A thin rod lies along the positive x-axis with one end at the origin and the other end at $(L,0,0)$. It has a linear charge density given by $\lambda(x) = Ax$, where A is a constant (figure 7.8).
    (a) Find an expression for A if the total charge on the rod is Q.
    (b) Point P is located at $(-b,0,0)$ from the origin along the negative x-axis. Find an expression for the electrostatic potential at P.

Solution:
    (a) The total charge distributed on the thin rod can be written as

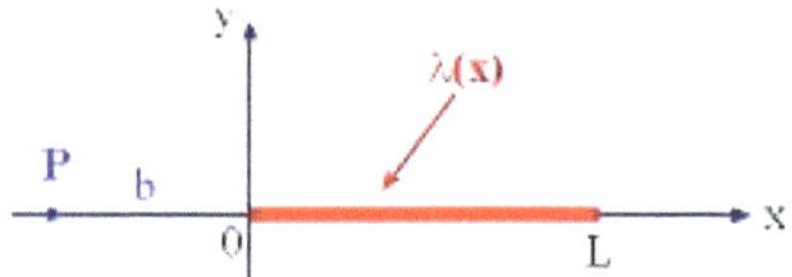

Figure 7.8. A thin rod of length L on the x-axis.

$$Q = \int_{Line} \lambda(\vec{r}')dr' = \int_0^L Axdx = \frac{AL^2}{2} \Rightarrow A = \frac{2Q}{L^2}. \tag{7.70}$$

(b) For a line charge distribution, the potential is given by

$$V(r) = \frac{1}{4\pi\epsilon_0} \int_0^L \frac{\lambda(\vec{r}')}{|\vec{r} - \vec{r}'|}dr'. \tag{7.71}$$

Noting that

$$\begin{aligned}
\vec{r} &= -b\hat{x}, \\
\vec{r}' &= x'\hat{x} \Rightarrow r' = x' \Rightarrow dr' = dx', \\
|\vec{r} - \vec{r}'| &= |-b\hat{x} - x'\hat{x}| = b + x',
\end{aligned} \tag{7.72}$$

and using the charge density $\lambda(\vec{r}') = Ax'$, the potential becomes

$$\begin{aligned}
V(b) &= \frac{A}{4\pi\epsilon_0} \int_0^L \frac{x'}{b + x'}dx' = \frac{A}{4\pi\epsilon_0} \int_0^L \left[1 - \frac{b}{b + x'}\right]dx' \\
&= \frac{A}{4\pi\epsilon_0}\left[\int_0^L dx' - b\int_0^L \frac{dx'}{b + x'}\right] = \frac{A}{4\pi\epsilon_0}[L - b\ln(b + x)]\Big|_{x=0}^{x=L} \\
\Rightarrow V(r) &= \frac{A}{4\pi\epsilon_0}\left[L - b\ln\left(\frac{b + L}{b}\right)\right].
\end{aligned} \tag{7.73}$$

Using the result in part b for A, we may express the electrostatic potential as

$$V(b) = \frac{Q}{2\pi\epsilon_0 L}\left[1 - \frac{b}{L}\ln\left(\frac{b + L}{b}\right)\right]. \tag{7.74}$$

Example 7.4. Consider a semi-circular plate of radius R on the upper-half of the x–y plane with its center of curvature at the origin (see figure 7.9). The plate has a surface charge density given by $\sigma(x, y) = Ax^2y$, where A is a constant.

(a) What is the total charge on the plate in terms of A?
(b) Re-do part (a), switching the order of integration.

Solution:
(a) The total charge is given by the surface integral

$$Q = \int_{\text{Area}} \sigma(\vec{r}')da' = \int_{x_1'}^{x_2'} \left[\int_{y_1(x')}^{y'(x')} \sigma(x', y')dy' \right]dx'$$

$$= \int_{x_1'}^{x_2'} Ax'^2 \left[\int_{y_1(x')}^{y'(x')} y'dy' \right]dx' \tag{7.75}$$

where we replaced $\sigma(x', y') = Ax'^2 y'$. Using the equation for a circle

$$x'^2 + y'^2 = R^2 \tag{7.76}$$

for the semi-circle in figure 7.9, we can write

$$y'(x') = \sqrt{R^2 - x'^2}. \tag{7.77}$$

For the region shown in the figure, the limits for the integrations are: $x_1' = -R$ to $x_2' = R$ and $y_1'(x') = 0$ to $y_2'(x') = \sqrt{R^2 - x'^2}$. Then the integral in equation (7.75) becomes

$$Q = A \int_{-R}^{R} \left[x'^2 \left(\int_0^{\sqrt{R^2-x'^2}} y'dy' \right) \right]dx' = A \int_{-R}^{R} x'^2 \left[\frac{y'^2}{2} \Big|_0^{\sqrt{R^2-x'^2}} \right]dx',$$

$$= \frac{A}{2} \int_{-R}^{R} [R^2 x'^2 - x'^4]dx' = \frac{A}{2} \left[\frac{R^2 x'^3}{3} - \frac{x'^5}{5} \Big|_{-R}^{R} \right] \tag{7.78}$$

$$\Rightarrow Q = \frac{2AR^5}{15}.$$

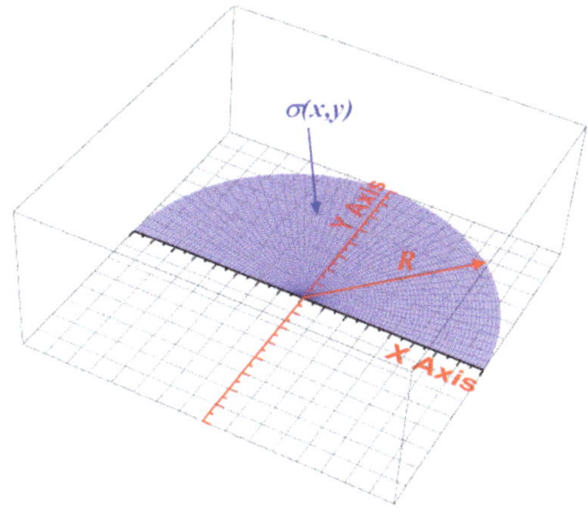

Figure 7.9. A semi-circular plate of radius R with a surface charge density, $\sigma(x, y) = Ax^2 y$.

(b) If we switch the order of integration, we have

$$Q = \int_{\text{Area}} \sigma(\vec{r}\,')da' = \int_{y_1}^{y_2} \int_{x_1(y')}^{x_2(y')} [\sigma(x', y')dy']dx'$$

$$= A \int_{y_1}^{y_2} y' \left[\int_{x_1(y')}^{x_2(y')} x'^2 dx' \right] dy'. \tag{7.79}$$

Referring to figure 7.9, the limits of integrations become: $y_1 = 0$ to $y_2 = R$ and $x_1(y') = -\sqrt{R^2 - y'^2}$ to $x_2(y') = \sqrt{R^2 - y'^2}$. This leads to

$$Q = A \int_0^R y' \left[\int_{-\sqrt{R^2-y'^2}}^{\sqrt{R^2-y'^2}} x'^2 dx' \right] dy' = A \int_0^R y' \left[\frac{x'^3}{3} \Big|_{-\sqrt{R^2-y'^2}}^{\sqrt{R^2-y'^2}} \right] dy' \tag{7.80}$$

$$\Rightarrow Q = A \int_0^R \frac{2(R^2 - y'^2)^{\frac{3}{2}}}{3} y' dy'.$$

The above integration can be carried out using transformation of variables defined by

$$u = (R^2 - y'^2) \Rightarrow -\frac{du}{2} = y' dy'. \tag{7.81}$$

Using these relation, one can write

$$\int \left[\frac{2(R^2 - y'^2)^{\frac{3}{2}}}{3} \right] y' dy' = -\int \frac{u^{\frac{3}{2}}}{3} du = -\frac{2}{15} u^{\frac{5}{2}}$$

$$\Rightarrow \int \left[\frac{2(R^2 - y'^2)^{\frac{3}{2}}}{3} \right] y' dy' = -\frac{2}{15} (R^2 - y^2)^{\frac{5}{2}}. \tag{7.82}$$

Therefore, the charge becomes

$$Q = A \int_0^R \frac{2\left(R^2 - y'^2\right)^{\frac{3}{2}}}{3} y' dy' = -\frac{2A}{15} (R^2 - y^2)^{\frac{5}{2}} \Big|_0^R \tag{7.83}$$

$$\Rightarrow Q = \frac{2AR^5}{15}.$$

Example 7.5. A cylinder has a height H, a radius R, and a mass M distributed uniformly throughout its volume (see figure 7.10). Find the moment of inertia of the cylinder about its axis of symmetry.

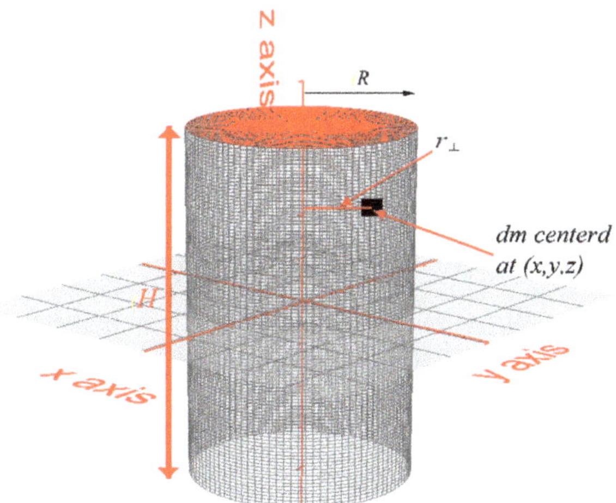

Figure 7.10. A cylinder with radius R and height H. It has a mass M distributed uniformly.

Solution: Since the mass is distributed uniformly in the cylinder, the volume density, $\rho(r)$, is a constant

$$\rho(r) = \frac{dm}{dv} = \frac{M}{V} = \frac{M}{\pi R^2 H}, \tag{7.84}$$

where we used the volume of a cylinder $V = \pi R^2 H$. The axis of symmetry is the z-axis that passes through the center of the cylinder (see figure 7.10). Thus using equation (7.60), the moment of inertia about the z-axis is given by the triple integral (in Cartesian coordinates)

$$I = \int_M dm\, r_\perp^2 = \iiint_V \rho(r) r_\perp^2 dx\,dy\,dz = \frac{M}{\pi R^2 H} \iiint_V r^2 dx\,dy\,dz. \tag{7.85}$$

Noting that $r_\perp$ is the perpendicular distance of the mass, dm, (in the volume, $dv = dx\,dy\,dz$, centered at a point with coordinates (x, y, z)) from the axis of symmetry (z-axis), we can write

$$r_\perp = \sqrt{x^2 + y^2}, \tag{7.86}$$

where

$$0 \leqslant r_\perp^2 \leqslant R. \tag{7.87}$$

Furthermore, since the cylinder, with height H, is centered about the z-axis,

$$-\frac{H}{2} \leqslant z \leqslant \frac{H}{2}. \tag{7.88}$$

Thus the moment of inertia in equation (7.85) becomes

$$I = \frac{M}{\pi R^2 H} \int_{z_1=-H/2}^{z_2=H/2} \left[\int_{y_1=-R}^{y_2=R} \left[\int_{x_1=-\sqrt{R^2-y^2}}^{x_2=\sqrt{R^2-y^2}} (x^2 + y^2) dx \right] dy \right] dz. \tag{7.89}$$

Since the integrand is independent of z we can carry out integration with respect to z followed by integration with respect to y. This leads to

$$\begin{aligned}
I &= \frac{M}{\pi R^2 H} \left[\int_{-R}^{R} \left(\frac{x^3}{3} + y^2 x \right) \Bigg|_{-\sqrt{R^2-y^2}}^{\sqrt{R^2-y^2}} dy \right] H \\
&= \frac{M}{\pi R^2} \int_{-R}^{R} 2 \left(\frac{R^2 + 2y^2}{3} \right) \sqrt{R^2 - y^2} \, dy.
\end{aligned} \tag{7.90}$$

Introducing the transformation of variable defined by

$$y = R \sin(\theta) \Rightarrow dy = R \cos(\theta) d\theta, \tag{7.91}$$

we may write

$$\begin{aligned}
I &= \frac{M}{\pi} \int_{-\pi/2}^{\pi/2} 2 \left(\frac{R^2 + 2R^2 \sin^2(\theta)}{3} \right) \cos^2(\theta) d\theta \\
&= \frac{2MR^2}{3\pi} \int_{-\pi/2}^{\pi/2} [\cos^2(\theta) + 2\sin^2(\theta)\cos^2(\theta)] d\theta.
\end{aligned} \tag{7.92}$$

Note that we have re-expressed the limits of integration in terms of θ (i.e., $-R \leqslant x \leqslant R \Rightarrow -\pi/2 \leqslant \theta \leqslant \pi/2$). Using the result obtained by Mathematica (which you can also easily show),

$$\begin{aligned}
\int_{-\pi/2}^{\pi/2} \mathrm{Cos}[\theta]^2 \, d\theta &= \frac{\pi}{2}, \\
\int_{-\pi/2}^{\pi/2} \mathrm{Cos}[\theta]^2 \mathrm{Sin}[\theta]^2 d\theta &= \frac{\pi}{8},
\end{aligned} \tag{7.93}$$

the integral becomes

$$\int_{-\pi/2}^{\pi/2} [\cos^2(\theta) + 2\sin^2(\theta)\cos^2(\theta)] = \frac{3\pi}{4}. \tag{7.94}$$

The moment of inertia is then found to be

$$I = \frac{MR^2}{2}. \tag{7.95}$$

Example 7.6. Consider a wire of constant linear mass density λ bent into the arc of a circle of radius R, as shown in figure 7.11.

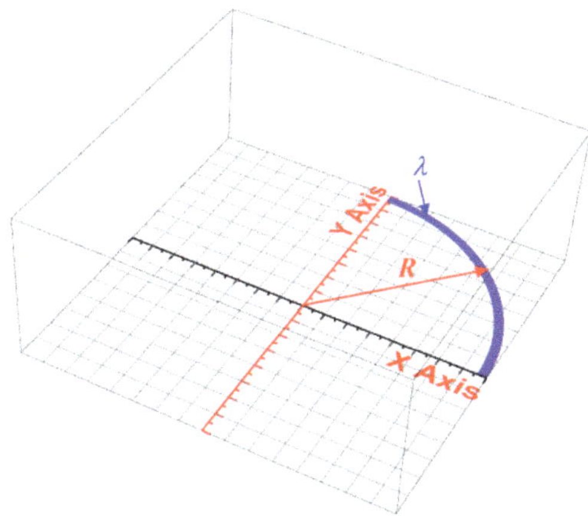

Figure 7.11. A wire of constant linear mass density λ bent into the arc of a circle of radius R.

(a) What is the total mass of the wire?
(b) Find the wire's center of mass.
(c) Find the area under the arc of the wire.
(d) The wire is now rotated about the x-axis. Find the volume inside the surface thus generated.
(e) Find the surface area of this curved surface of rotation.

Solution:

(a) The total mass of the wire is given by

$$M = \int dm. \tag{7.96}$$

For uniformly distributed linear mass over curved length the linear mass density $\lambda(l)$ is given by

$$\lambda(l) = \frac{dm}{dl} = \lambda = \text{constant} \Rightarrow dm = \lambda\,dl, \tag{7.97}$$

the total mass becomes

$$M = \lambda \int dl = \lambda l. \tag{7.98}$$

Using the result in Example 1(b) for the arc length of the curve

$$l = \frac{R\pi}{2}, \tag{7.99}$$

the total mass becomes

$$M = \frac{\lambda R\pi}{2}. \tag{7.100}$$

(b) The mass is distributed uniformly on a curved line on the x–y plane. Thus coordinates of the center of mass are given by

$$\bar{x} = \frac{1}{M} \int x\,dm, \quad \bar{y} = \frac{1}{M} \int y\,dm, \quad \text{and } \bar{z} = \frac{1}{M} \int z\,dm = 0. \tag{7.101}$$

where

$$M = \int dm.$$

From the result in (a)

$$M = \int dm = \frac{\lambda R\pi}{2}, \tag{7.102}$$

one can rewrite the x- and y-coordinates of the center of mass as

$$\bar{x} = \frac{\displaystyle\int x\,dm}{M} = \frac{2}{\lambda R\pi} \int x\,dm = \frac{2}{R\pi} \int x\,dl,$$

$$\bar{y} = \frac{\displaystyle\int y\,dm}{\displaystyle\int dm} = \frac{2}{\lambda R\pi} \int y\,dm = \frac{2}{R\pi} \int y\,dl, \tag{7.103}$$

where we replaced $dm = \lambda\,dl$ and take into account that λ is constant. From equation (7.12), we recall

$$dl = \sqrt{\left[\left(\frac{dy}{dx}\right)^2 + 1\right]}\,dx = \sqrt{1 + \left(\frac{dx}{dy}\right)^2}\,dy \tag{7.104}$$

so that equation (7.103) becomes

$$\bar{x} = \frac{2}{R\pi} \int_0^R x\sqrt{\left[\left(\frac{dy}{dx}\right)^2 + 1\right]}\,dx,$$

$$\bar{y} = \frac{2}{R\pi} \int_0^R y\sqrt{1 + \left(\frac{dx}{dy}\right)^2}\,dy. \tag{7.105}$$

The curve is defined by the equation of a circle

$$x^2 + y^2 = R^2 \Rightarrow y = \sqrt{R^2 - y^2} \text{ or } x = \sqrt{R^2 - y^2}, \tag{7.106}$$

we chose the positive roots for both x and y since in the first quadrant both x and y are positives. Using the expressions for x and y, we have

$$\frac{dx}{dy} = -\frac{y}{\sqrt{R^2 - y^2}}, \quad \frac{dy}{dx} = -\frac{x}{\sqrt{R^2 - x^2}}, \tag{7.107}$$

so that substituting these equations into equation (7.105), we find

$$\bar{x} = \frac{2}{R\pi} \int_0^R x \sqrt{\left[\left(\frac{x}{\sqrt{R^2 - x^2}}\right)^2 + 1\right]}\, dx = \frac{2}{\pi} \int_0^R \frac{x\, dx}{\sqrt{R^2 - x^2}} x\, dx,$$

$$\bar{y} = \frac{2}{R\pi} \int_0^R y \sqrt{\left[\left(\frac{y}{\sqrt{R^2 - y^2}}\right)^2 + 1\right]}\, dx = \frac{2}{\pi} \int_0^R \frac{y\, dy}{\sqrt{R^2 - y^2}}.$$

$$(7.108)$$

Upon carrying out the integrations, one can show that and one finds

$$\bar{x} = \frac{2}{\pi}\left(-\sqrt{R^2 - x^2}\right)\Big|_0^R,\ \bar{y} = \frac{2}{\pi}\left(-\sqrt{R^2 - y^2}\right)\Big|_0^R,$$

$$\Rightarrow \bar{x} = \bar{y} = \frac{2R}{\pi}.$$

$$(7.109)$$

Therefore the coordinates of the center of mass is $\left(\frac{2R}{\pi}, \frac{2R}{\pi}, 0\right)$.

(c) The area can be determined using

$$A = \int_{x_1}^{x_2}\left[\int_{y_1(x)}^{y_2(x)} dy\right] dx, \tag{7.110}$$

or

$$A = \int_{x_1}^{x_2}\left[\int_{x_1(y)}^{x_2(y)} dx\right] dy. \tag{7.111}$$

Here we chose to use equation (7.110). For the boundaries shown in figure 7.11, using the limits of integrations

$$x_1 = 0,\ x_2 = R,\ y_1(x) = 0,\ y_2(x) = \sqrt{R^2 - x^2}, \tag{7.112}$$

the area becomes

$$A = \int_0^R\left[\int_0^{\sqrt{R^2 - x^2}} dy\right] dx = \int_0^R \sqrt{R^2 - x^2}\, dx. \tag{7.113}$$

Introducing the transformation of variable defined by

$$x = R\sin(\theta) \Rightarrow \begin{cases} dx = R\cos(\theta)d\theta \\ x = 0 \Rightarrow \theta = 0, \\ x = R \Rightarrow \theta = \frac{\pi}{2} \end{cases}, \tag{7.114}$$

where we used $\theta = \sin^{-1}(x/R)$, equation (7.113) can be rewritten as

$$A = \int_0^R \sqrt{R^2 - R^2 \sin^2(\theta)}\, R \cos(\theta) d\theta = R^2 \int_0^{\pi/2} \cos^2(\theta) d\theta$$

$$= R^2 \int_0^{\pi/2} \left[\frac{\cos(2\theta) + 1}{2} \right] d\theta = \left. \frac{\sin(2\theta)}{4} + \frac{\theta}{2} \right|_0^{\pi/2} \tag{7.115}$$

$$\Rightarrow A = \frac{\pi R^2}{4},$$

where we used the relation

$$\cos(2\theta) = \cos^2(\theta) - \sin^2(\theta) \Rightarrow \cos^2(\theta) = \frac{\cos(2\theta) + 1}{2}. \tag{7.116}$$

(d) The rotation of the wire about the x-axis creates a hemisphere of radius R, shown in figure 7.12. A point on the surface of the sphere with coordinates (x, y, z) is defined by the equation of a sphere

$$x^2 + y^2 + z^2 = R^2 \Rightarrow x = \sqrt{R^2 - (y^2 + z^2)}, \tag{7.117}$$

and you can determine the volume by evaluating the integral in equation (7.24), which takes the form for the hemisphere in figure 7.12,

$$V = \int_{-R}^R \left[\int_{-\sqrt{R^2-x^2}}^{\sqrt{R^2-x^2}} \left(\int_0^{\sqrt{R^2-(y^2+z^2)}} dx \right) dy \right] dz. \tag{7.118}$$

However, here we will use a shortcut. Let the infinitesimal volume dv be a cylinder of radius y and thickness (i.e., height) dx so that we can write

$$dv = \pi y^2 dx \tag{7.119}$$

where y can be related to the radius of the sphere by

$$y = \sqrt{R^2 - x^2}. \tag{7.120}$$

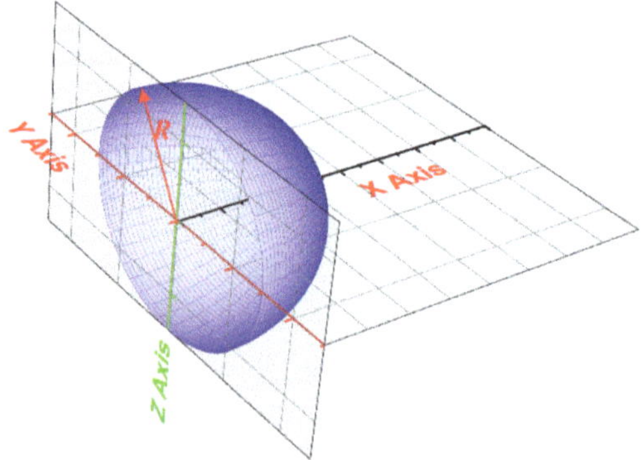

Figure 7.12. A rotation of an arc on the x–y plane rotated about the x-axis forms a hemisphere.

The volume will then be

$$V = \int_V dv = \int_0^R \pi y^2 dx = \pi \int_0^R (R^2 - x^2)dx = \pi\left(R^3 - \frac{R^3}{3}\right)$$
$$\Rightarrow V = \frac{2\pi R^3}{3}.$$

(7.121)

In the following sections, we will consider the methods of integrations in curvilinear coordinates. We first consider one-dimensional and two-dimensional curvilinear coordinates. For three-dimensional case, we consider only cylindrical and spherical coordinates. These sections will demonstrate the simplicity of carrying out the integrations in curvilinear instead of Cartesian coordinates for curved geometries by re-considering some of the problems we have already solved.

7.3 1-D and 2-D curvilinear coordinates

Let us consider a point (in pink) along a curved path on the x–y plane shown in figure 7.13. The coordinates of this point are (x, y) in Cartesian and (r, θ) in polar coordinates. These two system of coordinates are related by

$$x = r \cos \theta, \, y = r \sin (\theta).$$

(7.122)

We recall the infinitesimal arc length dl on this curved path in Cartesian coordinates given by

$$dl = \sqrt{dx^2 + dy^2}.$$

(7.123)

Using the relations in equation (7.122), we can rewrite dl as

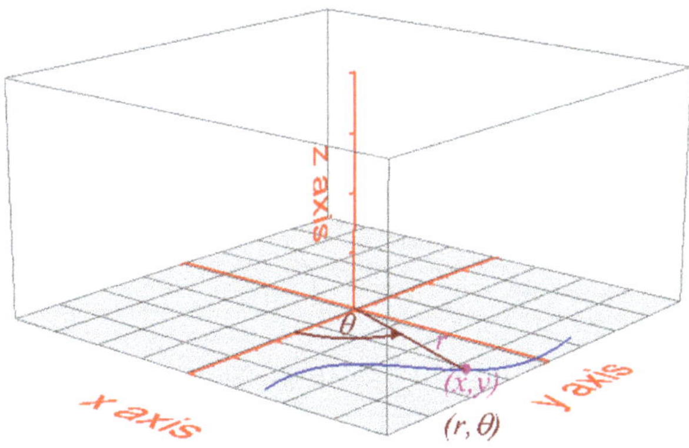

Figure 7.13. A point on a curved path on the x–y plane described by Cartesian (x, y) and polar (r, θ) coordinates.

$$
\begin{aligned}
dl &= \sqrt{[d(r\cos\theta)]^2 + [d(r\sin(\theta))]^2} \\
&= \sqrt{[\cos\theta\, dr - r\sin(\theta)d\theta]^2 + [\sin(\theta)dr + r\cos\theta\, d\theta]^2} \\
&= \sqrt{(\cos^2\theta + \sin^2\theta)dr^2 + r^2(\sin^2\theta + \cos^2\theta)d\theta^2} \\
\Rightarrow dl &= \sqrt{dr^2 + r^2 d\theta^2}.
\end{aligned}
\tag{7.124}
$$

By factoring dr or $rd\theta$, one can establish the relations

$$
ds = \sqrt{1 + \left(r\frac{d\theta}{dr}\right)^2}\, dr = \sqrt{\left(\frac{1}{r}\frac{dr}{d\theta}\right)^2 + 1}\, rd\theta.
\tag{7.125}
$$

Similarly, for the infinitesimal area da, shown in figure 7.14, one can write

$$
da = dxdy = rdrd\theta,
\tag{7.126}
$$

where we replaced Δr and $\Delta\theta$ with dr and $d\theta$ for the infinitesimal area da.

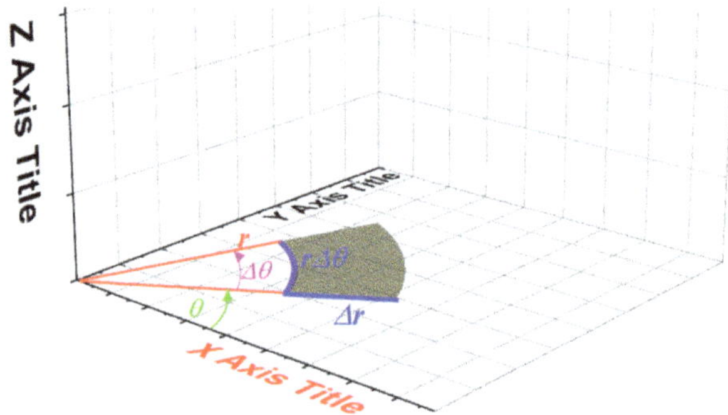

Figure 7.14. An infinitesimal area on the x–y plane in polar coordinates (r, θ).

Example 7.7. Let us reconsider the wire of constant linear mass density λ bent into the arc of a circle of radius R, as shown in figure 7.15. Using polar coordinates, find
    (a) the total mass of the wire,
    (b) the area under the arc of the wire.

Solution:
    (a) The total mass of the wire is given by

$$
M = \int dm = \int \lambda ds.
\tag{7.127}
$$

In polar coordinates this can be expressed as

$$
M = \int_0^{\pi/2} \lambda \sqrt{\left(\frac{1}{r}\frac{dr}{d\theta}\right)^2 + 1}\, rd\theta.
\tag{7.128}
$$

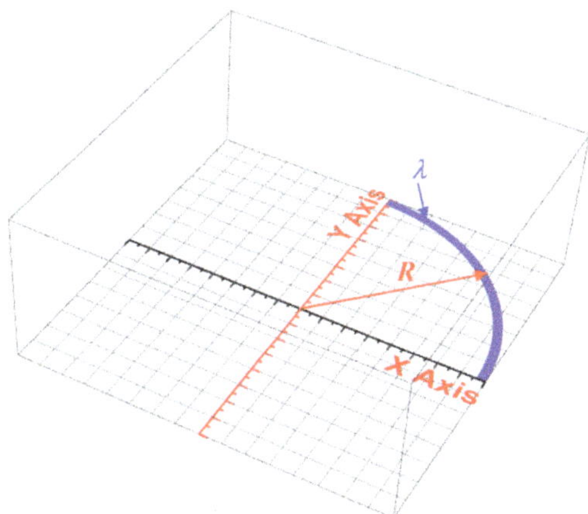

Figure 7.15. A wire of constant linear mass density λ bent into the arc of a circle of radius R.

Referring to figure 7.15, we have

$$r = R \Rightarrow \frac{dr}{d\theta} = 0, \tag{7.129}$$

so that the total mass becomes

$$M = \int_0^{\pi/2} \lambda R d\theta = \frac{\lambda R \pi}{2}. \tag{7.130}$$

(b) The area in polar coordinates is given by

$$A = \int da = \int \int r dr d\theta = \int_0^{\pi/2} \int_0^R r dr d\theta = \frac{\pi R^2}{4}. \tag{7.131}$$

Example 7.8. Consider a semi-circular plate of radius R on the upper-half of the x–y plane with its center of curvature at the origin (see figure 7.16). The plate has a surface charge density given by $\sigma(x, y) = Ax^2 y$, where A is a constant. Determine the total charge on the plate using polar coordinates. (*Note: this is the same as Example 7.4.*)

Solution: In polar coordinates, the total charge is expressible as

$$Q = \int_{Area} \sigma(\vec{r}) da = \int \int \sigma(\vec{r}) r dr d\theta. \tag{7.132}$$

Using the charge density, in polar coordinates

$$\sigma(\vec{r}) = Ax^2 y = Ar^3 \cos^2(\theta) \sin(\theta) \tag{7.133}$$

one finds

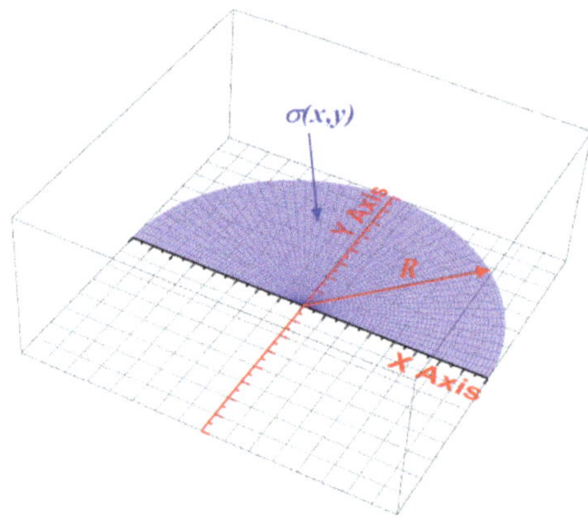

Figure 7.16. A semi-circular plate of radius R with a surface charge density, $\sigma(x, y) = Ax^2y$.

$$Q = A \int_0^\pi \int_0^R r^3 \cos^2(\theta) \sin(\theta) r dr d\theta = A\left[\int_0^R r^4 dr\right]\left[\int_0^\pi \cos^2(\theta) \sin(\theta) d\theta\right] \quad (7.134)$$

Noting that

$$\int_0^R r^4 dr = \left[\frac{r^5}{5}\right]_0^R = \frac{R^5}{5}, \quad (7.135)$$

and

$$u = \cos(\theta) \Rightarrow \begin{cases} -du = \sin(\theta)d\theta \\ \theta = 0 \Rightarrow u = 1, \\ \theta = \pi \Rightarrow u = -1 \end{cases} \quad (7.136)$$

$$\Rightarrow \int_0^\pi \cos^2(\theta) \sin(\theta) d\theta = -\int_1^{-1} u^2 du = -\left[\frac{u^3}{3}\right]_1^{-1} = \frac{2}{3},$$

one finds for equation (7.134)

$$Q = \frac{2AR^5}{15}, \quad (7.137)$$

which is the same as the result we found in Example 7.4.

7.4 3-D curvilinear coordinates: *cylindrical*

Let us consider a point in three-dimensional space. We describe this point by the Cartesian coordinates (x, y, z) and cylindrical coordinates (s, φ, z) as shown in figure 7.17. The coordinates (x, y) and (s, φ) are related by

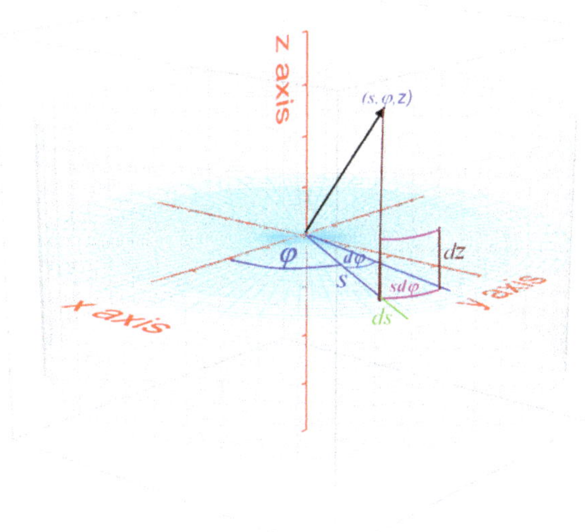

Figure 7.17. A point in space described by spherical coordinates, s, φ, and z.

$$x = s\cos(\varphi), \quad y = s\sin(\varphi). \tag{7.138}$$

In cylindrical coordinates, an infinitesimal volume dv with length ds, width $sd\varphi$, and height dz can be expressed as

$$dv = ds(sd\varphi)dz = s\,ds\,d\varphi\,dz. \tag{7.139}$$

Infinitesimal area da on the curved surface of the cylinder, with height dz and width $sd\varphi$, given by

$$da = dz(sd\varphi) = s\,dz\,d\varphi, \tag{7.140}$$

whereas on the top or bottom plane surface of the cylinder is

$$da = (ds)(sd\varphi) = s\,ds\,d\varphi. \tag{7.141}$$

Example 7.9. (*This is the same problem considered in Example 7.5*) A right-circular cylinder has a height H, a radius R, and a mass M distributed uniformly throughout its volume. Find the moment of inertia of the cylinder about its axis of symmetry using cylindrical coordinates. For the sake of convenience we have re-displayed the cylinder in figure 7.10 below in figure 7.18.

Solution: We recall the moment of inertia is given by

$$I = \int_{\text{Volume}} r_\perp^2 \, dm = \int_{\text{Volume}} r_\perp^2 \rho(\vec{r}) \, dv \tag{7.142}$$

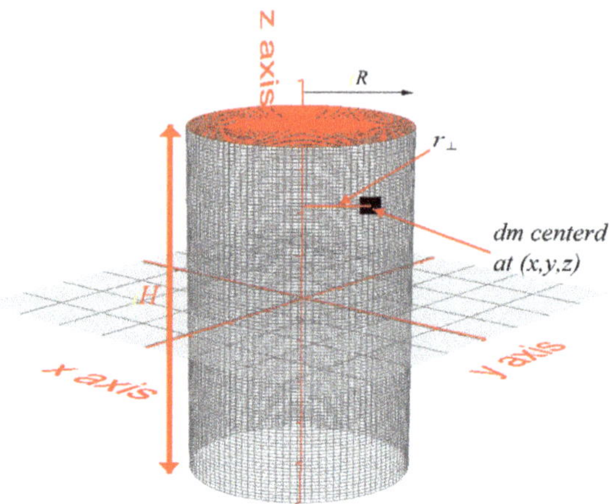

Figure 7.18. A cylinder with radius R and height H. It has a mass M distributed uniformly.

where $r_\perp$ is the perpendicular distance of the infinitesimal mass dm from the axis of rotation which is the z-axis. The mass is distributed uniformly and therefore the density is constant. Using

$$\rho(\vec{r}) = \rho = \frac{M}{\pi R^2 H}, \tag{7.143}$$

and in cylindrical coordinates

$$r_\perp = \sqrt{x^2 + y^2}, \, = s, \, dv = s\,ds\,d\varphi\,dz, \tag{7.144}$$

one can show that the moment of inertia of the cylinder is,

$$\begin{aligned}
I &= \rho \int\int\int r_\perp^2 s\,ds\,d\varphi\,dz = \frac{M}{\pi R^2 H} \int_{-H/2}^{H/2} \int_0^R \int_0^{2\pi} s^2 s\,ds\,d\varphi\,dz \\
&= \frac{M}{\pi R^2 H}\left[\int_{-H/2}^{H/2} dz\right]\left[\int_0^R s^3 ds\right]\left[\int_0^{2\pi} d\varphi\right] \Rightarrow I = \frac{1}{2} M R^2.
\end{aligned} \tag{7.145}$$

Example 7.10. Using cylindrical coordinates prove the volume of a right-circular cone of base radius R and height H shown in figure 7.19 to be

$$V = \frac{\pi R^2 H}{3}. \tag{7.146}$$

Solution: In cylindrical coordinates an infinitesimal volume dv is given by

$$dv = s\,ds\,d\varphi\,dz \Rightarrow V = \int\int\int s\,ds\,d\varphi\,dz. \tag{7.147}$$

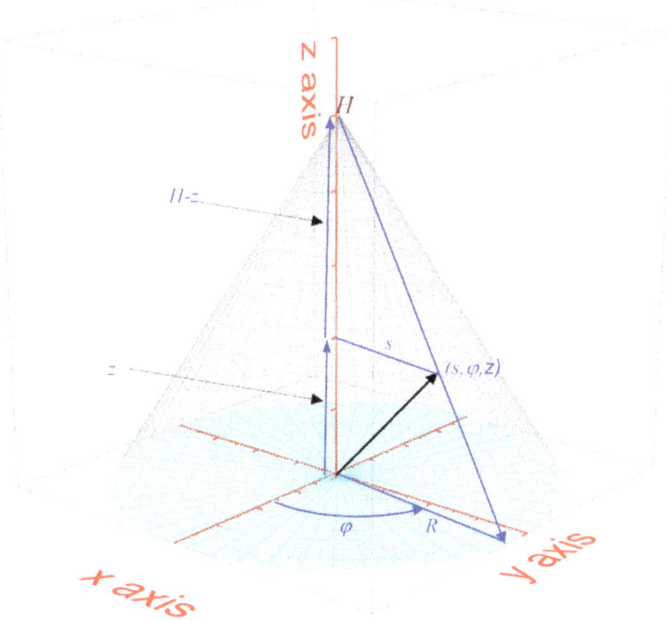

Figure 7.19. A right-circular cone of base radius R and height H.

Using the properties of two similar triangles, from figure 7.19, one can establish the relation

$$\frac{s}{H - z} = \frac{R}{H} \Rightarrow s = \frac{(H - z)R}{H}. \tag{7.148}$$

As we cover the entire volume of the cone, the variables s, φ, and z varies as

$$0 < \varphi < 2\pi, \ 0 < s < \frac{(H - z)R}{H}, \ 0 < z < H, \tag{7.149}$$

and the volume can be written as

$$V = \int_0^{2\pi} \int_0^H \int_0^{\frac{(H-z)R}{H}} s\,ds\,dz\,d\varphi = \left[\int_0^{2\pi} d\varphi\right] \int_0^H \left[\int_0^{\frac{(H-z)R}{H}} s\,ds\right] dz. \tag{7.150}$$

Upon integrating with respect to φ and s, one finds

$$V = 2\pi \int_0^H \frac{s^2}{2} \Big|_0^{\frac{(H-z)R}{H}} dz = \frac{\pi R^2}{H^2} \int_0^H (H^2 - 2Hz + z^2)\,dz$$

$$\Rightarrow V = \frac{\pi R^2}{H^2}\left[H^2 z - Hz^2 + \frac{z^3}{3}\Big|_0^H\right] = \frac{\pi R^2 H}{3}. \tag{7.151}$$

7.5 3-D curvilinear coordinate: *spherical*

Let us reconsider a point in three-dimensional space, but this time, we describe this point by the Cartesian coordinates (x, y, z) and spherical coordinates (r, θ, φ) as shown in figure 7.20, which are related by

$$x = r \cos(\varphi) \sin(\theta), \ y = r \sin(\varphi) \sin(\theta), \ z = r \cos(\theta). \tag{7.152}$$

Suppose we walk on the surface of the sphere shown in figure 7.20 along $\hat{\theta}$ for an infinitesimal displacement $rd\theta$ and then along $\hat{\varphi}$ for an infinitesimal displacement $r \sin(\theta)d\varphi$. Then the infinitesimal area da of the region with length $rd\theta$ and width $r \sin(\theta)d\varphi$ can be expressed as

$$da = (rd\theta)(r \sin(\theta)d\varphi) = r^2 \sin(\theta)d\theta d\varphi. \tag{7.153}$$

If now we continue walking for another infinitesimal displacement dr, but this time in the radial direction $\hat{r}$. Then one can express the volume of the infinitesimal 3-D space with base area da and height dr as

$$dv = (da)dr = r^2 \sin(\theta)d\theta d\varphi dr. \tag{7.154}$$

The equations in (7.153) and (7.154) are infinitesimal area and volume in spherical coordinates. We will see how we will apply these equations for problems with spherical symmetry in the following examples.

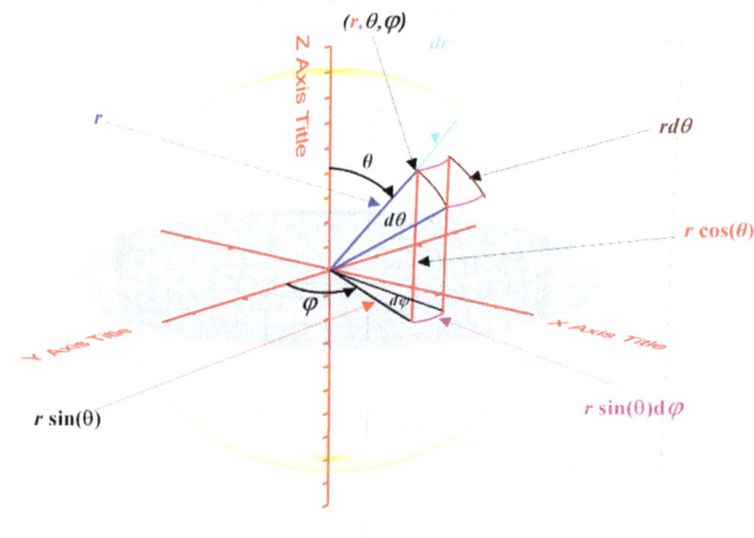

Figure 7.20. Infinitesimal displacements in spherical coordinates.

Example 7.11. A thin spherical dielectric shell of radius R (see figure 7.21) has a surface charge density given by

$$\sigma(\theta, \varphi) = \sigma_0 \sin(\theta) \cos\left(\frac{\varphi}{4}\right) \tag{7.155}$$

where σ_0 is a constant. Find the total charge.

Solution: In view of the relation in equation (7.153), an infinitesimal area on the surface of a sphere with radius R can be expressed as

$$da = R^2 \sin(\theta) d\theta d\varphi. \tag{7.156}$$

Thus the total surface charge given by the surface integral

$$Q = \int \int \sigma(\theta, \varphi) da, \tag{7.157}$$

becomes

$$Q = \int \int \sigma_0 \sin(\theta) \cos\left(\frac{\varphi}{4}\right) R^2 \sin(\theta) d\theta d\varphi.$$

$$= \sigma_0 R^2 \int_0^{2\pi} \int_0^{\pi} \sin^2(\theta) \cos\left(\frac{\varphi}{4}\right) d\theta d\varphi \tag{7.158}$$

$$= \sigma_0 R^2 \left(\int_0^{\pi} \sin^2(\theta) d\theta \right) \left(\int_0^{2\pi} \cos\left(\frac{\varphi}{4}\right) d\varphi \right),$$

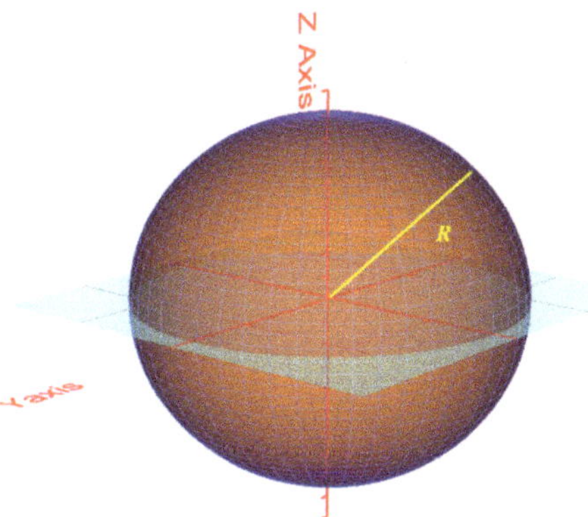

Figure 7.21. A spherical shell with surface charge density, $\sigma(\theta, \varphi) = \sigma_0 \sin(\theta) \cos\left(\frac{\varphi}{4}\right)$.

where we used for a full sphere the limits of integration for θ is $(0 \leqslant \theta \leqslant \pi)$ and for φ is $(0 \leqslant \varphi \leqslant 2\pi)$. Using the results for the integrals

$$\int_0^\pi \sin^2(\theta)d\theta = \frac{\theta}{2} - \frac{\sin 2\theta}{4} \bigg|_0^\pi = \frac{\pi}{2},$$

$$\int_0^{2\pi} \cos\left(\frac{\varphi}{4}\right)d\varphi = 4\sin\frac{\varphi}{4} \bigg|_0^{2\pi} = 4,$$

(7.159)

one finds for total charge,

$$Q = 2\pi\sigma_0 R^2.$$

(7.160)

Example 7.12. A zone of a sphere is defined to be that section of the sphere that lies between two parallel planes intersecting the sphere (see figure 7.22). Consider two planes separated by a distance h that intersect a sphere of radius R. Prove this somewhat surprising result (surprising because this area does not depend on where we slice the sphere—toward the equator, closer to the north, or south of the sphere!)

Solution: We recall that an infinitesimal area da on the surface of a sphere with radius R is given by

$$da = R^2 \sin(\theta)d\theta d\varphi.$$

(7.161)

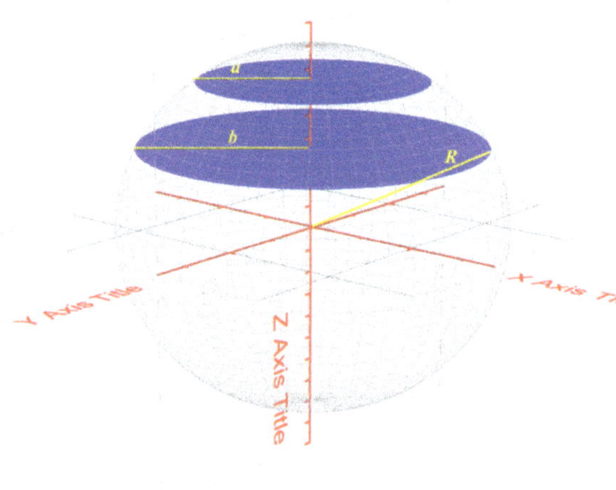

Figure 7.22. A zone of a sphere is defined to be that section of the sphere that lies between two parallel planes intersecting the sphere.

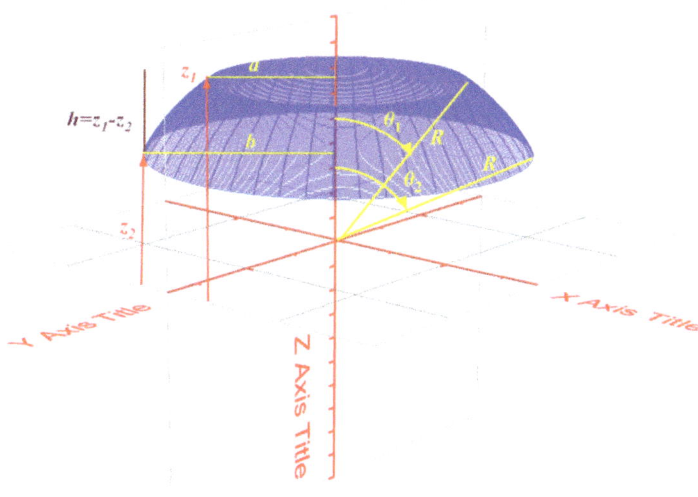

Figure 7.23. The surface of a zone.

For the surface area of the zone, one can write

$$A = \int_0^{2\pi} \int_{\theta_1}^{\theta_2} R^2 \sin(\theta)\, d\theta\, d\varphi = -2\pi R^2 \cos(\theta)\, \big|_{\theta_1}^{\theta_2} \tag{7.162}$$
$$= 2\pi R^2 [\cos(\theta_1) - \cos(\theta_2)],$$

where the limits of integrations θ_1 and θ_2 represent the angular positions of the two planes forming the zone as shown in figure 7.23. Suppose in Cartesian coordinates, the upper plane is defined by $z = z_1$ and the bottom by $z = z_2$, we can easily see that

$$\cos(\theta_1) = \frac{z_1}{R}, \ \cos(\theta_2) = \frac{z_2}{R}. \tag{7.163}$$
$$\Rightarrow R[\cos(\theta_1) - \cos(\theta_2)] = z_1 - z_2 = h,$$

where h is the height of the zone. Thus the surface area of the zone becomes

$$A = 2\pi R h, \tag{7.164}$$

which is indeed independent of the position of the two planes forming the zone. Surprising indeed! The volume sliced out near the equator is greater than the volume sliced out near one of the poles. The two slices, however, have the same lateral surface areas. Moreover, if we replace h by $2R$, the result gives the total surface area for a sphere. This is the case when one of the planes is tangent to the sphere at the north pole and the other tangent at the south pole.

7.6 Scalar integrals and Mathematica

These are the basic commands for carrying out integrations to scalar function. These integrations could be used for single or multiple integrations.

- Integrate $[f, x]$ gives the indefinite integral.
- Integrate $[f, \{x, x_{\min}, x_{\max}\}]$ gives the definite integral.
- Integrate $\left[f, \{x, x_{\min}, x_{\max}\}, \{y, y_{\min}, y_{\max}\}, \ldots\right]$ gives the multiple integral.
- Integrate $[f, \{x, y, \ldots\} \in reg]$ integrates over the geometric region *reg*.

In the following, we will see how we integrate some of the examples we considered.

- Example 7.1

$$I = \text{Integrate}$$

$$\left[R \Big/ \sqrt{R^2 - x^2}, \{x, 0, R\}, \text{ Assumptions} \to R > 0,\right.$$

$$\left.\text{Assumptions} \to x < R\right]$$

$$= \frac{\pi R}{2}$$

$$I = \text{Integrate}$$

$$\left[1, \{x, 0, R\}, \left\{y, 0, \sqrt{R^2 - x^2}\right\}, \text{ Assumptions} \to R > 0,\right.$$

$$\left.\text{Assumptions} \to x < R\right]$$

$$= \frac{\pi R^2}{4}$$

- Example 7.2

$$A = \text{Integrate}[1, \{x, 0, 1\}, \{y, 0, -2x + 2\}, \{z, 0, 2 + x + y\}] = 3$$

- Example 7.4

$$Q = A \text{ Integrate}$$

$$\left[x^2 y, \{x, -R, R\}, \left\{y, 0, \sqrt{R^2 - x^2}\right\},\right.$$

$$\left.\text{Assumptions} \to R > 0, \text{ Assumptions} \to x < R\right]$$

$$= \frac{2AR^5}{15}$$

- Example 7.8

$$Q = A \text{ Integrate}[r^4 \text{Cos}[\theta]^2 \text{Sin}[\theta], \{r, 0, R\}, \{\theta, 0, \pi\}, \text{ Assumptions} \to r > 0]$$

$$= \frac{2AR^5}{15}$$

- Example 7.10

$$V = \text{Integrate}[[s, \{z, 0, H\}, \{\phi, 0, 2\pi\}, \{s, 0, (H - z)R/H\},$$

$$\text{Assumptions} \to H > 0, \text{ Assumptions} \to R > 0]$$

$$= \frac{1}{3} H \pi R^2$$

7.7 Homework assignment

Problem 1. Consider an area on the x–y plane bounded by the x-axis, the line defined by the equation, $x = 2$, and a curve defined by the equation, $y = \sqrt{x}$, as shown in figure 7.24. *Using integration in Cartesian coordinates*, find
    (a) the length of the curved side of the area,
    (b) the area.

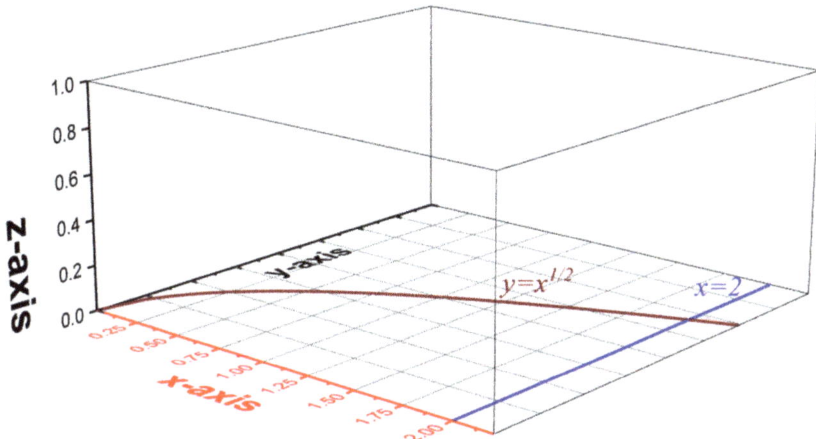

Figure 7.24. An area bounded by the x-axis, the line $x = 2$, and the curve $y = \sqrt{x}$.

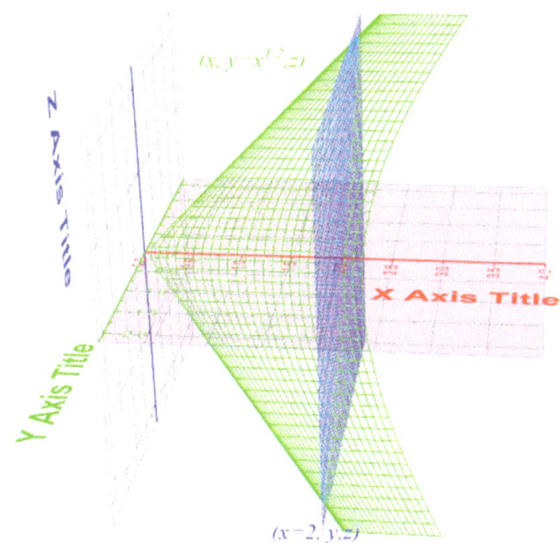

Figure 7.25. The area in the previous problem rotated about the z-axis.

Problem 2. The area in the previous problem is rotated about the x-axis to form a volume as shown in figure 7.25. *Using integration in Cartesian coordinates,* find the surface area for
    (a) the curved surface,
    (b) the plane surface.

Problem 3. For an area (A) bounded by the x-axis $(y = 0)$, $y = \sin(x)$, and $x = \pi/2$ evaluate the integral

$$I = \int\int_A 6y^2 \cos x \, dx dy. \tag{7.165}$$

Problem 4. For the area (A) bounded by the parabola $y = x^2$ and the straight line $2x - y + 8 = 0$, evaluate the integral

$$I = \int\int_A x \, dx dy.$$

Problem 5. Using double integrals determine the volume bounded above the square plane formed by the vertices $(0, 0)$, $(2, 0)$, $(0, 2)$, and $(2, 2)$ and below the plane $z = 8 - x + y$.

Problem 6. A rope of length, $L = 10$ cm, is stretched along the positive y-axis with one end at the origin. The linear mass density, $\lambda(y)$, over this rope increases uniformly from 4 g cm^{-1} at its one end at the origin to 24 g cm^{-1} at the other end. Find
    (a) the expression of the linear mass density in terms of y,
    (b) the total mass of the rope,
    (c) the center of mass,
    (d) the moment of inertia about an axis passing through the center of mass,
    (e) the moment of inertia about an axis passing through the heavy end.

Problem 7. A solid disk with radius, $r = a$, and negligible thickness is sitting on the x–y plane centered at the origin. It has a uniform surface mass density, σ_0. *Using integration in Cartesian coordinates,* find
    (a) the circumference of the circle,
    (b) the area of the disk,
    (c) the centroid (or center of mass) for the portion of *the disk* in the first quadrant,
    (d) the moment of inertia of the disk about an axis along the diameter of the disk,
    (e) the centroid for the portion of *the side* of the disk in the first quadrant,

Problem 8. The surface charge density σ on the area bounded by the x-axis $(y = 0)$ and curve $(y = 16 - x^2)$ is proportional to y. That means

$$\sigma(y) = ky.$$

Find the total charge Q on the area (figure 7.26).

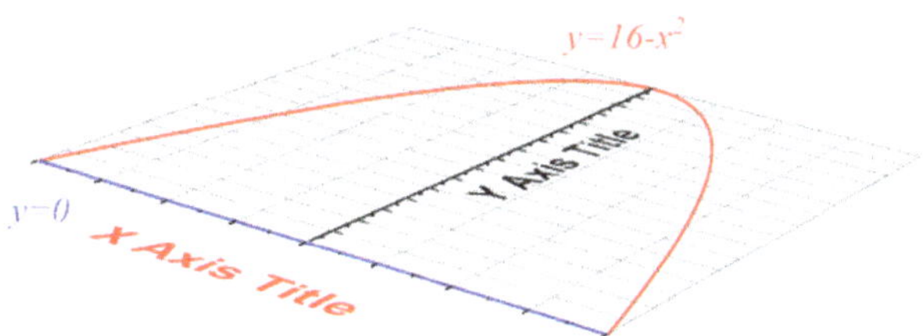

Figure 7.26. The region bounded by the x-axis ($y = 0$) and curve ($y = 16 - x^2$).

Problem 9. A light with an intensity, $I_0 = 42$ J m^{-2}, is incident on a square silvered mirror sitting on the x–y plane. The square has side length 2 m and is centered about the origin. The fraction of the incident light that is reflected at a point on the mirror (x, y) is $(x - y)^2/4$. In other words, the intensity of the reflected light at the point is

$$I_R(x, y) = I_0(x - y)^2/4.$$

Find the fraction of the reflected light from the entire mirror.

Problem 10. Using double integrals determine the volume formed below the surface defined by $z = y(x + 2)$ and above the surface bounded by $x + y = 0$, $y = 1$, and $y = \sqrt{x}$.

Problem 11. Repeat Problem 7 using polar coordinates.

Problem 12. Consider the volume of the space bounded by the curved surface of the cylinder defined $x^2 + y^2 = 4$, $z = 2x^2 + y^2$, and the x–y plane. Find the volume of the cylinder using integrations in cylindrical coordinates (figure 7.27).

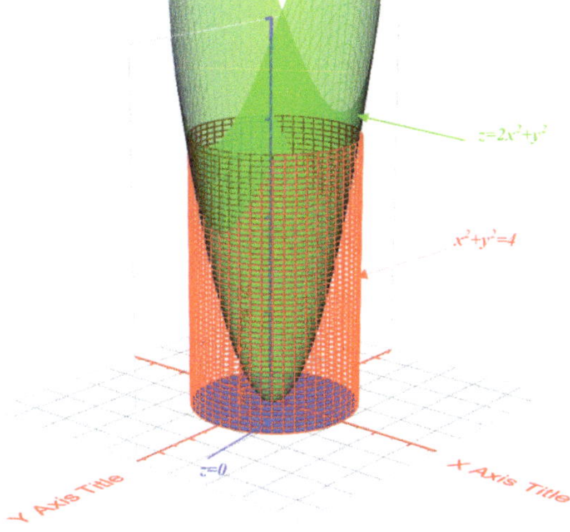

Figure 7.27. The curved surface of the cylinder defined $x^2 + y^2 = 4$, $z = 2x^2 + y^2$, and the x–y plane.

Problem 13. Consider a solid sphere of radius, $r = a$ and mass, M. The mass is distributed uniformly over the volume of the sphere. Using integrations in spherical coordinates, find

    (a) the centroid for the northern hemisphere of the ball,
    (b) the moment of inertia of the sphere about an axis passing through the diameter (along the z-axis),
    (c) the volume of the sphere.

Problem 14. In Problem 3 we considered a solid sphere of radius $r = a$ and mass M that is distributed uniformly over the volume of the sphere. Now let's imagine the sphere is hollow and the same mass is distributed over the surface of the sphere. Using integrations in spherical coordinates, find

    (a) the surface area of the sphere,
    (b) the centroid of the curved area of the upper hemisphere,
    (c) the moment of inertia of the of the sphere about an axis passing through the diameter (along the z-axis).

Problem 15. Consider a sphere with constant mass density ρ_0 and radius $r = 2a$. Determine the gravitational attraction force on a unit mass ($m = 1$) at the center of the sphere due to the portion of the mass of the sphere occupying the volume above the plane defined by $z = a$ (figure 7.28).

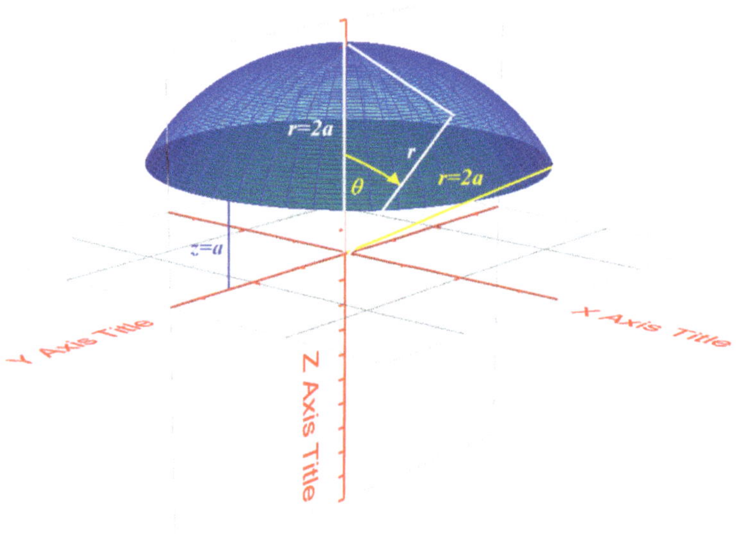

Figure 7.28. Portion of a sphere with radius $r = 2a$ occupying the volume above the plane defined by $z = a$.

Problem 16. Using the expression

$$l = \int_{y_1}^{y_2} \sqrt{1 + \left(\frac{dx}{dy}\right)^2}\, dy, \tag{7.166}$$

show that you will find the same result as in Example 7.1(b).

IOP Publishing

Studies in Theoretical Physics, Volume 1
Fundamental mathematical methods
Daniel Erenso and Victor Montemayor

Chapter 8

Vector calculus

The last two chapters introduced us to scalar functions' differential and integral calculus. This chapter will present differential and integral calculus for vector functions. We begin with a review of vector algebra involving scalar and vector product of two vectors, followed by an introduction to triple scalar and vector multiplications. We then study how to produce a vector function from a scalar function by introducing the gradient operator applying the differential of a scalar function of three variables and the properties for the scalar product for two vectors. Following this, we study how to determine the divergence of a vector, the curl of a vector, and the Laplacian for a scalar (or a vector) function using the gradient operator in Cartesian, cylindrical, and spherical coordinate systems. The relationship between a vector function and its curl with a line and surface integrals under Green's and Stokes' theorems is introduced and applied to various physical problems. We then introduce the divergence theorem that relates volume and surface integrals of vector functions and studies its applications in solving physical problems. Finally, we present the basic commands in Mathematica for vector calculus and resolve some examples that we already considered involving gradient, divergence, curl, and the Laplacian.

8.1 Review of vector products

In chapter 3, we saw that two vectors

$$\vec{A} = A_x\hat{x} + A_y\hat{y} + A_z\hat{z}, \ \vec{B} = B_x\hat{x} + B_y\hat{y} + B_z\hat{z} \tag{8.1}$$

can be multiplied in two different ways: one resulting in a scalar and the other resulting in another vector. Let these vectors be the vectors shown in figure 8.1.

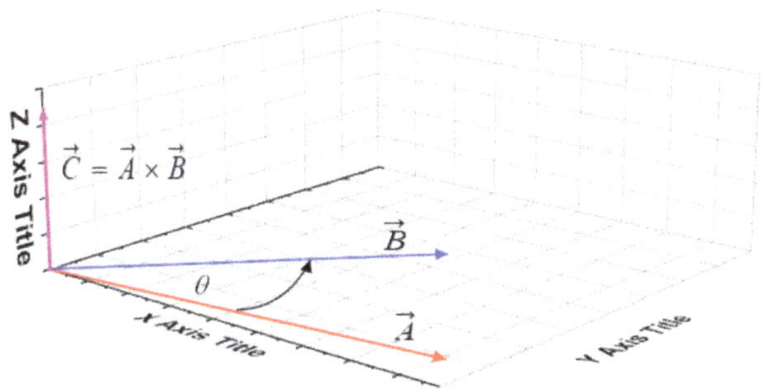

Figure 8.1. The cross (vector) product of two vectors.

Scalar product (dot product)

The scalar product of $\vec{A}$ and $\vec{B}$ is given by

$$\vec{A} \cdot \vec{B} = A_x B_x + A_y B_y + A_z B_z = |\vec{A}| |\vec{B}| \cos\theta, \tag{8.2}$$

where θ is the angle between the two vectors (see figure 8.1).

Cross product (vector product)

The vector product of $\vec{A}$ and $\vec{B}$ results in another vector $\vec{C}$ normal to the plane formed by the vectors $\vec{A}$ and $\vec{B}$ as shown in figure 8.1 and is given by

$$\vec{C} = \vec{A} \times \vec{B} = \begin{vmatrix} \hat{x} & \hat{y} & \hat{z} \\ A_x & A_y & A_z \\ B_x & B_y & B_z \end{vmatrix} \tag{8.3}$$

$$\Rightarrow \vec{C} = (A_y B_z - A_z B_y)\hat{x} + (A_z B_x - A_x B_z)\hat{y} + (A_x B_y - A_y B_x)\hat{z}.$$

When the angle between the two vectors is θ, the magnitude of vector $\vec{C}$ is given by

$$|\vec{C}| = |\vec{A} \times \vec{B}| = |\vec{A}| |\vec{B}| \sin(\theta) \tag{8.4}$$

and the direction is determined using the right-hand rule (see section 3.3). Vector $\vec{C}$ is always perpendicular to the plane formed by the two vectors.

Triple scalar and vector products

Consider the three vectors $\vec{A}$, $\vec{B}$, and $\vec{C}$ in Cartesian coordinates

$$\vec{A} = A_x \hat{x} + A_y \hat{y} + A_z \hat{z}, \quad \vec{B} = B_x \hat{x} + B_y \hat{y} + B_z \hat{z}, \tag{8.5}$$

$$\vec{C} = C_x \hat{x} + C_y \hat{y} + C_z \hat{z}.$$

We can multiply these vectors in two different ways: one leads to a scalar known as triple scalar product, and the other results in another vector known as triple vector product. The scalar triple product is given by

$$\vec{C} \cdot \left(\vec{A} \times \vec{B}\right) = \begin{vmatrix} C_x & C_y & C_z \\ A_x & A_y & A_z \\ B_x & B_y & B_z \end{vmatrix}. \tag{8.6}$$

We recall that exchanging two rows changes the sign of the determinant. Applying this property, one can write

$$\begin{vmatrix} C_x & C_y & C_z \\ A_x & A_y & A_z \\ B_x & B_y & B_z \end{vmatrix} = - \begin{vmatrix} B_x & B_y & B_z \\ A_x & A_y & A_z \\ C_x & C_y & C_z \end{vmatrix} = \begin{vmatrix} A_x & A_y & A_z \\ B_x & B_y & B_z \\ C_x & C_y & C_z \end{vmatrix}, \tag{8.7}$$

so that in view of the expression for the triple scalar product, on can write

$$\vec{C} \cdot \left(\vec{A} \times \vec{B}\right) = -\vec{B} \cdot \left(\vec{A} \times \vec{C}\right) = \vec{A} \cdot \left(\vec{B} \times \vec{C}\right). \tag{8.8}$$

Taking into account the relation

$$\vec{A} \times \vec{C} = -\vec{C} \times \vec{A}, \tag{8.9}$$

one can establish the relation

$$\vec{C} \cdot \left(\vec{A} \times \vec{B}\right) = \vec{B} \cdot \left(\vec{C} \times \vec{A}\right) = \vec{A} \cdot \left(\vec{B} \times \vec{C}\right). \tag{8.10}$$

The triple vector product of the three vectors is given by

$$\vec{A} \times \left(\vec{B} \times \vec{C}\right) = \vec{B}\left(\vec{A} \cdot \vec{C}\right) - \vec{C}\left(\vec{A} \cdot \vec{B}\right). \tag{8.11}$$

The easy way to memorize this expression for the triple vector product is to read the phrase '*BAC-CAB.*'

8.2 Vectors product physical applications

Work done

Consider a constant force $\vec{F}$ applied on an object. This force displaces the object from a position $\vec{r}_1$ to a position $\vec{r}_2$ for a displacement

$$\vec{d} = \Delta \vec{r} = \vec{r}_2 - \vec{r}_1, \tag{8.12}$$

(see figure 8.2) the work done by this force W is the scalar product of the force vector $\vec{F}$ and the displacement vector $\vec{d}$,

$$\begin{aligned} W = |\vec{F}| |\vec{d}| \cos(\theta) = \vec{F} \cdot \vec{d} = \vec{F} \cdot (\vec{r}_2 - \vec{r}_1) \\ \Rightarrow W = \vec{F} \cdot \Delta \vec{r}. \end{aligned} \tag{8.13}$$

For non-constant force that depends on the position of the object $\vec{F} = \vec{F}(\vec{r})$, the work done is determined from an integral form of the scalar product for the force and displacement vectors given by

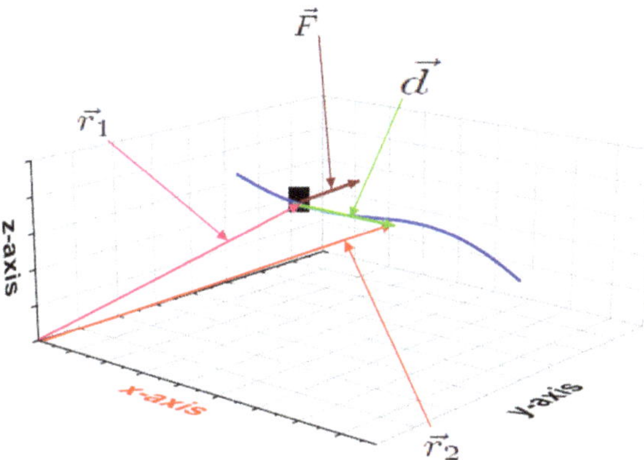

Figure 8.2. An object displaced from an initial position $\vec{r}_1$ to a new position $\vec{r}_2$ by a constant force $\vec{F}$.

$$W = \int_{r_1}^{r_2} \vec{F} \cdot d\vec{r}.$$ (8.14)

Angular and linear dynamics

Consider an object under a rotational motion. Suppose the object is rotating with a constant angular velocity $\vec{\omega}$, about some axis of rotation. Over a time interval $\Delta t = t_2 - t_1$, the position of the object has changed from $\vec{r}_1$ to $\vec{r}_2$ and the linear displacement becomes $\vec{d} = \Delta\vec{r} = \vec{r}_2 - \vec{r}_1$. The linear velocity $\vec{v}$ of the object which is a constant for a constant angular velocity, $\vec{\omega}$, can be determined using the vector product of the two vectors expressed as

$$\vec{v} = \vec{\omega} \times \vec{d} = \vec{\omega} \times (\vec{r}_2 - \vec{r}_1) = \vec{\omega} \times \Delta\vec{r}.$$ (8.15)

However, when the angular velocity is not a constant and depends on the relative position of the object, $\vec{\omega} = \vec{\omega}(\vec{r})$, the average linear velocity is determined using the line integral,

$$\vec{v} = \int_{\vec{r}_1}^{\vec{r}_2} \vec{\omega} \times d\vec{r}.$$ (8.16)

Angular momentum: Suppose the object for which we described the linear velocity, $\vec{v}$, and constant angular velocity, $\vec{\omega}$ has a mass, m (figure 8.3). This mass has a linear momentum, $\vec{p} = m\vec{v}$. The angular momentum, $\vec{J}$, of this object, is given by the vector product,

$$\vec{J} = \vec{r} \times \vec{p}.$$ (8.17)

When the linear momentum changes with the position of the object, the average angular momentum becomes

$$\vec{J} = \int d\vec{r} \times \vec{p}.$$ (8.18)

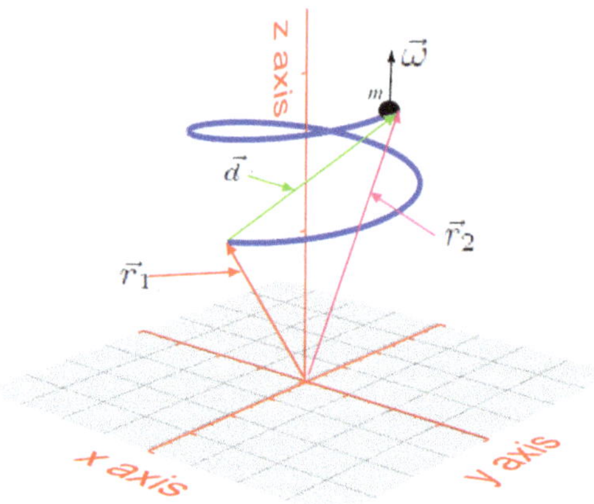

Figure 8.3. A mass m rotating about the z-axis in a spiral path with constant angular velocity $\vec{\omega}$. The initially position $\vec{r}_1$ has changed to new position $\vec{r}_2$.

Torque: The mechanical torque, $\vec{\tau}$, is a physical vector quantity that we often use to describe the rotational motion of an object acted by force, $\vec{F}$. Suppose the force acted on the object has rotated the object about an axis such that the position relative to the axis has changed from $\vec{r}_1$ to $\vec{r}_2$ for a total displacement, $\vec{d} = \Delta\vec{r} = \vec{r}_2 - \vec{r}_1$. For a constant force, the torque about the axis of rotation is given by the vector product,

$$\vec{\tau} = \vec{d} \times \vec{F}. \tag{8.19}$$

However, when the force is not constant and depends on the position of the object, the torque is given by the integral,

$$\vec{\tau} = \int_{\vec{r}_1}^{\vec{r}_2} d\vec{r} \times \vec{F}. \tag{8.20}$$

Replace the linear velocity $\vec{v} = \vec{\omega} \times \vec{r}$ in the expression for the linear momentum,

$$\vec{p} = m\vec{v} = m\vec{v} = m(\vec{\omega} \times \vec{r}), \tag{8.21}$$

the angular momentum can be expressed as a triple vector product

$$\vec{J} = m \int d\vec{r} \times (\vec{\omega} \times \vec{r}). \tag{8.22}$$

Magnetic field and force

Suppose an object with charge q moves with a constant velocity $\vec{v}$ in a region with a uniform magnetic field $\vec{B}$. The magnetic force experienced by the object is given by

$$\vec{F}_B = q\vec{v} \times \vec{B}. \tag{8.23}$$

For a non-uniform magnetic field $\vec{B}(\vec{r})$, such as the magnetic field due to a wire of length l that carries a current I where the magnetic field is given by

$$\vec{B}(r) = \frac{\mu_0 I}{4\pi}\left(\frac{\vec{l} \times \vec{r}}{r^3}\right), \tag{8.24}$$

the magnetic force can be expressed using a triple vector product

$$\vec{F}_B(\vec{r}) = \frac{q\mu_0 I}{4\pi}\left[\frac{\vec{v} \times (\vec{l} \times \vec{r})}{r^3}\right]. \tag{8.25}$$

Solid state physics

In solid state physics, we often see what is known as lattice vectors $\{\vec{a}_i\}$ and reciprocal lattice vectors $\{\vec{b}_i\}$. These vectors are related by

$$\vec{b}_i = \frac{\vec{a}_j \times \vec{a}_k}{\vec{a}_1 \cdot (\vec{a}_2 \times \vec{a}_3)}, \quad \text{where } (i, j, k) = (1, 2, 3). \tag{8.26}$$

The lattice vectors are used to describe crystal structure in solids. The reciprocal lattice vectors are useful to study waves in solids (lattice waves) known as phonons. These waves are described by the wave vector $\vec{k}$ expressed in terms of the reciprocal vectors.

8.3 Vectors derivatives

Next, we shall see how we carry out the derivatives of a vector function. To this end, let us consider the time-dependent scalar function $c(t)$, vector functions $\vec{A}(t)$ and $\vec{B}(t)$ that can be expressed in Cartesian coordinates as

$$\begin{aligned}
\vec{A}(t) &= A_x(t)\hat{x} + A_y(t)\hat{y} + A_z(t)\hat{z}, \\
\vec{B}(t) &= B_x(t)\hat{x} + B_y(t)\hat{y} + B_z(t)\hat{z}.
\end{aligned} \tag{8.27}$$

Using these vectors, we can establish the following derivatives with respect to time.

Derivative of a vector:

$$\frac{d\vec{A}}{dt} = \frac{dA_x}{dt}\hat{x} + \frac{dA_y}{dt}\hat{y} + \frac{dA_z}{dt}\hat{z}. \tag{8.28}$$

Derivative of a vector multiplied by a scalar that is also function of t:

$$\frac{d}{dt}\left(c\vec{A}\right) = \vec{A}\frac{dc}{dt} + \frac{d\vec{A}}{dt}c. \tag{8.29}$$

Derivative of scalar vector product:

$$\frac{d}{dt}\left(\vec{A} \cdot \vec{B}\right) = \vec{A} \cdot \frac{d\vec{B}}{dt} + \frac{d\vec{A}}{dt} \cdot \vec{B}. \tag{8.30}$$

Derivative of vector product:

$$\frac{d}{dt}\left(\vec{A} \times \vec{B}\right) = \vec{A} \times \frac{d\vec{B}}{dt} + \frac{d\vec{A}}{dt} \times \vec{B} = \vec{A} \times \frac{d\vec{B}}{dt} - \vec{B} \times \frac{d\vec{A}}{dt}. \qquad (8.31)$$

Example 8.1. The position of an object at time t is described by the two-dimensional vector $\vec{r}(t)$. This vector using polar coordinates (r, θ) can be expressed as

$$\vec{r}(t) = r(t) \cos [\theta(t)]\hat{x} + r(t) \sin [\theta(t)]\hat{y}. \qquad (8.32)$$

Find the general expressions for the velocity $\vec{v}(t)$ and the acceleration $\vec{a}(t)$ and express your answer in polar coordinates (figure 8.4).

Solution: The velocity $\vec{v}(t)$ is given by the derivative of the position vector

$$\vec{v}(t) = \frac{d\vec{r}}{dt} = \frac{d}{dt}\{r(t) \cos [\theta(t)]\}\hat{x} + \frac{d}{dt}\{r(t) \sin [\theta(t)]\}\hat{y}. \qquad (8.33)$$

Assuming the general case where both r and θ are time-dependent upon differentiating with respect to time, we find

$$\vec{v}(t) = \frac{dr}{dt} \cos (\theta)\hat{x} + r\frac{d \cos (\theta)}{dt}\hat{x} + \frac{dr}{dt} \sin (\theta)\hat{y} + r\frac{d \sin (\theta)}{dt}\hat{y}$$
$$\Rightarrow \vec{v}(t) = \frac{dr}{dt} \cos (\theta)\hat{x} - r \sin (\theta)\frac{d\theta}{dt}\hat{x} + \frac{dr}{dt} \sin (\theta)\hat{y} + r \cos (\theta)\frac{d\theta}{dt}\hat{y}, \qquad (8.34)$$

which can be put in the form

$$\vec{v}(t) = \frac{dr}{dt}(\cos (\theta)\hat{x} + \sin (\theta)\hat{y}) + \frac{d\theta}{dt}r(-\sin (\theta)\hat{x} + \cos (\theta)\hat{y}). \qquad (8.35)$$

Noting that in two-dimensional polar coordinates, the unit vector along the radial direction $\hat{r}$ (i.e., along $\vec{r}$) is

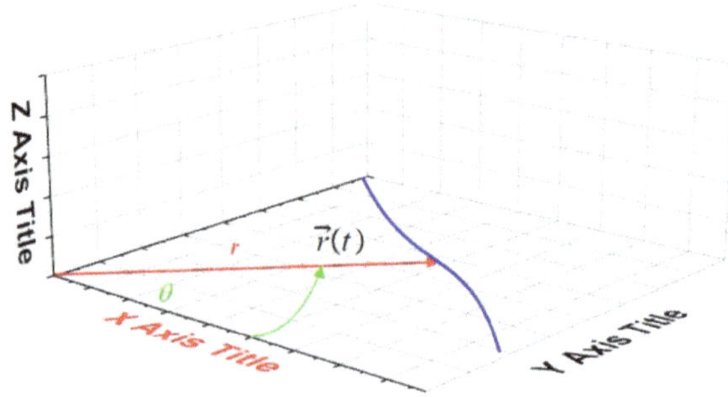

Figure 8.4. Curvilinear trajectory on the x–y plane.

$$\hat{r} = \frac{d\vec{r}}{dr} \bigg/ \left| \frac{d\vec{r}}{dr} \right| = \cos(\theta)\hat{x} + \sin(\theta)\hat{y}, \tag{8.36}$$

and along the angular direction $\hat{\theta}$ (i.e., along $\vec{\theta}$) is

$$\hat{\theta} = \frac{d\vec{r}}{d\theta} \bigg/ \left| \frac{d\vec{r}}{d\theta} \right| = -\sin(\theta)\hat{x} + \cos(\theta)\hat{y}; \tag{8.37}$$

one can express the velocity as

$$\vec{v}(t) = \frac{dr}{dt}\hat{r} + \frac{d\theta}{dt}r\hat{\theta} = \frac{dr}{dt}\hat{r} + r\omega\hat{\theta}, \tag{8.38}$$

where ω

$$\omega = \frac{d\theta}{dt} \tag{8.39}$$

is the angular speed. Introducing the radial speed v_r and tangential speed v_t, defined by

$$v_r = \frac{dr}{dt}, \; v_t = \omega r, \tag{8.40}$$

one can also write the velocity as

$$\vec{v}(t) = v_r\hat{r} + v_t\hat{\theta}. \tag{8.41}$$

When the magnitude of the radial displacement is constant (i.e., the motion is in a circular orbit), we have

$$v_r = \frac{dr}{dt} = 0, \tag{8.42}$$

and the velocity becomes

$$\vec{v}(t) = v_t(t)\hat{\theta} = \omega(t)r\hat{\theta}. \tag{8.43}$$

The acceleration $\vec{a}(t)$ is given by the time derivative of the instantaneous velocity,

$$\vec{a}(t) = \frac{d\vec{v}}{dt} = \frac{d}{dt}(v_r\hat{r} + v_t\hat{\theta})$$
$$\Rightarrow \vec{a}(t) = \frac{dv_r}{dt}\hat{r} + v_r\frac{d\hat{r}}{dt} + \frac{dv_t}{dt}\hat{\theta} + v_t\frac{d\hat{\theta}}{dt}. \tag{8.44}$$

As shown in the expressions, both unit vectors $\hat{r}$ and $\hat{\theta}$ depend on the angle θ that changes with time. Differentiating these unit vectors with respect to time, we have

$$\frac{d\hat{r}}{dt} = \frac{d}{dt}(\cos(\theta)\hat{x} + \sin(\theta)\hat{y}) = \frac{d\theta}{dt}(-\sin(\theta)\hat{x} + \cos(\theta)\hat{y})$$
$$= \frac{d\theta}{dt}\hat{\theta} \Rightarrow \frac{d\hat{r}}{dt} = \omega\hat{\theta}, \tag{8.45}$$

and

$$\frac{d\hat{\theta}}{dt} = \frac{d}{dt}(-\sin(\theta)\hat{x} + \cos(\theta)\hat{y}) = \frac{d\theta}{dt}(-\cos(\theta)\hat{x} - \sin(\theta)\hat{y})$$

$$= -\frac{d\theta}{dt}\hat{r} \Rightarrow \frac{d\hat{\theta}}{dt} = -\omega\hat{r},$$
(8.46)

so that the acceleration becomes

$$\vec{a}(t) = \frac{dv_r}{dt}\hat{r} + v_r\omega\hat{\theta} + \frac{dv_t}{dt}\hat{\theta} - v_t\omega\hat{r}$$

$$= \left(\frac{dv_r}{dt} - v_t\omega\right)\hat{r} + \left(v_r\omega + \frac{dv_t}{dt}\right)\hat{\theta}.$$
(8.47)

In terms of the magnitude of the acceleration in the radial a_r and tangential a_t directions,

$$a_r = \frac{dv_r}{dt}, \; a_t = \frac{dv_t}{dt},$$
(8.48)

we can rewrite the acceleration as

$$\vec{a}(t) = (a_r - v_t\omega)\hat{r} + (v_r\omega + a_t)\hat{\theta}.$$
(8.49)

Using

$$a_r = \frac{d^2r}{dt^2}, \; v_t = \omega r, \; v_r = \frac{dr}{dt}, \; \alpha = \frac{d\omega}{dt} = \frac{d^2\theta}{dt^2}$$

$$a_t = \frac{dv_t}{dt} = \frac{d(r\omega)}{dt} = \omega\frac{dr}{dt} + \frac{d\omega}{dt}r = \omega v_r + \alpha r,$$
(8.50)

the acceleration can also be put in the form

$$\vec{a}(t) = \left(\frac{d^2r}{dt^2} - r\omega^2\right)\hat{r} + \left(2\frac{dr}{dt}\omega + r\alpha\right)\hat{\theta}.$$
(8.51)

For a circular motion, where the radius is constant, $r = R$, we have

$$\frac{dr}{dt} = 0, \; \frac{d^2r}{dt^2} = 0,$$
(8.52)

the acceleration becomes

$$\vec{a}(t) = -R\omega^2\hat{r} + R\alpha\hat{\theta}.$$
(8.53)

Using

$$v_t = \omega R, \Rightarrow -R\omega^2\hat{r} = -\frac{v_t^2}{R}\hat{r} = \vec{a}_c,$$

$$R\alpha\hat{\theta} = R\frac{d\omega}{dt}\hat{\theta} = \vec{a}_{\parallel},$$
(8.54)

which are the centripetal (radial) $\vec{a}_c$ and tangential $\vec{a}_{\parallel}$ accelerations, equation (8.53) can be rewritten as.

$$\vec{a}(t) = \vec{a}_c + \vec{a}_{\parallel}, \tag{8.55}$$

which is the vector sum of the radial and tangential accelerations.

8.4 The gradient operator and directional derivative

In the previous section, we considered the derivative of scalar and vector functions of one variable, time t. In most problems in physics, we often encounter products of scalars and vectors that are functions of two or more variables. Let us consider the scalar function of three variables in Cartesian coordinates $f(x, y, z)$. We recall that the infinitesimal change in this function df (the total differential) is given by

$$df = \frac{\partial f}{\partial x}dx + \frac{\partial f}{\partial y}dy + \frac{\partial f}{\partial z}dz. \tag{8.56}$$

In view of the expression for scalar vector product of $\vec{A}$ and $\vec{B}$ in Cartesian coordinates

$$\vec{A} \cdot \vec{B} = A_x B_x + A_y B_y + A_z B_z, \tag{8.57}$$

one can rewrite df as

$$df = (dx\hat{x} + dy\hat{y} + dz\hat{z}) \cdot \left(\frac{\partial f}{\partial x}\hat{x} + \frac{\partial f}{\partial y}\hat{y} + \frac{\partial f}{\partial z}\hat{z}\right). \tag{8.58}$$

Noting that the differential for the position vector in Cartesian coordinates,

$$\vec{r} = x\hat{x} + y\hat{y} + z\hat{z}, \tag{8.59}$$

is given by

$$d\vec{r} = dx\hat{x} + dy\hat{y} + dz\hat{z}, \tag{8.60}$$

so that one rewrites df in the form

$$df = d\vec{r} \cdot \left(\frac{\partial f}{\partial x}\hat{x} + \frac{\partial f}{\partial y}\hat{y} + \frac{\partial f}{\partial z}\hat{z}\right) = d\vec{r} \cdot \left(\hat{x}\frac{\partial}{\partial x} + \hat{y}\frac{\partial}{\partial y} + \hat{z}\frac{\partial}{\partial z}\right)f$$
$$\Rightarrow df = \nabla f \cdot d\vec{r}, \tag{8.61}$$

where

$$\nabla = \hat{x}\frac{\partial}{\partial x} + \hat{y}\frac{\partial}{\partial y} + \hat{z}\frac{\partial}{\partial z}, \tag{8.62}$$

is known as *the gradient operator*. Note that when the gradient operator acts on a scalar function $f(x, y, z)$

$$\nabla f = \frac{\partial f}{\partial x}\hat{x} + \frac{\partial f}{\partial y}\hat{y} + \frac{\partial f}{\partial z}\hat{z} = \vec{V}(x, y, z), \tag{8.63}$$

it generates a vector function $\vec{V}(x, y, z)$.

The directional derivative
The function $f(x, y, z)$ defines a surface. Suppose this function is a constant $f(x, y, z) = C$, then the total differential for the function, df, becomes

$$df = \nabla f \cdot d\vec{r} = \vec{V} \cdot d\vec{r} = 0, \tag{8.64}$$

indicating that the scalar product of the vectors $\vec{V}(x, y, z) = \nabla f(x, y, z)$ and $d\vec{r}$ is zero and the two vectors are orthogonal at the point (x, y, z). Since

$$d\vec{r} = dx\hat{x} + dy\hat{y} + dz\hat{z}, \tag{8.65}$$

is an infinitesimal displacement on the surface defined by the function $f(x, y, z)$, the vector $d\vec{r}$ is tangent to the surface at the point (x, y, z). Therefore, $\nabla f(x, y, z)$ must be a vector normal to the surface at the point (x, y, z).

Now let's consider a point with the coordinates (x_0, y_0, z_0) on the surface defined by the function $f(x, y, z)$. We then moved a distance s and reach a point described by the coordinates (x, y, z), which could be on or off the surface (see figure 8.5). Let $\hat{n}$ be a unit vector along the line joining the two points as shown in figure 8.5. Using the parametric equation of a line, one can write

$$x(s) = x_0 + sn_x,\ y(s) = y_0 + sn_y,\ z(s) = z_0 + sn_z. \tag{8.66}$$

For the function $f(x(s), y(s), z(s))$, one can then write

$$\frac{df}{ds} = \frac{\partial f}{\partial x}\frac{dx}{ds} + \frac{\partial f}{\partial y}\frac{dy}{ds} + \frac{\partial f}{\partial z}\frac{dz}{ds}. \tag{8.67}$$

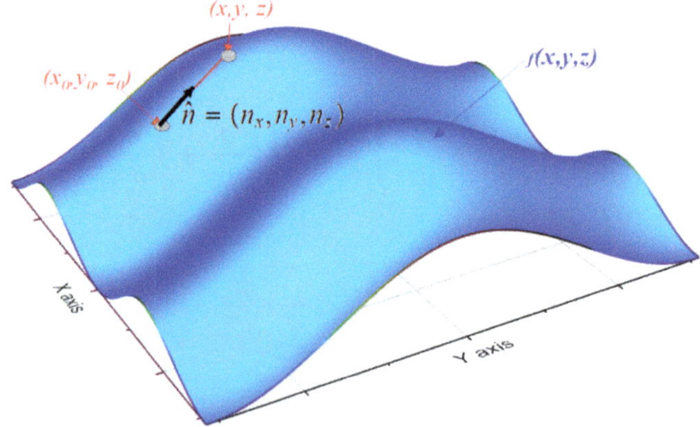

Figure 8.5. A surface defined by the function $f(x, y, z)$.

Using equation (8.66), we have

$$\frac{dx}{ds} = n_x, \frac{dy}{ds} = n_y, \frac{dz}{ds} = n_z,$$ (8.68)

so that equation (8.67) becomes,

$$\frac{df}{ds} = \frac{\partial f}{\partial x}n_x + \frac{\partial f}{\partial y}n_y + \frac{\partial f}{\partial z}n_z$$

$$= \left(\frac{\partial f}{\partial x}\hat{x} + \frac{\partial f}{\partial y}\hat{y} + \frac{\partial f}{\partial z}\hat{z}\right) \cdot \left(n_x\hat{x} + n_y\hat{y} + n_z\hat{z}\right)$$ (8.69)

which can be put in the form

$$\frac{df}{ds} = \nabla f \cdot \hat{n},$$ (8.70)

where $\hat{n}$ is the unit vector along the vector $\vec{s}$ joining the two points in figure 8.5. Equation (8.70) is the directional derivative, which describes how the function $f(x, y, z)$ changes along the direction $\hat{n}$.

The gradient operator in curvilinear coordinates
In chapter 7, we saw the two curvilinear coordinate systems, cylindrical and spherical coordinates systems. Using the equations relating these coordinate systems,

$$\vec{r} = x\hat{x}, +y\hat{y} + z\hat{z} \text{ (Cartesian)}$$
$$\vec{r} = s \cos{(\varphi)}\hat{x}, +s \sin{(\varphi)}\hat{y} + z\hat{z} \text{ (Cylindrical)},$$
$$\vec{r} = r \sin{(\theta)} \cos{(\varphi)}\hat{x} + r \sin{(\theta)} \sin{(\varphi)}\hat{y} +,$$
$$+ r \cos{(\theta)}\hat{z} \text{ (Spherical)},$$ (8.71)

we can write the unit vectors in cylindrical coordinates,

$$\hat{s} = \frac{d\vec{r}}{d\hat{s}} \bigg/ \left|\frac{d\vec{r}}{d\hat{s}}\right| = \cos{(\varphi)}\hat{x}, +\sin{(\varphi)}\hat{y},$$

$$\hat{\varphi} = \frac{d\vec{r}}{d\varphi} \bigg/ \left|\frac{d\vec{r}}{d\varphi}\right| = -s \sin{(\varphi)}\hat{x}, +s \cos{(\varphi)}\hat{y},$$ (8.72)

and in spherical coordinates,

$$\hat{r} = \frac{d\vec{r}}{dr} \Big/ \left| \frac{d\vec{r}}{dr} \right|$$
$$= \sin(\theta)\cos(\varphi)\hat{x} + \sin(\theta)\sin(\varphi)\hat{y} + \cos(\theta)\hat{z},$$
$$\hat{\theta} = \frac{d\vec{r}}{d\theta} \Big/ \left| \frac{d\vec{r}}{d\theta} \right|$$
$$= r\cos(\theta)\cos(\varphi)\hat{x} + r\cos(\theta)r\sin(\varphi)\hat{y} - r\sin(\theta)\hat{z},$$
$$\hat{\varphi} = \frac{d\vec{r}}{d\varphi} \Big/ \left| \frac{d\vec{r}}{d\varphi} \right|$$
$$= -r\sin(\theta)\sin(\varphi)\hat{x} + r\sin(\theta)\cos(\varphi)\hat{y}. \tag{8.73}$$

Using equations (8.71)–(8.73) and the methods we introduced in section 5.6 for change of variables and partial differentiation, we can show that the gradient is given by

$$\nabla f = \frac{\partial f}{\partial x}\hat{x} + \frac{\partial f}{\partial y}\hat{y} + \frac{\partial f}{\partial z}\hat{z} \text{ (in Cartesian)}, \tag{8.74}$$

$$\nabla f = \hat{r}\frac{\partial f}{\partial r} + \frac{\hat{\varphi}}{r}\frac{\partial f}{\partial \varphi} + \hat{z}\frac{\partial f}{\partial z} \text{ (in Cylindrical)}, \tag{8.75}$$

$$\nabla f = \hat{r}\frac{\partial f}{\partial r} + \frac{\hat{\theta}}{r}\frac{\partial f}{\partial \theta} + \frac{\hat{\varphi}}{r\sin(\theta)}\frac{\partial f}{\partial \varphi} \text{ (in Spherical)}. \tag{8.76}$$

Note that the unit vectors in cylindrical and spherical coordinates are not constant unlike the Cartesian unit vectors.

Example 8.2. A point charge Q is located at the origin (see figure 8.6). The electrostatic potential at a point P a distance r from the point charge is given by

$$\phi(r) = \frac{kQ}{r}, \tag{8.77}$$

where k is the electrostatic constant expressed in terms the electrical permittivity of a free space ϵ_0 as

$$k = \frac{1}{4\pi\epsilon_0}. \tag{8.78}$$

In classical electrodynamics, the electric field is related to the electrostatic potential $\phi(r)$ by the equation

$$\vec{E}(\vec{r}) = -\nabla\phi(r). \tag{8.79}$$

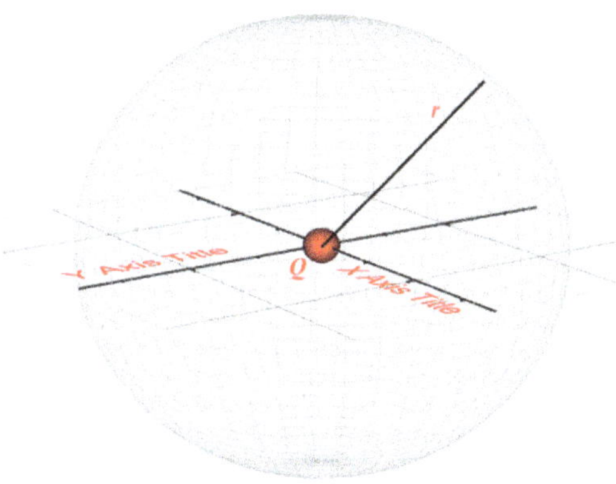

Figure 8.6. A point charge placed at the origin.

(a) Find an expression for the electric field at point P using the gradient in spherical coordinates,

(b) and Cartesian coordinates.

(c) Find the directional derivative of the electrostatic potential at $(1, 2, -1)$ in the direction,

$$\vec{S} = 2\hat{x} - 2\hat{y} + \hat{z}. \tag{8.80}$$

Solution:

(a) The electric potential depends only on r. As shown in figure 8.7 for points equidistant from the charge at the origin, the potential is constant. These points form a spherical surface with radius r which is known as an equipotential surface. Using the gradient in spherical coordinates, the electric field can be written as

$$\vec{E}(\vec{r}) = -\nabla\phi(r)$$

$$= -\hat{r}\frac{\partial}{\partial r}\left(\frac{kQ}{r}\right) + \hat{\theta}\frac{1}{r}\frac{\partial}{\partial\theta}\left(\frac{kQ}{r}\right) + \hat{\varphi}\frac{1}{r\sin(\theta)}\frac{\partial}{\partial\varphi}\left(\frac{kQ}{r}\right) \tag{8.81}$$

$$\Rightarrow \vec{E}(\vec{r}) = -\hat{r}\frac{\partial}{\partial r}\left(\frac{kQ}{r}\right) = \frac{kQ}{r^2}\hat{r}.$$

(b) In Cartesian coordinates, the electric potential is given by

$$\phi(r) = \frac{kQ}{r(x, y, z)} = \frac{kQ}{\sqrt{x^2 + y^2 + z^3}}, \tag{8.82}$$

where we used

$$\vec{r} = x\hat{x} + y\hat{y} + z\hat{z} \Rightarrow r = \sqrt{x^2 + y^2 + z^3}. \tag{8.83}$$

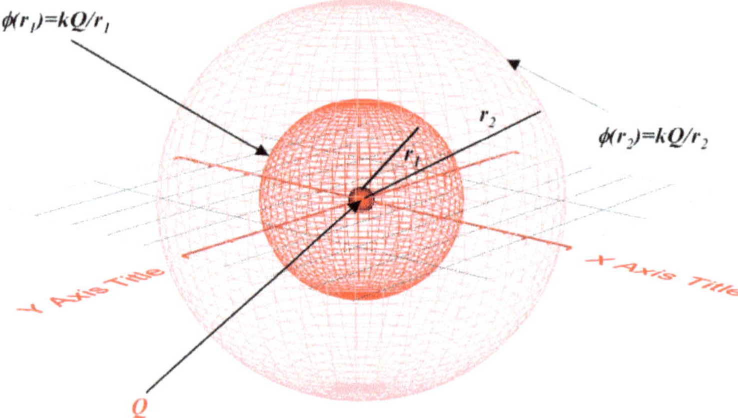

Figure 8.7. Equipotential surfaces for a point charge Q placed at the origin.

Noting that the partial derivative with respect to x,

$$\frac{\partial \phi(r)}{\partial x} = \frac{\partial}{\partial x}\left[\frac{kQ}{\sqrt{x^2 + y^2 + z^3}}\right]$$
$$= \left(-\frac{1}{2}\right)\frac{kQ}{(x^2 + y^2 + z^3)^{3/2}}\frac{\partial}{\partial x}[x^2 + y^2 + z^3] \tag{8.84}$$
$$\Rightarrow \frac{\partial \phi(r)}{\partial x} = -\frac{kQ}{(x^2 + y^2 + z^3)^{3/2}}x = -\frac{kQ}{r^2}\frac{x}{r} = -\frac{kQ}{r^3}x.$$

Similarly, for the partial derivatives with respect to y and z, we find

$$\frac{\partial \phi(\vec{r})}{\partial y} = -\frac{kQ}{r^3}y, \quad \frac{\partial \phi(\vec{r})}{\partial z} = -\frac{kQ}{r^3}z. \tag{8.85}$$

Then using equations (8.84) and (8.85), the electric field

$$\vec{E}(\vec{r}) = -\nabla\phi(r) = -\left[\frac{\partial \phi(r)}{\partial x}\hat{x} + \frac{\partial \phi(r)}{\partial y}\hat{y} + \frac{\partial \phi(r)}{\partial z}\hat{z}\right], \tag{8.86}$$

becomes

$$\vec{E}(\vec{r}) = \frac{kQ}{r^3}x\hat{x} + \frac{kQ}{r^3}y\hat{y} + \frac{kQ}{r^3}z\hat{z} = \frac{kQ}{r^3}(x\hat{x} + y\hat{y} + z\hat{z})$$
$$\Rightarrow \vec{E}(\vec{r}) = \frac{kQ}{r^3}\vec{r}. \tag{8.87}$$

In terms of the unit vector in the direction of $\vec{r}$

$$\hat{r} = \frac{\vec{r}}{r}, \tag{8.88}$$

equation (8.87) can be rewritten as

$$\vec{E}(\vec{r}) = -\nabla\phi(\vec{r}) = \frac{kQ}{r^2}\hat{r}.$$ (8.89)

In order to see how the electric field vectors changes both in magnitude and direction by making a vector plot, let's find the three components at a point (x, y, z). Using

$$\vec{E}(\vec{r}) = \frac{kQ}{r^3}\vec{r} = \frac{kQ}{(x^2 + y^2 + z^2)^{3/2}}(x\hat{x} + y\hat{y} + z\hat{z})$$ (8.90)

one can write

$$E_x(x, y, z) = \frac{x}{(x^2 + y^2 + z^2)^{3/2}},$$
$$E_y(x, y, z) = \frac{y}{(x^2 + y^2 + z^2)^{3/2}},$$
$$E_z(x, y, z) = \frac{z}{(x^2 + y^2 + z^2)^{3/2}},$$ (8.91)

where we assume the charge is positive and normalized the electric field by kQ. The figure shown in figure 8.8 displays the electric field vector of a positive point charge placed at the origin.

(c) We are given

$$\vec{S} = 2\hat{x} - 2\hat{y} + \hat{z}$$ (8.92)

and we want to find

$$\frac{d\phi(r)}{ds} = \nabla\phi(r) \cdot \hat{n}\bigg|_{(1,\,2,-1)}.$$ (8.93)

Here $\hat{n}$ is the unit vector along the vector $\vec{S}$ tangent to the surface defined by the electrostatic potential function $\phi(r)$ at a point (x, y, z). This unit vector is given by

$$\hat{n} = \frac{\vec{S}}{|\vec{S}|} = \frac{2}{3}\hat{x} - \frac{2}{3}\hat{y} + \frac{1}{3}\hat{z}.$$ (8.94)

We are interested in the change in the potential $\phi(r)$ along this direction at a point $(1, 2, -1)$ on the surface. This is given by the directional derivative

$$\frac{d\phi(r)}{ds} = \nabla\phi(r) \cdot \hat{n}\bigg|_{(1,\,2,-1)}.$$ (8.95)

Using the result we obtained in part (a) or (b) and the unit vector $\hat{n}$, one can write

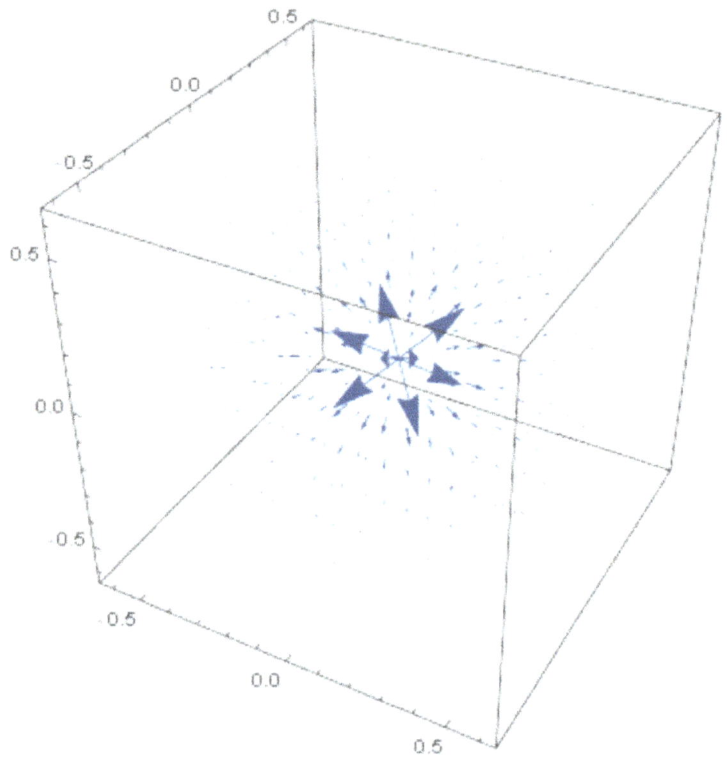

Figure 8.8. The electric field vector for positive point charge placed at the origin.

$$\frac{d\phi(\vec{r})}{ds} = -\frac{kQ}{(x^2 + y^2 + z^2)^{3/2}}(x\hat{x} + y\hat{y} + z\hat{z}) \cdot \left(\frac{2}{3}\hat{x} - \frac{2}{3}\hat{y} + \frac{1}{3}\hat{z}\right)\Bigg|_{(1,\,2,-1)}$$

$$\Rightarrow \frac{d\phi(\vec{r})}{ds} = -\frac{kQ}{(x^2 + y^2 + z^2)^{3/2}}\left(\frac{2}{3}x - \frac{2}{3}y + \frac{1}{3}z\right)\Bigg|_{(1,\,2,-1)} \tag{8.96}$$

$$\Rightarrow \frac{d\phi(\vec{r})}{ds} = -\frac{kQ}{\sqrt{6}}\left(\frac{2}{3} - \frac{4}{3} - \frac{1}{3}\right) = \frac{kQ}{\sqrt{6}}.$$

For the scalar product of the electric field $\vec{E}(\vec{r})$ and the unit vector $\hat{n}$, we have

$$\vec{E}(\vec{r}) \cdot \hat{n} = |\,\vec{E}(\vec{r})\,|\,|\,\hat{n}\,|\,\cos(\theta) \Rightarrow \cos(\theta) = \frac{\vec{E}(\vec{r}) \cdot \hat{n}}{|\,\vec{E}(\vec{r})\,|}, \tag{8.97}$$

where θ is the angle between the electric field and the unit vector $\hat{n}$. For $KQ = 1$, the electric field

$$\vec{E}(\vec{r}) = \frac{kQ}{r^3}\vec{r}, \tag{8.98}$$

at point $(1, 2, -1)$ becomes,

$$\vec{E}(\vec{r}) = \frac{1}{\left(\sqrt{6}\right)^3}(x + 2\hat{y} - \hat{z}).$$ (8.99)

Using this result and equation (8.94), one finds for equation (8.97)

$$\cos(\theta) = \frac{\dfrac{2}{3\left(\sqrt{6}\right)^3} - \dfrac{4}{3\left(\sqrt{6}\right)^3} - \dfrac{1}{3\left(\sqrt{6}\right)^3}}{\left(1/\sqrt{6}\right)^2} = -\frac{1}{\sqrt{6}},$$ (8.100)

8.5 The divergence, the curl, and the Laplacian

We have seen the result of the action the gradient operator on a scalar function. It can provide information regarding how the scalar function changes along a given direction. This section will consider how the gradient operator acts on vector functions, such as the electric field due to a point charge, gravitational field due to a given mass, or magnetic field due to a current-carrying wire. For vector functions, the gradient operator acts in two different ways to describe *the divergence* and *the curl* (rotation) of the vector function. For a vector function

$$\vec{V}(x, y, z) = V_x(x, y, z)\hat{x} + V_y(x, y, z)\hat{y} + V_z(x, y, z)\hat{z},$$ (8.101)

the divergence of $\vec{V}(x, y, z)$ $(\nabla \cdot \vec{V})$ is given by

$$\nabla \cdot \vec{V} = \frac{\partial V_x}{\partial x} + \frac{\partial V_y}{\partial y} + \frac{\partial V_z}{\partial z}.$$ (8.102)

Note that the divergence of a vector function results in a scalar function. For *the curl* of the vector $\vec{V}(x, y, z)$ $(\nabla \times \vec{V})$, in Cartesian coordinates is given by the determinant

$$\nabla \times \vec{V} = \begin{vmatrix} \hat{x} & \hat{y} & \hat{z} \\ \dfrac{\partial}{\partial x} & \dfrac{\partial}{\partial y} & \dfrac{\partial}{\partial z} \\ V_x & V_y & V_z \end{vmatrix},$$ (8.103)

that leads to

$$\nabla \times \vec{V} = \left(\frac{\partial V_z}{\partial y} - \frac{\partial V_y}{\partial z}\right)\hat{x} + \left(\frac{\partial V_x}{\partial z} - \frac{\partial V_z}{\partial x}\right)\hat{y} + \left(\frac{\partial V_y}{\partial x} - \frac{\partial V_x}{\partial y}\right)\hat{z}.$$ (8.104)

Suppose the vector field $\vec{V}(x, y, z)$ is expressible as a gradient of some scalar function $\phi(x, y, z)$,

$$\vec{V}(x, y, z) = \nabla\phi(x, y, z),$$ (8.105)

the divergence of this vector becomes

$$\nabla \cdot \vec{V} = \nabla \cdot [\nabla \phi(x, y, z)] = (\nabla \cdot \nabla)\phi(x, y, z) = \nabla^2 \phi(x, y, z), \qquad (8.106)$$

where

$$\nabla^2 = \frac{\partial^2}{\partial x^2} + \frac{\partial^2}{\partial y^2} + \frac{\partial^2}{\partial z^2}, \qquad (8.107)$$

is called the *Laplacian operator*. As an example, let's consider the electric field vector which is given by

$$\vec{E}(\vec{r}) = -\nabla \phi(\vec{r}). \qquad (8.108)$$

The divergence of $\vec{E}(\vec{r})$ leads to

$$\nabla \cdot \vec{E}(\vec{r}) = -\nabla \cdot \nabla \phi(x, y, z) = -\left(\frac{\partial^2 \phi}{\partial x^2} + \frac{\partial^2 \phi}{\partial y^2} + \frac{\partial^2 \phi}{\partial z^2}\right) = -\nabla^2 \phi(x, y, z). \quad (8.109)$$

In classical electrodynamics the divergence of the electric field is related to the volume charge density $\rho(x, y, z)$ by the differential form of Gauss' law,

$$\nabla \cdot \vec{E}(\vec{r}) = \frac{\rho(x, y, z)}{\epsilon_0}, \qquad (8.110)$$

that can be expressed using the Laplacian as

$$\nabla \cdot \vec{E}(\vec{r}) = -\nabla \cdot \nabla \phi(x, y, z) \Rightarrow \frac{\partial^2 \phi}{\partial x^2} + \frac{\partial^2 \phi}{\partial y^2} + \frac{\partial^2 \phi}{\partial z^2} = -\frac{\rho(x, y, z)}{\epsilon_0} \qquad (8.111)$$

$$\Rightarrow \nabla^2 \phi(x, y, z) = F(x, y, z)$$

where

$$F(x, y, z) = -\frac{\rho(x, y, z)}{\epsilon_0}. \qquad (8.112)$$

Equation (8.111) is known as *Poison's equation*. In a free space where there is no charge, $\rho(x, y, z) = 0$, we find

$$\nabla^2 \phi(x, y, z) = 0, \qquad (8.113)$$

which is known as *Laplace's equation*. We often see *Poison's* and *Laplace's* equations in classical mechanics and electromagnetism. However, generally, in most branches of theoretical physics, we frequently see equations involving the gradient, the divergence, and the curl of a scalar or a vector function. Thus it is worth remembering the following relations involving these operators:

(a) *The divergence of a vector and a scalar*: For a vector function $\vec{V}$ multiplied by a scalar functions f, the divergence can be expressed as

$$\nabla \cdot \left(f\vec{V}\right) = \vec{V} \cdot (\nabla f) + f\left(\nabla \cdot \vec{V}\right). \qquad (8.114)$$

(b) *The curl of the curl*: In view of the relation for triple vector product

$$\vec{A} \times \left(\vec{B} \times \vec{C} \right) = \vec{B}\left(\vec{A} \cdot \vec{C} \right) - \left(\vec{A} \cdot \vec{B} \right)\vec{C}, \tag{8.115}$$

for the curl of the curl of the vector function $\vec{V}$, one can establish the relation

$$\nabla \times \left(\nabla \times \vec{V} \right) = \nabla\left(\nabla \cdot \vec{V} \right) - \nabla^2 \vec{V}, \tag{8.116}$$

that involves the divergence and the Laplacian of the vector function. There follows that for *the Laplacian of a vector function,*

$$\nabla^2 \vec{V} = \nabla\left(\nabla \cdot \vec{V} \right) - \nabla \times \left(\nabla \times \vec{V} \right). \tag{8.117}$$

(c) *The curl of a vector and a scalar*: For a vector function $\vec{V}$ multiplied by a scalar functions f, the curl can be expressed as

$$\nabla \times \left(f\vec{V} \right) = \left(\nabla f \right) \times \vec{V} + f\left(\nabla \times \vec{V} \right) = f\left(\nabla \times \vec{V} \right) - \vec{V} \times \left(\nabla f \right). \tag{8.118}$$

Example 8.3. Consider two functions—a scalar function and a vector function—defined as follows:

$$f(\vec{r}) = xyz, \quad \vec{V}(x,\, y,\, z) = (xy,\, yz,\, zx). \tag{8.119}$$

At the point $(1,\, -1,\, 1)$, evaluate the following quantities :
  (a) ∇f
  (b) $\nabla \cdot \vec{V}$
  (c) $\nabla \times \vec{V}$
  (d) $\nabla^2 f$
  (e) $\nabla^2 \vec{V}$
  (f) $\nabla^2(f\vec{V})$
  (g) graph the equipotential surface assuming that $f(\vec{r})$ represents the electric potential function of some charge distribution.

Solution:
  (a) The gradient of the scalar function f is given by

$$\nabla f = \frac{\partial f}{\partial x}\hat{x} + \frac{\partial f}{\partial y}\hat{y} + \frac{\partial f}{\partial z}\hat{z}.$$

$$\nabla f = \frac{\partial(xyz)}{\partial x}\hat{x} + \frac{\partial(xyz)}{\partial y}\hat{y} + \frac{\partial(xyz)}{\partial z}\hat{z} \tag{8.120}$$

$$\Rightarrow \nabla f = yz\hat{x} + xz\hat{y} + xy\hat{z}.$$

Upon evaluating this at $(x,\, y,\, z) = (1,\, -1,\, 1)$, we find

$$\nabla f = -\hat{x} + \hat{y} - \hat{z}. \tag{8.121}$$

(b) The divergence of the vector function,

$$\vec{V} = \left(V_x,\ V_y,\ V_z\right) = (xy,\ yz,\ zx),$$

is given by

$$\nabla \cdot \vec{V} = \frac{\partial V_x}{\partial x} + \frac{\partial V_y}{\partial y} + \frac{\partial V_z}{\partial z} = \frac{\partial(xy)}{\partial x} + \frac{\partial(yz)}{\partial y} + \frac{\partial(xz)}{\partial z} \tag{8.122}$$

$$\Rightarrow \nabla \cdot \vec{V} = y + z + x \Rightarrow \nabla \cdot \vec{V} = 1,$$

where we substitute $(x,\ y,\ z) = (1,\ -1,\ 1)$.

(c) For $\nabla \times \vec{V}$, we have

$$\nabla \times \vec{V} = \begin{vmatrix} \hat{x} & \hat{y} & \hat{z} \\ \dfrac{\partial}{\partial x} & \dfrac{\partial}{\partial y} & \dfrac{\partial}{\partial z} \\ xy & yz & xz \end{vmatrix}$$

$$\nabla \times \vec{V} = \left(\frac{\partial(xz)}{\partial y} - \frac{\partial(yz)}{\partial z}\right)\hat{x} + \left(\frac{\partial(xy)}{\partial z} - \frac{\partial(xz)}{\partial x}\right)\hat{y} \tag{8.123}$$

$$+ \left(\frac{\partial(yz)}{\partial x} - \frac{\partial(xy)}{\partial y}\right)\hat{z}$$

$$\Rightarrow \nabla \times \vec{V} = (0 - y)\hat{x} + (0 - z)\hat{y} + (0 - x)\hat{z} = -y\hat{x} - z\hat{y} - x\hat{z}, \tag{8.124}$$

so that evaluating this at the point $(1,\ -1,\ 1)$, one finds

$$\nabla \times \vec{V} = \hat{x} - \hat{y} - \hat{z}. \tag{8.125}$$

(d) For the Laplacian of the scalar function f,

$$\nabla^2 f = \frac{\partial^2 f}{\partial x^2} + \frac{\partial^2 f}{\partial y^2} + \frac{\partial^2 f}{\partial z^2}, \tag{8.126}$$

we find

$$\nabla^2 f = \frac{\partial^2(xyz)}{\partial x^2} + \frac{\partial^2(xyz)}{\partial y^2} + \frac{\partial^2(xyz)}{\partial z^2} = 0. \tag{8.127}$$

Noting that

$$\nabla^2 f = \nabla \cdot \nabla f, \tag{8.128}$$

one can also find the same result

$$\nabla^2 f = \nabla \cdot (yz\hat{x} + xz\hat{y} + xy\hat{z}) = \frac{\partial(yz)}{\partial x} + \frac{\partial(xz)}{\partial y} + \frac{\partial(xy)}{\partial z} = 0, \qquad (8.129)$$

using the expression for the gradient of the function determined in (a).

(e) Here we want to determine the Laplacian of a vector function instead of a scalar function. In such cases, one should evaluate

$$\nabla^2 \vec{V} = \frac{\partial^2 \vec{V}}{\partial x^2} + \frac{\partial^2 \vec{V}}{\partial y^2} + \frac{\partial^2 \vec{V}}{\partial z^2}, \qquad (8.130)$$

where

$$\vec{V} = V_x\hat{x} + V_y\hat{y} + V_z\hat{z}. \qquad (8.131)$$

Thus for

$$\vec{V} = xy\hat{x} + yz\hat{y} + zx\hat{z} \qquad (8.132)$$

one finds

$$\frac{\partial^2 \vec{V}}{\partial x^2} = \frac{\partial^2(xy\hat{x} + yz\hat{y} + zx\hat{z})}{\partial x^2} = 0,$$
$$\frac{\partial^2 \vec{V}}{\partial y^2} = \frac{\partial^2(xy\hat{x} + yz\hat{y} + zx\hat{z})}{\partial y^2} = 0, \qquad (8.133)$$
$$\frac{\partial^2 \vec{V}}{\partial z^2} = \frac{\partial^2(xy\hat{x} + yz\hat{y} + zx\hat{z})}{\partial z^2} = 0.$$

The Laplacian becomes

$$\nabla^2 \vec{V} = 0. \qquad (8.134)$$

We can also apply the relation

$$\nabla^2 \vec{V} = \nabla\left(\nabla \cdot \vec{V}\right) - \nabla \times \left(\nabla \times \vec{V}\right). \qquad (8.135)$$

Using the results we obtained for the divergence and the curl

$$\nabla \cdot \vec{V} = y + z + x, \quad \nabla \times \vec{V} = -y\hat{x} - z\hat{y} - x\hat{z}, \qquad (8.136)$$

we find

$$\nabla\left(\nabla \cdot \vec{V}\right) = \hat{x}\frac{\partial}{\partial x}(y + z + x) + \hat{y}\frac{\partial}{\partial y}(y + z + x) + \hat{z}\frac{\partial f}{\partial z}(y + z + x)$$
$$\Rightarrow \nabla\left(\nabla \cdot \vec{V}\right) = \hat{x} + \hat{y} + \hat{z}, \qquad (8.137)$$

and

$$\nabla \times \left(\nabla \times \vec{V} \right) = \nabla \times \vec{U} = \begin{vmatrix} \hat{x} & \hat{y} & \hat{z} \\ \dfrac{\partial}{\partial x} & \dfrac{\partial}{\partial y} & \dfrac{\partial}{\partial z} \\ -y & -z & -x \end{vmatrix}$$

$$= \left(\frac{\partial(y-z)}{\partial y} - \frac{\partial(-x)}{\partial z} \right)\hat{x} + \left(\frac{\partial(-y)}{\partial z} - \frac{\partial(-x)}{\partial x} \right)\hat{y}$$

$$+ \left(\frac{\partial(-z)}{\partial x} - \frac{\partial(-y)}{\partial y} \right)\hat{z} \Rightarrow \nabla \times \left(\nabla \times \vec{V} \right) = \hat{x} + \hat{y} + \hat{z}. \tag{8.138}$$

Substituting equations (8.137) and (8.138) into equation (8.135), one also finds the same result for the Laplacian, $\nabla^2 \vec{V} = 0$.

(f) Noting that

$$\nabla^2 \left(f\vec{V} \right) = \hat{x}\nabla^2 \left(fV_x \right) + \hat{y}\nabla^2 \left(fV_z \right) + \hat{z}\nabla^2 \left(fV_z \right) \tag{8.139}$$

and

$$\nabla^2 \left(fV_x \right) = \frac{\partial^2((xy)^2 z)}{\partial x^2} + \frac{\partial^2((xy)^2 z)}{\partial y^2} + \frac{\partial^2((xy)^2 z)}{\partial z^2} = 2y^2 z + 2x^2 z,$$

$$\nabla^2 \left(fV_y \right) = \frac{\partial^2((yz)^2 x)}{\partial x^2} + \frac{\partial^2((yz)^2 x)}{\partial y^2} + \frac{\partial^2((yz)^2 x)}{\partial z^2} = 2z^2 x + 2y^2 x, \tag{8.140}$$

$$\nabla^2 \left(fV_z \right) = \frac{\partial^2((zx)^2 y)}{\partial x^2} + \frac{\partial^2((zx)^2 y)}{\partial y^2} + \frac{\partial^2((zx)^2 y)}{\partial z^2} = 2z^2 y + 2x^2 y,$$

we find

$$\nabla^2 \left(f\vec{V} \right) = \hat{x}(2y^2 z + 2x^2 z) + \hat{y}(2z^2 x + 2y^2 x) + \hat{z}(2z^2 y + 2x^2 y). \tag{8.141}$$

(g) Figure 8.9 displays the equipotential surface for $f(x, y, z) = xyz = $ constant and figure 8.10 shows the magnitude and direction for the vector field defined by $\nabla f = yz\hat{x} + xz\hat{y} + xy\hat{z}$.

The gradient, the divergence, the Laplacian, and the curl that we have seen so far are for scalar and vector functions expressed in Cartesian coordinates. Depending on the problem we want to solve, sometimes it is convenient to evaluate these quantities in curvilinear coordinates by expressing the functions using the corresponding transformations from the Cartesian to the convenient curvilinear coordinates. Thus it is essential to know the expression for these quantities in the two curvilinear coordinates we studied (cylindrical and spherical coordinates). We can derive these expressions from their corresponding expressions in Cartesian.

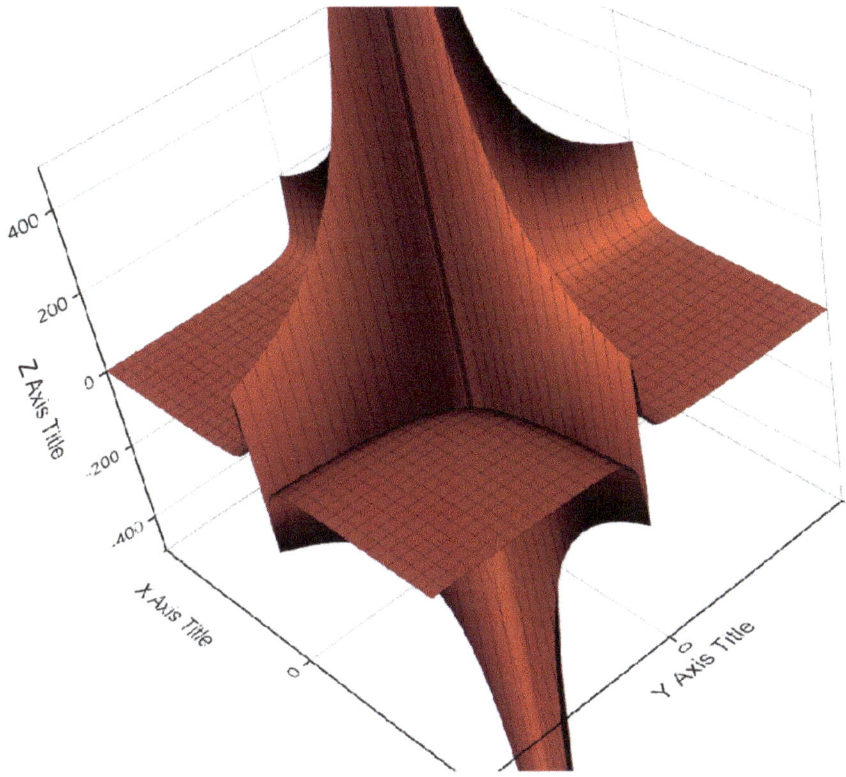

Figure 8.9. An equipotential surface for $f(x, y, z) = xyz$.

However, here, we list these expressions in cylindrical and spherical coordinates instead of deriving them.

(a) **Cylindrical coordinates** (r, φ, z): For a scalar function $f(r, \varphi, z)$ the gradient and Laplacian are given by

$$\nabla f = \hat{s}\frac{\partial f}{\partial s} + \hat{\varphi}\frac{1}{s}\frac{\partial f}{\partial \varphi} + \hat{z}\frac{\partial f}{\partial z}, \tag{8.142}$$

$$\nabla^2 f = \frac{1}{s}\frac{\partial}{\partial s}\left(s\frac{\partial f}{\partial s}\right) + \frac{1}{s^2}\frac{\partial^2 f}{\partial \varphi^2} + \frac{\partial^2 f}{\partial z^2}. \tag{8.143}$$

For a vector function

$$\vec{V}(s, \varphi, z) = V_s(s, \varphi, z)\hat{s} + V_\varphi(s, \varphi, z)\hat{\varphi} + V_z(s, \varphi, z)\hat{z}, \tag{8.144}$$

the divergence and the curl can be determined using

$$\nabla \cdot \vec{V} = \frac{1}{s}\frac{\partial}{\partial s}(sV_s) + \frac{1}{s}\frac{\partial}{\partial \varphi}(V_\varphi) + \frac{\partial V_z}{\partial z},$$

$$\nabla \times \vec{V} = \hat{s}\left(\frac{1}{s}\frac{\partial V_z}{\partial \varphi} - \frac{\partial V_\varphi}{\partial z}\right) + \hat{\varphi}\left(\frac{\partial V_s}{\partial z} - \frac{\partial V_z}{\partial s}\right) \tag{8.145}$$

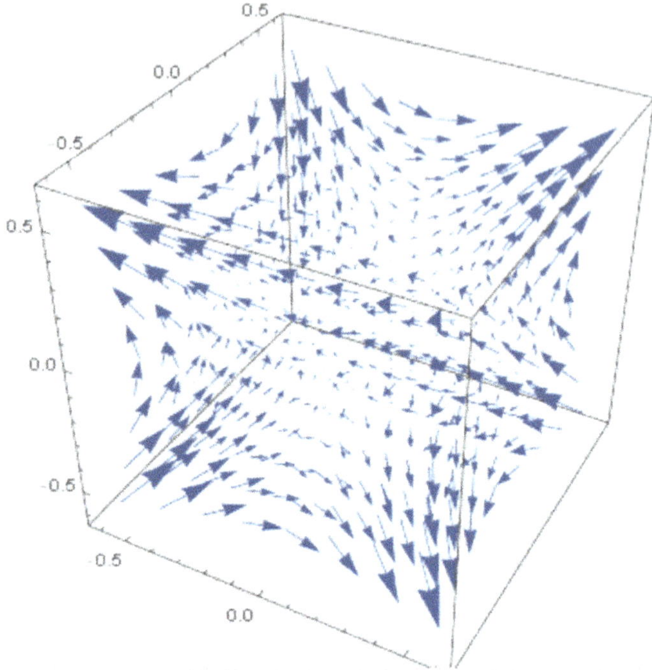

Figure 8.10. The vector defined by ∇f.

$$+\hat{z}\frac{1}{s}\left(\frac{\partial}{\partial s}(sV_\varphi) - \frac{\partial V_s}{\partial\varphi}\right).\tag{8.146}$$

(b) **Spherical coordinates** (r, θ, φ): For a scalar function $f(r, \theta, \varphi)$ the gradient and the Laplacian:

$$\nabla f = \hat{r}\frac{\partial f}{\partial r} + \hat{\theta}\frac{1}{r}\frac{\partial f}{\partial\theta} + \hat{\varphi}\frac{1}{r\sin(\theta)}\frac{\partial f}{\partial\varphi},$$

$$\nabla^2 f = \frac{1}{r^2}\frac{\partial}{\partial r}\left(r^2\frac{\partial f}{\partial r}\right) + \frac{1}{r^2\sin(\theta)}\frac{\partial}{\partial\theta}\left(\sin(\theta)\frac{\partial f}{\partial\theta}\right)\tag{8.147}$$

$$+ \frac{1}{r^2\sin^2(\theta)}\frac{\partial^2 f}{\partial\varphi^2}.\tag{8.148}$$

For a vector function

$$\vec{V}(r, \varphi, \theta) = V_r(r, \varphi, \theta)\hat{r} + V_\theta(r, \varphi, \theta)\hat{\theta} + V_\varphi(r, \varphi, \theta)\hat{\varphi},\tag{8.149}$$

the divergence and the curl are given by

$$\nabla \cdot \vec{V} = \frac{1}{r^2}\frac{\partial}{\partial r}(r^2 V_r) + \frac{1}{r\sin(\theta)}\frac{\partial}{\partial \theta}(\sin(\theta) V_\theta) + \frac{1}{r\sin(\theta)}\frac{\partial V_\varphi}{\partial \varphi},$$

(8.150)

$$\nabla \times \vec{V} = \hat{r}\frac{1}{r\sin(\theta)}\left(\frac{\partial}{\partial \theta}(\sin(\theta) V_\varphi) - \frac{\partial V_\theta}{\partial \varphi}\right)$$

$$+ \hat{\theta}\left(\frac{1}{r\sin(\theta)}\frac{\partial V_r}{\partial \varphi} - \frac{1}{r}\frac{\partial}{\partial r}(r V_\varphi)\right) + \hat{\varphi}\frac{1}{r}\left(\frac{\partial}{\partial r}(r V_\theta) - \frac{\partial V_r}{\partial \theta}\right).$$

(8.151)

Example 8.4. Consider an infinitely long cylindrical wire of radius a, carrying a current I in the positive z-direction, and this current is uniformly distributed (see figure 8.11).

The magnetic field for a point inside the wire is given by

$$\vec{B} = \left(\frac{\mu_0 I s}{2\pi a^2}\right)\hat{\varphi},$$

(8.152)

where $s < a$ is the distance of the point with coordinates (s, φ, z) from the axis of the cylindrical wire and μ_0 is magnetic permeability of a free space.

(a) Using vector plot show that the magnetic field lines form a concentric circle.

(b) Show that magnetic field vector is a rotational vector field (i.e., the $\nabla \times \vec{B} \neq 0$) inside the cylinder.

(c) Show that $\nabla \cdot \vec{B} = 0$.

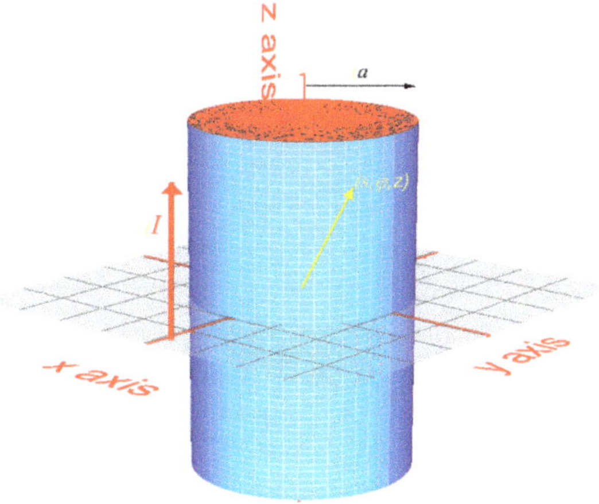

Figure 8.11. A current flowing I in the positive z-axis in a cylinder of radius a.

Solution:

(a) Recalling that the unit vector $\hat{\varphi}$ in cylindrical coordinates is

$$\hat{\varphi} = -\sin(\varphi)\hat{x} + \cos(\hat{\varphi})\hat{y}, \tag{8.153}$$

we can express the magnetic field in Cartesian coordinates as

$$\vec{B} = \frac{\mu_0 I}{2\pi a^2}(-s\sin(\varphi)\hat{x} + s\cos(\hat{\varphi})\hat{y}) = \frac{\mu_0 I}{2\pi a^2}(-y\hat{x} + x\hat{y}). \tag{8.154}$$

By normalizing the magnetic field by $\dfrac{\mu_0 I}{2\pi a^2}$, one can write

$$\vec{B} = -y\hat{x} + x\hat{y}. \tag{8.155}$$

The vector plot of equation (8.155) is shown in figure 8.12.

(b) We note that the components of the magnetic field in cylindrical coordinates are

$$B_s = 0, \quad B_\varphi = \frac{\mu_0 I s}{2\pi a^2}, \quad B_z = 0. \tag{8.156}$$

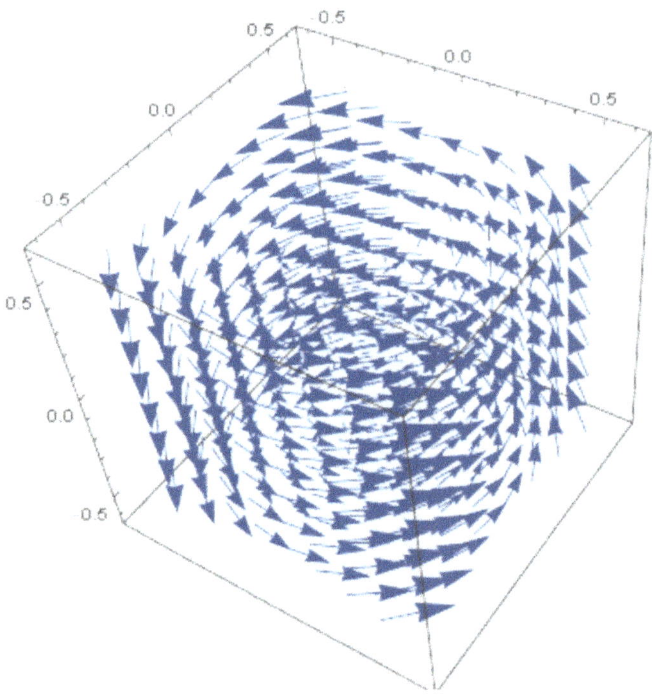

Figure 8.12. The magnetic field vector inside the infinitely long cylindrical wire carrying a current I.

Then using the expression for the curl of a vector in cylindrical coordinates,

$$\nabla \times \vec{B} = \hat{s}\left(\frac{1}{s}\frac{\partial B_z}{\partial \varphi} - \frac{\partial B_\varphi}{\partial z}\right) + \hat{\varphi}\left(\frac{\partial B_s}{\partial z} - \frac{\partial B_z}{\partial s}\right)$$
$$+ \hat{z}\frac{1}{s}\left(\frac{\partial}{\partial s}(sB_\varphi) - \frac{\partial B_s}{\partial \varphi}\right),$$

(8.157)

we find

$$\nabla \times \vec{B} = -\hat{s}\left(\frac{\partial B_\varphi}{\partial z}\right) + \hat{z}\frac{1}{s}\frac{\partial}{\partial s}(sB_\varphi) = \frac{\hat{z}}{s}\frac{\partial}{\partial s}\left(s\frac{\mu_0 Is}{2\pi a^2}\right)$$
$$= \frac{\hat{z}}{s}\frac{\partial}{\partial s}\left(\frac{\mu_0 Is^2}{2\pi a^2}\right) \Rightarrow \nabla \times \vec{B} = \frac{\mu_0 I}{\pi a^2}\hat{z}.$$

(8.158)

Using the current density defined by

$$J = \frac{\text{Current}}{\text{Area}} = \frac{I}{\pi a^2},$$

(8.159)

one can write

$$\nabla \times \vec{B} = \mu_0 J\hat{z} = \mu_0 \vec{J},$$

(8.160)

which is *Ampere's law*.

(c) The divergence in cylindrical coordinates is given by

$$\nabla \cdot \vec{B} = \frac{1}{s}\frac{\partial}{\partial s}(sB_s) + \frac{1}{s}\frac{\partial}{\partial \varphi}(B_\varphi) + \frac{\partial B_z}{\partial z}.$$

(8.161)

Using equation (8.156), we find

$$\nabla \cdot \vec{B} = \frac{1}{s}\frac{\partial}{\partial \varphi}(B_\varphi) = \frac{1}{s}\frac{\partial}{\partial \varphi}\left(\frac{\mu_0 Is}{2\pi a^2}\right) = 0.$$

(8.162)

Example 8.5. In example 8.2 we have shown that the electric field of a point charge Q positioned at the origin, is given

$$\vec{E}(\vec{r}) = -\nabla\phi(\vec{r}) = \frac{kQ}{r^2}\hat{r}.$$

(8.163)

Show that this electric field vector is irrotational,

$$\nabla \times \vec{E} = 0.$$

(8.164)

Solution: The electric field vector in spherical coordinates has components

$$E_r = \frac{kQ}{r^2}, \ E_\theta = E_\varphi = 0,$$

(8.165)

and the curl of the electric field

$$\nabla \times \vec{E} = \hat{r}\frac{1}{r\sin(\theta)}\left[\frac{\partial}{\partial\theta}(\sin(\theta)E_\varphi) - \frac{\partial E_\theta}{\partial\varphi}\right]$$
$$+ \hat{\theta}\left[\frac{1}{r\sin(\theta)}\frac{\partial E_r}{\partial\varphi} - \frac{1}{r}\frac{\partial}{\partial r}(rE_\varphi)\right] + \hat{\varphi}\frac{1}{r}\left[\frac{\partial}{\partial r}(rE_\theta) - \frac{\partial E_r}{\partial\theta}\right] \tag{8.166}$$

becomes

$$\nabla \times \vec{E} = 0, \tag{8.167}$$

which shows the electric field vector is irrotational.

8.6 Line vector integrals

The gradient, the divergence, and the curl of vector functions introduced in the last sections are essentially the derivatives of vector functions. In this and the following sections, we study line, surface, and volume integrals of vector functions and establish relationships between these integrals through the divergence and the curl of the vector function. We begin with the line vector integral of the form

$$I = \oint_C \vec{A}(\vec{r}) \cdot d\vec{r}, \tag{8.168}$$

over a closed curve C. As we will see, the result of such integral could tell us an essential property of vector functions, in particular vector fields such as electric and gravitational fields. However, before we get into this property of vectors, we want to be familiar with how we carry out line integrals over a closed curve.

Example 8.6. Consider the vector field

$$\vec{A}(\vec{r}) = x^2y\hat{x} + xy^2\hat{y} + a^3e^{-\beta y}x\cos(\alpha)\hat{z}, \tag{8.169}$$

where a, α, and β are known constants. Evaluate the line integral

$$I = \oint_C \vec{A}(\vec{r}) \cdot d\vec{r}, \tag{8.170}$$

around the curve C shown in figure 8.13.

Solution: We can carry out this closed line integral by dividing the integral into three integrals corresponding to the three sides (C_1, C_2, and C_3),

$$\oint_C \vec{A}(\vec{r}) \cdot d\vec{r} = \int_{C_1} \vec{A}(\vec{r}) \cdot d\vec{r} + \int_{C_2} \vec{A}(\vec{r}) \cdot d\vec{r} + \int_{C_3} \vec{A}(\vec{r}) \cdot d\vec{r}. \tag{8.171}$$

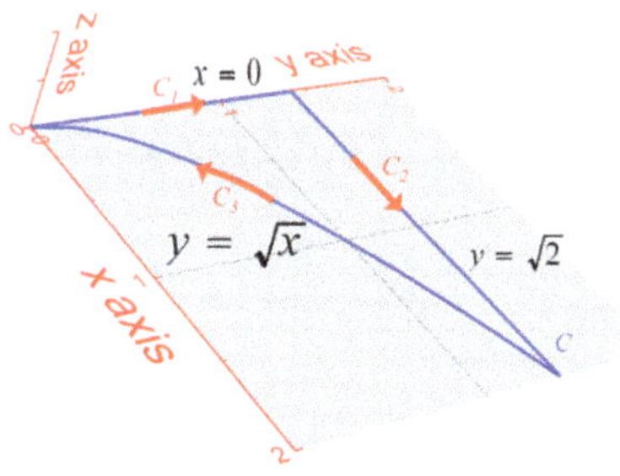

Figure 8.13. A closed path on the x–y plane defined by $x = 0$, $y = \sqrt{2}$, and $y = \sqrt{x}$.

Let us represent a point on any of these line segments by a vector

$$\vec{r} = x\hat{x} + y\hat{y}, \tag{8.172}$$

so that

$$d\vec{r} = dx\hat{x} + dy\hat{y}. \tag{8.173}$$

Referring to figure 8.13, for a point

$$\text{on } C_1, \ x = 0 \Rightarrow \vec{r} = y\hat{y} \Rightarrow d\vec{r} = dy\hat{y},$$
$$\text{on } C_2, \ y = \sqrt{2} \Rightarrow \vec{r} = x\hat{x} + \sqrt{2}\,\hat{y} \Rightarrow d\vec{r} = dx\hat{x},$$
$$\text{on } C_3, \ y = \sqrt{x} \Rightarrow \vec{r} = x\hat{x} + \sqrt{x}\,\hat{y} \Rightarrow d\vec{r} = dx\hat{x} + \frac{dx}{2\sqrt{x}}\hat{y}. \tag{8.174}$$

Furthermore on these lines, the function

$$\vec{A}(\vec{r}) = x^2 y\hat{x} + xy^2\hat{y} + a^3 e^{-\beta y} x \cos{(\alpha)}\hat{z}, \tag{8.175}$$

also becomes

$$\text{on } C_1, \ x = 0 \Rightarrow \vec{A}(\vec{r}) = 0,$$
$$\text{on } C_2, \ y = \sqrt{2}, z = 0$$
$$\Rightarrow \vec{A}(\vec{r}) = x^2\sqrt{2}\,\hat{x} + 2x\hat{y} + a^3 e^{-\sqrt{2}\beta} x \cos{(\alpha)}\hat{z}, \tag{8.176}$$
$$\text{on } C_3, \ y = \sqrt{x}, z = 0$$
$$\Rightarrow \vec{A}(\vec{r}) = x^{5/2}\hat{x} + x^2\hat{y} + a^3 e^{-\beta\sqrt{x}} x \cos{(\alpha)}\hat{z}.$$

Therefore, using the results in equations (8.174) and (8.176), one finds for the integrals on C_1 and C_2

$$\int_{C_1} \vec{A}(\vec{r}) \cdot d\vec{r} = 0, \tag{8.177}$$

$$\int_{C_2} \vec{A}(\vec{r}) \cdot d\vec{r} = \int_0^2 (x^2\sqrt{2}\,\hat{x} + 2x\hat{y}) \cdot dx\hat{x}$$
$$= \sqrt{2} \int_0^2 x^2 dx = \frac{\sqrt{2}}{3}x^3 \Big|_0^2 \Rightarrow \int_{C_2} \vec{A}(\vec{r}) \cdot d\vec{r} = \frac{8\sqrt{2}}{3}, \tag{8.178}$$

and on C_3

$$\int_{C_3} \vec{A}(\vec{r}) \cdot d\vec{r} = \int_2^0 (x^{5/2}\hat{x} + x^2\hat{y}) \cdot \left(dx\hat{x} + \frac{dx}{2\sqrt{x}}\hat{y}\right)$$

$$= \int_2^0 \left[x^{5/2} + \frac{x^2}{2\sqrt{x}}\right]dx = \int_2^0 \left[x^{5/2} + \frac{1}{2}x^{3/2}\right]dx$$

$$= \frac{2}{7}x^{7/2} + \frac{1}{5}x^{5/2}\bigg|_2^0$$

$$\Rightarrow \int_{C_3} \vec{A}(\vec{r}) \cdot d\vec{r} = -\frac{16\sqrt{2}}{7} - \frac{4\sqrt{2}}{5}.$$

(8.179)

Upon substituting the results in equations (8.177)–(8.179) into equation (8.171), one finds

$$\oint_C \vec{A}(\vec{r}) \cdot d\vec{r} = -\frac{44\sqrt{2}}{105}.$$

(8.180)

Example 8.7. Consider the Moon and the Earth. Imagine the origin of a spherical coordinate system coincides with the center of the Earth. Then the position of the Moon can be described by (r, θ, φ) in spherical coordinates. The gravitational potential energy of the Earth and the Moon is given by

$$U = \frac{GMm}{r},$$

(8.181)

where r is the distance between the Earth and the Moon, M is mass of the Earth, and m is mass of the Moon (see figure 8.14).

(a) Find the gravitational force

$$\vec{F} = \nabla U$$

(8.182)

on the Moon using the gravitational potential energy.

(b) Show the work done by the gravitational force when the Moon makes one complete revolution around the Earth

$$W = \oint_C \vec{F} \cdot d\vec{r} = 0.$$

(8.183)

Solution:

(a) The gravitational force is given by

$$\vec{F} = \nabla U.$$

(8.184)

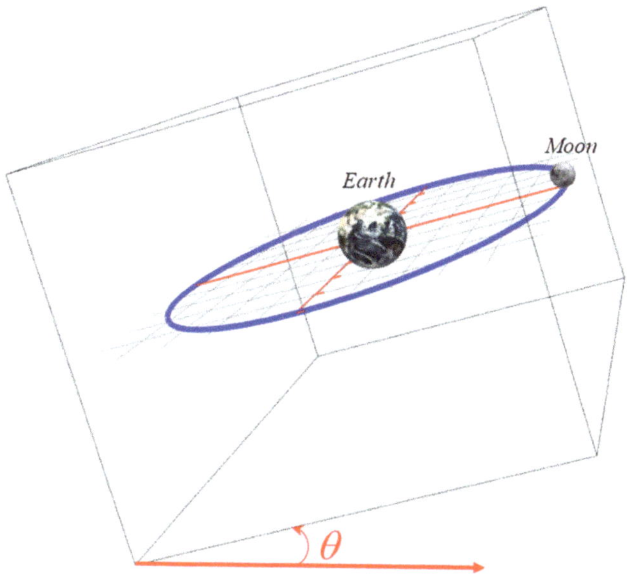

Figure 8.14. Moon's orbit around the Earth.

Using the gradient in spherical coordinates, we find

$$\nabla U = \hat{r}\frac{\partial U}{\partial r} + \hat{\theta}\frac{1}{r}\frac{\partial U}{\partial \theta} + \hat{\varphi}\frac{1}{r\sin(\theta)}\frac{\partial U}{\partial \varphi}$$

$$\Rightarrow \vec{F} = \hat{r}\frac{\partial U}{\partial r} = \hat{r}\frac{\partial}{\partial r}\left[\frac{GMm}{r}\right] \tag{8.185}$$

$$\Rightarrow \vec{F} = -\frac{GMm}{r^2}\hat{r}.$$

(b) The work done is given by

$$W = \oint_C \vec{F}(x, y, z) \cdot d\vec{r} = -\oint_C \frac{GMm}{a^2}\hat{r} \cdot d\vec{r}, \tag{8.186}$$

for a circular orbit r is a constant. That means in spherical coordinates, the position vector

$$\vec{r} = a\hat{r} \Rightarrow d\vec{r} = ad\hat{r}. \tag{8.187}$$

Recalling that in spherical coordinates

$$\vec{r} = r\sin(\theta)\cos(\varphi)\hat{x} + r\sin(\theta)\sin(\varphi)\hat{y} + r\cos(\theta)\hat{z}, \tag{8.188}$$

and the unit vector

$$\hat{r} = \frac{\partial\vec{r}}{\partial r}\bigg/\left|\frac{\partial\vec{r}}{\partial r}\right| \tag{8.189}$$

$$= \sin(\theta)\cos(\varphi)\hat{x} + \sin(\theta)\sin(\varphi)\hat{y} + \cos(\theta)\hat{z},$$

one can write

$$d\hat{r} = [\cos(\theta)\cos(\varphi)d\theta - \sin(\theta)\sin(\varphi)d\varphi]\hat{x}$$
$$+ [\cos(\theta)\sin(\varphi)d\theta + \sin(\theta)\cos(\varphi)d\varphi]\hat{y} - \sin(\theta)d\theta\hat{z}. \tag{8.190}$$

There follows that

$$\hat{r} \cdot d\vec{r} = \sin(\theta)\cos(\theta)\cos^2(\varphi)d\theta - \sin^2(\theta)\sin(\varphi)\cos(\varphi)d\varphi$$
$$+ \sin(\theta)\cos(\theta)\sin^2(\varphi)d\theta + \sin^2(\theta)\sin(\varphi)\cos(\varphi)d\varphi \tag{8.191}$$
$$- \sin(\theta)\cos(\theta)d\theta. \Rightarrow \hat{r} \cdot d\vec{r} = 0.$$

Then the work done becomes

$$W = \oint_C \vec{F}(x, y, z) \cdot d\vec{r} = -\oint_C \frac{GMm}{a^2}\hat{r} \cdot d\vec{r} = 0. \tag{8.192}$$

Note that for this particular vector (the force), the line integral over a closed curve is zero. Note that in this problem, the Earth and the Moon are considered as point particles.

Conservative vector: In the previous two examples, we saw one for which the line integral over a closed curve is zero, and the other is not. Generally, the vector function

$$\vec{V}(x, y, z) = V_x(x, y, z)\hat{x} + V_y(x, y, z)\hat{y} + V_z(x, y, z)\hat{z}, \tag{8.193}$$

is called a conservative vector field if any one of the following conditions is satisfied:
- the line integral over a closed curve is zero independent of the path,

$$\oint_C \vec{V}(x, y, z) \cdot d\vec{r} = 0, \tag{8.194}$$

- the curl of the vector is zero,

$$\nabla \times \vec{V}(x, y, z) = 0, \tag{8.195}$$

- the vector field can be expressed as a gradient of some scalar function $U(x, y, z)$,

$$\vec{V}(x, y, z) = \nabla U(x, y, z). \tag{8.196}$$

Note that as we will see in the following sections, if any one of these is true, then the other two are automatically true.

8.7 Conservative vectors and exact differentials

Exact vs. inexact differentials: Consider the scalar function $U(x, y, z)$. We recall that the differential for this function is given by

$$dU = \frac{\partial U}{\partial x}dx + \frac{\partial U}{\partial y}dy + \frac{\partial U}{\partial z}dz$$

$$dU = \left(\frac{\partial U}{\partial x}\hat{x} + \frac{\partial U}{\partial y}\hat{y} + \frac{\partial U}{\partial z}\hat{z}\right) \cdot (dx\hat{x} + dy\hat{y} + dz\hat{z}) = \nabla U \cdot d\vec{r}. \tag{8.197}$$

Let us assume that the scalar potential $U(x, y, z)$ be a potential function for some conservative vector field,

$$\vec{V}(x, y, z) = V_x(x, y, z)\hat{x} + V_y(x, y, z)\hat{y} + V_z(x, y, z)\hat{z},$$

$$\Rightarrow \vec{V}(x, y, z) = \nabla U(x, y, z), \tag{8.198}$$

then we can write

$$dU = \nabla U \cdot d\vec{r} = \vec{V} \cdot d\vec{r}$$

$$\Rightarrow V_x = \frac{\partial U}{\partial x}, \; V_y = \frac{\partial U}{\partial y}, \; V_z = \frac{\partial U}{\partial z}. \tag{8.199}$$

For a conservative vector field, we know that

$$\nabla \times \vec{V}(x, y, z) = \begin{vmatrix} \hat{x} & \hat{y} & \hat{z} \\ \frac{\partial}{\partial x} & \frac{\partial}{\partial y} & \frac{\partial}{\partial z} \\ V_x & V_y & V_z \end{vmatrix} = \left(\frac{\partial V_z}{\partial y} - \frac{\partial V_y}{\partial z}\right)\hat{x} + \left(\frac{\partial V_x}{\partial z} - \frac{\partial V_z}{\partial x}\right)\hat{y}$$

$$+ \left(\frac{\partial V_y}{\partial x} - \frac{\partial V_x}{\partial y}\right)\hat{z} = 0, \tag{8.200}$$

which leads to

$$\frac{\partial V_z}{\partial y} - \frac{\partial V_y}{\partial z} = \frac{\partial V_x}{\partial z} - \frac{\partial V_z}{\partial x} = \frac{\partial V_y}{\partial x} - \frac{\partial V_x}{\partial y} = 0$$

$$\Rightarrow \frac{\partial V_z}{\partial y} = \frac{\partial V_y}{\partial z}, \; \frac{\partial V_x}{\partial z} = \frac{\partial V_z}{\partial x}, \; \frac{\partial V_y}{\partial x} = \frac{\partial V_x}{\partial y}. \tag{8.201}$$

Based on our assumption, the scalar function U is related to a conservative vector field V by equation (8.199). Thus using equation (8.199), we can rewrite equation (8.201) as

$$\frac{\partial^2 U}{\partial y \partial z} = \frac{\partial^2 U}{\partial z \partial y}, \; \frac{\partial^2 U}{\partial z \partial x} = \frac{\partial^2 U}{\partial x \partial z}, \; \frac{\partial^2 U}{\partial x \partial y} = \frac{\partial^2 U}{\partial y \partial x}. \tag{8.202}$$

These relations are known as *reciprocity relations*. When the reciprocity relations are completely satisfied, the differential

$$dU = \frac{\partial U}{\partial x}dx + \frac{\partial U}{\partial y}dy + \frac{\partial U}{\partial z}dz, \tag{8.203}$$

is said to be an exact differential. We derived the *reciprocity relations* assuming the scalar function U is a function that its gradient gives a conservative vector field V (i.e., $\vec{V}(x, y, z) = \nabla U(x, y, z)$). Therefore, we can conclude that a differential

$$dU = \frac{\partial U}{\partial x}dx + \frac{\partial U}{\partial y}dy + \frac{\partial U}{\partial z}dz$$
$$= f(x, y, z)dx + g(x, y, z)dy + h(x, y, z)dz, \tag{8.204}$$

is *exact* when the vector field

$$\vec{V}(x, y, z) = f(x, y, z)\hat{x} + g(x, y, z)\hat{y} + h(x, y, z)\hat{z} \tag{8.205}$$

is a *conservative vector field*. Which means, we can also say, when

$$\oint_C \vec{V}(x, y, z) \cdot d\vec{r} = \oint_C f(x, y, z)dx + \oint_C g(x, y, z)dy$$
$$+ \oint_C h(x, y, z)dz = 0, \tag{8.206}$$

the differential is *exact* otherwise it is *inexact*.

8.8 Double integral and Green's theorem

Let us consider two functions $P(x, y)$ and $Q(x, y)$. When these two functions and their first derivatives are continuous for all x and y in the region bounded (shaded pink) by the curve C of area A (see figure 8.15), Green's theorem states

$$\iint_A \left[\frac{\partial Q(x, y)}{\partial x} - \frac{\partial P(x, y)}{\partial y} \right]dxdy = \oint_C [P(x, y)dx + Q(x, y)dy], \tag{8.207}$$

where the line integral is in a counterclockwise direction around the boundary of area A (i.e., on the curve C). Next, we shall prove this theorem. To this end, we note that

$$\iint_A \left[\frac{\partial Q(x, y)}{\partial x} - \frac{\partial P(x, y)}{\partial y} \right]dxdy = \iint_A \frac{\partial Q(x, y)}{\partial x}dxdy - \iint_A \frac{\partial P(x, y)}{\partial y}dydx. \tag{8.208}$$

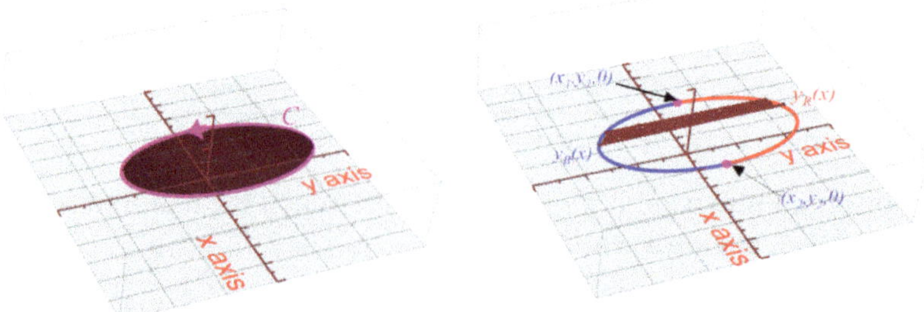

Figure 8.15. A region on the x–y plane bounded by two curves $y_B(x)$ and $y_R(x)$.

For the area shown in figure 8.15, we may write

$$\iint_A \frac{\partial P(x, y)}{\partial y}\, dy\, dx = \int_{x_1}^{x_2}\left[\int_{y_B(x)}^{y_R(x)} \frac{\partial P(x, y)}{\partial y}\, dy\right] dx, \tag{8.209}$$

so that applying the relation

$$\int_a^b \frac{df(t)}{dt}\, dt = f(b) - f(a), \tag{8.210}$$

one can write

$$\iint_A \frac{\partial P(x, y)}{\partial y}\, dy\, dx = \int_{x_1}^{x_2} [P(x, y_R(x)) - P(x, y_B(x))]\, dx$$

$$= \int_{x_1}^{x_2} P(x, y_R(x))\, dx - \int_{x_1}^{x_2} P(x, y_B(x))\, dx$$

$$= -\int_{x_2}^{x_1} P(x, y_R(x))\, dx - \int_{x_1}^{x_2} P(x, y_B(x))\, dx = -\oint_C P(x, y) \tag{8.211}$$

$$dx$$

$$\Rightarrow -\iint_A \frac{\partial P(x, y)}{\partial y}\, dy\, dx = \oint_C P(x, y)\, dx.$$

Note that the line integral over the closed curve C is the region's boundary with area A and must be integrated with a counterclockwise direction. The limits of integration in the two integrals indicate that we perform the integration first on the upper part of the curve from $x_1 \to x_2$ and then from $x_2 \to x_1$. Similarly, when we perform the integration in the reverse order (i.e., first with respect to x and then with respect to y) for the integral

$$\iint_A \frac{\partial Q(x, y)}{\partial x}\, dx\, dy = \int_{y_1}^{y_2}\left[\int_{x_B(y)}^{x_R(y)} \frac{\partial Q(x, y)}{\partial x}\, dx\right] dy, \tag{8.212}$$

as shown in figure 8.16, for the same area, we can write

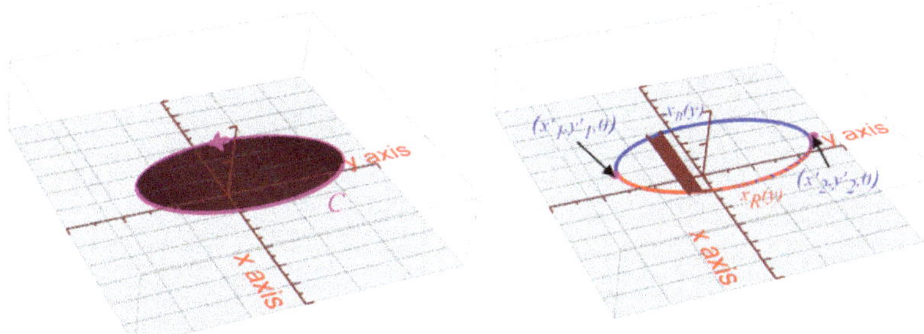

Figure 8.16. A region on the x–y plane bounded by two curves $y_B(x)$ and $y_R(x)$.

$$\iint_A \frac{\partial Q(x, y)}{\partial x} dx dy = \int_{y_1'}^{y_2'} \left[\int_{x_B(y)}^{x_R(y)} \frac{\partial Q(x, y)}{\partial x} dx \right] dy$$

$$= \int_{y_1'}^{y_2'} [Q(x_R(y), y) - Q(x_B(y), y)] dy$$

$$= \int_{y_1'}^{y_2'} Q(x_R(y), y) dy - \int_{y_1'}^{y_2'} Q(x_B(y), y) dy \qquad (8.213)$$

$$= \int_{y_1'}^{y_2'} Q(x_R(y), y) dy + \int_{y_2'}^{y_1'} Q(x_B(y), y) dy = \oint_C Q(x, y) dy$$

$$\Rightarrow \iint_A \frac{\partial Q(x, y)}{\partial x} dx dy = \oint_C Q(x, y) dy.$$

The line integral over the same curve C is also carried out in a counterclockwise direction. Therefore, combining equations (8.211) and (8.213), one finds

$$\iint_A \left[\frac{\partial Q(x, y)}{\partial x} - \frac{\partial P(x, y)}{\partial y} \right] dx dy = \oint_C [P(x, y) dx + Q(x, y) dy], \qquad (8.214)$$

which is Green's theorem.

Example 8.8. Let the area A be the region bounded by the curves $y = x^2$ and $x = y^2$. Verify Green's theorem for the integral

$$I = \oint_C [(2xy - x^2) dx + (x + y^2) dy]. \qquad (8.215)$$

Solution: We want to verify that the result for

$$I_1 = \oint_C [P(x, y) dx + Q(x, y) dy] = \oint_C [(2xy - x^2) dx + (x + y^2) dy], \qquad (8.216)$$

is equal to

$$I_2 = \iint_A \left[\frac{\partial Q(x, y)}{\partial x} - \frac{\partial P(x, y)}{\partial y} \right] dx dy$$

$$= \iint_A \left[\frac{\partial (x + y^2)}{\partial x} - \frac{\partial (2xy - x^2)}{\partial y} \right] dx dy, \qquad (8.217)$$

for the area shown in figure 8.17. To this end, the line integral can be rewritten as

$$I_1 = \int_{C_{green}} [(2xy - x^2)dx + (x + y^2)dy]$$
$$+ \int_{C_{blue}} [(2xy - x^2)dx + (x + y^2)dy].$$

(8.218)

For the green curve,

$$y = x^2 \Rightarrow dy = 2xdx,$$

(8.219)

for the blue curve,

$$x = y^2 \Rightarrow dx = 2ydy,$$

(8.220)

and noting that the two curves intersect at the two points $(0, 0)$ and $(1, 1)$, we find for equation (8.218)

$$I_1 = \int_0^1 [(2x^3 - x^2)dx + (x + x^4)2xdx] + \int_1^0 [(2y^3 - y^4)2ydy + 2y^2dy]$$
$$= \int_0^1 [2x^5 + 2x^3 + x^2]dx + \int_1^0 [4y^4 - 2y^5 + 2y^2]dy$$

(8.221)

$$\Rightarrow I_1 = \frac{1}{30}.$$

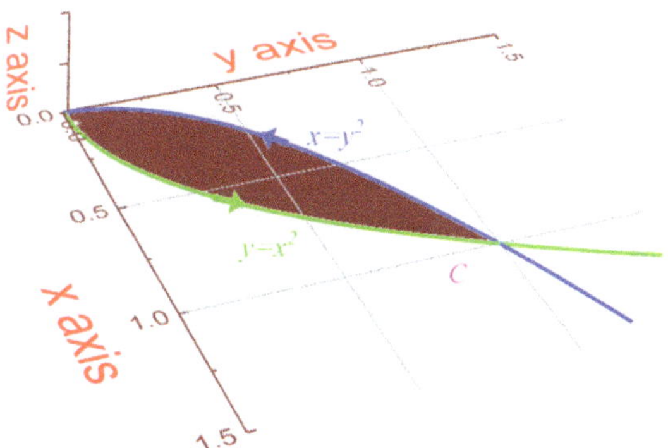

Figure 8.17. A region bounded by the curves $y = x^2$ and $x = y^2$ on the x–y plane.

For the area integral

$$I_2 = \iint_A \left[\frac{\partial(x + y^2)}{\partial x} - \frac{\partial(2xy - x^2)}{\partial y} \right] dx\,dy = \int_0^1 \left[\int_{y^2}^{\sqrt{y}} (1 - 2x) dx \right] dy$$

$$= \int_0^1 (x - x^2) \Big|_{y^2}^{\sqrt{y}} dy = \int_0^1 \left(\sqrt{y} - y - y^2 + y^4 \right) dy \qquad (8.222)$$

$$= \left[\frac{2}{3} y^{3/2} - \frac{y^2}{2} - \frac{y^3}{3} + \frac{y^5}{5} \right]_0^1 \Rightarrow I_2 = \frac{1}{30}.$$

Therefore, according to the results in equation (8.221) and (8.222),

$$I_1 = I_2, \qquad (8.223)$$

which verifies Green's theorem.

8.9 The Stokes' theorem

Suppose we are given a vector function,

$$\vec{D}(\vec{r}) = D_x(\vec{r})\hat{x} + D_y(\vec{r})\hat{y} + D_z(\vec{r})\hat{z}. \qquad (8.224)$$

and a closed curve C on the x–y plane as shown in figure 8.18. Applying Green's theorem,

$$\oint_C [P(x, y)dx + Q(x, y)dy] = \iint_A \left[\frac{\partial Q(x, y)}{\partial x} - \frac{\partial P(x, y)}{\partial y} \right] dx\,dy, \qquad (8.225)$$

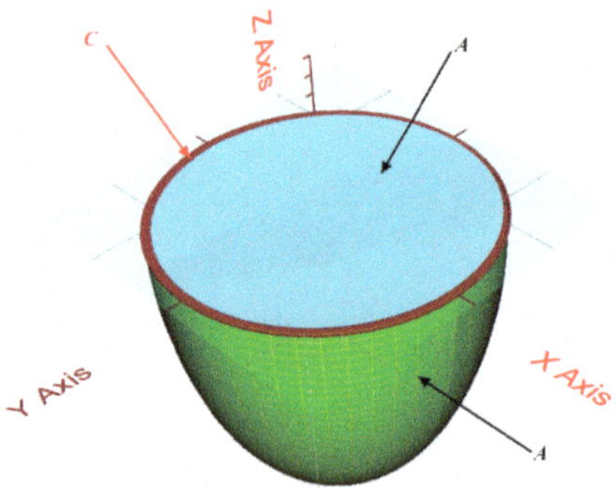

Figure 8.18. A closed curve C (red) on the x–y plane. This curve is a boundary to a 2-D circular surface on the x–y plane (cyan) and also a 3-D balloon like surface (green).

for

$$P(x, y) = D_x(x, y, z), \quad Q(x, y) = D_y(x, y, z), \tag{8.226}$$

and the closed curve shown in figure 8.18, one can write

$$\oint_C \left[D_x(\vec{r})dx + D_y(\vec{r})dy \right] = \iint_A \left[\frac{\partial D_y(\vec{r})}{\partial x} - \frac{\partial D_x(\vec{r})}{\partial y} \right] dxdy, \tag{8.227}$$

which can be put in the form

$$\oint_C \left[D_x(\vec{r})\hat{x} + D_y(\vec{r})\hat{y} + D_z(\vec{r})\hat{z} \right] \cdot (dx\hat{x} + dy\hat{y} + 0\hat{z})$$
$$= \iint_A \left[\frac{\partial D_y(\vec{r})}{\partial x} - \frac{\partial D_x(\vec{r})}{\partial y} \right] dxdy. \tag{8.228}$$

An infinitesimal displacement on the curve C can be written as

$$d\vec{r} = dx\hat{x} + dy\hat{y} + 0\hat{z}. \tag{8.229}$$

For a vector $\vec{D}$ the curl in Cartesian coordinates is given by

$$\nabla \times \vec{D}(\vec{r}) = \begin{vmatrix} \hat{x} & \hat{y} & \hat{z} \\ \dfrac{\partial}{\partial x} & \dfrac{\partial}{\partial y} & \dfrac{\partial}{\partial z} \\ D_x & D_y & D_z \end{vmatrix} = \left(\frac{\partial D_z}{\partial y} - \frac{\partial D_y}{\partial z} \right)\hat{x} + \left(\frac{\partial D_x}{\partial z} - \frac{\partial D_z}{\partial x} \right)\hat{y}$$
$$+ \left(\frac{\partial D_y}{\partial x} - \frac{\partial D_x}{\partial y} \right)\hat{z}, \tag{8.230}$$

so that one can write

$$(\nabla \times \vec{D}(\vec{r})) \cdot \hat{z} = \frac{\partial D_y}{\partial x} - \frac{\partial D_x}{\partial y}. \tag{8.231}$$

Now substituting equations (8.229) and (8.231) into equation (8.228), one finds

$$\oint_C \vec{D}(\vec{r}) \cdot d\vec{r} = \iint_A (\nabla \times \vec{D}(\vec{r})) \cdot \hat{z}(dxd)y. \tag{8.232}$$

Noting that $\hat{z}$ is a unit vector normal to the infinitesimal area $dxdy$, which is on the x–y plane, one can define an infinitesimal area vector $d\vec{a}$ in Cartesian coordinates

$$d\vec{a} = da_x\hat{x} + da_y\hat{y} + da_z\hat{z} = \hat{x}(dydz) + \hat{y}(dzdx) + \hat{z}(dxdy), \tag{8.233}$$

so that instead of the area bounded by the curve C on the x–y plane, for an infinitesimal area in 3-D (for example, for the balloon surface in figure 8.18), one can write

$$\hat{z}dxdy \rightarrow \hat{n}da, \tag{8.234}$$

where $\hat{n}$ is the unit vector normal to the infinitesimal area da in 3-D. Therefore equation (8.232) can be rewritten as

$$\oint_C \vec{D}(\vec{r}) \cdot d\vec{r} = \int_A \nabla \times \vec{D}(\vec{r}) \cdot \hat{n}\,da. \qquad (8.235)$$

Equation (8.235) is *Stokes' theorem*. The following are some important notes to make when we apply Stokes' theorem.

1. The curve C bounds the open area A. The integral must therefore be the same for any area A bounded by the curve C. This means that you can pick an easy area A if you wish to make life simpler!

2. The integral

$$\Phi = \int_A \left(\nabla \times \vec{D}(\vec{r}) \right) \cdot \hat{n}\,dA, \qquad (8.236)$$

   is just the flux of the curl of the vector $\vec{D}$ through the open area A.

3. The integral

$$\Phi = \oint_C \vec{D}(\vec{r}) \cdot d\vec{r} \qquad (8.237)$$

   is the line integral of the vector $\vec{D}$ around the closed curve C. If $\vec{D}$ is a force, then this integral is just the work done by that force around the curve C. If the force is a *conservative force*, then the integral is zero. Thus for a conservative vector $\vec{D}$

$$\int_A \nabla \times \vec{D}(\vec{r}) \cdot \hat{n}\,da = \oint_C \vec{D}(\vec{r}) \cdot d\vec{r} = 0$$
$$\Rightarrow \nabla \times \vec{D}(\vec{r}) = 0. \qquad (8.238)$$

This proves one of the conditions we stated in section 8.6 regarding conservative vector functions.

Example 8.9. Consider the vector field

$$\vec{D}(\vec{r}) = (3y, \, -xz, \, yz^2), \qquad (8.239)$$

and the paraboloid

$$2z = x^2 + y^2, \qquad (8.240)$$

bounded by the plane $z = 2$ as shown in figure 8.19.

    (a) Evaluate the flux Φ directly from

$$\Phi = \int_A \nabla \times \vec{D} \cdot \hat{n}\,da. \qquad (8.241)$$

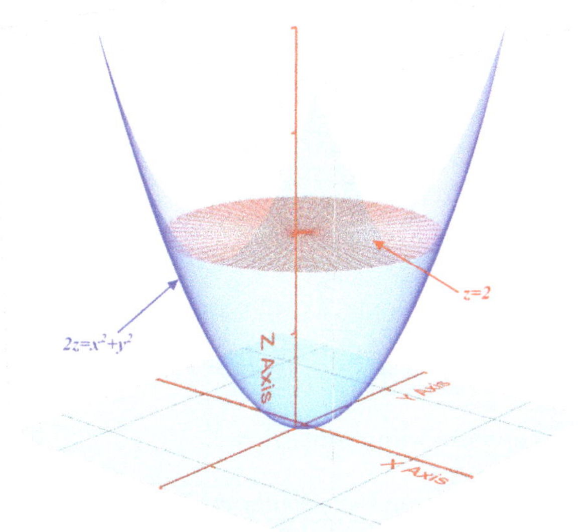

Figure 8.19. A paraboloid surface defined by $2z = x^2 + y^2$ and bounded by a plane defined by $z = 2$.

(b) Verify the result in (a) by calculating the flux Φ applying Stokes' theorem. In other words, evaluate the integral

$$\oint_C \vec{D} \cdot d\vec{r}, \tag{8.242}$$

and show that the result is the same as in (a).

Solution:

(a) The curl of the vector

$$\vec{D}(\vec{r}) = 3y\hat{x} - xz\hat{y} + yz^2\hat{z}, \tag{8.243}$$

is given by

$$\nabla \times \vec{D}(\vec{r}) = \begin{vmatrix} \hat{x} & \hat{y} & \hat{z} \\ \dfrac{\partial}{\partial x} & \dfrac{\partial}{\partial y} & \dfrac{\partial}{\partial z} \\ D_x & D_y & D_z \end{vmatrix} \tag{8.244}$$

$$= \left(\frac{\partial(yz^2)}{\partial y} - \frac{\partial(-xz)}{\partial z}\right)\hat{x} + \left(\frac{\partial(3y)}{\partial z} - \frac{\partial(yz^2)}{\partial x}\right)\hat{y} + \left(\frac{\partial(-xz)}{\partial x} - \frac{\partial(3y)}{\partial y}\right)\hat{z}$$

$$\Rightarrow \nabla \times \vec{D}(\vec{r}) = (z^2 + x)\hat{x} - (3 + z)\hat{z}.$$

On the plane defined by $z = 2$, equation (8.244) becomes

$$\nabla \times \vec{D}(\vec{r}) = (4 + x)\hat{x} - 5\hat{z}, \tag{8.245}$$

and for the flux, one can write

$$\Phi = \int_A \nabla \times \vec{D}(\vec{r}) \cdot \hat{n}dA = \int_A [(4 + x)\hat{x} - 5\hat{z}] \cdot \hat{n}da$$

$$= \int_{-2}^{2} \int_{-\sqrt{4-x^2}}^{\sqrt{4-x^2}} [(4 + x)\hat{x} - 5\hat{z}] \cdot \hat{z}dxdy \tag{8.246}$$

$$\Rightarrow \Phi = -5 \int_{-2}^{2} \int_{-\sqrt{4-x^2}}^{\sqrt{4-x^2}} dxdy = -20\pi,$$

where we used

$$\int_{-2}^{2} \int_{-\sqrt{4-x^2}}^{\sqrt{4-x^2}} dxdy = \pi R^2,$$

for an area of a circle with radius $R = 2$.

(b) Noting that for a point on the circular curve on the plane defined by $z = 2$,

$$\vec{D}(\vec{r}) \cdot d\vec{r} = (3y\hat{x} - 2x\hat{y} + 4y\hat{z}) \cdot (dx\hat{x} + dy\hat{y} + 0\hat{z})$$
$$\Rightarrow \vec{D}(\vec{r}) \cdot d\vec{r} = 3ydx - 2xdy, \tag{8.247}$$

applying Stokes' theorem

$$\Phi = \int_A \nabla \times \vec{D}(\vec{r}) \cdot \hat{n}da = \oint_C \vec{D}(\vec{r}) \cdot d\vec{r}, \tag{8.248}$$

one can write for the flux

$$\Phi = \oint_C \vec{D}(\vec{r}) \cdot d\vec{r} = \oint_C (3ydx - 2xdy). \tag{8.249}$$

For a circle with radius $R = 2$, we have

$$x = 2\cos(\varphi), \; y = 2\sin(\varphi), \; z = 2$$
$$\Rightarrow dx = -2\sin(\varphi)d\varphi, \; dy = 2\cos(\varphi)d\varphi, \tag{8.250}$$

and equation (8.249) becomes

$$\oint_C \vec{D}(\vec{r}) \cdot d\vec{r} = \int_0^{2\pi} (-12\sin^2(\varphi)d\varphi - 8\cos^2(\varphi)d\varphi)$$

$$= \int_0^{2\pi} (-12 + 4\cos^2(\varphi))d\varphi \Rightarrow \oint_C \vec{D}(\vec{r}) \cdot d\vec{r} = -20\pi, \tag{8.251}$$

where we used the integral

$$\int_0^{2\pi} \cos^2(\varphi)d\varphi = \pi. \tag{8.252}$$

The results in equation (8.251) shows the flux is the same as in (a) as expected.

8.10 The divergence theorem

Consider a three-dimensional region occupying a volume V and a surface S enclosing this region, such as the one shown in figure 8.20. The divergence theorem states that for a vector function $\vec{D}(\vec{r})$,

$$\iiint_V \left(\nabla \cdot \vec{D}(\vec{r}) \right) dv = \iint_S \vec{D}(\vec{r}) \cdot \hat{n} \, da, \qquad (8.253)$$

where dv is an infinitesimal volume inside the region with total volume V, and da is an infinitesimal area on the enclosing surface S, and $\hat{n}$ is an outward unit vector normal to the area da.

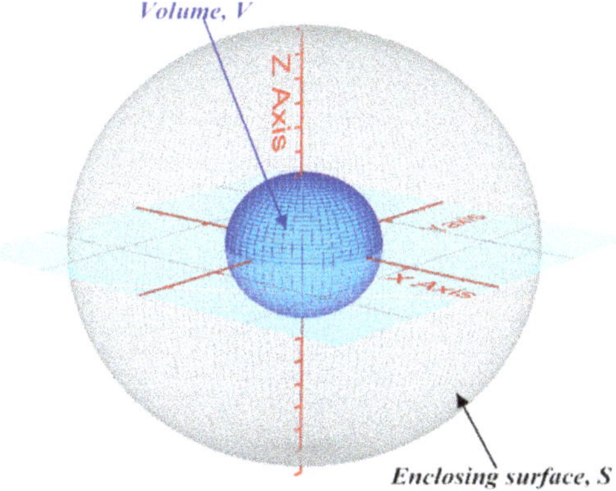

Figure 8.20. A region occupying a volume V enclosed by a surface S.

Example 8.10. Consider the vector field

$$\vec{D}(\vec{r}) = xy\hat{x} + yz\hat{y} + zx\hat{z}, \qquad (8.254)$$

shown in figure 8.21.

    (a) Compute the flux Φ_v through the closed rectangular surface of sides a, b, and c shown in figure 8.22 by direct application of the definition of flux,

$$\Phi = \iint_S \vec{D}(\vec{r}) \cdot \hat{n} \, da. \qquad (8.255)$$

    (b) Compute Φ through the same surface indirectly by applying the divergence theorem,

$$\Phi = \iint_S \vec{D}(\vec{r}) \cdot \hat{n} \, da = \iiint_V \left(\nabla \cdot \vec{D}(\vec{r}) \right) dv. \qquad (8.256)$$

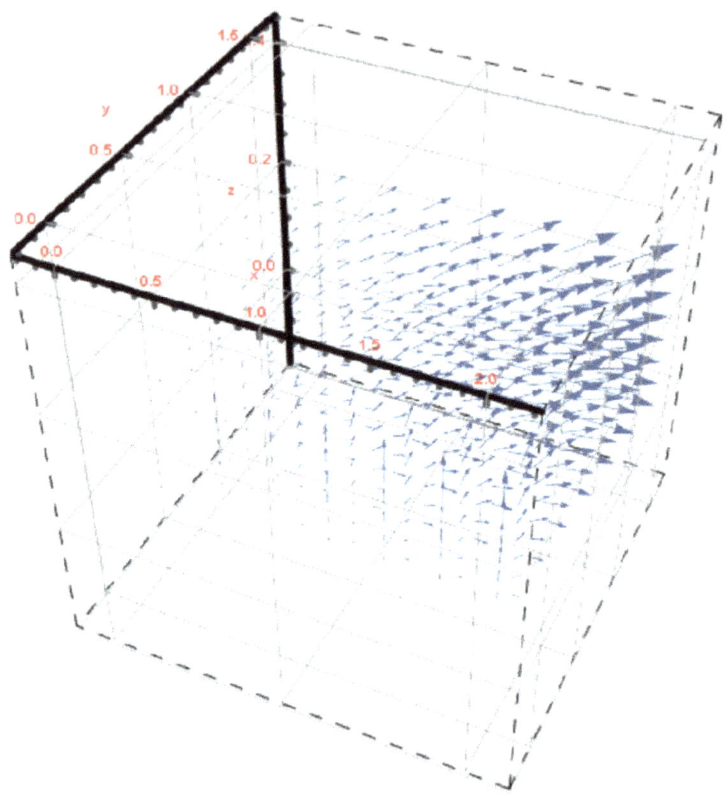

Figure 8.21. A 3-D vector plot for a vector field $\vec{D}(x, y, z) = xy\hat{x} + yz\hat{y} + zx\hat{z}$.

Solution:

(a) The flux Φ for the vector field $\vec{D}(x, y, z)$ over a given surface area A is defined as

$$\Phi_v = \int\int \vec{D}(x, y, z) \cdot \hat{n}\,da. \tag{8.257}$$

Figure 8.22 has three pairs of surfaces normal to the three axis. For the flux through the surfaces normal to the x-axis (i.e., the plane surfaces defined by $x = 0$ and $x = a$), we have

$$\Phi_x = \int_0^c \int_0^b \vec{D}(a, y, z) \cdot \hat{x}\,dydz + \int_0^c \int_0^b \vec{D}(0, y, z) \cdot (-\hat{x})\,dydz$$
$$= \int_0^c \int_0^b D_x(a, y, z)\,dydz - \int_0^c \int_0^b D_x(0, y, z)\,dydz \tag{8.258}$$

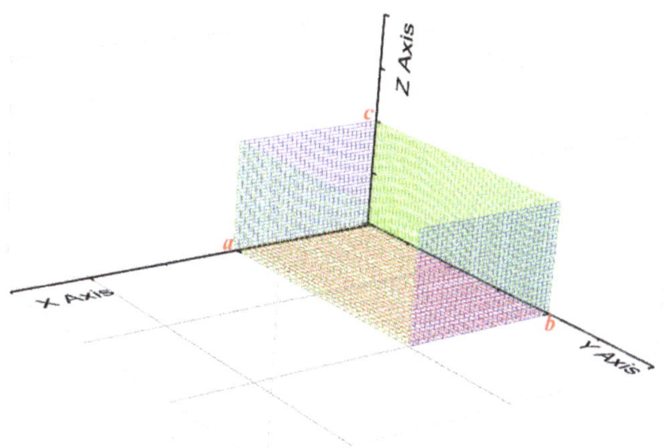

Figure 8.22. Rectangular closed surface of length a, width b, and height c.

and using

$$D_x(x, y, z) = xy \Rightarrow D_x(0, y, z) = 0, \ D_x(a, y, z) = ay, \tag{8.259}$$

we find

$$\phi_x = \int_0^c \int_0^b ay\,dy\,dz = \frac{ab^2c}{2}. \tag{8.260}$$

For the surfaces normal to the y-axis, (i.e., the plane surfaces defined by $y = 0$ and $y = b$), one can write

$$\Phi_y = \int\int \vec{D}(x, b, z) \cdot \hat{y}\,dx\,dz + \int\int \vec{D}(x, 0, z) \cdot (-\hat{y})\,dx\,dz \tag{8.261}$$

$$= \int_0^c \int_0^a D_y(x, b, z)\,dx\,dz - \int_0^c \int_0^a D_y(x, 0, z)\,dx\,dz, \tag{8.262}$$

so that using

$$D_y(x, y, z) = yz \Rightarrow D_y\big(x, 0, z\big) = 0, \ D_y\big(x, b, z\big) = bz, \tag{8.263}$$

one finds

$$\Phi_y = \int_0^c \int_0^a D_z(x, b, z)\,dx\,dz - \int_0^c \int_0^a D_y(x, 0, z)\,dx\,dz = \frac{abc^2}{2}. \tag{8.264}$$

Following a similar procedure for the surfaces normal to the z-axis (i.e., the plane surfaces defined by $z = 0$ and $z = c$), we can show that

$$\Phi_z = \int_0^b \int_0^a D_z(x, y, c)dxdy - \int_0^b \int_0^a D_y(x, y, 0)dxdy = \frac{a^2bc}{2}. \qquad (8.265)$$

Therefore, the total flux becomes

$$\Phi = \Phi_x + \Phi_y + \Phi_z = \frac{a^2bc}{2} + \frac{ab^2c}{2} + \frac{abc^2}{2}. \qquad (8.266)$$

(b) According to the divergence theorem

$$\Phi = \iint_S \vec{D}(x, y, z) \cdot \hat{n}da = \iiint_V (\nabla \cdot \vec{D}(x, y, z))dxdydz. \qquad (8.267)$$

Thus the flux can be expressed as

$$\begin{aligned}
\Phi &= \iiint_V \left(\nabla \cdot \vec{D}(x, y, z)\right)dxdydz \\
&= \int_0^c \int_0^b \int_0^a \left(\frac{\partial D_x}{\partial x} + \frac{\partial D_y}{\partial y} + \frac{\partial D_z}{\partial z}\right)dxdydz \\
&= \int_0^c \int_0^b \int_0^a \left(\frac{\partial(xy)}{\partial x} + \frac{\partial(yz)}{\partial y} + \frac{\partial(zx)}{\partial z}\right)dxdydz \qquad (8.268) \\
&= \int_0^c \int_0^b \int_0^a (y + z + x)dxdydz \\
\Rightarrow \Phi &= \frac{a^2bc}{2} + \frac{ab^2c}{2} + \frac{abc^2}{2},
\end{aligned}$$

which is the same as the result we found in (a).

Example 8.11. *The divergence theorem in cylindrical coordinates*: Verify the divergence theorem

$$\iiint_V \left(\nabla \cdot \vec{D}(\vec{r})\right)dv = \iint_S \vec{D}(\vec{r}) \cdot \hat{n}da, \qquad (8.269)$$

for the vector field

$$\vec{D} = as\hat{s} + b\hat{\varphi} + cz\hat{z} \qquad (8.270)$$

(where a, b, and c are constants) for a cylinder with radius R and height H centered about the z-axis (see figure 8.23).

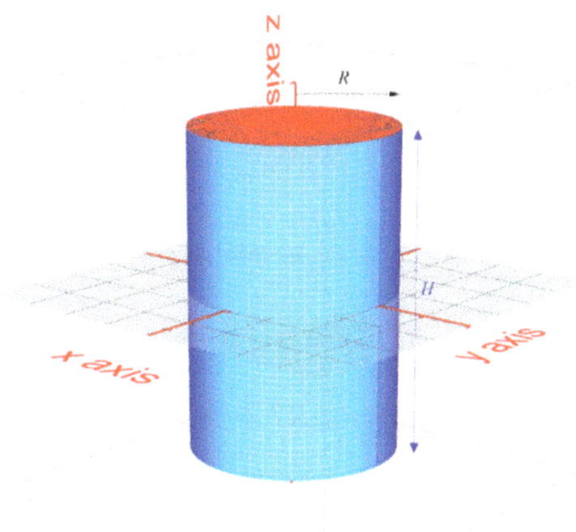

Figure 8.23. A cylinder with radius R and height H centered about the z-axis.

Solution: For the components of the vector $\vec{V}$ in cylindrical coordinates, we have

$$V_s = as, \ V_\varphi = b, \ Vz = cz \tag{8.271}$$

and the divergence

$$\nabla \cdot \vec{D} = \frac{1}{s}\frac{\partial}{\partial s}(sV_s) + \frac{1}{s}\frac{\partial}{\partial \varphi}(V_\varphi) + \frac{\partial}{\partial z}(V_z) \tag{8.272}$$

becomes

$$\nabla \cdot \vec{D} = \frac{1}{s}\frac{\partial}{\partial s}(s^2 a) + \frac{1}{s}\frac{\partial}{\partial \varphi}(b) + \frac{\partial}{\partial z}(cz) = 2a + c. \tag{8.273}$$

Using equation (8.273), for the volume integral in cylindrical coordinates, we find

$$\iiint_V (\nabla \cdot \vec{D}(\vec{r}))dv = (2a + c)\int_0^R \int_0^{2\pi} \int_{-H/2}^{H/2} sdsd\varphi dz$$
$$\Rightarrow \iiint_V (\nabla \cdot \vec{D}(\vec{r}))dv = (2a + c)\pi R^2 H. \tag{8.274}$$

For the surface integral over cylindrical surface enclosing this volume, one can write

$$\iint_S \vec{D}(\vec{r}) \cdot \hat{n}da = \iint_{\text{bottom}} \vec{D}_b(\vec{r}) \cdot \hat{n}_b da$$
$$+ \iint_{\text{top}} \vec{D}_t(\vec{r}) \cdot \hat{n}_t da + \iint_{\text{side}} \vec{D}_s(\vec{r}) \cdot \hat{n}_s da. \tag{8.275}$$

We note that an outward unit vectors normal to the bottom ($\hat{n}_b$), top ($\hat{n}_t$), and side ($\hat{n}_s$) surfaces are

$$\hat{n}_b = -\hat{z}, \ \hat{n}_t = \hat{z}, \ \hat{n}_s = \hat{s}. \tag{8.276}$$

Furthermore, using the vector

$$\vec{D} = as\hat{s} + b\hat{\varphi} + cz\hat{z}, \tag{8.277}$$

we can write

$$\vec{D}_b(\vec{r}) = \vec{D}(s, \varphi, -H/2) = as\hat{s} + b\hat{\varphi} - \frac{cH}{2}\hat{z},$$

$$\vec{D}_t(\vec{r}) = \vec{D}(s, \varphi, H/2) = as\hat{s} + b\hat{\varphi} + \frac{cH}{2}\hat{z}, \tag{8.278}$$

$$\vec{D}_s(\vec{r}) = \vec{D}(R, \varphi, z) = aR\hat{s} + b\hat{\varphi} + cz\hat{z}.$$

Using the results in equations (8.276) and (8.278), one finds

$$\iint_{\text{bottom}} \vec{D}_b(\vec{r}) \cdot \hat{n}_b da = \int_0^R \int_0^{2\pi} \left(as\hat{s} + b\hat{\varphi} - \frac{cH}{2}\hat{z}\right) \cdot (-\hat{z})s\,ds\,d\varphi$$

$$\Rightarrow \iint_{\text{bottom}} \vec{D}_b(\vec{r}) \cdot \hat{n}_b da = \frac{\pi cHR^2}{2}, \tag{8.279}$$

$$\iint_{\text{top}} \vec{D}_t(\vec{r}) \cdot \hat{n}_t da = \int_0^R \int_0^{2\pi} \left(as\hat{s} + b\hat{\varphi} + \frac{cH}{2}\hat{z}\right) \cdot (\hat{z})s\,ds\,d\varphi$$

$$\Rightarrow \iint_{\text{top}} \vec{D}_t(\vec{r}) \cdot \hat{n}_t da = \frac{\pi cHR^2}{2}, \tag{8.280}$$

$$\iint_{\text{side}} \vec{D}_s(\vec{r}) \cdot \hat{n}_s da = \int_0^{2\pi} \int_{-H/2}^{H/2} (aR\hat{s} + b\hat{\varphi} + cz\hat{z}) \cdot (\hat{s})R\,d\varphi\,dz$$

$$\Rightarrow \iint_{\text{side}} \vec{D}_s(r, \varphi, z) \cdot \hat{n}_s da = 2a\pi R^2 H. \tag{8.281}$$

Upon substituting equations (8.279)–(8.281) into equation (8.275), we find for the surface integral

$$\iint_S \vec{D}(r, \varphi, z) \cdot \hat{n} da = c\pi R^2 H + 2a\pi R^2 H. = (2a + c)\pi R^2 H$$

$$\Rightarrow \iint_S \vec{D}(r, \varphi, z) \cdot \hat{n} da = \iiint_V \left(\nabla \cdot \vec{D}(x, y, z)\right)dx\,dy\,dz, \tag{8.282}$$

and this verifies the divergence theorem.

Example 8.12. *The divergence theorem in spherical coordinates*: Verify the divergence theorem for the vector field

$$\vec{D}(\vec{r}) = \vec{r} \tag{8.283}$$

using the sphere shown in figure 8.24.

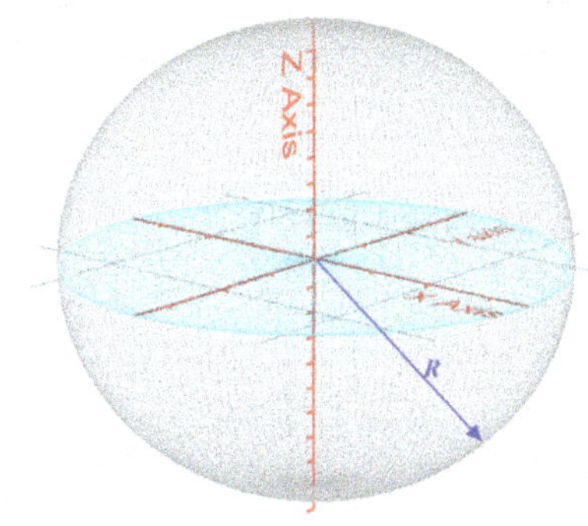

Figure 8.24. A sphere with radius R centered about the origin.

Solution: In spherical coordinates the vector $\vec{D}(\vec{r})$ can be rewritten as

$$\vec{D}(\vec{r}) = \vec{r} = r\hat{r}. \tag{8.284}$$

Then noting that on the surface of the sphere with radius R, this vector becomes

$$\vec{D}(\vec{r}) = R\hat{r}, \tag{8.285}$$

and the outward unit vector normal to an infinitesimal area on the surface of the sphere (i.e., $da = R^2 \sin(\theta)d\theta d\varphi$) is $\hat{n} = \hat{r}$, for the closed surface integral, one finds

$$\iint_S \vec{D}(\vec{r}) \cdot \hat{n}da = \int_0^{2\pi} \int_0^{\pi} (R\hat{r}) \cdot \hat{r} R^2 \sin(\theta)d\theta d\varphi = 4\pi R^3. \tag{8.286}$$

Using the divergence in spherical coordinates

$$\nabla \cdot \vec{D}(\vec{r}) = \frac{1}{r^2}\frac{\partial}{\partial r}(r^2 D_r) + \frac{1}{r \sin(\theta)}\frac{\partial}{\partial \theta}(\sin(\theta)D_\theta) + \frac{1}{r \sin(\theta)}\frac{\partial}{\partial \varphi}(D_\varphi), \tag{8.287}$$

for the vector $\vec{D}(\vec{r}) = r\hat{r}$, we have

$$\nabla \cdot \vec{D} = \frac{1}{r^2}\frac{\partial}{\partial r}(r^2 V_r) = \frac{1}{r^2}\frac{\partial}{\partial r}(r^3) = 3 \tag{8.288}$$

so that the volume integral becomes

$$\iiint_V \left(\nabla \cdot \vec{D}(\vec{r})\right)dV = \int_0^R 3r^2 dr \int_0^\pi \sin(\theta)d\theta \int_0^{2\pi} d\varphi = 4\pi R^3, \tag{8.289}$$

where we used $dV = r^2 dr \sin(\theta)d\theta d\varphi$ in spherical coordinates. The results in equations (8.286) and (8.289) verify the divergence theorem.

Example 8.13. *Stokes' theorem in spherical coordinates*: Verify Stokes' theorem for the magnetic field vector

$$\vec{B} = 4\hat{r} + 3\hat{\theta} - 2\hat{\varphi} \tag{8.290}$$

and the closed curve C shown in the figure 8.25.

Solution: We are given the magnetic field in spherical coordinates with components

$$B_r = 4, \; B_\theta = 3, \; B_\varphi = -2, \tag{8.291}$$

so that the curl of the magnetic field

$$\nabla \times \vec{B} = \frac{1}{r \sin(\theta)}\left(\frac{\partial}{\partial \theta}(\sin(\theta)B_\varphi) - \frac{\partial}{\partial \varphi}(B_\theta)\right)\hat{r}$$
$$+ \left(\frac{1}{r \sin(\theta)}\frac{\partial}{\partial \varphi}(B_r) - \frac{1}{r}\frac{\partial}{\partial r}(rB_\varphi)\right)\hat{\theta} + \frac{1}{r}\left(\frac{\partial}{\partial r}(rB_\theta) - \frac{\partial}{\partial \theta}(B_r)\right)\hat{\varphi}, \tag{8.292}$$

becomes

$$\nabla \times \vec{B} = \frac{1}{r \sin(\theta)}\frac{\partial}{\partial \theta}(-2 \sin(\theta))\hat{r} - \frac{1}{r}\frac{\partial}{\partial r}(-2r)\hat{\theta} + \frac{1}{r}\frac{\partial}{\partial r}(3r)\hat{\varphi}$$
$$= -\frac{2 \cos(\theta)}{r \sin(\theta)}\hat{r} + \frac{2}{r}\hat{\theta} + \frac{3}{r}\hat{\varphi} \Rightarrow \nabla \times \vec{B} = -\frac{2}{r}\cot(\theta)\hat{r} + \frac{2}{r}\hat{\theta} + \frac{3}{r}\hat{\varphi}. \tag{8.293}$$

Noting that the infinitesimal area on the x–y plane inside the curve shown in figure 8.25, in spherical coordinates, can be expressed as

$$\hat{n}da = \hat{z}r \sin(\theta)drd\varphi = [\cos(\theta)\hat{r} - \sin(\theta)\hat{\theta}]r \sin(\theta)drd\varphi, \tag{8.294}$$

one can write

$$\nabla \times \vec{B} \cdot \hat{n}da = \left(-\frac{2}{r}\cot(\theta)\hat{r} + \frac{2}{r}\hat{\theta} + \frac{3}{r}\hat{\varphi}\right)[\cos(\theta)\hat{r} - \sin(\theta)\hat{\theta}]r \sin(\theta)drd\varphi$$
$$\Rightarrow \nabla \times \vec{B} \cdot \hat{n}da = \left(-\frac{2}{r}\cot(\theta)\cos(\theta) - \frac{2}{r}\sin(\theta)\right)r \sin(\theta)drd\varphi. \tag{8.295}$$

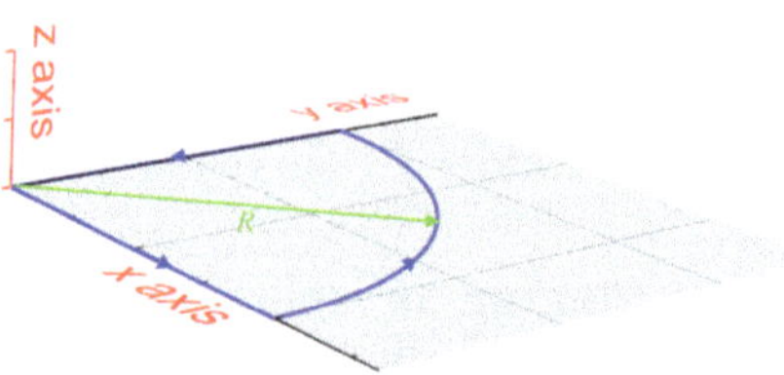

Figure 8.25. A quarter of a circle with radius R on the x–y plane.

Since on the x–y plane $\theta = \pi/2$, we have

$$\cot(\theta) = \cos(\theta) = 0,\ \sin(\theta) = 1 \Rightarrow \nabla \times \vec{B} \cdot \hat{n} da = -2dr d\varphi, \tag{8.296}$$

and for the surface bounded by the curve shown in figure 8.25, one can write

$$\int_A \nabla \times \vec{B} \cdot \hat{n} dA = -\int_0^R \int_0^{\pi/2} 2dr d\varphi = -R\pi. \tag{8.297}$$

We recall for the position vector $\vec{r}$ in spherical coordinates

$$\vec{r} = r\hat{r} \Rightarrow d\vec{r} = dr\hat{r} + rd\hat{r}. \tag{8.298}$$

When $\vec{r}$ represent a point on the x–y plane, where $\theta = \pi/2$, we find

$$\vec{r} = r\sin(\pi/2)\cos(\varphi)\hat{x} + r\sin(\pi/2)\sin(\varphi)\hat{y} + r\cos(\pi/2)\hat{z}$$

$$\Rightarrow \vec{r} = r\cos(\varphi)\hat{x} + r\sin(\varphi)\hat{y},\ \hat{r} = \frac{\partial \vec{r}}{\partial r} \Big/ \left| \frac{\partial \vec{r}}{\partial r} \right| = \cos(\varphi)\hat{x} + \sin(\varphi)\hat{y} \tag{8.299}$$

$$\Rightarrow d\hat{r} = [-\sin(\varphi)\hat{x} + \cos(\varphi)\hat{y}]d\varphi = \hat{\varphi}d\varphi.$$

In equation (8.299), we have used the relation

$$\hat{\varphi} = \frac{\partial \vec{r}}{\partial \varphi} \Big/ \left| \frac{\partial \vec{r}}{\partial \varphi} \right| = -\sin(\varphi)\hat{x} + \cos(\varphi)\hat{y}. \tag{8.300}$$

In view of equation (8.299), one can rewrite equation (8.298) as

$$d\vec{r} = dr\hat{r} + r\hat{\varphi}d\varphi, \tag{8.301}$$

so that

$$\vec{B} \cdot d\vec{r} = B_r dr + B_\varphi r d\varphi = 4dr - 2rd\varphi. \tag{8.302}$$

Then using the result in equation (8.302) the integral for the closed curve shown in figure 8.25, one can write

$$\oint_C \vec{B} \cdot d\vec{r} = \int_{l_1} (4dr - 2rd\varphi) + \int_{l_2} (4dr - 2rd\varphi) + \int_{l_3} (4dr - 2rd\varphi). \tag{8.303}$$

Noting that l_1 and l_3 are the straight lines (where $d\varphi = 0$) and l_2 is the curved line (where $r = R$ and $dr = 0$), equation (8.303) becomes

$$\oint_C \vec{B} \cdot d\vec{r} = \int_0^R 4dr - \int_0^{\pi/2} 2Rd\varphi + \int_R^0 4dr = -R\pi. \tag{8.304}$$

This results in equations (8.297) and (8.304) verify Stokes' theorem.

8.11 Vector calculus and Mathematica

The following are the basic commands that we can use in vector calculus:
- Grad $[f, \{x_1, \ldots, x_n\}]$ gives the gradient $(\partial f/\partial x_1, \ldots, \partial f/\partial x_n)$.

- Grad [f, $\{x_1, \ldots, x_n\}$, *chart*] gives the gradient in the coordinates *chart*.
- Div $\left[\{f_1, \ldots, f_n\}, \{x_1, \ldots, x_n\}\right]$ gives the divergence.
- Div $\left[\{f_1, \ldots, f_n\}, \{x_1, \ldots, x_n\}, \textit{chart}\right]$ gives the divergence in the coordinates *chart*.
- Curl [$\{f_1, f_2\}$, $\{x_1, x_2\}$] gives the curl.
- Curl $\left[\{f_1, f_2, f_3\}, \{x_1, x_2, x_3\}\right]$ gives the curl.
- Curl [f, $\{x_1, \ldots, x_n\}$] gives the curl of the ××...× array f with respect to the-dimensional vector $\{x_1, \ldots, x_n\}$.
- Curl [f, x, *chart*] gives the curl in the coordinates *chart*.
- Laplacian [f, $\{x_1, \ldots, x_n\}$] gives the Laplacian.
- Laplacian [f, $\{x_1, \ldots, x_n\}$, *chart*] gives the Laplacian in the given coordinates *chart*.

The following are some of the examples we considered in the previous sections solved using Mathematica.

- Example 8.3: The gradient, divergence, curl, and Laplacian in three-dimensional Cartesian coordinates:

$$\text{Grad}[f[x, y, z] = xyz, \{x, y, z\}]$$
$$\{yz, xz, xy\}$$

$$\text{Div}\left[\{V_x[x, y, z] = xy, V_y[x, y, z] = yz, V_z[x, y, z] = zx\}, \{x, y, z\}\right]$$
$$x + y + z$$

$$\text{Curl}\left[\{V_x[x, y, z] = xy, V_y[x, y, z] = yz, V_z[x, y, z] = zx\}, \{x, y, z\}\right]$$
$$\{-y, -z, -x\}$$

$$\text{Laplacian}[f[x, y, z] = xyz, \{x, y, z\}]$$
$$0$$

- Example 8.4: The gradient, divergence, curl, and Laplacian using an orthonormal basis for three-dimensional cylindrical coordinates:

$$\text{Curl}\left[\left\{B_s[s, \phi, z] = 0, B_\phi[s, \phi, z] = \frac{\mu_o I s}{2\pi a^2}, B_z[s, \phi, z] = 0\right\}, \{s, \phi, z\}, \text{'Cylindrical'}\right]$$
$$\left\{0, 0, \frac{i\mu_o}{a^2 \pi}\right\}$$

$$\text{Div}\left[\left\{B_s[s, \phi, z] = 0, B_\phi[s, \phi, z] = \frac{\mu_o I s}{2\pi a^2}, B_z[s, \phi, z] = 0\right\}, \{s, \phi, z\}, \text{'Cylindrical'}\right]$$
$$0$$

$$\mathrm{Grad}\left[B_\phi[s,\,\phi,\,z] = \frac{\mu_o Is}{2\pi a^2},\,\{s,\,\phi,\,z\},\,\text{`Cylindrical'}\right]$$

$$\left\{\frac{i\mu_o}{2a^2\pi},\,0,\,0\right\}$$

$$\mathrm{Div}\left[\left\{\frac{i\mu_o}{2a^2\pi},\,0,\,0\right\},\,\{s,\,\phi,\,z\},\,\text{`Cylindrical'}\right]$$

$$\frac{i\mu_o}{2a^2\pi s}$$

$$\mathrm{Laplacian}\left[B_\phi[s,\,\phi,\,z] = \frac{\mu_o Is}{2\pi a^2},\,\{s,\,\phi,\,z\},\,\text{`Cylindrical'}\right]\!/\!/\mathrm{Expand}$$

$$\frac{i\mu_o}{2a^2\pi s}$$

- Examples 8.2 and 8.5: The gradient, divergence, curl, and Laplacian in three-dimensional spherical coordinates:

$$E_Q[r,\,\theta,\,\phi] = -\mathrm{Grad}\left[\Phi[r,\,\theta,\,\phi] = \frac{KQ}{r},\,\{r,\,\theta,\,\phi\},\,\text{`Spherical''}\right]$$

$$\left\{\frac{KQ}{r^2},\,0,\,0\right\}$$

$$\mathrm{Curl}\left[\left\{E_r[r,\,\theta,\,\phi] = \frac{KQ}{r^2},\,E_\theta[r,\,\theta,\,\phi] = 0,\,E_\phi[r,\,\theta,\,\phi] = 0\right\},\,\{r,\,\theta,\,\phi\},\,\text{`Spherical'}\right]$$

$$\{0,\,0,\,0\}$$

8.12 Homework assignment

Problem 1. Consider an object with a mass m rotating about the z-axis with angular velocity ω as shown in figure 8.26. The position of the object is described by the position vector $\vec{r}$. The angular momentum of this mass is given by the triple vector product

$$\vec{L} = m\vec{r} \times (\vec{\omega} \times \vec{r}). \tag{8.305}$$

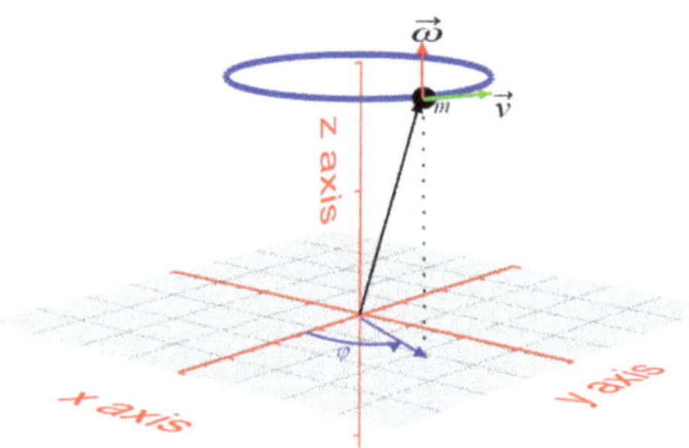

Figure 8.26. An object with a mass m rotating about z-axis.

(a) Expand this triple vector product applying the relation

$$\vec{A} \times \left(\vec{B} \times \vec{C}\right) = \left(\vec{A} \cdot \vec{C}\right)\vec{B} - \left(\vec{A} \cdot \vec{B}\right)\vec{C}, \tag{8.306}$$

and show that when the position vector $\vec{r}$ is perpendicular to the angular velocity

$$\vec{\omega} = \omega\hat{z},$$

the angular momentum is given by

$$L = mvr, \tag{8.307}$$

where

$$v = r\omega = r\frac{d\varphi}{dt}, \tag{8.308}$$

is the tangential speed.

(b) Show that for the angular momentum of the object

$$\vec{L} = m\vec{r} \times \vec{v} = m\vec{r} \times \left(\frac{d\vec{r}}{dt}\right), \tag{8.309}$$

the time derivative is

$$\frac{d\vec{L}}{dt} = m\vec{r} \times \frac{d^2\vec{r}}{dt^2}. \tag{8.310}$$

Problem 2. The position of a particle at a time t is given by

$$\vec{r} = \cos(t)\hat{x} + \sin(t)\hat{y} + (t)\hat{z}. \tag{8.311}$$

Show that both the speed and the magnitude of the acceleration are constant. Make a graph and describe the motion of the particle.

Problem 3. Consider two functions—a scalar function and a vector function—defined as follows:

$$f(\vec{r}) = xy^2z, \quad \vec{V}(x, y, z) = (x^2y, y^2z, z^2x). \tag{8.312}$$

At the point $(1, -1, 1)$, evaluate the following quantities :

(a) ∇f

(b) $\nabla \cdot \vec{V}$

(c) $\nabla \times \vec{V}$

(d) $\nabla^2 f$

(e) $\nabla^2 \vec{V}$

(f) $\nabla^2(f\vec{V})$

(g) graph the equipotential surface assuming that $f(\vec{r})$ represents the electric potential function of some charge distribution.

(h) Reproduce the results in (a)–(f) using Mathematica.

Problem 4.

(a) Find the gradient of the function

$$f(x, y, z) = x^2y^3z, \tag{8.313}$$

at $(1, 2, -1)$.

(b) Starting from the point $(1, 1)$, in what direction does the function

$$f(x, y) = x^2 - y^2 + 2xy \tag{8.314}$$

decrease most rapidly. Verify this direction by making 3-D plot for the function.

Problem 5. Find the derivative of the function

$$f(x, y, z) = ze^x \cos y, \tag{8.315}$$

at $(1, 0, \pi/3)$ in the direction of the vector

$$\vec{s} = i + 2j. \tag{8.316}$$

Make a 3-D graph for the equipotential surface defined by this function,

$$ze^x \cos y = c \Rightarrow z = c/e^x \cos y, \tag{8.317}$$

where c is a constant.

Problem 6. Find the direction of the line normal to the surface defined by the function

$$f(x, y, z) = x^2y + y^2z + z^2x + 1 = 0, \tag{8.318}$$

at the point $(1, 2, -1)$. Find the equation of the tangent plane and the normal line at this point. Make a 3-D plot to show the surface, the tangent plane, and the normal line.

Problem 7. Find the divergence and curl of the vectors
(a)

$$\vec{v}(x, y, z) = x \sin (y)\hat{x} + \cos (y)\hat{y} + xy\hat{z}, \tag{8.319}$$

(b)

$$\vec{v}(x, y, z) = \sinh (z)\hat{x} + 2y\hat{y} + x \cosh (z)\hat{z}. \tag{8.320}$$

Problem 8. Find the gradient and the Laplacian for the scalar functions
(a)

$$f(x, y, z) = xy(x^2 + y^2 - 5z^2), \tag{8.321}$$

(b)

$$f(x, y, z) = \ln (x^2 + y^2 + z^2). \tag{8.322}$$

Problem 9. For the position vector in Cartesian coordinates

$$\vec{r} = x\hat{x} + y\hat{y} + z\hat{z} \Rightarrow \hat{r} = \frac{\vec{r}}{|\vec{r}|} = \frac{x\hat{x} + y\hat{y} + z\hat{z}}{\sqrt{x^2 + y^2 + z^2}}, \tag{8.323}$$

evaluate $\nabla \times (\hat{z} \times \vec{r})$, $\nabla \cdot \hat{r}$, and $\nabla \times \hat{r}$.

Problem 10. Find the work done by the force

$$\vec{F} = (2xy - 3)\hat{x} + x^2\hat{y} + z\hat{z}. \tag{8.324}$$

in moving an object from the position $\vec{r}_1 = \hat{x}$ to $\vec{r}_2 = \hat{y}$ along each of the three paths (green, pink, and blue) shown in figure 8.27.

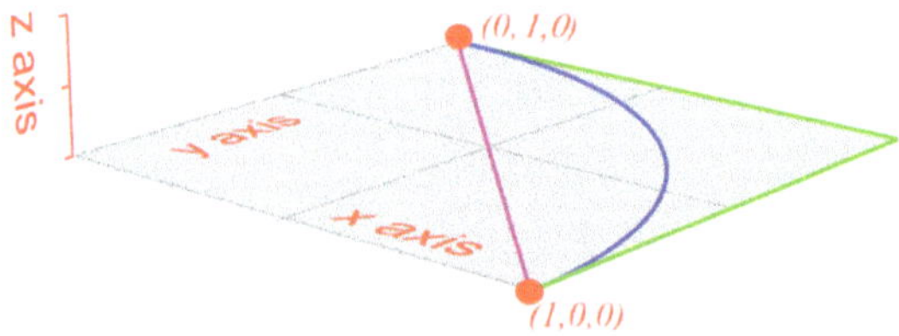

Figure 8.27. The three paths.

Problem 11. Verify that the vectors

(a)

$$\vec{F} = \hat{x} - z\hat{y} - y\hat{z}, \tag{8.325}$$

(b)

$$\vec{F} = y\sin(2x)\hat{x} + \sin^2(x)\hat{y}, \tag{8.326}$$

(c)

$$\vec{F} = (3x^2yz - 3y)\hat{x} + (x^3z - 3x)\hat{y} + (x^3y + 2z)\hat{z}, \tag{8.327}$$

are conservative and then find for each vector a scalar function $\phi(x, y, z)$ such that

$$\vec{F} = -\nabla\phi.$$

Problem 12. For the force vector

$$\vec{F} = -y\hat{x} + x\hat{y} + z\hat{z},$$

acting on an object moving it from the position $(1, 0, 0)$ to $(-1, 0, \pi)$, calculate the work done

(a) along the helix path defined by $x = \cos(t)$, $y = \sin(t)$, $z = t$,
(b) along the straight line joining the points $(1, 0, 0)$ and $(-1, 0, \pi)$.
    The two paths are shown in figure 8.28.

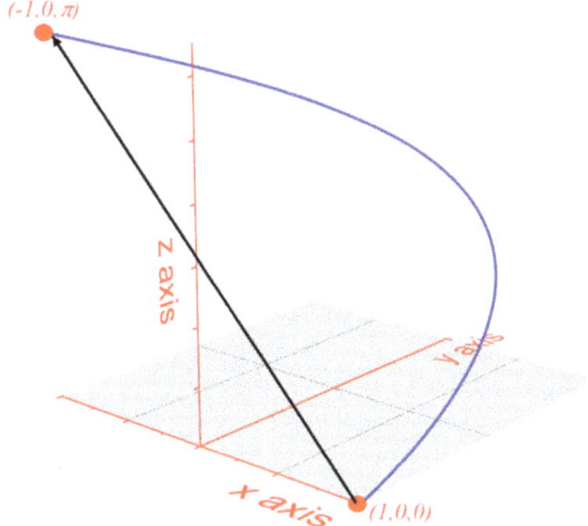

Figure 8.28. The two paths.

Problem 13. Evaluate the line integral

$$I = \oint (2x\,dy - 3y\,dx), \tag{8.328}$$

around the square with vertices $(0, 2, 0)$, $(2, 0, 0)$, $(-2, 0, 0)$, and $(0, -2, 0)$ (figure 8.29).

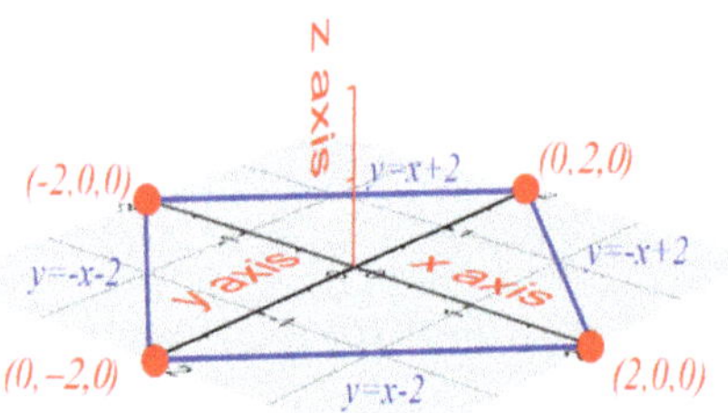

Figure 8.29. The square.

Problem 14.

    (a) For a simple closed curve C on the x–y plane shown in figure 8.30, using Green's theorem show that the area enclosed is

$$A = \iint_A dx\,dy = \frac{1}{2} \oint_C (x\,dy - y\,dx). \tag{8.329}$$

    (b) Using the result in (a), for an ellipse defined by $x = a\cos(\theta)$, $y = b\sin(\theta)$, $0 \leqslant \theta \leqslant 2\pi$, show that the area $A = \pi ab$.

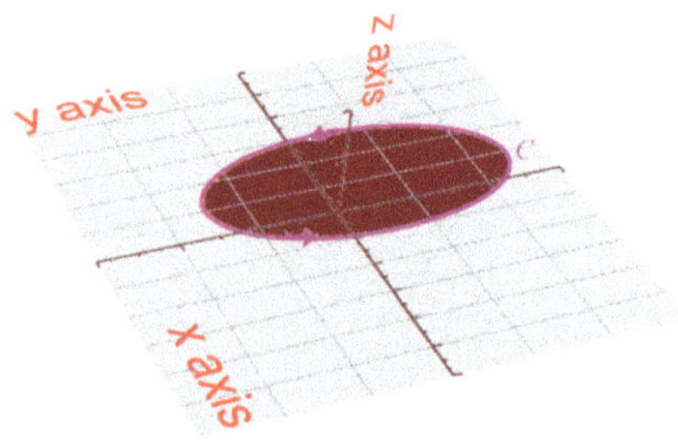

Figure 8.30. A region bounded by a curve C.

Problem 15. Consider the triangular region on the x–y plane with vertices $(0, 0, 0)$, $(0, 3, 0)$, and $(3, 0, 0)$ as shown in figure 8.31.

(a) Apply Green's theorem

$$\iint_A \left[\frac{\partial Q(x, y)}{\partial x} - \frac{\partial P(x, y)}{\partial y} \right] dxdy = \oint_C [P(x, y)dx + Q(x, y)dy], \qquad (8.330)$$

show that for

$$P = 0, \; Q = \frac{1}{2}x^2, \qquad (8.331)$$

we find

$$\iint_A xdxdy = \oint_C \frac{1}{2}x^2dy. \qquad (8.332)$$

And for

$$P = -\frac{1}{2}y^2, \; Q = 0, \qquad (8.333)$$

we find

$$\iint_A ydxdy = -\oint_C \frac{1}{2}y^2dx. \qquad (8.334)$$

(b) Using the results in (a) and Problem 14

$$A = \iint_A dxdy = \frac{1}{2} \oint_C [xdy - ydx], \qquad (8.335)$$

find the centroid

$$\bar{x} = \frac{\displaystyle\iint_A xdxdy}{\displaystyle\iint_A dxdy}, \; \bar{y} = \frac{\displaystyle\iint_A ydxdy}{\displaystyle\iint_A dxdy}, \qquad (8.336)$$

for the triangle shown in figure 8.31.

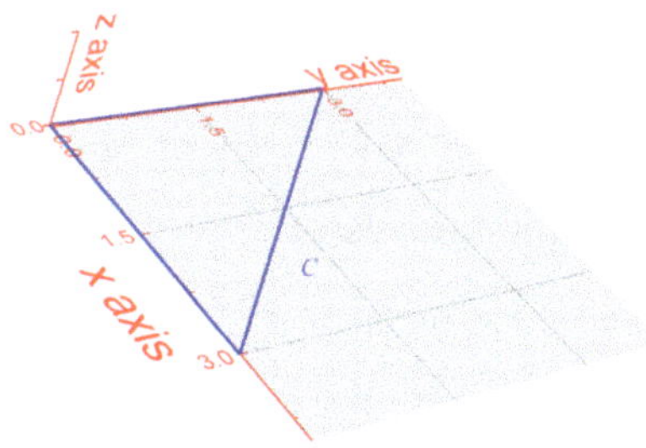

Figure 8.31. The triangular region on the x–y plane.

Problem 16. Using Green's theorem, evaluate the integral

$$I = \oint_C [(x\sin(x) - 3xdy)dx + (x - y^2)dy], \tag{8.337}$$

where C is the triangle on the x–y plane with vertices $(0, 0, 0)$, $(1, 1, 0)$, and $(2, 0, 0)$ shown in figure 8.32.

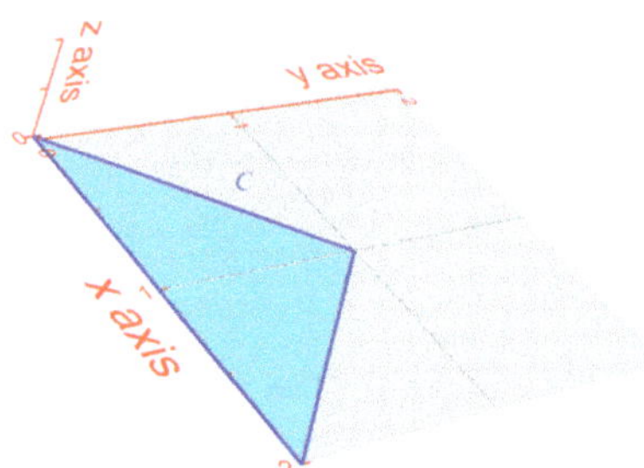

Figure 8.32. A triangular region.

Problem 17. Using the a cubic box with unit length, verify the divergence theorem

$$\iiint_V (\nabla \cdot \vec{V}(x, y, z))dxdydz = \iint_S \vec{V}(x, y, z) \cdot \hat{n}da, \tag{8.338}$$

for the vector

$$\vec{V}(x, y, z) = x^2\hat{x} + y^2\hat{y} + z^2\hat{z}. \tag{8.339}$$

Problem 18. Using a sphere with radius $r = 3$ units, verify the divergence theorem for the vector

$$\vec{V}(x, y, z) = x\cos^2 y\hat{x} + xz\hat{y} + z\sin^2 y\hat{z}. \tag{8.340}$$

Problem 19. For the volume of a sphere defined by

$$x^2 + y^2 + z^3 \leqslant 25, \tag{8.341}$$

verify the divergence theorem for the vector

$$\vec{F}(x, y, z) = (x^2 + y^2 + z^3)(x\hat{x} + y\hat{y} + z\hat{z}), \tag{8.342}$$

in spherical coordinates.

Problem 20. For the volume of a hemisphere defined by

$$x^2 + y^2 + z^3 \leqslant 9, \tag{8.343}$$

verify the divergence theorem for the vector

$$\vec{E}(x, y, z) = y\hat{x} + xz\hat{y} + (2x - 1)\hat{z}, \tag{8.344}$$

in spherical coordinates.

Problem 21. Consider part of the surface defined by $z = 4 - x^2 - y^2$ above the x–y plane (see figure 8.33). Using the vector

$$\vec{B}(x, y, z) = x^2\hat{x} + z^2\hat{y} - y^2\hat{z}, \tag{8.345}$$

verify Stokes' theorem.

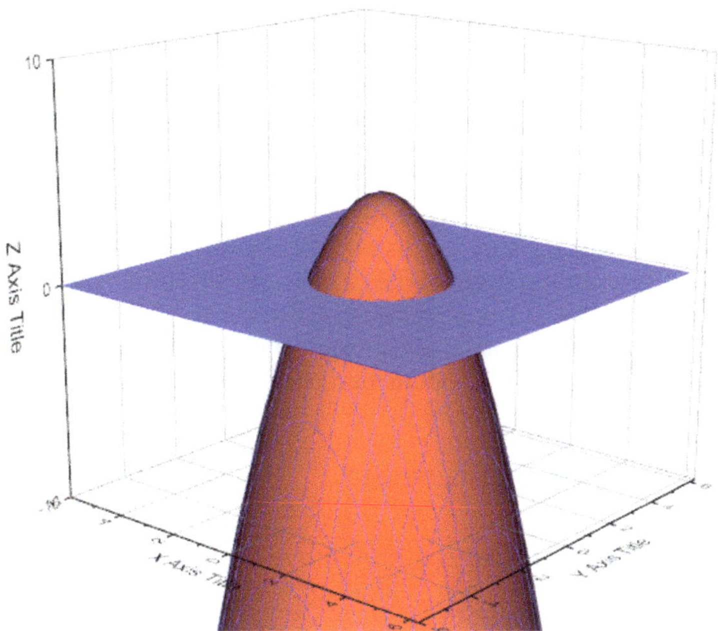

Figure 8.33. The x–y plane and the surface defined by $z(x, y) = 4 - x^2 - y^2$.

IOP Publishing

Studies in Theoretical Physics, Volume 1
Fundamental mathematical methods
Daniel Erenso and Victor Montemayor

Chapter 9

Introduction to the calculus of variations

Finding the shortest path between two points in Euclidean space, phase space, or even in Minkowski space-time are some of the challenging problems of physics. The interest to solve these challenging problems was the main factor for the invention of the calculus of variations, and it is the objective of this chapter to introduce this novel mathematical method. We begin with an introduction to some basic terminologies that we often encounter in the calculus of variations and determine the equation of a line for the shortest distance between two points on a plane surface. We then solve the general problem of the calculus of variation by deriving the Euler–Lagrange equation and see its application to solve the brachistochrone and coupled-harmonic oscillator problems in classical mechanics. In the final section, we provide some of the basic commands in Mathematica that we can use to solve problems using the calculus of variations.

9.1 Stationary points and geodesic

Stationary point

A point with coordinates $(x_0, f(x_0))$ on a curve defined by the function $f(x)$ is said to be a stationary point when

$$\left. \frac{df(x)}{dx} \right|_{x=x_0} = 0. \tag{9.1}$$

It means the point described by the coordinate $(x_0, f(x_0))$ is either a maximum, a minimum, or an inflection point. The points where the three people stand in figure 9.1 are stationary.

Geodesic

The curve along an arbitrary surface which marks the shortest path between two neighboring points.

doi:10.1088/978-0-7503-3135-7ch9   9-1

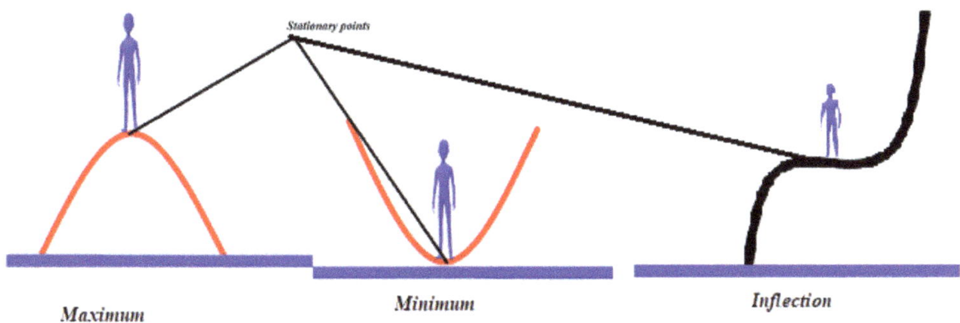

Figure 9.1. Stationary points.

Example 9.1. A ball of mass, m, is kicked from the top of a building with height, y_0, with an initial speed, v_o, at an angle, θ_o, above the horizontal. Find the time, t_0, at which the height of the ball, $y(t)$, becomes stationary.

Solution: We recall that from kinematics, the projectile motion of the ball along the y-direction is determined by Newton's second law

$$m\frac{dv_y}{dt} = -mg \Rightarrow \frac{dv_y}{dt} = -g \Rightarrow v_y(t) = v_{0y} - gt. \tag{9.2}$$

From the component of the velocity of the ball in the y-direction, we can show that

$$v_y(t) = \frac{dy(t)}{dt} = v_{0y} - gt \Rightarrow \int_{y_0}^{y(t)} dy = \int_0^t (v_{0y} - gt)dt,$$

$$\Rightarrow y(t) = y_0 + v_{0y}t - \frac{1}{2}gt^2. \tag{9.3}$$

The time at which the height of the ball $y(t)$ becomes stationary is determined from

$$v_y(t) = \frac{dy(t)}{dt}\bigg|_{t=t_0} = 0 \Rightarrow v_{0y} - gt\,\big|_{t=t_0} = 0$$

$$\Rightarrow t_0 = \frac{v_{0y}}{g} = \frac{v_0 \sin(\theta_o)}{g}. \tag{9.4}$$

Example 9.2. *Geodesic:* Consider two points on the x–y plane, P_1 and P_2 (see figure 9.2). Prove that the geodesic (shortest path) between the two points is the distance measured along a straight line (i.e., show that the geodesic is given by an equation of a straight line, $y(x) = mx + b$)

Solution: Let the coordinates for P_1 be (x_1, y_1) and P_2 be (x_2, y_2). Then the distance between these points is given by the integral

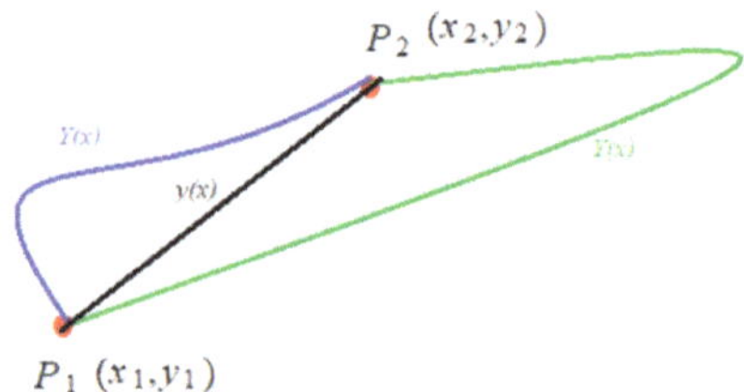

Figure 9.2. Geodesic and non-geodesic paths.

$$L = \int_{(1)}^{(2)} ds, \tag{9.5}$$

where

$$ds = \sqrt{dx^2 + dy^2} = \sqrt{1 + \left(\frac{dy}{dx}\right)^2}\, dx = \sqrt{1 + \left(\frac{dx}{dy}\right)^2}\, dy. \tag{9.6}$$

Let us express this distance as

$$L = \int_{x_1}^{x_2} \sqrt{1 + \left(\frac{dy}{dx}\right)^2}\, dx. \tag{9.7}$$

Out of the infinitely many paths that connect the two points, we want to determine the one that would give the minimum distance (the geodesic). Let us denote these functions by $Y(x, \epsilon)$. From these infinite number of functions, only one function gives the minimum distance between the two points. Let us denote this function by $y(x)$. Then we may express $Y(x)$ in terms of $y(x)$ as

$$Y(x, \epsilon) = y(x) + \epsilon \eta(x), \tag{9.8}$$

where $\eta(x)$ is an arbitrary function which must satisfy the condition

$$\eta(x_1) = \eta(x_2) = 0, \tag{9.9}$$

so that at the two end points ($P_1 = (x_1, y_1)$ and $P_2 = (x_2, y_2)$), we find

$$Y(x_1, \epsilon) = y(x_1) = y_1, \quad Y(x_2, \epsilon) = y(x_2) = y_2. \tag{9.10}$$

This means all varied paths must pass through the two points P_1 and P_2. Here ϵ is the scale factor for the magnitude of the variation. It is this constant that determines by how much $Y(x, \epsilon)$ differs from $y(x)$. We can easily see this from figure 9.3, where we considered three different paths. Thus for any arbitrary path that connects the two points, the length, L, would depend on the variational parameter and can be expressed as,

$$L(\epsilon) = \int_{x_1}^{x_2} \sqrt{1 + Y'^2}\, dx, \tag{9.11}$$

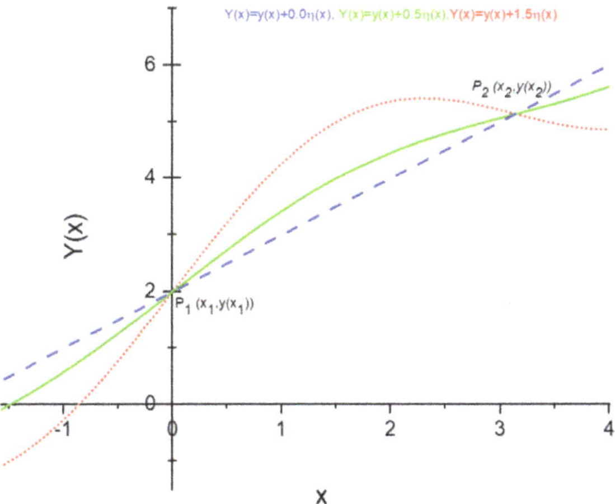

Figure 9.3. The functions are $\eta(x) = \sin(x)$ and $y(x) = x + 2$. Note that the smaller the variational parameter ϵ $(0.5 < 1.5)$, the closer the function $Y(x)$ to $y(x)$.

where

$$Y' = \frac{dY(x, \epsilon)}{dx}. \tag{9.12}$$

We are interested in the path that gives the minimum distance between the two points (i.e., the geodesic). The necessary condition for the distance $L(\epsilon)$, to be minimum is that the length function, $L(\epsilon)$, must have a stationary point at $(\epsilon = 0, L(\epsilon = 0))$. This requires

$$\left. \frac{dL(\epsilon)}{d\epsilon} \right|_{\epsilon=0} = 0, \tag{9.13}$$

which leads to

$$\left. \frac{dL(\epsilon)}{d\epsilon} \right|_{\epsilon=0} = \int_{x_1}^{x_2} \left. \frac{d}{d\epsilon}\left(\sqrt{1 + Y'^2}\right) \right|_{\epsilon=0} dx$$

$$= \int_{x_1}^{x_2} \frac{Y'}{\sqrt{1 + Y'^2}} \frac{dY'}{d\epsilon} dx \Bigg|_{\epsilon=0} = 0. \tag{9.14}$$

Using equation (9.8), we can write

$$Y' = \frac{dY(x, \epsilon)}{dx} = \frac{dy}{dx} + \epsilon\frac{d\eta}{dx} \Rightarrow Y' = y'(x) + \epsilon\eta'(x),$$

$$\frac{dY'(\epsilon)}{d\epsilon} = \frac{d}{d\epsilon}[y'(x) + \epsilon\eta'(x)] = \eta'(x), \tag{9.15}$$

so that upon evaluating these equations at $\epsilon = 0$, we find

$$Y'(x, \epsilon)\,\big|_{\epsilon=0} = y'(x), \qquad (9.16)$$

$$\frac{dY'(\epsilon)}{d\epsilon}\bigg|_{\epsilon=0} = \eta'(x). \qquad (9.17)$$

Substituting equations (9.16) and (9.17) into equation (9.14), for the stationary point (the geodesic), one finds

$$\frac{dL(\epsilon)}{d\epsilon}\bigg|_{\epsilon=0} = \int_{x_1}^{x_2} \frac{y'\eta'(x)}{\sqrt{1+y'^2}}dx = 0. \qquad (9.18)$$

Using integration by parts

$$\int u\,dv = uv - \int v\,du, \qquad (9.19)$$

for

$$dv = \eta'(x) \Rightarrow v = \eta(x),$$

$$u = \frac{y'}{\sqrt{1+y'^2}} \Rightarrow du = \frac{d}{dx}\left(\frac{y'}{\sqrt{1+y'^2}}\right)dx, \qquad (9.20)$$

equation (9.18) can be rewritten as

$$\int_{x_1}^{x_2} \frac{y'\eta'(x)}{\sqrt{1+y'^2}}dx = \frac{y'\eta(x)}{\sqrt{1+y'^2}}\bigg|_{x_1}^{x_2} - \int_{x_1}^{x_2} \eta(x)\frac{d}{dx}\left(\frac{y'}{\sqrt{1+y'^2}}\right)dx = 0. \qquad (9.21)$$

In view of the condition set in equation (9.9), the first term in equation (9.21) becomes zero and equation (9.21) becomes

$$\int_{x_1}^{x_2} \frac{y'\eta'(x)}{\sqrt{1+y'^2}}dx = -\int_{x_1}^{x_2} \eta(x)\frac{d}{dx}\left(\frac{y'}{\sqrt{1+y'^2}}\right)dx = 0. \qquad (9.22)$$

Thus substituting equation (9.22) into equation (9.18), one can write

$$\frac{dL(\epsilon)}{d\epsilon}\bigg|_{\epsilon=0} = -\int_{x_1}^{x_1} \eta(x)\frac{d}{dx}\left(\frac{y'}{\sqrt{1+y'^2}}\right)dx = 0. \qquad (9.23)$$

Since $\eta(x)$ is an arbitrary function, for the integral to be zero, we must have

$$\frac{d}{dx}\left(\frac{y'(x)}{\sqrt{1+y'^2(x)}}\right) = 0 \Rightarrow \frac{y'(x)}{\sqrt{1+y'^2(x)}} = c, \qquad (9.24)$$

where c is a constant. Upon solving for $y'(x)$, we find

$$y' = \left[\frac{c^2}{1 - c^2} \right]^{1/2} = m. \tag{9.25}$$

Note that we have introduced another constant m in terms of the constant c. From equation (9.25), one can show that

$$\frac{dy}{dx} = m \Rightarrow y(x) = mx + b, \tag{9.26}$$

which is the equation of a straight line.

9.2 The general problem of the calculus of variations

In the previous section, we saw the application of the calculus of variations to show that the shortest path connecting two points on a plane (the geodesic) is a straight line

$$y(x) = mx + b. \tag{9.27}$$

Next, we shall consider the application of the calculus of variations to the general problem. In a Euclidean space a surface is defined by the function, $F(x, y, z)$, where it depends on the Cartesian coordinates x, y, and z (see figure 9.4(a)). Instead of the Euclidean space let's consider a surface defined by the function, $F\left(x, y(x), y'(x) = \frac{dy}{dx}\right)$. This surface could, for example, be a surface in phase space (in classical mechanics) if we replace

$$x \rightarrow t, \, y(x) \rightarrow y(t), \, y'(x) \rightarrow y'(t) = \frac{dy}{dt} = v_y(t) = \frac{p_y}{m} \tag{9.28}$$

describing the dynamics of a particle mass, m, moving along the y-direction in terms of the parameters (time $= t$, position $= y(t)$, velocity $= v_y(t)\hat{y} = \frac{p_y\hat{y}}{m}$), where $p_y\hat{y}$ is the momentum (see figure 9.4(b)). In classical mechanics, the dynamics of a particle is determined by an equation derived from Newton's second law. As we shall see, this equation can be derived from a more general equation known as *the Euler–Lagrange equation*

$$\frac{\partial}{\partial x}\left(\frac{\partial F}{\partial y'}\right) - \frac{\partial F}{\partial y} = 0, \tag{9.29}$$

where $F = F(x, y(x), y'(x))$ is the function that defines the surface constructed by the set of points with coordinates, $(x, y(x), y'(x))$. *The Euler–Lagrange equation is derived by applying the calculus of variations. In general, in the problem that we want to solve applying the calculus of variations, we know the coordinates of two different points $(x_1, y(x_1), y'(x_1))$ and $(x_2, y(x_2), y'(x_2))$ on the surface defined by $F = F(x, y(x), y'(x))$* (figure 9.5). From the infinitely many trajectories that can connect these two points, there is only one trajectory on this surface that is the

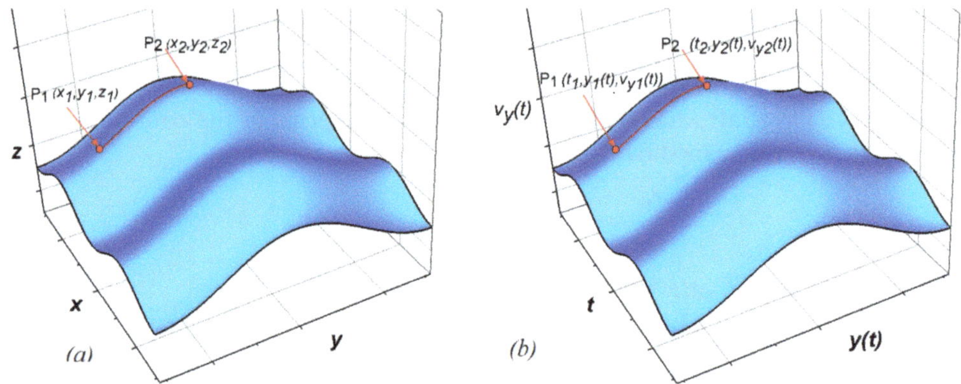

Figure 9.4. (a) A surface defined by the function $F(x, y, z)$ in Euclidean space and (b) a surface defined by the function $F(t, y(t), v_y(t) = \frac{p_y(t)}{m})$ in a phase space.

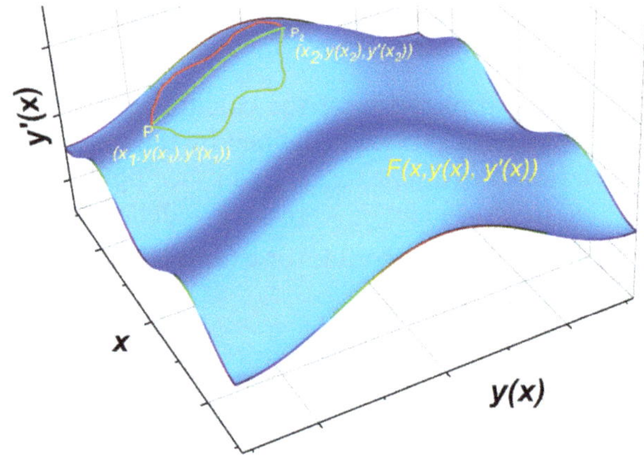

Figure 9.5. A surface defined by the function $F(x, y(x), y'(x) = \frac{dy}{dx})$.

shortest (the geodesic). Finding such geodesic is the general problem that can be solved by applying the calculus of variations.

The surface is defined by the function $F(x, y(x), y'(x))$. The distance between these two points is determined by evaluating the integral

$$I = \int_{x_1}^{x_2} F(x, y(x), y'(x)) \, dx. \tag{9.30}$$

To determine the equation that the function F is governed by, so that we find the shortest length joining the two points, let the function for any path connecting the two points be $Y(x)$. From these infinite number of functions, only one function gives the minimum distance between the two points. If this function is $y(x)$, then we may write $Y(x)$ in terms of $y(x)$ as

$$Y(x, \epsilon) = y(x) + \epsilon\eta(x), \tag{9.31}$$

where $\eta(x)$ is an arbitrary function that must satisfy the condition

$$\eta(x_1) = \eta(x_2) = 0, \tag{9.32}$$

so that at the two points ($P_1 = (x_1, y_1)$ and $P_2 = (x_2, y_2)$), we find

$$Y(x_1, \epsilon) = y(x_1) = y_1, \; Y(x_2, \epsilon) = y(x_2) = y_2. \tag{9.33}$$

Using equation (9.31), we can also establish the relations

$$Y(x, \epsilon)\,|_{\epsilon=0} = y(x), \tag{9.34}$$

$$\frac{dY(x, \epsilon)}{d\epsilon} = \eta(x) \Rightarrow \frac{dY(\epsilon)}{d\epsilon}\bigg|_{\epsilon=0} = \eta(x), \tag{9.35}$$

$$\frac{dY}{dx} = \frac{dy}{dx} + \epsilon\frac{d\eta}{dx} \text{ or } Y'(x, \epsilon) = y'(x) + \epsilon\eta'(x) \Rightarrow Y'(x, \epsilon)\,|_{\epsilon=0} = y'(x), \tag{9.36}$$

$$\frac{dY'(\epsilon)}{d\epsilon} = \frac{d}{d\epsilon}[y'(x) + \epsilon\eta'(x)] = \eta'(x) \Rightarrow \frac{dY'(\epsilon)}{d\epsilon}\bigg|_{\epsilon=0} = \eta'(x). \tag{9.37}$$

For the geodesic the integral

$$I(\epsilon) = \int_{x_1}^{x_2} F(x, Y(x, \epsilon), Y'(x, \epsilon))\, dx, \tag{9.38}$$

must be stationary, that means

$$\frac{dI(\epsilon)}{d\epsilon}\bigg|_{\epsilon=0} = \int_{x_1}^{x_2} \frac{d}{d\epsilon}[F(x, Y(x, \epsilon), Y'(x, \epsilon))]\bigg|_{\epsilon=0} dx = 0. \tag{9.39}$$

Noting that

$$\frac{d}{d\epsilon}[F(x, Y(x, \epsilon), Y'(x, \epsilon))]\bigg|_{\epsilon=0} = \frac{\partial F}{\partial Y}\frac{dY(\epsilon)}{d\epsilon} + \frac{\partial F}{\partial Y'}\frac{dY'(\epsilon)}{d\epsilon}\bigg|_{\epsilon=0}$$

$$= \frac{\partial F}{\partial Y}\bigg|_{\epsilon=0}\frac{dY(\epsilon)}{d\epsilon}\bigg|_{\epsilon=0} + \frac{\partial F}{\partial Y'}\bigg|_{\epsilon=0}\frac{dY'(\epsilon)}{d\epsilon}\bigg|_{\epsilon=0}, \tag{9.40}$$

and using equations (9.34)–(9.37), we find

$$\frac{d}{d\epsilon}[F(x, Y(x, \epsilon), Y'(x, \epsilon))]\bigg|_{\epsilon=0} = \left[\frac{\partial}{\partial y}F(x, y(x), y'(x))\right]\eta(x)$$
$$+ \left[\frac{\partial}{\partial y'}F(x, y(x), y'(x))\right]\eta'(x). \tag{9.41}$$

Then in view of equation (9.41), the integral for the geodesic line in equation (9.39) becomes

$$\frac{dI(\epsilon)}{d\epsilon}\bigg|_{\epsilon=0} = \int_{x_1}^{x_2}\left[\frac{\partial}{\partial y}F(x, y(x), y'(x))\right]\eta(x)dx+$$
$$+ \int_{x_1}^{x_2}\left[\frac{\partial}{\partial y'}F(x, y(x), y'(x))\right]\eta'(x)dx. \tag{9.42}$$

Using integration by parts the second integral can be rewritten as

$$\int_{x_1}^{x_2}\left[\frac{\partial}{\partial y'}F(x, y(x), y'(x))\right]\eta'(x)dx = \eta(x)\frac{\partial}{\partial x}\left[\frac{\partial F}{\partial y'}\right]\bigg|_{x_1}^{x_2}$$
$$- \int_{x_1}^{x_2}\eta(x)\frac{\partial}{\partial x}\left[\frac{\partial F}{\partial y'}\right]\eta(x)dx, \tag{9.43}$$

so that using equation (9.32), we find

$$\int_{x_1}^{x_2}\left[\frac{\partial}{\partial y'}F(x, y(x), y'(x))\right]\eta'(x)dx = -\int_{x_1}^{x_2}\eta(x)\frac{\partial}{\partial x}\left[\frac{\partial F}{\partial y'}\right]\eta(x)dx. \tag{9.44}$$

In view of the result in equation (9.44), equation (9.42) can be put in the form

$$\frac{dI(\epsilon)}{d\epsilon}\bigg|_{\epsilon=0} = \int_{x_1}^{x_2}\left[\frac{\partial}{\partial y}F(x, y(x), y'(x))\right]\eta(x)dx+$$
$$- \int_{x_1}^{x_2}\eta(x)\frac{\partial}{\partial x}\left[\frac{\partial F}{\partial y'}\right]\eta(x)dx = 0. \tag{9.45}$$
$$\Rightarrow \frac{dI(\epsilon)}{d\epsilon}\bigg|_{\epsilon=0} = \int_{x_1}^{x_2}\left[\frac{\partial}{\partial x}\left[\frac{\partial F}{\partial y'}\right] - \frac{\partial F}{\partial y}\right]\eta(x)dx = 0.$$

Then from which follows

$$\frac{\partial}{\partial x}\left(\frac{\partial F}{\partial y'}\right) - \frac{\partial F}{\partial y} = 0, \tag{9.46}$$

where $F = F(x, y(x), y'(x))$. Equation (9.46) is known as the Euler–Lagrange equation, which makes the integral in equation (9.30).

Euler–Lagrange equation for multivariable function

Suppose we are given a function $F\left(x,\, y(x),\, z(x),\, \frac{dy(x)}{dx},\, \frac{dz(x)}{dx}\right)$ and we want to find $y = y(x)$ and $z = z(x)$, that make the integral

$$I = \int_{x_1}^{x_2} F(x,\, y,\, z,\, \frac{dy(x)}{dx},\, \frac{dz(x)}{dx})\, dx, \tag{9.47}$$

stationary. Following a procedure similar to the one used earlier, one can show that the integral in equation (9.47) becomes stationary when

$$\frac{\partial}{\partial x}\left(\frac{\partial F}{\partial y'}\right) - \frac{\partial F}{\partial y} = 0, \quad \frac{\partial}{\partial x}\left(\frac{\partial F}{\partial z'}\right) - \frac{\partial F}{\partial z} = 0. \tag{9.48}$$

For the generalized coordinates

$$t,\, q_1(t),\, q_2(t),\, ...q_N(t),\, q_1'(t),\, q_2'(t),\, ...q_N'(t), \tag{9.49}$$

and a function, $F(t,\, q_1(t),\, q_2(t),\, ...q_N(t),\, q_1'(t),\, q_2'(t),\, ...q_N'(t))$, then the integral

$$I = \int_{x_1}^{x_2} F(t,\, q_1(t),\, q_2(t),\, ...q_N(t),\, q_1'(t),\, q_2'(t),\, ...q_N'(t))dt, \tag{9.50}$$

becomes stationary, when

$$\frac{\partial}{\partial t}\left(\frac{\partial F}{\partial q_i'}\right) - \frac{\partial F}{\partial q_i} = 0, \tag{9.51}$$

for all $i = 1, 2, 3...N$.

9.3 The Brachistochrone problem

The brachistochrone problem: If two points A and B are given, at different heights but not laying one above the other (see figure 9.6), it is required to find among all possible curves connecting them the one along which a bead slides from A to B under the influence of gravity (neglecting friction) in the shortest possible time. This curve is called a brachistochrone curve (Gr. $\beta\rho\alpha\chi\acute{\iota}\sigma\tau\circ\varsigma$, brachistos—the shortest, $\chi\rho\acute{o}\nu\circ\varsigma$, chronos—time), or curve of fastest descent.

This problem occupied the time of the leading mathematicians in the whole of Europe: Newton, Leibniz, Bernoulli, L'Hospital, and others. From then on, the calculus of variations developed as a particular mathematical discipline.

Example 9.3. Using the Euler–Lagrange equation solve the brachistochrone problem, assuming the 'material point' starts from rest.

Solution: We are given the two points $(x_1,\, y_1)$ and $(x_2,\, y_2)$; we choose axes through point 1 with the positive y-axis downward as shown in figure 9.7. We want to find the

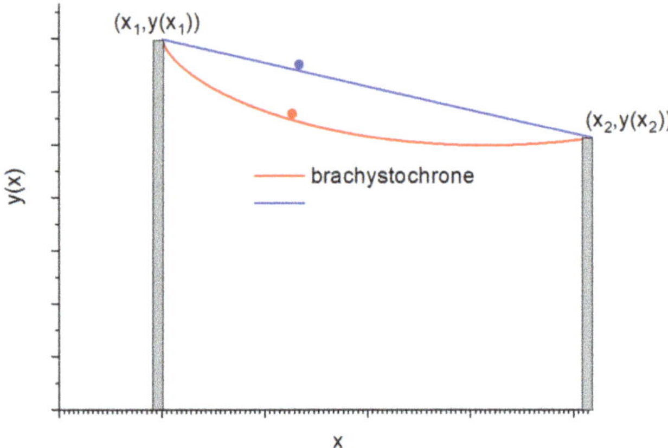

Figure 9.6. Two beads released along two different frictionless paths. The paths connect two points at two different elevation. The curved path is the fastest path and it is known as the brachistochrone path.

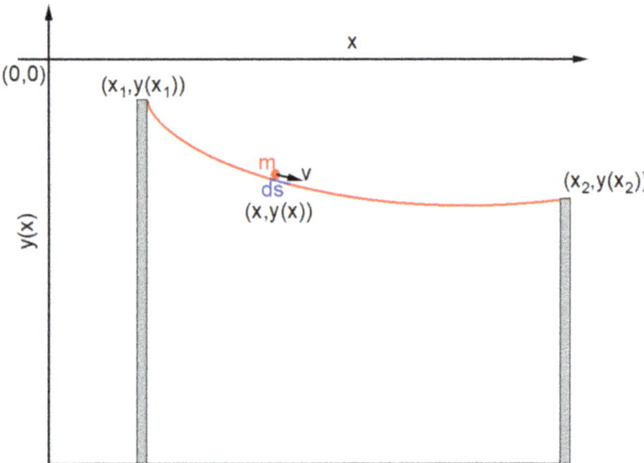

Figure 9.7. A bead of mass m sliding on the *brachistochrone* path after a time, t, released from the initial position $(x_1, y(x_1))$. Let this position be $(x(t), y(t))$.

curve joining the two points, down which a bead will slide (from rest) in the least time. In other words, we want to minimize time t by finding the stationary value for the integral

$$I = \int_1^2 dt = \int_1^2 \frac{ds}{v}. \tag{9.52}$$

The total energy of the bead is zero since it starts from rest assuming the zero energy level is the origin. If there is no friction, then from conservation of energy, we can write the energy at any point (x, y) below the origin as

$$\frac{1}{2}mv^2 - mgy = 0 \Rightarrow v = \sqrt{2gy}, \tag{9.53}$$

and equation (9.52) becomes

$$I = \int_1^2 \frac{ds}{v} = \int_1^2 \frac{ds}{\sqrt{2gy}}.$$ (9.54)

Using

$$ds = \sqrt{dx^2 + dy^2} = \sqrt{1 + \left(\frac{dx}{dy}\right)^2}\, dy,$$ (9.55)

we can put equation (9.54) in the form

$$I = \int_1^2 \frac{\sqrt{1 + \left(\dfrac{dx}{dy}\right)^2}}{\sqrt{2gy}}\, dy = \frac{1}{\sqrt{2g}} \int_{y_1}^{y_2} \frac{\sqrt{1 + x'^2}}{\sqrt{y}}\, dy$$ (9.56)

$$\Rightarrow I = \frac{1}{\sqrt{2g}} \int_{y_1}^{y_2} F(y, x(y), x'(y))dy,$$

where

$$F(y, x(y), x'(y)) = \frac{1}{\sqrt{2g}} \frac{\sqrt{1 + x'^2}}{\sqrt{y}}.$$ (9.57)

In view of the integral in equation (9.30) and the corresponding Euler–Lagrange equation (9.46), the integral in equation (9.56) becomes stationary when

$$\frac{\partial}{\partial y}\left(\frac{\partial F}{\partial x'}\right) - \frac{\partial F}{\partial x} = 0,$$ (9.58)

where

$$x'(y) = \frac{dx(y)}{dy}.$$ (9.59)

Using equation (9.57), one can show that

$$\frac{\partial F}{\partial x} = 0, \quad \frac{\partial F}{\partial x'} = \frac{1}{\sqrt{2g}} \frac{x'}{\sqrt{1 + x'^2}\,\sqrt{y}}.$$ (9.60)

Upon substituting the results in equation (9.60) into equation (9.58), one finds

$$\frac{\partial}{\partial y}\left(\frac{1}{\sqrt{2g}} \frac{x'}{\sqrt{1 + x'^2}\,\sqrt{y}}\right) = 0 \Rightarrow \frac{x'}{\sqrt{1 + x'^2}\,\sqrt{y}} = \sqrt{c},$$ (9.61)

where c is a constant. Solving for x',

$$\frac{x'}{\sqrt{1 + x'^2}\,\sqrt{y}} = \sqrt{c} \Rightarrow x'^2(1 - cy) = cy \Rightarrow \frac{dx}{dy} = \sqrt{\frac{cy}{1 - cy}}$$

$$\Rightarrow x = \int_0^{-y} \sqrt{\frac{cy}{1 - cy}}\,dy. \tag{9.62}$$

Introducing the transformation of variable defined by

$$cy = \sin^2\left(\frac{\theta}{2}\right) = \frac{1}{2}(1 - \cos(\theta)) \Rightarrow dy = \frac{1}{c}\sin\left(\frac{\theta}{2}\right)\cos\left(\frac{\theta}{2}\right)d\theta, \tag{9.63}$$

equation (9.60) can be rewritten as

$$x = \frac{1}{c}\int_0^{-\theta} \sqrt{\frac{\sin^2\left(\frac{\theta}{2}\right)}{1 - \sin^2\left(\frac{\theta}{2}\right)}}\,\sin\left(\frac{\theta}{2}\right)\cos\left(\frac{\theta}{2}\right)d\theta = \frac{1}{c}\int_0^{-\theta} \sin^2\left(\frac{\theta}{2}\right)d\theta \tag{9.64}$$

$$\Rightarrow x = \frac{1}{c}\int_0^{-\theta} \frac{1}{2}(1 - \cos(\theta))d\theta \Rightarrow x = \frac{1}{2c}(\theta - \sin(\theta)).$$

Therefore, the trajectory of the bead that takes the smallest possible time is defined by

$$x = \frac{1}{2c}(\theta - \sin(\theta)), \; y = \frac{1}{2c}(1 - \cos(\theta)). \tag{9.65}$$

Cycloid: Consider a circle of radius r rolling along the positive x-axis with a constant angular velocity starting from the origin. If you mark the point on the circle coinciding with the origin at the initial time (as shown in figure 9.8) and follow the trajectory of this point, its x and y coordinates are given by

$$x = r(\theta - \sin(\theta)), \; y = r(1 - \cos(\theta)), \tag{9.66}$$

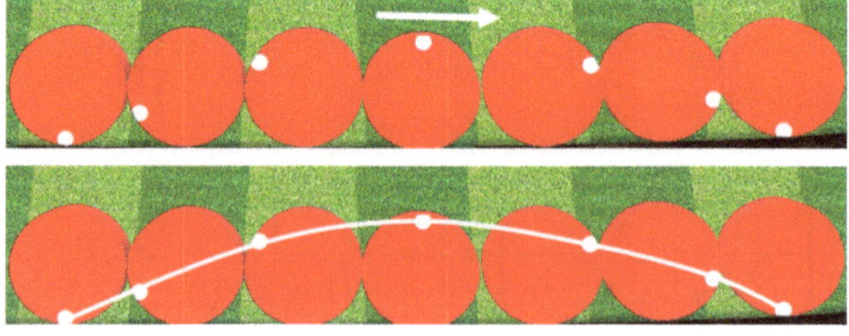

Figure 9.8. Cycloid motion.

where θ is the angle that the circle (the point) rotated. For example, figure 9.9 shows this trajectory for a point on a circle of unit radius ($r = 1$). For a given θ, the circle's center lies at

$$x = r\theta, \ y = r. \tag{9.67}$$

On the other hand, if the circles center becomes

$$x = r\theta, \ y = -r, \tag{9.68}$$

the trajectory would be the one shown in figure 9.10. The first quarter of this trajectory is the trajectory of the bead that takes the smallest possible time.

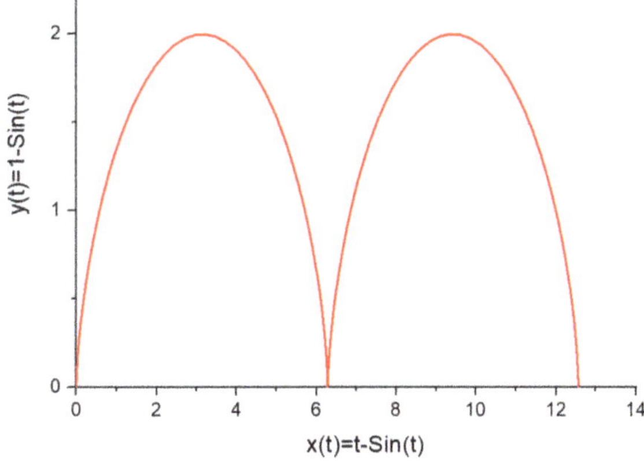

Figure 9.9. Parametric plot for $x(t) = r(t - \sin(t))$ and $y(t) = r(1 - \sin(t))$ for $r = 1$.

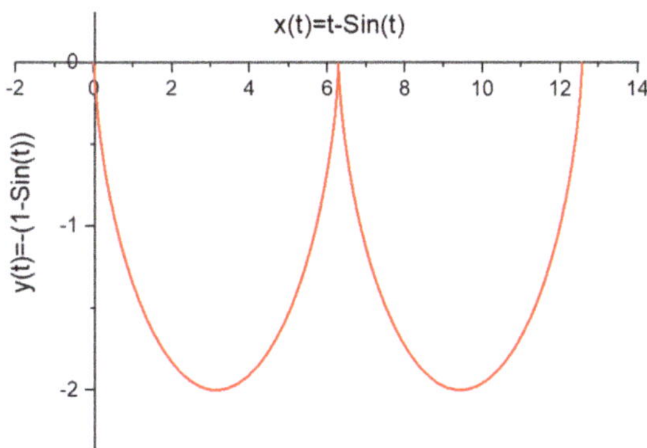

Figure 9.10. Parametric plot for $x(t) = r(t - \sin(t))$ and $y(t) = r(1 - \sin(t))$ for $r = 1$.

9.4 The Euler–Lagrange equation in classical mechanics

In this section, we will see the applications of the Euler–Lagrange equation in classical mechanics. The following are essential quantities we should be familiar with before we see these applications.

1. *The Hamiltonian (H)*: The sum of the kinetic energy (*KE*) and potential energy (*V*)

$$H = KE + V.$$
(9.69)

2. *The Lagrangian ($\mathcal{L}$)*: The kinetic energy minus the potential energy

$$\mathcal{L} = KE - V.$$
(9.70)

3. *The classical action (S)*: The integral of the Lagrangian with respect to time over a given interval of time

$$S = \int_{t_1}^{t_2} \mathcal{L} dt.$$
(9.71)

4. *Hamilton's principle*: The motion of a given system from time t_1 to time t_2 is such that the classical action

$$S = \int_{t_1}^{t_2} \mathcal{L} dt,$$
(9.72)

has a stationary value for the correct path of the motion. The path actually followed by a system, as specified in terms of the generalized coordinates q_i, is the path that makes the action integral stationary,

$$\delta I = \delta \int_{t_1}^{t_2} \mathcal{L}(q_i, \dot{q}_i, t)\, dt = 0,$$
(9.73)

for $i = 1, 2, 3...N$. It means that the Lagrangian must satisfy the set of equations

$$\frac{\partial}{\partial t}\left(\frac{\partial \mathcal{L}}{\partial \dot{q}_i}\right) - \frac{\partial \mathcal{L}}{\partial q_i} = 0$$
(9.74)

for $i = 1, 2, 3...N$.

Example 9.4. Use Lagrange's equations to find the equation of motion for a particle traveling along the x–y plane where its potential energy depends on the x-coordinate $U(x)$.

Solution: The kinetic energy of a particle moving in the x–y plane can be expressed as

$$KE = \frac{1}{2}m\left(v_x^2 + v_y^2\right) = \frac{1}{2}m(\dot{x}^2 + \dot{y}^2). \tag{9.75}$$

Then the Lagrangian becomes

$$\mathcal{L} = KE - U = \frac{1}{2}m(\dot{x}^2 + \dot{y}^2) - U(x). \tag{9.76}$$

Since $\mathcal{L} = \mathcal{L}(x, y, \dot{x}, \dot{y}, t)$ is a function of two variables, applying equation (9.74), one can write

$$\frac{\partial}{\partial t}\left(\frac{\partial \mathcal{L}}{\partial \dot{x}}\right) - \frac{\partial \mathcal{L}}{\partial x} = 0, \quad \frac{\partial}{\partial t}\left(\frac{\partial \mathcal{L}}{\partial \dot{y}}\right) - \frac{\partial \mathcal{L}}{\partial y} = 0. \tag{9.77}$$

Therefore, using the Lagrangian, we find

$$\frac{\partial}{\partial t}\left(\frac{\partial \mathcal{L}}{\partial \dot{x}}\right) - \frac{\partial \mathcal{L}}{\partial x} = 0 \Rightarrow \frac{\partial}{\partial t}(m\dot{x}) + \frac{\partial U(x)}{\partial x} = 0$$

$$\Rightarrow m\ddot{x} = -\frac{\partial U(x)}{\partial x}, \tag{9.78}$$

$$\frac{\partial}{\partial t}\left(\frac{\partial \mathcal{L}}{\partial \dot{y}}\right) - \frac{\partial \mathcal{L}}{\partial y} = 0 \Rightarrow m\ddot{y} = 0 \text{ (Zero acceleration)}.$$

Example 9.5. *Atwood's machine*: A string of length, l, passes over a frictionless pulley connecting two masses m_1 and m_2. Find an expression for the acceleration of the masses in the system.

Solution: Let us define the origin of the y-axis on the ground's surface, also assume $m_1 > m_2$, and the length of the string is l. At a given time t let the position of m_1 and m_2 be y_1 and y_2, as shown in figure 9.11. Then the kinetic and the potential energies of the system can be expressed as

$$KE = \frac{1}{2}m_1\dot{y}_1^2 + \frac{1}{2}m_2\dot{y}_2^2, \quad V = m_1 g y_1 + m_2 g y_2. \tag{9.79}$$

Then the Lagrangian becomes

$$\mathcal{L} = KE - V = \frac{1}{2}m_1\dot{y}_1^2 + \frac{1}{2}m_2\dot{y}_2^2 - m_1 g y_1 - m_2 g y_2. \tag{9.80}$$

This equation appears to be a function of two variables. However, because of the constraint

$$y_1 + y_2 + l = C, \tag{9.81}$$

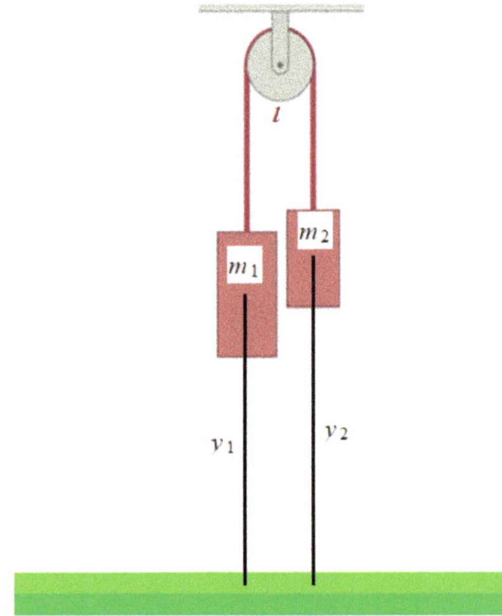

Figure 9.11. Atwood's machine.

where C is a constant, we end up with a Lagrangian that depends only on one variable. This can be achieved by substituting

$$y_2 = C - y_1 - l \Rightarrow \dot{y}_2 = -\dot{y}_1 \tag{9.82}$$

into equation (9.80) that results in

$$\mathcal{L} = \frac{1}{2}(m_1 + m_2)\dot{y}_1^2 - m_1 g y_1 - m_2 g(C - y_1 - l)$$

$$\Rightarrow \mathcal{L} = \frac{1}{2}(m_1 + m_2)\dot{y}_1^2 - (m_1 - m_2)g y_1 - C_1, \tag{9.83}$$

where we replaced, $C_1 = m_2 g(C - l)$. Now using the Euler–Lagrange equation

$$\frac{\partial}{\partial t}\left(\frac{\partial \mathcal{L}}{\partial \dot{q}_i}\right) - \frac{\partial \mathcal{L}}{\partial q_i} = 0 \tag{9.84}$$

for $q_i = y_1$, we find

$$\frac{\partial \mathcal{L}}{\partial y_1} = -(m_1 - m_2)g, \quad \frac{\partial \mathcal{L}}{\partial \dot{y}_1} = (m_1 + m_2)\dot{y}_1$$

$$\Rightarrow \frac{\partial}{\partial t}\left[(m_1 + m_2)\dot{y}_1\right] = -(m_1 - m_2)g \tag{9.85}$$

$$\Rightarrow a_1 = \ddot{y}_1 = -\frac{m_1 - m_2}{m_1 + m_2}g.$$

Using the result $\dot{y}_2 = -\dot{y}_1$, from equation (9.82), the acceleration of the second mass becomes

$$a_2 = \frac{m_1 - m_2}{m_1 + m_2}g = -a_1. \tag{9.86}$$

The minus sign for a_1 indicates the first mass is accelerating in the negative y-direction.

Example 9.6. *Central forces*: Describe the properties of the motion of a mass, m, moving under the influence of a central force (that is, a force acting only along the radial direction) given by

$$\vec{F} = -f(r)\hat{r}, \tag{9.87}$$

for some function $f(r)$. Assume that the motion is confined to a plane (see figure 9.12).

Solution: The kinetic energy

$$KE = \frac{1}{2}mv^2. \tag{9.88}$$

Using spherical coordinates, the mass position can be described by the coordinates $(r, \theta = \pi/2, \varphi)$ so that the position and the velocity of the mass become

$$\vec{r} = r\sin\left(\frac{\pi}{2}\right)\cos(\varphi)\hat{x} + r\sin\left(\frac{\pi}{2}\right)\sin(\varphi)\hat{y} + r\cos\left(\frac{\pi}{2}\right)\hat{z},$$

$$\vec{v} = \frac{d\vec{r}}{dt} = \frac{d}{dt}\left[r\sin\left(\frac{\pi}{2}\right)\cos(\varphi)\hat{x} + r\sin\left(\frac{\pi}{2}\right)\sin(\varphi)\hat{y} + r\cos\left(\frac{\pi}{2}\right)\hat{z}\right] \tag{9.89}$$

$$\vec{v} = [\dot{r}\cos(\varphi) - r\dot{\varphi}\sin(\varphi)]\hat{x} + [\dot{r}\sin(\varphi) + r\dot{\varphi}\cos(\varphi)]\hat{y}.$$

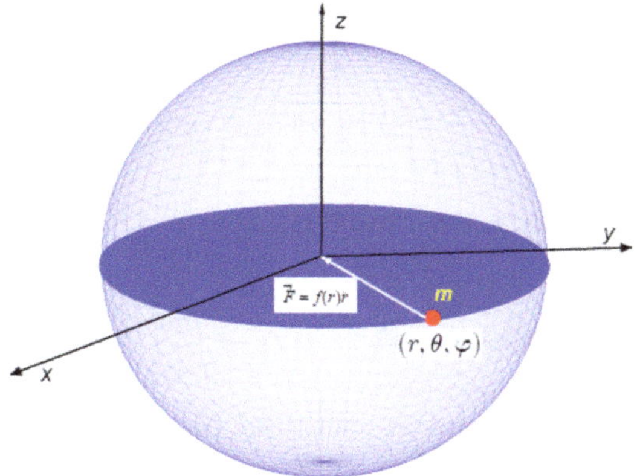

Figure 9.12. A mass m under a central force motion confined to the x–y plane.

From equation (9.89), we can determine the magnitude of the velocity and write the kinetic energy as

$$v^2 = \dot{r}^2 + r^2\dot{\varphi}^2 \Rightarrow T = \frac{1}{2}m(\dot{r}^2 + r^2\dot{\varphi}^2). \tag{9.90}$$

The potential energy $U(r)$ is related to the central force $\vec{F}$ by

$$\vec{F} = -\nabla \cdot U(r). \tag{9.91}$$

Since the force is a central force, the direction is radially inward and depends on r only. Therefore, using equation (9.87), the potential energy can be expressed as

$$U(r) = \int f(r)\, dr. \tag{9.92}$$

Employing the results in equations (9.87) and (9.92), for the Lagrangian one finds,

$$L(t, r, \dot{r}, \varphi, \dot{\varphi}) = T - U = \frac{1}{2}m(\dot{r}^2 + r^2\dot{\varphi}^2) - \int f(r)\, dr. \tag{9.93}$$

Then applying the Euler–Lagrange equation in equation (9.74) for $q_1 = r$ and $q_2 = \varphi$, we have

$$\frac{\partial}{\partial t}\left(\frac{\partial \mathcal{L}}{\partial \dot{r}}\right) - \frac{\partial \mathcal{L}}{\partial r} = 0, \quad \frac{\partial}{\partial t}\left(\frac{\partial \mathcal{L}}{\partial \dot{\varphi}}\right) - \frac{\partial \mathcal{L}}{\partial \varphi} = 0, \tag{9.94}$$

and using the Lagrangian in equation (9.93),

$$\frac{\partial \mathcal{L}}{\partial \varphi} = 0, \quad \frac{\partial \mathcal{L}}{\partial \dot{\varphi}} = mr^2\dot{\varphi}, \quad \frac{\partial \mathcal{L}}{\partial r} = mr\dot{\varphi}^2 - f(r), \quad \frac{\partial \mathcal{L}}{\partial \dot{r}} = m\dot{r}, \tag{9.95}$$

so that, one finds

$$\frac{\partial}{\partial t}\left(\frac{\partial \mathcal{L}}{\partial \dot{\varphi}}\right) = 0 \Rightarrow \frac{\partial \mathcal{L}}{\partial \dot{\varphi}} = \text{const} \Rightarrow mr^2\dot{\varphi} = \text{const} \Rightarrow I\omega = \text{const.} \tag{9.96}$$

$$\frac{\partial}{\partial t}\left(\frac{\partial \mathcal{L}}{\partial \dot{r}}\right) - \frac{\partial \mathcal{L}}{\partial r} = 0 \Rightarrow m\ddot{r} = mr\dot{\varphi}^2 - f(r). \tag{9.97}$$

For a circular motion, where the radius is fixed, $\dot{r} = 0$ and $\ddot{r} = 0$. Thus equation (9.97) becomes

$$mr\dot{\varphi}^2 = f(r) \Rightarrow \frac{m(r\dot{\varphi})^2}{r} = f(r) \Rightarrow \frac{mv^2}{r} = -f(r), \tag{9.98}$$

where $a_c = v^2/r$ is the centripetal acceleration that you know from introductory physics.

9.5 The calculus of variations and Mathematica

The basic problem of the calculus of variations is to determine the functions that maximize or minimize a functional

$$I = \int_{x_1}^{x_2} F(x, y(x), y'(x))\, dx. \tag{9.99}$$

In general, there can be more than one independent variable and the integrand can depend on several functions and their higher derivatives. The extremal functions are solutions of the Euler–Lagrange equations that are obtained by setting the first variational derivatives of the functional, with respect to each function, equal to zero. Since many ordinary and partial differential equations that occur in physics and engineering can be derived as the Euler equations for appropriate functionals, variational methods are of general utility.

- VariationalD $[f, u[x], x]$ returns the variational derivative of the integral with respect to $u[x]$, where the integrand f is a function of u, its derivatives, and x.
- VariationalD $[f, u[x, y, ...], x, y, ...]$ returns the variational derivative of the multiple integral with respect to $u[x, y, ...]$, where f is a function of u, its derivatives, and the coordinates $x, y, ...$.
- VariationalD $[f, u[x, y, ...], v[x, y, ...], ..., \{x, y, ...\}]$ gives a list of variational derivatives with respect to $u, v, ...$.

9.6 Homework assignment

Problem 1. For each of the following line integrals, solve the Euler–Lagrangian equation and discuss the results.

(a)

$$I = \int_{t_1}^{t_2} \sqrt{t}\,\sqrt{1 + \dot{q}^2}\, dt, \tag{9.100}$$

where t is the independent variable and $\dot{q} = \frac{dq}{dt}$.

(b)

$$I = \int_{y_1}^{y_2} (x^2 + x'^2)\, dy \tag{9.101}$$

here y is the independent variable and $x' = \frac{dx}{dy}$.

(c)

$$I = \int_{\theta_1}^{\theta_2} \theta\left(\sqrt{1 - r'^2}\right) d\theta, \tag{9.102}$$

θ is the independent variable and $r' = \frac{dr}{d\theta}$.

Problem 2. From introductory physics you know that light travels with a speed, $c = 3 \times 10^8 \text{ms}^{-1}$, in a vacuum. When it travels in a medium with a refractive index, $n > 1$ the light slows down and the speed, v, is given by

$$v = \frac{c}{n}. \tag{9.103}$$

You also know that when light travels from one medium with refractive index, n_1, to another with refractive index, n_2, the light could bend toward or away from the normal depending on which refractive index is greater or less (figure 9.13). Using calculus of variations show that:

(a) the angle of incidence, θ_1, and angle of transmission, θ_2, are related by the law of refraction (Snell's law)

$$n_1 \sin \theta_1 = n_2 \sin \theta_2. \tag{9.104}$$

Hint: minimize the time taken by the light

$$T = \int_1^2 dt. \tag{9.105}$$

Note that dt is the infinitesimal time for the light to travel an infinitesimal distance ds along the path of the light.

(b) the angle of incidence, θ_1, and the angle of reflection, θ_1', are related by the law of reflection

$$\theta_1 = \theta_1'. \tag{9.106}$$

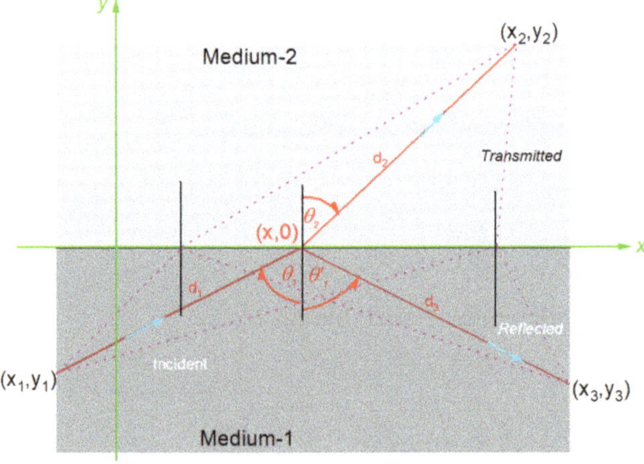

Figure 9.13. Different straight-line paths for the light traveling from point 1 to points 2 and 3. No total internal reflection. The incident light partly gets reflected and partly gets refracted. The shortest paths are defined by the law of reflection and law of refraction.

Hint: minimize the time taken by the light

$$T = \int_1^3 dt. \tag{9.107}$$

N.B. In either case the light travels in a straight line in each medium.

Problem 3. Starting from

$$I(\epsilon) = \int_{x_1}^{x_2} F(x, Y, Y') \, dx, \tag{9.108}$$

where

$$Y(x) = y(x) + \epsilon\eta(x), \tag{9.109}$$

derive the Euler–Lagrange equation

$$\frac{\partial}{\partial x}\left(\frac{\partial F}{\partial y'}\right) - \frac{\partial F}{\partial y} = 0, \tag{9.110}$$

by applying the calculus of variations.

Problem 4. Redo Example 9.2 applying the Euler–Lagrange equation,

$$\frac{\partial}{\partial x}\left(\frac{\partial F}{\partial y'}\right) - \frac{\partial F}{\partial y} = 0, \tag{9.111}$$

$$\text{or } \frac{\partial}{\partial y}\left(\frac{\partial F}{\partial x'}\right) - \frac{\partial F}{\partial x} = 0. \tag{9.112}$$

whichever is easier (figure 9.14).

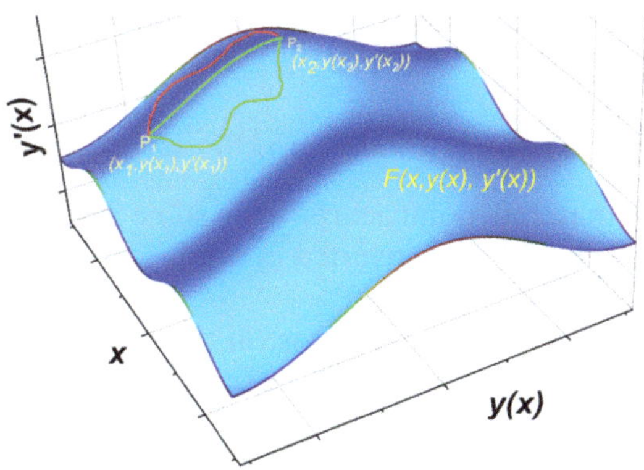

Figure 9.14. A surface defined by the function $F(x, y(x), y'(x) = \frac{dy}{dx})$.

Problem 5. Consider a surface of revolution generated by revolving a curve $y(x)$ about the x-axis. The curve is required to pass through fixed end points (x_1, y_1) and (x_2, y_2) as shown in figure 9.15. Find the curve $y(x)$ that gives the minimum area for the resulting surface using the *Euler–Lagrange equation.*

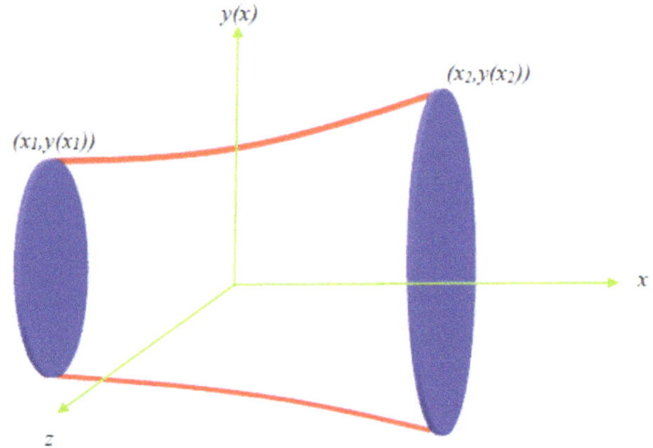

Figure 9.15. A curve rotated about the x-axis.

Problem 6. The speed of an electromagnetic wave in an atmosphere increases in proportion to the height, $v(y) = y/b$, where $b > 0$ is some parameter describing the speed of the electromagnetic wave. This shows that, for example, the speed becomes zero when the height is zero (i.e., $y = 0 \Rightarrow v = 0$), which simulates the condition at the surface of a black hole, called its *event horizon*, where the gravitational force is so strong that the velocity of light goes to zero, thus even trapping the light. Find the optical path in an atmosphere. *Hint: the optical path must be the shortest path and do not expect a straight line.*

Problem 7. In Example 9.3 using the Euler–Lagrange equation we solved the brachistochrone problem, assuming the bead started from rest. We showed that in fact the shortest time is defined by the equation of an inverted cycloid

$$x = \frac{1}{2c}(\theta - \sin(\theta)), \quad y = -\frac{1}{2c}(1 - \cos(\theta)). \tag{9.113}$$

Show that when the material point starts with some initial velocity, v_0, still the shortest time path is defined by an inverted cycloid.

Problem 8. Consider the motion of a mass, m, moving under the influence of a central force (that is, a force acting only along the radial direction) given by

$$\vec{F} = -f(r)\hat{r} \tag{9.114}$$

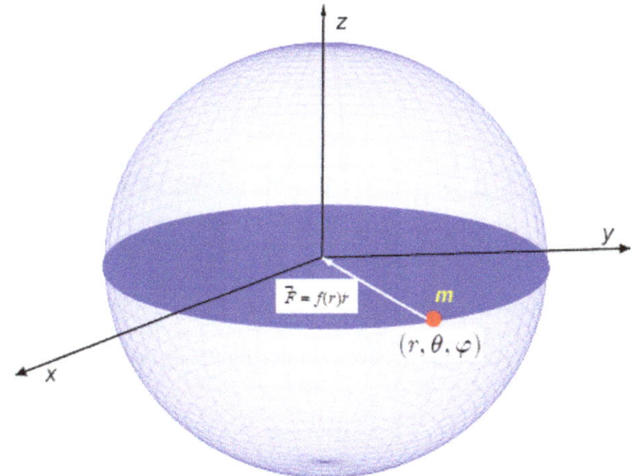

Figure 9.16. A mass m under a central force motion confined to the x–y plane.

for some function $f(r)$. Assume that the motion is confined to a plane and the position of the mass can be described using polar coordinates (r, φ)

$$\vec{r} = r\cos(\varphi)\hat{x} + r\sin(\varphi)\hat{y} \tag{9.115}$$

where $\theta = \pi/2$ as shown in figure 9.16.

    (a) Find the Lagrangian.

    (b) Using the Euler–Lagrange equation find the equation of motion for the mass for the radial and angular coordinates (i.e., r and φ). Show that one of these equations gives you the law of conservation of angular momentum, $\vec{L}$,

$$\frac{d\vec{L}}{dt} = I\frac{d\vec{\omega}}{dt} = 0 \Rightarrow \vec{L} = I\vec{\omega} = \text{Constant}, \tag{9.116}$$

where $I = mr^2$ is the moment of inertia.

    (c) For the case in which the radial distance is constant

$$\dot{r} = \frac{dr}{dt} = 0 \tag{9.117}$$

you will find the equation of motion for a mass, m, moving in a circle

$$mr\dot{\theta}^2 = -f(r) \Rightarrow \frac{m(r\dot{\theta})^2}{r} = -f(r) \Rightarrow \frac{mv^2}{r} = -f(r). \tag{9.118}$$

Problem 9. *A one-dimensional harmonic oscillator*: Consider a mass, m, attached to one end of a spring with spring constant, k. Find the Lagrangian and the equation of motion for the mass using the Euler–Lagrange equation for the following two cases.

    (a) The other end of the spring is attached to a wall as shown in figure 9.17 and the mass is oscillating on a frictionless table.

(b) The other end of the spring is attached to a ceiling as shown in figure 9.18 and the mass is oscillating in a vertical plane.

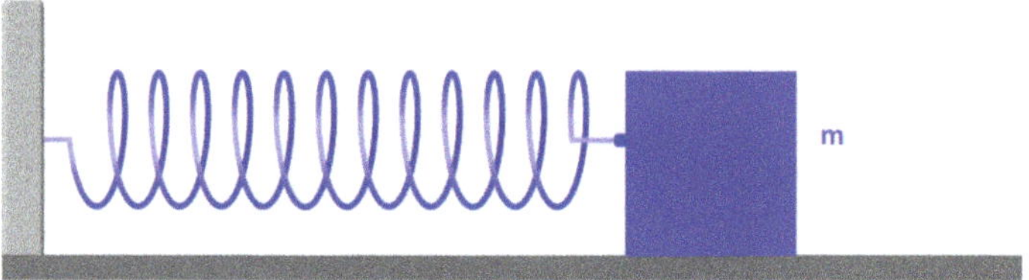

Figure 9.17. A harmonic oscillator in a horizontal plane.

Figure 9.18. A harmonic oscillator in a vertical plane.

Problem 10. Consider two masses, m_1 and m_2, connected by a string of length, l, with negligible mass. The string passes through a hole at the center of a table. The mass m_1 is on the table and it can move on the table. The surface of the table is frictionless. The second mass, m_2, hanging from the other end of the string can move up or down on a vertical plane (see figure 9.19).

(a) Using cylindrical coordinates (r, φ, z), find the Lagrangian.

(b) Find the equation of motion for the two masses using the Euler–Lagrange equation.

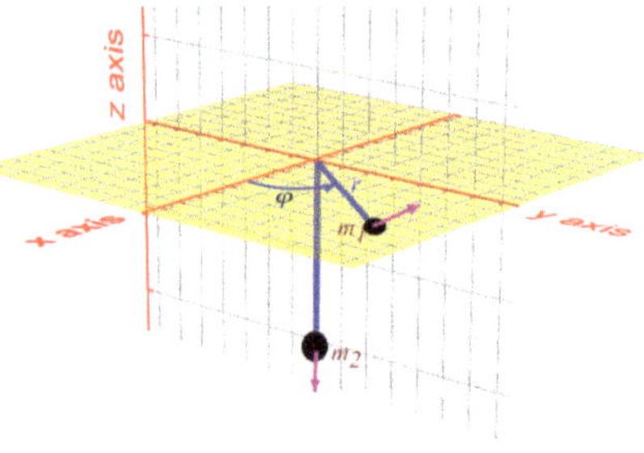

Figure 9.19. Two masses connected by a string of length l. In cylindrical coordinates the position for m_1 is $(r, \varphi, 0)$ and for m_2 it is $(0, 0, -z_2)$. Note that $z_2 + r = l$.

IOP Publishing

Studies in Theoretical Physics, Volume 1
Fundamental mathematical methods
Daniel Erenso and Victor Montemayor

Chapter 10

Introduction to the eigenvalue problem

We often encounter eigenvalue problems in most branches of physics. Such problems involve a square matrix, an eigenvalue, and eigenvectors. This chapter introduces how we solve eigenvalue problems. We begin with a brief review of matrices and determinants that we studied in chapter 4. We then introduce the Dirac notation for vectors and orthogonal transformations of vectors. Following this, we discuss how we find eigenvalues and eigenvectors for square matrices and see the unique properties of eigenvalue problems involving Hermitian square matrices. Finally, after introducing the similarity transformation, we see how Mathematica can determine the eigenvalues and eigenvectors of a square matrix by resolving the same problems that we considered as an example.

10.1 Eigenvalue problem in physics

In most branches of physics, we often find equations that involve matrices with non-zero off-diagonal elements. Solving these equations requires transforming the equations into a different form where the off-diagonal elements of the matrices vanish (diagonalized matrix). In classical mechanics, one good example is the angular momentum, $\vec{L}$, equation

$$\vec{L} = I\vec{\omega}, \tag{10.1}$$

where $\vec{\omega}$ is the angular velocity, and I is the moment of inertia matrix,

$$I = \begin{pmatrix} I_{xx} & I_{xy} & I_{xz} \\ I_{yx} & I_{yy} & I_{yz} \\ I_{zx} & I_{zy} & I_{zz} \end{pmatrix}, \tag{10.2}$$

which is a symmetric square matrix (i.e., $I_{ij} = I_{ji}$). In quantum mechanics, we often need to solve the time-independent Schrödinger equation

$$\hat{H}\phi_n = E_n\phi_n, \tag{10.3}$$

where ϕ_n and E_n are the energy eigenfunctions and eigenvalues, respectively, and $\hat{H}$ is the energy operator. For example, for a two-level atom, with two energy eigenfunctions and eigenvalues, equation (10.3) can be put in a matrix form

$$\begin{bmatrix} H_{11} & H_{12} \\ H_{21} & H_{22} \end{bmatrix}\begin{bmatrix} \phi_1 \\ \phi_2 \end{bmatrix} = \begin{bmatrix} E_1 \\ E_2 \end{bmatrix}\begin{bmatrix} \phi_1 \\ \phi_2 \end{bmatrix}. \tag{10.4}$$

To find the energy eigenvalues and eigenfunctions, one may have to diagonalize the energy matrix

$$H = \begin{bmatrix} H_{11} & H_{12} \\ H_{21} & H_{22} \end{bmatrix}. \tag{10.5}$$

This chapter introduces how to diagonalize and solve such problems described by equation (10.3), which are commonly known as eigenvalue problems. We begin with matrix and determinant review that we introduced in chapter 4.

10.2 Matrix review

Matrix arithmetic: Consider the following matrices:

$$A = \begin{pmatrix} 2 & 3 & 1 \\ 2 & 1 & 0 \end{pmatrix}, \quad B = \begin{pmatrix} 2 & 2i \\ i & -1 \\ 3 & 2i \end{pmatrix},$$

$$C = \begin{pmatrix} 2 & 1 & 3 \\ 4 & -1 & -2 \\ -1 & 0 & 1 \end{pmatrix}, \quad D = \begin{pmatrix} -2 & 0 & 1 \\ 1 & -1 & 2 \\ 3 & 1 & 0 \end{pmatrix}. \tag{10.6}$$

A. **Addition and subtraction:** We can add or subtract matrices if the matrices have the same number of rows and columns. From matrices A, B, C, and D we can add/subtract only matrices C and D

$$C + D = \begin{pmatrix} 2 & 1 & 3 \\ 4 & -1 & -2 \\ -1 & 0 & 1 \end{pmatrix} + \begin{pmatrix} -2 & 0 & 1 \\ 1 & -1 & 2 \\ 3 & 1 & 0 \end{pmatrix}$$

$$= \begin{pmatrix} 2-2 & 1+0 & 3+1 \\ 4+1 & -1-1 & -2+2 \\ -1+3 & 0+1 & 1+0 \end{pmatrix} = \begin{pmatrix} 0 & 1 & 4 \\ 5 & -2 & 0 \\ 2 & 1 & 1 \end{pmatrix}. \tag{10.7}$$

B. **Matrix multiplication:** we can multiply two matrices when the number of columns of the first matrix equals the number of rows of the second matrix. From the above matrices, we can make the multiplications AB and CD but we cannot make the multiplications BC or BD. The element in row i and column j of the product matrix AB (i.e., $(AB)_{ij}$) is given by

$$(AB)_{ij} = \sum_{k=1}^{n} A_{ik} B_{kj} = A_{i1} B_{1j} + A_{i2} B_{2j} + \dots A_{in} B_{nj}. \tag{10.8}$$

C. **Commutativities:** For any two multipliable matrices C and D,

$$CD \neq DC. \tag{10.9}$$

D. **The commutator:** For square matrices C and D the commutator $[C, D]$ is defined as

$$[C, D] = CD - DC. \tag{10.10}$$

E. **The identity matrix I:** A square matrix, when multiplied with another square matrix A, keeps the matrix unchanged,

$$IA = AI = A. \tag{10.11}$$

F. **Transpose of a matrix A^T:** A matrix constructed by changing the ith row of the matrix A into the ith column ($i.e.,(A^T)_{ji} = A_{ij}$),

$$A = \begin{pmatrix} 2 & 3 & 1 \\ 2 & 1 & 0 \end{pmatrix} \Rightarrow A^T = \begin{pmatrix} 2 & 2 \\ 3 & 1 \\ 1 & 0 \end{pmatrix}.$$

G. **Transpose conjugate of a matrix $A^\dagger$:** The transpose and conjugate of the matrix A (i.e., $(A^\dagger)_{ji} = \left((A^T)_{ji}\right)^* = A_{ij}^*$),

$$A = \begin{pmatrix} 2 & 2i \\ i & -1 \\ 3 & 2i \end{pmatrix} \Rightarrow A^\dagger = (A^T)^* = \begin{pmatrix} 2 & -i & 3 \\ -2i & -1 & -2i \end{pmatrix}. \tag{10.12}$$

For a product of two matrices, A and B,

$$(AB)^\dagger = B^\dagger A^\dagger. \tag{10.13}$$

Note that this relation applies for a product of more than two matrices.

H. **Adjoint of a matrix:** The adjoint of a square matrix A is given by

$$\mathrm{adj}(A) = [\mathrm{cof}(A)]^T, \tag{10.14}$$

where $\mathrm{cof}(A)$ is the cofactor of the matrix A. We recall that the minor of matrix A, (M_{ij}), is the determinant of the matrix formed from matrix A by removing the ith row and jth column. For the cofactor matrix the elements are given by

$$[\text{cof}(A)]_{ij} = (-1)^{i+j} M_{ij}.$$

(10.15)

I. **Inverse of a (square) matrix A^{-1}:**

$$A^{-1}A = AA^{-1} = I.$$

(10.16)

We can determine the inverse of an invertible matrix (det $|A| \neq 0$) using *row reduction* or the *adjoint matrix*. In terms of the adjoint matrix, the inverse is given by

$$A^{-1} = \frac{[\text{cof}(A)]^T}{\det |A|}.$$

(10.17)

10.3 Orthogonal transformations and Dirac's notation

Vectors, matrices, and Dirac notation: Consider a vector $\vec{r}$ describing a point in space as shown in figure 10.1. In Cartesian coordinates this vector is given by

$$\vec{r} = x\hat{x} + y\hat{y} + z\hat{z},$$

(10.18)

and it can be expressed using a column matrix as

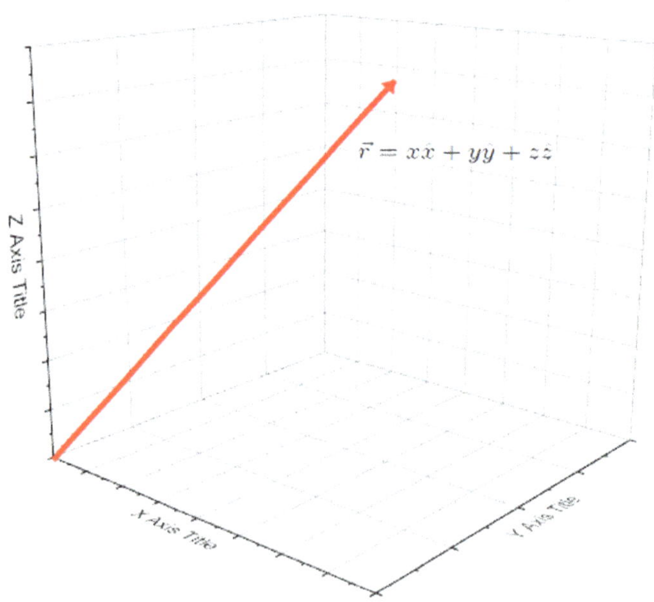

Figure 10.1. A vector, $\vec{r}$, representing a point in spaces in terms of the Cartesian coordinates, x, y, and z.

$$\vec{r} = x\hat{x} + y\hat{y} + z\hat{z} = \begin{pmatrix} x \\ y \\ z \end{pmatrix}, \tag{10.19}$$

where each row corresponds to the component along each unit vector in Cartesian coordinates. For an N-dimensional vector $\vec{r}$, one can then write

$$\vec{r} = x_1\hat{e}_1 + x_2\hat{e}_2 + \dots x_N\hat{e}_N = \sum_{n=1}^{N} x_n\hat{e}_n = \begin{pmatrix} x_1 \\ x_2 \\ \vdots \\ x_N \end{pmatrix},$$

where x_n is the component along the unit vector $\hat{e}_n$. A vector, using Dirac's notation is represented by a ket vector $|\,\rangle$, for example, the vector $\vec{r}$ is represented as $|\,r\rangle$,

$$\vec{r} = \sum_{n=1}^{N} x_n\hat{e}_n = |r\rangle = \begin{pmatrix} x_1 \\ x_2 \\ \vdots \\ x_N \end{pmatrix}. \tag{10.20}$$

The transpose conjugate of a vector $\vec{r}$, which we represent by a bra vector $\langle r\,|$ can be written as

$$\langle r| = \left(|r\rangle\right)^{\dagger} = \begin{pmatrix} x_1 \\ x_2 \\ \vdots \\ x_N \end{pmatrix}^{\dagger} = \begin{pmatrix} x_1^* & x_2^* & \dots & x_N^* \end{pmatrix} \tag{10.21}$$

$$= x_1^*\hat{e}_1 + x_2^*\hat{e}_2 + \dots x_N^*\hat{e}_N.$$

Recalling that the magnitude of a vector is determined using a scalar product

$$|\vec{r}|^2 = \vec{r}^* \cdot \vec{r} = (x_1^*\hat{e}_1 + x_2^*\hat{e}_2 + \dots x_N^*\hat{e}_N) \cdot (x_1\hat{e}_1 + x_2\hat{e}_2 + \dots x_N\hat{e}_N)$$

$$= x_1^*x_1 + x_2^*x_2 + \dots x_N^*x_N = \sum_{n=1}^{N} x_n^*x_n, \tag{10.22}$$

using the bra–ket notation, one can write

$$|\vec{r}|^2 = \langle r|r\rangle = x_1^*x_1 + x_2^*x_2 + \dots x_N^*x_N = \sum_{n=1}^{N} x_n^*x_n. \tag{10.23}$$

Suppose we make some operation on the vector $|\,r\rangle$ and found another vector $|\,r'\rangle$ and we represent this operation by a matrix, M. Then one can describe this action using the equation

$$|r'\rangle = M|r\rangle. \tag{10.24}$$

For example, a 3-D vector equation (10.24), using matrices, can be expressed as

$$\begin{pmatrix} x' \\ y' \\ z' \end{pmatrix} = \begin{pmatrix} M_{11} & M_{12} & M_{13} \\ M_{21} & M_{22} & M_{23} \\ M_{31} & M_{32} & M_{33} \end{pmatrix} \begin{pmatrix} x \\ y \\ z \end{pmatrix}. \tag{10.25}$$

For the transpose conjugate of equation (10.24), one can write

$$\left(|r'\rangle \right)^{\dagger} = \left(M \, |r\rangle \right)^{\dagger} \Rightarrow \langle r'| = \left(|r\rangle \right)^{\dagger} M^{\dagger}$$
$$\Rightarrow \langle r'| = \langle r| \, M^{\dagger}. \tag{10.26}$$

Noting that

$$\begin{pmatrix} x' \\ y' \\ z' \end{pmatrix}^{\dagger} = \begin{pmatrix} x'^{*} & y'^{*} & z'^{*} \end{pmatrix},$$

$$\begin{pmatrix} M_{11} & M_{12} & M_{13} \\ M_{21} & M_{22} & M_{23} \\ M_{31} & M_{32} & M_{33} \end{pmatrix}^{\dagger} = \begin{pmatrix} M_{11}^{*} & M_{21}^{*} & M_{31}^{*} \\ M_{12}^{*} & M_{22}^{*} & M_{32}^{*} \\ M_{13}^{*} & M_{23}^{*} & M_{33}^{*} \end{pmatrix}, \tag{10.27}$$

in view of equation (10.26), for the transpose conjugate of equation (10.25), one can write

$$\langle r'| = \langle r| \, M^{\dagger}$$

$$\Rightarrow \begin{pmatrix} x'^{*} & y'^{*} & z'^{*} \end{pmatrix} = \begin{pmatrix} x^{*} & y^{*} & z^{*} \end{pmatrix} \begin{pmatrix} M_{11}^{*} & M_{21}^{*} & M_{31}^{*} \\ M_{12}^{*} & M_{22}^{*} & M_{32}^{*} \\ M_{13}^{*} & M_{23}^{*} & M_{33}^{*} \end{pmatrix}. \tag{10.28}$$

Then the magnitude of the new vector $|\,r'\rangle$ can be determined from the inner product

$$|\vec{r}'|^{2} = \langle r'|r'\rangle = \langle r| \, M^{\dagger}M \, |r\rangle. \tag{10.29}$$

When the operation on the vector $|\,r\rangle$, leaves the magnitude of the vector unchanged,

$$|\vec{r}'|^{2} = |\vec{r}|^{2} \Rightarrow \langle r'|r'\rangle = \langle r|r\rangle \Rightarrow \langle r| \, M^{\dagger}M \, |r\rangle = \langle r|r\rangle, \tag{10.30}$$

for the transformation matrix M, we must have

$$M^{\dagger}M = I = M^{-1}M \Rightarrow M^{-1} = M^{\dagger}, \tag{10.31}$$

where I is the identity matrix. Such a matrix, where its inverse M^{-1} is equal to its transpose conjugate $M^{\dagger}$, is known as an *orthogonal matrix*.

One example of an orthogonal matrix is the rotation matrix (representing the 'rotation operation') that we considered in chapter four. For a counter-clockwise rotation about the z-axis by an angle, θ, we found a vector $\vec{r}'$ described by the coordinates $(\hat{x}', \hat{y}', \hat{z}')$. As shown in figure 4.2 the length of the vector has not changed by the rotation. In terms of the unit vectors in the rotated coordinate, the vector $\vec{r}'$ can be expressed as

$$\left|r'\right\rangle = \vec{r}' = x'\hat{x}' + y'\hat{y}' + z'\hat{z}' = \begin{pmatrix} x' \\ y' \\ z' \end{pmatrix}. \tag{10.32}$$

figure 4.3 shows the top view of the x–y–z and the rotated x'-y'-z' coordinate systems. From this diagram, we have shown that the equation that relates the components in the two coordinates system is given by

$$\left|r'\right\rangle = R\left|r\right\rangle \Rightarrow \begin{pmatrix} x' \\ y' \\ z' \end{pmatrix} = \begin{pmatrix} \cos\theta & \sin(\theta) & 0 \\ -\sin(\theta) & \cos\theta & 0 \\ 0 & 0 & 1 \end{pmatrix} \begin{pmatrix} x \\ y \\ z \end{pmatrix}, \tag{10.33}$$

where we denote the rotation matrix by $R_z(\theta) = R$. The rotation matrix, R, is orthogonal as it has not changed the magnitude of the vector, $r' = r$. One can easily verify that by checking, $R^{-1} = R^\dagger$.

10.4 Eigenvalues and eigenvectors

In general, we can express a linear transformation of a vector, $\vec{A}$, into another vector, $\vec{A}'$, as

$$\vec{A}' = M\,\vec{A} \quad \text{or} \quad \left|A'\right\rangle = M\left|A\right\rangle. \tag{10.34}$$

Using matrices, for example for a 3-D vector, this can be put in the form

$$\begin{pmatrix} A_1' \\ A_2' \\ A_3' \end{pmatrix} = \begin{pmatrix} M_{11} & M_{12} & M_{13} \\ M_{21} & M_{22} & M_{23} \\ M_{31} & M_{32} & M_{33} \end{pmatrix} \begin{pmatrix} A_1 \\ A_2 \\ A_3 \end{pmatrix}. \tag{10.35}$$

Under this transformation, when the vector, $\vec{A}'$, is expressible as,

$$\vec{A}' = M\vec{A} = \lambda\vec{A}, \quad \text{or} \quad \left|A'\right\rangle = M\left|A\right\rangle = \lambda\left|A\right\rangle, \tag{10.36}$$

where λ is a constant, then $\vec{A}$ and λ are called *the eigenvector* (characteristic vector) and *the eigenvalue* (characteristic value) of the matrix M, respectively. For example for the 3-D vector in equation (10.35), the eigenvalue equation can be written as

$$\begin{pmatrix} M_{11} & M_{12} & M_{13} \\ M_{21} & M_{22} & M_{23} \\ M_{31} & M_{32} & M_{33} \end{pmatrix} \begin{pmatrix} A_1 \\ A_2 \\ A_3 \end{pmatrix} = \lambda \begin{pmatrix} A_1 \\ A_2 \\ A_3 \end{pmatrix}. \tag{10.37}$$

Using a 3×3 identity matrix I,

$$I = \begin{pmatrix} 1 & 0 & 0 \\ 0 & 1 & 0 \\ 0 & 0 & 1 \end{pmatrix}, \tag{10.38}$$

one can put equation (10.37) in the form

$$\begin{pmatrix} M_{11} & M_{12} & M_{13} \\ M_{21} & M_{22} & M_{23} \\ M_{31} & M_{32} & M_{33} \end{pmatrix} \begin{pmatrix} A_1 \\ A_2 \\ A_3 \end{pmatrix} = \lambda \begin{pmatrix} 1 & 0 & 0 \\ 0 & 1 & 0 \\ 0 & 0 & 1 \end{pmatrix} \begin{pmatrix} A_1 \\ A_2 \\ A_3 \end{pmatrix}$$

$$= \begin{pmatrix} \lambda & 0 & 0 \\ 0 & \lambda & 0 \\ 0 & 0 & \lambda \end{pmatrix} \begin{pmatrix} A_1 \\ A_2 \\ A_3 \end{pmatrix}, \tag{10.39}$$

which can be rewritten as

$$\left[\begin{pmatrix} M_{11} & M_{12} & M_{13} \\ M_{21} & M_{22} & M_{23} \\ M_{31} & M_{32} & M_{33} \end{pmatrix} - \begin{pmatrix} \lambda & 0 & 0 \\ 0 & \lambda & 0 \\ 0 & 0 & \lambda \end{pmatrix} \right] \begin{pmatrix} A_x \\ A_y \\ A_z \end{pmatrix} = 0$$

$$\Rightarrow \begin{pmatrix} M_{11} - \lambda & M_{12} & M_{13} \\ M_{21} & M_{22} - \lambda & M_{23} \\ M_{31} & M_{32} & M_{33} - \lambda \end{pmatrix} \begin{pmatrix} A_x \\ A_y \\ A_z \end{pmatrix} = 0. \tag{10.40}$$

The eigenvalues are determined from the condition

$$\begin{vmatrix} M_{11} - \lambda & M_{12} & M_{13} \\ M_{21} & M_{22} - \lambda & M_{23} \\ M_{31} & M_{32} & M_{33} - \lambda \end{vmatrix} = 0, \tag{10.41}$$

which is known as the characteristic equation of matrix M. The roots of this equation are the eigenvalues of M. To find the eigenvectors, we substitute the eigenvalues and solve the resulting equations. For example, from equation (10.41), we find three eigenvalues, $\lambda = \lambda_1$, λ_2, and λ_3 and we determine the corresponding eigenvectors

$$|A_1\rangle = \begin{pmatrix} A_{11} \\ A_{21} \\ A_{31} \end{pmatrix}, \quad |A_2\rangle = \begin{pmatrix} A_{12} \\ A_{22} \\ A_{32} \end{pmatrix}, \quad |A_3\rangle = \begin{pmatrix} A_{13} \\ A_{23} \\ A_{33} \end{pmatrix}, \tag{10.42}$$

by solving the set of linear equations resulting from

$$\begin{pmatrix} M_{11} - \lambda_i & M_{12} & M_{13} \\ M_{21} & M_{22} - \lambda_i & M_{23} \\ M_{31} & M_{32} & M_{33} - \lambda_i \end{pmatrix} \begin{pmatrix} A_{1i} \\ A_{2i} \\ A_{3i} \end{pmatrix} = 0. \tag{10.43}$$

for $i = 1, 2$, and 3. When we solve these equations, we can determine only two components of the eigenvectors in terms of the third component. We determine this third component by imposing the condition that eigenvectors must be normalized, $\langle A_i \mid A_i \rangle = 1$.

From *The VNR Concise Encyclopedia of Mathematics* (*New York: Van Nostrand Reinhold Co., 1977*):

Eigenvalues: Eigenvalue problems are important in many branches of physics. They make it possible to find coordinate systems in which the transformations in question take on their simplest forms. In mechanics for instance, the principal moments of a rigid body are found with the help of the eigenvalues of the symmetric matrix representing the inertia tensor.... Eigenvalues are of central importance in quantum mechanics, in which the measured values of physical 'observables' appear as the eigenvalues of certain operators. The term 'transformation' is used predominantly in pure mathematical (geometrical) context, whereas 'operator' is more customary in applications (physics, technology).

Example 10.1. Find the eigenvalues and the eigenvectors for the matrix

$$M = \begin{pmatrix} 0 & 1 & 0 \\ 1 & 0 & 0 \\ 0 & 0 & 0 \end{pmatrix}. \tag{10.44}$$

Solution: We determine the eigenvalues from the eigenvalue equation

$$\begin{vmatrix} 0-\lambda & 1 & 0 \\ 1 & 0-\lambda & 0 \\ 0 & 0 & 0-\lambda \end{vmatrix} = 0 \Rightarrow -\lambda \begin{vmatrix} -\lambda & 0 \\ 0 & -\lambda \end{vmatrix} - \begin{vmatrix} 1 & 0 \\ 0 & -\lambda \end{vmatrix} = 0 \tag{10.45}$$

$$\Rightarrow -\lambda^3 + \lambda = 0 \Rightarrow \lambda_1 = 0, \lambda_2 = 1, \lambda_3 = -1.$$

The corresponding eigenvectors are determined from

$$\begin{pmatrix} M_{11}-\lambda_i & M_{12} & M_{13} \\ M_{21} & M_{22}-\lambda_i & M_{23} \\ M_{31} & M_{32} & M_{33}-\lambda_i \end{pmatrix} \begin{pmatrix} A_{1i} \\ A_{2i} \\ A_{3i} \end{pmatrix} = 0. \tag{10.46}$$

For $i = 1$, $\lambda_1 = 0$, we find

$$\begin{pmatrix} 0 & 1 & 0 \\ 1 & 0 & 0 \\ 0 & 0 & 0 \end{pmatrix} \begin{pmatrix} A_{11} \\ A_{21} \\ A_{31} \end{pmatrix} = 0 \Rightarrow A_{11} = 0, A_{21} = 0 \tag{10.47}$$

$$\Rightarrow |A_1\rangle = \begin{pmatrix} 0 \\ 0 \\ A_{31} \end{pmatrix}.$$

From equation (10.47), the corresponding bra vector can be written as

$$\langle A_1 | = (0 \quad 0 \quad A_{31}), \tag{10.48}$$

so that applying the normalization condition for the eigenvectors, one finds

$$\langle A_1 | A_1 \rangle = 1 \Rightarrow (0 \quad 0 \quad A_{31}) \begin{pmatrix} 0 \\ 0 \\ A_{31} \end{pmatrix} = 1 \Rightarrow A_{31} = 1$$

$$\Rightarrow |A_1\rangle = \begin{pmatrix} 0 \\ 0 \\ 1 \end{pmatrix}. \tag{10.49}$$

Similarly, for $\lambda_2 = 1$, we have

$$\begin{pmatrix} -1 & 1 & 0 \\ 1 & -1 & 0 \\ 0 & 0 & -1 \end{pmatrix} \begin{pmatrix} A_{12} \\ A_{22} \\ A_{32} \end{pmatrix} = 0$$

$$\Rightarrow -A_{12} + A_{22} = 0, \; A_{12} - A_{22} = 0, \; A_{32}$$

$$= 0 \Rightarrow A_{12} = A_{22}, \; A_{32} = 0, \tag{10.50}$$

that leads to

$$|A_2\rangle = A_{21}\begin{pmatrix} 1 \\ 1 \\ 0 \end{pmatrix} \Rightarrow \langle A_2| = A_{21}(1 \quad 1 \quad 0),$$

$$\langle A_2 | A_2 \rangle = 1 \Rightarrow A_{21}^2 (1 \quad 1 \quad 0) \begin{pmatrix} 1 \\ 1 \\ 0 \end{pmatrix} = 1 \Rightarrow A_{21} = \frac{1}{\sqrt{2}} \tag{10.51}$$

$$\Rightarrow |A_2\rangle = \frac{1}{\sqrt{2}} \begin{pmatrix} 1 \\ 1 \\ 0 \end{pmatrix}.$$

Once again, following a similar procedure for $\lambda_3 = -1$, one can also show that

$$\begin{pmatrix} 1 & 1 & 0 \\ 1 & 1 & 0 \\ 0 & 0 & 1 \end{pmatrix} \begin{pmatrix} A_{13} \\ A_{23} \\ A_{33} \end{pmatrix} = 0 \Rightarrow A_{13} + A_{23} = 0, \; A_{13} + A_{23} = 0, \; A_{33} = 0$$

$$\Rightarrow A_{13} = -A_{23}, \; A_{33} = 0 \Rightarrow |A_3\rangle = A_{23}\begin{pmatrix} -1 \\ 1 \\ 0 \end{pmatrix} \Rightarrow |A_3\rangle = \frac{1}{\sqrt{2}} \begin{pmatrix} -1 \\ 1 \\ 0 \end{pmatrix}. \tag{10.52}$$

We now can say we have completely solved the eigenvalue equation.

10.5 Eigenvalue equation and Hermitian matrices

A matrix M is said to be Hermitian when it is equal to its transpose conjugate matrix $M^\dagger$

$$M^\dagger = M \Rightarrow (M^*)^T = M. \tag{10.53}$$

The eigenvalues for a Hermitian matrix are real, and the corresponding eigenvectors form an orthonormal set of vectors. We can easily show this as follows:

$$M\left|A_i\right\rangle = \lambda_i\left|A_i\right\rangle \Rightarrow \left\langle A_j\right|M\left|A_i\right\rangle = \left\langle A_j\right|\lambda_i\left|A_i\right\rangle = \lambda_i\left\langle A_j\middle|A_i\right\rangle. \tag{10.54}$$

Now let us take the transpose conjugate of equation (10.54),

$$\left\langle A_j\right|M\left|A_i\right\rangle^\dagger = \left(\lambda_i\left\langle A_j\right|A_i\right\rangle\right)^\dagger \Rightarrow \left\langle A_i\right|M^\dagger\left|A_j\right\rangle = \lambda_i^*\left\langle A_i\right|A_j\right\rangle. \tag{10.55}$$

For a Hermitian matrix, $M^\dagger = M$, one can rewrite equation (10.55) as

$$\left\langle A_i\right|M\left|A_j\right\rangle = \lambda_i^*\left\langle A_i\right|A_j\right\rangle. \tag{10.56}$$

Upon switching i and j, we can also write equation (10.56) as

$$\left\langle A_j\right|M\left|A_i\right\rangle = \lambda_j^*\left\langle A_j\right|A_i\right\rangle. \tag{10.57}$$

Subtracting equation (10.57) from equation (10.54), one finds

$$\left(\lambda_i - \lambda_j^*\right)\left\langle A_j\right|A_i\right\rangle = 0. \tag{10.58}$$

In order for equation (10.58) to be true, when $i \neq j$, we must have $\left\langle A_j\middle|A_i\right\rangle = 0$ and when $i = j$ since $\left\langle A_i\middle|A_i\right\rangle \geq 1$, we must have $\lambda_i = \lambda_i^*$. Therefore, for a Hermitian square matrix,

$$\lambda_i = \lambda_i^*, \ \left\langle A_j\middle|A_i\right\rangle = \begin{cases} 0 & i \neq j \\ 1 & i = j \end{cases} = \delta_{ij}. \tag{10.59}$$

This proves that the eigenvalues for a Hermitian matrix must be real and the corresponding eigenvectors must form an orthonormal set of vectors.

10.6 The similarity transformation

We recall from the previous section for the matrix M, the eigenvalue equation,

$$M\left|A\right\rangle = \lambda\left|A\right\rangle, \tag{10.60}$$

where λ is the eigenvalue and $|A\rangle$ is the eigenvector. For example for a 3×3 Hermitian matrix

$$M = \begin{pmatrix} M_{11} & M_{12} & M_{13} \\ M_{21} & M_{22} & M_{23} \\ M_{31} & M_{32} & M_{33} \end{pmatrix}, \tag{10.61}$$

where the three eigenvalues λ_1, λ_2, and λ_3 are real and the three eigenvectors are given by

$$|A_1\rangle = \begin{pmatrix} A_{11} \\ A_{21} \\ A_{31} \end{pmatrix}, \quad |A_2\rangle = \begin{pmatrix} A_{12} \\ A_{22} \\ A_{32} \end{pmatrix}, \quad |A_3\rangle = \begin{pmatrix} A_{13} \\ A_{23} \\ A_{33} \end{pmatrix},$$

$$\Rightarrow \langle A_1| = (A_{11} \ A_{21} \ A_{31}), \quad \langle A_2| = (A_{12} \ A_{22} \ A_{32}),$$

$$\langle A_1| = (A_{13} \ A_{23} \ A_{33}), \tag{10.62}$$

since these eigenvectors form an orthonormal set of vectors

$$\langle A_i|A_j\rangle = \delta_{ij} = \begin{cases} 1, & i = j \\ 0, & i \neq j \end{cases}, \tag{10.63}$$

one can then write

$$\langle A_i| M |A_j\rangle = \lambda_j \langle A_i|A_j\rangle = \lambda_i \delta_{ij}. \tag{10.64}$$

Using equations (10.61) and (10.62), one can rewrite equation (10.64) in a matrix form as

$$\begin{pmatrix} A_{11} & A_{21} & A_{31} \\ A_{12} & A_{22} & A_{32} \\ A_{13} & A_{23} & A_{33} \end{pmatrix} \begin{pmatrix} M_{11} & M_{12} & M_{13} \\ M_{21} & M_{22} & M_{23} \\ M_{31} & M_{32} & M_{33} \end{pmatrix} \begin{pmatrix} A_{11} & A_{12} & A_{13} \\ A_{21} & A_{22} & A_{23} \\ A_{31} & A_{32} & A_{33} \end{pmatrix}$$

$$= \begin{pmatrix} \lambda_1 & 0 & 0 \\ 0 & \lambda_2 & 0 \\ 0 & 0 & \lambda_3 \end{pmatrix} \Rightarrow C^T M C = \Lambda, \tag{10.65}$$

where

$$C = \begin{pmatrix} A_{11} & A_{12} & A_{13} \\ A_{21} & A_{22} & A_{23} \\ A_{31} & A_{32} & A_{33} \end{pmatrix}, \quad \Lambda = \begin{pmatrix} \lambda_1 & 0 & 0 \\ 0 & \lambda_2 & 0 \\ 0 & 0 & \lambda_3 \end{pmatrix}, \tag{10.66}$$

and C^T is the transpose matrix for C. Note that C is a matrix whose columns are the eigenvectors, and the Λ is a diagonal matrix constructed from the eigenvalues for matrix M. For a more general case, where the matrix M is a non-Hermitian, equation (10.71) becomes

$$C^\dagger M C = \Lambda \tag{10.67}$$

where $C^\dagger$ is the transpose conjugate of the matrix C. For an orthogonal transformation, we recall that

$$C^\dagger = C^{-1} \tag{10.68}$$

and one can write

$$C^{-1} M C = \Lambda \tag{10.69}$$

and such transformation is known as the similarity transformation of the matrix M.

Example 10.2. *The normal modes of vibration for coupled harmonic oscillators*: Consider a system consisting of two equal masses m connected by three identical springs with spring constant, k.

The masses can slide on a horizontal, frictionless surface. The springs are at their unstretched/uncompressed lengths when the masses are at their equilibrium positions. At $t = 0$, the masses are displaced by x_{10} and x_{20} from their equilibrium positions and then released. For these coupled harmonic oscillators, provide a complete description of the resulting motion (figure 10.2). It means,

    (a) Find the equations of motion.
    (b) Make a similarity transformation: determine the eigenvalues and eigenvectors.
    (c) Solve the de-coupled transformed equations of motion.
    (d) Determine the propagator matrix U.
    (e) Find the normal modes of vibration.

Solution:

    (a) *The equations of motion*: To find the equations of motion, we use the Euler–Lagrange equation. Of course, we can use Newton's second law if we want to. To use the Euler–Lagrange equation, we need to find the kinetic energy and the potential energy. Suppose at a given instant of time the position of the first mass is x_1, and the second mass is x_2 as measured from their corresponding equilibrium positions. Then the corresponding kinetic energy can be expressed as

$$KE_1 = \frac{1}{2}m\dot{x}_1^2, \quad KE_2 = \frac{1}{2}m\dot{x}_2^2, \tag{10.70}$$

so that the total kinetic energy is

$$KE = \frac{1}{2}m(\dot{x}_1^2 + \dot{x}_2^2). \tag{10.71}$$

The potential energy (elastic) is due to the displacement of the springs from their equilibrium position. When the masses are displaced by x_1 and x_2, the left and right springs stretch by x_1 and compresses by x_2, respectively. As a result, the net stretching (or compression) of the middle spring will be $x_2 - x_1$. Therefore, the total potential energy can be written as

$$U = \frac{1}{2}kx_1^2 + \frac{1}{2}kx_2^2 + \frac{1}{2}k(x_2 - x_1)^2. \tag{10.72}$$

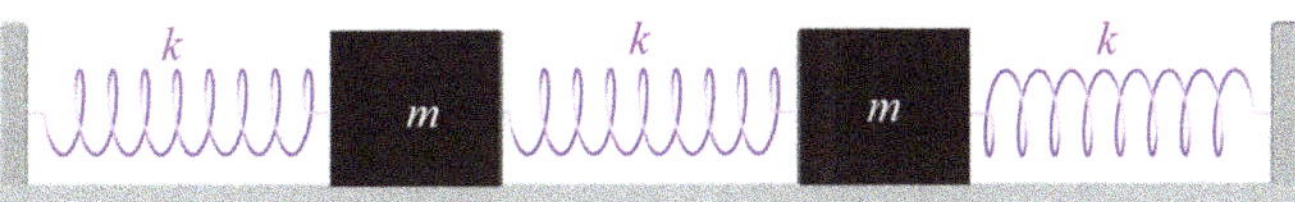

Figure 10.2. Coupled harmonic oscillators.

Then the Lagrangian

$$\mathcal{L}(t, x_1, x_2, \dot{x}_1, \dot{x}_2) = KE - U$$

$$\Rightarrow \mathcal{L}(t, x_1, x_2, \dot{x}_1, \dot{x}_2) = \frac{1}{2}m(\dot{x}_1^2 + \dot{x}_2^2) - \frac{1}{2}kx_1^2 - \frac{1}{2}kx_2^2 \tag{10.73}$$

$$- \frac{1}{2}k(x_2 - x_1)^2,$$

which leads to

$$\frac{d\mathcal{L}}{dx_1} = -kx_1 + k(x_2 - x_1), \quad \frac{\partial\mathcal{L}}{\partial x_2} = -kx_2 - k(x_2 - x_1),$$

$$\frac{d}{dt}\left(\frac{\partial\mathcal{L}}{\partial\dot{x}_1}\right) = m\ddot{x}_1, \quad \frac{d}{dt}\left(\frac{\partial\mathcal{L}}{\partial\dot{x}_2}\right) = m\ddot{x}_2. \tag{10.74}$$

Using the Euler–Lagrange equation

$$\frac{d}{dt}\left(\frac{\partial\mathcal{L}}{\partial\dot{x}_i}\right) - \frac{\partial\mathcal{L}}{\partial x_i} = 0, \quad \text{for } i = 1, 2, \tag{10.75}$$

we find

$$\begin{aligned}
m\ddot{x}_1 - [-kx_1 + k(x_2 - x_1)] = 0 &\Rightarrow m\ddot{x}_1 = -2kx_1 + kx_2, \\
m\ddot{x}_2 - [-kx_2 - k(x_2 - x_1)] = 0 &\Rightarrow m\ddot{x}_2 = -2kx_2 + kx_1.
\end{aligned} \tag{10.76}$$

Introducing a constant,

$$\omega = \sqrt{\frac{k}{m}}, \tag{10.77}$$

the equations in equation (10.76) can be put in the form

$$\ddot{x}_1 = -2\omega^2 x_1 + \omega^2 x_2, \quad \ddot{x}_2 = \omega^2 x_1 - 2\omega^2 x_2. \tag{10.78}$$

These two equations, which are second-order differential equations, describe the motion for the two masses.

(b) *Similarity transformation, eigenvalues, and eigenvectors*: Using matrices and Dirac's bra–ket representation, equation (10.78) can be rewritten as,

$$\begin{pmatrix} \ddot{x}_1 \\ \ddot{x}_2 \end{pmatrix} = \begin{pmatrix} -2\omega^2 & \omega^2 \\ \omega^2 & -2\omega^2 \end{pmatrix}\begin{pmatrix} x_1 \\ x_2 \end{pmatrix}$$

$$\frac{d^2}{dt^2}\begin{pmatrix} x_1 \\ x_2 \end{pmatrix} = \begin{pmatrix} -2\omega^2 & \omega^2 \\ \omega^2 & -2\omega^2 \end{pmatrix}\begin{pmatrix} x_1 \\ x_2 \end{pmatrix} \Rightarrow \frac{d^2}{dt^2}\,|r\rangle = M\,|r\rangle, \tag{10.79}$$

where

$$|r\rangle = x_1\hat{e}_1 + x_2\hat{e}_2 = \begin{pmatrix} x_1 \\ x_2 \end{pmatrix}, \quad M = \begin{pmatrix} -2\omega^2 & \omega^2 \\ \omega^2 & -2\omega^2 \end{pmatrix}. \tag{10.80}$$

Note that $\hat{e}_1$ and $\hat{e}_2$ are column matrices given by

$$\hat{e}_1 = \begin{pmatrix} 1 \\ 0 \end{pmatrix}, \quad \hat{e}_2 = \begin{pmatrix} 0 \\ 1 \end{pmatrix}. \tag{10.81}$$

Suppose if one solved the eigenvalue equation for the matrix M,

$$M\,|A\rangle = \lambda\,|A\rangle, \tag{10.82}$$

and found the normalized eigenvectors

$$|A_1\rangle = \begin{pmatrix} A_{11} \\ A_{21} \end{pmatrix}, \quad |A_2\rangle = \begin{pmatrix} A_{12} \\ A_{22} \end{pmatrix} \tag{10.83}$$

corresponding to the eigenvalues, $\lambda = \lambda_1$ and λ_2. Then we can write the matrix

$$C = \begin{pmatrix} A_{11} & A_{12} \\ A_{21} & A_{22} \end{pmatrix}, \tag{10.84}$$

so that we can make *the similarity transformation* for the matrix M,

$$\begin{aligned}
C^{-1}MC &= \Lambda \\
&\Rightarrow \begin{pmatrix} a_{11} & a_{12} \\ a_{21} & a_{22} \end{pmatrix}^{-1} \begin{pmatrix} M_{11} & M_{12} \\ M_{21} & M_{22} \end{pmatrix} \begin{pmatrix} a_{11} & a_{12} \\ a_{21} & a_{22} \end{pmatrix} \\
&= \begin{pmatrix} \lambda_1 & 0 \\ 0 & \lambda_2 \end{pmatrix}.
\end{aligned} \tag{10.85}$$

Noting that

$$C^{-1}C = CC^{-1} = I, \tag{10.86}$$

one can write

$$\frac{d^2}{dt^2}\,|r\rangle = M\,|r\rangle \Rightarrow \frac{d^2}{dt^2}\,|r\rangle = MCC^{-1}\,|r\rangle, \tag{10.87}$$

and multiplying from the left by the matrix C^{-1}, we find

$$C^{-1}\frac{d^2}{dt^2}\,|r\rangle = C^{-1}MCC^{-1}\,|r\rangle \Rightarrow \frac{d^2}{dt^2}C^{-1}\,|r\rangle = (C^{-1}MC)C^{-1}\vec{r}. \tag{10.88}$$

Introducing a new vector defined by

$$|R\rangle = C^{-1}\,|r\rangle = C^{-1}\begin{pmatrix} x_1 \\ x_2 \end{pmatrix} = \begin{pmatrix} y_1 \\ y_2 \end{pmatrix}, \tag{10.89}$$

and noting that

$$C^{-1}MC = \Lambda = \begin{pmatrix} \lambda_1 & 0 \\ 0 & \lambda_2 \end{pmatrix}, \tag{10.90}$$

one can rewrite equation (10.88) as

$$\frac{d^2}{dt^2}\begin{pmatrix} y_1 \\ y_2 \end{pmatrix} = \begin{pmatrix} \ddot{y}_1 \\ \ddot{y}_2 \end{pmatrix} = \begin{pmatrix} \lambda_1 & 0 \\ 0 & \lambda_2 \end{pmatrix}\begin{pmatrix} y_1 \\ y_2 \end{pmatrix}. \tag{10.91}$$

From equation (10.91), we find the following homogenous second-order linear differential equations,

$$\frac{d^2 y_1}{dt^2} - \lambda_1 y_1 = 0, \quad \frac{d^2 y_2}{dt^2} - \lambda_2 y_2 = 0. \tag{10.92}$$

Using the techniques introduced in chapter 6, we can determine the solutions to equation (10.92). What we need is the eigenvalues λ_1 and λ_2. For the moment let's pretend that we have determined λ_1 and λ_2 and the corresponding eigenvectors in equation (10.83) so that we can construct the matrix C in equation (10.84) and find the inverse $C^{-1}(= C^{\dagger})$. Multiplying equation (10.89) by the matrix C, one can write

$$CC^{-1}\begin{pmatrix} x_1(t) \\ x_2(t) \end{pmatrix} = C\begin{pmatrix} y_1(t) \\ y_2(t) \end{pmatrix} \Rightarrow \begin{pmatrix} x_1(t) \\ x_2(t) \end{pmatrix} = C\begin{pmatrix} y_1(t) \\ y_2(t) \end{pmatrix}, \tag{10.93}$$

where we used $CC^{-1} = I$. Note that equation (10.93) is the solution to equation (10.78). Next we solve the eigenvalue equation

$$M\,|A\rangle = \lambda\,|A\rangle, \tag{10.94}$$

for the matrix

$$M = \begin{pmatrix} -2\omega^2 & \omega^2 \\ \omega^2 & -2\omega^2 \end{pmatrix} \tag{10.95}$$

and determine the eigenvalues and the eigenvectors. To this end, we note that for the eigenvalues, we find

$$\begin{vmatrix} -2\omega^2 - \lambda & \omega^2 \\ \omega^2 & -2\omega^2 - \lambda \end{vmatrix} = 0 \Rightarrow (2\omega^2 + \lambda)^2 - (\omega^2)^2 = 0 \tag{10.96}$$

$$\Rightarrow (2\omega^2 + \lambda - \omega^2)(2\omega^2 + \lambda + \omega^2) = 0 \Rightarrow \lambda_1 = -\omega^2,\ \lambda_2 = -3\omega^2.$$

We then find the corresponding eigenvectors for $\lambda_1 = -\omega^2$,

$$\begin{pmatrix} -2\omega^2 - \lambda_1 & \omega^2 \\ \omega^2 & -2\omega^2 - \lambda_1 \end{pmatrix}\begin{pmatrix} a_{11} \\ a_{21} \end{pmatrix} = \begin{pmatrix} -\omega^2 & \omega^2 \\ \omega^2 & -\omega^2 \end{pmatrix}\begin{pmatrix} a_{11} \\ a_{21} \end{pmatrix} = 0$$

$$\Rightarrow -a_{11} + a_{21} = 0 \Rightarrow a_{11} = a_{21} \Rightarrow |A_1\rangle \tag{10.97}$$

$$= \begin{pmatrix} a_{11} \\ a_{21} \end{pmatrix} = a_{11}\begin{pmatrix} 1 \\ 1 \end{pmatrix},$$

and $\lambda_2 = -3\omega^2$.

$$\begin{pmatrix} -2\omega^2 - \lambda_2 & \omega^2 \\ \omega^2 & -2\omega^2 - \lambda_2 \end{pmatrix}\begin{pmatrix} a_{12} \\ a_{22} \end{pmatrix} = \begin{pmatrix} \omega^2 & \omega^2 \\ \omega^2 & \omega^2 \end{pmatrix}\begin{pmatrix} a_{12} \\ a_{22} \end{pmatrix} = 0$$

$$\Rightarrow a_{12} + a_{22} = 0 \Rightarrow a_{22} = -a_{12} \Rightarrow |A_2\rangle \tag{10.98}$$

$$= \begin{pmatrix} a_{12} \\ a_{22} \end{pmatrix} = a_{12}\begin{pmatrix} 1 \\ -1 \end{pmatrix}.$$

Upon normalizing equations (10.97) and (10.100), one finds for the eigenvectors,

$$|A_1\rangle = \frac{1}{\sqrt{2}}\begin{pmatrix} 1 \\ 1 \end{pmatrix}, \ |A_2\rangle = \frac{1}{\sqrt{2}}\begin{pmatrix} 1 \\ -1 \end{pmatrix}. \tag{10.99}$$

Using these eigenvectors as columns, we can construct the transformation matrix C and also find C^{-1}

$$C = \begin{pmatrix} \frac{1}{\sqrt{2}} & \frac{1}{\sqrt{2}} \\ \frac{1}{\sqrt{2}} & -\frac{1}{\sqrt{2}} \end{pmatrix}, \ \Rightarrow C^{-1} = C^{\dagger} = (C^*)^T = \begin{pmatrix} \frac{1}{\sqrt{2}} & \frac{1}{\sqrt{2}} \\ \frac{1}{\sqrt{2}} & -\frac{1}{\sqrt{2}} \end{pmatrix}. \tag{10.100}$$

We note that

$$CC^{-1} = \begin{pmatrix} \frac{1}{\sqrt{2}} & \frac{1}{\sqrt{2}} \\ \frac{1}{\sqrt{2}} & -\frac{1}{\sqrt{2}} \end{pmatrix}\begin{pmatrix} \frac{1}{\sqrt{2}} & \frac{1}{\sqrt{2}} \\ \frac{1}{\sqrt{2}} & -\frac{1}{\sqrt{2}} \end{pmatrix} = \begin{pmatrix} 1 & 0 \\ 0 & 1 \end{pmatrix} \Rightarrow C^{-1}C = I. \tag{10.101}$$

(c) *Solving the de-coupled transformed equations of motion*: Upon substituting the eigenvalues we determined $\lambda_1 = -\omega^2$, $\lambda_2 = -3\omega^2$, into equation (10.92), we find the homogenous linear second-order differential equations

$$\frac{d^2y_1}{dt^2} + \omega^2 y_1 = 0, \ \frac{d^2y_2}{dt^2} + 3\omega^2 y_2 = 0, \tag{10.102}$$

the solutions of which are given by

$$y_1(t) = A_1 \cos(\omega t) + A_2 \sin(\omega t), \ y_2(t) = B_1 \cos(\sqrt{3}\,\omega t) + B_2 \sin(\sqrt{3}\,\omega t). \tag{10.103}$$

Now substituting the results in equations (10.100) and (10.103) into equation (10.93), we find for the positions of the masses

$$\begin{pmatrix} x_1(t) \\ x_2(t) \end{pmatrix} = \begin{pmatrix} \frac{1}{\sqrt{2}} & \frac{1}{\sqrt{2}} \\ \frac{1}{\sqrt{2}} & -\frac{1}{\sqrt{2}} \end{pmatrix}\begin{pmatrix} A_1 \cos(\omega t) + A_2 \sin(\omega t) \\ B_1 \cos(\sqrt{3}\,\omega t) + B_2 \sin(\sqrt{3}\,\omega t) \end{pmatrix}, \tag{10.104}$$

and for the corresponding velocity

$$\frac{d}{dt}\begin{pmatrix} x_1(t) \\ x_2(t) \end{pmatrix} = \begin{pmatrix} \frac{1}{\sqrt{2}} & \frac{1}{\sqrt{2}} \\ \frac{1}{\sqrt{2}} & -\frac{1}{\sqrt{2}} \end{pmatrix} \frac{d}{dt}\begin{pmatrix} A_1 \cos(\omega t) + A_2 \sin(\omega t) \\ B_1 \cos(\sqrt{3}\,\omega t) + B_2 \sin(\sqrt{3}\,\omega t) \end{pmatrix}$$

$$\Rightarrow \begin{pmatrix} \dot{x}_1(t) \\ \dot{x}_2(t) \end{pmatrix} = \begin{pmatrix} -\omega A_1 \sin(\omega t) + \omega A_2 \cos(\omega t) \\ -\sqrt{3}\,\omega B_1 \sin(\sqrt{3}\,\omega t) + \sqrt{3}\,\omega B_2 \cos(\sqrt{3}\,\omega t) \end{pmatrix}.$$

$$(10.105)$$

Initially ($t = 0$) the first mass is displaced, $x_1(0) = x_{10}$, and the second mass is displaced, $x_2(0) = x_{20}$, and both masses are released from rest, which means the speed for each mass at $t = 0$ is zero, $\dot{x}_1(0) = \dot{x}_2(0) = 0$. Using these initial conditions in equations (10.104) and (10.105), we find

$$\begin{pmatrix} x_{10} \\ x_{20} \end{pmatrix} = \begin{pmatrix} \frac{1}{\sqrt{2}} & \frac{1}{\sqrt{2}} \\ \frac{1}{\sqrt{2}} & -\frac{1}{\sqrt{2}} \end{pmatrix}\begin{pmatrix} A_1 \\ B_1 \end{pmatrix} = \frac{1}{\sqrt{2}}\begin{pmatrix} A_1 + B_1 \\ A_1 - B_1 \end{pmatrix}$$

$$\Rightarrow A_1 = \frac{1}{\sqrt{2}}(x_{10} + x_{20}), \quad B_1 = \frac{1}{\sqrt{2}}(x_{10} - x_{20}),$$

$$(10.106)$$

and

$$\begin{pmatrix} \dot{x}_1(0) \\ \dot{x}_2(0) \end{pmatrix} = \begin{pmatrix} 0 \\ 0 \end{pmatrix} = \begin{pmatrix} \frac{1}{\sqrt{2}} & \frac{1}{\sqrt{2}} \\ \frac{1}{\sqrt{2}} & -\frac{1}{\sqrt{2}} \end{pmatrix}\begin{pmatrix} \omega A_2 \\ \sqrt{3}\,\omega B_2 \end{pmatrix}$$

$$= \frac{\omega}{\sqrt{2}}\begin{pmatrix} A_2 + B_2 \\ \sqrt{3}(A_2 - B_2) \end{pmatrix} \Rightarrow A_2 = B_2 = 0.$$

$$(10.107)$$

In view of the results in equations (10.106) and (10.107), one finds for the positions in equation (10.104)

$$x_1(t) = \frac{1}{2}(x_{10} + x_{20})\cos(\omega t) + \frac{1}{2}(x_{10} - x_{20})\cos(\sqrt{3}\,\omega t),$$

$$x_2(t) = \frac{1}{2}(x_{10} + x_{20})\cos(\omega t) - \frac{1}{2}(x_{10} - x_{20})\cos(\sqrt{3}\,\omega t),$$

$$(10.108)$$

and the speeds in equation (10.105),

$$\dot{x}_1(t) = -\frac{\omega}{2}[(x_{10} + x_{20})\sin(\omega t) + \sqrt{3}(x_{10} - x_{20})\sin(\sqrt{3}\,\omega t)],$$

$$\dot{x}_2(t) = \frac{\omega}{2}[-(x_{10} + x_{20})\sin(\omega t) + \sqrt{3}(x_{10} - x_{20})\sin(\sqrt{3}\,\omega t)].$$

$$(10.109)$$

(d) *The propagator matrix U:* We can rewrite equation (10.108), as

$$x_1(t) = \frac{1}{2}[\cos(\omega t) + \cos(\sqrt{3}\,\omega t)]x_{10} + \frac{1}{2}[\cos(\omega t) - \cos(\sqrt{3}\,\omega t)]x_{20},$$

$$x_2(t) = \frac{1}{2}[\cos(\omega t) - \cos(\sqrt{3}\,\omega t)]x_{10} + \frac{1}{2}[\cos(\omega t) + \cos(\sqrt{3}\,\omega t)]x_{20},$$

(10.110)

which one can put, using matrices, in the form

$$\begin{pmatrix} x_1(t) \\ x_2(t) \end{pmatrix} = \frac{1}{2}\begin{pmatrix} \cos(\omega t) + \cos(\sqrt{3}\,\omega t) & \cos(\omega t) - \cos(\sqrt{3}\,\omega t) \\ \cos(\omega t) - \cos(\sqrt{3}\,\omega t) & \cos(\omega t) + \cos(\sqrt{3}\,\omega t) \end{pmatrix}\begin{pmatrix} x_{10} \\ x_{20} \end{pmatrix},$$

(10.111)

$$\Rightarrow \begin{pmatrix} x_1(t) \\ x_2(t) \end{pmatrix} = U(t)\begin{pmatrix} x_1(0) \\ x_2(0) \end{pmatrix} \Rightarrow |r(t)\rangle = U(t)|r(0)\rangle,$$

where

$$|r(0)\rangle = \begin{pmatrix} x_1(0) \\ x_2(0) \end{pmatrix} = \begin{pmatrix} x_{10} \\ x_{20} \end{pmatrix},$$

(10.112)

$$U(t) = \frac{1}{2}\begin{pmatrix} \cos(\omega t) + \cos(\sqrt{3}\,\omega t) & \cos(\omega t) - \cos(\sqrt{3}\,\omega t) \\ \cos(\omega t) - \cos(\sqrt{3}\,\omega t) & \cos(\omega t) + \cos(\sqrt{3}\,\omega t) \end{pmatrix}.$$

The matrix $U(t)$ in equation (10.112) is the *propagator matrix.*

(e) *The normal modes of vibration:* Suppose the two masses initially are equally displaced in the same direction (i.e., $x_{10} = x_{20}$), then the initial state of the two masses can be described by the first eigenvector,

$$|\,r(0)\rangle = |\,A_1\rangle \Rightarrow \begin{pmatrix} x_{10} \\ x_{20} \end{pmatrix} = \frac{1}{\sqrt{2}}\begin{pmatrix} 1 \\ 1 \end{pmatrix} \Rightarrow x_{10} = x_{20} = \frac{1}{\sqrt{2}}.$$

(10.113)

and equation (10.111) becomes

$$\begin{pmatrix} x_1(t) \\ x_2(t) \end{pmatrix} = \frac{1}{2\sqrt{2}}\begin{bmatrix} \cos(\omega t) + \cos(\sqrt{3}\,\omega t) & \cos(\omega t) - \cos(\sqrt{3}\,\omega t) \\ \cos(\omega t) - \cos(\sqrt{3}\,\omega t) & \cos(\omega t) + \cos(\sqrt{3}\,\omega t) \end{bmatrix}\begin{bmatrix} 1 \\ 1 \end{bmatrix}$$

(10.114)

$$= \frac{1}{\sqrt{2}}\begin{pmatrix} \cos(\omega t) \\ \cos(\omega t) \end{pmatrix} \Rightarrow x_1(t) = x_2(t).$$

The result in equation (10.114) shows that two masses oscillate with a frequency ω with the same amplitude in the same direction. Note that when we set, $x_{10} = x_{20} = \frac{1}{\sqrt{2}}$, in equation (10.110), we also find the same result. On the other hand, if the two masses, initially displaced with the same amount but in opposite directions (i.e., $x_{10} = -x_{20}$), we can describe the initial state of the two masses by the second eigenvector

$$|\,r(0)\rangle = |\,A_2\rangle = \begin{pmatrix} x_{10} \\ x_{20} \end{pmatrix} = \frac{1}{\sqrt{2}}\begin{pmatrix} 1 \\ -1 \end{pmatrix} \Rightarrow x_1(t) = x_2(t) = \frac{1}{\sqrt{2}},$$

(10.115)

and one finds from equation (10.111)

$$\begin{pmatrix} x_1(t) \\ x_2(t) \end{pmatrix} = \frac{1}{2\sqrt{2}} \begin{bmatrix} \cos(\omega t) + \cos(\sqrt{3}\,\omega t) & \cos(\omega t) - \cos(\sqrt{3}\,\omega t) \\ \cos(\omega t) - \cos(\sqrt{3}\,\omega t) & \cos(\omega t) + \cos(\sqrt{3}\,\omega t) \end{bmatrix} \begin{bmatrix} 1 \\ -1 \end{bmatrix},$$

$$= \frac{1}{\sqrt{2}} \begin{pmatrix} \cos(\sqrt{3}\,\omega t) \\ -\cos(\sqrt{3}\,\omega t) \end{pmatrix} = \frac{1}{\sqrt{2}} \begin{pmatrix} \cos(\sqrt{3}\,\omega t) \\ \cos(\sqrt{3}\,\omega t + \pi) \end{pmatrix} \qquad (10.116)$$

$$\Rightarrow x_1(t) = -x_2(t).$$

The two masses oscillate with a frequency $\sqrt{3}\,\omega$ out of phase by π. *The two modes of vibrations described by equations* (10.114) *and* (10.116) *are known as normal modes of vibration.*

10.7 Eigenvalue equation and Mathematica

The following are the basic commands in Mathematica for finding eigenvalues and eigenvectors for a square matrix.

- Eigenvalues[m] gives a list of the eigenvalues of the square matrix m.
- Eigenvalues[m][$\{m, a\}$] gives the generalized eigenvalues of m with respect to a.
- Eigenvalues[m][m, k] gives the first k eigenvalues of m.
- Eigenvalues[m][$\{m, a\}, k$] gives the first k generalized eigenvalues.
- Eigenvectors[m] gives a list of the eigenvectors of the square matrix m.
- Eigenvectors[$\{m, a\}$] gives the generalized eigenvectors of m with respect to a.
- Eigenvectors[$\{m, a\}, k$] gives the first k generalized eigenvectors.
- Eigensystem[$\{m, a\}$] gives the generalized eigenvalues and eigenvectors of m with respect to a.
- Eigensystem[m, k] gives the eigenvalues and eigenvectors for the first k eigenvalues of m.
- Eigensystem[$\{m, a\}, k$] gives the first k generalized eigenvalues and eigenvectors.

In the following we verify the eigenvalues and eigenvectors of the matrices in Example 10.1 and Example 10.2.

- Example 10.1

$$m1 = \text{MatrixForm}[\{\{0, 1, 0\}, \{1, 0, 0\}, \{0, 0, 0\}\}]$$

$$\begin{pmatrix} 0 & 1 & 0 \\ 1 & 0 & 0 \\ 0 & 0 & 0 \end{pmatrix}$$

$$\text{Eigenvalues}\begin{bmatrix} \begin{pmatrix} 0 & 1 & 0 \\ 1 & 0 & 0 \\ 0 & 0 & 0 \end{pmatrix} \end{bmatrix}$$

$$\{-1, 1, 0\}$$

$$\text{Eigenvectors}\left[\begin{pmatrix}0&1&0\\1&0&0\\0&0&0\end{pmatrix}\right]$$

$$\{\{-1,\,1,\,0\},\,\{1,\,1,\,0\},\,\{0,\,0,\,1\}\}$$

$$\text{Eigensystem}\left[\begin{pmatrix}0&1&0\\1&0&0\\0&0&0\end{pmatrix}\right]$$

$$\{\{-1,\,1,\,0\},\,\{\{-1,\,1,\,0\},\,\{1,\,1,\,0\},\,\{0,\,0,\,1\}\}\}$$

- Example 10.2

$$\text{m2} = \text{MatrixForm}[\{\{-2\omega^2,\,\omega^2\},\,\{\omega^2,\,-2\omega^2\}\}]$$

$$\begin{pmatrix}-2\omega^2 & \omega^2 \\ \omega^2 & -2\omega^2\end{pmatrix}$$

$$\text{Eigenvalues}\left[\begin{pmatrix}-2\omega^2 & \omega^2 \\ \omega^2 & -2\omega^2\end{pmatrix}\right]$$

$$\{-3\omega^2,\,-\omega^2\}$$

$$\text{Eigenvectors}\left[\begin{pmatrix}-2\omega^2 & \omega^2 \\ \omega^2 & -2\omega^2\end{pmatrix}\right]$$

$$\{\{-1,\,1\},\,\{1,\,1\}\}$$

$$\text{Eigensystem}\left[\begin{pmatrix}-2\omega^2 & \omega^2 \\ \omega^2 & -2\omega^2\end{pmatrix}\right]$$

$$\{\{-3\omega^2,\,-\omega^2\},\,\{\{-1,\,1\},\,\{1,\,1\}\}\}$$

Note that Mathematica does not give normalized eigenvectors.

10.8 Homework assignment

Problem 1. Consider the vectors

$$\vec{A}_1 = a_1(\hat{e}_1 - \sqrt{2}\,\hat{e}_2 + \hat{e}_3),\ \vec{A}_2 = a_2(\hat{e}_1 + \sqrt{2}\,\hat{e}_2 + \hat{e}_3),$$
$$\vec{A}_3 = a_3(-\hat{e}_1 + \hat{e}_3), \tag{10.117}$$

where $\hat{e}_1$, $\hat{e}_2$, and $\hat{e}_3$ form a complete orthonormal set of basis vectors,

$$\hat{e}_i \cdot \hat{e}_2 = \delta_{ij},$$

and a_1, a_2, a_3 are real constants.

(a) Write the vectors $\vec{A}_1$, $\vec{A}_2$, $\vec{A}_3$ and the corresponding complex conjugates using matrices and Dirac's bra–ket notation.

(b) Assuming that the magnitude of these vectors is one, find the constant a_1, a_2, and a_3.

(c) Construct a 3×3 matrix C with the normalized vectors $\vec{A}_1$, $\vec{A}_2$, $\vec{A}_3$ making the columns of this matrix.

(d) Find C^{-1}, C^T, and determine the relationship between these two matrices.

(e) Suppose you are given the matrix, L_x,

$$L_x = \frac{\hbar}{\sqrt{2}} \begin{pmatrix} 0 & 1 & 0 \\ 1 & 0 & 1 \\ 0 & 1 & 0 \end{pmatrix}. \tag{10.118}$$

[L_x is the matrix representation for x-component of the orbital angular momentum operator for spin-1 particle. You will see this matrix in quantum mechanics]. Find the matrix

$$M = C^{-1}L_x C. \tag{10.119}$$

(f) Based on the results in (d), (e), and the properties of the normalized vectors $(\vec{A}_1, \vec{A}_2, \vec{A}_3)$, comment on the relationship between the matrices C, L_x, M; and the transformation from matrix L_x to M.

Problem 2. Find the magnitude of the vector directed from P to Q, when

(a) $P = (4, -1, 2, 7)$ and $Q = (2, 3, 1, 9)$.

(b) $P = (-1, 5, -3, 2, 4)$ and $Q = (2, 6, 2, 7, 6)$.

(c) $P = (x_1, y_1, z_1, ct_1)$ and $Q = (x_2, y_2, z_2, ct_2)$ where c is the speed of light in free space.

For the matrices listed in problem 3–6

(a) Find the eigenvalues and eigenvectors.

(b) Construct the matrix C that diagonalizes each these matrices, determine its inverse matrix, $C^{-1} = C^\dagger$.

(c) Compute $C^{-1}MC$ for each matrices. (*this part may be done using Mathematica, but appropriate output must be provided*).

Problem 3.

$$M = \begin{bmatrix} 1 & 3 \\ 2 & 2 \end{bmatrix}. \tag{10.120}$$

Problem 4.

$$M = \begin{pmatrix} 0 & 1 \\ 1 & 0 \end{pmatrix}, \quad M = \begin{pmatrix} 0 & -i \\ i & 0 \end{pmatrix}, \quad M = \begin{pmatrix} 1 & 0 \\ 0 & -1 \end{pmatrix}. \tag{10.121}$$

These matrices are usually represented as σ_x, σ_y, and σ_z and are known as the Pauli Spin-1/2 matrices. You will also see these matrices in quantum mechanics.

Problem 5.

$$M = \begin{bmatrix} 2 & 3 & 0 \\ 3 & 2 & 0 \\ 0 & 0 & 1 \end{bmatrix}. \tag{10.122}$$

Problem 6.

$$M = \begin{pmatrix} 0 & -i & 0 \\ i & 0 & -i \\ 0 & i & 0 \end{pmatrix}. \tag{10.123}$$

This matrix is also related to spin-matrices (but for spin-1 particles).

Problem 7. Show that the matrices in Problems 4 and 6 are Hermitian.

Problem 8. Consider a system consisting of two masses m_1 and m_2 connected by three springs with spring constant k_1, k_2, and k_2 as shown in figure 10.3. The masses can slide on a horizontal, frictionless surface. The springs are at their unstretched/uncompressed lengths when the masses are at its equilibrium positions. At $t = 0$, the masses are displaced from its equilibrium positions by the amounts x_{10} and x_{20} and released from rest.

(a) Find the kinetic energy, the potential energy, and the Lagrangian. **Using the Euler–Lagrange equation** derive the equations of motion for each of the masses and express the equations using matrices

$$\begin{bmatrix} \ddot{x}_1 \\ \ddot{x}_2 \end{bmatrix} = M \begin{bmatrix} x_1 \\ x_2 \end{bmatrix}. \tag{10.124}$$

(b) Let us assume that two atoms have nearly the same mass (i.e., $m_1 \simeq m_2 = m$) and,

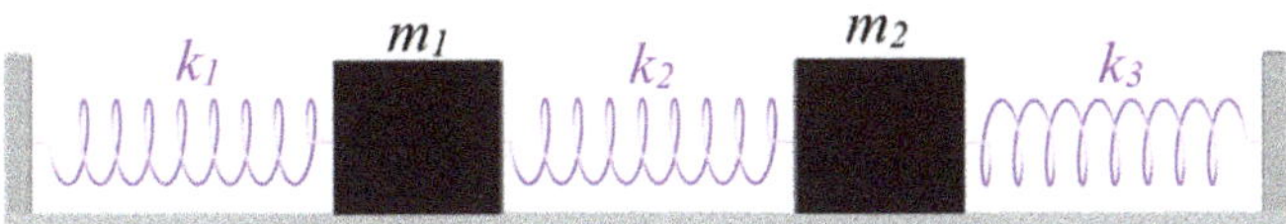

Figure 10.3. Two masses and three different springs.

$$k_1 = 5k, \; k_2 = 2k, \; k_3 = 2k. \tag{10.125}$$

Using the similarity transformation find the eigenvalues and eigenvectors for the matrix M.

(c) For the two masses find the displacements ($x_1(t)$ and $x_2(t)$) and speeds ($\dot{x}_1(t)$ and $\dot{x}_2(t)$).

(d) Find the propagator matrix.

(e) Describe the normal modes of vibration of the atoms.

IOP Publishing

Studies in Theoretical Physics, Volume 1
Fundamental mathematical methods
Daniel Erenso and Victor Montemayor

Chapter 11

Special functions

This chapter will introduce the essential, special functions we see in advanced physics courses such as statistical mechanics, classical mechanics, electromagnetism, and quantum mechanics. These special functions are integral equations that we often cannot find symbolic, analytical expressions. We will introduce these functions, including the gamma function, the beta function, the error function, the elliptical integrals, and the Dirac delta function. We begin with the gamma function and introduce how the factorial function is related to the gamma function. Furthermore, after introducing the beta function, we see how it is related to the gamma function. Following that, we see the error function and some of the elliptic integral functions. Finally, we introduce the Dirac delta function and see how we use Mathematica to generate all the special functions considered in this chapter. We illustrate the application of these special functions using physical problems, which we will solve analytically and numerically using Mathematica.

11.1 The factorial, the gamma function, and Stirling's formula

The factorial $n!$:

For a positive integer n, the factorial $n!$ is defined as

$$n! = n \times (n-1) \times (n-2) \times (n-3) \times (n-4) \dots 3 \times 2 \times 1 \times 0!, \qquad (11.1)$$

where by definition

$$0! = 1. \qquad (11.2)$$

We most likely have seen $n!$ in introductory algebra courses but probably not how and when one ends up with a factorial that involves such multiplication of positive integers. Next, we will see the 'ancestry' integral function known as the gamma function from which the factorial function is derived. To this end, consider the integral function given by

$$F(\alpha) = \int_0^\infty e^{-\alpha x} dx. \tag{11.3}$$

For any real number, $\alpha > 0$, the value of this integral is

$$F(\alpha) = \int_0^\infty e^{-\alpha x} dx = -\frac{e^{-\alpha x}}{\alpha} \Big|_0^\infty = \frac{1}{\alpha}. \tag{11.4}$$

Now let us differentiate this integral with respect to α as many as we can, say n times,

$$\frac{\partial^n F(\alpha)}{\partial \alpha^n} = \frac{\partial^n}{\partial \alpha^n} \int_0^\infty e^{-\alpha x} dx = \frac{\partial^n}{\partial \alpha^n} \frac{1}{\alpha}. \tag{11.5}$$

For $n = 0$, one can write

$$\frac{\partial^0}{\partial \alpha^0} \int_0^\infty e^{-\alpha x} dx = \frac{\partial^0}{\partial \alpha^0} \frac{1}{\alpha} \Rightarrow \int_0^\infty x^0 e^{-\alpha x} dx = \frac{0!}{\alpha^1}, \tag{11.6}$$

where we used $0! = 1$. For the first derivative, $n = 1$,

$$\frac{\partial}{\partial \alpha} \int_0^\infty e^{-\alpha x} dx = \frac{\partial}{\partial \alpha} \frac{1}{\alpha} \Rightarrow \int_0^\infty \frac{\partial}{\partial \alpha}(e^{-\alpha x}) dx = \frac{\partial}{\partial \alpha}\left(\frac{1}{\alpha}\right)$$

$$\Rightarrow \int_0^\infty - xe^{-\alpha x} dx = -\frac{1}{\alpha^2} \Rightarrow \int_0^\infty x^1 e^{-\alpha x} dx = \frac{1!}{\alpha^2}. \tag{11.7}$$

For the second derivative, $n = 2$,

$$\frac{\partial^2}{\partial \alpha^2}\left(\int_0^\infty e^{-\alpha x} dx\right) = \frac{\partial^2}{\partial \alpha^2}\left(\frac{1}{\alpha}\right) \Rightarrow \frac{\partial}{\partial \alpha}\left(\int_0^\infty xe^{-\alpha x} dx\right) = \frac{\partial}{\partial \alpha}\left(\frac{1!}{\alpha^2}\right)$$

$$\Rightarrow \int_0^\infty x^2 e^{-\alpha x} dx = \frac{2 \times 1!}{\alpha^3} = \frac{2!}{\alpha^3}. \tag{11.8}$$

For the third derivative, $n = 3$,

$$\frac{\partial^3}{\partial \alpha^3}\left(\int_0^\infty e^{-\alpha x} dx\right) = \frac{\partial^3}{\partial \alpha^3}\left(\frac{1}{\alpha}\right) \Rightarrow \frac{\partial}{\partial \alpha}\left(\int_0^\infty x^2 e^{-\alpha x} dx\right) = \frac{\partial}{\partial \alpha}\left(\frac{2!}{\alpha^3}\right)$$

$$\Rightarrow \int_0^\infty x^3 e^{-\alpha x} dx = \frac{3 \times 2!}{\alpha^3} \Rightarrow \int_0^\infty x^3 e^{-\alpha x} dx = \frac{3!}{\alpha^3}. \tag{11.9}$$

Therefore, it is not difficult to see for the nth derivative, we find

$$\frac{\partial^n F(\alpha)}{\partial \alpha^n} = \frac{\partial^n}{\partial \alpha^n} \int_0^\infty e^{-\alpha x} dx = \int_0^\infty x^n e^{-\alpha x} dx = \frac{n!}{\alpha^n}. \tag{11.10}$$

Upon setting $\alpha = 1$, we find

$$n! = \int_0^\infty x^n e^{-x} dx, \tag{11.11}$$

which is the integral form of the factorial which we defined in equation (11.1) and is valid only for a positive integer or zero, $n \geqslant 0$.

The Gamma function

For any positive real number p, $(p > 0)$, replacing n by $p - 1$ in the integral in equation (11.11), one can define the function $\Gamma(p)$, known as *the gamma function*

$$\Gamma(p) = \int_0^\infty x^{p-1}e^{-x}dx. \tag{11.12}$$

Replacing p by $p + 1$ in equation (11.12), we have

$$\Gamma(p + 1) = \int_0^\infty x^p e^{-x}dx, \tag{11.13}$$

so that introducing the transformation of variables defined by

$$u = x^p, \; dv = e^{-x}dx \Rightarrow du = px^{p-1}, \; v = -e^{-x}, \tag{11.14}$$

and applying integration by parts

$$\int u\,dv = uv - \int v\,du, \tag{11.15}$$

one can rewrite equation (11.13) as

$$\Gamma(p + 1) = -e^{-x}x^p \big|_0^\infty + \int_0^\infty px^{p-1}e^{-x}dx = p\int_0^\infty x^{p-1}e^{-x}dx, \tag{11.16}$$

so that using equation (11.12), one can easily find

$$\Gamma(p + 1) = p\Gamma(p). \tag{11.17}$$

Equation (11.17) is the recursion relation for the gamma function. Note that for $p = n$, with $n \geqslant 0$ (positive integer or zero), equation (11.13) becomes $n! = \Gamma(n + 1)$, which is the factorial in terms of the gamma function and equation (11.17) gives $n! = n\Gamma(n)$.

Example 11.1. Show that

$$\Gamma(1/2) = \sqrt{\pi}, \tag{11.18}$$

and

$$\int_0^\infty e^{-u^2}du = \frac{\Gamma(1/2)}{2}. \tag{11.19}$$

Solution: Using the definition of the gamma function in equation (11.12) for $p = 1/2$, we have

$$\Gamma(1/2) = \int_0^\infty x^{-1/2}e^{-x}dx, \tag{11.20}$$

and introducing a variable defined by

$$u^2 = x \Rightarrow 2udu = dx, \qquad (11.21)$$

we can write

$$\Gamma(1/2) = 2 \int_0^\infty e^{-u^2} du. \qquad (11.22)$$

Upon squaring both sides of equation (11.22), we find

$$\Gamma^2\left(\frac{1}{2}\right) = 4\left(\int_0^\infty e^{-u^2} du\right)\left(\int_0^\infty e^{-v^2} dv\right) = 4 \int_0^\infty \int_0^\infty e^{-(u^2+v^2)} dudv. \qquad (11.23)$$

Using the polar coordinates (r, θ) (see figure 11.1) defined by

$$u = r\cos(\theta), \, v = r\sin(\theta) \Rightarrow dudv = rdrd\theta, \, u^2 + v^2 = r^2, \qquad (11.24)$$

one can rewrite equation (11.23) as

$$\Gamma^2\left(\frac{1}{2}\right) = 4 \int_0^\infty e^{-r^2} rdr \int_0^{\pi/2} d\theta. \qquad (11.25)$$

Here the upper limit of integration for θ is $\pi/2$ since both u and v are positive, and we must integrate only in the first quadrant (the region shown in green). Therefore, integrating with respect to r and θ leads to

$$\Gamma^2\left(\frac{1}{2}\right) = -4\frac{e^{-r^2}}{2}\bigg|_0^\infty \frac{\pi}{2} = \pi \Rightarrow \Gamma\left(\frac{1}{2}\right) = \sqrt{\pi}. \qquad (11.26)$$

Upon substituting this result into equation (11.22), we find

$$\sqrt{\pi} = 2 \int_0^\infty e^{-u^2} du \Rightarrow \int_0^\infty e^{-u^2} du = \frac{\sqrt{\pi}}{2}. \qquad (11.27)$$

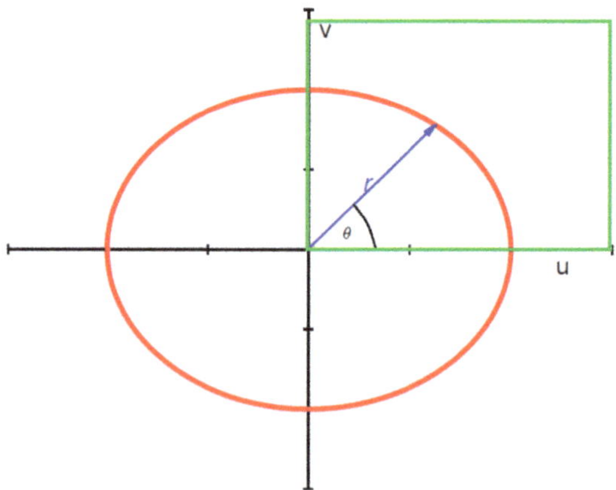

Figure 11.1. A point described by polar coordinates (r, θ) and Cartesian coordinates (u, v).

Stirling's formula:

We recall for positive integer or zero values for p, from the gamma function in equation (11.12) we find the factorial

$$p!\,\Gamma(p + 1) = \int_0^\infty x^p e^{-x} dx. \tag{11.28}$$

We often see the factorial in statistical mechanics where we need to deal with a large number of nuclei, electrons, atoms, or molecules usually at the order of Avogadro's number, $N_A = 6.02 \times 10^{23}$. The statistical analyses of such physical systems require an approximation for the factorial function. Therefore, it is essential to find the approximate expression for the factorial for a large number like Avogadro's number. The approximate expression for the factorial that we find in such limit is known as Stirling's formula and is given by

$$p! = \Gamma(p + 1) \cong p^p e^{-p} \sqrt{2\pi p}. \tag{11.29}$$

Stirling's formula is often expressed using a logarithm function. Taking the logarithm of both sides of equation (11.29), we have

$$\ln(p!) \cong \ln\left(p^p e^{-p}\sqrt{2\pi p}\right) = \ln\left(p^p\right) + \ln\left(e^{-p}\right) + \ln\left((2\pi p)^{\frac{1}{2}}\right)$$

$$= p\ln(p) - p\ln(e) + \frac{1}{2}\ln(2\pi p) \Rightarrow \ln(p!) \cong p\ln p - p + \frac{1}{2}\ln(2\pi p), \tag{11.30}$$

for an extremely large value of p, the last term is very small as compared to the first two terms and Stirling's formula can be rewritten as

$$\ln(p!) \cong p\ln p - p. \tag{11.31}$$

Next, we shall derive Stirling's formula. To this end, we note that introducing a new variable defined by

$$x = p + y\sqrt{p} \Rightarrow \begin{cases} dx = \sqrt{p}\,dy, \\ x = 0 \Rightarrow y = -\sqrt{p}, \\ x \to \infty \Rightarrow y \to \infty, \end{cases} \tag{11.32}$$

the Γ function in equation (11.13) can be put in the form

$$\Gamma(p + 1) = \int_0^\infty x^p e^{-x} dx = \int_{-\sqrt{p}}^\infty \left(p + y\sqrt{p}\right)^p e^{-(p+y\sqrt{p})} \sqrt{p}\,dy. \tag{11.33}$$

Using the relation

$$\left(p + y\sqrt{p}\right)^p = e^{\ln\left[(p+y\sqrt{p})^p\right]} = e^{p\ln(p+y\sqrt{p})}, \tag{11.34}$$

equation (11.33) can be rewritten as

$$\Gamma(p+1) = \int_{-\sqrt{p}}^{\infty} e^{p\ln(p+y\sqrt{p})}e^{-(p+y\sqrt{p})}\sqrt{p}\,dy = \sqrt{p}\int_{-\sqrt{p}}^{\infty} e^{p\ln(p+y\sqrt{p})-p-y\sqrt{p}}dy \tag{11.35}$$

We recall that the Taylor series expansion for $f(y)$ about $y = 0$ is given by

$$f(y) = f(0) + \frac{1}{1!}\frac{df(y)}{dy}\bigg|_{y=0} y + \frac{1}{2!}\frac{d^2f(y)}{dy^2}\bigg|_{y=0} y^2 + ..., \tag{11.36}$$

so that for $f(y) = \ln\left(p + y\sqrt{p}\right)$, using

$$f(0) = \ln(p), \quad \frac{df(y)}{dy}\bigg|_{y=0} = \frac{\sqrt{p}}{p+y\sqrt{p}}\bigg|_{y=0} = \frac{\sqrt{p}}{p},$$

$$\frac{d^2f(y)}{dy^2}\bigg|_{y=0} = \frac{d}{dy}\left[\frac{\sqrt{p}}{p+y\sqrt{p}}\right]\bigg|_{y=0} = -\frac{\sqrt{p}\sqrt{p}}{\left(p+y\sqrt{p}\right)^2}\bigg|_{y=0} = -\frac{1}{p}, \tag{11.37}$$

for extremely large values of p, we find an approximate expression

$$\ln\left(p + y\sqrt{p}\right) \cong \ln(p) + \frac{\sqrt{p}\,y}{p} - \frac{y^2}{2p}. \tag{11.38}$$

Then upon substituting equation (11.38) into equation (11.35), we find

$$\Gamma(p+1) \cong \sqrt{p}\int_{-\sqrt{p}}^{\infty} e^{\left\{p\left(\ln(p)+\frac{\sqrt{p}\,y}{p}-\frac{y^2}{2p}\right)-p-\sqrt{p}\,y\right\}}dy = \sqrt{p}\int_{-\sqrt{p}}^{\infty} e^{\left(p\ln(p)-p-\frac{y^2}{2}\right)}dy \tag{11.39}$$

$$\Rightarrow p! \cong \sqrt{p}\,e^{p\ln(p)-p}\int_{-\sqrt{p}}^{\infty} e^{-\frac{y^2}{2}}dy.$$

Again for extremely large values of p ($p \to \infty \Rightarrow \sqrt{p} \to \infty$), applying the result in equation (11.27), one can write

$$\int_{-\sqrt{p}}^{\infty} e^{-\frac{y^2}{2}}dy \simeq \int_{-\infty}^{\infty} e^{-\frac{y^2}{2}}dy = \sqrt{2\pi}, \tag{11.40}$$

and equation (11.39) can further be approximated as

$$p! = \Gamma(p+1) \simeq \sqrt{2p\pi}\,e^{p\ln(p)-p} = \sqrt{2p\pi}\,e^{\ln(p^p)}e^{-p}$$

$$\Rightarrow p! = \Gamma(p+1) \simeq \sqrt{2p\pi}\,p^p e^{-p}. \tag{11.41}$$

Taking the logarithm of equation (11.41)

$$\ln(p!) \simeq \ln\left(\sqrt{2p\pi}\, p^{p} e^{-p}\right) = \frac{1}{2}\ln(2p\pi) + p(\ln p - \ln e) \tag{11.42}$$

$$\Rightarrow \ln(p!) \simeq p \ln p - p,$$

where we dropped the first term as it is negligible compared to the second term.

Example 11.2. Consider a classroom full of gas molecules. There are approximately $N = 5000N_A = 3 \times 10^{27}$ molecules in the room, where $N_A = 6.02 \times 10^{23}$ is Avogadro's number. From binomial theorem, it can be shown that the probability for n of the molecules to be in the front half and $n' = N - n$ molecules to be in the back half of the room is given by

$$P(n) = \binom{N}{n} p^n q^{n'} = \frac{N!}{n!(N-n)!} p^n q^{N-n}, \tag{11.43}$$

p and q are the probability that a molecule is found in the front and the back half, respectively. From the symmetry of the problem, we must have

$$p = \frac{1}{2}, \quad q = 1 - p = \frac{1}{2} = p. \tag{11.44}$$

On average, we would expect to find half of the molecules in the front half of the room and the other half of the molecules in the back half of the room. Find the probability that 0.1% of the molecules in the room shifted from the front to the back half.

Solution: Since $q = p$, we can write

$$P(n) = \frac{N!}{n!(N-n)!} p^n p^{N-n} = \frac{N!}{n!(N-n)!} p^N. \tag{11.45}$$

On average, there are $n_{\text{ave}} = 0.5N$ of molecules in the front half of the room and $n'_{\text{ave}} = N - n_{\text{ave}} = 0.5N$ in the back half of the room. Here we want to find the probability that 0.1% of the molecules shifted to the back half of the room. In other words, we want to determine the probability that the number of molecules in the front half (n_{ave}) is reduced by 0.1%. It means we want to find $P(n)$ for

$$n = 0.5N - (N \times 0.1\%) \Rightarrow n = 0.499N = 0.499 \times 3 \times 10^{27} \tag{11.46}$$

$$\Rightarrow n = 1.497 \times 10^{27}.$$

Both N and n are huge numbers, therefore we can make Stirling's approximation

$$\ln n! \cong n \ln n - n, \quad \ln N! \cong N \ln N - N. \tag{11.47}$$

Thus for

$$\ln\left[P(n)\right] = \ln\left[\frac{N!}{n!(N-n)!}p^N\right] = \ln N! - \ln n! - \ln(N-n)! + \ln p^N, \qquad (11.48)$$

we can make the approximation

$$\ln\left[P(n)\right] \cong \left[N \ln N - N\right] - \left[n \ln n - n\right]$$
$$- \left[(N-n)\ln(N-n) - (N-n)\right]$$
$$+ N \ln p = N \ln N - N - n \ln n + n - N \ln(N-n)$$
$$+ n \ln(N-n) + N - n$$
$$+ N \ln p = N[\ln N - \ln(N-n) + \ln p] + n[\ln(N-n) - \ln n]$$
$$= N \ln\left(\frac{N}{N-n}p\right) + n \ln\left(\frac{N-n}{n}\right) n \ln\left(\frac{N-n}{n}\right)$$
$$- N \ln\left(\frac{N-n}{Np}\right) \qquad\qquad (11.49)$$
$$= \ln\left(\frac{N-n}{n}\right)^n + \ln\left(\frac{N-n}{Np}\right)^{-N}$$
$$= \ln\left[\left(\frac{N-n}{n}\right)^n\left(\frac{N-n}{Np}\right)^N\right]$$
$$\Rightarrow P(n) \simeq \left(\frac{N-n}{n}\right)^n\left(\frac{N-n}{Np}\right)^N.$$

Substituting the values $N = 3 \times 10^{27}$, $n = 1.497 \times 10^{27}$, $p = 0.5$ and evaluating the expression using Mathematica, we find

$$P(n) \cong \exp[-6 \times 10^{21}]. \qquad (11.50)$$

which is almost zero.

11.2 The beta function

The beta function, $B(p, q)$, is defined by the definite integral,

$$B(p, q) = \int_0^1 x^{p-1}(1-x)^{q-1}dx, \qquad (11.51)$$

where p and q are positive real numbers ($p > 0$ and $q > 0$). There are different forms of the beta function that one can derive from equation (11.51) using transformation of variables. Next, we shall derive a few of them.

(a) Introducing the variables defined by

$$x = y/a \Rightarrow \begin{cases} dx = dy/a, \\ x = 0 \Rightarrow y = 0, \\ x = 1 \Rightarrow y = a, \end{cases} \tag{11.52}$$

the beta function in equation (11.51) can be rewritten as

$$B(p, q) = \frac{1}{a} \int_0^a \left(\frac{y}{a}\right)^{p-1} \left(1 - \frac{y}{a}\right)^{q-1} dy = \frac{1}{a^{p+q-1}} \int_0^a y^{p-1}(a - y)^{q-1} dy. \tag{11.53}$$

(b) Introducing the variables defined by

$$x = \sin^2(\theta) \Rightarrow \begin{cases} dx = 2 \sin(\theta) \cos(\theta) d\theta, \\ x = 0 \Rightarrow \theta = 0, \\ x = 1 \Rightarrow \theta = \pi/2, \end{cases} \tag{11.54}$$

the beta function in equation (11.51) becomes

$$B(p, q) = \int_0^{\pi/2} \sin^{2p-2}(\theta)(1 - \sin^2(\theta))^{q-1} 2 \sin(\theta) \cos(\theta) d\theta$$
$$\Rightarrow B(p, q) = 2 \int_0^{\pi/2} \sin^{2p-1}(\theta) \cos^{2q-1}(\theta) d\theta. \tag{11.55}$$

(c) Introducing the variables defined by

$$x = \frac{y}{1 + y} \Rightarrow y = \frac{x}{1 - x} \Rightarrow \begin{cases} dx = \dfrac{dy}{1 + y} - \dfrac{y dy}{(1 + y)^2} = \dfrac{dy}{(1 + y)^2}, \\ x = 0 \Rightarrow y = 0, \\ x = 1 \Rightarrow y = \lim\limits_{x \to 1}\left(\dfrac{x}{1 - x}\right) = \infty, \end{cases} \tag{11.56}$$

for the beta function in equation (11.51), we find

$$B(p, q) = \int_0^\infty \left(\frac{y}{1 + y}\right)^{p-1} \left(1 - \frac{y}{1 + y}\right)^{q-1} \frac{dy}{(1 + y)^2}$$
$$= \int_0^\infty \left(\frac{y}{1 + y}\right)^{p-1} \left(\frac{1}{1 + y}\right)^{q-1} \frac{dy}{(1 + y)^2} \tag{11.57}$$
$$\Rightarrow B(p, q) = \int_0^\infty \frac{y^{p-1} dy}{(1 + y)^{p+q}}.$$

Example 11.3. Prove that the gamma and beta functions are related by

$$B(p, q) = \frac{\Gamma(p)\Gamma(q)}{\Gamma(p + q)}.$$ (11.58)

Solution: For the gamma functions

$$\Gamma(q) = \int_0^\infty x^{q-1}e^{-x}dx, \quad \Gamma(p) = \int_0^\infty y^{p-1}e^{-y}dy,$$ (11.59)

introducing the transformation of variables defined by

$$u^2 = x \Rightarrow 2udu = dx, \quad v^2 = y \Rightarrow 2vdv = dy,$$ (11.60)

one can write

$$\Gamma(q) = 2\int_0^\infty u^{2q-1}e^{-u^2}du \ , \quad \Gamma(p) = 2\int_0^\infty v^{2p-1}e^{-v^2}dv.$$ (11.61)

Multiplying $\Gamma(q)$ with $\Gamma(p)$, we have

$$\Gamma(p)\Gamma(q) = 4\int_0^\infty \int_0^\infty u^{2q-1}v^{2p-1}e^{-(u^2+v^2)}du\ dv,$$ (11.62)

so that using the polar coordinates defined in equation (11.24) (see figure 11.1)

$$u = r\cos(\theta), \ v = r\sin(\theta) \Rightarrow dudv = rdrd\theta, \ u^2 + v^2 = r^2$$ (11.63)

we find

$$\Gamma(p)\Gamma(q) = 4\int_0^\infty \int_0^{\pi/2} (r\cos(\theta))^{2q-1} (r\sin(\theta))^{2p-1} e^{-r^2}rdrd\theta,$$
$$= 4\int_0^\infty r^{2(p+q-1)}e^{-r^2}rdr \int_0^{\pi/2} (\sin(\theta))^{2p-1}(\cos(\theta))^{2q-1}d\theta.$$ (11.64)

Noting that for the first integral in equation (11.64), one can write

$$\int_0^\infty r^{2(p+q-1)}e^{-r^2}rdr = \frac{1}{2}\int_0^\infty (r^2)^{(p+q-1)}e^{-r^2}d(r^2)$$
$$= \frac{1}{2}\int_0^\infty R^{(p+q-1)}e^{-R}dR = \frac{\Gamma(p+q)}{2},$$ (11.65)

where we replaced r^2 by R and used the definition for the gamma function in equation (11.12). For the second integral in equation (11.64), applying the second form of the beta function in equation (11.55), one can write

$$\int_0^{\pi/2} \sin^{2p-1}(\theta)\cos^{2q-1}(\theta)d\theta = \frac{B(p, q)}{2}.$$ (11.66)

Upon substituting equations (11.65) and (11.66) into equation (11.64), one finds

$$B(p, q) = \frac{\Gamma(p)\Gamma(q)}{\Gamma(p + q)}. \tag{11.67}$$

This equation relates the beta and gamma functions.

11.3 The error function

Let us consider the function,

$$f(t) = e^{-t^2}. \tag{11.68}$$

In Example 11.1, we have shown that the integral,

$$\int_0^\infty e^{-t^2}dt = \frac{\sqrt{\pi}}{2}, \tag{11.69}$$

which is the area of the region shown in figure 11.2(a). This region is bounded by the functions $f(t) = e^{-t^2}$, $t = 0$, and $f(t) = 0$ for $0 < t < \infty$. However, sometimes it is necessary to find the area when $0 < t < x$, where $x < \infty$, like for the areas of the regions shown in figure 11.2(b) or (c). In such cases, one must evaluate the integral,

$$A(x) = \int_0^x e^{-t^2}dt, \tag{11.70}$$

which does not have an explicit expression. The error function is the function that is defined in terms of this area as

$$\text{erf}(x) = \frac{2}{\sqrt{\pi}}A(x) = \frac{2}{\sqrt{\pi}}\int_0^x e^{-t^2}dt. \tag{11.71}$$

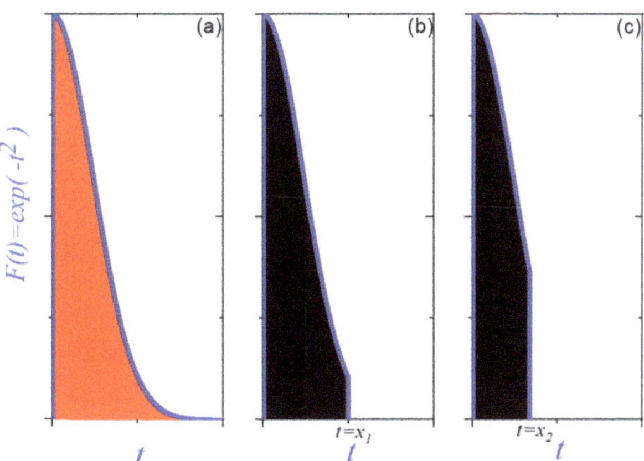

Figure 11.2. Area under the curve defined by the function $f(t) = e^{-t^2}$, $t = 0$, and $f(t) = 0$ for (a) $0 < t < \infty$ (b) and (c) are for case $0 < t < x$.

There are also other related integrals which are sometimes referred to as error functions. These include

(a) The standard normal or Gaussian cumulative distribution function, $\Phi(x)$,

$$\Phi(x) = \frac{1}{\sqrt{2\pi}} \int_{-\infty}^{x} e^{-t^2/2}dt = \frac{1}{2} + \frac{1}{2}\operatorname{erf}(x/\sqrt{2}). \tag{11.72}$$

Noting that the integral in equation (11.72) can be split into two integrals,

$$\begin{aligned}
\Phi(x) &= \frac{1}{\sqrt{2\pi}}\left[\int_{-\infty}^{0} e^{-t^2/2}dt + \int_{0}^{x} e^{-t^2/2}dt\right]\\
&= \frac{1}{\sqrt{2\pi}}\left[-\int_{0}^{-\infty} e^{-t^2/2}dt + \int_{0}^{x} e^{-t^2/2}dt\right]\\
\Rightarrow \Phi(x) &= \frac{1}{\sqrt{2\pi}}\left[\int_{0}^{\infty} e^{-t^2/2}dt + \int_{0}^{x} e^{-t^2/2}dt\right],
\end{aligned} \tag{11.73}$$

using transformation of variables

$$t = \sqrt{2}\,t' \Rightarrow \begin{cases} dt = \sqrt{2}\,dt'\\ t = 0 \Rightarrow t' = 0,\\ t = \infty \Rightarrow t' = \infty,\\ t = x \Rightarrow t' = x/\sqrt{2}, \end{cases} \tag{11.74}$$

one can write

$$\Phi(x) = \frac{1}{\sqrt{\pi}} \int_{0}^{\infty} e^{-t'^2}dt' + \frac{1}{\sqrt{\pi}} \int_{0}^{x/\sqrt{2}} e^{-t'^2}dt'. \tag{11.75}$$

Using equation (11.68) and equation (11.71), we can rewrite equation (11.75) as

$$\begin{aligned}
\Phi(x) &= \frac{1}{\sqrt{2\pi}} \int_{-\infty}^{x} e^{-t^2/2}dt = \frac{1}{2} + \frac{1}{2}\operatorname{erf}(x/\sqrt{2})\\
\Rightarrow \operatorname{erf}(x/\sqrt{2}) &= 2\Phi(x) - 1.
\end{aligned} \tag{11.76}$$

(b) *The complementary error function* $\operatorname{erf} c(x)$: the complementary error function is defined by

$$\begin{aligned}
\operatorname{erf} c(x) &= \frac{2}{\sqrt{\pi}} \int_{x}^{\infty} e^{-t^2}dt = 1 - \operatorname{erf}(x),\\
\text{or } \operatorname{erf} c(x/\sqrt{2}) &= \frac{2}{\sqrt{\pi}} \int_{x}^{\infty} e^{-t^2/2}dt = 1 - \operatorname{erf}(x/\sqrt{2}).
\end{aligned} \tag{11.77}$$

Example 11.4. Consider a criterion that either is or is not satisfied. We look at a system with many elements, each of which satisfies or does not satisfy the criterion.

Criterion: the answer to a test question is correct. For example: Consider a test with many multiple-choice questions. Each answer on the test is either correct (satisfies criterion) or is not correct (does not satisfy the criterion).

We then look at a large number of these systems (for example, a large number of tests consisting of multiple-choice questions). We let x represent the number of elements within a given system satisfying the criterion. We then define the following:

$\bar{x} \equiv$ the average number of elements satisfying the criterion.

$\sigma \equiv$ the standard deviation from the mean number of elements satisfying the criterion.

The probability that any one system will have x to $x + dx$ elements satisfying the criterion is then given by the Gaussian distribution (see figure 11.3):

$$a(x)dx = \frac{1}{\sqrt{2\pi}\,\sigma}e^{-\frac{(x-\bar{x})^2}{2\sigma^2}}dx. \tag{11.78}$$

Find an expression in terms of the error function for the probability that the number of elements of a given system satisfying the criterion, x, will be in the range,

$$\bar{x} - n\sigma \leqslant x \leqslant \bar{x} + n\sigma \tag{11.79}$$

for some real value of n.

Solution: The probability that one system will have x to $x + dx$ elements satisfying the criterion is

$$a(x)dx = \frac{1}{\sqrt{2\pi}\,\sigma}e^{-\frac{(x-\bar{x})^2}{2\sigma^2}}dx, \tag{11.80}$$

then the probability that the number of elements in the range $\bar{x} - n\sigma \leqslant x \leqslant \bar{x} + n\sigma$ satisfying the criterion is given by the integral

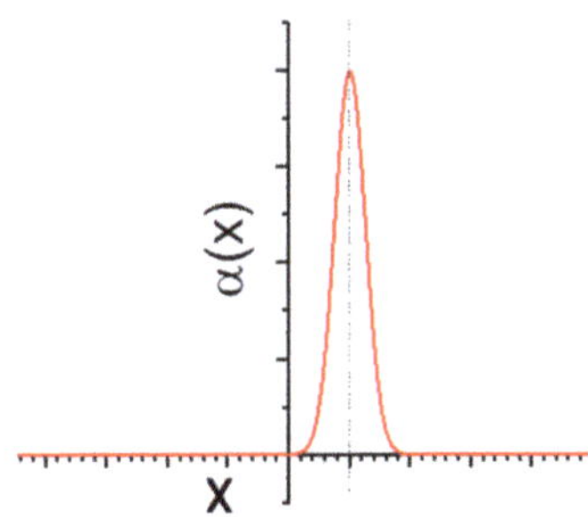

Figure 11.3. Gaussian distribution.

$$P_n(\bar{x}) = \int_{\bar{x}-n\sigma}^{\bar{x}+n\sigma} \alpha(x)dx = \int_{\bar{x}-n\sigma}^{\bar{x}+n\sigma} \frac{1}{\sqrt{2\pi}\,\sigma} e^{-\frac{(x-\bar{x})^2}{2\sigma^2}} dx. \tag{11.81}$$

Introducing a new variable defined by

$$\frac{x-\bar{x}}{\sqrt{2}\,\sigma} = y \Rightarrow \begin{cases} dx = \sqrt{2}\,\sigma dy, \\[4pt] x = \bar{x} - n\sigma \Rightarrow y = \dfrac{\bar{x} - n\sigma - \bar{x}}{\sqrt{2}\,\sigma} = \dfrac{-n}{\sqrt{2}}, \\[6pt] x = \bar{x} + n\sigma \Rightarrow y = \dfrac{\bar{x} + n\sigma - \bar{x}}{\sqrt{2}\,\sigma} = \dfrac{n}{\sqrt{2}}, \end{cases} \tag{11.82}$$

we can rewrite equation (11.81) as

$$P_n(\bar{x}) = \int_{\frac{-n}{\sqrt{2}}}^{\frac{n}{\sqrt{2}}} \frac{1}{\sqrt{2\pi}\,\sigma} e^{-y^2} \sqrt{2}\,\sigma dy = \frac{1}{\sqrt{\pi}} \int_{\frac{-n}{\sqrt{2}}}^{\frac{n}{\sqrt{2}}} e^{-y^2} dy. \tag{11.83}$$

Upon splitting the integral into two intervals, $(-n/\sqrt{2} < y < 0)$ and $(0 < y < n/\sqrt{2})$, we have

$$P_n(\bar{x}) = \frac{1}{\sqrt{\pi}} \left[\int_{-n/\sqrt{2}}^{0} e^{-y^2} dy + \int_{0}^{n/\sqrt{2}} e^{-y^2} dy \right], \tag{11.84}$$

so that using

$$\int_{-n/\sqrt{2}}^{0} e^{-y^2} dy = \int_{n/\sqrt{2}}^{0} e^{-y^2} d(-y) = \int_{0}^{n/\sqrt{2}} e^{-y^2} dy, \tag{11.85}$$

we find

$$P_n(\bar{x}) = \frac{2}{\sqrt{\pi}} \int_{0}^{n/\sqrt{2}} e^{-y^2} dy. \tag{11.86}$$

In terms of the error function

$$\mathrm{erf}(x) = \frac{2}{\sqrt{\pi}} \int_{0}^{x} e^{-t^2} dt, \tag{11.87}$$

equation (11.86) can be expressed as

$$P_n(\bar{x}) = \mathrm{erf}\left(\frac{n}{\sqrt{2}}\right). \tag{11.88}$$

We can find numerical values for the probability for different values of n using Mathematica, for example,

$$P_n(\bar{x}) = \mathrm{erf}\left(\frac{n}{\sqrt{2}}\right) = \begin{cases} 0.682\,6, & n = 1, \\ 0.954\,4, & n = 2, \\ 0.997\,4, & n = 3. \end{cases} \tag{11.89}$$

11.4 Elliptic integrals

Elliptic integrals do often appear in many physical problems. Here we shall consider one real physical problem in classical mechanics. However, before we do that, we shall consider the kinds of elliptical integrals we encounter in real physical problems. There are two types of elliptical integral that we will consider here.

(a) **The elliptic integral of the first kind** is denoted using the notation $K(\alpha/\xi)$ and is defined by the integral expression

$$K(\alpha/\xi) = \int_0^\alpha \frac{d\varphi}{\sqrt{1 - \sin^2(\xi)\sin^2(\varphi)}}. \tag{11.90}$$

Introducing the transformation defined by

$$m = \sin^2(\xi),$$

$$t = \sin(\varphi) \Rightarrow \begin{cases} \cos(\varphi) = \sqrt{1 - \sin^2(\varphi)} = \sqrt{1 - t^2}, \\[2mm] dt = \cos(\varphi)d\varphi \Rightarrow d\varphi = \dfrac{dt}{\sqrt{1 - t^2}}, \\[2mm] \varphi = \alpha \Rightarrow \sin(\alpha) = x, \end{cases} \tag{11.91}$$

we can put equation (11.90) in the form

$$K(x/m) = \int_0^x \frac{dt}{\sqrt{(1 - t^2)(1 - mt^2)}}, \tag{11.92}$$

for $0 \leqslant m \leqslant 1$. When $\varphi = \alpha = \pi/2$, we have

$$\sin(\varphi) = x \Rightarrow x = \sin(\pi/2) = 1, \tag{11.93}$$

and equation (11.92) becomes

$$K(m) = \int_0^1 \frac{dt}{\sqrt{(1 - t^2)(1 - mt^2)}}, \tag{11.94}$$

which is *the complete elliptic integral of the first kind.*

(b) **The elliptic integral of the second kind:** the elliptic integral of the second kind is defined by the integral expression

$$E(\alpha/\xi) = \int_0^\alpha \sqrt{1 - \sin^2(\xi)\sin^2(\varphi)}\,d\varphi, \tag{11.95}$$

and using the transformation in equation (11.91), one can write

$$E(x/m) = \int_0^x \sqrt{\frac{1 - mt^2}{1 - t^2}}\,dt, \tag{11.96}$$

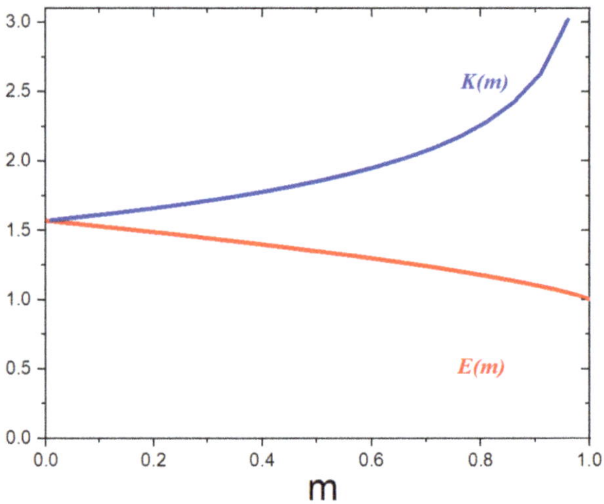

Figure 11.4. The complete elliptical integrals of the first and second kind.

for $0 \leqslant m \leqslant 1$. Similarly, for $\alpha = \pi/2$ and $x = 1$, one finds *the complete elliptic integral of the second kind,*

$$E(m) = \int_0^{\pi/2} \sqrt{1 - m \sin^2{(\varphi)}}\, d\varphi = \int_0^1 \sqrt{\frac{1 - mt^2}{1 - t^2}}\, dt. \tag{11.97}$$

The graphs for the complete elliptic integral of the first kind (equation (11.94)) and the second kind (equation (11.97)) are shown in figure 11.4.

Example 11.5. Consider a simple pendulum with a mass M suspended from the end of a light rigid rod of length l. We pull the pendulum to the side by an angle, α, and release it from rest. Let $\theta = -\alpha$ and $d\theta/dt = 0$ at $t = 0$, where θ is the pendulum angle from the vertical (see figure 11.5). Find an integral expression for the period for the pendulum, T. Using a series expansion for the period, find an approximate expression for small amplitudes of motion.

Solution: Using conservation of mechanical energy,

$$ME_I = ME_\theta, \tag{11.98}$$

where ME_I is the initial mechanical energy (when the pendulum is pulled to the side by an angle α) which is just only the gravitational potential energy given by

$$ME_I = Mgh_{\text{max}} = Mgl(1 - \cos{(\alpha)}), \tag{11.99}$$

and ME_θ is the mechanical energy at some instant of time (i.e., at an angle θ see figure 11.5) which is the sum of the kinetic and the potential energy given by

$$ME_\theta = Mgh + \frac{1}{2}Mv^2 = Mgl(1 - \cos{(\theta)}) + \frac{1}{2}Ml^2\left(\frac{d\theta}{dt}\right)^2. \tag{11.100}$$

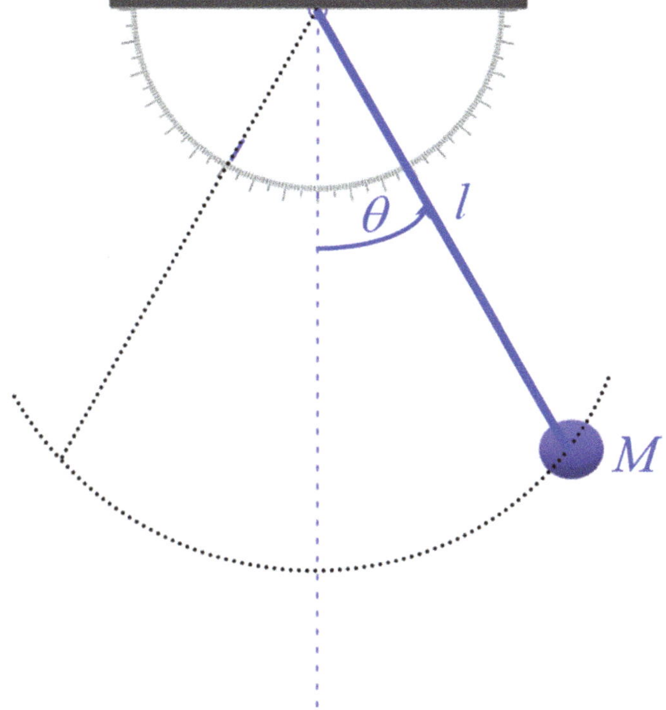

Figure 11.5. A simple pendulum. At the initial time, $t = 0$, the mass m was displaced by an angle, α, from the vertical.

Combining equations (11.98)–(11.100), one finds

$$Mgl(1 - \cos(\alpha)) = Mgl(1 - \cos(\theta)) + \frac{1}{2}Ml^2\left(\frac{d\theta}{dt}\right)^2$$

$$\Rightarrow -g\cos(\alpha) = -g\cos(\theta) + \frac{1}{2}l\left(\frac{d\theta}{dt}\right)^2 \Rightarrow \frac{d\theta}{dt} \qquad (11.101)$$

$$= \sqrt{\frac{2g}{l}(\cos(\theta) - \cos(\alpha))}.$$

Noting that the period is the time for one complete oscillation, which we can express as

$$T = 2\int_0^{t'} dt, \qquad (11.102)$$

where $t = 0$ is the initial time, the pendulum at the maximum displacement from the vertical ($\theta = -\alpha$) and $t = t'$ is the time at which the pendulum reached the position ($\theta = \alpha$). Using equation (11.101), we can rewrite equation (11.102) as

$$\frac{d\theta}{dt} = \sqrt{\frac{2g}{l}(\cos(\theta) - \cos(\alpha))} \Rightarrow dt = \frac{d\theta}{\sqrt{\frac{2g}{l}(\cos(\theta) - \cos(\alpha))}},$$

$$\Rightarrow T = 2\int_0^{t'} dt = 2\int_{-\alpha}^{\alpha} \frac{d\theta}{\sqrt{\frac{2g}{l}(\cos(\theta) - \cos(\alpha))}}. \tag{11.103}$$

Using the relation

$$\cos(\theta) = 1 - 2\sin^2\left(\frac{\theta}{2}\right), \quad \cos(\alpha) = 1 - 2\sin^2\left(\frac{\alpha}{2}\right), \tag{11.104}$$

one can put equation (11.103) in the form

$$T = \sqrt{\frac{l}{g}} \int_{-\alpha}^{\alpha} \frac{d\theta}{\sqrt{\sin^2\left(\frac{\alpha}{2}\right) - \sin^2\left(\frac{\theta}{2}\right)}}. \tag{11.105}$$

Introducing the transformation of variable defined by

$$\sin\left(\frac{\theta}{2}\right) = \sin\left(\frac{\alpha}{2}\right)\sin(\varphi)$$

$$\cos(\varphi) = \sqrt{1 - \frac{\sin^2\left(\frac{\theta}{2}\right)}{\sin^2\left(\frac{\alpha}{2}\right)}},$$

$$\Rightarrow \begin{cases} \frac{1}{2}\cos\left(\frac{\theta}{2}\right)d\theta = \sin\left(\frac{\alpha}{2}\right)\cos(\varphi)d\varphi \Rightarrow d\theta = \frac{2\sin\left(\frac{\alpha}{2}\right)\cos(\varphi)d\varphi}{\sqrt{1 - \sin^2\left(\frac{\alpha}{2}\right)\sin^2(\varphi)}}, \\[4mm] \sqrt{\sin^2\left(\frac{\alpha}{2}\right) - \sin^2\left(\frac{\theta}{2}\right)} = \sin\left(\frac{\alpha}{2}\right)\cos(\varphi) = \sin\left(\frac{\alpha}{2}\right)\cos(\varphi), \\[4mm] \theta = \pm\alpha \Rightarrow \sin\left(\pm\frac{\alpha}{2}\right) = \sin\left(\pm\frac{\alpha}{2}\right)\sin(\varphi) \Rightarrow \sin(\varphi) = \pm 1 \Rightarrow \varphi = \pm\frac{\pi}{2}, \end{cases} \tag{11.106}$$

we can show that equation (11.105) can be rewritten as

$$T = 2\sqrt{\frac{l}{g}} \int_{-\pi/2}^{\pi/2} \frac{d\varphi}{\sqrt{1 - \sin^2\left(\frac{\alpha}{2}\right)\sin^2(\varphi)}}. \tag{11.107}$$

Taking into consideration the fact that $\sin^2(\varphi)$ is an even function, we find

$$T = 4\sqrt{\frac{l}{g}} \int_0^{\pi/2} \frac{d\varphi}{\sqrt{1 - m\sin^2(\varphi)}} = 4\sqrt{\frac{l}{g}} K(m), \qquad (11.108)$$

where

$$K(m) = \int_0^{\pi/2} \frac{d\varphi}{\sqrt{1 - m\sin^2(\varphi)}} = \int_0^{\pi/2} (1 - m\sin^2(\varphi))^{-\frac{1}{2}} d\varphi, \qquad (11.109)$$

is the *complete elliptic integral of the first kind*. We recall from chapter 1, the series expansion

$$(1 + x)^p = \sum_{n=0}^{\infty} \binom{p}{n} x^n = 1 + px + \frac{p(p-1)}{2!} x^2 \dots \qquad (11.110)$$

Applying equation (11.110), equation (11.109) can be expressed as

$$T = 4\sqrt{\frac{l}{g}} \sum_{n=0}^{\infty} \binom{-1/2}{n} (-m)^n \int_0^{\pi/2} \sin^{2n}(\varphi) d\varphi$$

$$= 4\sqrt{\frac{l}{g}} \left[\int_0^{\pi/2} d\varphi + \frac{1}{2}m \int_0^{\pi/2} \sin^2(\varphi) d\varphi \right. \qquad (11.111)$$

$$\left. + \frac{3}{8}m^2 \int_0^{\pi/2} \sin^4(\varphi) d\varphi \dots \right].$$

For small amplitudes of motion (i.e., when α is very small), we have

$$m = \sin^2\left(\frac{\alpha}{2}\right) << 1, \qquad (11.112)$$

so that one can only keep the zero-order term and find

$$T \simeq 2\pi\sqrt{\frac{l}{g}}, \qquad (11.113)$$

which is the well-known expression for the period of a simple pendulum.

11.5 The Dirac delta function

Some physical properties of some objects or systems may require some description confined to a point or a surface. For example, how one can describe the density of the charge of the nucleus or a point mass in space using some function. The Dirac delta function is a function used to describe such kinds of quantities in space. The Dirac delta function does not exist in the usual sense of functions, but we can use several functions to approximate the Dirac delta function. Before we see these functions, we first define the Dirac delta function. To this end, let us consider a point

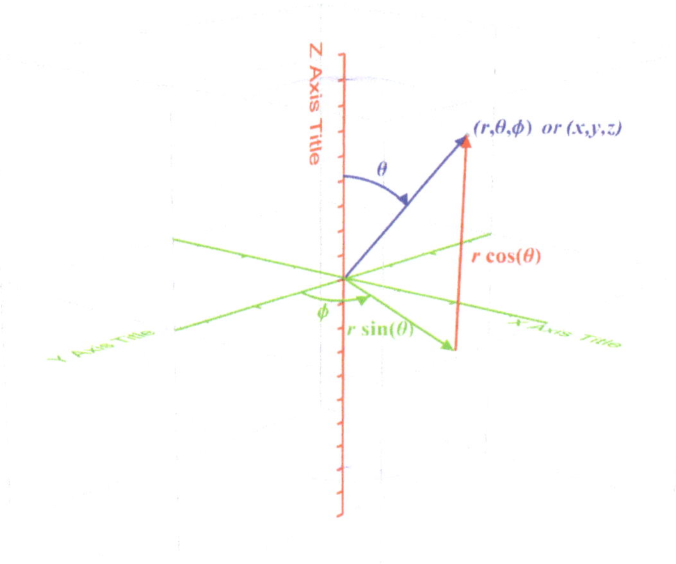

Figure 11.6. A point in space in spherical and Cartesian coordinates.

in space (see figure 11.6) described by the Cartesian coordinates (x, y, z) or the spherical coordinates (r, θ, φ)

$$\begin{aligned}
\vec{r} &= x\hat{x} + y\hat{y} + z\hat{z} \\
&= r\sin(\theta)\cos(\varphi)\hat{x} + r\sin(\theta)\sin(\varphi)\hat{y} + r\cos(\theta)\hat{z} = r\hat{r},
\end{aligned} \tag{11.114}$$

where

$$\hat{r} = \sin(\theta)\cos(\varphi)\hat{x} + \sin(\theta)\sin(\varphi)\hat{y} + \cos(\theta)\hat{z} = \frac{\dfrac{\partial\vec{r}}{\partial r}}{\left|\dfrac{\partial\vec{r}}{\partial r}\right|}, \tag{11.115}$$

is the unit vector along the radial direction. In spherical coordinates, we recall the gradient and the Laplacian for a scalar function $f(r, \theta, \varphi)$ are given by

$$\nabla f = \hat{r}\frac{\partial f}{\partial r} + \hat{\theta}\frac{1}{r}\frac{\partial f}{\partial \theta} + \hat{\varphi}\frac{1}{r\sin(\theta)}\frac{\partial f}{\partial \varphi}, \tag{11.116}$$

and

$$\nabla^2 f = \frac{1}{r^2}\frac{\partial}{\partial r}\left(r^2\frac{\partial f}{\partial r}\right) + \frac{1}{r^2 \sin(\theta)}\frac{\partial}{\partial \theta}\left(\sin(\theta)\frac{\partial f}{\partial \theta}\right) + \frac{1}{r^2 \sin^2(\theta)}\frac{\partial^2 f}{\partial \varphi^2}, \tag{11.117}$$

respectively. On the other hand, for a vector field,

$$\vec{V}(r, \varphi, \theta) = V_r(r, \varphi, \theta)\hat{r} + V_\theta(r, \varphi, \theta)\hat{\theta} + V_\varphi(r, \varphi, \theta)\hat{\varphi}, \tag{11.118}$$

instead of the gradient, in most physical problems, we often need to find the divergence of the vector field which can be expressed, in spherical coordinates, as

$$\nabla \cdot \vec{V} = \frac{1}{r^2}\frac{\partial}{\partial r}(r^2 V_r) + \frac{1}{r \sin(\theta)}\frac{\partial}{\partial \theta}(\sin(\theta) V_\theta) + \frac{1}{r \sin(\theta)}\frac{\partial V_\varphi}{\partial \varphi}. \tag{11.119}$$

Suppose we have some physical quantity that depends on the inverse distance between two points in space. An excellent example of such physical quantity is electrical and gravitational potentials. Let us describe these quantities using a function in spherical coordinates. If we set the first point at the origin and the second point at some point in space described in spherical coordinates as shown in figure 11.6, then we may write this function as

$$f(r, \theta, \varphi) = \frac{1}{r}. \tag{11.120}$$

Then one can construct a vector field by taking the gradient of this function,

$$\vec{V} = \nabla\left(\frac{1}{r}\right) = \hat{r}\frac{\partial}{\partial r}\left(\frac{1}{r}\right) = -\frac{\hat{r}}{r^2}, \quad \text{for } r > 0. \tag{11.121}$$

The divergence of the vector $\vec{V}$ becomes

$$\nabla \cdot \vec{V} = \nabla \cdot \left(-\frac{\hat{r}}{r^2}\right) = \frac{1}{r^2}\frac{\partial}{\partial r}\left[r^2\left(-\frac{1}{r^2}\right)\right] = \begin{cases} 0, & \text{for } r > 0, \\ \infty, & \text{for } r = 0 \end{cases}. \tag{11.122}$$

There follows that

$$\nabla^2\left(\frac{1}{r}\right) = \nabla \cdot \left[\nabla\left(\frac{1}{r}\right)\right] = \nabla \cdot \left(-\frac{\hat{r}}{r^2}\right) = \begin{cases} 0, & \text{for } r > 0, \\ \infty, & \text{for } r = 0, \end{cases} \tag{11.123}$$

which leads to

$$\iiint_V (\nabla \cdot \vec{V})d\tau = \iiint_V \nabla \cdot \left(-\frac{\hat{r}}{r^2}\right)d\tau = 0, \quad \text{for } r > 0$$

$$\Rightarrow \int_0^\infty \int_0^\pi \int_0^{2\pi} \nabla \cdot \left(-\frac{\hat{r}}{r^2}\right)d\tau = 0, \quad \text{for } r > 0. \tag{11.124}$$

On the other hand, if one evaluates this integral using the divergence theorem,

$$\iiint_V \left(\nabla \cdot \vec{V}\right)d\tau = \iint_S \vec{V}(x, y, z) \cdot d\vec{a},$$

$$\Rightarrow \iint_S \vec{V} \cdot d\vec{a} = \int_0^\pi \int_0^{2\pi} \left(-\frac{\hat{r}}{r^2}\right)$$

$$\cdot \hat{r} r^2 \sin(\theta) \cos(\varphi) d\theta d\varphi,$$

$$\Rightarrow \iint_S \vec{V} \cdot d\vec{a} = -\int_0^\pi \int_0^{2\pi} \sin(\theta) \cos(\varphi) d\theta d\varphi$$

$$= -4\pi.$$

(11.125)

This non-zero value must be a result when $r = 0$ is included. Therefore, one can then write

$$\int_0^\infty \int_0^\pi \int_0^{2\pi} \left[\nabla \cdot \left(\frac{\hat{r}}{r^2}\right)\right] d\tau = \begin{cases} 0, & \text{for } r > 0, \\ 4\pi, & \text{for } r = 0. \end{cases}$$

(11.126)

The function $\delta(\vec{r})$, defined as

$$4\pi\sigma(\vec{r}) = 4\pi\delta(x)\delta(y)\delta(z) = \nabla \cdot \left(\frac{\hat{r}}{r^2}\right) = \nabla \cdot \left(\frac{\vec{r}}{r^3}\right) = -\nabla \cdot \nabla\left(\frac{1}{r}\right)$$

$$= -\nabla^2\left(\frac{1}{r}\right),$$

(11.127)

is known as the *Dirac delta function*, where

$$\delta(x) = \begin{cases} 0, & \text{for } x \neq 0 \\ \infty, & \text{for } x = 0 \end{cases}, \quad \delta(y) = \begin{cases} 0, & \text{for } y \neq 0 \\ \infty, & \text{for } y = 0 \end{cases},$$

$$\text{and } \delta(z) = \begin{cases} 0, & \text{for } z \neq 0 \\ \infty, & \text{for } z = 0 \end{cases}.$$

(11.128)

Then for the *Dirac delta function*

$$\int_0^\infty \int_0^\pi \int_0^{2\pi} \delta(\vec{r}) d\tau = 1 \Rightarrow f(0) \int_0^\infty \int_0^\pi \int_0^{2\pi} \delta(\vec{r}) d\tau$$

$$= f(0) \Rightarrow \int_0^\infty \int_0^\pi \int_0^{2\pi} f(\vec{r})\delta(\vec{r}) d\tau = f(0).$$

(11.129)

For the one-dimensional case,

$$\int_{-\infty}^\infty \delta(x) dx = 1, \quad \int_{-\infty}^\infty f(x)\delta(x) dx = f(0).$$

(11.130)

There are more properties of the *Dirac delta function* that can be derived from equation (11.130). Here are a couple

$$\delta(x) = \delta(-x), \quad \delta(ax) = \frac{1}{a}\delta(x).$$

(11.131)

As we mentioned initially, the Dirac delta function does not exist in the usual sense of functions. However, in the limiting case, various forms of functions display the properties of the Dirac delta function. These functions include: *a Gaussian function*

$$\lim_{n\to\infty}\sigma_n(x) = \lim_{n\to\infty}\frac{n}{\sqrt{\pi}}\exp(-n^2x^2) \simeq \delta(x), \tag{11.132}$$

(Figure 11.7(a)), a *Lorentz function*

$$\lim_{n\to\infty}\sigma_n(x) = \lim_{n\to\infty}\frac{n}{\pi}\frac{1}{1+n^2x^2} \simeq \delta(x), \tag{11.133}$$

(Figure 11.7(b)), and *a sinc function*

$$\lim_{n\to\infty}\sigma_n(x) = \lim_{n\to\infty}\frac{1}{2\pi}\int_{-n}^{n}e^{ixt}dt = \lim_{n\to\infty}\frac{\sin(nx)}{\pi x} \simeq \delta(x) \tag{11.134}$$

(figure 11.7(c)). The general form of the Dirac delta function for 1-D, about $x = a$, is defined by

$$\delta(x-a) = \begin{cases} 0, & \text{for } x \neq a, \\ \infty, & \text{for } x = a. \end{cases}$$

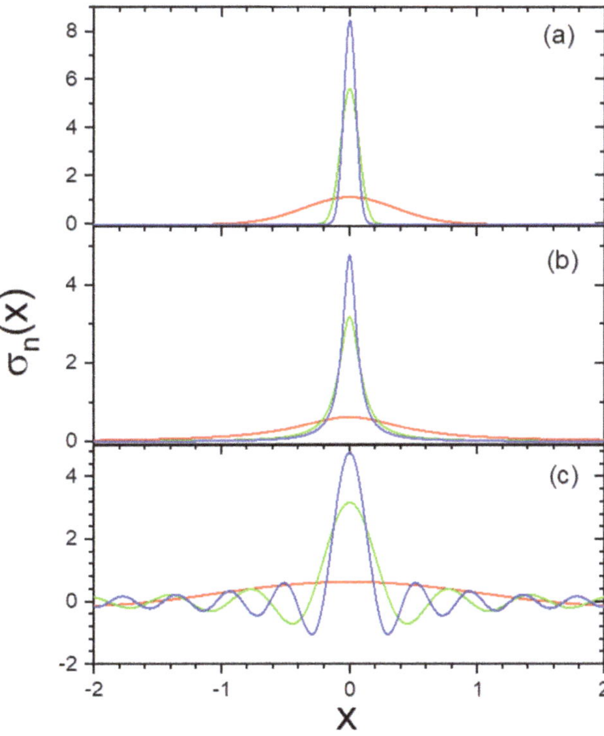

Figure 11.7. Different forms of Dirac delta function: (a) Gaussian, $\sigma_n(x) = \frac{n}{\sqrt{\pi}}\exp(-n^2x^2)$ (b) Lorentz, $\sigma_n(x) = \frac{n}{\pi}\frac{1}{1+n^2x^2}$, and (c) sinc function, $\sigma_n(x) = \frac{\sin(nx)}{\pi x}$. Note that $n = 2$ (red), $n = 10$ (green), and $n = 15$ (blue).

In terms of an integral expression it can also expressed as

$$\int_{-\infty}^{\infty} \delta(x-a)dx = 1, \quad \int_{-\infty}^{\infty} f(x)\delta(x-a)dx = f(a). \tag{11.135}$$

For the 3-D case, the 1-D integral expression in equation (11.135) becomes

$$\int_{0}^{\infty}\int_{0}^{\pi}\int_{0}^{2\pi} \delta(\vec{r})r^2 \sin(\theta)\,dr\,d\theta\,d\varphi = 1,$$

$$\int_{0}^{\infty}\int_{0}^{\pi}\int_{0}^{2\pi} f(\vec{r})\delta(\vec{r})r^2 \sin(\theta)\,dr\,d\theta\,d\varphi = f(0). \tag{11.136}$$

$$\int_{0}^{\infty}\int_{0}^{\pi}\int_{0}^{2\pi} f(\vec{r}-\vec{r_0})\delta(\vec{r}-\vec{r_0})r^2 \sin(\theta)\,dr\,d\theta\,d\varphi = f(\vec{r_0}).$$

Example 11.6. From introductory physics, the electric potential, $V(\vec{r})$, due to a point charge located at the origin (0, 0, 0) (i.e., $r = 0$) at a point in space described by the position vector, $\vec{r}$, (see figure 11.8) is given by

$$V(\vec{r}) = \frac{1}{4\pi\epsilon_0}\frac{q}{r}. \tag{11.137}$$

Show that the volume charge density, $\rho(\vec{r})$, for this point charge can be expressed in terms of the Dirac delta function,

$$\rho(\vec{r}) = \frac{dq}{d\tau} = q\delta(\vec{r}) = q\delta(x)\delta(y)\delta(z), \tag{11.138}$$

where dq is an infinitesimal charge in an infinitesimal volume $d\tau$.

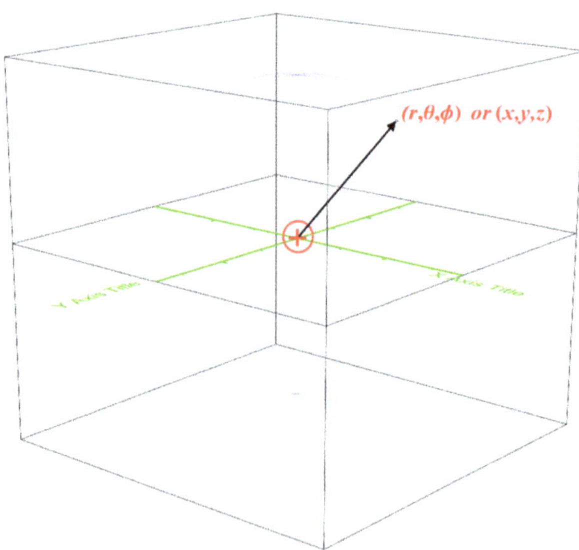

Figure 11.8. A positive point charge, q, at the origin.

Solution: The electric potential, $dV(\vec{r})$ of an infinitesimal charge dq' in a volume $d\tau'$ can be expressed as

$$dV(\vec{r}) = \frac{1}{4\pi\epsilon_0} \frac{dq'}{|\vec{r} - \vec{r}'|} = \frac{1}{4\pi\epsilon_0} \frac{\rho(\vec{r}')d\tau'}{|\vec{r} - \vec{r}'|} \Rightarrow V(\vec{r}) = \frac{1}{4\pi\epsilon_0} \iiint_V \frac{\rho(\vec{r}')d\tau'}{|\vec{r} - \vec{r}'|}. \quad (11.139)$$

Using spherical coordinates, we can write

$$V(\vec{r}) = \frac{1}{4\pi\epsilon_0} \int_0^\infty \int_0^\pi \int_0^{2\pi} \frac{\rho(\vec{r}')r'^2 \sin(\theta)dr'd\theta'd\varphi'}{|\vec{r} - \vec{r}'|}. \quad (11.140)$$

This potential for a point charge becomes

$$\frac{1}{4\pi\epsilon_0} \int_0^\infty \int_0^\pi \int_0^{2\pi} \frac{\rho(\vec{r}')r'^2 \sin(\theta)dr'd\theta'd\varphi'}{|\vec{r} - \vec{r}'|} = \frac{1}{4\pi\epsilon_0} \frac{q}{r}$$

$$\Rightarrow \int_0^\infty \int_0^\pi \int_0^{2\pi} \frac{1}{|\vec{r} - \vec{r}'|} \frac{\rho(\vec{r}')}{q} r'^2 \sin(\theta)dr'd\theta'd\varphi' = \frac{1}{r}. \quad (11.141)$$

From the property of the Dirac delta function

$$\int_0^\infty \int_0^\pi \int_0^{2\pi} f(\vec{r}')\delta(\vec{r}' - \vec{r}_0)r'^2 \sin(\theta')dr'd\theta'd\varphi' = f(\vec{r}_0), \quad (11.142)$$

one can easily find

$$f(\vec{r}') = \frac{1}{|\vec{r} - \vec{r}'|}, \quad \delta(\vec{r}' - \vec{r}_0) = \frac{\rho(\vec{r}')}{q}$$

$$\Rightarrow f(\vec{r}_0) = \frac{1}{|\vec{r} - \vec{r}_0|} = \frac{1}{r} \Rightarrow \vec{r}_0 = 0, \quad (11.143)$$

which leads to

$$\frac{\rho(\vec{r}')}{q} = \delta(\vec{r}') \Rightarrow \rho(\vec{r}') = q\delta(\vec{r}'). \quad (11.144)$$

Example 11.7. The volume charge density, $\rho(\vec{r})$, of a point charge, q, placed at a point on the z-axis, $\vec{r}_0 = a\hat{z}$, is given by

$$\rho(\vec{r}) = q\delta(\vec{r} - \vec{r}_0) = q\delta(x)\delta(y)\delta(z - a), \quad (11.145)$$

where $\delta(\vec{r})$ is the Dirac delta function. Show that the electric potential, $V(\vec{r})$, due to this point charge is given by

$$V(\vec{r}) = \frac{1}{4\pi\epsilon_0} \frac{q}{|\vec{r} - \vec{r}_0|} = \frac{q}{4\pi\epsilon_0} \frac{1}{\sqrt{x^2 + y^2 + (z - a)^2}}. \quad (11.146)$$

The electric potential for a volume charge distribution is given by

$$V(\vec{r}) = \frac{1}{4\pi\epsilon_0} \iiint_V \frac{\rho(\vec{r}')d\tau'}{|\vec{r} - \vec{r}'|},$$
(11.147)

where $\vec{r}'$ is the position of the infinitesimal charge $dq' = \rho(\vec{r}')d\tau'$, in an infinitesimal volume $d\tau'$, and $\rho(\vec{r}')$ is the charge density in the volume V.

Solution: Using the given charge density and the expression for the potential, one can write

$$V(\vec{r}) = \frac{q}{4\pi\epsilon_0} \iiint_V \frac{\delta(\vec{r}' - \vec{r}_0)d\tau'}{|\vec{r} - \vec{r}'|} = \frac{q}{4\pi\epsilon_0} \iiint_V \frac{\delta(\vec{r}' - \vec{r}_0)d\tau'}{|\vec{r} - \vec{r}'|}.$$
(11.148)

In Cartesian coordinates, using

$$\vec{r}' = x'\hat{x} + y'\hat{y} + z'\hat{z}, \; \vec{r} = x\hat{x} + y\hat{y} + z\hat{z}, \; \vec{r}_0 = a\hat{z}$$
$$\Rightarrow |\vec{r} - \vec{r}'| = \sqrt{(x - x')^2 + (y - y')^2 + (z - z')^2}$$
(11.149)

and

$$\delta(\vec{r}' - \vec{r}_0) = \delta(x' - x_0)\delta(y' - y_0)\delta(z' - z_0)$$
$$= \delta(x')\delta(y')\delta(z' - a)$$
(11.150)

one can rewrite equation (11.148) as

$$V(\vec{r}) = \frac{q}{4\pi\epsilon_0} \int_{-\infty}^{\infty} \int_{-\infty}^{\infty} \int_{-\infty}^{\infty} \frac{\delta(x')\delta(y')\delta(z' - a)dx'dy'dz'}{\sqrt{(x - x')^2 + (y - y')^2 + (z - z')^2}}$$
$$= \frac{q}{4\pi\epsilon_0} \int_{-\infty}^{\infty} \delta(z' - a)dz' \int_{-\infty}^{\infty} \delta(y')dy' \int_{-\infty}^{\infty} f(x', y', z')\delta(x')dx',$$
(11.151)

where

$$f(x', y', z') = \frac{1}{\sqrt{(x - x')^2 + (y - y')^2 + (z - z')^2}}.$$
(11.152)

Now applying the property of the Dirac delta function

$$\int_{-\infty}^{\infty} f(x)\delta(x - a)dx = f(a),$$
(11.153)

we find

$$\int_{-\infty}^{\infty} f(x', y', z')\delta(x')dx' = \int_{-\infty}^{\infty} f(x', y', z')\delta(x' - 0)dx'$$
$$= f(0, y', z') = \frac{1}{\sqrt{x^2 + (y - y')^2 + (z - z')^2}}.$$
(11.154)

Substituting equation (11.154) into equation (11.151), we find for the electric potential

$$V(\vec{r}) = \frac{q}{4\pi\epsilon_0} \int_{-\infty}^{\infty} \delta(z' - a)dz' \int_{-\infty}^{\infty} f(0, y', z')\delta(y')dy'. \tag{11.155}$$

Once again using the property of the Dirac delta function, we can write

$$\int_{-\infty}^{\infty} f(0, y', z')\delta(y')dy' = \int_{-\infty}^{\infty} f(0, y', z')\delta(y' - 0)dy'$$

$$= f(0, 0, z') = \frac{1}{\sqrt{x^2 + y^2 + (z - z')^2}}. \tag{11.156}$$

In view of the result in equation (11.156), equation (11.155) becomes

$$V(\vec{r}) = \frac{q}{4\pi\epsilon_0} \int_{-\infty}^{\infty} f(0, 0, z')\delta(z' - a)dz'. \tag{11.157}$$

One last time using the Dirac delta function property, we find for the potential to be

$$V(\vec{r}) = \frac{q}{4\pi\epsilon_0} \int_{-\infty}^{\infty} f(0, 0, z')\delta(z' - a)dz' = \frac{q}{4\pi\epsilon_0}f(0, 0, a)$$

$$\Rightarrow V(\vec{r}) = \frac{q}{4\pi\epsilon_0} \frac{1}{\sqrt{x^2 + y^2 + (z - a)^2}}. \tag{11.158}$$

11.6 Mathematica and special functions

The following are the basic commands in Mathematica we can use to generate and evaluate the special functions we studied in the previous sections.
- Gamma[z] is the Euler gamma function $\Gamma(z)$.
- Gamma[a,z] is the incomplete gamma function $\Gamma(a, z)$.
- Gamma [a, z_0, z_1] is the generalized incomplete gamma function $\Gamma(a, z_0) - \Gamma(a, z_1)$.
- Beta [a, b] gives the Euler beta function $B(a, b)$.
- Beta [z, a, b] gives the incomplete beta function $B_z(a, b)$.
- Erf[z] gives the error function erf (z).
- Erfc [z] gives the complementary error function erfc (z).
- Erf [z_0, z_1] gives the generalized error function erf $(z_1) -$ erf (z_0).
- Erfi [z] gives the imaginary error function erf$(iz)/i$.
- EllipticF [ϕ, m] gives the elliptic integral of the first kind $F(\phi|m)$.
- EllipticK [m] gives the complete elliptic integral of the first kind $K(m)$.
- EllipticE [m] gives the complete elliptic integral of the second kind $E(m)$.
- EllipticE [ϕ, m] gives the elliptic integral of the second kind $E(\phi|m)$.
- EllipticPi [n, m] gives the complete elliptic integral of the third kind $\Pi(n|m)$.

- EllipticPi $[n, \phi, m]$ gives the incomplete elliptic integral of the third kind $\Pi(n; \phi|m)$.
- DiracDelta $[x]$ represents the Dirac delta function $\delta(x)$.
- DiracDelta $[x_1, x_2, \ldots]$ represents the multidimensional Dirac delta function $\delta(x_1, x_2, \ldots)$.

Next we see how some of the examples we considered are resolved using Mathematica.

- Example 11.7

$$\frac{q}{4\pi\epsilon_0}\text{Integrate}\left[\frac{\text{DiracDelta}[x'-a]\text{DiracDelta}[y']\text{DiracDelta}[z']}{\sqrt{(x-x')^2+(y-y')^2+(z-z')^2}}, \{x', -\text{Infinity}, \text{Infinity}\},\right.$$

$$\{y', -\text{Infinity}, \text{Infinity}\}, \{z', -\text{Infinity}, \text{Infinity}\}\right]$$

$$\text{ConditionalExpression}\left[\frac{q}{4\sqrt{(a-x)^2+y^2+z^2}\,\pi\epsilon_0}, a \in \mathbb{R}\right]$$

11.7 Homework assignments

Problem 1.

(a) Prove that

$$B(q, p) = B(p, q). \tag{11.159}$$

(b) Express the integrals

$$I_1 = \int_0^1 \frac{x^4}{\sqrt{1-x^2}}dx, \quad I_2 = \int_0^\pi \sin^3(\theta)\cos(\theta)d\theta \tag{11.160}$$

as beta functions and then write each beta functions in terms of gamma functions using the relation we derived in Example 11.3,

$$B(p, q) = \frac{\Gamma(p)\Gamma(q)}{\Gamma(p+q)}. \tag{11.161}$$

When possible use the gamma function formulas such as

$$\Gamma(p) = \int_0^\infty x^{p-1}e^{-x}dx, \quad \Gamma(p+1) = p\Gamma(p), \quad \Gamma(1/2) = \sqrt{\pi} \tag{11.162}$$

to write an exact answer in terms of π, $\sqrt{2}$, etc.

(c) Applying the result in Example 11.1 show that the integral

$$\int_{-\infty}^\infty e^{-y^2/a}dy = \sqrt{a\pi}, \tag{11.163}$$

for $a > 0$.

Problem 2. Using Stirling's formula evaluate

(a)

$$\lim_{n \to \infty} \left[\frac{\Gamma\left(n + \frac{3}{2}\right)}{\sqrt{n}\,\Gamma(n + 1)} \right] \tag{11.164}$$

(b)

$$\lim_{n \to \infty} \left[\frac{(2n)!\,\sqrt{n}}{2^{2n}(n!)^2} \right]. \tag{11.165}$$

Problem 3. The integral

$$\int_x^\infty u^{p-1} e^{-u}\,du = \Gamma(p, x) \tag{11.166}$$

is called an *incomplete gamma function*. Note that for $x = 0$, you find the gamma function

$$\int_0^\infty u^{p-1} e^{-u}\,du = \Gamma(p). \tag{11.167}$$

By repeated integration find several terms of the asymptotic series for $\Gamma(p, x)$ and you would find

$$\begin{aligned}
\Gamma(p, x) &= \int_x^\infty u^{p-1} e^{-u}\,du \\
&= x^{p-1} e^{-x}[1 + (p - 1)x^{-1} + (p - 1)(p - 2)x^{-2} \\
&\quad + (p - 1)(p - 2)(p - 3)x^{-3}...].
\end{aligned} \tag{11.168}$$

Problem 4. Using the gamma and beta function formulas show that

$$\int_0^\infty \frac{dy}{(1 + y)\sqrt{y}} = \pi. \tag{11.169}$$

Problem 5.

(a) Prove that the error function

$$\operatorname{erf}(x) = \frac{2}{\sqrt{\pi}} \int_0^x e^{-t^2}\,dt \tag{11.170}$$

is an odd function.

(b) Show that

$$\Phi(x) = \frac{1}{\sqrt{2\pi}} \int_{-\infty}^{x} e^{-t^2/2} dt = \frac{1}{2} + \frac{1}{2} \operatorname{erf}(x/\sqrt{2}), \tag{11.171}$$

where

$$\operatorname{erf}(x/\sqrt{2}) = \frac{2}{\sqrt{\pi}} \int_{0}^{x/\sqrt{2}} e^{-t^2} dt \tag{11.172}$$

is the error function.

(c) Derive the relation between the error function and the complementary error function given in equation (11.77).

Problem 6.

For the same pendulum in Example 11.5

    (a) Use the Euler–Lagrange equation to find the equation of motion for the mass, m.

    (b) The resulting equation is a non-linear differential equation. Show that this equation for small amplitudes of oscillation gives a homogenous linear second-order differential equation. By solving this equation show that the period of oscillation is given by same expression

$$T = 2\pi \sqrt{\frac{l}{g}}. \tag{11.173}$$

Problem 7.

Prove the relations

$$\delta(x) = \delta(-x), \quad \delta(ax) = \frac{1}{a}\delta(x). \tag{11.174}$$

Problem 8.

Show that

$$\int_{-\infty}^{\infty} \frac{d\delta(x)}{dx} f(x) dx = -\frac{df(x)}{dx}\bigg|_{x=0}. \tag{11.175}$$

Problem 9.

From introductory physics, the electric potential, $V(\vec{r})$, due to a point charge located at the origin $(0, 0, 0)$ (i.e., $r = 0$) is given by

$$V(\vec{r}) = \frac{1}{4\pi\epsilon_0} \frac{q}{r}. \tag{11.176}$$

Show that the volume charge density, $\rho(\vec{r})$, for this point charge can be expressed in terms of the Dirac delta function

$$\rho(\vec{r}) = \frac{dq}{d\tau} = q\sigma(\vec{r}) = q\sigma(x)\sigma(y)\sigma(z), \tag{11.177}$$

where dq is an infinitesimal charge in an infinitesimal volume $d\tau$.

Problem 10.
The volume charge density, $\rho(\vec{r})$, of a point charge, q, placed at a point on the x-axis, $\vec{r}_0 = a\hat{x}$, can be expressed as

$$\rho(\vec{r}) = q\sigma(\vec{r} - \vec{r}_0), \tag{11.178}$$

where $\sigma(\vec{r})$ is the Dirac delta function. Show that the electric potential, $V(\vec{r})$, due to this point charge is given by

$$V(\vec{r}) = \frac{1}{4\pi\epsilon_0} \frac{q}{|\vec{r} - \vec{r}_0|} = \frac{q}{4\pi\epsilon_0} \frac{1}{\sqrt{(x-a)^2 + y^2 + z^2}}. \tag{11.179}$$

The electric potential for a volume charge distribution is given by

$$V(\vec{r}) = \frac{1}{4\pi\epsilon_0} \iiint_V \frac{\rho(\vec{r}')d\tau'}{|\vec{r} - \vec{r}'|}, \tag{11.180}$$

where $\vec{r}'$ is the position of the infinitesimal charge $dq' = \rho(\vec{r}')d\tau'$, in an infinitesimal volume $d\tau'$, and $\rho(\vec{r}')$ is the charge density in the volume V.

Problem 11.
Show that

$$\delta[(x - x_1)(x - x_2)] = \frac{\delta(x - x_1) + \delta(x - x_2)}{|x_1 - x_2|}. \tag{11.181}$$

IOP Publishing

Studies in Theoretical Physics, Volume 1
Fundamental mathematical methods
Daniel Erenso and Victor Montemayor

Chapter 12

Power series and differential equations

Chapter 6 introduced the various methods for solving homogeneous and non-homogeneous linear differential equations with constant coefficients. This chapter introduces the more general method, the power-series substitution method, for solving homogeneous linear differential equations with constant and non-constant coefficients. The power-series substitution method, essentially, is based on the assumption that the solution to the linear differential equation is expressible as convergent power series. By substituting this series into the differential equation and deriving a recursion relation for the expansion coefficients in the power series, we can determine the solution to the linear differential equation. We introduce this method by resolving the homogeneous second-order linear differential equation with constant coefficients. Specifically we will use the differential equation describing the motion of a non-damped harmonic oscillator, from chapter six. We then demonstrate how the series substitution method is used in homogeneous linear differential equations with non-constant coefficients by solving the Legendre and the Bessel differential equations. Unlike the Legendre, the Bessel differential equation has singular points where the differential equation is not defined. In such cases, we need what is known as the Method of Frobenius that we can apply to determine the solution to the Bessel differential equation, which are known as the Bessel functions. The Legendre Polynomials and the Bessel functions form what is known as a complete set of orthonormal functions. The properties of these functions are explored in detail after we introduce the conditions for any orthonormal set of vectors and functions. We then introduce the associated Legendre polynomials, the solutions to the associated Legendre differential equation that we will derive from the Legendre differential equation and discuss completeness of these polynomials. Following this, the Spherical Harmonics, which are related to the associated Legendre polynomials, and the Addition theorem will be introduced with applications to solve real physical problems. Finally, after we discussed Fuch's theorem, which is the condition that determines when we can or can not use the Method of

Frobenius, we introduce the essential commands in Mathematica to solve the differential equations of the type considered in this chapter.

12.1 Power series substitution

The differential equations we seek to find the solution are linear differential equations like those we studied in chapter 6. However, unlike those equations here, the coefficients in the differential equations we shall consider are not constants and depend on the variable, x, like the following differential equation,

$$\frac{d^2y(x)}{dx^2} + f(x)\frac{dy(x)}{dx} + g(x)y(x) = 0. \tag{12.1}$$

Such differential equations or even those with constant coefficients can be solved using the series substitution method. The method involves a simple procedure with a bit of algebra. We assume the solution to the differential equation expressed as a power series,

$$y(x) = \sum_{n=0}^{\infty} a_n x^n = a_0 + a_1 x + a_2 x^2 + \ldots a_n x^n + \ldots. \tag{12.2}$$

We then substitute this series into the differential equation and try to establish the general relationship between the expansion coefficients. We will demonstrate the application of this method using the following example which we already determined the solution using a different method (see section 6.4).

Example 12.1. A mass, m, is attached to one end of a horizontal spring of spring constant, k. The other end is attached to a rigid vertical wall. The mass slides on a horizontal and frictionless surface. Find the position of the mass as a function of time, $x(t)$. At the time, $t = 0$, the mass is stretched from its equilibrium position by a distance, $x = x_{\max}$, and released from rest (figure 12.1).

Solution: Using Newton's second law, the differential equation of motion for the mass, m, can be written as

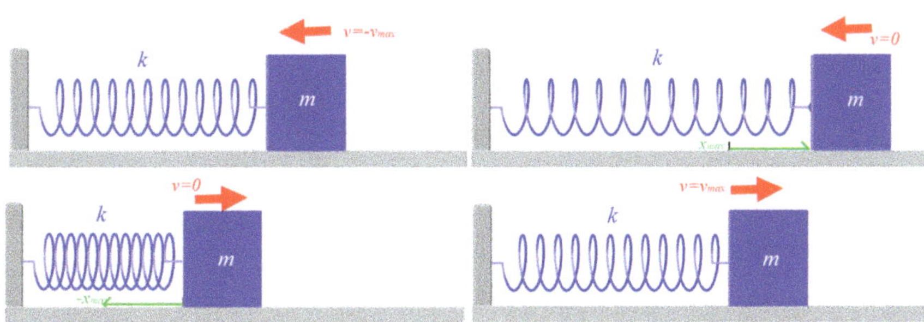

Figure 12.1. A simple harmonic oscillator.

$$F_{net} = ma \Rightarrow m\frac{d^2x(t)}{dt^2} = -kx(t), \Rightarrow \frac{d^2x(t)}{dt^2} + \omega^2 x(t) = 0, \tag{12.3}$$

where

$$\omega = \sqrt{\frac{k}{m}}, \tag{12.4}$$

is the angular frequency. Let us assume that the solution to this differential equation can be written as

$$x(t) = \sum_{n=0}^{\infty} a_n t^n = a_0 + a_1 t + a_2 t^2 + \ldots a_n t^n + \ldots, \tag{12.5}$$

so that

$$\frac{dx(t)}{dt} = \sum_{n=0}^{\infty} a_n \frac{d(t^n)}{dt} = \sum_{n=0}^{\infty} n a_n t^{n-1} = a_1 + 2a_2 t + 3a_3 t^2 +$$
$$\ldots n a_n t^{n-1} + \ldots \Rightarrow \frac{dx(t)}{dt} = \sum_{n=0}^{\infty} (n+1)a_{n+1} t^n, \tag{12.6}$$

$$\frac{d^2x(t)}{dt^2} = \sum_{n=0}^{\infty} a_n \frac{d^2(t^n)}{dt^2} = \sum_{n=0}^{\infty} n(n-1)a_n t^{n-2}$$
$$= 2a_2 + 3 \cdot 2a_3 t + 4 \cdot 3a_4 t^2 + \ldots n(n-1)a_n t^{n-2} + \ldots \tag{12.7}$$
$$\Rightarrow \frac{d^2x(t)}{dt^2} = \sum_{n=0}^{\infty} (n+2)(n+1)a_{n+2} t^n.$$

Now substituting equations (12.5) and (12.7) into the differential equation in equation (12.3), one finds

$$\sum_{n=0}^{\infty} (n+2)(n+1)a_{n+2} t^n + \omega^2 \sum_{n=0}^{\infty} a_n t^n = 0$$
$$\Rightarrow \sum_{n=0}^{\infty} [(n+2)(n+1)a_{n+2} + \omega^2 a_n] t^n = 0. \tag{12.8}$$

Equation (12.8) holds provided

$$(n+2)(n+1)a_{n+2} + \omega^2 a_n = 0 \Rightarrow a_{n+2} = -\frac{a_n}{(n+2)(n+1)}\omega^2. \tag{12.9}$$

Next we shall examine the first few terms resulting from this recursion relation. First we consider when n is even. For $n = 0$, we find

$$n = 0 \Rightarrow a_2 = -\frac{a_0}{2 \cdot 1}\omega^2 \Rightarrow a_{2\times 1} = \frac{a_0(-1)^1}{2!}\omega^{2\times 1}, \tag{12.10}$$

$$n = 2 \Rightarrow a_4 = -\frac{a_2}{4 \cdot 3}\omega^2 = \frac{a_0(-1)^2\omega^4}{4 \cdot 3 \cdot 2!} = \frac{a_0(-1)^2\omega^{2\times2}}{4!}$$
$$\Rightarrow a_{2\times2} = \frac{a_0}{(2 \times 2)!}(-1)^2\omega^{2\times2}, \tag{12.11}$$

$$n = 4 \Rightarrow a_6 = -\frac{a_4\omega^2}{6 \cdot 5} = \frac{a_0(-1)^3\omega^{2\times3}}{6 \cdot 5 \cdot 4 \cdot 3 \cdot 2 \cdot 1} = \frac{a_0(-1)^3\omega^{2\times3}}{6!}$$
$$\Rightarrow a_{2\times3} = \frac{a_0(-1)^3\omega^{2\times3}}{(2 \times 3)!}. \tag{12.12}$$

From the results described in equations (12.10)–(12.12), one can easily establish the relation

$$a_{2m} = a_0\frac{(-1)^m\omega^{2m}}{(2m)!}, \quad \text{for } m = 0, 1, 2, \ldots \tag{12.13}$$

that generates the values for the even term coefficients in the series in equation (12.5). Next we shall consider the odd terms using the recursion relation in equation (12.9). For the first three odd terms, we have

$$n = 1 \Rightarrow a_3 = \frac{a_1(-1)^1}{3 \cdot 2}\omega^2 = \frac{a_1(-1)^1}{3!}\omega^2$$
$$\Rightarrow a_{(2\times1+1)} = \frac{a_1(-1)^1\omega^{2\times1}}{[(2 \times 1) + 1]!} \tag{12.14}$$

$$n = 3 \Rightarrow a_5 = -\frac{a_3}{5 \cdot 4}\omega^2 = \frac{a_1(-1)^2\omega^4}{5 \cdot 4 \cdot 3 \cdot 2 \cdot 1} = \frac{a_1(-1)^2\omega^4}{5!}$$
$$\Rightarrow a_{(2\times3-1)} = \frac{a_1(-1)^2\omega^{2\times2}}{(2 \times 2 + 1)!} \tag{12.15}$$

$$n = 5 \Rightarrow a_7 = -\frac{a_5\omega^2}{7 \cdot 6} = \frac{a_1(-1)^3\omega^6}{7 \cdot 6 \cdot 5 \cdot 4 \cdot 3 \cdot 2 \cdot 1} = \frac{a_1(-1)^3\omega^{2\times3}}{7!}$$
$$\Rightarrow a_{(2\times4-1)} = \frac{a_1(-1)^3\omega^{2\times3}}{(2 \times 3 + 1)!} \tag{12.16}$$

so that one can write

$$a_{2m+1} = \frac{a_1(-1)^m\omega^{2m}}{(2m + 1)!}, \quad \text{for } m = 0, 1, 2, 3\ldots \tag{12.17}$$

This generates the odd term coefficients in the series in equation (12.5). Thus the series in equation (12.5), that we put in the form,

$$x(t) = \sum_{m=0}^{\infty} a_{2m}t^{2m} + \sum_{m=0}^{\infty} a_{2m+1}t^{2m+1}, \tag{12.18}$$

using the results in equations (12.13) and (12.17), can be expressed as

$$x(t) = \sum_{m=0}^{\infty} \frac{a_0(-1)^m \omega^{2m}}{(2m)!} t^{2m} + \sum_{m=0}^{\infty} \frac{a_1(-1)^m \omega^{2m}}{(2m+1)!} t^{2m+1}$$

$$\Rightarrow x(t) = a_0 \sum_{m=0}^{\infty} \frac{(-1)^m}{(2m)!} (\omega t)^{2m} + \frac{a_1}{\omega} \sum_{m=0}^{\infty} \frac{(-1)^m}{(2m+1)!} (\omega t)^{2m+1}. \tag{12.19}$$

We recall that the Taylor series expansions for $\sin(x)$ and $\cos(x)$ from chapter 3,

$$\sin(x) = \sum_{k=0}^{\infty} \frac{(-1)^k x^{2k+1}}{(2k+1)!}, \quad \cos(x) = \sum_{n=0}^{\infty} \frac{(-1)^n x^{2n}}{(2n)!} \tag{12.20}$$

so that we can rewrite equation (12.19) as

$$x(t) = a_0 \cos(\omega t) + \frac{a_1}{\omega} \sin(\omega t) = A \cos(\omega t) + B \sin(\omega t), \tag{12.21}$$

where $A = a_0$ and $B = a_1/\omega$ are constants determined by the initial conditions. Since initially ($t = 0$) the spring is stretched by $x_{\max}$ and the mass is released from rest, we have

$$x(0) = a_0 \cos(0) - a_1 \omega \sin(0) = x_{\max} \Rightarrow a_0 = x_{\max} \tag{12.22}$$

$$v_x(t) = \frac{dx(t)}{dt} = -a_0 \sin(\omega t) - a_1 \omega^2 \cos(\omega t), \ v_x(0) = 0 \tag{12.23}$$

$$\Rightarrow - a_0 \sin(0) - a_1 \omega^2 \cos(0) = 0 \Rightarrow a_1 \omega^2 = 0 \Rightarrow a_1 = 0.$$

In view of equations (12.22) and (12.23), equation (12.21) becomes

$$x(t) = x_{\max} \cos(\omega t). \tag{12.24}$$

Equation (12.24) describes the position of the mass as a function of time.

12.2 Orthonormal set of vectors and functions

In the following sections, we will solve differential equations that lead to forming a complete set of functions. Before we talk about such functions, it is essential to know about scalar or vectors functions that form an orthonormal set. We recall that if the scalar product of two real vectors, $\vec{A}$ and $\vec{B}$, is zero,

$$\vec{A} \cdot \vec{B} = \sum_{i=1}^{3} A_i B_i = 0, \tag{12.25}$$

the two vectors are said to be *orthogonal*. For orthogonal complex vectors equation (12.25) becomes

$$\vec{A}^* \cdot \vec{B} = \sum_{i=1}^{3} A_i^* B_i = 0, \tag{12.26}$$

where $\vec{A}^*$ is the complex conjugate of $\vec{A}$. Using Dirac notation we introduced in chapter 10, we represent any vector $\vec{A}$ by a ket vector $| A \rangle$ and its complex conjugate $\vec{A}^*$ by a bra vector $\langle A |$. equation (12.26) can then be rewritten as

$$\langle A \mid B \rangle = \sum_{i=1}^{3} A_i^* B_i. \tag{12.27}$$

Orthonormal sets of vectors
An orthonormal set of vectors is a set of N−dimensional vectors

$$\{\vec{A}_1, \vec{A}_2, \vec{A}_3...\vec{A}_n...\}, \tag{12.28}$$

that meet the requirement

$$\langle A_n \mid A_m \rangle = \delta_{nm}, \tag{12.29}$$

where

$$\delta_{nm} = \begin{cases} 0, & n \neq m, \\ 1, & n = m, \end{cases} \tag{12.30}$$

is the Kronecker delta.

Orthonormal sets of functions
Two different functions $A_1(x)$ and $A_2(x)$ are said to be orthogonal for all $x \, \epsilon \, (a, b)$ when

$$\int_a^b A_1^*(x)A_2(x)dx = 0, \tag{12.31}$$

where $A_1^*(x)$ is the complex conjugate of the function $A_1(x)$. A set of functions

$$\{A_1(x), A_2(x), A_3(x)...A_n(x)...\} \tag{12.32}$$

that meet the requirement

$$\int_a^b A_n^*(x)A_m(x)dx = \delta_{nm} \tag{12.33}$$

are said to form an *orthonormal set of functions*. Like vectors, we can also express functions using Dirac notation,

$$A_m(x) = \langle x \mid A_m \rangle \Rightarrow A_m^*(x) = (\langle x \mid A_m \rangle)^* = \langle A_m \mid x \rangle. \tag{12.34}$$

Applying equation (12.34), one can then rewrite equation (12.33) as

$$\int_a^b A_n^*(x)A_m(x)dx = \int_a^b \langle A_n \mid x \rangle \langle x \mid A_m \rangle dx = \delta_{nm}. \tag{12.35}$$

When the functions form an orthonormal set for all $x \, \epsilon \, (-\infty, \infty)$, equation (12.35) becomes

$$\int_{-\infty}^{\infty} A_n^*(x)A_m(x)dx = \int_{-\infty}^{\infty} \langle A_n \mid x \rangle \langle x \mid A_m \rangle dx = \delta_{nm},$$

$$\Rightarrow \langle A_n \mid \left(\int_{-\infty}^{\infty} dx \mid x \rangle \langle x \mid \right) \mid A_m \rangle = \delta_{nm}. \qquad (12.36)$$

Since for orthonormal set of vectors,

$$\langle A_n \mid A_m \rangle = \delta_{nm}, \qquad (12.37)$$

one must have

$$\int_{-\infty}^{\infty} dx \mid x \rangle \langle x \mid = 1. \qquad (12.38)$$

The relation in equation (12.38) is known as *the completeness relation*. We often see such kind of relations in quantum mechanics for continuum Eigenstates such as position and momentum eigenstates.

Example 12.2. Show that the set of functions defined by

$$A_n(x) = \frac{1}{\sqrt{\pi}} \sin(nx), \qquad (12.39)$$

for $n = 1, 2, 3 \ldots.$ form an orthonormal set of functions for all $x \in (-\pi, \pi)$.

Solution: Using Euler's formula, we can express

$$A_m(x) = \frac{1}{\sqrt{\pi}} \sin(mx) = \frac{e^{imx} - e^{-imx}}{2i\sqrt{\pi}} = A_m^*(x)$$

$$\Rightarrow A_n^*(x) = \frac{e^{inx} - e^{-inx}}{2i\sqrt{\pi}}, \qquad (12.40)$$

so that

$$\int_a^b A_n^*(x)A_m(x)dx = \int_{-\pi}^{\pi} \left(\frac{e^{inx} - e^{-inx}}{2i\sqrt{\pi}} \right)\left(\frac{e^{imx} - e^{-imx}}{2i\sqrt{\pi}} \right)dx$$

$$= \frac{1}{4\pi} \int_{-\pi}^{\pi} [e^{i(m-n)x} + e^{-i(m-n)x} - e^{-i(m+n)x} - e^{i(m+n)x}]dx$$

$$= \frac{1}{4\pi i} \left[\frac{e^{i(m-n)x}}{m-n} + \frac{e^{-i(m+n)x}}{m+n} - \frac{e^{i(m+n)x}}{m+n} - \frac{e^{-i(m-n)x}}{m-n} \right]_{-\pi}^{\pi} \qquad (12.41)$$

$$\Rightarrow \int_a^b A_n^*(x)A_m(x)dx = \frac{1}{\pi}\left[\frac{\sin[(m-n)\pi]}{m-n} - \frac{\sin[(m+n)\pi]}{m+n} \right].$$

In equation (12.41), we note that the first term is zero only when $m \neq n$. The second term, however, is always zero whether $m = n$ or $m \neq n$. This leads to

$$\int_a^b A_n^*(x)A_m(x)dx = \frac{\sin\left[(m-n)\pi\right]}{(m-n)\pi} = \delta_{nm}. \tag{12.42}$$

Note that we have applied L'Hospital's rule to show that the result in equation (12.42) is one when $m \neq n$. We can find the result in equation (12.42) applying the double angle relations

$$\begin{aligned}
\cos\left(\alpha + \beta\right) &= \cos\left(\alpha\right)\cos\left(\beta\right) - \sin\left(\alpha\right)\sin\left(\beta\right), \\
\cos\left(\alpha - \beta\right) &= \cos\left(\alpha\right)\cos\left(\beta\right) + \sin\left(\alpha\right)\sin\left(\beta\right),
\end{aligned} \tag{12.43}$$

that gives

$$\sin\left(\alpha\right)\sin\left(\beta\right) = \frac{1}{2}\left[\cos\left(\alpha - \beta\right) - \cos\left(\alpha + \beta\right)\right]. \tag{12.44}$$

In view of these relation, one can show that

$$\begin{aligned}
\int_a^b A_n^*(x)A_m(x)dx &= \int_{-\pi}^{\pi} \sin(nx)\sin(mx)dx \\
&= \frac{1}{2\pi}\left\{\int_{-\pi}^{\pi}\cos[(n-m)x]dx - \int_{-\pi}^{\pi}\cos[(n+m)x]dx\right\}.
\end{aligned}$$

Noting that the integral

$$\int_{-\pi}^{\pi}\cos[(n+m)x]dx = -\frac{\sin[(m+n)\pi]}{(m+n)\pi} = 0,$$

one finds

$$\begin{aligned}
\int_a^b A_n^*(x)A_m(x)dx &= \frac{1}{2\pi}\int_{-\pi}^{\pi}\cos[(n-m)x]dx \\
&= \left[\begin{array}{ll} \dfrac{1}{2\pi}\displaystyle\int_{-\pi}^{\pi}dx & \text{for } n = m, \\[2ex] \dfrac{1}{2\pi}\displaystyle\int_{-\pi}^{\pi}\cos[(n-m)x]dx, & \text{for } n \neq m \end{array}\right] \\
&= \left[\begin{array}{ll} 1 & \text{for } n = m, \\[2ex] -\dfrac{\sin[(m-n)\pi]}{(m-n)\pi} = 0, & \text{for } n \neq m \end{array}\right] \\
&\Rightarrow \int_a^b A_n^*(x)A_m(x)dx = \delta_{nm},
\end{aligned} \tag{12.45}$$

which is the same as the result in equation (12.41).

12.3 Complete set of functions

We recall a set of functions, $A_n(x)$, form an orthonormal set for all $x \in (a, b)$, when

$$\int_a^b A_n^*(x)A_m(x)dx = \delta_{nm}. \tag{12.46}$$

Next, we shall study when such a set of functions form a complete set. However, before we do that, we want to review the complete set of vectors in vector space.

Complete sets of vectors in vector space: Consider a three dimensional real space $\mathbb{R}^3$ in Cartesian coordinate system. We recall the unit vectors $\hat{x}$, $\hat{y}$, and $\hat{z}$ (or $\hat{x}_1$, $\hat{x}_2$, and $\hat{x}_3$) form an orthonormal set of vectors in 3-D vector space

$$\hat{x}_i \cdot \hat{x}_j = \delta_{ij} \Rightarrow \langle x_i \mid x_j \rangle = \delta_{ij}. \tag{12.47}$$

Any vector, $\vec{r} \in \mathbb{R}^3$, can be expressed in terms of these three unit vectors

$$\vec{r} = \sum_{i=1}^{3} r_i \hat{x}_i \Rightarrow \mid r \rangle = \sum_{i=1}^{3} r_i \mid x_i \rangle. \tag{12.48}$$

Multiplying both sides from the left by $\langle x_j \mid$,

$$\langle x_j \mid r \rangle = \sum_{i=1}^{3} r_i \langle x_j \mid x_i \rangle = \sum_{i=1}^{3} r_i \delta_{ji} \Rightarrow r_j = \langle x_j \mid r \rangle, \tag{12.49}$$

where we applied the relation for an orthonormal set of vectors in equation (12.47). Note that if one of the vectors, $\hat{x}_i$, is missing, then we cannot express the vector, $\vec{r} \in \mathbb{R}^3$, in terms of the remaining two vectors only. However, as long as we have all the three orthonormal vectors in the set, any vector $\vec{r} \in \mathbb{R}^3$ can be expressed as the linear sum of these vectors. Then we say that the set of vectors $\{\hat{x}_i\} = \{\hat{x}_1, \hat{x}_2, \hat{x}_3\}$ form a complete orthonormal set in $\mathbb{R}^3$.

Now let's make a transition from a vector space to a function space. Suppose we have an infinite complete orthonormal set of vectors

$$\{A_i\} = \{A_1, A_2, A_3...\} = \{\mid A_1 \rangle, \mid A_2 \rangle, \mid A_3 \rangle ...\}, \tag{12.50}$$

and a vector g ($\mid g \rangle$). Since the set of vectors $\{A_i\}$ for a complete set, one can write

$$\mid g \rangle = \sum_{n=0}^{\infty} c_n \mid A_1 \rangle \Rightarrow \langle x \mid g \rangle = \sum_{n=0}^{\infty} c_n \langle x \mid A_1 \rangle,$$

$$\Rightarrow g(x) = \sum_{n=0}^{\infty} c_n A_n(x), \tag{12.51}$$

where

$$g(x) = \langle x \mid g \rangle, \quad A_n(x) = \langle x \mid A_1 \rangle,$$
$$\{A_n(x)\} = \{A_0(x), A_1(x), A_2(x), A_3(x), ...\}. \tag{12.52}$$

In equation (12.51), the infinite set of functions $\{A_n(x)\}$ are an orthonormal set of functions for all $x \in (a, b)$. Furthermore, we note that the function $g(x)$ which is also defined for all $x \in (a, b)$ is expressed as a linear combination of the orthonormal set of functions $\{A_n(x)\}$. Therefore, the set of functions $\{A_n(x)\}$ form *a complete set* of functions. The expansion coefficient c_n is determined using the property for orthonormal set of functions

$$\int_a^b A_n^*(x)A_m(x)dx = \delta_{nm}. \tag{12.53}$$

Multiplying equation (12.51) by $A_m^*(x)$, we have

$$A_m^*(x)g(x) = \sum_{n=0}^{\infty} c_n A_m^*(x)A_n(x), \tag{12.54}$$

and integrating with respect to x for $x \in (a, b)$,

$$\int_a^b A_m^*(x)g(x)dx = \sum_{n=0}^{\infty} c_n \int_a^b A_m^*(x)A_n(x)dx, \tag{12.55}$$

and using equation (12.53), we find

$$\int_a^b A_m^*(x)g(x)dx = \sum_{n=0}^{\infty} c_n \delta_{nm} = c_m \Rightarrow c_m = \int_a^b A_m^*(x)g(x)dx. \tag{12.56}$$

Example 12.3. Any periodic function $g(x)$ defined in the interval $(-\pi, \pi)$ can be expressed in terms of the complete set of orthonormal functions

$$A_n(x) = \frac{1}{\sqrt{\pi}} \sin(nx) = \left\{ \frac{1}{\sqrt{\pi}} \sin(x), \frac{1}{\sqrt{\pi}} \sin(2x), \frac{1}{\sqrt{\pi}} \sin(3x), ... \right\}, \tag{12.57}$$

as

$$g(x) = \sum_{n=1}^{\infty} c_n A_n(x) \Rightarrow g(x) = \sum_{n=1}^{\infty} \frac{c_n}{\sqrt{\pi}} \sin(nx), \tag{12.58}$$

where

$$c_n = \int_{-\pi}^{\pi} A_n^*(x)g(x)dx = \frac{1}{\sqrt{\pi}} \int_{-\pi}^{\pi} \sin(nx)g(x)dx. \tag{12.59}$$

Thus the set of functions $\{A_n(x)\}$ form a complete orthonormal set of functions. In chapter 16, we will see that equation (12.58) is the Fourier series expansion. As an example, let us consider the step function defined by

$$g(x) = \begin{cases} 1, & 0 < x < \pi, \\ -1, & -\pi < x < 0. \end{cases} \tag{12.60}$$

We will show in chapter 16 the series (Fourier series) expansion for this function

$$g(x) = \sum_{n=1}^{\infty} \frac{c_n}{\sqrt{\pi}} \sin(nx), \tag{12.61}$$

is given by

$$g(x) = \frac{4}{\pi} \sum_{n=0}^{\infty} \frac{\sin[(2n+1)x]}{2n+1}, \tag{12.62}$$

by evaluating the expansion coefficient

$$c_n = \int_{-\pi}^{\pi} A_n^*(x)g(x)dx = \frac{1}{\sqrt{\pi}} \int_{-\pi}^{\pi} \sin{(nx)}g(x)dx,$$ (12.63)

for $g(x)$ in equation (12.60). This series expansion for different number of terms n in the series is shown in figure 12.2. It shows the higher is n in the series, the closer the series function is to the exact function in equation (12.60).

12.4 The Legendre differential equation

We can use the series substitution method to solve a differential equation of the form

$$a_n(x)\frac{d^n y}{dx^n} + a_{n-1}(x)\frac{d^{n-1}y}{dx^n} + a_{n-2}(x)\frac{d^{n-2}y}{dx^{n-2}}$$
$$+ \dots a_1(x)\frac{dy}{dx} + a_0(x) = 0.$$ (12.64)

For the case where the coefficients $a_n(x)$ are independent of x, we have seen the application of this method by solving the harmonic oscillator problem. Next, we shall see, using the series substitution method, how we determine the solution to the Legendre differential equation in which $a_n(x)$ depends on x. The Legendre differential equation is given by

$$(1 - x^2)y'' - 2xy' + l(l + 1)y = 0,$$ (12.65)

where $l = 0, 1, 2, 3, \dots$. We are interested in the solution of this differential equation for all $x \in [-1, 1]$. As we will see later on, the Legendre differential equation is derived from Laplace's equation in spherical coordinates. For now, we focus on how to find the solution to this differential equation. To this end, we may want to rewrite equation (12.65) in the form

$$l(l + 1)y - 2xy' - x^2y'' + y'' = 0.$$ (12.66)

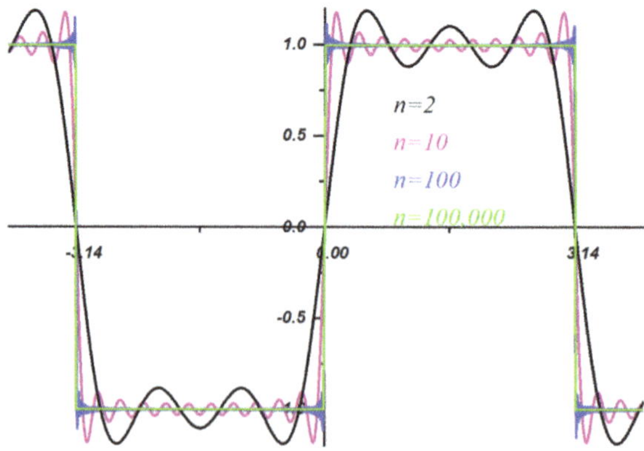

Figure 12.2. Fourier series expansion of the step function.

Suppose the solution to this differential equation, $y(x)$, exists; one must be able to express the function, $y(x)$, *as a convergent* power series for all $x \in [-1, 1]$. Let this series be,

$$y(x) = \sum_{n=0}^{\infty} a_n x^n = a_0 + a_1 x + a_2 x^2 + a_3 x^3 + \ldots a_n x^n + \ldots$$

$$\Rightarrow l(l + 1)y = \sum_{n=0}^{\infty} l(l + 1)a_n x^n. \tag{12.67}$$

so that

$$y'(x) = \sum_{n=0}^{\infty} a_n \frac{d}{dx}(x^n) = \sum_{n=0}^{\infty} a_n n x^{n-1} \Rightarrow 2xy'(x) = \sum_{n=0}^{\infty} 2na_n x^n, \tag{12.68}$$

$$y''(x) = \sum_{n=0}^{\infty} a_n n \frac{d}{dx}(x^{n-1}) = \sum_{n=0}^{\infty} a_n n(n - 1)x^{n-2} \Rightarrow x^2 y''(x) = \sum_{n=0}^{\infty} a_n n(n - 1)x^n$$

$$y''(x) = \sum_{n=0}^{\infty} a_n n(n - 1)x^{n-2} = 0 + 0 + (2 \times 1)a_2 x^0 + (3 \times 2)a_3 x^1 + \ldots \tag{12.69}$$

$$= \sum_{m=2}^{\infty} a_m m(m - 1)x^{m-2} \Rightarrow y''(x) = \sum_{n=0}^{\infty} a_{n+2}(n + 2)(n + 1)x^n,$$

where we replaced m by $n + 2$ in Eq. (12.69). Now substituting Eqs. (12.67), (12.68) and (12.69) into Eq. (12.66), we find

$$l(l + 1)y - 2xy' - x^2 y'' + y'' = 0,$$

$$\sum_{n=0}^{\infty}(n + 2)(n + 1)a_{n+2}x^n + \sum_{n=0}^{\infty}[l(l + 1) - n(n - 1) - 2n]a_n x^n = 0$$

$$\Rightarrow \sum_{n=0}^{\infty}\{a_{n+2}(n + 2)(n + 1) + a_n[l(l + 1) - n(n + 1)]\}x^n = 0 \tag{12.70}$$

$$\Rightarrow a_{n+2}(n + 2)(n + 1) + a_n[l(l + 1) - n(n + 1)] = 0$$

$$\Rightarrow a_{n+2} = -\frac{l(l + 1) - n(n + 1)}{(n + 2)(n + 1)}a_n.$$

Noting that

$$l(l + 1) - n(n + 1) = l^2 - n^2 + l - n = (l - n)(l + n + 1), \tag{12.71}$$

one can rewrite equation (12.70) as

$$a_{n+2} = -\frac{(l - n)(l + n + 1)}{(n + 2)(n + 1)}a_n. \tag{12.72}$$

Equation (12.72) is the *the recursion relation* that we use to determine a_n, when n is even and odd. To this end, using equation (12.72) for even $n = 0, 2, 4, 6\ldots$ we find

$$n = 0 \Rightarrow a_2 = -\frac{(l - 0)(l + 0 + 1)}{(0 + 2)(0 + 1)}a_0 \Rightarrow a_2 = -\frac{l(l + 1)}{2!}a_0 \qquad (12.73)$$

$$n = 2 \Rightarrow a_4 = -\frac{(l - 2)(l + 2 + 1)}{(2 + 2)(2 + 1)}a_2 = -\frac{(l - 2)(l + 3)}{4 \cdot 3}a_2$$

$$\Rightarrow a_4 = \frac{l(l + 1)(l - 2)(l + 3)}{4!}a_0 \qquad (12.74)$$

$$n = 4 \Rightarrow a_6 = -\frac{(l - 4)(l + 4 + 1)}{(4 + 2)(4 + 1)}a_4 = -\frac{(l - 4)(l + 5)}{6 \cdot 5}a_4$$

$$\Rightarrow a_6 = -\frac{l(l + 1)(l - 2)(l + 3)(l - 4)(l + 5)}{6!}a_0. \qquad (12.75)$$

Similarly for odd $n = 1, 3, 5 \ldots$, one can also show

$$n = 1 \Rightarrow a_3 = -\frac{(l - 1)(l + 1 + 1)}{(1 + 2)(1 + 1)}a_1 \Rightarrow a_3 = -\frac{(l - 1)(l + 2)}{3!}a_1, \qquad (12.76)$$

$$n = 3 \Rightarrow a_5 = -\frac{(l - 3)(l + 3 + 1)}{(3 + 2)(3 + 1)}a_3 = -\frac{(l - 3)(l + 4)}{5 \cdot 4}a_3$$

$$\Rightarrow a_5 = \frac{(l - 1)(l + 2)(l - 3)(l + 4)}{5!}a_1, \qquad (12.77)$$

$$a_7 = -\frac{(l - 5)(l + 5 + 1)}{(5 + 2)(5 + 1)}a_5 = -\frac{(l - 5)(l + 6)}{7 \cdot 6}a_5$$

$$\Rightarrow a_7 = -\frac{(l - 1)(l + 2)(l - 3)(l + 4)(l - 5)(l + 6)}{7!}a_1. \qquad (12.78)$$

The power series for the function $y(x)$ can be expressed as a sum of even and odd series

$$y(x) = \sum_{n=0}^{\infty} a_n x^n = \sum_{m=0}^{\infty} a_{2m} x^{2m} + \sum_{m=0}^{\infty} a_{2m+1} x^{2m+1} = y_e(x) + y_o(x). \qquad (12.79)$$

Using the results in equations (12.73)–(12.75), the even series can be written as

$$y_e(x) = a_0 + a_2 x^2 + a_4 x^4 + a_6 x^6 + \ldots$$
$$= a_0 \left\{ 1 - \frac{l(l + 1)}{2!}x^2 + \frac{l(l + 1)(l - 2)(l + 3)}{4!}x^4 \right.$$
$$\left. - \frac{l(l + 1)(l - 2)(l + 3)(l - 4)(l + 5)}{6!}x^6 + \ldots \right\}. \qquad (12.80)$$

Similarly, using equations (12.76)–(12.78), one can also write for the odd series

$$y_o(x) = y_e(x) = a_1 x^1 + a_3 x^3 + a_5 x^5 + \dots$$

$$= a_1 \left\{ x - \frac{(l-1)(l+2)}{3!} x^3 + \frac{(l-1)(l+2)(l-3)(l+4)}{5!} x^5 \right. \tag{12.81}$$

$$\left. - \frac{(l-1)(l+2)(l-3)(l+4)(l-5)(l+6)}{7!} x^7 + \dots \right\}.$$

Note that a_0 and a_1 are constants determined from the boundary conditions. Furthermore, we do not know if this series $y(x) = y_e(x) + y_o(x)$ is convergent or divergent at this stage. For this function to be a solution to the differential equation, it must be a convergent series for all $x \in [-1, 1]$. The following section focuses on the conditions that must be fulfilled for the series to be a convergent series for all $x \in [-1, 1]$.

12.5 The Legendre polynomials

The Legendre polynomials are the solution to the Legendre differential equation. These polynomials are determined from, $y(x)$, in equation (12.79) by imposing the condition for convergence to the series for all $x \in [-1, 1]$. Any function that is a solution to any differential equation must be finite. It means the series form of the function must be convergent in the given domain. We use the ratio test and the integral test (see section 1.2) for both the even and odd series to determine any additional conditions so that the series be convergent for all $x \in [-1, 1]$. To this end, using the recursion relation in equation (12.72), which generates both the even and odd series, we have

$$a_{n+2} = -\frac{(l-n)(l+n+1)}{(n+2)(n+1)} a_n \Rightarrow \frac{a_{n+2}}{a_n} = -\frac{(l-n)(l+n+1)}{(n+2)(n+1)}, \tag{12.82}$$

so that according to the ratio test, both the odd and even series are convergent for all $x \in [-1, 1]$ that satisfy the condition

$$\lim_{n \to \infty} \left| \frac{a_{n+2}}{a_n} x^2 \right| < 1. \tag{12.83}$$

Therefore, the interval of convergence for the function,

$$y(x) = y_e(x) + y_o(x), \tag{12.84}$$

must be the same for both series. Then substituting equation (12.82) into equation (12.83), one finds

$$\lim_{n \to \infty} \left| \frac{a_{n+2}}{a_n} x^2 \right| = \lim_{n \to \infty} \left| -\frac{(l-n)(l+n+1)}{(n+2)(n+1)} x^2 \right| < 1$$

$$\Rightarrow \lim_{n \to \infty} \left| -\frac{(l-n)(l+n+1)}{(n+2)(n+1)} x^2 \right| = \lim_{n \to \infty} | x | < 1. \tag{12.85}$$

Equation (12.85) shows that the series (both the odd and even series) are convergent for $-1 < x < 1$. However, since the ratio test fails at $x = \pm 1$, we need to examine the convergence of the series at these values using a different test of convergence. We note that for $x = \pm 1$, the even series in equation (12.80) and the odd series in equation (12.81) become

$$y_e(\pm 1) = a_0\left\{1 - \frac{l(l+1)}{2!} + \frac{l(l+1)(l-2)(l+3)}{4!}\right.$$
$$\left. - \frac{l(l+1)(l-2)(l+3)(l-4)(l+5)}{6!} + \dots\right\}, \tag{12.86}$$

$$y_o(\pm 1) = a_1\left\{1 - \frac{(l-1)(l+2)}{3!} + \frac{(l-1)(l+2)(l-3)(l+4)}{5!}\right.$$
$$\left. - \frac{(l-1)(l+2)(l-3)(l+4)(l-5)(l+6)}{7!} + \dots\right\}, \tag{12.87}$$

where we included the factor -1 in the odd series in the constant a_1. In the Legendre differential equation, we recall that, $l = 0, 1, 2, \dots$. Let us examine each of these series for the first few values of l.

(a) $l = 0$ at $x = \pm 1$: The even series terminate

$$y_e(\pm 1) = a_0, \tag{12.88}$$

but for the odd series, we find

$$y_o(\pm 1) = a_1\left\{1 + \frac{1 \times 2}{3!} + \frac{1 \times 2 \times 3 \times 4}{5!} + \frac{1 \times 2 \times 3 \times 4 \times 5 \times 6}{7!} + \dots\right\}$$
$$= a_1\left\{1 + \frac{1}{3} + \frac{1}{5} + \frac{1}{7} \dots + \frac{1}{2n+1} + \dots\right\} \tag{12.89}$$
$$\Rightarrow y_o(\pm 1) = a_1 \sum_{n=0}^{\infty} \frac{1}{2n+1}.$$

The even series is clearly a convergent series but the odd series is not. We can easily show this using the integral test,

$$\int_0^\infty \frac{dn}{2n+1} = \frac{1}{2}\ln(2n+1)\Big|_0^\infty = \infty. \tag{12.90}$$

Since we are looking for a well-defined function for which the series

$$y(x) = y_e(x) + y_o(x), \tag{12.91}$$

is convergent for all x in the interval, $-1 \leqslant x \leqslant 1$, we must set, $a_1 = 0$, so that

$$y_o(\pm 1) = a_1 \sum_{n=0}^{\infty} \frac{1}{2n + 1} = 0, \tag{12.92}$$

and the solution to the Legendre differential equation, for $l = 0$, is

$$y(x) = y_0(x) = a_0. \tag{12.93}$$

(b) $l = 1$ *at* $x = \pm 1$: For the even series, from equation (12.86), we find

$$y_e(\pm 1) = a_0\left\{1 - \frac{1(1 + 1)}{2!} + \frac{1(1 + 1)(1 - 2)(1 + 3)}{4!}\right.$$
$$\left. - \frac{1(1 + 1)(1 - 2)(1 + 3)(1 - 4)(1 + 5)}{6!} + ...\right\}$$
$$= a_0\left\{1 - \frac{2}{2} - \frac{1 \times 2 \times 4}{4 \times 3 \times 2} - \frac{1 \times 2 \times 4 \times 3 \times 6}{6 \times 5 \times 4 \times 3 \times 2} + ...\right\} \tag{12.94}$$
$$= -a_0\left\{\frac{1}{3} + \frac{1}{5} + \frac{1}{7}...+\frac{1}{2n + 3} + ...\right\}$$
$$\Rightarrow y_e(\pm 1) = -a_0 \sum_{n=0}^{\infty} \frac{1}{2n + 3}.$$

On the other hand, for the odd series from equation (12.87),

$$y_o(\pm 1) = a_1\left\{1 - \frac{(1 - 1)(1 + 2)}{3!} + \frac{(1 - 1)(1 + 2)(1 - 3)(1 + 4)}{5!}\right.$$
$$\left. - \frac{(1 - 1)(1 + 2)(l - 3)(1 + 4)(1 - 5)(1 + 6)}{7!} + ...\right\} \tag{12.95}$$
$$y_o(\pm 1) = a_1,$$

This time at $x = \pm 1$, we easily see that the odd series is a convergent and the even series is a divergent series. Therefore, the only way out from having a divergent series in the solution to the Legendre differential equation,

$$y(x) = y_e(x) + y_o(x), \tag{12.96}$$

is to set, $a_0 = 0$, such that

$$y_e(\pm 1) = a_0 \sum_{n=0}^{\infty} \frac{1}{2n + 3} = 0. \tag{12.97}$$

Therefore the solution for, $l = 1$, should be

$$y(x) = y_1(x) = a_1 x. \tag{12.98}$$

Following similar procedure for $l = 2, 3, 4...$, one can reach the conclusion that the solution to the Legendre differential equation for all $x \in [-1, 1]$ must be given by

$$y(x) = \begin{cases} y_e(x), & l = \text{even}, \\ y_o(x), & l = \text{odd}, \end{cases}$$

(12.99)

where

$$y_e(x) = a_0\left\{1 - \frac{l(l+1)}{2!}x^2 + \frac{l(l+1)(l-2)(l+3)}{4!}x^4 \right.$$
$$\left. - \frac{l(l+1)(l-2)(l+3)(l-4)(l+5)}{6!}x^6 + ...\right\},$$

(12.100)

$$y_o(x) = a_1\left\{x - \frac{(l-1)(l+2)}{3!}x^3 + \frac{(l-1)(l+2)(l-3)(l+4)}{5!}x^5 \right.$$
$$\left. - \frac{(l-1)(l+2)(l-3)(l+4)(l-5)(l+6)}{7!}x^7 + ...\right\}.$$

(12.101)

For example, one can easily find for $l = 2$,

$$y(x) = y_2(x) = a_0(1 - 3x^2),$$

(12.102)

and $l = 3$,

$$y(x) = y_3(x) = a_1\left(x - \frac{5}{3}x^3\right).$$

(12.103)

When one chooses the values for the constants, a_n, by imposing the condition that $y(x)$ should be one at $x = 1$, one finds what is known as *the Legendre polynomials*, $P_n(x)$, For example,

$$l = 0 \Rightarrow y_0(x = 1) = 1 \Rightarrow a_0 = 1 \Rightarrow P_0(x) = 1,$$
$$l = 1 \Rightarrow y_1(x = 1) = 1 \Rightarrow a_1 = 1 \Rightarrow P_1(x) = x,$$
$$l = 2 \Rightarrow y_2(x = 1) = 1 \Rightarrow a_0(1 - 3) = 1 \Rightarrow a_0 = -\frac{1}{\sqrt{2}}$$
$$\Rightarrow P_2(x) = \frac{1}{2}(3x^2 - 1),$$

(12.104)

$$l = 3 \Rightarrow y_3(x = 1) = 1 \Rightarrow a_1\left(1 - \frac{5}{3}\right) = 1 \Rightarrow a_1 = -\frac{3}{2}$$
$$\Rightarrow P_3(x) = \frac{1}{2}(5x^3 - 3x).$$

Here are some more Legendre Polynomials generated using Mathematica

Table 12.1. Legendre polynomials generated using Mathematica.

LegendreP[2, x]	$\frac{1}{2}(3x^2 - 1)$
LegendreP[3, x]	$\frac{1}{2}(5x^3 - 3x)$
LegendreP[4, x]	$\frac{1}{8}(35x^4 - 30x^2 + 3)$
LegendreP[5, x]	$\frac{1}{8}(63x^5 - 70x^3 + 15x)$
LegendreP[6, x]	$\frac{1}{16}(231x^6 - 315x^4 + 105x^2 - 5)$
LegendreP[7, x]	$\frac{1}{16}(429x^7 - 693x^5 + 315x^3 - 35x)$
LegendreP[8, x]	$\frac{1}{128}(6435x^8 - 12\,012x^6 + 6930x^4 - 1260x^2 + 35)$
LegendreP[9, x]	$\frac{1}{128}(12\,155x^9 - 25\,740x^7 + 18\,018x^5 - 4620x^3 + 315x)$

Note that the number of zero values for the polynomials is equal to the degree of the polynomials. For example, the polynomials $P_1(x)$, $P_2(x)$, $P_3(x)$, and $P_4(x)$ have 1, 2, 3, and 4 zero values, respectively.

12.6 The generating function for the Legendre polynomials

In the previous section, we have determined the solutions to the Legendre differential equation, which are the Legendre polynomials, $P_l(x)$. We recall these polynomials for $l = 0, 1, 2, 3$ are

$$P_0(x) = 1, \; P_1(x) = x, \; P_2(x) = \frac{1}{2}(3x^2 - 1), \; P_3(x) = \frac{1}{2}(5x^3 - 3x). \quad (12.105)$$

In this section, we are interested in finding a function that generates these polynomials. To this end, we consider the function,

$$\phi(x, h) = (1 - 2xh + h^2)^{-1/2}. \quad (12.106)$$

We want to examine the Taylor series expansion for this function about, $h = 0$, given by

$$\phi(x, h) = \sum_{l=0}^{\infty} \frac{1}{l!} \frac{d^l}{dh^l} \phi(x, h) \Big|_{h=0} h^l$$
$$= \phi(x, 0) + \left[\frac{d}{dh}\phi(x, 0)\right]\frac{h}{1!} + \left[\frac{d^2}{dh^2}\phi(x, 0)\right]\frac{h^2}{2!} + \dots, \quad (12.107)$$

for $l = 0, 1, 2, 3$ and predict the relationship between this function and the Legendre polynomials.

$$l = 0 \Rightarrow f_0(x) = \frac{1}{0!}\frac{d^0}{dh^0}\phi(x, h)\Big|_{h=0} = (1 - 2xh + h^2)^{-1/2}\,|_{h=0} \quad (12.108)$$
$$\Rightarrow f_0(x) = 1 = P_0(x),$$

$$l = 1 \Rightarrow f_1(x) = \frac{1}{1!}\frac{d}{dh}\phi(x, h)\bigg|_{h=0} = \frac{x - h}{(1 - 2xh + h^2)^{3/2}}\bigg|_{h=0} \tag{12.109}$$

$$\Rightarrow f_1(x) = x = P_1(x),$$

$$\begin{aligned}
l = 2 \Rightarrow f_2(x) &= \frac{1}{2!}\frac{d^2}{dh^2}\phi(x, h)\bigg|_{h=0} = \frac{1}{2}\frac{d}{dh}\left[\frac{x - h}{(1 - 2xh + h^2)^{3/2}}\right]\bigg|_{h=0} \\
&= \frac{1}{2}\left[-\frac{1}{(1 - 2xh + h^2)^{3/2}} + \frac{3(x - h)^2}{(1 - 2xh + h^2)^{5/2}}\right]_{h=0}
\end{aligned} \tag{12.110}$$

$$\Rightarrow f_2(x) = \frac{1}{2}(3x^2 - 1) = P_2(x),$$

$$\begin{aligned}
l = 3 \Rightarrow f_3(x) &= \frac{1}{3!}\frac{d^3}{dh^3}\phi(x, h)\bigg|_{h=0} \\
&= \frac{1}{6}\frac{d}{dh}\left[-\frac{1}{(1 - 2xh + h^2)^{3/2}} + \frac{3(x - h)^2}{(1 - 2xh + h^2)^{5/2}}\right]\bigg|_{h=0} \\
&= \frac{1}{6}\left[\frac{-3(x - h)}{(1 - 2xh + h^2)^{5/2}} - \frac{3 \times 2(x - h)}{(1 - 2xh + h^2)^{5/2}} + \frac{3 \times 5(x - h)^3}{(1 - 2xh + h^2)^{7/2}}\right]_{h=0}
\end{aligned} \tag{12.111}$$

$$\Rightarrow f_3(x) = \frac{1}{6}(15x^3 - 9x) = \frac{1}{2}(5x^3 - 3x) = P_3(x).$$

From the results listed above, we can then conclude that

$$f_l(x) = \frac{1}{l!}\frac{d^l}{dh^l}\phi(x, h)\bigg|_{h=0} = P_l(x). \tag{12.112}$$

Therefore, the Taylor series expansion for the function

$$\phi(x, h) = \frac{1}{\sqrt{1 - 2xh + h^2}}, \tag{12.113}$$

about $h = 0$ can be written, in terms of the Legendre polynomials, as

$$\phi(x, h) = \frac{1}{\sqrt{1 - 2xh + h^2}} = \sum_{l=0}^{\infty} P_l(x)h^l. \tag{12.114}$$

The function $\phi(x, h)$ is *the generating function* for the Legendre polynomials.

Example 12.4. Consider a point charge, Q, located at the position, $\vec{r}' = (a, 0, 0)$, in spherical coordinates or $\vec{r}' = (0, 0, a)$, in Cartesian coordinates (the source point). Find an expression for the electrostatic potential, $V(\vec{r})$, at the point located at the position, $\vec{r} = (r, \theta, \varphi) = (x, y, z)$, (the field point) (see figure 12.3).

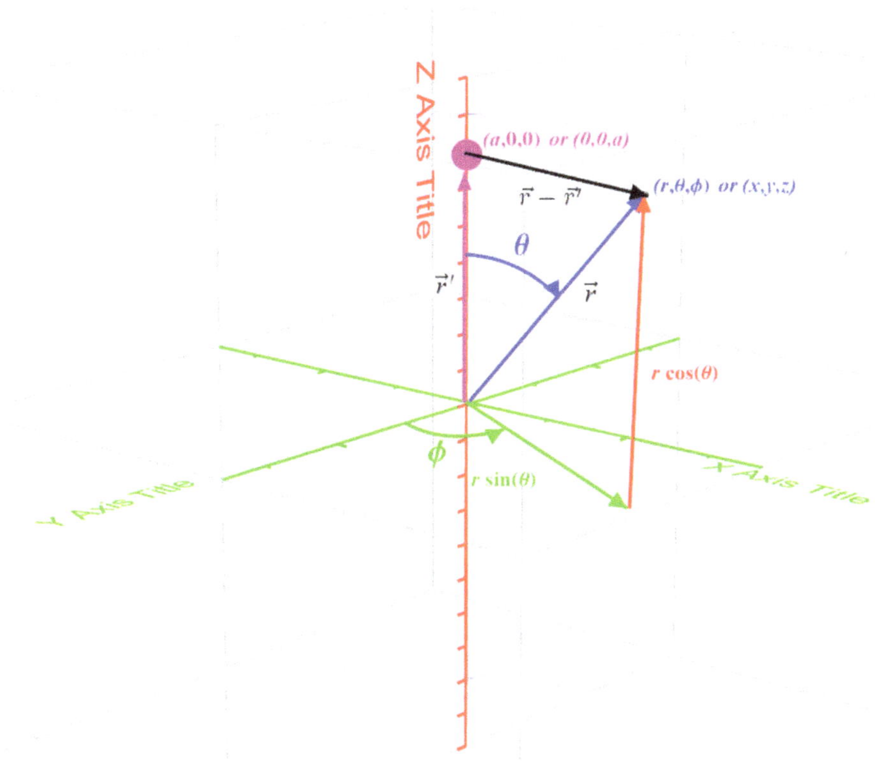

Figure 12.3. A point charge, Q, placed on the z-axis at a distance $r' = a$.

Solution: For a point charge located at the position described by the vector $\vec{r}'$, the electric potential at the point p described by the position $\vec{r}$ is inversely proportional to the distance between the charge and the point p ($|\vec{r} - \vec{r}'|$) and directly proportional to the charge. It can be expressed as

$$V(\vec{r}) = \frac{Q}{4\pi\epsilon_0} \frac{1}{|\vec{r} - \vec{r}'|}. \tag{12.115}$$

For $a < r$, one can write

$$\frac{1}{|\vec{r} - \vec{r}'|} = \frac{1}{\sqrt{r^2 + a^2 - 2ra\cos(\theta)}} = \frac{1}{r}\frac{1}{\sqrt{1 - 2\frac{a}{r}\cos(\theta) + \left(\frac{a}{r}\right)^2}} \tag{12.116}$$

$$= \frac{1}{r}\frac{1}{\sqrt{1 - 2xh + h^2}},$$

where we introduced the variables

$$h = \frac{a}{r}, \quad x = \cos(\theta). \tag{12.117}$$

Using the relation in equation (12.114) and equation (12.117), one can express equation (12.116) as

$$\frac{1}{|\vec{r} - \vec{r}'|} = \frac{1}{\sqrt{1 - 2xh + h^2}} = \sum_{l=0}^{\infty} P_l(\cos(\theta))\left(\frac{a}{r}\right)^l. \tag{12.118}$$

Upon substituting equation (12.118) into equation (12.115), one finds

$$V(\vec{r}) = \frac{Q}{4\pi\epsilon_0}\frac{1}{|\vec{r} - \vec{r}'|} = \frac{Q}{4\pi\epsilon_0 r}\sum_{l=0}^{\infty} P_l(\cos(\theta))\left(\frac{a}{r}\right)^l$$

$$\Rightarrow V(\vec{r}) = \frac{Q}{4\pi\epsilon_0}\sum_{l=0}^{\infty} P_l(\cos(\theta))\frac{a^l}{r^{l+1}}. \tag{12.119}$$

On the other hand when $a > r$, since the series expansion in equation (12.118) is convergent when $h < 1$, we must write

$$\frac{1}{|\vec{r} - \vec{r}'|} = \frac{1}{a\sqrt{1 - 2\frac{r}{a}\cos(\theta) + \left(\frac{r}{a}\right)^2}} = \frac{1}{a\sqrt{1 - 2xh + h^2}}, \tag{12.120}$$

where

$$h = \frac{r}{a}, \ x = \cos(\theta). \tag{12.121}$$

In this case the potential in equation (12.115) becomes

$$V(r) = \frac{Q}{4\pi\epsilon_0}\frac{1}{|\vec{r} - \vec{r}'|} = \frac{Q}{4\pi\epsilon_0 a}\sum_{l=0}^{\infty} P_l(\cos(\theta))\left(\frac{r}{a}\right)^l$$

$$\Rightarrow V(r) = \frac{Q}{4\pi\epsilon_0}\sum_{l=0}^{\infty} P_l(\cos(\theta))\frac{r^l}{a^{l+1}}. \tag{12.122}$$

Rodrigues' Formula

Rodrigues' formula is a formula that can be used to generate the Legendre polynomials. It is given by

$$P_l(x) = \frac{1}{2^l l!}\frac{d^l}{dx^l}(x^2 - 1)^l. \tag{12.123}$$

Example 12.5. Use Rodrigues' formula to find the equation for $P_2(x)$.

Solution: Using Rodrigues' formula $P_2(x)$ can be expressed as

$$P_2(x) = \frac{1}{2^2 2!}\frac{d^2}{dx^2}(x^2 - 1)^2 \tag{12.124}$$

which can be simplified into

$$
P_2(x) = \frac{1}{2^3}\frac{d^2}{dx^2}[x^4 - 2x^2 + 1] = \frac{1}{2^3}\frac{d}{dx}[4x^3 - 4x]
$$
$$
= \frac{1}{2^3}[12x^2 - 4] = \frac{4}{2^3}[3x^2 - 1] \Rightarrow P_2(x) = \frac{1}{2}[3x^2 - 1].
$$

$$(12.125)$$

Leibniz' Rule

Leibniz' rule is useful when we need to find higher order derivatives of a product (for example when we apply *Rodrigues' formula*). It is given by

$$
\frac{d^N}{dx^N}(uv) = \sum_{n=0}^{N}\binom{N}{n}\left(\frac{d^n u}{dx^n}\right)\left(\frac{d^{N-n}v}{dx^{N-n}}\right).
$$

$$(12.126)$$

Example 12.6. Use Leibniz' rule to find

$$
\frac{d^8}{dx^8}(x^2 e^{2x}).
$$

$$(12.127)$$

Solution: For $N = 8$ Leibniz' rule can be expressed as

$$
\frac{d^8}{dx^8}(uv) = \sum_{n=0}^{8}\binom{8}{n}\left(\frac{d^n u}{dx^n}\right)\left(\frac{d^{8-n}v}{dx^{8-n}}\right).
$$

$$(12.128)$$

For $u = x^2$, we have

$$
\frac{d^0 u}{dx^0} = x^2, \ \frac{d^1 u}{dx^1} = 2x, \ \frac{d^2 u}{dx^2} = 2, \ \frac{d^n u}{dx^n} = 0 \text{ for } n \geq 3,
$$

$$(12.129)$$

so that

$$
\frac{d^8}{dx^8}(uv) = \sum_{n=0}^{8}\binom{8}{n}\left(\frac{d^n u}{dx^n}\right)\left(\frac{d^{8-n}v}{dx^{8-n}}\right) = \binom{8}{0}\left(\frac{d^0 u}{dx^0}\right)\left(\frac{d^8 v}{dx^8}\right)
$$
$$
+ \binom{8}{1}\left(\frac{d^1 u}{dx^1}\right)\left(\frac{d^7 v}{dx^7}\right) + \binom{8}{2}\left(\frac{d^2 u}{dx^2}\right)\left(\frac{d^6 v}{dx^6}\right)
$$
$$
= \left(\frac{8!}{8!0!}\right)x^2\left(\frac{d^8 v}{dx^8}\right) + \left(\frac{8!}{7!1!}\right)2x\left(\frac{d^7 v}{dx^7}\right) + \left(\frac{8!}{6!2!}\right)2\left(\frac{d^6 v}{dx^6}\right)
$$
$$
\Rightarrow \frac{d^8}{dx^8}(uv) = x^2\left(\frac{d^8 v}{dx^8}\right) + 16x\left(\frac{d^7 v}{dx^7}\right) + 56\left(\frac{d^6 v}{dx^6}\right).
$$

$$(12.130)$$

Noting that for $v = e^{2x}$

$$\frac{d^n v}{dx^n} = 2^n e^{2x}, \tag{12.131}$$

we find

$$\frac{d^8}{dx^8}(uv) = x^2(2^8 e^{2x}) + 16x(2^7 e^{2x}) + 56(2^6 e^{2x}) = 2^8(x^2 + 8x + 14)e^{2x}$$

$$\Rightarrow \frac{d^8}{dx^8}(uv) = (256x^2 + 2048x + 3584)e^{2x}. \tag{12.132}$$

The Legendre polynomials recursion relation
We recall the generating function

$$(1 - 2xh + h^2)^{-1/2} = \sum_{l=0}^{\infty} P_l(x)h^l, \tag{12.133}$$

so that

$$\frac{d}{dh}(1 - 2xh + h^2)^{-1/2} = \frac{d}{dh}\sum_{l=0}^{\infty} P_l(x)h^l$$

$$\Rightarrow \frac{(x - h)}{(1 - 2xh + h^2)}\frac{1}{\sqrt{1 - 2xh + h^2}} = \sum_{l=0}^{\infty} lP_l(x)h^{l-1}. \tag{12.134}$$

Using the relation in equation (12.114), one can rewrite equation (12.134) as

$$\frac{(x - h)}{1 - 2xh + h^2}\sum_{l=0}^{\infty} P_l(x)h^l = \sum_{l=0}^{\infty} lP_l(x)h^{l-1}$$

$$\Rightarrow (x - h)\sum_{l=0}^{\infty} P_l(x)h^l = (1 - 2xh + h^2)\sum_{l=0}^{\infty} lP_l(x)h^{l-1}$$

$$\Rightarrow x\sum_{l=0}^{\infty} P_l(x)h^l - \sum_{l=0}^{\infty} P_l(x)h^{l+1} \tag{12.135}$$

$$= \sum_{l=0}^{\infty} lP_l(x)h^{l-1} - 2x\sum_{l=0}^{\infty} lP_l(x)h^l + \sum_{l=0}^{\infty} lP_l(x)h^{l+1}$$

$$\Rightarrow \sum_{l=0}^{\infty} lP_l(x)h^{l-1} - \sum_{l=0}^{\infty} x(1 + 2l)P_l(x)h^l + \sum_{l=0}^{\infty} (1 + l)P_l(x)h^{l+1} = 0.$$

Upon expanding the first and the third series in equation (12.135), one finds,

$$\sum_{l=0}^{\infty} l P_l(x) h^{l-1} = 1 P_1(x) + 2 P_2(x) h^1 + 2 P_3(x) h^2 + 4 P_4(x) h^3 + \ldots$$

$$\Rightarrow \sum_{l=0}^{\infty} l P_l(x) h^{l-1} = \sum_{l=0}^{\infty} (l+1) P_{l+1}(x) h^l,$$

(12.136)

$$\sum_{l=0}^{\infty} (1+l) P_l(x) h^{l+1} = P_0(x) h + 2 P_1(x) h^2 + 3 P_2(x) h^3 + 4 P_3(x) h^4$$

$$+ 5 P_4(x) h^5 + \ldots = 0 \times P_{-1}(x) + P_0(x) h + 2 P_1(x) h^2 + 3 P_2(x) h^3 \quad (12.137)$$

$$+ 4 P_3(x) h^4 \Rightarrow \sum_{l=0}^{\infty} (1+l) P_l(x) h^{l+1} = \sum_{l=0}^{\infty} l P_{l-1}(x) h^l.$$

Upon substituting equations (12.136) and (12.137) into equation (12.135), one finds

$$\sum_{l=0}^{\infty} (l+1) P_{l+1}(x) h^l - \sum_{l=0}^{\infty} x(1+2l) P_l(x) h^l + \sum_{l=0}^{\infty} l P_{l-1}(x) h^l = 0$$

$$\Rightarrow \sum_{l=0}^{\infty} [(l+1) P_{l+1}(x) - x(1+2l) P_l(x) + l P_{l-1}(x)] h^l = 0.$$

(12.138)

There follows that

$$(l+1) P_{l+1}(x) - x(1+2l) P_l(x) + l P_{l-1}(x) = 0$$
$$\Rightarrow (l+1) P_{l+1}(x) = x(1+2l) P_l(x) - l P_{l-1}(x).$$

(12.139)

Equation (12.139) is the recursion relation for the Legendre polynomials.

12.7 Legendre series

Orthogonality of the Legendre polynomials
The Legendre polynomials form an orthogonal set of functions. The orthogonality condition is given by

$$\langle P_l \mid P_m \rangle = \langle P_l \mid \int_{-1}^{1} dx \mid x \rangle\langle x \mid P_m \rangle = \int_{-1}^{1} \langle P_l \mid x \rangle\langle x \mid P_m \rangle dx$$

$$\Rightarrow \langle P_l \mid P_m \rangle = \int_{-1}^{1} P_l^*(x) P_m(x) dx = \frac{2}{2l+1} \delta_{lm},$$

(12.140)

for $x \in [-1, 1]$.

Completeness of the Legendre polynomials

The Legendre polynomials also form a complete set of orthonormal functions in the interval $[-1, 1]$. That means any function $f(x)$ can be expressed as a linear combination of the Legendre polynomials

$$f(x) = \sum_{m=0}^{\infty} a_m P_m(x). \tag{12.141}$$

The expression for the expansion coefficients is determined using the orthogonality property of the Legendre polynomials. Multiplying both sides by $P_l(x)$ and then integrating over x, we can write

$$P_l(x)f(x) = \sum_{m=0}^{\infty} a_m P_l(x) P_m(x) \Rightarrow \int_{-1}^{1} P_l(x)f(x)dx$$

$$= \sum_{m=0}^{\infty} a_m \int_{-1}^{1} P_l(x) P_m(x)dx \Rightarrow \int_{-1}^{1} P_l(x)f(x)dx = \sum_{m=0}^{\infty} a_m \frac{2}{2l+1}\delta_{lm} \tag{12.142}$$

$$\Rightarrow \int_{-1}^{1} P_l(x)f(x)dx = a_l \frac{2}{2l+1} \Rightarrow a_l = \frac{2l+1}{2} \int_{-1}^{1} P_l(x)f(x)dx.$$

Recalling that $P_l(x)$ is real, the expansion coefficients are given by

$$a_l = \frac{2l+1}{2} \int_{-1}^{1} P_l(x)f(x)dx. \tag{12.143}$$

Example 12.7. Expand the step function,

$$f(x) = \begin{cases} -1, & -1 \le x \le 0, \\ +1, & 0 \le x \le 1, \end{cases} \tag{12.144}$$

in terms of the Legendre polynomials. The function $f(x)$ is shown in figure 12.4.

Solution: The Legendre series is given by

$$f(x) = \sum_{l=0}^{\infty} a_l P_l(x), \tag{12.145}$$

where

$$a_l = \frac{2l+1}{2} \int_{-1}^{1} P_l^*(x)f(x)dx. \tag{12.146}$$

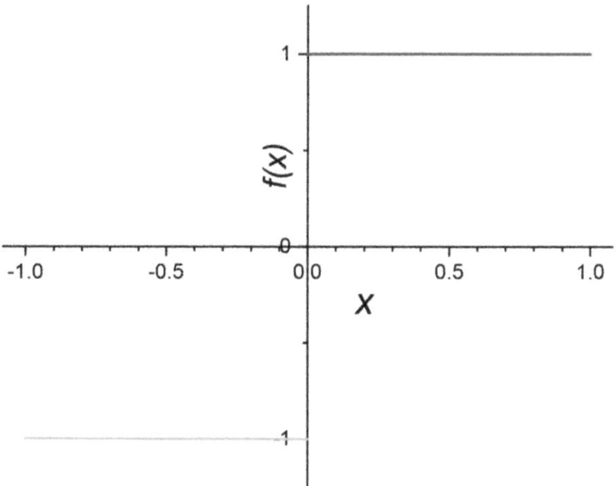

Figure 12.4. The step function.

The function has different values in the interval $(-1, 0)$, $(f(x) = -1)$ and $(0, 1)$, $(f(x) = 1)$. Hence the expansion coefficients, a_l becomes

$$
\begin{aligned}
a_l &= \frac{2l + 1}{2} \left[\int_0^1 P_l(x) f(x) dx + \int_{-1}^0 P_l(x) f(x) dx \right] \\
&= \frac{2l + 1}{2} \left[\int_0^1 P_l(x) dx - \int_{-1}^0 P_l(x) dx \right] = \frac{2l + 1}{2} \left[\int_0^1 P_l(x) dx \right. \\
&\quad \left. - \int_1^0 P_l(-x) d(-x) = \frac{2l + 1}{2} \left[\int_0^1 P_l(x) dx + \int_1^0 P_l(-x) dx \right] \right. \quad (12.147) \\
&= \frac{2l + 1}{2} \left[\int_0^1 P_l(x) dx - \int_0^1 P_l(-x) dx \right] \\
\Rightarrow a_l &= \frac{2l + 1}{2} \left[\int_0^1 [P_l(x) - P_l(-x)] dx \right].
\end{aligned}
$$

Using Rodrigues' formula for the Legendre polynomials

$$
P_l(x) = \frac{1}{2^l l!} \frac{d^l}{dx^l} (x^2 - 1)^l, \tag{12.148}
$$

by replacing x by $-x$, one finds,

$$
P_l(-x) = \frac{1}{2^l l!} \frac{d^l}{d(-x)^l} (x^2 - 1)^l = (-1)^l \frac{1}{2^l l!} \frac{d^l}{dx^l} (x^2 - 1)^l
$$

$$
\Rightarrow P_l(-x) = (-1)^l P_l(-x) = \begin{cases} P_l(x), & l = \text{even}, \\ -P_l(x), & l = \text{odd}. \end{cases} \tag{12.149}
$$

Applying equation (12.149), equation (12.147) can be rewritten as

$$a_l = \begin{cases} (2l + 1) \displaystyle\int_0^1 P_l(x)dx, & l = \text{odd}, \\[2ex] 0, & l = \text{even}. \end{cases} \tag{12.150}$$

Using the Legendre polynomials for $l = 1$ and 3,

$$P_1(x) = x, \quad P_3(x) = \frac{1}{2}(5x^3 - 3x), \tag{12.151}$$

one finds for equation (12.150),

$$a_1 = 3 \int_0^1 x\,dx = \frac{3}{2}, \quad a_3 = \frac{7}{2} \int_0^1 (5x^3 - 3x)dx = -\frac{7}{8}, \tag{12.152}$$

and the Legendre series expansion for the step function becomes

$$f(x) = \frac{3}{2}P_1(x) - \frac{7}{8}P_3(x) + \dots. \tag{12.153}$$

Noting that the series expansion exists only for the odd terms, one can write

$$f(x) = a_1 P_1(x) + a_3 P_3(x) + a_5 P_5(x)\dots = \sum_{m=0}^{\infty} a_{2m+1} P_{2m+1}(x), \tag{12.154}$$

where

$$a_{2m+1} = (4m + 3) \int_0^1 P_{2m+1}(x)dx, \tag{12.155}$$

is the expansion coefficient

Least-squares fit and the Legendre series

Suppose we are given a curve or a graph that is described by a function $f(x)$ defined in the interval $(-1, 1)$ and we want to find a polynomial

$$g_n(x) = a_n x^n + a_{n-1}x^{n-1} + \dots a_0 = \sum_{k=0}^{n} a_k x^k \tag{12.156}$$

that can best fit to the curve defined by the function $f(x)$. Then the polynomial $g_n(x)$ can best fit the curve described by $f(x)$ when the integral

$$\int_{-1}^1 [f(x) - g_n(x)]^2 dx \tag{12.157}$$

is a minimum. This polynomial is nothing but the nth order Legendre polynomial.

12.8 The associated Legendre differential equation

The associated Legendre differential equation is given by

$$(1 - x^2)y'' - 2xy' + \left[l(l + 1) - \frac{m^2}{1 - x^2} \right]y = 0, \tag{12.158}$$

where $m^2 \leq l^2$. When $m = 0$, we find the Legendre differential equation

$$(1 - x^2)y'' - 2xy' + l(l + 1)y = 0, \tag{12.159}$$

the solution of which we found to be the Legendre polynomials

$$y(x) = P_l(x). \tag{12.160}$$

We can solve this differential equation using the series substituting method. However, here we use a different approach. To this end, we introduce a transformation of variable defined by

$$y(x) = (1 - x^2)^{\frac{m}{2}}u(x), \tag{12.161}$$

$$\Rightarrow \left[l(l + 1) - \frac{m^2}{1 - x^2} \right]y = \left[l(l + 1) - \frac{m^2}{1 - x^2} \right](1 - x^2)^{\frac{m}{2}}u(x). \tag{12.162}$$

Upon differentiating equation (12.161) with respect to x

$$y'(x) = (1 - x^2)^{\frac{m}{2}}u'(x) - \frac{m(1 - x^2)^{\frac{m}{2}}}{1 - x^2}xu(x) \tag{12.163}$$

$$\Rightarrow 2xy'(x) = 2x(1 - x^2)^{\frac{m}{2}}\left[u'(x) - \frac{mx}{1 - x^2}u(x) \right]. \tag{12.164}$$

Differentiating equation (12.163) with respect to x, we have

$$\begin{aligned}
y''(x) = {}&(1 - x^2)^{\frac{m}{2}}u''(x) - \frac{mx(1 - x^2)^{\frac{m}{2}}}{1 - x^2}u'(x) \\
&- \frac{m(1 - x^2)^{\frac{m}{2}}}{1 - x^2}u(x) + \frac{m(m - 2)(1 - x^2)^{\frac{m}{2}}}{(1 - x^2)^2}x^2u(x) \\
&- \frac{m(1 - x^2)^{m/2}}{1 - x^2}xu'(x).
\end{aligned} \tag{12.165}$$

There follows that

$$\begin{aligned}
(1 - x^2)y''(x) = {}&(1 - x^2)^{\frac{m}{2}}\{(1 - x^2)u''(x) - 2mxu'(x) \\
&+ \frac{m(m - 1)x^2 - m}{1 - x^2}u(x)\}.
\end{aligned} \tag{12.166}$$

Now substituting equations (12.162), (12.164), (12.166) into equation (12.158), we find

$$+ (1 - x^2)^{\frac{m}{2}} \left\{ (1 - x^2)u''(x) - 2mxu'(x) + \frac{m(m - 1)x^2 - m}{1 - x^2}u(x) \right.$$

$$\left. -2xu'(x) + \frac{2mx^2}{1 - x^2}u(x) + \left[l(l + 1) - \frac{m^2}{1 - x^2} \right]u(x) \right\} = 0 \qquad (12.167)$$

$$\Rightarrow (1 - x^2)u''(x) - 2(m + 1)u'(x) + [l(l + 1) - m(m + 1)]u(x).$$

To find the general solution for all values of m, we shall examine equation (12.167) for the first three values of m. To this end, we note that for $m = 0$, equation (12.167) gives

$$(1 - x^2)\frac{d^2u(x)}{dx^2} - 2x\frac{du(x)}{dx} + l(l + 1)u(x) = 0, \qquad (12.168)$$

which is the Legendre differential equation and the solution is given by the Legendre polynomials

$$u(x) = P_l(x) = \frac{d^0 P_l(x)}{dx^0} = P_l^0(x). \qquad (12.169)$$

In view of equation (12.169), we can rewrite the differential equation in equation (12.168) as

$$(1 - x^2)\frac{d^2 P_l(x)}{dx^2} - 2x\frac{dP_l(x)}{dx} + l(l + 1)P_l(x) = 0. \qquad (12.170)$$

Let us differentiate equation (12.170) with respect to x,

$$\frac{d}{dx}\left\{ (1 - x^2)\frac{d^2 P_l(x)}{dx^2} - 2x\frac{dP_l(x)}{dx} + l(l + 1)P_l(x) = 0 \right\}$$

$$\Rightarrow (1 - x^2)\frac{d^3 P_l(x)}{dx^3} - 2x\frac{d^2 P_l(x)}{dx^2} - 2x\frac{d^2 P_l(x)}{dx^2} - 2\frac{dP_l(x)}{dx}$$

$$+ l(l + 1)\frac{dP_l(x)}{dx} = 0 \qquad (12.171)$$

$$\Rightarrow (1 - x^2)\frac{d^3 P_l(x)}{dx^3} - 4x\frac{d^2 P_l(x)}{dx^2} + [l(l + 1) - 2]\frac{dP_l(x)}{dx} = 0.$$

Suppose we represent the first derivative of the Legendre polynomials by a function

$$P_l^1(x) = \frac{d^1 P_l(x)}{dx^1}, \qquad (12.172)$$

the differential equation in equation (12.171) can be rewritten as

$$(1 - x^2)\frac{d^2 P_l^1(x)}{dx^2} - 4x\frac{dP_l^1(x)}{dx} + (l(l + 1) - 2)P_l^1(x) = 0. \qquad (12.173)$$

Now let us consider equation (12.167) for $m = 1$,

$$(1 - x^2)\frac{d^2u(x)}{dx^2} - 4x\frac{du(x)}{dx} + [l(l + 1) - 2]u(x) = 0. \tag{12.174}$$

So that comparing this with equation (12.173), one can easily see that

$$u(x) = \frac{d^1 P_l(x)}{dx^1} \text{ for } m = 1. \tag{12.175}$$

Differentiating equation (12.172) with respect to x, we have

$$\frac{d}{dx}\left\{(1 - x^2)\frac{d^2 P_l(x)}{dx^2} - 4x\frac{dP_l(x)}{dx} + (l(l + 1) - 2)P_l(x) = 0\right\}$$
$$\Rightarrow (1 - x^2)\frac{d^3 P_l(x)}{dx^3} - 6x\frac{d^2 P_l(x)}{dx^2} + (l(l + 1)u'(x) - 6)\frac{dP_l(x)}{dx} = 0, \tag{12.176}$$

and introducing the function

$$u(x) = \frac{d^2 P_l(x)}{dx^2}, \tag{12.177}$$

we find

$$(1 - x^2)\frac{d^2 P_l(x)}{dx^2} - 6x\frac{dP_l(x)}{dx} + [l(l + 1) - 6]P_l(x) = 0. \tag{12.178}$$

For $m = 2$ equation (12.167) becomes

$$(1 - x^2)u''(x) - 6xu'(x) + [l(l + 1) - 6]u(x) = 0, \tag{12.179}$$

so that comparing this result with equation (12.178), we also find

$$u(x) = \frac{d^2 P_l(x)}{dx^2} \text{ for } m = 2. \tag{12.180}$$

Following the same procedure one can easily establish that the solution to equation (12.167) for a given m is given by

$$u(x) = \frac{d^m P_l(x)}{dx^m}. \tag{12.181}$$

In view of equation (12.181), one finds for equation (12.161),

$$y(x) = (1 - x^2)^{m/2}\frac{d^m}{dx^m}P_l(x), \tag{12.182}$$

which the solution to the associated Legendre differential equation

$$(1 - x^2)y'' - 2xy' + \left[l(l + 1) - \frac{m^2}{1 - x^2}\right]y = 0. \tag{12.183}$$

The associated Legendre functions:
The solutions to the associated Legendre differential equation

$$(1 - x^2)\frac{d^2 P_l^m(x)}{dx^2} - 2x\frac{dP_l^m(x)}{dx} + \left[l(l + 1) - \frac{m^2}{1 - x^2} \right]P_l^m(x) = 0, \qquad (12.184)$$

are the associated Legendre functions given by

$$P_l^m(x) = (1 - x^2)^{m/2}\frac{d^m P_l(x)}{dx^m}. \qquad (12.185)$$

Note that when $m = 0$, we find the Legendre polynomials $P_l^0(x) = P_l(x)$.

Orthogonality of the associated Legendre functions
Using the Dirac bra-ket notation, we can express that the orthogonality condition to the associated Legendre polynomials as

$$\begin{aligned}
\langle P_l^m \mid P_{l'}^m \rangle &= \int_{-1}^{1} \langle P_l^m \mid x \rangle\langle x \mid P_{l'}^m \rangle dx = \int_{-1}^{1} P_l^{*m}(x)P_{l'}^m(x)dx \\
\Rightarrow \langle P_l^m \mid P_{l'}^m \rangle &= \int_{-1}^{1} P_l^m(x)P_{l'}^m(x)dx = \frac{2}{2l + 1}\frac{(l + m)!}{(l - m)!}\delta_{ll'},
\end{aligned} \qquad (12.186)$$

where we used the fact that the associated Legendre polynomials are real functions.

12.9 Spherical harmonics and the addition theorem

Consider the second-order differential equation

$$\frac{d^2\Phi(\varphi)}{d\varphi^2} + m^2\Phi(\varphi) = 0, \qquad (12.187)$$

for $0 \le \varphi \le 2\pi$. The solution is given by

$$\Phi(\varphi) = A(m)e^{im\varphi} = \frac{1}{\sqrt{2\pi}}e^{im\varphi}, \qquad (12.188)$$

where we determine the constant $A(m)$ by imposing the normalization condition

$$\int_0^{2\pi} \mid \Phi(\varphi) \mid^2 d\varphi = 1.$$

Let us also consider the associated Legendre differential equation

$$(1 - x^2)\frac{d^2 y(x)}{dx^2} - 2x\frac{dy(x)}{dx} + \left[l(l + 1) - \frac{m^2}{1 - x^2} \right]y(x) = 0, \qquad (12.189)$$

which we determined the solution to be

$$y(x) = P_l^m(x) = (1 - x^2)^{m/2} \frac{d^m}{dx^m} P_l(x). \tag{12.190}$$

Now introducing the transformation

$$x = \cos(\theta) \Rightarrow \sin^2(\theta) = 1 - x^2, \ dx = -\sin(\theta)d\theta, \ P_l^m(x) = \Theta(\cos(\theta)), \tag{12.191}$$

one finds

$$\frac{dy(x)}{dx} = \frac{d\Theta}{d\theta}\frac{d\theta}{dx} = -\frac{1}{\sin(\theta)}\frac{d\Theta}{d\theta}, \tag{12.192}$$

$$\Rightarrow \frac{d^2y(x)}{dx^2} = -\frac{d}{dx}\left[\frac{1}{\sin(\theta)}\frac{d\Theta}{d\theta}\right]$$

$$= -\frac{d\Theta}{d\theta}\frac{d}{d\theta}\left[\frac{1}{\sin(\theta)}\right]\left(\frac{d\theta}{dx}\right) - \frac{1}{\sin(\theta)}\frac{d^2\Theta}{d\theta^2}\left(\frac{d\theta}{dx}\right) \tag{12.193}$$

$$\Rightarrow \frac{d^2y(x)}{dx^2} = -\frac{\cos(\theta)}{\sin^3(\theta)}\frac{d\Theta(\theta)}{d\theta} + \frac{1}{\sin^2(\theta)}\frac{d^2\Theta(\theta)}{d\theta^2}.$$

Then substituting the results in equations (12.191)–(12.193) into (12.189), the associated Legendre differential equation can be expressed as

$$\sin^2(\theta)\left[-\frac{\cos(\theta)}{\sin^3(\theta)}\frac{d\Theta(\theta)}{d\theta} + \frac{1}{\sin^2(\theta)}\frac{d^2\Theta(\theta)}{d\theta^2}\right] + 2\frac{\cos(\theta)}{\sin(\theta)}\frac{d\Theta}{d\theta}$$

$$+ \left[l(l+1) - \frac{m^2}{\sin^2(\theta)}\right]\Theta = 0$$

$$\Rightarrow \frac{d^2\Theta(\theta)}{d\theta^2} + \frac{\cos(\theta)}{\sin(\theta)}\frac{d\Theta}{d\theta} + \left[l(l+1) - \frac{m^2}{\sin^2(\theta)}\right]\Theta = 0 \tag{12.194}$$

$$\Rightarrow \frac{1}{\sin(\theta)}\frac{d}{d\theta}\left[\sin(\theta)\frac{d\Theta(\theta)}{d\theta}\right] + \left[l(l+1) - \frac{m^2}{\sin^2\theta}\right]\Theta(\theta) = 0.$$

The solution to this form of the associated Legendre differential equation can then be written as

$$\Theta(\theta) = P_l^m(\cos(\theta)) = \sin^m(\theta)\frac{d^m}{d(\cos(\theta))^m}P_l(\cos(\theta)). \tag{12.195}$$

For the orthogonality relation for the associated Legendre polynomials in equation (12.186), we have

$$\int_{-1}^{1} P_l^m(x)P_l^m(x)dx = \int_0^{\pi} [P_l^m(\cos(\theta))]^2 \sin(\theta)d\theta = \frac{2}{2l+1}\frac{(l+m)!}{(l-m)!}, \tag{12.196}$$

so that one can write a normalized function $\Theta(\theta)$

$$\Theta(\theta) = \sqrt{\frac{2l+1}{2}\frac{(l-m)!}{(l+m)!}}\, P_l^m(\cos(\theta)). \tag{12.197}$$

The functions $Y_{lm}(\theta, \varphi)$, known as the *spherical harmonics, are functions* defined in terms of the product of the functions $\Phi(\varphi)$ in equation (12.188) and $\Theta(\theta)$ in equation (12.197)

$$Y_{lm}(\theta, \varphi) = \Phi(\varphi)\Theta(\theta) = C(l, m)P_l^m(\cos(\theta))e^{im\varphi}, \tag{12.198}$$

where

$$C(l, m) = (-1)^m\sqrt{\frac{2l+1}{4\pi}\frac{(l-m)!}{(l+m)!}},$$

$$m^2 \leq l^2 \Rightarrow m = -l, -l+1, -l+2 \ldots 0, 1, , 3\ldots l,$$
$$l = 0, 1, 2, 3\ldots \tag{12.199}$$

The spherical harmonics form an orthonormal set of vectors/functions and *the orthonormality condition is given by*

$$\int_{\theta=0}^{\pi}\int_{\varphi=0}^{2\pi} Y_{lm}^*(\theta, \varphi)\,Y_{l'm'}(\theta, \varphi)\sin(\theta)d\theta d\varphi = \delta_{ll'}\delta_{mm'}. \tag{12.200}$$

Spherical harmonics form a complete set of vectors, which we may express, using Dirac's notion, as

$$\sum_{l=0}^{\infty}\sum_{m=-l}^{l} |\,Y_{lm}\rangle\langle Y_{lm}\,| = 1, \tag{12.201}$$

or form a complete set of functions,

$$\sum_{l=0}^{\infty}\sum_{m=-l}^{l}\int_0^{\pi}\int_0^{2\pi} \langle\theta', \varphi'|\,Y_{lm}\rangle\langle Y_{lm}\,|\,\theta', \varphi'\rangle\sin(\theta')d\theta'd\varphi' = 1$$

$$\Rightarrow \sum_{l=0}^{\infty}\sum_{m=-l}^{l}\int_0^{\pi}\int_0^{2\pi} Y_{lm}^*(\theta', \varphi')\,Y_{lm}(\theta', \varphi')\sin(\theta')d\theta'd\varphi' = 1. \tag{12.202}$$

Thus any finite function, $f(\theta, \varphi)$

$$f(\theta, \varphi) = \langle\theta, \varphi\,|f\rangle \tag{12.203}$$

that is defined in the domain $\theta \in [0, \pi]$ and $\varphi \in [0, 2\pi]$ can be expressed as a linear combination of the spherical harmonics

$$f(\theta, \varphi) = \langle\theta, \varphi\,|f\rangle = \sum_{l=0}^{\infty}\sum_{m=-l}^{l} \langle\theta, \varphi\,|\,Y_{lm}\rangle\langle Y_{lm}\,|f\rangle. \tag{12.204}$$

Using the completeness relation for a continuous space

$$\int_0^\pi \int_0^{2\pi} |\theta', \varphi'\rangle\langle\theta', \varphi'| \, \sin(\theta')d\theta' d\varphi' = 1, \tag{12.205}$$

where

$$\langle\theta \mid \theta'\rangle = \delta(\theta - \theta'), \quad \langle\varphi \mid \varphi'\rangle = \delta(\varphi - \varphi'), \tag{12.206}$$

we can rewrite equation (12.204) as

$$f(\theta, \varphi) = \langle\theta, \varphi \mid f\rangle$$

$$= \sum_{l=0}^\infty \sum_{m=-l}^l \int_0^\pi \int_0^{2\pi} \langle\theta, \varphi \mid Y_{lm}\rangle\langle Y_{lm} \mid\mid \theta', \varphi'\rangle\langle\theta', \varphi'\mid f\rangle \sin(\theta')d\theta' d\varphi'$$

$$= \sum_{l=0}^\infty \sum_{m=-l}^l \int_0^\pi \int_0^{2\pi} \langle\theta, \varphi \mid Y_{lm}\rangle\langle Y_{lm} \mid \theta', \varphi'\rangle\langle\theta', \varphi'\mid f\rangle \sin(\theta')d\theta' d\varphi' \tag{12.207}$$

$$\Rightarrow f(\theta, \varphi) = \sum_{l=0}^\infty \sum_{m=-l}^l \left[\int_0^\pi \int_0^{2\pi} Y_{lm}^*(\theta', \varphi')f(\theta', \varphi') \sin(\theta')d\theta' d\varphi'\right] Y_{lm}(\theta, \varphi)$$

where we used Dirac's notation

$$Y_{lm}(\theta, \varphi) = \langle\theta, \varphi \mid Y_{lm}\rangle, \quad Y_{lm}^*(\theta', \varphi') = \langle Y_{lm} \mid \theta', \varphi'\rangle,$$

$$f(\theta', \varphi') = \langle\theta', \varphi'\mid f\rangle. \tag{12.208}$$

Introducing

$$A_{lm} = \int_0^\pi \int_0^{2\pi} Y_{lm}^*(\theta', \varphi')f(\theta', \varphi') \sin(\theta')d\theta' d\varphi'. \tag{12.209}$$

Equation (12.207) can be rewritten as

$$f(\theta, \varphi) = \sum_{l=0}^\infty \sum_{m=-l}^l A_{lm} Y_{lm}(\theta, \varphi), \tag{12.210}$$

Note that A_{lm} are the expansion coefficients.

The inverse distance between two points
Consider two points described by the vectors, $\vec{r}$ and $\vec{r}'$, as shown in figure 12.5. In spherical coordinates, these two points are described by (r, θ, φ) and (r', θ', φ'), respectively. The angle between these two vectors, γ (which is related to the angles describing each of the two vectors), can be used to determine the inverse distance between these two points. Given the equation we determined that relates the inverse distance and the Legendre polynomials, one can write

$$\frac{1}{|\vec{r} - \vec{r}'|} = \frac{1}{\sqrt{r^2 + r'^2 - 2rr' \cos(\gamma)}} = \sum_{l=0}^\infty \frac{r_<^l}{r_>^{l+1}} P_l(\cos(\gamma)), \tag{12.211}$$

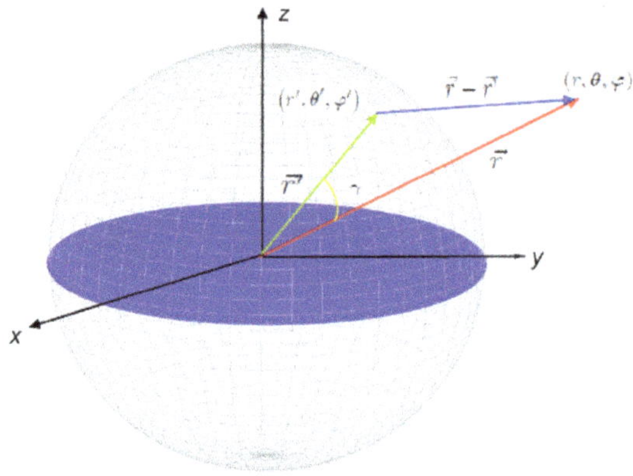

Figure 12.5. Two points described by two vectors, $\vec{r}$ and $\vec{r}'$. In spherical coordinates the two points have coordinates (r, θ, φ) and (r', θ', φ'), respectively. The angle between these two vectors is γ.

where

$$\frac{r_<^l}{r_>^{l+1}} = \begin{cases} \dfrac{r'^l}{r^{l+1}}, & \text{for } r' < r, \\[2mm] \dfrac{r^l}{r'^{l+1}}, & \text{for } r < r'. \end{cases} \tag{12.212}$$

The addition theorem for spherical harmonics
The addition theorem states that the Legendre polynomial $P_l(\cos(\gamma))$ can be expressed in terms of the spherical harmonics,

$$P_l(\cos(\gamma)) = \frac{4\pi}{2l+1} \sum_{m=-l}^{l} Y_{lm}(\theta, \varphi) Y_{lm}^*(\theta', \varphi'), \tag{12.213}$$

where

$$\cos(\gamma) = \cos(\theta)\cos(\theta') + \sin(\theta)\sin(\theta')\cos(\varphi - \varphi'). \tag{12.214}$$

Let us derive equation (12.214). To this end, we describe the two points in figure 12.5, using spherical coordinates by

$$\begin{aligned} \vec{r} &= r\sin(\theta)\cos(\varphi)\hat{x} + r\sin(\theta)\sin(\varphi)\hat{y} + r\cos(\theta)\hat{z}, \\ \vec{r}' &= r'\sin(\theta')\cos(\varphi')\hat{x} + r'\sin(\theta')\sin(\varphi')\hat{y} + r'\cos(\theta')\hat{z}, \end{aligned} \tag{12.215}$$

so that

$$|\vec{r} - \vec{r}'| = \sqrt{(x - x')^2 + (y - y')^2 + (z - z')^2}$$

$$= \{(r\sin(\theta)\cos(\varphi) - r'\sin(\theta')\cos(\varphi'))^2$$

$$+(r\sin(\theta)\sin(\varphi) - r'\sin(\theta')\sin(\varphi'))^2 \tag{12.216}$$

$$+ (r\cos(\theta) - r'\cos(\theta'))^2 \},$$

which leads to

$$|\vec{r} - \vec{r}'| = r^2\{\sin^2(\theta)[\cos^2(\varphi) + \sin^2(\varphi)] + \cos^2(\theta)\}$$

$$+ r'^2\{\sin^2(\theta')[\cos^2(\varphi') + \sin^2(\varphi')] + \cos^2(\theta')\}$$

$$- 2rr'\{\sin(\theta)\sin(\theta')[\cos(\varphi)\cos(\varphi') + \sin(\varphi)\sin(\varphi')] \tag{12.217}$$

$$+ \cos(\theta)\cos(\theta')\}.$$

Applying the relations

$$\cos(\varphi - \varphi') = \cos(\varphi)\cos(\varphi') + \sin(\varphi)\sin(\varphi'),$$

$$\cos^2(\varphi) + \sin^2(\varphi) = 1, \ \sin^2(\theta) + \cos^2(\theta) = 1, \tag{12.218}$$

one finds

$$|\vec{r} - \vec{r}'| = r^2 + r'^2 - 2rr'[\cos(\theta)\cos(\theta') + \sin(\theta)\sin(\theta')\cos(\varphi - \varphi')]. \tag{12.219}$$

This can be put in the form

$$|\vec{r} - \vec{r}'| = r^2 + r'^2 - 2rr'\cos(\gamma), \tag{12.220}$$

where

$$\cos(\gamma) = \cos(\theta)\cos(\theta') + \sin(\theta)\sin(\theta')\cos(\varphi - \varphi'), \tag{12.221}$$

and γ is the angle between the two vectors, $\vec{r}$ and $\vec{r}'$, shown in figure 12.5.

Example 12.8. A solid sphere of radius R has a constant volume charge density ρ. The sphere is centered at the origin of coordinates. The point P is a distance $d > R$ from the center of the sphere. Find an expression for the electrostatic potential at the point P, ϕ_P, due to the charged sphere (figure 12.6).

Solution: The electrostatic potential at point P is given by the integral expression

$$V(\vec{r}) = \frac{1}{4\pi\epsilon_0} \int_{vol} \frac{dq'}{|\vec{r} - \vec{r}'|}. \tag{12.222}$$

For a uniform charge density, ρ, we have

$$\frac{dq'}{d\tau'} = \rho \Rightarrow dq' = \rho d\tau' = \rho r'^2 dr' \sin(\theta') d\theta' d\varphi', \tag{12.223}$$

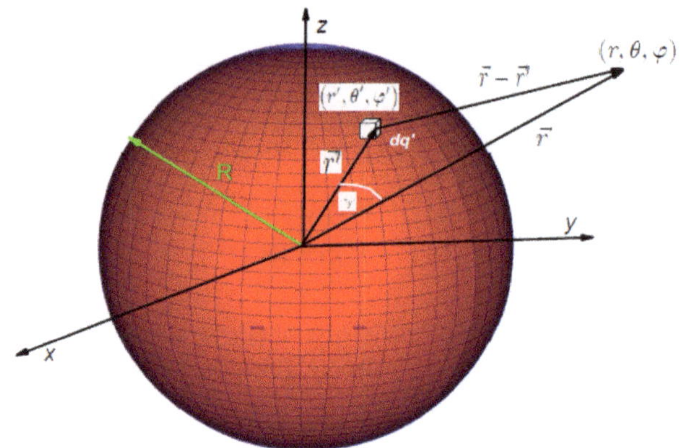

Figure 12.6. A charged solid sphere with uniform charge density, ρ.

where we used spherical coordinates for the infinitesimal value $d\tau'$. Using equation (12.223) the potential can then be expressed as

$$V(\vec{r}) = \frac{\rho}{4\pi\epsilon_0} \int_0^R \int_{\theta=0}^{\pi} \int_{\varphi=0}^{2\pi} \frac{1}{|\vec{r} - \vec{r}'|} r'^2 dr' \sin(\theta') d\theta' d\varphi'. \tag{12.224}$$

The relation

$$\frac{1}{|\vec{r} - \vec{r}'|} = \sum_{l=0}^{\infty} \frac{r_<^l}{r_>^{l+1}} P_l(\cos\gamma), \tag{12.225}$$

for a point outside the sphere ($r = d > r'$) gives

$$\frac{1}{|\vec{r} - \vec{r}'|} = \sum_{l=0}^{\infty} \frac{r'^l}{d^{l+1}} P_l(\cos\gamma). \tag{12.226}$$

Upon substituting equation (12.226) into equation (12.224), we find

$$V(\vec{r}) = \frac{\rho}{4\pi\epsilon_0} \sum_{l=0}^{\infty} \frac{r'^l}{d^{l+1}} \int_0^R \int_{\theta=0}^{\pi} \int_{\varphi=0}^{2\pi} P_l(\cos\gamma) r'^2 dr' \sin(\theta') d\theta' d\varphi'. \tag{12.227}$$

Now applying the addition theorem in equation (12.213), one can rewrite equation (12.227) as

$$V(\vec{r}) = \frac{\rho}{4\pi\epsilon_0} \sum_{l=0}^{\infty} \sum_{m=-l}^{l} \left(\frac{4\pi}{2l+1}\right)$$
$$\times \int_0^R \int_{\theta=0}^{\pi} \int_{\varphi=0}^{2\pi} \frac{r'^l}{d^{l+1}} Y_{lm}(\theta, \varphi) Y_{lm}^*(\theta', \varphi') r'^2 dr' \sin(\theta') d\theta' d\varphi'. \tag{12.228}$$

Since the integration is with respect to the prime variables, equation (12.228) can be put in the form

$$V(\vec{r}) = \frac{\rho}{\epsilon_0} \sum_{l=0}^{\infty} \sum_{m=-l}^{l} \frac{r'^l}{d^{l+1}} \left(\frac{1}{2l+1}\right) Y_{lm}(\theta, \varphi)$$
$$\times \int_0^R \frac{r'^{l+2}}{d^{l+1}} dr' \int_{\theta=0}^{\pi} \int_{\varphi=0}^{2\pi} Y_{lm}^*(\theta', \varphi') \sin(\theta') d\theta' d\varphi'. \tag{12.229}$$

We note that for, $l = m = 0$, the spherical harmonics,

$$Y_{lm}(\theta', \varphi') = (-1)^m \sqrt{\frac{2l+1}{4\pi} \frac{(l-m)!}{(l+m)!}} P_l^m(\cos(\theta')) e^{im\varphi'}, \tag{12.230}$$

become

$$Y_{00}(\theta, \varphi) = \sqrt{\frac{1}{4\pi}} \Rightarrow \sqrt{4\pi} \, Y_{00}(\theta, \varphi) = 1. \tag{12.231}$$

Substituting equation (12.231) into equation (12.229),

$$V(\vec{r}) = \frac{\rho}{\epsilon_0} \sum_{l=0}^{\infty} \sum_{m=-l}^{l} \left(\frac{1}{2l+1}\right) Y_{lm}(\theta, \varphi)$$
$$\times \sqrt{4\pi} \int_0^R \frac{r'^{l+2}}{d^{l+1}} dr' \int_{\theta=0}^{\pi} \int_{\varphi=0}^{2\pi} Y_{00}(\theta', \varphi') Y_{lm}^*(\theta', \varphi') \sin(\theta') d\theta' d\varphi'. \tag{12.232}$$

In view of the orthonormality relation for the spherical harmonics,

$$\langle Y_{lm} \mid Y_{l'm'} \rangle = \int_{\theta=0}^{\pi} \int_{\varphi=0}^{2\pi} Y_{lm}^*(\theta', \varphi') Y_{l'm'}(\theta', \varphi') \sin(\theta') d\theta' d\varphi',$$
$$= \delta_{ll'} \delta_{mm'} \tag{12.233}$$

we have

$$\int_{\theta=0}^{\pi} \int_{\varphi=0}^{2\pi} Y_{00}(\theta', \varphi') Y_{lm}^*(\theta', \varphi') \sin(\theta') d\theta' d\varphi' = \delta_{0l} \delta_{0m}, \tag{12.234}$$

so that equation (12.232) becomes

$$V(\vec{r}) = \frac{\rho\sqrt{4\pi}}{\epsilon_0} \sum_{l=0}^{\infty} \sum_{m=-l}^{l} \left(\frac{1}{2l+1}\right) \delta_{0l} \delta_{0m} Y_{lm}(\theta, \varphi) \int_0^R \frac{r'^{l+2}}{d^{l+1}} dr'. \tag{12.235}$$

Because of the Kronecker delta function, all the terms in the summation in equation (12.235) becomes zero except when $l = 0$ and $m = 0$. Thus one finds

$$V(\vec{r}) = \frac{\rho\sqrt{4\pi}}{\epsilon_0} \frac{Y_{00}(\theta, \varphi)}{d} \int_0^R r'^2 dr' = \frac{\rho\sqrt{4\pi}}{\epsilon_0} \frac{Y_{00}(\theta, \varphi) R^3}{3d}$$
$$\Rightarrow V(\vec{r}) = \frac{\rho R^3}{3\epsilon_0 d}, \tag{12.236}$$

where we replaced $\sqrt{4\pi}\, Y_{00}(\theta, \varphi) = 1$. In terms of the total charge, Q, for a uniform distribution, the charge density is given by

$$\frac{Q}{\frac{4}{3}\pi R^3} = \rho. \tag{12.237}$$

Equation (12.236) can be rewritten as

$$V(\vec{r}) = \frac{Q}{4\pi\epsilon_0 d}, \tag{12.238}$$

which is the potential of a point charge.

12.10 The method of Frobenius and the Bessel equation

When the standard power series solutions fail:

The power series solution

$$y(x) = \sum_{n=0}^{\infty} a_n x^n = a_0 + a_1 x + a_2 x^2 + a_3 x^3 \ldots, \tag{12.239}$$

is applicable to solve a differential equation when the differential equation has no singular point for all $x \in \mathbb{R}$. In some physical problems, we find differential equations that may not be defined for all $x \in \mathbb{R}$. For example, differential equations involving a factor with negative or fractions as an exponent for x are not defined for all $x \in \mathbb{R}$. In such cases, the standard power series solution fails, and we need to use a slightly modified method known as *the method of Frobenius*, which we will see next. However, before we do so, we first consider some simple differential equations where the solutions do not satisfy the standard power series solution.

Example 12.9. Solve the differential equation

$$y' + \frac{y}{x} = 0. \tag{12.240}$$

Solution: First we note that this differential equation is not defined at $x = 0$. But it can be solved easily as follows:

$$y' + \frac{y}{x} = 0 \Rightarrow \frac{dy}{dx} = -\frac{1}{x}y \Rightarrow \frac{dy}{y} = -\frac{dx}{x}$$

$$\Rightarrow \int_{y_0}^{y} \frac{dy}{y} = -\int_{x_0}^{x} \frac{dx}{x} \Rightarrow \ln\left(\frac{y}{y_0}\right) = -\ln\left(\frac{x}{x_0}\right) = \ln\left(\frac{x}{x_0}\right)^{-1} \tag{12.241}$$

$$\Rightarrow \ln\left(\frac{y}{y_0}\right) = \ln\left(\frac{x_0}{x}\right) \Rightarrow \frac{y}{y_0} = \frac{x_0}{x} \Rightarrow y = (x_0 y_0)x^{-1}.$$

Noting that $y_0 x_0 = a_1$ (constant), the solution can be written as

$$y = a_1 x^{-1}, \tag{12.242}$$

which clearly show that both the differential equation and solution involve x with a *negative exponent* and the standard power series expansion does not work!

Example 12.10. Solve the differential equation

$$y' - \frac{3y}{2x} = 0. \tag{12.243}$$

Solution: Here also the differential equation is singular at $x = 0$. The solution is given by

$$y' - \frac{3y}{2x} = 0 \Rightarrow \frac{dy}{dx} = \frac{3}{2}\frac{1}{x}y \Rightarrow \frac{dy}{y} = \frac{3}{2}\frac{dx}{x} \Rightarrow \int_{y_0}^{y} \frac{dy}{y} = \frac{3}{2}\int_{x_0}^{x} \frac{dx}{x}$$

$$\Rightarrow \ln\left(\frac{y}{y_0}\right) = \frac{3}{2}\ln\left(\frac{x}{x_0}\right) = \ln\left(\frac{x}{x_0}\right)^{3/2} \Rightarrow \frac{y}{y_0} = \left(\frac{x}{x_0}\right)^{3/2} \tag{12.244}$$

$$\Rightarrow y = \left(\frac{y_0}{x_0^{3/2}}\right)x^{3/2}.$$

Noting that $y_0/x_0^{3/2} = a_1$ (constant), the solution can be written as

$$y = a_1 x^{3/2} = x^{1/2}(a_1 x), \tag{12.245}$$

which also involves x with a fraction exponent.

The method of Frobenius
When we face differential equations not defined for all $x \in \mathbb{R}$ like the examples we saw above, we use the method of Frobenius. In this method, we use a generalized power series of the form

$$y(x) = x^s \sum_{n=0}^{\infty} a_n x^n = \sum_{n=0}^{\infty} a_n x^{n+s}, \tag{12.246}$$

where s is a number to be determined along with the expansion coefficients, a_n. It may be positive, negative, fraction, or even complex, although we do not consider the complex number case here.

Example 12.11. *The Bessel differential equation*: As an application of the method of Frobenius, we will solve the Bessel differential equation,

$$x^2 y'' + xy' + (x^2 - p^2)y = 0, \tag{12.247}$$

where p is a constant parameter characterizing the differential equation. The Bessel differential equation is derived from the Laplace equation in cylindrical coordinates, which we will see in the next chapter.

Solution: Using a generalized power series in equation (12.246), one can write

$$(x^2 - p^2)y = \sum_{n=0}^{\infty} a_n x^{n+s+2} - \sum_{n=0}^{\infty} p^2 a_n x^{n+s}, \tag{12.248}$$

$$y'(x) = \sum_{n=0}^{\infty} (n+s)a_n x^{n+s-1} \Rightarrow xy'(x) = \sum_{n=0}^{\infty} (n+s)a_n x^{n+s}, \tag{12.249}$$

$$y''(x) = \sum_{n=0}^{\infty} (n+s)(n+s-1)a_n x^{n+s-2}$$

$$\Rightarrow x^2 y''(x) = \sum_{n=0}^{\infty} (n+s)(n+s-1)a_n x^{n+s}. \tag{12.250}$$

Upon substituting equations (12.248)–(12.250) into equation (12.247), we find

$$\sum_{n=0}^{\infty} \{(n+s)(n+s-1) + (n+s) - p^2\} a_n x^{n+s} + \sum_{n=0}^{\infty} a_n x^{n+s+2} = 0$$

$$\Rightarrow \sum_{n=0}^{\infty} [(n+s)^2 - p^2] a_n x^{n+s} + \sum_{n=0}^{\infty} a_n x^{n+s+2} = 0. \tag{12.251}$$

Expanding the first series in equation (12.251), we have

$$\sum_{n=0}^{\infty} [(n+s)^2 - p^2] a_n x^{n+s} = [s^2 - p^2] a_0 x^s + [(1+s)^2 - p^2] a_1 x^{s+1}$$

$$+ \sum_{n=2}^{\infty} [(n+s)^2 - p^2] a_n x^{n+s}, \tag{12.252}$$

so that using transformation of indices, $n = m + 2$ and noting that for $n = 2 \Rightarrow m = 0$, one can put equation (12.252) in the form

$$\sum_{n=0}^{\infty} [(n+s)^2 - p^2] a_n x^{n+s} = [s^2 - p^2] a_0 x^s + [(1+s)^2 - p^2] a_1 x^{s+1}$$

$$+ \sum_{m=0}^{\infty} [(m+2+s)^2 - p^2] a_{m+2} x^{m+s+2}. \tag{12.253}$$

Since m is a dummy variable, we can replace it with n and rewrite equation (12.253) as

$$\sum_{n=0}^{\infty} [(n+s)^2 - p^2] a_n x^{n+s} = [s^2 - p^2] a_0 x^s + [(1+s)^2 - p^2] a_1 x^{s+1}$$

$$+ \sum_{n=0}^{\infty} [(n+2+s)^2 - p^2] a_{n+2} x^{n+s+2}. \tag{12.254}$$

Now substituting equation (12.254) into equation (12.251), we find

$$[s^2 - p^2]a_0 x^s + [(1 + s)^2 - p^2]a_1 x^{s+1}$$
$$+ \sum_{n=0}^{\infty} \{[(n + 2 + s)^2 - p^2]a_{n+2} + a_n\} x^{n+s+2} = 0. \tag{12.255}$$

For equation (12.253) to be true, the coefficients in equation (12.255) must be zero

$$[s^2 - p^2]a_0 = 0 \Rightarrow s = \pm p \tag{12.256}$$

$$[(1 + s)^2 - p^2]a_1 = 0 \Rightarrow [(1 \pm p)^2 - p^2]a_1 = 0$$
$$\Rightarrow [1 \pm 2p]a_1 = 0 \Rightarrow a_1 = 0, \tag{12.257}$$

$$[(n + 2 + s)^2 - p^2]a_{n+2} + a_n = 0 \Rightarrow a_{n+2} = -\frac{1}{(n + 2 + s)^2 - p^2}a_n. \tag{12.258}$$

The result in equations (12.256)–(12.258) show that there are two different resulting functions when we substitute $s = p$ and $s = -p$ into the recursion relation in equation (12.258).

First solution ($s = p$): For this case, the recursion formula in equation (12.258) becomes

$$a_{n+2} = -\frac{a_n}{(n + 2 + p)^2 - p^2} = -\frac{a_n}{(n + 2)(n + 2 + 2p)}. \tag{12.259}$$

Since we found $a_1 = 0$ all the odd terms in the power series vanish and the recursion relation can be expressed, for the even terms, using $n + 2 = 2m \Rightarrow n = 2m - 2$, as

$$a_{2m} = -\frac{a_{2m-2}}{(2m)(2m + 2p)} = -\frac{a_{2m-2}}{2^2 m(m + p)}, \tag{12.260}$$

where $m = 1, 2, 3....$ Next, we shall evaluate equation (12.260) for the first few terms,

$$m = 1 \Rightarrow a_2 = -\frac{a_0}{2^2 1!(1 + p)},$$
$$m = 2 \Rightarrow a_4 = -\frac{a_2}{2^2 2(2 + p)} = \frac{a_0}{2^4 2!(1 + p)(2 + p)}, \tag{12.261}$$
$$m = 3 \Rightarrow a_6 = -\frac{a_4}{2^2 3(3 + p)} = -\frac{a_0}{2^6 3!(1 + p)(2 + p)(3 + p)}.$$

Recalling that the gamma function

$$\Gamma(p + 1) = p\Gamma(p),$$
$$\Gamma(p + 2) = (p + 1)\Gamma(p + 1) = p(p + 1)\Gamma(p), \tag{12.262}$$
$$\Gamma(p + 3) = (p + 2)\Gamma(p + 2) = p(p + 1)(p + 2)\Gamma(p),$$

we can rewrite equation (12.261) as

$$a_2 = -\frac{a_0}{2^2 1!(1+p)} = -\frac{a_0 p\Gamma(p)}{2^2 1! p(1+p)\Gamma(p)} \Rightarrow a_2 = -\frac{a_0\Gamma(p+1)}{2^2 1!\Gamma(p+2)}, \qquad (12.263)$$

$$a_4 = \frac{a_0}{2^4 2!(1+p)(2+p)} = \frac{a_0 p\Gamma(p)}{2^4 2! p(1+p)(2+p)\Gamma(p)}$$
$$\Rightarrow a_4 = \frac{a_0\Gamma(p+1)}{2^4 2!\Gamma(p+3)}, \qquad (12.264)$$

$$a_6 = -\frac{a_0}{2^6 3!(1+p)(2+p)(3+p)} = -\frac{a_0 p\Gamma(p)}{2^6 3! p(1+p)(2+p)(3+p)\Gamma(p)}$$
$$\Rightarrow a_6 = -\frac{a_0\Gamma(p+1)}{2^6 3!\Gamma(p+4)}. \qquad (12.265)$$

Therefore, the solution for *the Bessel differential equation*

$$y(x) = x^s \sum_{n=0}^{\infty} a_n x^n = x^s \sum_{m=0}^{\infty} a_{2m} x^{2m}, \qquad (12.266)$$

when $s = p$, using equations (12.263)–(12.265), can be put in the form

$$y(x) = x^p\left\{ a_0 - \frac{a_0\Gamma(p+1)}{2^2 1!\Gamma(p+2)}x^2 + \frac{a_0\Gamma(p+1)}{2^4 2!\Gamma(p+3)}x^4 - \frac{a_0\Gamma(p+1)}{2^6 3!\Gamma(p+4)}x^6 + \ldots \right\}$$
$$= a_0 x^p \Gamma(p+1)\left\{ \frac{1}{0!\Gamma(p+1)} - \frac{1}{1!\Gamma(p+2)}\left(\frac{x}{2}\right)^2 + \frac{1}{2!\Gamma(p+3)}\left(\frac{x}{2}\right)^4 \right. \qquad (12.267)$$
$$\left. - \frac{1}{3!\Gamma(p+4)}\left(\frac{x}{2}\right)^6 \ldots + \frac{(-1)^n}{n!\Gamma(p+n+1)}\left(\frac{x}{2}\right)^{2n} \ldots \right\}.$$

Recalling that the factorial and the gamma functions are related by

$$\Gamma(1) = \Gamma(2) = 1, \quad n! = \Gamma(n+1), \qquad (12.268)$$

equation (12.267) can be rewritten as

$$y(x) = a_0 2^p \Gamma(p+1)\left(\frac{x}{2}\right)^p \left\{ \frac{1}{\Gamma(1)\Gamma(p+1)} - \frac{1}{\Gamma(2)\Gamma(p+2)}\left(\frac{x}{2}\right)^2 \right.$$
$$+ \frac{1}{\Gamma(3)\Gamma(p+3)}\left(\frac{x}{2}\right)^4 - \frac{1}{\Gamma(4)\Gamma(p+4)}\left(\frac{x}{2}\right)^6 \ldots$$
$$\left. + \frac{(-1)^n}{n!\Gamma(p+n+1)}\left(\frac{x}{2}\right)^{2n} + \ldots \right\} \qquad (12.269)$$
$$\Rightarrow y(x) = a_0 2^p \Gamma(p+1) \sum_{n=0}^{\infty} \frac{(-1)^n}{\Gamma(n+1)\Gamma(n+1+p)}\left(\frac{x}{2}\right)^{2n+p}.$$

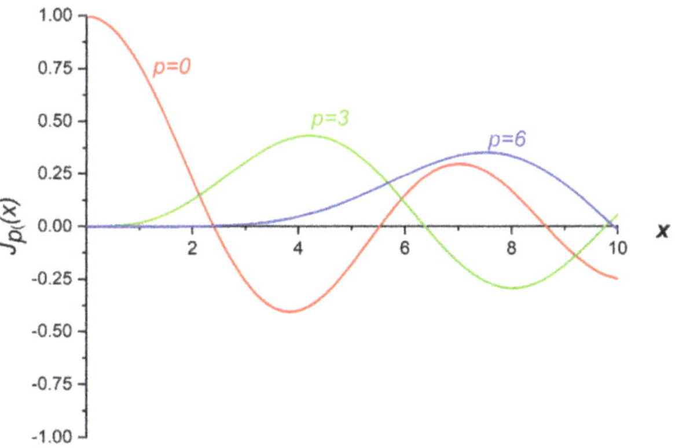

Figure 12.7. Bessel function, $J_p(x)$, for $p = 0, 3, 6$.

The function $J_p(x)$ defined by

$$J_p(x) = \frac{y(x)}{a_0 2^p \Gamma(p + 1)} = \sum_{n=0}^{\infty} \frac{(-1)^n}{\Gamma(n + 1)\Gamma(n + 1 + p)}\left(\frac{x}{2}\right)^{2n+p}, \qquad (12.270)$$

is known as the *Bessel function of the first kind* of order p. Figure 12.7 shows $J_p(x)$ when $p = 0, 3, 6$.

Second Solution $s = -p$: The second solution, when $s = -p$, can easily be obtained from the first solution by replacing p with $-p$,

$$J_{-p}(x) = \sum_{n=0}^{\infty} \frac{(-1)^n}{\Gamma(n + 1)\Gamma(n + 1 - p)}\left(\frac{x}{2}\right)^{2n-p}. \qquad (12.271)$$

If p is not an integer, $J_p(x)$ is a series starting with x^p and $J_{-p}(x)$ is a series starting with x^{-p}. Then $J_p(x)$ and $J_{-p}(x)$ are two independent solutions and the linear combination of these two functions is the general solution. The combination is called the *Neumann* or *Weber* function denoted by either $N_p(x)$ or $Y_p(x)$ and is given by

$$N_p(x) = Y_p(x) = \frac{\cos(\pi p)J_p(x) - J_{-p}(x)}{\sin(\pi p)}. \qquad (12.272)$$

However, if p is an integer, then the first few terms in $J_{-p}(x)$ are zero because $\Gamma(n + 1 - p)$ is the gamma function of a negative integer, which is infinite. As a result, one finds

$$J_{-p}(x) = (-1)^p J_p(x). \qquad (12.273)$$

This shows that $J_{-p}(x)$ and $J_p(x)$ are not independent solutions.

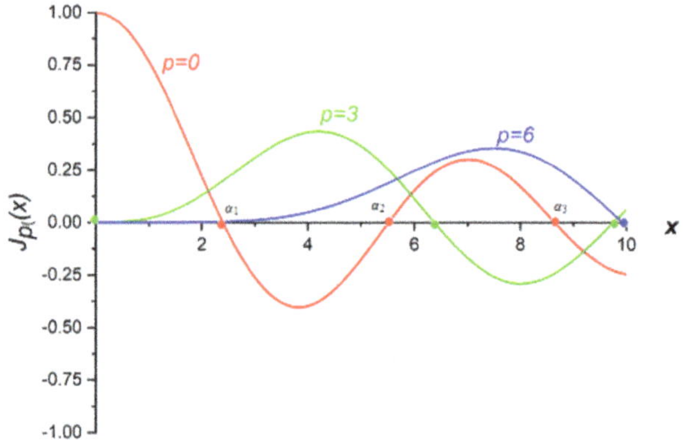

Figure 12.8. The zeroes of the Bessel function (the dots on the x-axis).

The zeroes of the Bessel function

The zeroes of the Bessel function $\{\alpha_1, \alpha_2, \alpha_3, ...\}$ are the values of x at which

$$J_p(x) = 0. \tag{12.274}$$

In figure 12.8 the points on the x-axis where the $J_p(x)$ intersects (i.e., $J_p(x) = 0$) are the zeroes of the Bessel function.

12.11 The orthogonality of the Bessel functions

In the previous section we solved the Bessel differential equation

$$x^2 y'' + xy' + (x^2 - p^2)y = 0. \tag{12.275}$$

We saw that the solution can be expressed in terms of the Bessel function $J_p(x)$. Next, we determine the orthonormality relation for the Bessel functions. To this end, noting that

$$x^2 y'' + xy' = x(xy'' + y') = x\frac{d}{dx}\left(x\frac{dy}{dx}\right), \tag{12.276}$$

we can rewrite the Bessel equation as

$$x\frac{d}{dx}\left(x\frac{dy}{dx}\right) + (x^2 - p^2)y = 0, \tag{12.277}$$

the solution of which is given by the Bessel functions, $y(x) = J_p(x)$. Thus for the Bessel function, $J_p(u)$, one can write

$$u\frac{d}{du}\left(u\frac{dJ_p(u)}{du}\right) + (u^2 - p^2)J_p(u) = 0. \tag{12.278}$$

We introduce the transformation of variable defined by

$$u = \alpha_i x \Rightarrow du = \alpha_i dx, \tag{12.279}$$

where

$$\alpha_i = \{\alpha_1, \alpha_2, \alpha_3, \ldots\}, \tag{12.280}$$

are the zeroes of the Bessel function. Using equation (12.279), one can rewrite equation (12.278) as

$$x\frac{d}{dx}\left(x\frac{dJ_p(\alpha_i x)}{dx}\right) + (\alpha_i^2 x^2 - p^2)J_p(\alpha_i x) = 0. \tag{12.281}$$

Replacing α_i with α_j, we can also write

$$x\frac{d}{dx}\left(x\frac{dJ_p(\alpha_j x)}{dx}\right) + \left(\alpha_j^2 x^2 - p^2\right)J_p(\alpha_j x) = 0. \tag{12.282}$$

Multiplying equation (12.281) by $J_p(\alpha_j x)$ and equation (12.282) by $-J_p(\alpha_i x)$, we have

$$xJ_p(\alpha_j x)\frac{d}{dx}\left(x\frac{dJ_p(\alpha_i x)}{dx}\right) + (\alpha_i^2 x^2 - p^2)J_p(\alpha_j x)J_p(\alpha_i x) = 0, \tag{12.283}$$

$$-xJ_p(\alpha_i x)\frac{d}{dx}\left(x\frac{dJ_p(\alpha_j x)}{dx}\right) - \left(\alpha_j^2 x^2 - p^2\right)J_p(\alpha_i x)J_p(\alpha_j x) = 0, \tag{12.284}$$

so that upon adding these two equations, one finds

$$J_p(\alpha_j x)\frac{d}{dx}\left(x\frac{d}{dx}J_p(\alpha_i x)\right) - J_p(\alpha_i x)\frac{d}{dx}\left(x\frac{d}{dx}J_p(\alpha_j x)\right)$$
$$+ \left(\alpha_i^2 - \alpha_j^2\right)J_p(\alpha_i x)J_p(\alpha_j x)x = 0. \tag{12.285}$$

Integrating equation (12.285) with respect to x from 0 to 1,

$$\int_0^1 J_p(\alpha_j x)\frac{d}{dx}\left(x\frac{d}{dx}J_p(\alpha_i x)\right)dx - \int_0^1 J_p(\alpha_i x)\frac{d}{dx}\left(x\frac{d}{dx}J_p(\alpha_j x)\right)dx$$
$$+ \left(\alpha_i^2 - \alpha_j^2\right)\int_0^1 J_p(\alpha_i x)J_p(\alpha_j x)xdx = 0. \tag{12.286}$$

Applying integration by parts,

$$\int_0^1 u\frac{dv}{dx}dx = uv\,|_0^1 - \int_0^1 v\frac{du}{dx}dx, \tag{12.287}$$

one can write

$$\int_0^1 J_p(\alpha_j x)\frac{d}{dx}\left(x\frac{d}{dx}J_p(\alpha_i x)\right)dx = J_p(\alpha_j x)x\frac{d}{dx}(J_p(\alpha_i x))\ \Big|_0^1 \tag{12.288}$$

$$-\int_0^1 x\frac{d}{dx}(J_p(\alpha_i x))\frac{d}{dx}(J_p(\alpha_j x))dx,$$

$$\int_0^1 J_p(\alpha_i x)\frac{d}{dx}\left(x\frac{d}{dx}J_p(\alpha_j x)\right) = J_p(\alpha_i x)x\frac{d}{dx}(J_p(\alpha_j x))\ \Big|_0^1 \tag{12.289}$$

$$-\int_0^1 x\frac{d}{dx}(J_p(\alpha_j x))\frac{d}{dx}(J_p(\alpha_i x))dx.$$

Noting that

$$\begin{aligned} x = 0 &\Rightarrow xJ_p(\alpha_i x) = 0, \\ x = 1 &\Rightarrow J_p(\alpha_i x)x = J_p(\alpha_i) = 0, \end{aligned} \tag{12.290}$$

since α_i & α_j are the zeroes of the Bessel function, equations (12.288) and (12.289) reduce to

$$\int_0^1 J_p(\alpha_j x)\frac{d}{dx}\left(x\frac{d}{dx}J_p(\alpha_i x)\right)dx = -\int_0^1 x\frac{d}{dx}(J_p(\alpha_i x))\frac{d}{dx}(J_p(\alpha_j x))dx, \tag{12.291}$$

$$\int_0^1 J_p(\alpha_i x)\frac{d}{dx}\left(x\frac{d}{dx}J_p(\alpha_j x)\right) = -\int_0^1 x\frac{d}{dx}(J_p(\alpha_j x))\frac{d}{dx}(J_p(\alpha_i x))dx. \tag{12.292}$$

Now substituting equations (12.291) and (12.292) into equation (12.286), one finds

$$\left(\alpha_i^2 - \alpha_j^2\right)\int_0^1 J_p(\alpha_i x)J_p(\alpha_j x)xdx = 0. \tag{12.293}$$

There follows that, for $i \neq j$, since

$$\alpha_i^2 - \alpha_j^2 \neq 0, \tag{12.294}$$

we must have

$$\int_0^1 J_p(\alpha_i x)J_p(\alpha_j x)xdx = 0. \tag{12.295}$$

But for $i = j$, since

$$\alpha_i^2 - \alpha_j^2 = \alpha_i^2 - \alpha_i^2 = 0, \tag{12.296}$$

we should generally expect that

$$\int_0^1 J_p(\alpha_i x)J_p(\alpha_j x)xdx = \int_0^1 J_p^2(\alpha_i x)xdx = \text{Constant.} \tag{12.297}$$

Therefore, the orthogonality relation for the Bessel functions can be written as

$$\int_0^1 J_p(\alpha_i x) J_p(\alpha_j x) x \, dx = \begin{cases} 0, & i \neq j \\ C, & i = j \end{cases} = \delta_{ij} C. \tag{12.298}$$

To find the constant C, we now consider the integral

$$I = \int_0^1 J_p(\alpha_i x) J_p(\alpha_j x) x \, dx, \tag{12.299}$$

where α_i (which is not the zeroes of the Bessel function) but can be expressed in terms of α_j (which is the zeroes of the Bessel function) by

$$\alpha_i = \alpha_j + \epsilon \Rightarrow \alpha_i^2 - \alpha_j^2 = (\alpha_j + \epsilon)^2 - \alpha_j^2 = 2\alpha_j\epsilon + \epsilon^2 \simeq 2\alpha_j\epsilon, \tag{12.300}$$

for $\epsilon < < 1$, is a small constant which we set to zero to make $\alpha_i = \alpha_j$ for $i = j$.

It is important to note that (see figure 12.9),

$$J_p(\alpha_j) = 0, \quad J_p(\alpha_j + \epsilon) \neq 0, \tag{12.301}$$

but when $\epsilon \to 0$,

$$\lim_{\epsilon \to 0} J_p(\alpha_j + \epsilon) = J_p(\alpha_j) = 0. \tag{12.302}$$

Then for $\alpha_i = \alpha_j + \epsilon$, equations (12.288) and (12.289) become

$$\begin{aligned}
\int_0^1 J_p(\alpha_j x) \frac{d}{dx}\left(x\frac{d}{dx}J_p((\alpha_j + \epsilon)x)\right) dx &= J_p(\alpha_j x) x \frac{d}{dx}\left(J_p((\alpha_j + \epsilon)x)\right)\Big|_0^1 \\
&\quad - \int_0^1 x\frac{d}{dx}\left(J_p((\alpha_j + \epsilon)x)\right)\frac{d}{dx}\left(J_p(\alpha_j x)\right) dx \\
&= -\int_0^1 x\frac{d}{dx}\left(J_p((\alpha_j + \epsilon)x)\right)\frac{d}{dx}\left(J_p(\alpha_j x)\right) dx
\end{aligned} \tag{12.303}$$

and

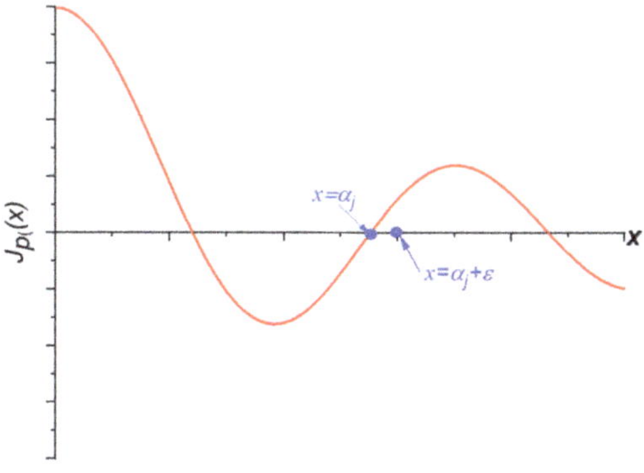

Figure 12.9. The Bessel function at two different values: $J_p(\alpha_j x)$ and $J_p((\alpha_j + \epsilon)x)$ where α_j is the zeroes of the Bessel function.

$$\int_0^1 J_p((\alpha_j + \epsilon)x)\frac{d}{dx}\left(x\frac{d}{dx}J_p(\alpha_j x)\right) = J_p((\alpha_j + \epsilon)x)x\frac{d}{dx}(J_p(\alpha_j x)) \Big|_0^1$$

$$- \int_0^1 x\frac{d}{dx}(J_p(\alpha_j x))\frac{d}{dx}(J_p((\alpha_j + \epsilon)x))dx$$

$$= J_p((\alpha_j + \epsilon))\left[\frac{d}{dx}(J_p(\alpha_j x))\right]_{x=1} - \int_0^1 x\frac{d}{dx}(J_p(\alpha_j x))\frac{d}{dx}$$

$$(J_p((\alpha_j + \epsilon)x))dx.$$

$$(12.304)$$

Noting that equation (12.286), for $\alpha_i = \alpha_j + \epsilon$, becomes

$$\int_0^1 J_p(\alpha_j x)\frac{d}{dx}\left(x\frac{d}{dx}J_p((\alpha_j + \epsilon)x)\right)dx$$

$$- \int_0^1 J_p((\alpha_j + \epsilon)x)\frac{d}{dx}\left(x\frac{d}{dx}J_p(\alpha_j x)\right)dx$$

$$+ ((\alpha_j + \epsilon)^2 - \alpha_j^2)\int_0^1 J_p((\alpha_j + \epsilon)x)J_p(\alpha_j x)xdx = 0,$$

$$(12.305)$$

and using the results in equations (12.300), (12.303) and (12.304), one finds

$$-J_p((\alpha_j + \epsilon))\left[\frac{d}{dx}(J_p(\alpha_j x))\right]_{x=1} + 2\alpha_j\epsilon\int_0^1 J_p((\alpha_j + \epsilon)x)J_p(\alpha_j x)xdx = 0. \quad (12.306)$$

Noting that

$$\frac{d}{dx}(J_p(\alpha_j x))\Big|_{x=1} = \alpha_j\frac{dJ_p(x)}{dx}\Big|_{x=\alpha_j} = \alpha_j J_p'(\alpha_j), \qquad (12.307)$$

we can rewrite equation (12.306) as

$$\int_0^1 J_p((\alpha_j + \epsilon)x)J_p(\alpha_j x)xdx = \frac{\alpha_j J_p((\alpha_j + \epsilon))J_p'(\alpha_j)}{2\epsilon\alpha_j}. \qquad (12.308)$$

We recall the Taylor series expansion for $f(\epsilon)$ about $\epsilon = 0$, is given by

$$f(\epsilon) = f(\epsilon)\,|_{\epsilon=0} + \epsilon\frac{df(\epsilon)}{d\epsilon}\Big|_{\epsilon=0} + \frac{\epsilon^2}{2!}\frac{d^2f(\epsilon)}{d\epsilon^2}\Big|_{\epsilon=0} + ..., \qquad (12.309)$$

and for $f(\epsilon) = J_p(\alpha_j + \epsilon)$, using

$$f(\epsilon)\,|_{\epsilon=0} = J_p(\alpha_j + \epsilon)\,\Big|_{\epsilon=0} = J_p(\alpha),$$

$$\frac{df(\epsilon)}{d\epsilon}\Big|_{\epsilon=0} = \frac{dJ_p(\alpha_j + \epsilon)}{d\epsilon}\Big|_{\epsilon=0} = \frac{dJ_p(x)}{dx}\Big|_{x=\alpha_j} = J_p'(\alpha_j),$$

$$(12.310)$$

one finds

$$J_p(\alpha_j + \epsilon) = J_p(\alpha_j) + \epsilon J_p'(\alpha_j) + \frac{\epsilon^2}{2!}J_p''\alpha_j + \ldots = \epsilon\left[J_p'(\alpha_j) + \frac{\epsilon}{2!}J_p''(\alpha_j) + \ldots\right], \quad (12.311)$$

where we used $J_p(\alpha_j) = 0$, since α_j is the zeroes of the Bessel function. Substituting equation (12.311) into equation (12.308), we have

$$\int_0^1 J_p\big((\alpha_j + \epsilon)x\big)J_p(\alpha_j x)x\,dx = \frac{\alpha_j J_p'(\alpha_j)\left[J_p'(\alpha_j) + \dfrac{\epsilon}{2!}J_p''(\alpha_j) + \ldots\right]}{2\alpha_j}, \quad (12.312)$$

so that in the limit as $\epsilon \to 0$, one finds

$$\int_0^1 J_p(\alpha_j x)J_p(\alpha_j x)x\,dx = \frac{J_p'(\alpha_j)J_p'(\alpha_j)}{2}. \quad (12.313)$$

Therefore, in view of the results in equations (12.298) and (12.313), one can establish, for the Bessel functions, the orthogonality condition

$$\int_0^1 J_p(\alpha_i x)J_p(\alpha_j x)x\,dx = \frac{J_p'(\alpha_i)J_p'(\alpha_j)}{2}\delta_{ij}. \quad (12.314)$$

Limiting (asymptotic) forms for the Bessel functions

The Bessel functions have different simpler approximate expressions for different limiting cases. Next we shall consider some of the approximations for the Bessel function for $x < <1$ and $x > >1$.

(a) $x < <1$: We recall the Bessel function,

$$J_p(x) = \left(\frac{x}{2}\right)^p\left\{\frac{1}{\Gamma(1)\Gamma(p+1)} - \frac{1}{\Gamma(2)\Gamma(p+2)}\left(\frac{x}{2}\right)^2 \right.$$
$$\left. + \frac{1}{\Gamma(3)\Gamma(p+3)}\left(\frac{x}{2}\right)^4 - \frac{1}{\Gamma(4)\Gamma(p+4)}\left(\frac{x}{2}\right)^6 + \ldots\right\}, \quad (12.315)$$

so that for $x < <1$, dropping all higher order terms starting from $\left(\frac{x}{2}\right)^2$, we have

$$J_p(x) \simeq \left(\frac{x}{2}\right)^p\frac{1}{\Gamma(1)\Gamma(p+1)}, \quad (12.316)$$

and using the relations

$$\Gamma(1) = 1, \ \Gamma(p+1) = p! \quad (12.317)$$

one finds

$$J_p(x) \simeq \frac{1}{p!}\left(\frac{x}{2}\right)^p.$$ (12.318)

Replacing p by $-p$, in equation (12.315), one can write for $J_{-p}(x)$

$$J_{-p}(x) = \left(\frac{2}{x}\right)^p \left\{ \frac{1}{\Gamma(1)\Gamma(1-p)} - \frac{1}{\Gamma(2)\Gamma(2-p)}\left(\frac{x}{2}\right)^2 \right.$$
$$+ \frac{1}{\Gamma(3)\Gamma(3-p)}\left(\frac{x}{2}\right)^4 - \frac{1}{\Gamma(4)\Gamma(4-p)}\left(\frac{x}{2}\right)^6 + \dots$$ (12.319)
$$\left. + \frac{(-1)^n}{\Gamma(n+1)\Gamma(n+1-p)}\left(\frac{x}{2}\right)^{2n} \dots \right\}.$$

Since, generally, the gamma function

$$\Gamma(p+1) = \int_0^\infty x^p e^{-x} dx,$$ (12.320)

for $p + 1 < 1$ $(p < 0)$ becomes infinity, for $x < <1$ in the series in equation (12.319) we must have

$$n + 1 - p = 1 \Rightarrow n = p \Rightarrow \Gamma(n+1-p) = \Gamma(1) = 0! = 1,$$

$$\Rightarrow J_{-p}(x) = \left(\frac{2}{x}\right)^p \left\{ \frac{1}{\Gamma(1)\Gamma(1-p)} - \frac{1}{\Gamma(2)\Gamma(2-p)}\left(\frac{x}{2}\right)^2 \right.$$
$$+ \frac{1}{\Gamma(3)\Gamma(3-p)}\left(\frac{x}{2}\right)^4 - \frac{1}{\Gamma(4)\Gamma(4-p)}\left(\frac{x}{2}\right)^6 + \dots$$ (12.321)
$$\left. \dots + \frac{(-1)^p}{\Gamma(p)}\left(\frac{x}{2}\right)^{2p} \right\}.$$

Using equation (12.321), for $x < <1$, one can show that

$$\begin{cases} J_{-p}(x) \simeq 1, & p = 0 \\ J_{-p}(x) \simeq \left(\frac{2}{x}\right)^p \frac{1}{\Gamma(p)}, & p > 0 \end{cases}$$ (12.322)

Then the *Neumann* function

$$N_p(x) = \frac{\cos(\pi p) J_p(x) - J_{-p}(x)}{\sin(\pi p)},$$ (12.323)

becomes

$$\begin{cases} N_p(x) \simeq \frac{2}{\pi} \ln x, & p = 0. \\ N_p(x) \simeq -\frac{(p-1)!}{\pi}\left(\frac{2}{x}\right)^p = \frac{\Gamma(p)}{\pi}\left(\frac{2}{x}\right)^p, & p > 0. \end{cases}$$ (12.324)

(b) For $x > > 1$

$$J_p(x) \simeq \sqrt{\frac{2}{\pi x}} \cos\left(x - \frac{2p+1}{4}\pi\right), \; N_p(x) \simeq \sqrt{\frac{2}{\pi x}} \sin\left(x - \frac{2p+1}{4}\pi\right). \qquad (12.325)$$

Many differential equations occur in practice that are not of the standard form of the Bessel differential equation in equation (12.247). However, the solution can be expressed in terms of the Bessel functions. For example, the differential equation

$$y'' + \frac{1-2a}{x}y' + \left[(bcx^{c-1})^2 + \frac{a^2 - p^2 c^2}{x^2}\right]y = 0, \qquad (12.326)$$

has the solution

$$y = x^a Z_p(bx^c), \qquad (12.327)$$

where Z_p stands for J_p or N_p or any linear combination, and a, b, c, p are constants.

Example 12.12. *The lengthening pendulum (Bessel functions, an application)*: Consider a simple pendulum that consists of a mass m attached to a string of length l_0 at the initial time $t = 0$. The length of the pendulum is increasing at a steady rate,

$$\dot{l} = \frac{dl}{dt} = v. \qquad (12.328)$$

Find the equation of motion and determine the solution for small oscillation.

Solution: At a given time, t, let the length of the pendulum be,

$$l(t) = \sqrt{x(t)^2 + y(t)^2}. \qquad (12.329)$$

Referring to figure 12.10, the kinetic (KE) and the gravitational potential (U) energies can be expressed as

$$U = -mg[l \cos(\theta(t)) - l_0], \; KE = \frac{1}{2}m(l^2\dot{\theta}^2 + \dot{l}^2\theta^2). \qquad (12.330)$$

Then the Lagrangian becomes

$$L = KE - U = \frac{1}{2}m(l^2\dot{\theta}^2 + \dot{l}^2) + mg[l \cos(\theta) - l_0]. \qquad (12.331)$$

Using Euler–Lagrange equations

$$\frac{d}{dt}\left(\frac{\partial L}{\partial \dot{\theta}}\right) - \frac{\partial L}{\partial \theta} = 0, \qquad (12.332)$$

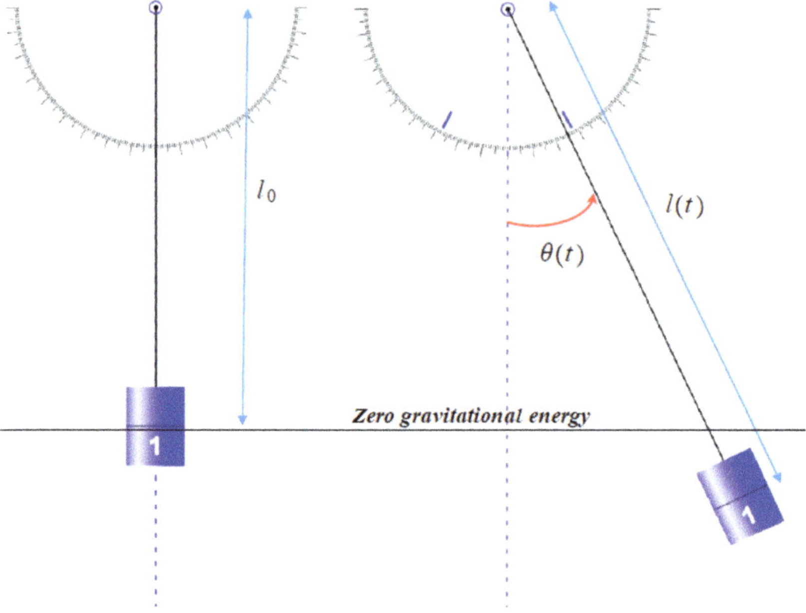

Figure 12.10. The lengthening pendulum.

we find

$$\frac{d}{dt}(ml^2\dot{\theta}) + mgl\,\sin(\theta) = 0 \Rightarrow \frac{d}{dt}(l^2\dot{\theta}) + gl\,\sin(\theta) = 0. \tag{12.333}$$

The length is increasing at a constant rate v,

$$\dot{l} = v \Rightarrow dt = \frac{dl}{v}, \tag{12.334}$$

then transforming the DE from t to l,

$$v\frac{d}{dl}\left(l^2 v\frac{d\theta}{dl}\right) + gl\,\sin(\theta) = 0 \Rightarrow l^2 v^2\frac{d^2\theta}{dl^2} + 2lv^2\frac{d\theta}{dl} + gl\,\sin(\theta) = 0 \tag{12.335}$$

$$\Rightarrow \frac{d^2\theta}{dl^2} + \frac{2}{l}\frac{d\theta}{dl} + \frac{g}{lv^2}\sin(\theta) = 0.$$

For small angle θ, we have $\sin(\theta) \simeq \theta$ and equation (12.335) becomes

$$\frac{d^2\theta}{dl^2} + \frac{2}{l}\frac{d\theta}{dl} + \frac{g}{lv^2}\theta = 0. \tag{12.336}$$

We now consider the differential equation (12.326)

$$\frac{d^2y}{dx^2} + \frac{1 - 2a}{x}\frac{dy}{dx} + \left[(bcx^{c-1})^2 + \frac{a^2 - p^2c^2}{x^2}\right]y = 0 \tag{12.337}$$

the solution of which is given, in terms of the Bessel function, by

$$y(x) = x^a Z_p(bx^c). \tag{12.338}$$

Replacing x by l and y by θ, one can rewrite equations (12.337) and (12.338) as

$$\frac{d^2\theta}{dl^2} + \frac{1-2a}{x}\frac{d\theta}{dl} + \left[(bcl^{c-1})^2 + \frac{a^2 - p^2 c^2}{l^2}\right]\theta = 0, \tag{12.339}$$

$$\theta(l) = l^a Z_p(bl^c). \tag{12.340}$$

Comparing equation (12.339) with equation (12.336), one can establish the relations

$$\frac{1-2a}{l} = \frac{2}{l}, \ (bcl^{c-1})^2 + \frac{a^2 - p^2 c^2}{l^2} = \frac{g}{lv^2}$$

$$a = -\frac{1}{2}, \ c = \frac{1}{2}, \ b = 2\sqrt{\frac{g}{v^2}}, \ p^2 = \frac{a^2}{c^2} \Rightarrow p = \pm 1. \tag{12.341}$$

Upon substituting the values in equation (12.341) into equation (12.340), one finds

$$\theta(l) = \frac{1}{\sqrt{l}} Z_1\left(2\frac{\sqrt{gl}}{v}\right), \tag{12.342}$$

which is the solution to the differential equation we found for the lengthening pendulum. Note that Z_1 is the linear superposition of the first and second kind of the Bessel function

$$Z_1\left(2\frac{\sqrt{gl}}{v}\right) = C_1 J_p\left(2\frac{\sqrt{gl}}{v}\right) + C_2 N_p\left(2\frac{\sqrt{gl}}{v}\right)$$

$$\Rightarrow \theta(l) = \frac{1}{\sqrt{l}}\left[C_1 J_p\left(2\frac{\sqrt{gl}}{v}\right) + C_2 N_p\left(2\frac{\sqrt{gl}}{v}\right)\right]. \tag{12.343}$$

The constants C_1 and C_2 are determined from the initial condition. In this case, at $t = 0$, we can take $\theta(0) = \theta_0$ and $\dot{\theta}(0) = 0$ and find the constants after some algebra that we will not do here.

12.12 Fuch's theorem

So far, what we have seen is how a second-order differential equation of the form,

$$y'' + f(x)y' + g(x)y = 0, \tag{12.344}$$

can be solved using the standard power series expansion method (e.g., the Legendre equation) or the more general Frobenius method (e.g., the Bessel equation). If we think that the Frobenius method can solve any second-order differential equation, then we are wrong. A necessary condition must be satisfied if the differential equation is solvable using *the Frobenius method. Fuch's theorem* determines the necessary and sufficient condition to apply the Frobenius method. It states that for the differential equation

$$y'' + f(x)y' + g(x)y = 0, \tag{12.345}$$

one can determine the solution using the Frobenius method when the functions defined by $F(x) = f(x)y'(x)$ and $G(x) = g(x)y(x)$ are expressible as power series (i.e., a convergent power series)

$$F(x) = f(x)y' = \sum_{n=0}^{\infty} a_n x^n, \quad G(x) = g(x)y = \sum_{n=0}^{\infty} b_n x^n. \tag{12.346}$$

When this condition is met, the solutions to the differential equation consist of either two independent Frobenius series

$$y_1(x) = \sum_{n=0}^{\infty} A_n x^{n+s}, \quad y_2(x) = \sum_{n=0}^{\infty} B_n x^{n+s}, \tag{12.347}$$

or only one Frobenius series, $y_1(x)$. The second solution can be expressed as

$$y_2(x) = y_1(x) \ln x + \sum_{n=0}^{\infty} B_n x^{n+s}, \tag{12.348}$$

when the values of s are equal or differ by an integer.

Example 12.13. Consider the differential equation

$$y'' + \frac{1}{x^2}y' - \frac{2}{x^3}y = 0, \tag{12.349}$$

the solution of which is

$$y(x) = e^{1/x}. \tag{12.350}$$

Discuss the applicability of a Frobenius-type generalized power-series solution to this differential equation.

Solution: We note that the differential equation has the form

$$y'' + f(x)y' + g(x)y = 0, \tag{12.351}$$

where

$$f(x) = \frac{1}{x^2}, \, g(x) = -\frac{2}{x^3}. \tag{12.352}$$

Using the given solution

$$y(x) = e^{1/x}, \tag{12.353}$$

we find

$$F(x) = f(x)y' = \frac{1}{x^2}\frac{d}{dx}(e^{1/x}) = -\frac{1}{x^4}e^{1/x}, \tag{12.354}$$

$$G(x) = g(x)y = -\frac{2}{x^3}e^{1/x}. \tag{12.355}$$

These two functions, $F(x)$ and $G(x)$, have a singular point at $x = 0$ and cannot be expressed as Frobenius series and therefore the Frobenius method cannot be used to find the solution of this differential equation.

12.13 Mathematica and serious substitution method

In this final section of this chapter, we are interested in basic commands that we can use to solve the Legendre, the associated Legendre, and the Bessel differential equations and the properties of the functions representing the solutions. These functions included the Legendre and the associated Legendre polynomials, Bessel functions, and spherical harmonics. We also see how we solve some of the problems we considered up to this point using Mathematica.

- *The Legendre differential equation*

    DSolve $[\{(1 - x^2)y''[x] - 2xy'[x] + l(l + 1)y[x] == 0\}, y[x], x]$
    $\{\{y[x] \rightarrow C[1]\text{LegendreP}[l, x] + C[2]LegendreQ[l, x]\}\}$

- *The associated Legendre differential equation*

    DSolve$\left[\left\{(1 - x^2)y''[x] - 2xy'[x] + \left(l(l + 1) - \frac{m^2}{1 - x^2}\right)y[x] == 0\right\}, y[x], x\right]$
    $\{\{y[x] \rightarrow C[1]\text{LegendreP}[l, m, x] + C[2]\text{LegendreQ}[l, m, x]\}\}$

- *Legendre and the associated Legendre polynomials*
    LegendreP[n,x] gives the Legendre polynomial $P_n(x)$.
    LegendreP[n,m,x] gives the associated Legendre polynomial $P_n^m(x)$.
    LegendreQ[n,z] gives the Legendre function of the second kind $Q_n(z)$.
    LegendreQ[n,m,z] gives the associated Legendre function of the second kind $Q_n^m(z)$.
- Example 12.7: The following is the Mathematica result for the series expansion for the step function

$$f(x) = \begin{cases} -1, & -1 \leqslant x \leqslant 0, \\ +1, & 0 \leqslant x \leqslant 1, \end{cases} \tag{12.356}$$

we did in Example 12.7 obtained by truncating the series for different values of m.

Input : $a[m] = (4m + 3) \int_0^1 \text{LegendreP}[2m + 1, x]\, dx$

Output : $\text{LegendreP}[2\,m, 0] - \text{LegendreP}\,[2 + 2\,m, 0]$

Input:Plot$\left[\left\{f_1[x] = \sum_{m=0}^1 a[m]\text{LegendreP}[2m + 1, x],\right.\right.$

$f_3[x] = \sum_{m=0}^3 a[m]\text{LegendreP}[2m + 1, x],$

$f_{10}[x] = \sum_{m=0}^{20} a[m]\text{LegendreP}[2m + 1, x], f_{100}[x]$

$= \sum_{m=0}^{100} a[m]\text{LegendreP}[2m + 1, x],$

$\{x, -1, 1\}, PlotLegends \rightarrow \text{'Expressions'}\}\big]$

Output:the result shown in figure 12.11

- *SphericalHarmonicY $[l, m, \theta, \phi]$ gives the spherical harmonic $Y_{lm}(\theta, \varphi)$. Here are a few examples:*

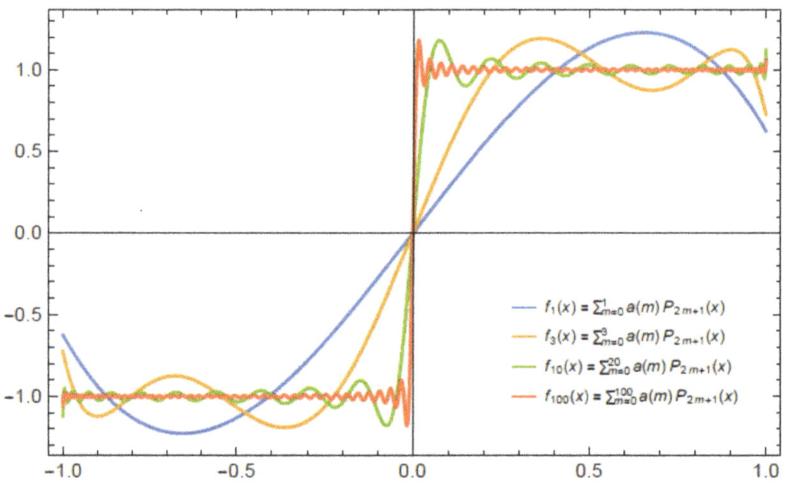

Figure 12.11. Plots for the Legendre series expansion for the step function.

SphericalHarmonicY[0, 0, θ, ϕ]	$\frac{1}{2\sqrt{\pi}}$
SphericalHarmonicY[1, -1, θ, ϕ]	$\frac{1}{2}e^{-i\phi}\sqrt{\frac{3}{2\pi}}\,\mathrm{Sin}[\theta]$
SphericalHarmonicY[1, 0, θ, ϕ]	$\frac{1}{2}\sqrt{\frac{3}{\pi}}\,\mathrm{Cos}[\theta]$
SphericalHarmonicY[1, 1, θ, ϕ]	$-\frac{1}{2}e^{i\phi}\sqrt{\frac{3}{2\pi}}\,\mathrm{Sin}[\theta]$
SphericalHarmonicY[2, -2, θ, ϕ]	$\frac{1}{4}e^{-2i\phi}\sqrt{\frac{15}{2\pi}}\,\mathrm{Sin}[\theta]^2$
SphericalHarmonicY[2, -1, θ, ϕ]	$\frac{1}{2}e^{-i\phi}\sqrt{\frac{15}{2\pi}}\,\mathrm{Cos}[\theta]\mathrm{Sin}[\theta]$
SphericalHarmonicY[2, 0, θ, ϕ]	$\frac{1}{4}\sqrt{\frac{5}{\pi}}(-1 + 3\mathrm{Cos}[\theta]^2)$
SphericalHarmonicY[2, 1, θ, ϕ]	$\frac{1}{2}e^{-i\phi}\sqrt{\frac{15}{2\pi}}\,\mathrm{Cos}[\theta]\mathrm{Sin}[\theta]$
SphericalHarmonicY[2, 2, θ, ϕ]	$\frac{1}{4}e^{2i\phi}\sqrt{\frac{15}{2\pi}}\,\mathrm{Sin}[\theta]^2$

- *The Bessel differential equation:*
  DSolve[$\{x^2 y''[x] + xy'[x] + (x^2 - p^2)y[x]==0\}$, $y[x]$, x]
  $\{\{y[x] \rightarrow \mathrm{BesselJ}[p, x]C[1] + \mathrm{BesselY}[p, x]C[2]\}\}$

- *Bessel functions:*
  BesselJ[n, z]gives the Bessel function of the first kind $J_n(z)$.
  BesselY[n, z]gives the Bessel function of the second kind $Y_n(z)$.

- Example 12.12
  $$\mathrm{DSolve}\left[\left\{\theta\,''[l] + \frac{2}{l}\theta'[l] + \frac{g}{lv^2}\theta[l]==0\right\}, \theta[l], l\right]$$
  $$\left\{\left\{\theta[l] \rightarrow \frac{v\mathrm{BesselJ}\left[1, \frac{2\sqrt{g}\sqrt{l}}{v}\right]C[1]}{\sqrt{g}\sqrt{l}} - \frac{2iv\mathrm{BesselY}\left[1, \frac{2\sqrt{g}\sqrt{l}}{v}\right]C[2]}{\sqrt{g}\sqrt{l}}\right\}\right\}$$ Note that this result is a
  bit different from the result in equation (12.342). However, taking into account $C[1]$ and $C[2]$ are constants determined from the initial conditions, this Mathematica result and the result in equation (12.342) must give the same result for the initial conditions $\theta(0) = \theta_0$ and $\dot{\theta}(0) = 0$.

12.14 Homework assignments

Problem 1. For the following differential equations find solutions using the series substitution method and elementary method (any of the methods you were introduced to in chapter 6).

(a)

$$\frac{dx(t)}{dt} - 3t^2 x(t) = 0 \tag{12.357}$$

(b)

$$\frac{d^2x(t)}{dt^2} - 2\frac{dx(t)}{dt} + x(t) = 0. \tag{12.358}$$

Problem 2. Show that the set of functions defined by

$$f_n(t) = \exp\left(\frac{int}{l}\right) \tag{12.359}$$

for $n = 0, \pm 1, \pm 2, \pm 3...$ form an orthogonal set of functions for all $t \,\varepsilon\,(-l, l)$. If we want to make this set of functions an orthonormal set, what must be the normalization factor.

Problem 3. We have shown that the solution to the Legendre differential equation

$$(1 - x^2)y'' - 2xy' + l(l + 1)y = 0, \tag{12.360}$$

is given by

$$y_e(x) = a_0\left\{1 - \frac{l(l + 1)}{2!}x^2 + \frac{l(l + 1)(l - 2)(l + 3)}{4!}x^4 \right.$$
$$\left. - \frac{l(l + 1)(l - 2)(l + 3)(l - 4)(l + 5)}{6!}x^6 + ...\right\} \tag{12.361}$$

for even l and by

$$y_o(x) = a_1\left\{x - \frac{(l - 1)(l + 2)}{3!}x^3 + \frac{(l - 1)(l + 2)(l - 3)(l + 4)}{5!}x^5 \right.$$
$$\left. - \frac{(l - 1)(l + 2)(l - 3)(l + 4)(l - 5)(l + 6)}{7!}x^7 + ...\right\} \tag{12.362}$$

odd l. Using these equations find the solutions for $l = 4$ and 5. By imposing the condition that $y(x)$ should be one at $x = 1$, determine the constants a_0 and a_1 in each case and write the expressions for *the Legendre polynomials*, $P_4(x)$ and $P_5(x)$. Compare the results with the results in the note.

Problem 4.
    (a) Using Leibniz' rule

$$\frac{d^N}{dt^N}(uv) = \sum_{n=0}^{N}\binom{N}{n}\left(\frac{d^n u}{dt^n}\right)\left(\frac{d^{N-n}v}{dt^{N-n}}\right) \tag{12.363}$$

find the derivative

$$f(t) = \frac{d^6}{dt^6}(t^2 \sin(t)). \tag{12.364}$$

(b) Use the Rodrigues formula

$$P_n(x) = \frac{1}{2^n n!} \frac{d^n}{dx^n} (x^2 - 1)^n \tag{12.365}$$

to determine $P_n(x)$ for $n = 0, 1, 2, 3$, and 4. Check your result with what we listed in table 12.1.

Problem 5.

(a) The Legendre polynomials can be generated using

$$P_l(x) = \frac{1}{l!} \frac{d^l}{dh^l} \phi(x, h) \bigg|_{h=0}, \tag{12.366}$$

for the function

$$\phi(x, h) = (1 - 2xh + h^2)^{-1/2}. \tag{12.367}$$

Show this is true for $l = 3$ and $l = 4$.

(b) Use the recursion relation

$$lP_l(x) = (2l - 1)xP_{l-1}(x) - (l - 1)P_{l-2}(x) \tag{12.368}$$

and the values of $P_0(x)$ and $P_1(x)$ to find $P_l(x)$ for $l = 2, 3, 4, 5$, and 6.

(c) You often see Legendre polynomials in electricity and magnetism. Such as in the description of charge distribution on aspherical surface. For example, a surface charge density could be expressed as

$$\sigma = \frac{q}{4\pi R^2}(\cos(\theta) - \cos^3(\theta)) \tag{12.369}$$

where q is the total charge on the surface of the sphere and R is the radius. Suppose we replace $x = \cos(\theta)$, the charge density can be expressed as

$$\sigma = \frac{q}{4\pi R^2}(x - x^3). \tag{12.370}$$

Express the charge density using Legendre polynomials.

Problem 6. Show that

$$\int_{-1}^{1} P_l(x)dx = 0 \text{ for } l > 0. \tag{12.371}$$

Problem 7. Consider the following functions

$$f(x) = \begin{cases} \cos(nx), & \text{for } x \, \varepsilon \, (-\infty, \infty), \\ e^{-x^2/2}, & \text{for } x \, \varepsilon \, (-\infty, \infty), \\ P_2(x) = \frac{1}{2}(3x^2 - 1), & \text{for } x \, \varepsilon \, (-1, 1), \end{cases} \tag{12.372}$$

(a) For the given intervals evaluate the integral

$$I = \int f^*(x)f(x)dx. \tag{12.373}$$

(b) Find the normalized functions for each.

Problem 8. For the function

$$f(x) = \begin{cases} 0, & -1 < x < 0, \\ x, & 0 < x < 1, \end{cases} \tag{12.374}$$

(a) find the first four non-zero terms in the Legendre series,
(b) using a computer (Mathematica) find these terms,
(c) plot the graph Legendre series as a function of x and considering a different number of terms show that as the number of terms increases the function becomes closer and closer to the exact function plot.

Problem 9. Using the transformation of variable defined by

$$x = \cos \theta, \tag{12.375}$$

show that the differential equation

$$\frac{1}{\sin \theta} \frac{d}{d\theta} \left(\sin \theta \frac{dy(\theta)}{d\theta} \right) + \left(l(l + 1) - \frac{m^2}{\sin^2 \theta} \right) y(\theta) = 0 \tag{12.376}$$

can be written as

$$(1 - x^2)\frac{d^2 y(x)}{dx^2} - 2x\frac{dy(x)}{dx} + \left(l(l + 1) - \frac{m^2}{1 - x^2} \right) y(x) = 0. \tag{12.377}$$

Problem 10. Using the differential equation

$$(1 - x^2)\frac{d^2 P_l^m(x)}{dx^2} - 2x\frac{dP_l^m(x)}{dx} + \left[l(l + 1) - \frac{m^2}{1 - x^2} \right] P_l^m(x) = 0 \tag{12.378}$$

and integration by parts, show that

$$\int_{-1}^{1} P_l^m(x)P_n^m(x)dx = 0, \quad \text{for } l \neq n. \tag{12.379}$$

Problem 11. Show that the length of the vector, $\vec{d} = \vec{r} - \vec{r}'$, (see figure 12.12) between the two points described by the vectors, $\vec{r}$ and $\vec{r}'$, is given by

$$|\vec{r} - \vec{r}'| = r^2 + r'^2 - 2rr' \cos(\gamma), \tag{12.380}$$

where

$$\cos(\gamma) = \cos(\theta)\cos(\theta') + \sin(\theta)\sin(\theta')\cos(\varphi - \varphi'). \tag{12.381}$$

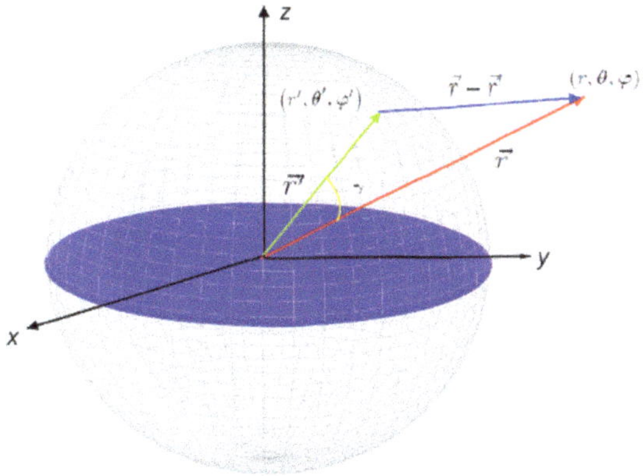

Figure 12.12. Two point described by two vectors, $\vec{r}$ and $\vec{r}'$. In spherical coordinates the two points have coordinates $(r,\ \theta,\ \varphi)$ and $(r',\ \theta',\ \varphi')$, respectively. The angle between these two vectors is γ.

Problem 12. A spherical shell of radius R has a constant surface charge density, σ. The sphere is centered at the origin of coordinates. The point P is a distance, $r = d < R$, from the center of the sphere. Find an expression for the electrostatic potential at the point P, due to the charged sphere. Note that

$$\sigma = \frac{dq'}{da} = \text{Constant}, \tag{12.382}$$

where da is an infinitesimal area on the surface of the sphere which we can express in spherical coordinates as

$$da = R^2 \sin(\theta)d\theta d\varphi, \tag{12.383}$$

R is the radius of the sphere (figure 12.13).

Problem 13. The spherical harmonics form a complete set and therefore any function $f(\theta,\ \varphi)$ defined for $0 < \theta < \pi$ and $0 < \varphi < 2\pi$, can be expressed as 'spherical harmonics series'

$$f(\theta,\ \varphi) = \sum_{l=0}^{\infty} \sum_{-l}^{l} A_{lm}\, Y_{lm}(\theta,\ \varphi), \tag{12.384}$$

where

$$A_{lm} = \int_0^{\pi} \int_0^{2\pi} Y_{lm}^*(\theta,\ \varphi) f(\theta,\ \varphi) \sin(\theta)d\theta d\varphi. \tag{12.385}$$

Applying this relation show that the function

$$f(\theta,\ \varphi) = P_l(\cos(\gamma)) = \frac{4\pi}{2l+1} \sum_{m=-l}^{l} Y_{lm}(\theta,\ \varphi) Y_{lm}^*(\theta',\ \varphi'), \tag{12.386}$$

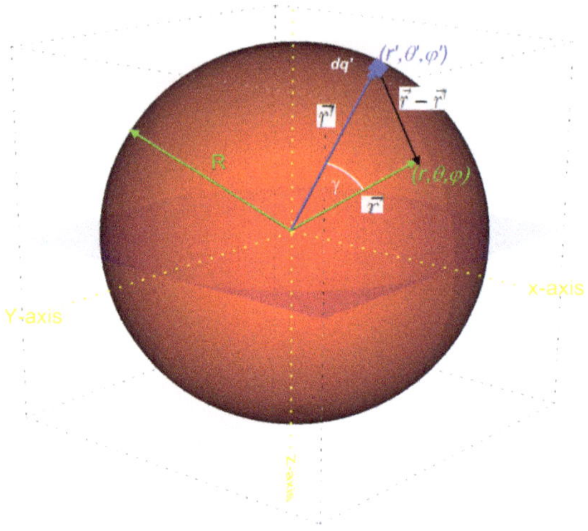

Figure 12.13. A spherical shell carrying a constant surface charge density, σ.

where

$$\cos(\gamma) = \cos(\theta)\cos(\theta') + \sin(\theta)\sin(\theta')\cos(\varphi - \varphi').$$ (12.387)

Problem 14. Using the ration test show that the infinite series

$$J_p(x) = \sum_{n=0}^{\infty} \frac{(-1)^n}{\Gamma(n+1)\Gamma(n+1+p)}\left(\frac{x}{2}\right)^{2n+p}$$ (12.388)

is a convergent series.

Problem 15. Using

$$J_p(x) = \sum_{n=0}^{\infty} \frac{(-1)^n}{\Gamma(n+1)\Gamma(n+1+p)}\left(\frac{x}{2}\right)^{2n+p},$$ (12.389)

and

$$J_{-p}(x) = \sum_{n=0}^{\infty} \frac{(-1)^n}{\Gamma(n+1)\Gamma(n+1-p)}\left(\frac{x}{2}\right)^{2n-p},$$ (12.390)

write out the first few terms of $J_0(x)$, $J_1(x)$, $J_2(x)$, $J_{-1}(x)$, and $J_{-2}(x)$.

Problem 16. The differential equation for transverse vibration of a string whose density increases linearly from one end to the other is

$$\frac{d^2y}{dx^2} + (\lambda x + \beta)y = 0,$$ (12.391)

where α and β are constants. Find the general solution of this equation in terms of Bessel functions. *Hint: make change of variable $\lambda x + \beta = \lambda u$.*

Problem 17. Using the relation

$$J_n(x) = \sqrt{\frac{\pi}{2x}} J_{(2n+1)/2}(x) = x^n \left(\frac{1}{x}\frac{d}{dx} \right)^n \left(\frac{\sin(x)}{x} \right) \tag{12.392}$$

and the orthogonality relation for the Bessel functions

$$\int_0^1 J_p(\alpha_i x) J_p(\alpha_j x) x\,dx = \frac{J_p'(\alpha_i) J_p'(\alpha_j)}{2} \delta_{ij} \tag{12.393}$$

find the orthogonality relation for the spherical Bessel functions.

Problem 18. Using the approximate relations for the Bessel functions find

$$\lim_{x \to 0} \frac{J_4(x)}{J_2^2(x)}. \tag{12.394}$$

Chapter 13

Partial differential equation

In the last chapter, we solved the associated Legendre and the Bessel differential equations, the solutions of which are the associated Legendre polynomials and the Bessel function, respectively. This chapter introduces how we generally solve homogeneous and non-homogeneous partial differential equations. We first introduce the standard partial differential equation (PDE) in different branches of advanced physics courses. From these equations, we focus on Laplace's PDE in Cartesian, spherical, and cylindrical coordinates and introduce the methods we use to determine the solutions to homogeneous PDEs. While doing so, we also see how the associated Legendre and Bessel differential equations derived from Laplace's equation in spherical and cylindrical coordinates. Finally, we see the method for solving non-homogeneous PDEs by solving Poisson's equation in spherical coordinates for real physical problems in electromagnetism.

13.1 PDE in physics

A partial differential equation (PDE) is an equation in some unknown function of more than one variable, say $f = V(x, y, z, t)$, involving possibly different orders of partial derivatives of that function. The following are some examples of PDEs that we often find in different branches of physics.

Gauss' law for the electric field

The differential form of Gauss' law states that the electric field, $\vec{E}(\vec{r})$ of a volume charge distribution $\rho(\vec{r})$ satisfies the PDE

$$\nabla \cdot \vec{E}(\vec{r}) = \frac{\rho(\vec{r})}{\epsilon_0}. \tag{13.1}$$

In Cartesian coordinates, using,

$$\vec{E}(x, y, z) = E_x(x, y, z)\hat{x} + E_y(x, y, z)\hat{y} + E_z(x, y, z)\hat{z},$$

$$\nabla = \frac{\partial}{\partial x}\hat{x} + \frac{\partial}{\partial y}\hat{y} + \frac{\partial}{\partial z}\hat{z}, \tag{13.2}$$

Gauss' law can be written as

$$\frac{\partial E_x(x, y, z)}{\partial x} + \frac{\partial E_y(x, y, z)}{\partial y} + \frac{\partial E_z(x, y, z)}{\partial z} = \frac{\rho(x, y, z)}{\epsilon_0}. \tag{13.3}$$

Poisson's equation

In terms of the electric potential in Cartesian coordinates $V(x, y, z)$, we can express the electric field, $\vec{E}(x, y, z)$, as

$$\vec{E}(x, y, z) = -\nabla V(x, y, z), \tag{13.4}$$

so that Gauss's Law in equation (13.3) can be rewritten as

$$\nabla \cdot [-\nabla V(x, y, z)] = \frac{\rho(x, y, z)}{\epsilon_0} \Rightarrow \nabla^2 V(x, y, z) = -\frac{\rho(x, y, z)}{\epsilon_0}$$

$$\Rightarrow \frac{\partial^2 V(x, y, z)}{\partial x^2} + \frac{\partial^2 V(x, y, z)}{\partial y^2} \tag{13.5}$$

$$+ \frac{\partial^2 V(x, y, z)}{\partial z^2} = -\frac{\rho(x, y, z)}{\epsilon_0}.$$

The PDE in equation (13.5) is known as *Poisson's equation.*

Laplace's equation

In regions where there is no charge, $\rho(x, y, z) = 0$, Poisson's equation becomes *Laplace's equation.* It is given by

$$\nabla^2 V(\vec{r}) = 0$$

$$\Rightarrow \nabla^2 V(x, y, z) = \frac{\partial^2 V(x, y, z)}{\partial x^2} + \frac{\partial^2 V(x, y, z)}{\partial y^2} \tag{13.6}$$

$$+ \frac{\partial^2 V(x, y, z)}{\partial z^2} = 0.$$

The diffusion or heat flow equation

Let us consider a function, $u(x, y, z, t)$, that describes the temperature at a given point in space with coordinates (x, y, z) time t. The temperature at this point changes with time as heat spreads throughout the space. The diffusion or heat equation governs how the temperature function, $u(x, y, z, t)$, changes at a given point in space over time. In a particular case of heat flow in an isotropic and homogeneous medium in a 3-dimensional space, the heat equation is given by

$$\frac{\partial u}{\partial t} = \alpha\left[\frac{\partial^2 u(x, y, z, t)}{\partial x^2} + \frac{\partial^2 u(x, y, z, t)}{\partial y^2} + \frac{\partial^2 u(x, y, z, t)}{\partial z^2}\right], \tag{13.7}$$

where α is the thermal diffusivity, a material-specific quantity depending on the thermal conductivity, the mass density, and the specific heat capacity.

One-dimensional wave equation

Consider a string with linear mass density, λ, under a tension force T. If one vibrates one end of the string with some frequency, as shown in figure 13.1, one can create a wave traveling on the string. Consider an infinitesimal segment of the string shown in figure 13.2. If the string is displaced from the equilibrium, the net transverse force on the segment between z and $z + \Delta z$ can be expressed as

$$\Delta F = T(\sin(\alpha) - \sin(\theta)). \tag{13.8}$$

For such a wave, since the displacement from the equilibrium position is small, the angular displacements (θ and α) are small. For small angles one can replace sin with tan and make the approximation

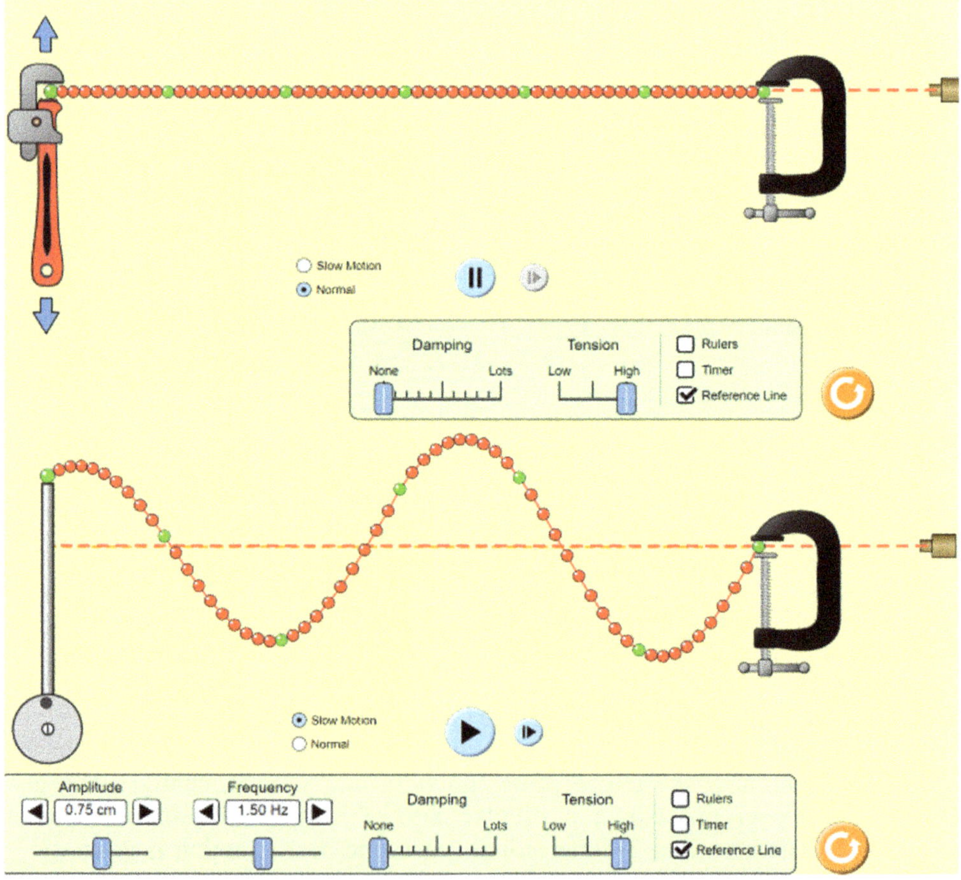

Figure 13.1. PhET standing wave simulation: The top is a snapshot of a tensioned string. The bottom is a snapshot of same string after a wave (a disturbance) is created on a strong. The incident and reflected wave at the ends form a standing wave.

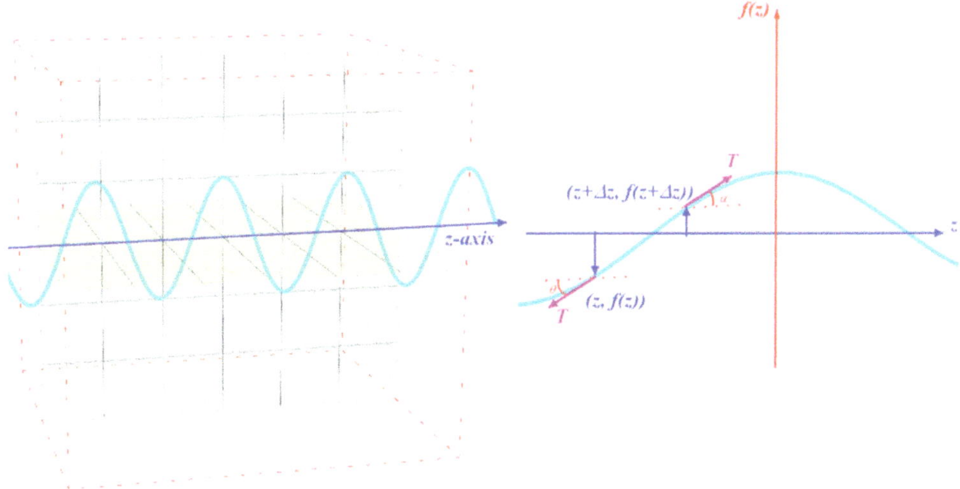

Figure 13.2. (a) A wave on a string oscillating on the y–z plane and (b) the displacement of the wave from the equilibrium position, $f(z)$ vs z.

$$\Delta F \simeq T(\tan(\alpha) - \tan(\theta)). \tag{13.9}$$

Applying Newton's second law, the net force can be related to the displacement from the equilibrium position $f(z, t)$ by

$$\Delta F = ma = m\frac{\partial^2 f}{\partial t^2} \Rightarrow T(\tan(\alpha) - \tan(\theta)) = m\frac{\partial^2 f}{\partial t^2}. \tag{13.10}$$

In chapter 6, we have shown that equation (13.10) leads to the one-dimensional wave equation given by

$$\frac{\partial^2 f}{\partial z^2} = \frac{1}{v^2}\frac{\partial^2 f}{\partial t^2}, \tag{13.11}$$

where

$$v = \sqrt{\frac{T}{\lambda}}, \tag{13.12}$$

is the speed of the wave.

Three-dimensional wave equation and Helmholtz equation

The three-dimensional wave equation governing a wave defined by the function $f(\vec{r}, t)$

$$\nabla^2 f(\vec{r}, t) = \frac{1}{v^2}\frac{\partial^2}{\partial t^2}f(\vec{r}, t), \tag{13.13}$$

that can be rewritten, in Cartesian coordinates, as

$$\frac{\partial^2 f}{\partial x^2} + \frac{\partial^2 f}{\partial y^2} + \frac{\partial^2 f}{\partial z^2} = \frac{1}{v^2}\frac{\partial^2 f}{\partial t^2}. \tag{13.14}$$

Using separation of variables we introduced in chapter 6,

$$f(x, y, z, t) = u(x, y, z)w(t) \tag{13.15}$$

one can derive the equation governing the special part $u(x, y, z)$, given by

$$\frac{1}{u}\left[\frac{\partial^2 u}{\partial x^2} + \frac{\partial^2 u}{\partial y^2} + \frac{\partial^2 u}{\partial z^2}\right] = \frac{1}{v^2 w}\frac{d^2 w}{dt^2} = -k^2$$

$$\Rightarrow \frac{\partial^2 u}{\partial x^2} + \frac{\partial^2 u}{\partial y^2} + \frac{\partial^2 u}{\partial z^2} + k^2 u = 0. \tag{13.16}$$

The PDE in equation (13.16) is known as *the Helmholtz equation*. We often see the Helmholtz equation in electromagnetism and quantum mechanics, where we deal with electromagnetic waves and their interaction at the macroscopic or atomic level.

The Schrödinger equation
In quantum mechanics, the wave function $\psi(\vec{r}, t)$ describes the state of a particle. The time-dependent Schrödinger equation governs the wave function which is given by,

$$V(\vec{r})\psi(\vec{r}, t) - \frac{\hbar^2}{2m}\nabla^2\psi(\vec{r}, t) = i\hbar\frac{\partial}{\partial t}\psi(\vec{r}, t), \tag{13.17}$$

where m is the mass of the particle, $V(\vec{r})$ is the potential energy, $\hbar = h/2\pi$ and h is Planck's constant. In Cartesian coordinates, the Schrödinger equation can be expressed as

$$-\frac{\hbar^2}{2m}\left[\frac{\partial^2\psi(x, y, z, t)}{\partial x^2} + \frac{\partial^2\psi(x, y, z, t)}{\partial y^2} + \frac{\partial^2\psi(x, y, z, t)}{\partial z^2}\right]$$

$$+ V(x, y, z, t)\psi(x, y, z, t) = i\hbar\frac{\partial}{\partial t}\psi(x, y, z, t). \tag{13.18}$$

This chapter uses the separation of variables we introduced in section 6.7 to separate these partial differential equation into three differential equations of single variables. Moreover, we use the solutions to these single-variable differential equations (most of which we determined in the previous chapter) to determine the solution to the partial differential equation. Now we will focus on Laplace's and Poison's equations in Cartesian, spherical and cylindrical coordinates.

13.2 Laplace's equation in Cartesian coordinates

Before we begin to solve Laplace's equation in Cartesian coordinates, it is essential to revise how we determine the solution for second-order HLDE with constant coefficients.

Example 13.1. Solve the differential equation

$$\frac{d^2 f(x)}{dx^2} = \pm k^2 f(x). \tag{13.19}$$

Solution: Assuming a solution of the form

$$f(x) = Ae^{\lambda x}, \tag{13.20}$$

we find the indicial equation

$$\lambda^2 \mp k^2 = 0. \tag{13.21}$$

For the minus case, the solutions are

$$\lambda^2 - k^2 = 0 \Rightarrow \lambda_1 = k, \ \lambda_2 = -k \tag{13.22}$$

and the general solution is given by

$$f(x) = A_1 e^{-kx} + A_2 e^{kx}. \tag{13.23}$$

For the plus case, the solution to the indicial equation are

$$\lambda^2 + k^2 = 0 \Rightarrow \lambda_1 = ik, \ \lambda_2 = -ik \tag{13.24}$$

and the general solution becomes

$$f(x) = B_1 \cos(kx) + B_2 \sin(kx). \tag{13.25}$$

For a point in space (see figure 13.3), described by the Cartesian coordinates

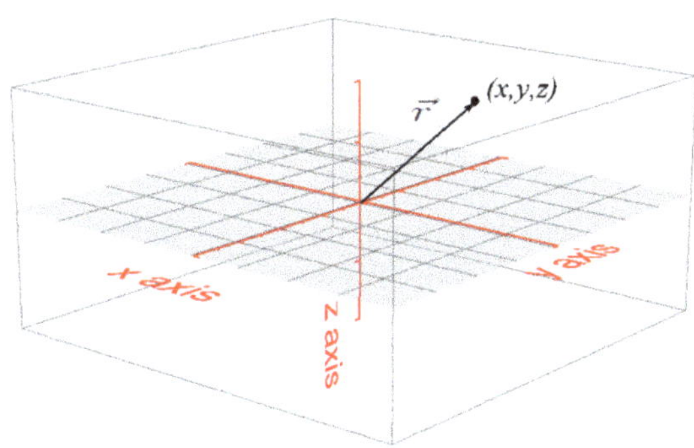

Figure 13.3. A point in space described by $\vec{r}$ using Cartesian coordinates (x, y, z).

$$\vec{r} = x\hat{x} + y\hat{y} + z\hat{z},\qquad(13.26)$$

Laplace's equation for the function $V(x, y, z)$ is given by

$$\frac{\partial^2 V(x, y, z)}{\partial x^2} + \frac{\partial^2 V(x, y, z)}{\partial y^2} + \frac{\partial^2 V(x, y, z)}{\partial z^2} = 0.\qquad(13.27)$$

Using separation of variables, one can express $V(x, y, z)$ as a product of three independent functions $X(x)$, $Y(y)$, and $Z(z)$

$$V(x, y, z) = X(x)Y(y)Z(z).\qquad(13.28)$$

Substituting equation (13.28) into equation (13.27), one can write

$$Y(y)Z(z)\frac{d^2 X(x)}{dx^2} + X(x)Z(z)\frac{d^2 Y(y)}{dy^2} + X(x)Y(y)\frac{d^2 Z(z)}{dz^2} = 0$$
$$\Rightarrow \frac{1}{X(x)}\frac{d^2 X(x)}{dx^2} + \frac{1}{Y(y)}\frac{d^2 Y(y)}{dy^2} + \frac{1}{Z(y)}\frac{d^2 Z(z)}{dz^2} = 0.$$
$$(13.29)$$

Equation (13.29) consists of three independent terms and the sum of which must be zero. Such a result can be possible if and only if each of these terms is a constant. We have to choose the signs of the constants so that their sum is zero. Equating the first two terms in equation (13.29) with negative real constants and the third term with a positive real constant, we obtain the three ordinary differential equations. Therefore, we can write

$$\frac{1}{X(x)}\frac{d^2 X(x)}{dx^2} = -k_x^2,\ \frac{1}{Y(y)}\frac{d^2 Y(y)}{dy^2} = -k_y^2,\ \frac{1}{Z(y)}\frac{d^2 Z(z)}{dz^2} = k_z^2,\qquad(13.30)$$

where k_x, k_y, and k_z are constants satisfying the equation

$$k_z^2 - k_y^2 - k_x^2 = 0.\qquad(13.31)$$

The equations in equation (13.30) are the second-order HLDE similar to what we saw in Example 13.1. These equations can be rewritten as

$$\frac{d^2 X(x)}{dx^2} + k_x^2 X(x) = 0,\ \frac{d^2 Y(y)}{dy^2} + k_y^2 Y(y) = 0$$
$$\frac{d^2 Z(z)}{dz^2} - k_z^2 Z(z) = 0.$$
$$(13.32)$$

Assuming k_x, k_y and k_z are real constants, in view of the results in Example 13.1, the solutions to the differential equations in equation (13.32) can be expressed in any one of the following forms

$$X_{k_x}(x) = \begin{cases} A_{k_x} e^{ik_x x} + B_{k_x} e^{-ik_x x}, \\ C_{k_x} \cos{(k_x x)} + D_{k_x} \sin{(k_x x)}, \\ \quad E_{k_x} \cos{\left(k_x x - \gamma_{k_x}\right)}, \\ \quad F_{k_x} \sin{\left(k_x x + \beta_{k_x}\right)}, \end{cases} \tag{13.33}$$

$$Y_{k_y}(y) = \begin{cases} A_{k_y} e^{ik_y y} + B_{k_y} e^{-ik_y y}, \\ C_{k_y} \cos{\left(k_y y\right)} + D_{k_y} \sin{\left(k_y y\right)}, \\ \quad E_{k_y} \cos{\left(k_y y - \gamma_{k_y}\right)}, \\ \quad F_{k_y} \sin{\left(k_y y + \beta_{k_y}\right)}, \end{cases} \tag{13.34}$$

$$Z_{k_z}(z) = \begin{cases} A_{k_z} e^{k_z z} + B_{k_z} e^{-k_z z}, \\ C_{k_z} \cosh{(k_z z)} + D_{k_z} \sinh{(k_z z)}, \\ \quad E_{k_z} \cosh{\left(k_z z + \gamma_{k_z}\right)}, \\ \quad F_{k_z} \sinh{\left(k_z z + \beta_{k_z}\right)}. \end{cases} \tag{13.35}$$

We choose the constants associated with the first two differential equations in equation (13.29) to be negative and the constant associated with the third term to be positive. There is nothing special about this choice. However, one should note that choosing a negative constant leads to oscillatory solutions having different phases. Choosing a positive constant leads to solutions involving exponentially decreasing and increasing terms. For example, if we had chosen

$$\frac{1}{X(x)}\frac{d^2X(x)}{dx^2} = \omega_x^2, \ \frac{1}{Y(y)}\frac{d^2Y(y)}{dy^2} = \omega_y^2, \ \frac{1}{Z(y)}\frac{d^2Z(z)}{dz^2} = -\omega_z^2,$$

$$\Rightarrow \frac{d^2X(x)}{dx^2} - \omega_x^2 X(x) = 0, \ \frac{d^2Y(y)}{dy^2} - \omega_y^2 Y(y) = 0, \tag{13.36}$$

$$\frac{d^2Z(z)}{dz^2} + \omega_z^2 Z(y) = 0,$$

where

$$\omega_x^2 + \omega_y^2 - \omega_z^2 = 0. \tag{13.37}$$

The solutions would have been

$$X_{\omega_x}(x) = \begin{cases} A_{\omega_x} e^{\omega_x x} + B_{\omega_x} e^{-\omega_x x}, \\ C_{\omega_x} \cosh{(\omega_x x)} + D_{\omega_x} \sinh{(\omega_x x)}, \\ \quad E_{\omega_x} \cosh{\left(\omega_x x - \gamma_{k_x'}\right)}, \\ \quad F_{\omega_x} \sinh{\left(\omega_x x + \beta_{k_x'}\right)}. \end{cases} \tag{13.38}$$

$$
Y_{\omega_y}(y) = \begin{cases}
A_{\omega_y} e^{\omega_y y} + B_{\omega_y} e^{-\omega_y y} \\
C_{\omega_y} \cosh(\omega_y y) + D_{\omega_y} \sinh(\omega_y y), \\
E_{\omega_y} \cosh\left(\omega_y y - \gamma_{\omega_y}\right), \\
F_{\omega_y} \sinh\left(\omega_y y + \beta_{\omega_y}\right),
\end{cases}
\tag{13.39}
$$

and

$$
Z_{\omega_z}(z) = \begin{cases}
A_{\omega_z} e^{i\omega_z z} + B_{\omega_z} e^{-i\omega_z z}, \\
C_{\omega_z} \cos(\omega_z z) + D_{\omega_z} \sin(\omega_z z), \\
E_{\omega_z} \cos(\omega_z z + \gamma_{\omega_z}), \\
F_{\omega_z} \sin\left(\omega_z z + \beta_{\omega_z}\right).
\end{cases}
\tag{13.40}
$$

Although the solutions in equations (13.38)–(13.40) appear to be different from equations (13.33)–(13.35), it is important to emphasize that these equations can be derived from one another. To this end, using

$$
\begin{aligned}
&\omega_z^2 = -k_z^2, \; -\omega_y^2 = k_y^2, \; -\omega_x^2 = k_x^2, \\
&\Rightarrow \omega_z = ik_z, \; \omega_y = ik_y, \; \omega_x = ik_x,
\end{aligned}
\tag{13.41}
$$

and the relations

$$
\begin{aligned}
\cosh(ik_x x) &= \frac{e^{ik_x x} + e^{-ik_x x}}{2} = \cos(k_x x), \\
\sinh(ik_x x) &= i\frac{e^{ik_x x} - e^{-ik_x x}}{2i} = i\sin(k_x x),
\end{aligned}
\tag{13.42}
$$

one can easily show that

$$
X_{\omega_x}(x) = \begin{cases}
A_{\omega_x} e^{\omega_x x} + B_{\omega_x} e^{-\omega_x x}, \\
C_{\omega_x} \cosh(\omega_x x) + D_{\omega_x} \sinh(\omega_x x), \\
E_{\omega_x} \cosh\left(\omega_x x - \gamma_{k_x'}\right), \qquad = X_{k_x}(x), \\
F_{\omega_x} \sinh\left(\omega_x x + \beta_{k_x'}\right),
\end{cases}
\tag{13.43}
$$

$$
Y_{\omega_y}(y) = \begin{cases}
A_{\omega_y} e^{\omega_y y} + B_{\omega_y} e^{-\omega_y y} \\
C_{\omega_y} \cosh(\omega_y y) + D_{\omega_y} \sinh(\omega_y y), \\
E_{\omega_y} \cosh\left(\omega_y y - \gamma_{\omega_y}\right), \qquad = Y_{k_y}(y), \\
F_{\omega_y} \sinh\left(\omega_y y + \beta_{\omega_y}\right),
\end{cases}
\tag{13.44}
$$

and

$$Z_{k_z}'(z) = \begin{cases} A_{\omega_z} e^{i\omega_z z} + B_{\omega_z} e^{-i\omega_z z}, \\ C_{\omega_z} \cos(\omega_z z) + D_{\omega_z} \sin(\omega_z z), \\ \quad E_{\omega_z} \cos(\omega_z z + \gamma_{\omega_z}), \\ \quad F_{\omega_z} \sin\left(\omega_z z + \beta_{\omega_z}\right), \end{cases} = Z_{k_z}(z). \tag{13.45}$$

The general solution to Laplace's equation in Cartesian coordinates in equation (13.27) is given by

$$V(x, y, z) = \sum_{k_x} \sum_{k_y} \sum_{k_z} X_{k_x}(x) Y_{k_y}(y) Z_{k_z}(z). \tag{13.46}$$

Note that the constants k_x, k_y, k_z along with the other constants are determined from the boundary condition set for the given problem and the condition

$$k_x^2 + k_y^2 - k_z^2 = 0. \tag{13.47}$$

We shall see how these constants are determined in the next example.

Example 13.2. Consider an infinitely long, rectangular waveguide of width, w, and height h with the top surface should be insulated from the rest held at the constant potential, $V(y = h) = Vo$, and with all other sides grounded. A vacuum pump has been connected to the waveguide to remove most of the air within it. Find the electrostatic potential, $V(x, y, z)$, everywhere within the waveguide. Inside a waveguide, the electric potential obeys Laplace's equation (figure 13.4),

$$\frac{\partial^2 V(x, y, z)}{\partial x^2} + \frac{\partial^2 V(x, y, z)}{\partial y^2} + \frac{\partial^2 V(x, y, z)}{\partial z^2} = 0. \tag{13.48}$$

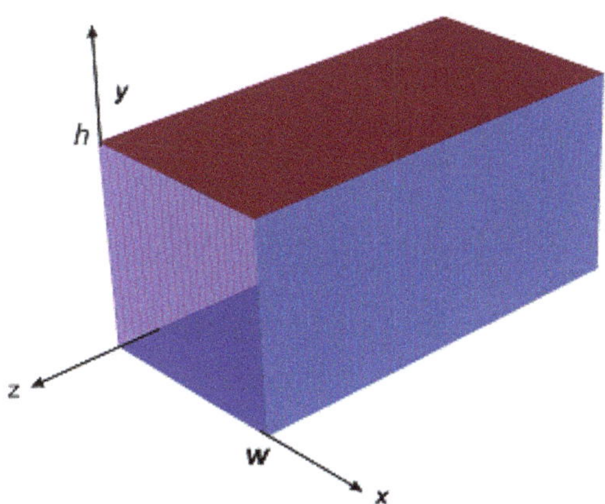

Figure 13.4. A rectangular waveguide with width w and height h. It is infinitely long along the z-direction.

Solution: We already determined the solution to Laplace's equation in Cartesian coordinates. To find the potential inside the waveguide, we need to state the given boundary conditions and impose these conditions on the general solution and find the unknown constants. To this end, we note that the general solution is given by

$$V(x, y, z) = \sum_{k_x, k_y, k_z} X_{k_x}(x) Y_{k_y}(y) Z_{k_z}(x), \tag{13.49}$$

where

$$k_x^2 + k_y^2 - k_z^2 = 0, \tag{13.50}$$

$X_{k_x}(x)$, $Y_{k_y}(y)$, and $Z_{k_z}(x)$ are given by equations (13.33)–(13.35). The boundary conditions are

$$V(0, y, z) = V(w, y, z) = V(x, 0, z) = 0, \ V(x, h, z) = V_0. \tag{13.51}$$

And also the potential must be finite everywhere inside the waveguide. Imposing the first three boundary conditions, we have

$$V(0, y, z) = 0 \Rightarrow \sum_{k_x, k_y, k_z} X_{k_x}(0) Y_{k_y}(y) Z_{k_z}(z) = 0 \Rightarrow X_{k_x}(0) = 0,$$

$$V(w, y, z) = 0 \Rightarrow \sum_{k_x, k_y, k_z} X_{k_x}(w) Y_{k_y}(y) Z_{k_z}(z)$$

$$= 0 \Rightarrow X_{k_x}(w) = 0,$$

$$V(x, 0, z) = 0 \Rightarrow \sum_{k_x, k_y, k_z} X_{k_x}(x) Y_{k_y}(0) Z_{k_z}(z)$$

$$= 0 \Rightarrow Y_{k_y}(0) = 0. \tag{13.52}$$

Using the solutions in equations (13.33) and (13.34), we find

$$X_{k_x}(0) = 0 \Rightarrow [C_{k_x} \cos (k_x x) + D_{k_x} \sin (k_x x)]_{x=0} = 0 \Rightarrow C_{k_x} = 0, \tag{13.53}$$

$$X_{k_x}(w) = 0 \Rightarrow [C_{k_x} \cos (k_x x) + D_{k_x} \sin (k_x x)]_{x=w} = 0$$
$$\Rightarrow D_{k_x} \sin (k_x w) = 0 \Rightarrow \sin (k_x w) = 0$$
$$\Rightarrow k_x = \frac{n\pi}{w}, \ \text{where } n = 1, 2, 3..., \tag{13.54}$$

$$Y_{k_y}(0) = 0 \Rightarrow \left[C_{k_y} \cos \left(k_y y\right) + D_{k_y} \sin \left(k_y y\right) \right]_{y=0} = 0 \Rightarrow C_{k_y} = 0. \tag{13.55}$$

Since the waveguide is infinitely long along the z-direction and the potential must be finite within the waveguide, we must have

$$\lim_{z \to \pm\infty} \left(\sum_{k_x, k_y, k_z} X_{k_x}(x) Y_{k_y}(y) Z_{k_z}(z) \right) \neq \pm\infty \Rightarrow \lim_{z \to \pm\infty} Z_{k_z}(z) \neq \pm\infty \tag{13.56}$$

and using the result in equation (13.35), one finds

$$\lim_{z \to \pm\infty} [A_{k_z} e^{k_z z} + B_{k_z} e^{-k_z z}] = \lim_{z \to \pm\infty} A_{k_z} e^{k_z z} \neq \pm\infty \Rightarrow k_z = 0$$

$$\Rightarrow Z_{k_z}(z) = A_{k_z} + B_{k_z} = \text{constant} = A.$$

(13.57)

Substituting $k_x = \frac{n\pi}{w}$ and $k_z = 0$ into

$$k_x^2 + k_y^2 - k_z^2 = 0,$$

(13.58)

we find

$$k_y = \pm i \frac{n\pi}{w}.$$

(13.59)

Taking into account all the results in equations (13.53)–(13.55), (13.57), and (13.59), we can write

$$X_{k_x}(x) = D_{k_x} \sin (k_x x) = D_n \sin \left(\frac{n\pi}{w} x \right),$$

(13.60)

$$Y_{k_y}(y) = D_{k_y} \sin \left(k_y y \right) = D_{k_y} \sin \left(\pm i \frac{n\pi}{w} y \right) = \pm i D_{k_y} \sinh \left(\frac{n\pi}{w} y \right)$$

$$\Rightarrow Y_{k_y}(y) = D_n' \sinh \left(\frac{n\pi}{w} y \right),$$

(13.61)

$$Z_{k_z}(z) = A,$$

(13.62)

where we used the relation

$$\sin (ix) = \frac{e^{i(ix)} - e^{-i(ix)}}{2i} = i \frac{e^x - e^{-x}}{2} \Rightarrow \sin (ix) = i \sinh (x)$$

(13.63)

in equation (13.61). Then the general solution can be expressed as

$$V(x, y, z) = \sum_{n=1}^{\infty} A D_n D_n' \sin \left(\frac{n\pi}{w} x \right) \sinh \left(\frac{n\pi}{w} y \right)$$

$$= \sum_{n=1}^{\infty} H_n \sin \left(\frac{n\pi}{w} x \right) \sinh \left(\frac{n\pi}{w} y \right)$$

(13.64)

where we introduced a constant $H_n = A D_n D_n'$. Now imposing the last boundary condition, $V(x, h, z) = V_0$, on equation (13.64), we find

$$V(x, h, z) = V_0 \Rightarrow \sum_{n=1}^{\infty} H_n \sin \left(\frac{n\pi}{w} x \right) \sinh \left(n\pi \frac{h}{w} \right) = V_0.$$

(13.65)

Multiplying equation (13.65) by $\sin\left(\frac{m\pi}{w}x\right)$ and integrating with respect to x over the width of the waveguide, one can write

$$\sum_{n=1}^{\infty} H_n \sinh\left(n\pi\frac{h}{w}\right) \int_0^W \sin\left(\frac{n\pi}{w}x\right)\sin\left(\frac{m\pi}{w}x\right)dx = V_0 \int_0^W \sin\left(\frac{m\pi}{w}x\right)dx. \quad (13.66)$$

Using

$$\int_0^W \sin\left(\frac{n\pi}{w}x\right)\sin\left(\frac{m\pi}{w}x\right)dx = \frac{w}{\pi}\int_0^W \sin\left(\frac{n\pi}{w}x\right)\sin\left(\frac{m\pi}{w}x\right)d\left(\frac{x\pi}{w}\right)$$

$$= \frac{w}{\pi}\int_0^\pi \sin(nu)\sin(mu)du = \frac{w}{\pi}\left(\frac{\pi}{2}\delta_{mn}\right)$$

$$= \frac{w}{2}\delta_{mn},$$

$$\int_0^W \sin\left(\frac{m\pi}{w}x\right)dx = -\frac{w}{m\pi}\cos\left(\frac{m\pi}{w}x\right)\Big|_0^w = \frac{w}{m\pi}[1-(-1)^m],$$

$\qquad\qquad(13.67)$

we rewrite equation (13.66) as

$$\sum_{n=1}^{\infty} H_n \sinh\left(n\pi\frac{h}{w}\right)\frac{w}{2}\delta_{mn} = V_0\frac{w}{m\pi}[1-(-1)^m]$$

$$\Rightarrow H_m \sinh\left(m\pi\frac{h}{w}\right) = V_0\frac{2}{m\pi}[1-(-1)^m]$$

$$\Rightarrow H_m = \begin{cases} 0, & m = \text{even}, \\[2mm] \dfrac{4V_0}{m\pi \sinh\left(m\pi\dfrac{h}{w}\right)}, & m = \text{odd}. \end{cases}$$

$\qquad\qquad(13.68)$

In view of the results in equation (13.68), equation (13.64) becomes

$$V(x,y,z) = \frac{4V_0}{\pi}\sum_{n=0}^{\infty} \frac{\sin\left[\dfrac{x}{w}(2n+1)\pi\right]\sinh\left[\dfrac{y}{w}(2n+1)\pi\right]}{(2n+1)\sinh\left[\dfrac{h}{w}(2n+1)\pi\right]}. \quad (13.69)$$

Equation (13.69) is the electric potential inside the waveguide.

It is important to emphasize that we will see such problems where we need to solve Laplace's equation in Cartesian coordinates in electromagnetism and quantum mechanics.

13.3 Laplace's equation in spherical coordinates

We have seen an example in the previous section illustrating how we solve Laplace's equation in Cartesian coordinates

$$\nabla^2 V(x, y, z) = \frac{\partial^2}{\partial x^2} V(x, y, z) + \frac{\partial^2}{\partial y^2} V(x, y, z) + \frac{\partial^2}{\partial z^2} V(x, y, z) = 0. \quad (13.70)$$

This same partial differential equation can also be expressed in spherical coordinates. In spherical coordinates, as shown in figure 13.5, a point in space is described by the coordinates r, θ, and φ. The Cartesian coordinates in terms of the spherical coordinates are given by

$$x = r \sin(\theta) \cos(\varphi), \ x = r \sin(\theta) \sin(\varphi), \ z = r \cos(\theta). \quad (13.71)$$

Using these relations, it can be shown that the Laplacian in spherical coordinates is given by

$$\nabla^2 = \frac{1}{r^2} \frac{\partial}{\partial r}\left(r^2 \frac{\partial}{\partial r}\right) + \frac{1}{r^2 \sin(\theta)} \frac{\partial}{\partial \theta}\left(\sin(\theta) \frac{\partial}{\partial \theta}\right) \quad (13.72)$$

$$+ \frac{1}{r^2 \sin^2(\theta)} \frac{\partial^2}{\partial \varphi^2}, \quad (13.73)$$

and Laplace's equation becomes

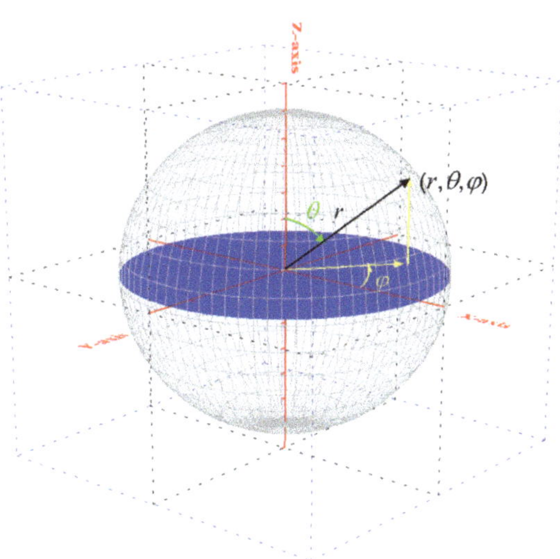

Figure 13.5. A point in space described by spherical coordinates, r, θ, and φ..

$$\nabla^2 V = \frac{1}{r^2} \frac{\partial}{\partial r}\left(r^2 \frac{\partial V}{\partial r}\right) + \frac{1}{r^2 \sin(\theta)} \frac{\partial}{\partial \theta}\left(\sin(\theta) \frac{\partial V}{\partial \theta}\right)$$
$$+ \frac{1}{r^2 \sin^2(\theta)} \frac{\partial^2 V}{\partial \varphi^2} = 0. \tag{13.74}$$

We now determine the general solution to equation (13.74) using the separation of variables. To this end, substituting

$$V(r, \theta, \varphi) = R(r)\Theta(\theta)\Phi(\varphi), \tag{13.75}$$

into equation (13.74), we have

$$\frac{1}{r^2} \frac{\partial}{\partial r}\left[r^2 \frac{\partial}{\partial r}(R(r)\Theta(\theta)\Phi(\varphi))\right] + \frac{1}{r^2 \sin(\theta)} \frac{\partial}{\partial \theta}\left[\right.$$
$$\left. \sin(\theta) \frac{\partial}{\partial \theta}(R(r)\Theta(\theta)\Phi(\varphi))\right]$$
$$+ \frac{1}{r^2 \sin^2(\theta)} \frac{\partial^2}{\partial \varphi^2}(R(r)\Theta(\theta)\Phi(\varphi)) = 0 \tag{13.76}$$
$$\Rightarrow \frac{\Theta(\theta)\Phi(\varphi)}{r^2} \frac{d}{dr}\left(r^2 \frac{dR(r)}{dr}\right) + \frac{R(r)\Phi(\varphi)}{r^2 \sin(\theta)} \frac{d}{d\theta}\left(\sin(\theta) \frac{d\Theta(\theta)}{d\theta}\right)$$
$$+ \frac{R(r)\Theta(\theta)}{r^2 \sin^2(\theta)} \frac{d^2\Phi(\varphi)}{d\varphi^2} = 0$$

so that multiplying equation (13.76) by $r^2 / R(r)\Theta(\theta)\Phi(\varphi)$, one finds

$$\frac{1}{R(r)} \frac{d}{dr}\left(r^2 \frac{dR(r)}{dr}\right)$$
$$+ \frac{1}{\sin(\theta)\Theta(\theta)} \frac{d}{d\theta}\left(\sin(\theta) \frac{d\Theta(\theta)}{d\theta}\right) + \frac{1}{\sin^2(\theta)\Phi(\varphi)} \frac{d^2\Phi(\varphi)}{d\varphi^2} = 0. \tag{13.77}$$

Equation (13.77) can be written as the summation of two functions

$$F_1(r) + F_2(\theta, \varphi) = 0, \tag{13.78}$$

where

$$F_1(r) = \frac{1}{R(r)} \frac{d}{dr}\left(r^2 \frac{dR(r)}{dr}\right),$$
$$F_2(\theta, \varphi) = \frac{1}{\sin(\theta)\Theta(\theta)} \frac{d}{d\theta}\left(\sin(\theta) \frac{d\Theta(\theta)}{d\theta}\right) + \frac{1}{\sin^2(\theta)\Phi(\varphi)} \frac{d^2\Phi(\varphi)}{d\varphi^2}. \tag{13.79}$$

We note that $F_1(r)$ depends only on r and $F_2(\theta, \varphi)$ depends only on θ and φ. But the three variables r, θ, and φ are independent. Therefore, each of these function must be some constant,

$$F_1(r) = k^2, \; F_2(\theta, \varphi) = -k^2, \tag{13.80}$$

such that

$$F_1(r) + F_2(\theta, \varphi) = 0. \tag{13.81}$$

There follows that

$$F_1(r) = k^2 \Rightarrow \frac{1}{R(r)} \frac{d}{dr}\left(r^2 \frac{dR(r)}{dr}\right) = k^2$$

$$\Rightarrow \frac{d}{dr}\left(r^2 \frac{dR(r)}{dr}\right) - k^2 R(r) = 0, \tag{13.82}$$

and

$$F_2(\theta, \varphi) = -k^2$$

$$\Rightarrow \frac{1}{\sin(\theta)\Theta(\theta)} \frac{d}{d\theta}\left(\sin(\theta)\frac{d\Theta(\theta)}{d\theta}\right) + \frac{1}{\sin^2(\theta)\Phi(\varphi)}\frac{d^2\Phi(\varphi)}{d\varphi^2} = -k^2$$

$$\Rightarrow \frac{\sin(\theta)}{\Theta(\theta)}\frac{d}{d\theta}\left(\sin(\theta)\frac{d\Theta(\theta)}{d\theta}\right) + \frac{1}{\Phi(\varphi)}\frac{d^2\Phi(\varphi)}{d\varphi^2} = -k^2 \sin^2\theta \tag{13.83}$$

$$\Rightarrow \frac{\sin(\theta)}{\Theta(\theta)}\frac{d}{d\theta}\left(\sin(\theta)\frac{d\Theta(\theta)}{d\theta}\right) + k^2 \sin^2(\theta) + \frac{1}{\Phi(\varphi)}\frac{d^2\Phi(\varphi)}{d\varphi^2} = 0.$$

In equation (13.83), the first two terms depend on θ, and the third term depends on φ. Thus for a similar reason, we explained earlier, we must have

$$F_2(\varphi) = \frac{1}{\Phi(\varphi)}\frac{d^2\Phi(\varphi)}{d\varphi^2} = -m^2,$$

$$F_4(\theta) = \frac{\sin(\theta)}{\Theta(\theta)}\frac{d}{d\theta}\left(\sin(\theta)\frac{d\Theta(\theta)}{d\theta}\right) + k^2 \sin^2(\theta) = m^2, \tag{13.84}$$

where m is a constant, so that one finds

$$F_2(\varphi) + F_4(\theta) = 0. \tag{13.85}$$

Using equations (13.82) and (13.84), the three second-order differential equations resulting from Laplace's equation in spherical coordinates can then be written as

$$\frac{d^2\Phi(\varphi)}{d\varphi^2} + m^2\Phi(\varphi) = 0,$$

$$\sin(\theta)\frac{d}{d\theta}\left(\sin(\theta)\frac{d\Theta(\theta)}{d\theta}\right) + \sin^2(\theta)\left[k^2 - \frac{m^2}{\sin^2(\theta)}\right]\Theta(\theta) = 0, \tag{13.86}$$

$$r^2\frac{d^2R(r)}{dr^2} + 2r\frac{dR(r)}{dr} - k^2 R(r) = 0.$$

Note that these differential equations are functions of single variables. We now proceed to find the solutions to these equations.

(a) *The function* $\Phi(\varphi)$: we can easily find the solution to the differential equation

$$\frac{d^2\Phi(\varphi)}{d\varphi^2} + m^2\Phi(\varphi) = 0, \tag{13.87}$$

can be expressed as

$$\Phi(\varphi) = A_m \cos(m\varphi) + B_m \sin(m\varphi). \tag{13.88}$$

(b) *The function* $\Theta(\theta)$: In chapter 12 we have solved the associated Legendre differential equation, the solutions of which are given by the associated Legendre polynomials. The differential equation

$$\sin(\theta)\frac{d}{d\theta}\left(\sin(\theta)\frac{d\Theta(\theta)}{d\theta}\right) + \sin^2(\theta)\left[k^2 - \frac{m^2}{\sin^2(\theta)}\right]\Theta(\theta) = 0, \tag{13.89}$$

has almost the same form as the associated Legendre differential equation and the solution can be expressed in terms of the Legendre polynomials. To this end, we introduce the transformation of variable $x = \cos(\theta)$, such that

$$\sin(\theta) = \sqrt{1 - x^2}, \; dx = -\sin(\theta)d\theta, \Rightarrow d\theta = -\frac{dx}{1 - x^2}, \tag{13.90}$$

and rewrite equation (13.89) as

$$(1 - x^2)\frac{d}{dx}\left[(1 - x^2)\frac{d\Theta(x)}{dx}\right] + (1 - x^2)\left[k^2 - \frac{m^2}{1 - x^2}\right]\Theta(x) = 0$$

$$\Rightarrow (1 - x^2)\frac{d^2\Theta(x)}{dx^2} - 2x\frac{d\Theta(x)}{dx} + \left[k^2 - \frac{m^2}{1 - x^2}\right]\Theta(x) = 0. \tag{13.91}$$

We recall the associated Legendre differential equation

$$(1 - x^2)\frac{d^2y(x)}{dx^2} - 2x\frac{dy(x)}{dx} + \left[l(l + 1) - \frac{m^2}{1 - x^2}\right]y = 0 \tag{13.92}$$

the solution of which we found to be the associate Legendre polynomials

$$y(x) = P_l^m(x), \tag{13.93}$$

where

$$l = 0, 1, 2, 3. \,.., \; m = -l....-2, -1, 0, 1, 2...l. \tag{13.94}$$

Note that we have not explained why m has the values given in equation (13.94). We will come to that later. For now, comparing equations (13.91) and (13.92), one finds,

$$k^2 = l(l + 1), \quad \Theta_{lm}(\theta) = P_l^m(\cos(\theta)). \tag{13.95}$$

(c) *The function $R(r)$*: Next we proceed to find the solution for the differential equation for the r-dependent part in equation (13.86),

$$r^2 \frac{d^2 R(r)}{dr^2} + 2r \frac{dR(r)}{dr} - k^2 R(r) = 0. \tag{13.96}$$

To this end, using the result $k^2 = l(l + 1)$, one can rewrite equation (13.96) as

$$\begin{aligned}
& r^2 \frac{d^2 R(r)}{dr^2} + 2r \frac{dR(r)}{dr} - l(l + 1)R(r) = 0 \\
\Rightarrow\ & \frac{d^2 R(r)}{dr^2} + \frac{2}{r} \frac{dR(r)}{dr} - \frac{l(l + 1)}{r^2} R(r) = 0.
\end{aligned} \tag{13.97}$$

This equation has a singularity at $r = 0$. In such cases, we determine the solution by using the method of Frobenius. In this method, we generally begin by assuming the solution is expressible by convergent Frobenius series of the form

$$R(r) = \sum_{n=0}^{\infty} a_n r^{n+s}, \tag{13.98}$$

and determine both the expansion coefficients a_n and the constant s. Noting that

$$-\frac{l(l + 1)}{r^2} R(r) = -\sum_{n=0}^{\infty} l(l + 1) a_n r^{n+s-2},$$

$$\frac{dR(r)}{dr} = \sum_{n=0}^{\infty} a_n(n + s)x^{n+s-1} \Rightarrow \frac{2}{r} \frac{dR(r)}{dr} = \sum_{n=0}^{\infty} 2a_n(n + s)x^{n+s-2}, \tag{13.99}$$

$$\frac{d^2 R(r)}{dr^2} = \sum_{n=0}^{\infty} a_n(n + s)(n + s - 1)x^{n+s-2}.$$

Upon substituting equation (13.99) into equation (13.97), we find

$$\begin{aligned}
& \frac{d^2 R(r)}{dr^2} + \frac{2}{r} \frac{dR(r)}{dr} - \frac{l(l + 1)}{r^2} R(r) \\
& = \sum_{n=0}^{\infty} a_n[(n + s)(n + s - 1) + 2(n + s) - l(l + 1)]x^{n+s-2} \\
& = \sum_{n=0}^{\infty} a_n[(n + s)(n + s + 1) - l(l + 1)]x^{n+s-2} = 0.
\end{aligned} \tag{13.100}$$

There follows that

$$(n + s)(n + s + 1) - l(l + 1) = 0$$
$$\Rightarrow s^2 + (2n + 1)s + n(n + 1) - l(l + 1) = 0, \tag{13.101}$$

which results in

$$s_1 = -(n + l + 1), \quad s_2 = l - n. \tag{13.102}$$

Noting that for $s_1 = -n - (l + 1)$,

$$R(r) = \sum_{n=0}^{\infty} a_n r^{n+s_1} = \frac{\left(\sum_{n=0}^{\infty} a_n\right)}{r^{l+1}} = \frac{A_l}{r^{l+1}}, \tag{13.103}$$

and for $s_2 = l - n$

$$\sum_{n=0}^{\infty} a_n r^{n+s_2} = \left(\sum_{n=0}^{\infty} a_n\right) r^l = C_l r^l, \tag{13.104}$$

one can write the general solution to equation (13.97) as

$$R_l(r) = B_l r^l + \frac{C_l}{r^{l+1}}. \tag{13.105}$$

Therefore, combining equations (13.88), (13.95), and (13.105), the solution to Laplace's equation in spherical coordinates can be written as

$$V(r,\, \theta,\, \varphi) = \sum_{l=0}^{\infty} \sum_{m=-l}^{l} R_l(r)\Theta_{lm}(\theta)\Phi_m(\varphi)$$
$$= \sum_{l=0}^{\infty} \sum_{m=-l}^{l} \left[B_l r^l + \frac{C_l}{r^{l+1}} \right] P_l^m(\cos(\theta)) \tag{13.106}$$
$$[A_m \cos(m\varphi) + B_m \sin(m\varphi)].$$

Note that the values assigned for l are

$$l = 0,\, 1,\, 2... \tag{13.107}$$

as we have already known from the solution to the Legendre differential equation in the previous chapter. But how did we know that

$$m = -l,\, l + 1...-1,\, 0,\, 1,\, ...l - 1? \tag{13.108}$$

We now proceed to prove that these are indeed the values for m. To this end, we note that for the maximum value of $x^2 = \cos^2(\theta)$, we have $x = \pm 1$ when $\theta = 0$ or π. Since in equation (13.92), the factor

$$l(l + 1) - \frac{m^2}{1 - x^2}, \tag{13.109}$$

diverges at $x = \pm 1$, we must have

$$m^2 = 0 \Rightarrow m = 0. \tag{13.110}$$

On the other hand, for the minimum value of $x^2 = \cos^2(\theta)$, we have $x = 0$ when $\theta = \pi/2$, we have

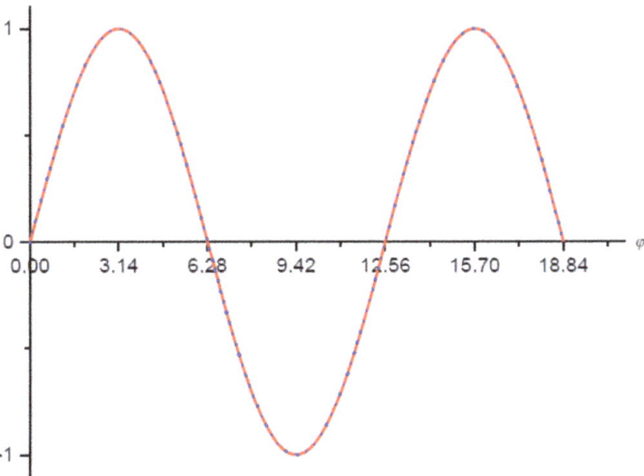

Figure 13.6. The functions $\sin(m\varphi)$ (blue dashed line) and $\sin(m(\varphi + 2\pi))$ (red dotted line) when m is an integer (in this case $m = 1$).

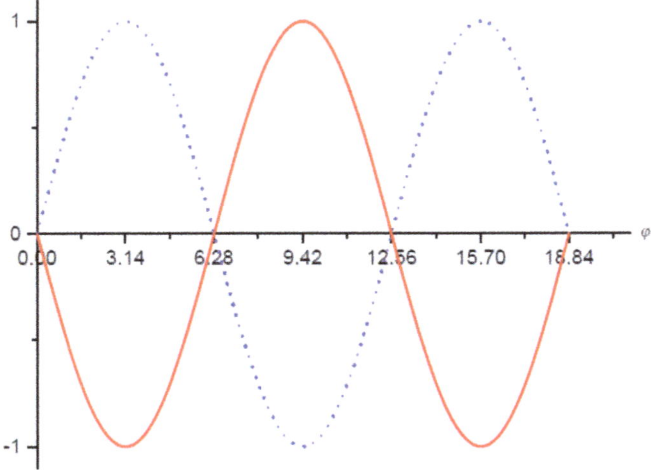

Figure 13.7. The functions $\sin(m\varphi)$ (blue dashed line) and $\sin(m(\varphi + 2\pi))$ (red dotted line) when m is not an integer (in this case $m = 0.5$).

$$l(l + 1) - \frac{m^2}{1 - x^2} = l(l + 1) - m^2, \tag{13.111}$$

which leads to

$$l(l + 1) - m^2 \geqslant 0 \Rightarrow m^2 \leqslant l(l + 1)$$
$$\Rightarrow -\sqrt{l(l + 1)} \leqslant m \leqslant \sqrt{l(l + 1)}. \tag{13.112}$$

However, we must expect m to be an integer. The reasoning behind this argument is shown in figures 13.6 and 13.7. We know that in spherical coordinates $0 \leqslant \varphi \leqslant 2\pi$, both the sine and cosine function,

$$\Phi_m(\varphi) = A_m \cos{(m\varphi)} + B_m \sin{(m\varphi)}, \tag{13.113}$$

must be a single-valued function. When m is an integer the sine function, as shown in figure 13.6, is indeed single valued but when it is non-integer, as shown in figure 13.7, the sine function is not single-valued function for all $\varphi \, \varepsilon \, [0, \, 2\pi]$. Therefore, the fact that m must be an integer set the upper and lower bounds for m to be

$$-\sqrt{l(l + 1)} \leqslant m \leqslant \sqrt{l(l + 1)} \Rightarrow -l \leqslant m \leqslant l. \tag{13.114}$$

Example 13.3. All of the space shown in figure 13.8 was initially filled with a uniform electric field of magnitude, E_o, pointing in the positive z-direction. A grounded conducting sphere of radius, a, with its center at the origin is placed in this uniform electric field. We wish to find the electrostatic potential at all points outside of this sphere.

Note: Inside a conductor, we have an equal number of positive and negative charges. The negative charges are free to move inside the conductor. Thus when we apply an external electric field, the negative charges migrate opposite to the direction

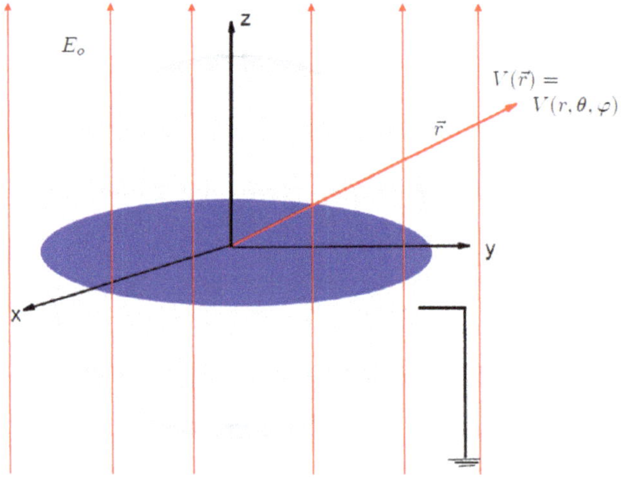

Figure 13.8. A grounded metallic sphere in a uniform electric field.

of the electric field. This results in a separation of positive and negative charges within the conductor (polarization). Furthermore, the electric field is a conservative vector field

$$\vec{E}(\vec{r}) = -\nabla V(\vec{r}).$$

(13.115)

For a uniform electric field, $\vec{E}(\vec{r}) = E_o\hat{z}$, using Cartesian coordinates, one can then write

$$\vec{E}(\vec{r}) = -\nabla V_0(\vec{r}) \Rightarrow -E_o\hat{z} = \frac{\partial V_0(\vec{r})}{\partial x}\hat{x} + \frac{\partial V_0(\vec{r})}{\partial y}\hat{y} + \frac{\partial V_0(\vec{r})}{\partial z}\hat{z}$$

$$\Rightarrow \frac{\partial V_0(\vec{r})}{\partial x} = 0, \ \frac{\partial V_0(\vec{r})}{\partial y} = 0, \ -E_o = \frac{\partial V_0(\vec{r})}{\partial z},$$

$$\Rightarrow V_0(\vec{r}) = -E_o z + C = -E_o r \cos(\theta) + C,$$

(13.116)

where C is a constant. For the sake of simplicity (for this particular problem), we chose this constant to be zero so that

$$V_0(\vec{r}) = -E_o r \cos(\theta).$$

(13.117)

Solution: Since there is no free charge outside the sphere, the electrical potential $V(\vec{r})$ satisfies Laplace's equation. In this case, because of the spherical symmetry, it is easy to solve Laplace's equation in spherical coordinates,

$$\nabla^2 V(\vec{r}) = 0$$

$$\Rightarrow \frac{1}{r^2}\frac{\partial}{\partial r}\left(r^2\frac{\partial}{\partial r}V\left(r, \theta, \varphi\right)\right) + \frac{1}{r^2 \sin(\theta)}$$

$$\frac{\partial}{\partial \theta}\left(\sin(\theta)\frac{\partial}{\partial \theta}V\left(r, \theta, \varphi\right)\right)$$

$$+ \frac{1}{r^2 \sin^2(\theta)}\frac{\partial^2}{\partial \varphi^2}V\left(r, \theta, \varphi\right) = 0.$$

(13.118)

While solving this partial differential equation, we will outline the basic steps we used in the previous example. It can be used as a general guideline to solve partial differential equations.

Step 1: *Identify and write down the boundary conditions for the given problem.*
    (a) *Boundary condition 1*: For this problem, we know that the sphere is grounded. Hence, the electric potential at any point on the surface of the sphere must be zero. That means

$$V(r, \theta, \varphi)|_{r=a} = V\left(a, \theta, \varphi\right) = 0.$$

(13.119)

(b) *Boundary condition 2*: In addition, we should expect that the presence of the conducting sphere far away from the sphere (like at infinity) is negligible. In other words, at infinity ($r \to \infty$), the electric potential is just the electric potential of a constant electric field pointing along the z-direction. Thus

$$\lim_{r \to \infty} V\left(r,\, \theta,\, \varphi\right) = V_0(\vec{r}) = -E_o r \cos(\theta). \tag{13.120}$$

Step 2: *Apply the boundary conditions to the general solution.* The general solution to Laplace's equation in spherical coordinates is given by

$$V(r,\, \theta,\, \varphi) = \sum_{l=0}^{\infty} \sum_{m=-l}^{l} \left[B_l r^l + \frac{C_l}{r^{l+1}} \right] P_l^m(\cos(\theta))[A_m \cos(m\varphi) \\ + B_m \sin(m\varphi)]. \tag{13.121}$$

Using *boundary condition 1*

$$V(a,\, \theta,\, \varphi) = 0, \tag{13.122}$$

we have

$$V(a,\, \theta,\, \varphi) = \sum_{l=0}^{\infty} \sum_{m=-l}^{l} \left(B_l a^l + \frac{C_l}{a^{l+1}} \right) P_l^m(\cos(\theta)) \\ \times (A_m \cos(m\varphi) + B_m \sin(m\varphi)) = 0 \Rightarrow B_l a^l + \frac{C_l}{a^{l+1}} \\ = 0 \Rightarrow C_l = -B_l a^{2l+1'}. \tag{13.123}$$

Using this result we may write the potential as

$$V(r,\, \theta,\, \varphi) = \sum_{l=0}^{\infty} \sum_{m=-l}^{l} B_l \left(r^l - \frac{a^{2l+1}}{r^{l+1}} \right) P_l^m(\cos(\theta)) \\ \times (A_m \cos(m\varphi) + B_m \sin(m\varphi)). \tag{13.124}$$

Now imposing *boundary condition 2* in equation (13.124), we find

$$\lim_{r \to \infty} V\left(r,\, \theta,\, \varphi\right) = \lim_{r \to \infty} \sum_{l=0}^{\infty} \sum_{m=-l}^{l} B_l r^l P_l^m(\cos(\theta)) \\ [A_m \cos(m\varphi) + B_m \sin(m\varphi)] \\ = -E_0 r \cos(\theta). \tag{13.125}$$

Since there is no φ dependence on the right-hand side, we must have

$$m = 0, \tag{13.126}$$

which leads to

$$\lim_{r \to \infty} \sum_{l=0}^{\infty} A_0 B_l r^l P_l(\cos(\theta)) = -E_0 r \cos(\theta) = -E_0 r P_1(\cos(\theta)), \qquad (13.127)$$

where we used

$$\cos(\theta) = P_1(\cos(\theta)). \qquad (13.128)$$

Expanding the series and absorbing the constant A_0 into B_l, we can write

$$B_0 + B_1 r P_1(\cos(\theta)) + \lim_{r \to \infty} \sum_{l=2}^{\infty} B_l r^l P_l(\cos(\theta)) = -E_0 r P_1(\cos(\theta)). \qquad (13.129)$$

Comparing the right and left-hand side of this equation, we find

$$B_1 = -E_0, \ B_l = 0 \text{ for } l \neq 1. \qquad (13.130)$$

Substituting the results in equations (13.124), (13.126) and (13.130) into equation (13.124), one finds

$$V(r, \theta, \varphi) = \left(\frac{a^3}{r^3} - 1\right) E_0 r \cos(\theta). \qquad (13.131)$$

Let us not expand the series; instead, let us apply what we know about the orthogonality of the Legendre polynomials from the previous chapter (equation (13.139))

$$\int_{-1}^{1} P_l(x) P_m(x) dx = \frac{2}{2l + 1} \delta_{lm}, \qquad (13.132)$$

which we rewrite as

$$\int_{0}^{\pi} P_l(\cos(\theta)) P_m(\cos(\theta)) \sin(\theta) d\theta = \frac{2}{2l + 1} \delta_{lm}. \qquad (13.133)$$

To this end, we can rewrite equation (13.127) as

$$\lim_{r \to \infty} \sum_{l=0}^{\infty} B_l r^l \int_{0}^{\pi} P_l(\cos(\theta)) P_m(\cos(\theta)) \sin(\theta) d\theta$$
$$= -E_0 r \int_{0}^{\pi} P_1(\cos(\theta)) P_m(\cos(\theta)) \sin(\theta) d\theta, \qquad (13.134)$$

where we absorbed the constant A_0 into B_l, multiplied equation (13.127) by $P_m(\cos(\theta)) \sin(\theta) d\theta$. Then using the orthogonality relation for the Legendre polynomials in equation (13.133), one finds

$$\lim_{r \to \infty} \sum_{l=0}^{\infty} B_l r^l \frac{2}{2l+1} \delta_{lm} = -E_0 r \frac{2}{2 \times 1 + 1} \delta_{1m}$$

$$\Rightarrow B_m r^m \frac{1}{2m+1} = -\frac{1}{3} E_0 r \delta_{1m}.$$

$$(13.135)$$

There follows that

$$B_m = -\frac{2m+1}{3} E_0 r^{1-m} \delta_{1m} = \begin{cases} 0, & m \neq 1, \\ -E_0 & m = 1, \end{cases} \qquad (13.136)$$

which is exactly the same as the result we found in equation (13.130).

13.4 Laplace's equation in cylindrical coordinates

One more important coordinate system for Laplace's equation to consider is cylindrical coordinates. A point in space in cylindrical coordinates is described by (s, φ, z) as shown in figure 13.9. The s and φ coordinates are related to the x and y Cartesian coordinates by

$$x = s \cos(\varphi), \, y = s \sin(\varphi) \Rightarrow s = \sqrt{x^2 + y^2}, \, \varphi = \tan^{-1}\left[\frac{y}{x}\right], \qquad (13.137)$$

and one can show that Laplace's equation in cylindrical coordinates is given by

$$\nabla^2 V = \frac{1}{s} \frac{\partial}{\partial s}\left(s \frac{\partial V}{\partial s}\right) + \frac{1}{s^2} \frac{\partial^2 V}{\partial \varphi^2} + \frac{\partial^2 V}{\partial z^2} = 0. \qquad (13.138)$$

Using separation of variables,

$$V(s, \varphi, z) = R(s)\phi(\varphi)Z(z), \qquad (13.139)$$

one can rewrite equation (13.138) as

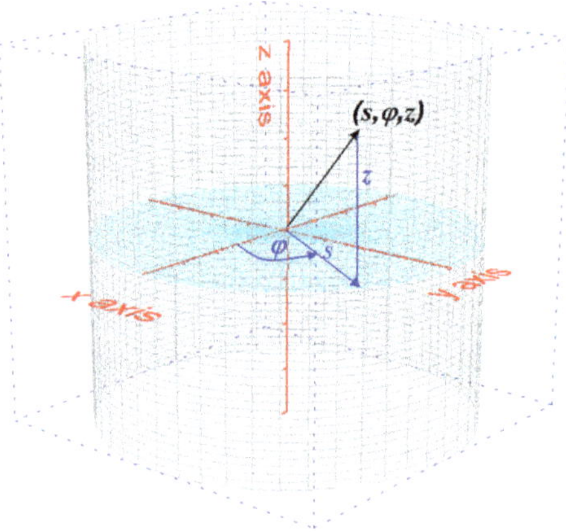

Figure 13.9. A point in space described by spherical coordinates, s, φ, and z..

$$\frac{\phi(\varphi)Z(z)}{s}\frac{d}{ds}\left(s\frac{dR(s)}{ds}\right) + \frac{R(s)Z(z)}{s^2}\frac{d^2\phi(\varphi)}{d\varphi^2} + R(s)\phi(\varphi)\frac{d^2Z(z)}{dz^2} = 0, \quad (13.140)$$

so that multiplying by

$$\frac{s^2}{R(s)\phi(\varphi)Z(z)}, \quad (13.141)$$

we find

$$s^2\left[\frac{1}{sR(s)}\frac{d}{ds}\left(s\frac{dR(s)}{ds}\right) + \frac{1}{Z(z)}\frac{d^2Z(z)}{dz^2}\right] + \frac{1}{\phi(\varphi)}\frac{d^2\phi(\varphi)}{d\varphi^2} = 0. \quad (13.142)$$

Using the similar reasoning we gave earlier in the case of Cartesian and spherical coordinates, one can write,

$$\frac{1}{\phi(\varphi)}\frac{d^2\phi(\varphi)}{d\varphi^2} = -p^2, \quad (13.143)$$

$$s^2\left[\frac{1}{sR(s)}\frac{d}{ds}\left(s\frac{dR(s)}{ds}\right) + \frac{1}{Z(z)}\frac{d^2Z(z)}{dz^2}\right] = p^2, \quad (13.144)$$

where p is a constant. For equation (13.143), the solution is given by

$$\phi(\varphi) = A_p \cos(p\varphi) + B_p \sin(p\varphi). \quad (13.145)$$

Upon rewriting equation (13.144) in the form

$$\frac{1}{sR(s)}\frac{d}{ds}\left(s\frac{dR(s)}{ds}\right) + \frac{1}{Z(z)}\frac{d^2Z(z)}{dz^2} = \frac{p^2}{s^2}$$
$$\Rightarrow \frac{1}{sR(s)}\frac{d}{ds}\left(s\frac{dR(s)}{ds}\right) - \frac{p^2}{s^2} + \frac{1}{Z(z)}\frac{d^2Z(z)}{dz^2} = 0, \quad (13.146)$$

using the same reasoning, one can establish the relation

$$\frac{1}{Z(z)}\frac{d^2Z(z)}{dz^2} = \left(\frac{q}{r_0}\right)^2 \Rightarrow \frac{d^2Z(z)}{dz^2} - \left(\frac{q}{r_0}\right)^2 Z(z) = 0, \quad (13.147)$$

$$\frac{1}{sR(s)}\frac{d}{ds}\left(s\frac{dR(s)}{ds}\right) - \frac{p^2}{s^2} = -\left(\frac{q}{r_0}\right)^2, \quad (13.148)$$

where q is a constant and r_0 is the radius of the cylinder which is also a constant. The solution to the differential equation in equation (13.147) can be rewritten as

$$Z(z) = C_q e^{\frac{q}{r_0}z} + D_q e^{-\frac{q}{r_0}z}. \quad (13.149)$$

For the differential equation in equation (13.148), we have

$$s\frac{d}{ds}\left(s\frac{dR(s)}{ds}\right) - p^2 R(s) = -\left(\frac{q}{r_0}\right)^2 s^2 R(s)$$

$$\Rightarrow s\frac{d}{ds}\left(s\frac{dR(s)}{ds}\right) + \left[\left(\frac{qs}{r_0}\right)^2 - p^2\right]R(s) = 0 \tag{13.150}$$

$$\Rightarrow s^2\frac{d^2 R(s)}{ds^2} + s\frac{dR(s)}{ds} + \left[\left(\frac{qs}{r_0}\right)^2 - p^2\right]R(s) = 0,$$

which one can put in the form

$$\left(\frac{qs}{r_0}\right)^2 \frac{d^2 R(s)}{d\left(\frac{qs}{r_0}\right)^2} + \left(\frac{qs}{r_0}\right)\frac{dR(s)}{d\left(\frac{qs}{r_0}\right)} + \left[\left(\frac{qs}{r_0}\right)^2 - p^2\right]R(s) = 0. \tag{13.151}$$

Introducing the dimensionless variable,

$$x = \frac{qs}{r_0}, \quad dx = d\left(\frac{qs}{r_0}\right), \quad R(s) = y(x), \tag{13.152}$$

we can rewrite equation (13.151) as

$$x^2\frac{d^2 y(x)}{dx^2} + x\frac{dy(x)}{dx} + (x^2 - p^2)y(x) = 0. \tag{13.153}$$

Inside the cylinder where $qs \leqslant r_0$, we have

$$0 \leqslant x = \frac{qs}{r_0} \leqslant 1, \tag{13.154}$$

and the solution to the differential equation

$$x^2\frac{d^2 y(x)}{dx^2} + x\frac{dy(x)}{dx} + (x^2 - p^2)y(x) = 0, \tag{13.155}$$

is the Bessel functions is given by

$$y(x) = a_p J_p(x) + b_p N_p(x). \tag{13.156}$$

Therefore, we can write the solution to equation (13.148) as

$$R(s) = a_p J_p\left(\frac{qs}{r_0}\right) + b_p N_p\left(\frac{qs}{r_0}\right).$$

Before we write the general solution, let us reconsider the solution to the φ dependent differential equation (equation (13.145)), which involves trigonometric functions that must be single-valued. By single-valued function, we mean one must have

$$\cos(p\varphi) = \cos[p(\varphi + 2\pi)], \ \sin(p\varphi) = \sin[p(\varphi + 2\pi)]. \tag{13.157}$$

As one can easily see from figures 13.10 and 13.11, we find a multi-valued function when p is not an integer but we find a single valued function when p is an integer. Consequently, p must be an integer, $p = n = \pm 1, \pm 2, \pm 3 \dots$. Therefore, the general solution to Laplace's equation in cylindrical coordinates can be expressed as

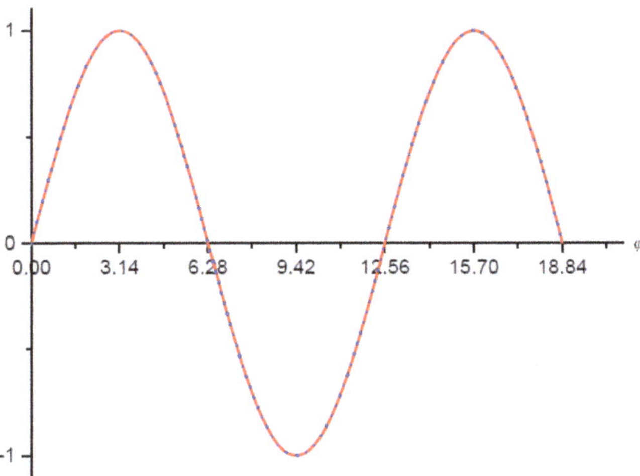

Figure 13.10. The functions $\sin(p\varphi)$ (blue dashed line) and $\sin(p(\varphi + 2\pi))$ (red dotted line), when p is an integer (in this case $p = 1$).

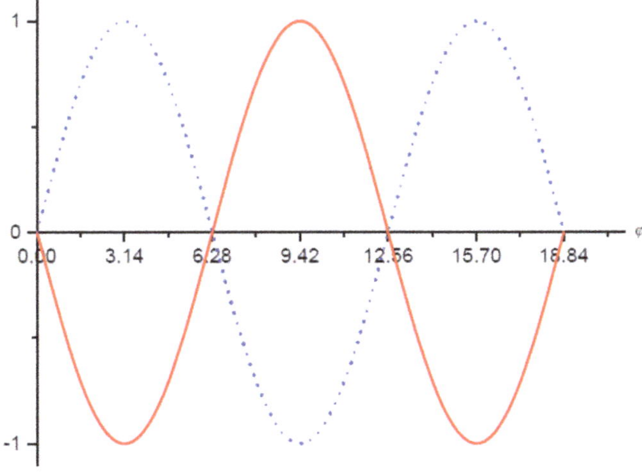

Figure 13.11. The functions $\sin(p\varphi)$ (blue dashed line) and $\sin(p(\varphi + 2\pi))$ (red dotted line), when p is not an integer (in this case $p = 0.5$).

$$V(s, \varphi, z) = \sum_{q \neq 0} \sum_{n=1} \left(C_q e^{\frac{q}{r_0}z} + D_q e^{-\frac{q}{r_0}z} \right)$$
$$\times \, (A_n \cos{(n\varphi)} + B_n \sin{(n\varphi)}) J_n\!\left(\frac{qs}{r_o}\right). \tag{13.158}$$

Note that in writing equation (13.158), from what we introduced in chapter 12, we have taken into account that when the constant p is an integer, $(p = n)$,

$$R(s) = a_p J_p\!\left(\frac{qs}{r_0}\right) + b_p N_p\!\left(\frac{qs}{r_0}\right) = a_n J_n\!\left(\frac{qs}{r_0}\right), \tag{13.159}$$

and also showed $J_n(x)$ and $J_{-n}(x)$ are not independent

$$J_{-n}(x) = (-1)^n J_n(x). \tag{13.160}$$

In the general solution in equation (13.158), we intentionally omitted $p = n = 0$ and $q = 0$ from the summation for a well-justified reason. In order to explain this reason and the solution to Laplace's equation for these values, we reconsider the three equations (13.143), (13.147), and (13.148), which we rewrite here

$$\frac{1}{Z(z)}\frac{d^2 Z(z)}{dz^2} = \left(\frac{q}{r_0}\right)^2, \quad \frac{1}{\phi(\varphi)}\frac{d^2\phi(\varphi)}{d\varphi^2} = -p^2$$
$$s^2\left[\frac{1}{sR(s)}\frac{d}{ds}\left(s\frac{dR(s)}{ds}\right) + \left(\frac{q}{r_0}\right)^2\right] - p^2 = 0. \tag{13.161}$$

For $p = 0$ and $q = 0$, these equations become

$$\frac{1}{Z(z)}\frac{d^2 Z(z)}{dz^2} = 0 \Rightarrow Z(z) = A_0 z + B_0, \tag{13.162}$$

$$\frac{1}{\phi(\varphi)}\frac{d^2\phi(\varphi)}{d\varphi^2} = 0 \Rightarrow \phi(\varphi) = C_0\varphi + D_0, \tag{13.163}$$

$$\frac{s}{R(s)}\frac{d}{ds}\left(s\frac{dR(s)}{ds}\right) = 0 \Rightarrow \frac{d}{ds}\left(s\frac{dR(s)}{ds}\right) = 0, \Rightarrow s\frac{dR(s)}{ds} = F_0$$
$$\Rightarrow dR(s) = F_0\frac{ds}{s} \Rightarrow R(s) = F_0 \ln{(s)}. \tag{13.164}$$

Since $Z(z)$ diverges for $z \to \pm\infty$, and $\phi(\varphi) = \phi(\varphi \pm 2\pi)$, we must have

$$Z(z) = B_0, \; \phi(\varphi) = D_0, \Rightarrow R(s)\phi(\varphi)Z(z) = A_0 \ln{(s)}, \tag{13.165}$$

where we included the constants B_0 and D_0 in A_0. Therefore, the general solution to Laplace's equation can be expressed as

$$V(s, \varphi, z) = A_0 \ln (s)$$

$$+ \sum_{q>0} \sum_{n=1} \left(C_q e^{\frac{q}{r_0} z} + D_q e^{-\frac{q}{r_0} z} \right) \tag{13.166}$$

$$\times (A_n \cos (n\varphi) + B_n \sin (n\varphi)) J_p \left(\frac{qs}{r_0} \right).$$

Example 13.4. A right, circular conducting cylindrical shell of radius r_0 and length L has its axis coincident with the z-axis and the plane bottom and top surfaces at $z = 0$ and $z = l$, respectively. The top and the curved surfaces of the cylinder are grounded,

$$V(s, \varphi, l) = V(r_0, \varphi, z) = 0. \tag{13.167}$$

But the bottom surface is maintained at a potential that depends on the coordinates s and φ given by

$$V(s, \varphi, 0) = s \cos (\varphi), \tag{13.168}$$

(see figure 13.12). Find the potential everywhere inside the cylinder.

Solution : The potential inside the cylinder satisfies Laplace's equation in cylindrical coordinates

$$\nabla^2 V = \frac{1}{s} \frac{\partial}{\partial s} \left(s \frac{\partial V}{\partial s} \right) + \frac{1}{s^2} \frac{\partial^2 V}{\partial \varphi^2} + \frac{\partial^2 V}{\partial z^2} = 0. \tag{13.169}$$

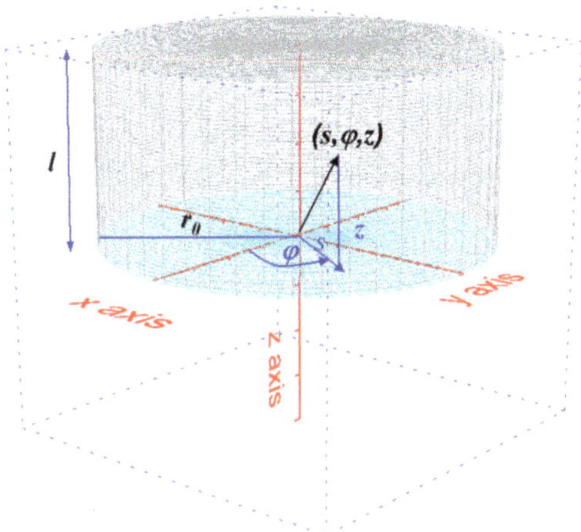

Figure 13.12. Conducting cylindrical shell.

The general solution to equation (13.169) is given by

$$V(s, \varphi, z) = A \ln(s)$$
$$+ \sum_{q>0} \sum_{n=1} \left(C_q e^{\frac{q}{r_0}z} + D_q e^{-\frac{q}{r_0}z} \right)$$
$$\times (A_n \cos(n\varphi) + B_n \sin(n\varphi)) J_n\left(\frac{qs}{r_o}\right). \tag{13.170}$$

This potential must satisfy the boundary conditions

$$V(s, \varphi, 0) = s\cos(\varphi), \quad V(s, \varphi, L) = 0, \quad \text{and} \quad V(r_0, \varphi, z) = 0. \tag{13.171}$$

Applying the boundary condition, $V(s, \varphi, 0) = s\cos(\varphi)$, we find

$$V(s, \varphi, 0) = A_0 \ln(s)$$
$$+ \sum_{q>0} \sum_{n=1} (C_q + D_q)(A_n \cos(n\varphi) + B_n \sin(n\varphi)) J_n\left(\frac{qs}{r_o}\right)$$
$$= s\cos(\varphi) \tag{13.172}$$
$$\Rightarrow A_n = 0, \quad \text{for } n \neq 1, \ B_n = 0, \quad \text{for } n \geqslant 1$$
$$\Rightarrow J_n\left(\frac{qs}{r_o}\right) = J_1\left(\frac{qs}{r_o}\right).$$

Therefore, in view of the results in equation (13.172), the potential in equation (13.170) reduced to

$$V(s, \varphi, z) = \sum_q \left(C_q e^{\frac{q}{r_0}z} + D_q e^{-\frac{q}{r_0}z} \right) \cos(\varphi) J_1\left(\frac{qs}{r_o}\right), \tag{13.173}$$

where we included A_1 into the constants, C_q and D_q. Imposing the second boundary condition, $V(s, \varphi, L) = 0$, on equation (13.173), one finds

$$V(s, \varphi, L) = \sum_{q>0} \left(C_q e^{\frac{q}{r_0}L} + D_q e^{\frac{q}{r_0}L} \right) \cos(\varphi) J_1\left(\frac{qs}{r_o}\right) = 0 \tag{13.174}$$
$$\Rightarrow C_q e^{\frac{q}{r_0}L} + D_q e^{-\frac{q}{r_0}L} = 0 \Rightarrow D_q = -C_q e^{2\frac{q}{r_0}L}.$$

Upon substituting equation (13.174) into equation (13.173), the potential can be rewritten as

$$V(s, \varphi, z) = \sum_{q>0} C_q \left(e^{\frac{q}{r_0}z} - e^{2\frac{q}{r_0}L}e^{-\frac{q}{r_0}z} \right) \cos(\varphi) J_1\left(\frac{qs}{r_o} \right)$$

$$= \sum_{q>0} 2C_q e^{\frac{q}{r_0}L} \left(\frac{e^{-\frac{q}{r_0}(L-z)} - e^{\frac{q}{r_0}(L-z)}}{2} \right) \cos(\varphi) J_p\left(\frac{qs}{r_o} \right) \tag{13.175}$$

$$\Rightarrow V(s, \varphi, z) = \sum_{q>0} F_q \sinh\left[\frac{q}{r_0}(L-z) \right] \cos(\varphi) J_1\left(\frac{qs}{r_o} \right).$$

Next, we apply the last boundary condition, $V(r_0, \varphi, z) = 0$,

$$V(r_0, \varphi, z) = \sum_{q>0} F_q \sinh\left[\frac{q}{r_0}(L-z) \right] \cos(\varphi) J_1(q) = 0 \tag{13.176}$$

$$\Rightarrow J_1(q) = 0, \Rightarrow q = \alpha_i, \text{ for } i = 1, 2, 3..., $$

where α_1, α_2, ... are *the zeroes* of the Bessel function. In view of the result in equation (13.176), the potential in equation (13.175) becomes

$$V(s, \varphi, z) = \sum_{i=1} F_i \sinh\left(\frac{\alpha_i}{r_0}(L-z) \right) J_1\left(\frac{\alpha_i s}{r_o} \right) \cos(\varphi). \tag{13.177}$$

To find the last constant F_i, first we reapply the first boundary condition

$$V(s, \varphi, 0) = s \cos(\varphi), \tag{13.178}$$

so that one can write

$$V(s, \varphi, 0) = \sum_i F_i \sinh\left(\frac{\alpha_i}{r_0}L \right) \cos(\varphi) J_1\left(\frac{\alpha_i s}{r_o} \right) = s \cos(\varphi)$$

$$\Rightarrow \sum_i F_i \sinh\left(\frac{\alpha_i}{r_0}L \right) J_1\left(\frac{\alpha_i s}{r_o} \right) = s. \tag{13.179}$$

We then employ the orthogonality relation for the Bessel functions

$$\int_0^1 J_p(\alpha_i x) J_p(\alpha_j x) x \, dx = \frac{J_p'(\alpha_i) J_p'(\alpha_j)}{2} \delta_{ij}. \tag{13.180}$$

To this end, multiplying both sides of equation (13.179) by

$$J_1\left(\frac{\alpha_j s}{r_o} \right) \frac{s}{r_o} d\left(\frac{s}{r_o} \right) \tag{13.181}$$

and integrating over s, we find

$$\sum_i F_i \sinh\left(\frac{\alpha_i}{r_0}L\right)\int_0^{r_0} J_1\left(\frac{\alpha_i s}{r_o}\right)J_1\left(\frac{\alpha_j s}{r_o}\right)\frac{s}{r_o}d\left(\frac{s}{r_o}\right)$$

$$= r_o \int_0^{r_0} \frac{s}{r_o}J_1\left(\frac{\alpha_j s}{r_o}\right)\frac{s}{r_o}d\left(\frac{s}{r_o}\right) \tag{13.182}$$

$$\Rightarrow \sum_i F_i \sinh\left(\frac{\alpha_i}{r_0}L\right)\int_0^1 J_1(\alpha_i x)J_1(\alpha_j x)x\,dx = r_0\int_0^1 x J_1(\alpha_j x)x\,dx,$$

where we replaced $x = s/r_o$. Applying the Bessel functions orthogonality relation in equation (13.180), one can put equation (13.182) in the form

$$\sum_i F_i \sinh\left(\frac{\alpha_i}{r_0}L\right)\frac{J_1'(\alpha_i)J_1'(\alpha_j)}{2}\delta_{ij} = r_0\int_0^1 x^2 J_1(\alpha_j x)dx.$$

$$\Rightarrow F_j \sinh\left(\frac{\alpha_j}{r_0}L\right)\frac{J_1'^2(\alpha_j)}{2} = r_0\int_0^1 x^2 J_1(\alpha_j x)dx \tag{13.183}$$

$$\Rightarrow F_j = \left[r_0\int_0^1 x^2 J_1(\alpha_j x)dx\right]\Big/\left[\sinh\left(\frac{\alpha_j}{r_0}L\right)\frac{J_1'^2(\alpha_j)}{2}\right].$$

Using the series for $J_n(\alpha_j x)$ or *Mathematica,* one can show that

$$\begin{cases} \displaystyle\int_0^1 x^2 J_1(\alpha_j x)dx = \frac{J_2(\alpha_j)}{\alpha_j} \\ J_p'(x) = J_{p+1}(x) \Rightarrow J_1'(\alpha_j) = J_2(\alpha_j) \end{cases}. \tag{13.184}$$

Upon substituting equation (13.184) into equation (13.183), we find

$$F_j = 2r_0/\alpha_j \sinh\left(\frac{\alpha_j}{r_0}L\right)J_2(\alpha_j) \tag{13.185}$$

and the potential in equation (13.177) can be rewritten as

$$V(s, \varphi, z) = \left[\sum_{j=1}^{\infty} \frac{2r_0 \sinh\left(\frac{\alpha_j}{r_0}(L-z)\right)}{\alpha_j \sinh\left(\frac{\alpha_j}{r_0}L\right)}\frac{J_1\left(\frac{\alpha_j s}{r_o}\right)}{J_2(\alpha_j)}\right]\cos(\varphi), \tag{13.186}$$

equation (13.186) is the potential inside the cylinder that satisfies the given three boundary conditions.

13.5 Poisson's equation

In the last three sections, we have seen several examples from electrostatics to determine the electric potential when there is no charge in the region with specific boundary conditions. In such cases, the electric potential satisfies Laplace's equation,

$$\frac{\partial^2}{\partial x^2} V(x, y, z) + \frac{\partial^2}{\partial y^2} V(x, y, z) + \frac{\partial^2}{\partial z^2} V(x, y, z) = 0, \tag{13.187}$$

$$\frac{1}{r}\frac{\partial}{\partial r}\left(r\frac{\partial}{\partial r} V(r, \varphi, z)\right) + \frac{1}{r^2}\frac{\partial^2}{\partial \varphi^2} V(r, \varphi, z) + \frac{\partial^2}{\partial z^2} V(r, \varphi, z) = 0, \tag{13.188}$$

$$\frac{1}{r^2}\frac{\partial}{\partial r}\left(r^2\frac{\partial}{\partial r} V(r, \theta, \varphi)\right)$$
$$+ \frac{1}{r^2 \sin(\theta)}\left[\sin(\theta)\frac{\partial}{\partial \theta}\left(\sin(\theta)\frac{\partial}{\partial \theta} V(r, \theta, \varphi)\right) + \frac{\partial^2}{\partial \varphi^2} V(r, \theta, \varphi)\right] \tag{13.189}$$
$$= 0,$$

and we have determined the general solutions in Cartesian, spherical, and cylindrical coordinates in the last three sections. We now see how we determine the electric potential in a region where there is some charge distribution, $\rho(\vec{r})$. In such cases the electric potential, $V(x, y, z)$, satisfies Poisson's equation given by

$$\nabla^2 V(x, y, z) = -\frac{\rho(\vec{r})}{\epsilon_0}, \tag{13.190}$$

where ϵ_0 is the electrical permittivity of a free space. We will discuss the basic procedures for solving Poisson's equation and applying the boundary conditions using an example in spherical coordinates.

Approach to solving Poisson's equation
Chapter 6 introduced the basic steps we can follow to solve second-order non-homogeneous linear differential equations. For example, for the differential equation of the form,

$$\frac{d^2y}{dx^2} + a\frac{dy}{dx} + y = f(x), \tag{13.191}$$

the solution is the sum of the homogenous, $y_h(x)$, and the particular, $y_p(x)$, solutions given by

$$y(x) = y_h(x) + y_p(x). \tag{13.192}$$

We determined the homogenous solution, $y_h(x)$, by solving

$$\frac{d^2 y_h}{dx^2} + a\frac{dy_h}{dx} + y_h = 0. \tag{13.193}$$

To find the particular solution, $y_p(x)$, we used different techniques that depends on the explicit form of the function $f(x)$. Poisson's equation

$$\nabla^2 V(\vec{r}) = -\frac{\rho(\vec{r})}{\epsilon_0}, \tag{13.194}$$

is a non-homogeneous second-order partial differential equation, and we can follow a similar approach to find its solutions. The homogenous solution, $V_h(\vec{r})$, is the solution to Laplace's equation,

$$\nabla^2 V_h(\vec{r}) = 0, \tag{13.195}$$

that we already know the general solutions in Cartesian, spherical, or cylindrical coordinates. If we determine the particular solution, $V_p(\vec{r})$, then the solution to Poisson's equation is given by

$$V(\vec{r}) = V_h(\vec{r}) + V_p(\vec{r}). \tag{13.196}$$

We can show that this is indeed the solution to Poisson's equation. Noting that

$$\nabla^2 V(\vec{r}) = \nabla^2\left[V_h(\vec{r}) + V_p(\vec{r})\right] = \nabla^2 V_h(\vec{r}) + \nabla^2 V_p(\vec{r}), \tag{13.197}$$

$V_h(\vec{r})$ is the solution to Laplace's equation, and $V_p(\vec{r})$ is the particular solution, one can show that

$$\nabla^2 V_h(\vec{r}) = 0, \ \nabla^2 V_p(\vec{r}) = -\frac{\rho(\vec{r})}{\epsilon_0},$$

$$\Rightarrow \nabla^2 V(\vec{r}) = \nabla^2\left[V_h(\vec{r}) + V_p(\vec{r})\right] = -\frac{\rho(\vec{r})}{\epsilon_0}. \tag{13.198}$$

After we determine the homogenous and the particular solutions, like we have done for linear non-homogeneous differential equations, we impose the boundary conditions to the general solution, $V(\vec{r}) = V_h(\vec{r}) + V_p(\vec{r})$.

Example 13.5. *Point charge and grounded conducting sphere*: A point charge Q has positioned a distance, a, from the center of a grounded, conducting sphere of radius R, where $R < a$. The center of the sphere is at the origin of coordinates, and the point charge is on the positive z-axis. Find the electrostatic potential outside the sphere (figure 13.13).

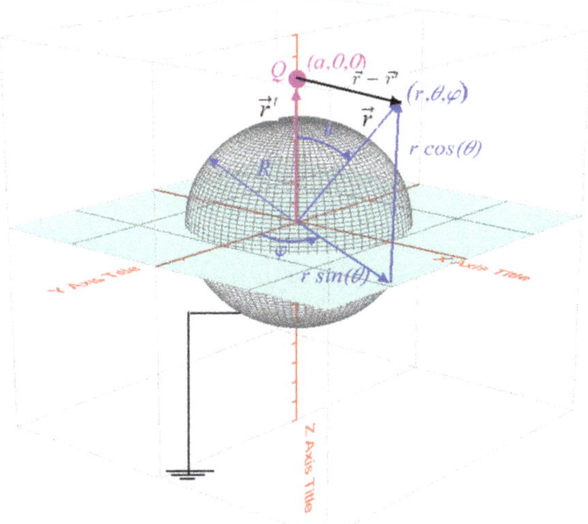

Figure 13.13. A point charge, Q, and a grounded conducting sphere of radius, R.

Solution: We want to find the electric potential at some point outside the sphere, $r > R$. Since there is a point charge along the z-axis at a distance, $a > R$, which is in the region outside the sphere, we cannot say that the charge density is zero everywhere outside the sphere (i.e., $\rho(\vec{r}) \neq 0$). As a result, Laplace's equation is not valid outside the sphere. Therefore, to find the electric potential, we need to find the solution to Poisson's equation. From the symmetry of the problem, it is better to use spherical coordinates

$$\nabla^2 V(r, \theta, \varphi) = -\frac{\rho(\vec{r})}{\epsilon_0}$$

$$\Rightarrow \frac{1}{r^2}\frac{\partial}{\partial r}\left(r^2\frac{\partial}{\partial r}V(r, \theta, \varphi)\right) + \frac{1}{r^2\sin(\theta)}\frac{\partial}{\partial \theta}\left(\sin(\theta)\frac{\partial}{\partial \theta}V(r, \theta, \varphi)\right) \tag{13.199}$$

$$+ \frac{1}{r^2\sin^2(\theta)}\frac{\partial^2}{\partial \varphi^2}V(r, \theta, \varphi) = -\frac{\rho(\vec{r})}{\epsilon_0}.$$

The homogenous solution for this equation is the general solution to Laplace's equation in spherical coordinates

$$\frac{1}{r^2}\frac{\partial}{\partial r}\left(r^2\frac{\partial}{\partial r}V_h(r, \theta, \varphi)\right) + \frac{1}{r^2\sin(\theta)}\frac{\partial}{\partial \theta}\left(\sin(\theta)\frac{\partial}{\partial \theta}V_h(r, \theta, \varphi)\right)$$

$$+ \frac{1}{r^2\sin^2(\theta)}\frac{\partial^2}{\partial \varphi^2}V_h(r, \theta, \varphi) = 0, \tag{13.200}$$

and the solution is given by

$$V_h(r, \theta, \varphi) = \sum_{l=0}^{\infty}\sum_{m=-l}^{l}\left(B_l r^l + \frac{C_l}{r^{l+1}}\right)P_l^m(\cos(\theta)) \tag{13.201}$$

$$\times (A_m\cos(m\varphi) + B_m\sin(m\varphi)).$$

We also know that for a point charge, Q, located at a position, $\vec{r}'$, the electric potential at a distance, $\vec{r}$, is given by

$$V_Q(r,\,\theta,\,\varphi) = \frac{1}{4\pi\epsilon_0}\frac{Q}{|\vec{r}-\vec{r}'|} = \frac{1}{4\pi\epsilon_0}\frac{Q}{\left(r^2 + a^2 - 2ra\cos(\theta)\right)^{1/2}}, \qquad (13.202)$$

where we used

$$\vec{r} = r\sin(\theta)\cos(\varphi)\hat{x} + r\sin(\theta)\sin(\varphi)\hat{y} + r\cos(\theta)\hat{z},\ \vec{r}' = a\hat{z},$$
$$\Rightarrow |\vec{r}-\vec{r}'| = \left(r^2 + a^2 - 2ra\cos(\theta)\right)^{1/2}. \qquad (13.203)$$

To find the particular solution, we note that the charge outside the sphere is zero except at the point where the point charge is positioned (the z-axis is distance a from the origin). Therefore, we can make an 'informed' choice to the particular solution for the Poisson's equation

$$\nabla^2 V(r,\,\theta,\,\varphi) = -\frac{\rho(\vec{r})}{\epsilon_0} \qquad (13.204)$$

is that of the electrical potential of this point charge

$$V_p(r,\,\theta,\,\varphi) = V_Q(r,\,\theta,\,\varphi) = \frac{1}{4\pi\epsilon_0}\frac{Q}{\left(r^2 + a^2 - 2ra\cos(\theta)\right)^{1/2}}. \qquad (13.205)$$

The general solution to the Poisson's equation can then be written as

$$V(r,\,\theta,\,\varphi) = V_h(r,\,\theta,\,\varphi) + V_p(r,\,\theta,\,\varphi) = \sum_{l=0}^{\infty}\sum_{m=-l}^{l}\left(B_l r^l + \frac{C_l}{r^{l+1}}\right)P_l^m(\cos(\theta))$$

$$\times (A_m\cos(m\varphi) + B_m\sin(m\varphi)) \qquad (13.206)$$

$$+ \frac{1}{4\pi\epsilon_0}\frac{Q}{\left(r^2 + a^2 - 2ra\cos(\theta)\right)^{1/2}}.$$

Next, we will apply the boundary conditions.

1. The electric potential must be finite anywhere outside including $r \to \infty$,

$$\lim_{r\to\infty} V(r,\,\theta,\,\varphi) = 0.$$

2. At any point on the surface of the sphere ($r = R$), the electric potential must be zero since the sphere is grounded,

$$V(R,\,\theta,\,\varphi) = 0. \qquad (13.207)$$

Applying the first boundary condition, we find

$$\lim_{r\to\infty} V(r,\,\theta,\,\varphi) = 0 \Rightarrow \lim_{n\to\infty} \sum_{l=0}^{\infty} \sum_{m=-l}^{l} \left(B_l r^l + \frac{C_l}{r^{l+1}} \right) P_l^m(\cos(\theta))$$

$$\times (A_m \cos(m\varphi) + B_m \sin(m\varphi))$$

$$+ \frac{1}{4\pi\epsilon_0} \frac{Q}{(r^2 + a^2 - 2ra\cos(\theta))^{1/2}} = 0. \qquad (13.208)$$

$$\Rightarrow \lim_{r\to\infty} \sum_{l=0}^{\infty} \sum_{m=-l}^{l} B_l r^l P_l^m(\cos(\theta))(A_m \cos(m\varphi) + B_m$$

$$\sin(m\varphi)) = 0,$$

equation (13.208) diverges as $r \to \infty$. The only way we can prevent this from happening is by setting $B_l = 0$ for all values of l. Therefore, the electric potential in equation (13.206) becomes

$$V(r,\,\theta,\,\varphi) = \sum_{l=0}^{\infty} \sum_{m=-l}^{l} \frac{C_l}{r^{l+1}} P_l^m(\cos(\theta))(A_m \cos(m\varphi) + B_m \sin(m\varphi))$$

$$+ \frac{1}{4\pi\epsilon_0} \frac{Q}{(r^2 + a^2 - 2ra\cos(\theta))^{1/2}}. \qquad (13.209)$$

Now imposing the second boundary condition $V(R,\,\theta,\,\varphi) = 0$, on equation (13.209), we find

$$V(R,\,\theta,\,\varphi) = \sum_{l=0}^{\infty} \sum_{m=-l}^{l} \frac{C_l}{R^{l+1}} P_l^m(\cos(\theta))(A_m \cos(m\varphi) + B_m \sin(m\varphi))$$

$$+ \frac{1}{4\pi\epsilon_0} \frac{Q}{(R^2 + a^2 - 2Ra\cos(\theta))^{1/2}} = 0. \qquad (13.210)$$

Applying the relation for the Legendre polynomials,

$$\frac{1}{|\vec{r} - \vec{r}\,'|} = \frac{1}{\sqrt{r^2 + r'^2 - 2rr'\cos(\theta)}} = \sum_{l=0}^{\infty} \frac{r_<^l}{r_>^{l+1}} P_l(\cos(\theta)), \qquad (13.211)$$

one can write

$$\frac{1}{4\pi\epsilon_0} \frac{Q}{(R^2 + a^2 - 2Ra\cos(\theta))^{1/2}} = \frac{Q}{4\pi\epsilon_0} \sum_{l=0}^{\infty} \frac{R^l}{a^{l+1}} P_l(\cos(\theta)), \qquad (13.212)$$

for $R < a$. In view of equation (13.212), equation (13.210) can be rewritten as

$$V(R, \theta, \varphi) = \sum_{l=0}^{\infty} \sum_{m=-l}^{l} \frac{C_l}{R^{l+1}}(A_m \cos(m\varphi) + B_m \sin(m\varphi))P_l^m(\cos(\theta))$$

$$+ \frac{Q}{4\pi\epsilon_0} \sum_{l=0}^{\infty} \frac{R^l}{a^{l+1}} P_l(\cos(\theta)) = 0. \tag{13.213}$$

The above equation to be zero independent of φ, at least the coefficients for $P_l^m(\cos(\theta))$ must be independent of $\cos(m\varphi)$ and $\sin(m\varphi)$. This requires

$$A_m = 0, \; B_m = 0 \text{ for } m \neq 0 \tag{13.214}$$

and equation (13.213) becomes

$$V(R, \theta, \varphi) = \sum_{l=0}^{\infty} \frac{C_l A_0}{R^{l+1}} P_l(\cos(\theta)) + \frac{Q}{4\pi\epsilon_0} \sum_{l=0}^{\infty} \frac{R^l}{a^{l+1}} P_l(\cos(\theta)) = 0$$

$$\Rightarrow \sum_{l=0}^{\infty} \left(\frac{C_l'}{R^{l+1}} + \frac{Q}{4\pi\epsilon_0} \frac{R^l}{a^{l+1}} \right) P_l(\cos(\theta)) = 0, \tag{13.215}$$

$$\Rightarrow \frac{C_l'}{R^{l+1}} + \frac{Q}{4\pi\epsilon_0} \frac{R^l}{a^{l+1}} = 0 \Rightarrow C_l' = -\frac{Q}{4\pi\epsilon_0} \frac{R^{2l+1}}{a^{l+1}}$$

where $C_l' = C_l A_0$. Using the results

$$C_l' = C_l A_0 = -\frac{Q}{4\pi\epsilon_0} \frac{R^{2l+1}}{a^{l+1}}, \; A_m = 0, \; B_m = 0 \text{ for } m \neq 0, \tag{13.216}$$

one finds for the potential in equation (13.210)

$$V(r, \theta, \varphi) = \sum_{l=0}^{\infty} \frac{C_l A_0}{r^{l+1}} P_l^0(\cos(\theta)) + \frac{1}{4\pi\epsilon_0} \frac{Q}{(r^2 + a^2 - 2ra\cos(\theta))^{1/2}}$$

$$= -\frac{Q}{4\pi\epsilon_0} \sum_{l=0}^{\infty} \frac{1}{r^{l+1}} \frac{R^{2l+1}}{a^{l+1}} P_l(\cos(\theta)) \tag{13.217}$$

$$+ \frac{1}{4\pi\epsilon_0} \frac{Q}{(R^2 + a^2 - 2Ra\cos(\theta))^{1/2}}.$$

The first term in equation (13.217) can be rewritten as

$$-\frac{Q}{4\pi\epsilon_0} \sum_{l=0}^{\infty} \frac{1}{r^{l+1}} \frac{R^{2l+1}}{a^{l+1}} P_l(\cos(\theta)) = -\frac{Q\frac{R}{a}}{4\pi\epsilon_0} \sum_{l=0}^{\infty} \frac{1}{r^{l+1}} \left(\frac{R^2}{a} \right)^l P_l(\cos(\theta))$$

$$= \frac{q'}{4\pi\epsilon_0} \sum_{l=0}^{\infty} \frac{r'^l}{r^{l+1}} P_l(\cos(\theta)), \tag{13.218}$$

where

$$r' = \frac{R^2}{a} < r, \; q' = -Q\frac{R}{a}. \tag{13.219}$$

Apply the relation in equation (13.211), one finds for equation (13.218)

$$-\frac{Q}{4\pi\epsilon_0}\sum_{l=0}^{\infty}\frac{1}{r^{l+1}}\frac{R^{2l+1}}{a^{l+1}}P_l(\cos(\theta)) = \frac{q'}{4\pi\epsilon_0}\frac{1}{\sqrt{r^2 + r'^2 - 2rr'\cos(\theta)}}, \tag{13.220}$$

and the electric potential in equation (13.217) can be expressed as

$$V(r,\,\theta,\,\varphi) = \frac{q'}{4\pi\epsilon_0}\frac{1}{\sqrt{r^2 + r'^2 - 2rr'\cos\theta}} + \frac{Q}{4\pi\epsilon_0}\frac{1}{\sqrt{r^2 + a^2 - 2ra\cos\theta}}, \tag{13.221}$$

where r' and q' are given by equation (13.219).

The result in equation (13.221) show the electric potential due to a point charge Q placed outside a conducting sphere is the sum of the potential due to the charge Q and its 'image charge' $q' = -Q\frac{R}{a}$. This image charge is inside the sphere positioned at a distance $r' = \frac{R^2}{a}$ from the sphere's center. In reality, this image charge is not a charge existing at this position. It represents the total induced charge on the grounded conducting sphere due to the electric field by the point charge Q. The method of replacing this induced charge with an image charge to find the electric potential is called the method of images. It is a valuable method in determining the electric potential of charges placed around conducting sphere, cylinder, or plate.

13.6 Homework assignment

Problem 1. In electromagnetism we often see electromagnetic wave equations. These equations are determined from Maxwell's equations in free space where there is no charge and current, given by

$$\nabla \cdot \vec{E} = 0,\ \nabla \cdot \vec{B} = 0,\ \nabla \times \vec{E} = -\frac{\partial \vec{B}}{\partial t},\ \nabla \times \vec{B} = \frac{1}{c^2}\frac{\partial \vec{E}}{\partial t}.$$

Note that c is the speed of light in vacuum. Using Maxwell's equation show that the electric field ($\vec{E}$) and the magnetic field ($\vec{B}$) satisfy the three-dimensional wave equations

$$\nabla^2\vec{E} = \frac{1}{c^2}\frac{\partial^2\vec{E}}{\partial t^2},\ \nabla^2\vec{B} = \frac{1}{c^2}\frac{\partial^2\vec{B}}{\partial t^2}. \tag{13.222}$$

Hint: take the curl of the curl of the electric field and apply the vector identity

$$\nabla \times \left(\nabla \times \vec{A}\right) = \nabla\left(\nabla \cdot \vec{A}\right) - \nabla^2\vec{A}. \tag{13.223}$$

Problem 2. Consider a string (for example a piano or violin or a guitar string) with linear mass density, λ, be stretched tightly by a tension force T and its ends fastened to supports at $x = 0$ and $x = l$. When the string is plucked (pulled aside a small distance and let go) and the string is vibrating (see figure 13.14), let's represent its vertical displacement from its equilibrium position at a given point on the string (x) and at a given time (t) be $f(x, t)$. Consider an infinitesimal segment of the string shown in figure 13.15. When the string is displaced from the equilibrium, the net transverse force on the segment between x and $x + \Delta x$ can be expressed as

$$\Delta F = T(\sin(\alpha) - \sin(\theta)). \tag{13.224}$$

We assume the equilibrium displacement $f(x, t)$ is always very small and that the slope at x and $x + \Delta x$, given by

$$\tan(\alpha) = \left.\frac{\partial f}{\partial x}\right|_x, \quad \tan(\theta) = \left.\frac{\partial f}{\partial x}\right|_{x+\Delta x}, \tag{13.225}$$

are very small and one can make the approximation

$$\sin(\alpha) \simeq \tan(\alpha), \; \sin(\theta) \simeq \tan(\theta). \tag{13.226}$$

Figure 13.14. The top is a snapshot of a tensioned string. The bottom is a snapshot of same string after a wave (a disturbance) is created on a strong. The incident and reflected wave at the ends form a standing wave.

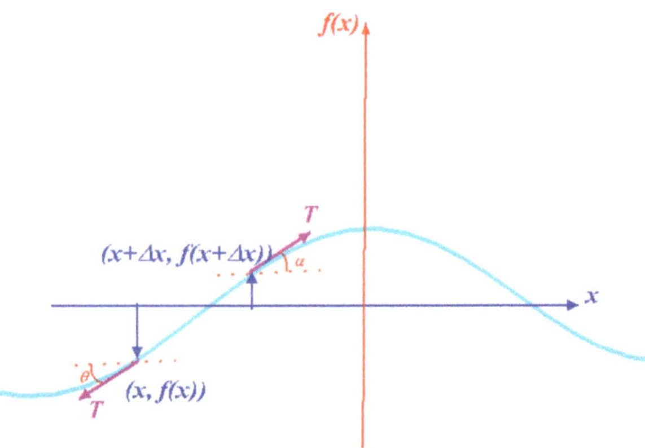

Figure 13.15. The displacement of the wave from the equilibrium position, $f(x)$ vs. x. Note that I did not show time dependence for the function $(f(x, t))$ in the graph as it represent a snapshot of the vibration.

Using this approximation and Newton's second law

$$\Delta F = ma = m\frac{\partial^2 f}{\partial t^2} \tag{13.227}$$

in the limit as $\Delta x \to 0$, show that the vibrating string obeys the wave equation

$$\frac{\partial^2 f}{\partial x^2} = \frac{1}{v^2}\frac{\partial^2 f}{\partial t^2}, \tag{13.228}$$

where v is the speed of the wave

$$v = \sqrt{\frac{T}{\lambda}}. \tag{13.229}$$

Hint: the mass is given by

$$m = \lambda \Delta x, \tag{13.230}$$

note that

$$\frac{\partial^2 f}{\partial x^2} = \lim_{\Delta z \to 0}\left[\frac{\left.\frac{\partial f}{\partial x}\right|_{x+\Delta x} - \left.\frac{\partial f}{\partial x}\right|_{x}}{\Delta x}\right]. \tag{13.231}$$

Problem 3. Using separation of variable

$$f(x, t) = X(x)T(t) \tag{13.232}$$

in the wave equation for the vibrating string

$$\frac{\partial^2 f}{\partial x^2} = \frac{1}{v^2}\frac{\partial^2 f}{\partial t^2}, \tag{13.233}$$

(a) Show that

$$\frac{d^2 X}{dx^2} + k^2 X = 0, \quad \frac{d^2 T}{dt^2} + k^2 v^2 T = 0, \tag{13.234}$$

where k is a positive constant.
(b) Write the solutions to the differential equations in part (a).
(c) The string is fastened at $x = 0$ and $x = l$. This means the displacement from the equilibrium position at these two points at any time t is zero, which means, conditions

$$f(0, t) = 0, f(l, t) = 0. \tag{13.235}$$

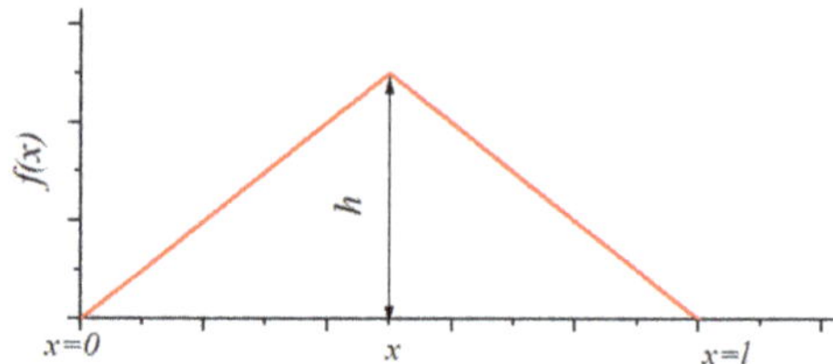

Figure 13.16. The shape of the string (the displacement from the equilibrium position) when it is plucked at the center at the initial time (i.e., $f(x, 0)$ vs x). Note that here $f(x)$ is $F_0(x) = f(x, 0)$.

Furthermore one begins the vibration by plucking the string a small distance, h, for example near center of the string as shown figure 13.16 This means at $t = 0$, the shape of the string is defined by some function $F_0(x) = f(x, 0)$ and none of the points are vibrating at this instant,

$$\left. \frac{df(x, t)}{dt} \right|_{t=0} = 0. \tag{13.236}$$

Using these boundary conditions and the solutions determined in part (b), show that the solution to the wave equation is given by

$$f(x, t) = \sum_{n=1}^{\infty} D_n \sin\left(\frac{n\pi x}{l}\right) \cos\left(\frac{n\pi v t}{l}\right). \tag{13.237}$$

Note: wave speed, v, is given by

$$v = \lambda f \tag{13.238}$$

where λ is the wavelength, and f is the frequency of the wave and also the wave number, k, is defined as

$$k = \frac{2\pi}{\lambda} = \frac{2\pi f}{\lambda f} = \frac{\omega}{v} \tag{13.239}$$

where $\omega = 2\pi f$ is the angular frequency.

(d) The shape of the string when it is plucked at the center at $t = 0$ (see figure 13.16) is given by

$$F_0(x) = f(x, 0) = \begin{cases} \dfrac{2h}{l}x, & 0 < x < l/2, \\[2mm] \dfrac{2h}{l}(l - x), & l/2 < x < l, \end{cases} \tag{13.240}$$

show that constants D_n in equation (13.237) takes the form

$$D_n = \frac{8h}{n^2\pi^2}\left[2\sin\left(\frac{n\pi}{2}\right)\sin^2\left(\frac{n\pi}{4}\right)\right]. \tag{13.241}$$

Problem 4. In Problem 3 we assume the string is plucked at the initial time ($t = 0$) at the center ($x = 0.5l$). Let's consider the case where you plucked the string at $x = 0.25l$ near the lower or left end of the string (like you do in a guitar).

(a) Write the function that defines the shape of the string at $t = 0$.

(b) Find the function describing the displacement of the string from the equilibrium position at any time t.

Problem 5. Find the steady-state temperature distribution in a plate with the boundary temperature $u = 30°$ for $x = 0$ and $y = 3$; $u = 20°$ for $x = 5$ and $y = 0$. For two-dimensional heat flow the temperature, $u(x, y, t)$, satisfies the partial differential equation

$$\frac{\partial u}{\partial t} - \frac{1}{\alpha^2}\left[\frac{\partial^2 u(x, y, t)}{\partial x^2} + \frac{\partial^2 u(x, y, t)}{\partial y^2}\right] = 0, \tag{13.242}$$

and at steady-state

$$\frac{\partial u(x, y, t)}{\partial t} = 0 \Rightarrow \frac{\partial^2 u(x, y)}{\partial x^2} + \frac{\partial^2 u(x, y)}{\partial y^2} = 0. \tag{13.243}$$

Also for a temperature function defined by

$$u'(x, y) = u(x, y) - 20° \tag{13.244}$$

at steady-state the partial differential equation is still

$$\frac{\partial^2 u'(x, y)}{\partial x^2} + \frac{\partial^2 u'(x, y)}{\partial y^2} = 0. \tag{13.245}$$

Problem 6. A bar of length l sitting on the x-axis with one end at the origin is initially at a temperature $u(x, 0) = 0°$. At this initial time, the ends are held at a temperature $u(l, t) = 20°$, find the temperature, $u(x, 0)$, for $t > 0$.

Problem 7. A spherical cavity has a radius R. The temperature on the surface of the sphere is given by

$$T(R, \theta) = 5° \cos^3 \theta - 3° \sin^2 \theta. \tag{13.246}$$

Find the temperature inside the sphere $T(r, \theta, \varphi)$ which is determined by Laplace's equation

$$\nabla^2 T = \frac{1}{r^2}\frac{\partial}{\partial r}\left(r^2\frac{\partial T}{\partial r}\right) + \frac{1}{r^2 \sin \theta}\frac{\partial}{\partial \theta}\left(\sin \theta\frac{\partial T}{\partial \theta}\right) + \frac{1}{r^2 \sin^2 \theta}\frac{\partial^2 T}{\partial \varphi^2} = 0. \tag{13.247}$$

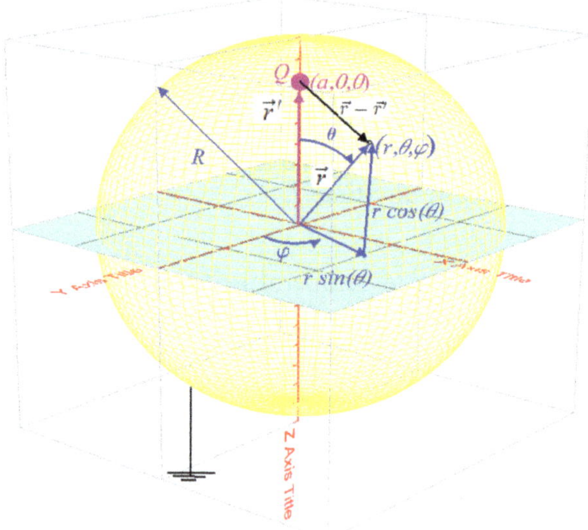

Figure 13.17. A point charge, Q, inside a grounded conducting sphere of radius, R. The position of the point charge is a distance, $a < R$, from the center of the sphere on the z-axis.

Problem 8. In Example 13.5 we have considered a *point charge and grounded conducting sphere* where the point charge Q is located outside a grounded sphere of radius R, (i.e., $a > R$). In this example we determined the potential outside the sphere. Now consider the case where the charge is located inside the sphere at distance $a < R$, from the center of the sphere on the z-axis as shown in figure 13.17. Find the electrostatic potential everywhere inside the sphere.

Problem 9. Show that the gravitational potential

$$V(\vec{r}) = -\frac{Gm}{r} \tag{13.248}$$

satisfies Laplace's equation

$$\nabla^2 V(\vec{r}) = 0. \tag{13.249}$$

Problem 10. A metallic spherical shell has a radius R. Somebody glued a charge on the northern hemisphere and grounded the southern hemisphere of the spherical shell such that the potential on the surface of the sphere is given by

$$V(R, \theta) = \begin{cases} V_0 \cos\theta, & 0 < \theta < \dfrac{\pi}{2} \\ 0, & \dfrac{\pi}{2} < \theta < \pi \end{cases}. \tag{13.250}$$

Find the electric potential inside the spherical shell, $V(r, \theta, \varphi)$, which is determined by Laplace's equation

$$\nabla^2 V = \frac{1}{r^2}\frac{\partial}{\partial r}\left(r^2\frac{\partial V}{\partial r}\right) + \frac{1}{r^2 \sin\theta}\frac{\partial}{\partial\theta}\left(\sin\theta\frac{\partial V}{\partial\theta}\right) + \frac{1}{r^2 \sin^2\theta}\frac{\partial^2 V}{\partial\varphi^2} = 0. \qquad (13.251)$$

Note: Make the assumption that the northern hemisphere is electrically insulated from the southern hemisphere.

Problem 11. Consider a semi-infinite solid cylinder shown in figure 13.18 with radius r_0. The base of the cylinder is held at a temperature $100°$ and the curved sides at $0°$. The temperature inside the cylinder is determined by the *diffusion or heat flow equation*

$$\frac{\partial u}{\partial t} = \alpha\left[\frac{\partial^2 u(x, y, z, t)}{\partial x^2} + \frac{\partial^2 u(x, y, z, t)}{\partial y^2} + \frac{\partial^2 V(x, y, z, t)}{\partial z^2}\right], \qquad (13.252)$$

where α is the thermal diffusivity and $u(x, y, z, t)$ is the temperature at a point (x, y, z, t) inside the cylinder at time t.

(a) For steady-state,

$$\frac{\partial u}{\partial t} = 0 \qquad (13.253)$$

find the temperature inside the cylinder.

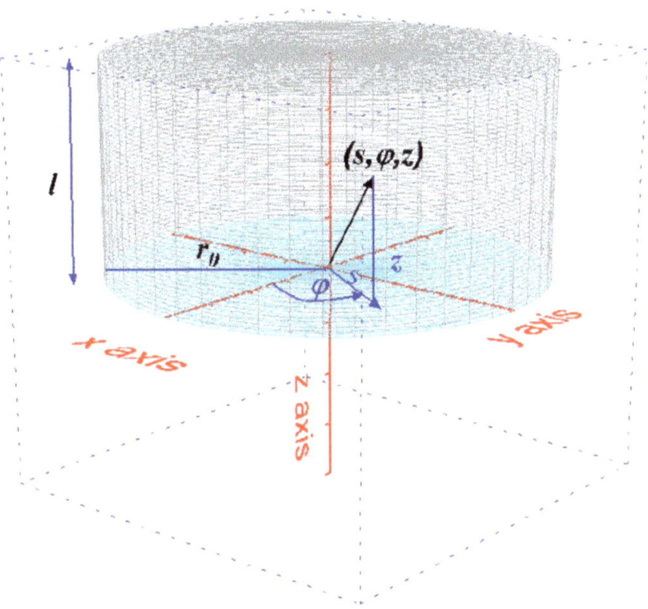

Figure 13.18. A semi-infinite cylinder (i.e., height of the cylinder extends to the positive infinity, $z \to \infty$).

(b) Compute numerically the coefficients

$$c_m = \frac{200}{\alpha_m J_1(\alpha_m)} \tag{13.254}$$

for $m = 1, 2, 3$ when $u = 0$, $s = 1$, and $u = 100$ at $z = 0$. Find u at $s = 1/2$, $z = 1$.

(c) In part (b) if $u = 0$ at $s = 10$ and $u = 100$ at $z = 0$, find u at $s = 5$, $z = 10$. What is the relation between parts (b) and (c)?

Problem 12. Find the steady-state temperature distribution in a solid semi-infinite cylinder if the boundary temperature are $u(s, \varphi, z) = 0°$ at $s = s_0$, and $u(s, \varphi, z) = y = s \sin(\varphi)$ at $z = 0$.

Problem 13. A plate in a shape of a quarter circle with radius a is sitting on the x–y plane as shown in figure 13.19. It has boundary temperatures $u(s, 0) = 0°$, $u(s, \pi/2) = 0°$, and $u(a, \pi/2) = 0°$. Find the interior steady-state temperature $u(s, \varphi)$.

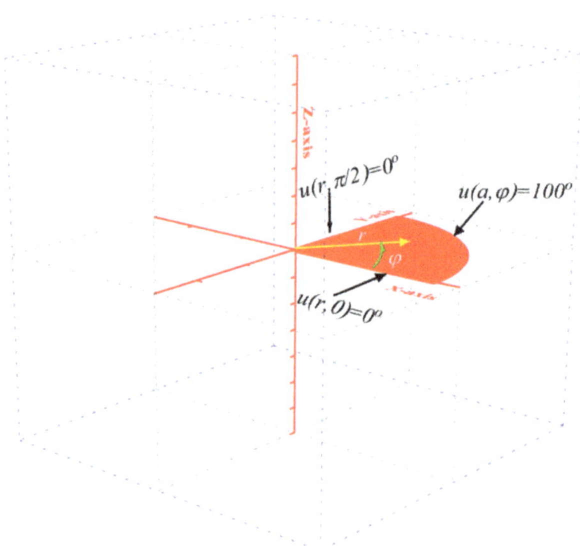

Figure 13.19. A quarter circular plate.

IOP Publishing

Studies in Theoretical Physics, Volume 1
Fundamental mathematical methods
Daniel Erenso and Victor Montemayor

Chapter 14

Functions of complex variables

This chapter begins with a review of the essential properties of complex functions from chapter 2, followed by an introduction to analytic and none analytic complex functions and associated fundamental terminologies to contour integrals involving complex functions. It then presents the methods for solving contour integrals involving complex functions and their application in solving real functions integrals. To this end, we introduce contour integration and Cauchy's theorem, Cauchy's integral formula, Laurent's series, the residue theorem, and methods for finding residues. We then see how the residue theorem is applied to find line integrals involving functions of single real variables using contour integrals on a complex plane. We then end the chapter by introducing the essential commands in Mathematica to resolve some of the examples considered.

14.1 Review of complex numbers

A complex number:
Numbers that have the form

$$z = a + ib, \qquad (14.1)$$

with a and b real numbers and

$$\pm i = \sqrt{-1} \Rightarrow i^2 = -1 \qquad (14.2)$$

are called *complex numbers*. In any complex number $z = a + ib$, a is the real part and b is the imaginary part

$$\mathrm{Re}\, z = a, \ \mathrm{Im}\, z = b. \qquad (14.3)$$

A complex variable in rectangular and polar coordinates:
In the rectangular complex plane the real part of a complex variable z is the x coordinate and its imaginary is the y coordinate (see figure 14.1)

$$z = x + iy \Rightarrow \mathrm{Re}(z) = x, \ \mathrm{Im}(z) = y. \qquad (14.4)$$

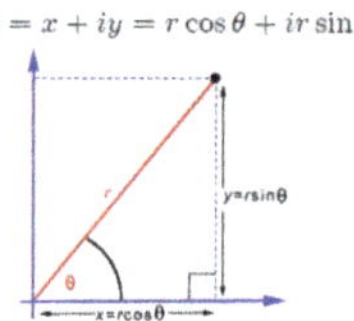

Figure 14.1. Rectangular and polar representation of a complex number $z = x + iy$.

In polar coordinate the same complex variable z is represented by

$$z = x + iy = r \cos(\theta) + ir \sin(\theta) \Rightarrow \mathrm{Re}(z) = r \cos(\theta),\ \mathrm{Im}(z) = r \sin(\theta), \quad (14.5)$$

where r is called *the modulus* or *the magnitude* and θ is called the phase angle of the complex number z given by

$$|z| = r = \sqrt{x^2 + y^2},\ \theta = \tan^{-1}\left(\frac{y}{x}\right). \quad (14.6)$$

Using Euler's equation, the complex variable z can be expressed in an exponential form as

$$z = x + iy = r[\cos(\theta) + i \sin(\theta)] = r \exp(i\theta). \quad (14.7)$$

Conjugate of complex variable:
The complex conjugate of a complex variable $z = x + iy$, denoted by z^*, is given by

$$z^* = x - iy. \quad (14.8)$$

The magnitude of a complex number z is square root of the product z and its complex conjugate,

$$|z| = \sqrt{zz^*} = \sqrt{x^2 + y^2}. \quad (14.9)$$

Example 14.1. Consider the complex function

$$f(z) = \frac{2z + 1}{z - i}. \quad (14.10)$$

Find the real and imaginary parts of this function.

Solution: To find the real and imaginary part of a complex function $f(z)$, one must substitute

$$z = x + iy, \quad (14.11)$$

and write the function $f(z)$ in the form

$$f(x + iy) = u(x, y) + iv(x, y),$$

(14.12)

where $u(x, y)$ is the real and $v(x, y)$ is the imaginary parts of the function $f(z)$. Then upon substituting equation (14.11) into equation (14.10), we find

$$\begin{aligned}
f(x + iy) &= \frac{2(x + iy) + 1}{(x + iy) - i} = \frac{2x + 1 + i2y}{x + i(y - 1)} \\
&= \frac{[2x + 1 + i2y][x - i(y - 1)]}{[x + i(y - 1)][x - i(y - 1)]} \\
&= \frac{(2x + 1)x + 2y(y - 1) + i[2xy - (2x + 1)(y - 1)]}{x^2 + (y - 1)^2} \\
&= \frac{(2x + 1)x + 2y(y - 1)}{x^2 + (y - 1)^2} \\
&\quad + i\frac{2x - y + 1}{x^2 + (y - 1)^2} \Rightarrow f(x + iy) = u(x, y) + iv(x, y),
\end{aligned}$$

(14.13)

where

$$u(x, y) = \frac{(2x + 1)x + 2y(y - 1)}{x^2 + (y - 1)^2}, \; v(x, y) = \frac{2x - y + 1}{x^2 + (y - 1)^2},$$

(14.14)

are the real and imaginary part of the function $f(z)$.

Example 14.2. Find the real and imaginary parts of the complex function

$$f(z) = z^{1/4}.$$

(14.15)

Solution: Whenever we are given exponential complex functions, expressing the function using polar coordinates is extremely useful. In polar coordinate we can write

$$z = x + iy = r \exp(i\theta),$$

(14.16)

where r and θ are related to the Cartesian coordinates x and y by

$$r = \sqrt{x^2 + y^2}, \; \theta = \tan^{-1}\left(\frac{y}{x}\right).$$

(14.17)

Substituting equation (14.16) into equation (14.15), one finds

$$f(z) = [r \exp(i\theta)]^{1/4} = r^{1/4} \exp\left(\frac{i\theta}{4}\right) = u(r, \theta) + iv(r, \theta),$$

(14.18)

where

$$u(r, \theta) = r^{1/4} \cos\left(\frac{\theta}{4}\right), \; v(r, \theta) = r^{1/4} \sin\left(\frac{\theta}{4}\right),$$

(14.19)

are the real and imaginary parts. Note that the polar coordinates r and θ are related to the Cartesian coordinates by equation (14.17).

Example 14.3. *Polarization of EM waves*: Consider an EM wave propagating along the positive z-direction. We can express the electric field of this wave using complex functions as

$$\vec{E} = \left(E_{0x}\hat{x} + E_{0y}\hat{y}\right)e^{i(kz-\omega t)}, \tag{14.20}$$

where k is the wave number, ω is the angular frequency, E_{0x} and E_{0y} are the complex amplitude of the electric field in the x- and y-direction, respectively.

(a) Express the complex electric field using Euler's formula.
(b) Assume the phase for the x-component to be $\varphi_x = 0$ and that of the y-component to be $\varphi_y = \varphi$, find the components of the electric field for $\varphi = 0, \pm\pi/2$.
(c) Find the real part of the electric field vector for $\varphi = 0, \pm\pi/2$ and plot the corresponding results for E_y versus E_x for a fixed value of z (for example $z = 0$) and $0 \leqslant \omega t \leqslant 2\pi$.

Solution:

(a) For complex electric field, using Euler's formula, we can write

$$E_{0x} = |E_{0x}|\, e^{i\varphi_x}, \quad E_{0y} = \left|E_{0y}\right| e^{i\varphi_y}, \tag{14.21}$$

so that

$$\vec{E} = \left(E_{0x}\hat{x} + E_{0y}\hat{y}\right)e^{i(kz-\omega t)}, \tag{14.22}$$

becomes

$$\vec{E} = \left(|E_{0x}|\, e^{i\varphi_x}\hat{x} + \left|E_{0y}\right| e^{i\varphi_y}\hat{y}\right)e^{i(kz-\omega t)}$$
$$= |E_{0x}|\, e^{i\left(kz-\omega t+\varphi_x\right)}\hat{x} + \left|E_{0y}\right| e^{i\left(kz-\omega t+\varphi_y\right)}\hat{y}. \tag{14.23}$$

(b) Let us assume the phase for the x-component $\varphi_x = 0$ and for the y-component $\varphi_y = \varphi$, then we can write equation (14.23) as

$$\vec{E} = |E_{0x}|\, e^{i(kz-\omega t)}\hat{x} + \left|E_{0y}\right| e^{i(kz-\omega t+\varphi)}\hat{y}. \tag{14.24}$$

When $\varphi = 0$, the electric field becomes

$$\vec{E} = |E_{0x}|\, e^{i(kz-\omega t)}\hat{x} + \left|E_{0y}\right| e^{i(kz-\omega t)}\hat{y}. \tag{14.25}$$

Noting that

$$E_x = |\,E_{0x}\,|\,e^{i(kz-\omega t)} = |\,E_{0x}\,|\,\cos{(kz - \omega t)}$$
$$+\, i\,|\,E_{0x}\,|\,\sin{(kz - \omega t)},$$
$$E_y = \left|\,E_{0y}\,\right|\,e^{i(kz-\omega t)} = \left|\,E_{0y}\,\right|\,\cos{(kz - \omega t)}$$
$$+\, i\,\left|\,E_{0y}\,\right|\,\sin{(kz - \omega t)},$$

(14.26)

so that electric field in equation (14.25) can be rewritten as

$$\vec{E} = E_x\hat{x} + E_y\hat{y} = [|\,E_{0x}\,|\,\cos{(kz - \omega t)} + i\,|\,E_{0x}\,|\,\sin{(kz - \omega t)}]\hat{x}$$
$$+ \left[\left|\,E_{0y}\,\right|\,\cos{(kz - \omega t)} + i\,\left|\,E_{0y}\,\right|\,\sin{(kz - \omega t)}\right]\hat{y}.$$

(14.27)

Then the real part of the electric field becomes

$$\mathrm{Re}\,[\vec{E}] = \left(|\,E_{0x}\,|\,\hat{x} + \left|\,E_{0y}\,\right|\,\hat{y}\right)\cos{(kz - \omega t)}.$$

(14.28)

For $\varphi = \pm\pi/2$, we still have the same form for the x-component

$$E_x = |\,E_{0x}\,|\,e^{i(kz-\omega t)} = |\,E_{0x}\,|\,\cos{(kz - \omega t)} + i\,|\,E_{0x}\,|\,\sin{(kz - \omega t)},$$

(14.29)

however for the y-component, we find

$$E_y = \left|\,E_{0y}\,\right|\,e^{i(kz-\omega t+\varphi)} = \left|\,E_{0y}\,\right|\,\cos{(kz - \omega t \pm \pi/2)}$$
$$+\, i\,\left|\,E_{0y}\,\right|\,\sin{(kz - \omega t \pm \pi/2)}.$$

(14.30)

Applying the relations

$$\cos{(\alpha \pm \pi/2)} = \cos{(\alpha)}\cos{(\pi/2)} \mp \sin{(\alpha)}\sin{(\pi/2)} = \mp\sin{(\alpha)},$$
$$\sin{(\alpha \pm \pi/2)} = \sin{(\alpha)}\cos{(\pi/2)} \pm \cos{(\alpha)}\sin{(\pi/2)} = \pm\cos{(\alpha)},$$

(14.31)

one can write equation (14.30) as

$$E_y = \mp\left|\,E_{0y}\,\right|\,\sin{(kz - \omega t)} \pm i\,\left|\,E_{0y}\,\right|\,\cos{(kz - \omega t)}.$$

(14.32)

Then using equations (14.29) and (14.32), we can express the electric field for such EM as

$$\vec{E} = E_x\hat{x} + E_y\hat{y} = [|\,E_{0x}\,|\,\cos{(kz - \omega t)} + i\,|\,E_{0x}\,|\,\sin{(kz - \omega t)}]\hat{x}$$
$$+ \left[\mp\left|\,E_{0y}\,\right|\,\sin{(kz - \omega t)} \pm i\,\left|\,E_{0y}\,\right|\,\cos{(kz - \omega t)}\right]\hat{y}.$$

(14.33)

Then for the real part of this electric field, one finds

$$\mathrm{Re}\,[\vec{E}] = |\,E_{0x}\,|\,\cos{(kz - \omega t)}\hat{x} \mp \left|\,E_{0y}\,\right|\,\sin{(kz - \omega t)}\hat{y}.$$

(14.34)

(c) For $\varphi = 0$ using the results in equation (14.28),

$$\mathrm{Re}\,[\vec{E}] = |\,E_{0x}\,|\,\cos{(kz - \omega t)}\hat{x} + \left|\,E_{0y}\,\right|\,\cos{(kz - \omega t)}\hat{y},$$

(14.35)

one can make the 3-D parametric plot shown in figure 14.2. Such an EM wave is called *linearly (plane) polarized*. On the other hand for $\varphi = \pm\pi/2$ using the results in equation (14.34),

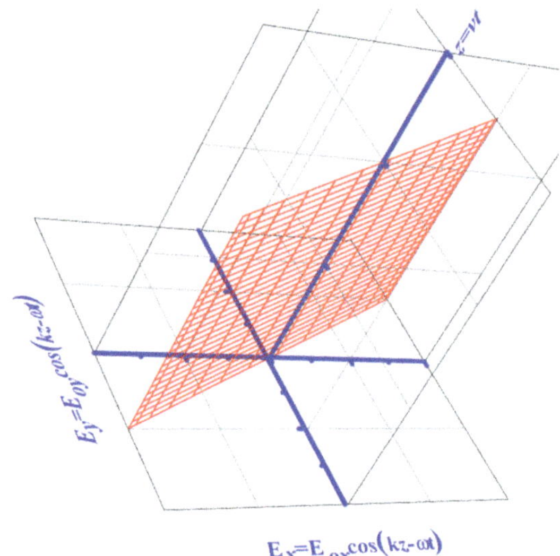

Figure 14.2. Linearly polarized EM wave.

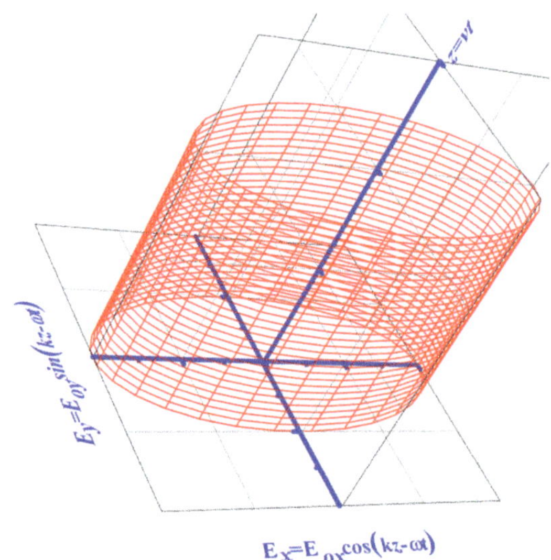

Figure 14.3. Elliptically polarized EM wave.

$$\mathrm{Re}\,[\vec{E}] = |\,E_{0x}\,|\,\cos\,(kz - \omega t)\hat{x} \mp \left|\,E_{0y}\,\right|\,\sin\,(kz - \omega t)\hat{y} \qquad (14.36)$$

we find the 3-D parametric plot shown in figure 14.3. Such an EM wave is known as *elliptically polarized*, when $|\,E_{0x}\,| \neq \left|\,E_{0y}\,\right|$. On the other hand, when $|\,E_{0x}\,| = \left|\,E_{0y}\,\right|$, we find a circularly polarized EM wave (see

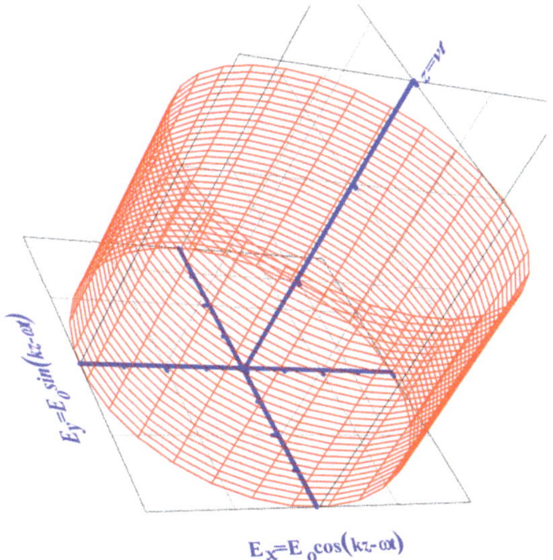

Figure 14.4. Circularly polarized EM wave.

figure 14.4). Note that in all the three polarization states, the EM wave is traveling along the positive z-direction with speed v. Also the plus and minus signs in the case of circular (elliptical) polarizations state lead to left and right circular (elliptical) polarization states, respectively.

14.2 Analytic functions

When a complex function $f(z)$ has a unique derivative at every point in a region on a complex plane, we say the function is analytic in the region. Sometimes, instead of *analytic*, the function is referred to as *regular*, *holomorphic*, or *monogenic*. Suppose we consider a point on a complex plane $z = z_0$. If the function $f(z)$ is analytic at this point, then the derivative at every point inside some small circle about $z = z_0$ given by,

$$\left. \frac{df}{dz} \right|_{z = z_0} = \lim_{\Delta z \to 0} \frac{f(z_0 + \Delta z) - f(z_0)}{\Delta z}, \tag{14.37}$$

must give a unique value. For example, while the functions

$$\frac{d(z^2)}{dz} = 2z, \quad \frac{d(\cos(z))}{dz} = -\sin z, \quad \frac{d}{dz}\left(\sqrt{1 + z^2}\right) = \frac{z}{\sqrt{1 + z^2}} \tag{14.38}$$

are *analytic*, the function

$$f(z) = |z| = \sqrt{x^2 + y^2}, \tag{14.39}$$

is *not analytic* for all z on the complex plane. We use *the Cauchy–Riemann conditions* to show these or any other complex function is analytic or not in a given region on the complex plane.

The Cauchy–Riemann conditions:
This is the condition that must be met by analytic function. It states that if the complex function

$$f(z) = u(x, y) + iv(x, y), \tag{14.40}$$

with $u(x, y)$ and $v(x, y)$ real functions, is analytic in a region, R, then in that region the following two conditions must be satisfied

$$\frac{\partial u}{\partial x} = \frac{\partial v}{\partial y}, \quad \frac{\partial v}{\partial x} = -\frac{\partial u}{\partial y}. \tag{14.41}$$

Next, we shall derive these conditions. To this end, for a complex function $f(z)$, where $z = x + iy$, one can write

$$\frac{\partial f}{\partial x} = \frac{\partial f}{\partial z}\frac{\partial z}{\partial x} = \frac{\partial f}{\partial z}\left[\frac{\partial x}{\partial x} + i\frac{\partial y}{\partial x}\right] = \frac{\partial f}{\partial z} \Rightarrow \frac{\partial f}{\partial z} = \frac{\partial f}{\partial x},$$

$$\frac{\partial f}{\partial y} = \frac{\partial f}{\partial z}\frac{\partial z}{\partial y} = \frac{\partial f}{\partial z}\left[\frac{\partial x}{\partial y} + i\frac{\partial y}{\partial y}\right] \Rightarrow \frac{\partial f}{\partial y} = i\frac{\partial f}{\partial z} \Rightarrow \frac{\partial f}{\partial z} = -i\frac{\partial f}{\partial y}. \tag{14.42}$$

Since for the analytic function, the derivative must be unique,

$$\frac{\partial f}{\partial z} = \frac{\partial f}{\partial x} = -i\frac{\partial f}{\partial y}. \tag{14.43}$$

Using the form of the function

$$f(z) = u(x, y) + iv(x, y), \tag{14.44}$$

one can also write for the derivatives

$$\frac{\partial f}{\partial x} = \frac{\partial u}{\partial x} + i\frac{\partial v}{\partial x}, \quad \frac{\partial f}{\partial y} = \frac{\partial u}{\partial y} + i\frac{\partial v}{\partial y}. \tag{14.45}$$

In view of equation (14.45), equation (14.43) can be put in the form

$$\frac{\partial f}{\partial z} = \frac{\partial f}{\partial x} = -i\frac{\partial f}{\partial y} \Rightarrow \frac{\partial u}{\partial x} + i\frac{\partial v}{\partial x} = -i\left(\frac{\partial u}{\partial y} + i\frac{\partial v}{\partial y}\right)$$

$$\Rightarrow \frac{\partial u}{\partial x} + i\frac{\partial v}{\partial x} = \frac{\partial v}{\partial y} - i\frac{\partial u}{\partial y}. \tag{14.46}$$

Two complex functions are equal when their real parts are equal and their imaginary parts are equal. Therefore, from equation (14.46) follows that

$$\frac{\partial u}{\partial x} = \frac{\partial v}{\partial y}, \frac{\partial v}{\partial x} = -\frac{\partial u}{\partial y}. \tag{14.47}$$

Equation (14.47) are *the Cauchy–Riemann conditions.* Now let us reconsider the function

$$f(z) = |z| = \sqrt{x^2 + y^2} = u(x, y) + iv(x, y) \tag{14.48}$$

which we stated is *not analytic* earlier. Clearly, one can see that

$$\Rightarrow u(x, y) = \sqrt{x^2 + y^2}, \, v(x, y) = 0$$

$$\Rightarrow \frac{\partial u}{\partial x} = \frac{x}{\sqrt{x^2 + y^2}}, \frac{\partial v}{\partial y} = 0, \frac{\partial v}{\partial x} = 0, \frac{\partial u}{\partial y} = \frac{y}{\sqrt{x^2 + y^2}} \tag{14.49}$$

$$\frac{\partial u}{\partial x} \neq \frac{\partial v}{\partial y}, \frac{\partial v}{\partial x} \neq -\frac{\partial u}{\partial y},$$

which is in violation of the *Cauchy–Riemann conditions.* On the other hand, for the analytic function $f(z) = z^2$, one finds

$$f(z) = (x + iy)^2 = x^2 - y^2 + i(2xy) \Rightarrow u(x, y)$$

$$= x^2 - y^2, \, v(x, y) = 2xy \tag{14.50}$$

$$\Rightarrow \frac{\partial u}{\partial x} = 2x = \frac{\partial v}{\partial y}, \frac{\partial v}{\partial x} = 2y = -\frac{\partial u}{\partial y},$$

which is in agreement with the *Cauchy–Riemann conditions.*

Example 14.4. Show that the function

$$f(z) = \cos(z), \tag{14.51}$$

is analytic.

Solution: In order to show that this function is analytic, it must satisfy the *Cauchy–Riemann conditions*

$$\frac{\partial u}{\partial x} = \frac{\partial v}{\partial y}, \frac{\partial v}{\partial x} = -\frac{\partial u}{\partial y}. \tag{14.52}$$

This requires writing the function $f(z)$ as

$$f(z) = u(x, y) + iv(x, y). \tag{14.53}$$

To this end, we note

$$f(z) = \cos(z) = \cos(x + iy) = \cos(x)\cos(iy) - \sin(x)\sin(iy), \tag{14.54}$$

so that using the relations

$$\cos(iy) = \cosh(x), \ \sin(iy) = i\sinh(y) \tag{14.55}$$

one can show that

$$f(z) = \cos(x)\cosh(y) - i\sin(x)\sinh(y)$$
$$\Rightarrow u(x, y) = \cos(x)\cosh(y), \ v(x, y) = -\sin(x)\sinh(y),$$
$$\Rightarrow \frac{\partial u}{\partial x} = -\sin(x)\cosh(y) = \frac{\partial v}{\partial y}, \tag{14.56}$$
$$\frac{\partial v}{\partial x} = -\cos(x)\sinh(y) = -\frac{\partial u}{\partial y}.$$

The result in equation (14.56) proves that the function $f(z)$ satisfies the *Cauchy–Riemann conditions* and therefore the function is analytic.

14.3 Essential terminologies

The following terminologies are essential in the description of functions of a complex variable.

(a) *A regular point:* The point z_0 is said to be a regular point of the function $f(z)$ if the function $f(z)$ is analytic at $z = z_0$.

(b) *Singular point:* If $f(z)$ is not analytic at a point, $z = z_0$, then z_0 is said to be a singular point or singularity of $f(z)$.

(c) *Isolated singularity:* When z_0 is the only singular point for the function $f(z)$ within an arbitrarily small region surrounding z_0, then z_0 is said to be isolated singularity.

(d) *Harmonic and conjugate harmonic functions:* Consider the first *Cauchy–Riemann condition*

$$\frac{\partial u}{\partial x} = \frac{\partial v}{\partial y}. \tag{14.57}$$

Upon differentiating with respect to x, we find

$$\frac{\partial^2 u}{\partial x^2} = \frac{\partial^2 v}{\partial x \partial y}. \tag{14.58}$$

Similarly, differentiate the second condition

$$\frac{\partial v}{\partial x} = -\frac{\partial u}{\partial y}, \tag{14.59}$$

with respect to y, one finds

$$\frac{\partial^2 v}{\partial x \partial y} = -\frac{\partial^2 u}{\partial y^2} \Rightarrow \frac{\partial^2 u}{\partial y^2} = -\frac{\partial^2 v}{\partial x \partial y}. \tag{14.60}$$

Then adding equations (14.58) and (14.60) results in

$$\frac{\partial^2 u}{\partial x^2} + \frac{\partial^2 u}{\partial y^2} = \frac{\partial^2 v}{\partial x \partial y} - \frac{\partial^2 v}{\partial x \partial y} = 0 \Rightarrow \nabla^2 u = 0, \tag{14.61}$$

which is Laplace's equation in two-dimensional Cartesian coordinates. Furthermore, upon differentiating equation (14.57) with respect to y

$$\frac{\partial^2 v}{\partial y^2} = \frac{\partial^2 u}{\partial x \partial y} \tag{14.62}$$

and equation (14.59) with respect to x,

$$\frac{\partial^2 v}{\partial x^2} = -\frac{\partial^2 u}{\partial y \partial x}, \tag{14.63}$$

one can show that

$$\frac{\partial^2 v}{\partial x^2} + \frac{\partial^2 v}{\partial y^2} = 0 \Rightarrow \nabla^2 v = 0, \tag{14.64}$$

which is also Laplace's equation. Functions satisfying Laplace's equation are called *harmonic functions*. The real and imaginary parts of the analytic function, $f(z)$, are called *conjugate harmonic functions*.

Complex function series expansion:
Let $f(z)$ is analytic within a region R, with one or more singular points on the borders (and possibly outside). Then since the derivatives of all orders for $f(z)$ at any point z_0 inside this region exist, $f(z)$ can be expanded in a Taylor series about the point z_0. This series converges for all points within the circle centered at z_0 that extends to the closest singularity of $f(z)$.

Example 14.5. Expand the complex function

$$f(z) = \sqrt{z + 2i}, \tag{14.65}$$

in a Taylor series about the origin, $z_0 = 0$, and find the circle of convergence for this series.

Solution: The Taylor series for a function, $f(z)$, about $z = z_0$, is given by

$$f(z) = \sum_{n=0}^{\infty} \frac{1}{n!} \frac{d^n f(z)}{dz^n} \bigg|_{z = z_0} (z - z_0)^n. \tag{14.66}$$

For the complex function, $f(z) = \sqrt{z + 2i}$, we have

$$\frac{d^0 f(z)}{dz^0} = f(z) = \sqrt{z + 2i} \Rightarrow \left.\frac{d^0 f(z)}{dz^0}\right|_{z = z_0} = \sqrt{z_0 + 2i}$$

$$\frac{df(z)}{dz} = \frac{1}{2}\frac{1}{\sqrt{z + 2i}} \Rightarrow \left.\frac{1}{1!}\frac{df(z)}{dz}\right|_{z = z_0} = \frac{1}{2}\frac{1}{\sqrt{z_0 + 2i}}$$

$$\frac{d^2 f(z)}{dz^2} = -\frac{1}{2^2}\frac{1}{(z + 2i)^{3/2}} \Rightarrow \left.\frac{1}{2!}\frac{d^2 f(z)}{dz^2}\right|_{z = z_0} = -\frac{1}{2^3}\frac{1}{(z_0 + 2i)^{3/2}}$$

$$\frac{d^3 f(z)}{dz^3} = \frac{3}{2^3}\frac{1}{(z + 2i)^{5/2}} \Rightarrow \left.\frac{1}{3!}\frac{d^3 f(z)}{dz^3}\right|_{z = z_0} = \frac{1}{2^4}\frac{1}{(z_0 + 2i)^{5/2}}.$$

$$(14.67)$$

and the series can be written as

$$f(z) = \sqrt{z_0 + 2i} + \frac{1}{2}\frac{1}{(z_0 + 2i)^{1/2}}(z - z_0) - \frac{1}{2^3}\frac{1}{(z_0 + 2i)^{3/2}}(z - z_0)^2$$
$$+ \frac{1}{2^4}\frac{1}{(z_0 + 2i)^{5/2}}(z - z_0)^3 \ldots$$

$$(14.68)$$

This series diverges at $z_0 = -2i$,

$$\lim_{z_0 \to -2i} \frac{1}{z_0 + 2i} \to \infty, \tag{14.69}$$

which shows that the function $f(z)$ has a singular point at $z_0 = -2i$. Therefore, the radius of convergence is determined by the condition

$$|z_0| < |-2i| = 2 \Rightarrow |z_0| < 2. \tag{14.70}$$

The series is convergent for all z inside the circle of radius, $R = 2$ centered about the origin, $z_0 = 0$. Then the series expansion in equation (14.68) about $z_0 = 0$ becomes

$$f(z) = \sqrt{2i} + \frac{1}{2}\frac{1}{(2i)^{1/2}}z - \frac{1}{2^3}\frac{1}{(2i)^{3/2}}z^2 + \frac{1}{2^4}\frac{1}{(2i)^{5/2}}z^3 \ldots, \tag{14.71}$$

which is convergent in the region on the complex plane defined by $|z| < 2$.

14.4 Contour integration and Cauchy's theorem

Cauchy's theorem: Consider a function, $f(z)$, that is analytic inside a region, R, on the complex plane. Within the region, R, if there is a closed curve, C, that does not cross itself, and that has a finite number of sharp corners, then Cauchy's theorem states that

$$\oint_C f(z)dz = 0. \tag{14.72}$$

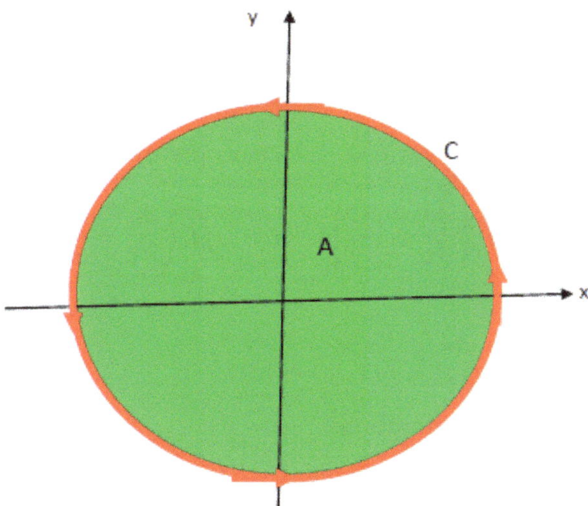

Figure 14.5. A circular contour on a complex plane. There is no singularity inside and on the contour.

The line integral on this curve is always zero. In the following, we prove Cauchy's theorem. To this end, let us consider a simple closed contour (a circular curve) shown in figure 14.5. Using

$$z = x + iy \Rightarrow dz = dx + idy, \tag{14.73}$$

for a complex function

$$f(z) = u(x, y) + iv(x, y), \tag{14.74}$$

the closed integral can be expressed as

$$\begin{aligned}
\oint_C f(z)dz &= \oint_C [(u(x, y) + iv(x, y))(dx + idy)] \\
&= \oint_C [(u(x, y)dx - v(x, y)dy)] \\
&\quad + i \oint_C [(v(x, y)dx + u(x, y)dy)].
\end{aligned} \tag{14.75}$$

Applying Green's theorem from chapter 8

$$\oint_C [(P(x, y)dx + Q(x, y)dy)] = \iint_A \left(\frac{\partial Q}{\partial x} - \frac{\partial P}{\partial y} \right) dxdy, \tag{14.76}$$

one can write equation (14.75) as

$$\oint_C f(z)dz = \oint_C [(u(x, y)dx - v(x, y)dy)]$$
$$+ i \oint_C [(v(x, y)dx + u(x, y)dy)] \tag{14.77}$$
$$= - \iint_A \left(\frac{\partial v}{\partial x} + \frac{\partial u}{\partial y}\right)dxdy + i \iint_A \left(\frac{\partial u}{\partial x} - \frac{\partial v}{\partial y}\right)dxdy.$$

For analytic function,

$$f(z) = u(x, y) + iv(x, y), \tag{14.78}$$

using the Cauchy–Riemann conditions

$$\frac{\partial u}{\partial x} = \frac{\partial v}{\partial y}, \frac{\partial v}{\partial x} = -\frac{\partial u}{\partial y}, \tag{14.79}$$

one can see that equation (14.77) becomes

$$\oint_C f(z)dz = - \iint_A \left(-\frac{\partial u}{\partial y} + \frac{\partial u}{\partial y}\right)dxdy + i \iint_A \left(\frac{\partial v}{\partial y} - \frac{\partial v}{\partial y}\right)dxdy = 0. \tag{14.80}$$

This proves Cauchy's theorem.

14.5 Cauchy's integral formula

Before we state and prove Cauchy's integral formula, it is essential to find the general equation describing a point on a circular contour on a complex plane.

The equation of a circle:

Consider a circle on a complex plane centered about a point, $z_0 = x_0 + iy_0$, shown in figure 14.6. Any point on this circle, $z = x + iy$, can be described by the equation

$$z - z_0 = (x - x_0) + i(y - y_0) = |z - z_0| e^{i\theta}, \tag{14.81}$$

where

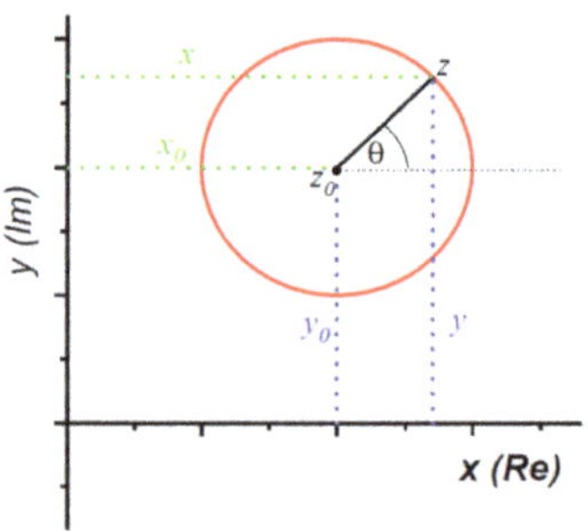

Figure 14.6. A circle centered about z_0 on a complex plane.

$$|z - z_0| = \sqrt{(x - x_0)^2 + (y - y_0)^2}, \quad \theta = \tan^{-1}\left(\frac{y - y_0}{x - x_0}\right). \tag{14.82}$$

Suppose R is the radius of the circle which is equal to $|z - z_0|$, one finds

$$R = |z - z_0| \Rightarrow R^2 = (x - x_0)^2 + (y - y_0)^2, \tag{14.83}$$

which is equation of a circle centered at (x_0, y_0).

Cauchy's integral formula
Suppose the complex function, $f(z)$, is analytic inside a region enclosed by a curve, C, and also on the curve itself (e.g., figure 14.7). The value for this function at the singular point, $z = z_0$, inside the curve C, (i.e., $f(z_0)$) can be determined using the line integral on the closed curve, C, given by

$$f(z_0) = \frac{1}{2\pi i} \oint_C \frac{f(z)}{z - z_0} dz, \tag{14.84}$$

where the integration being taken is in the counterclockwise sense. Equation (14.84) is *Cauchy's integral formula*. In the following we prove this formula. To this end, let us consider the function,

$$g(z) = \frac{f(z)}{z - z_0}, \tag{14.85}$$

which is not analytic everywhere inside the closed curve, C, in figure 14.7 due to the isolated singular point at $z = z_0$. Because of this singular point we cannot apply Cauchy's theorem,

$$\oint_C g(z) dz \neq 0. \tag{14.86}$$

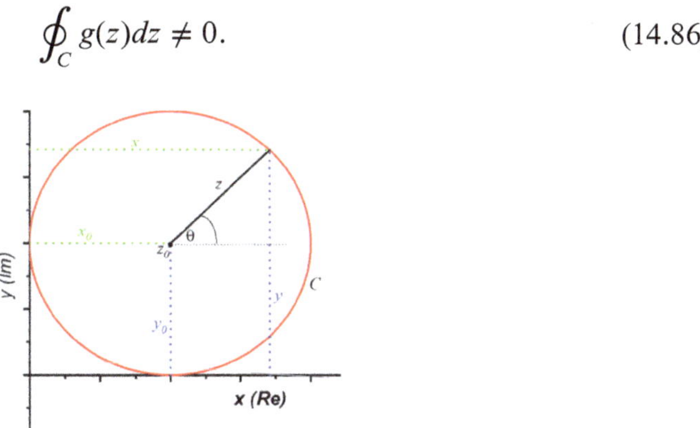

Figure 14.7. A circle centered about z_0. The function, $f(z)$, is analytic everywhere inside and on the closed curve, C.

However, if somehow we can exclude this isolated singular point from the region bounded by the curve, C, we can apply Cauchy's theorem. Let us consider the region bounded by the curve C' shown in figure 14.8 that keeps the singular point, z_0 outside the region bounded by C'. This closed curve, C', consist of two curves:

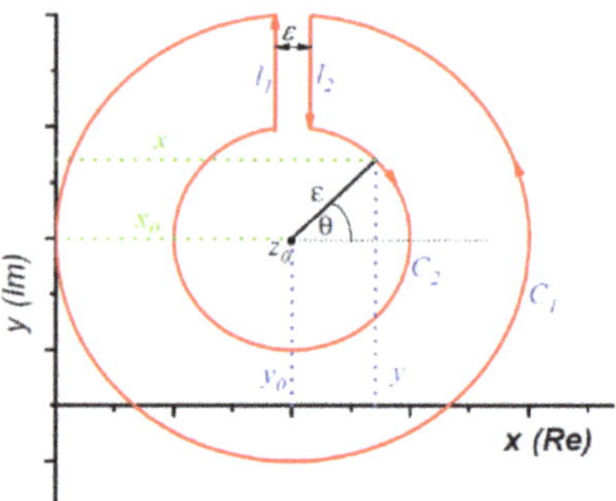

Figure 14.8. A region bounded by the curve C' that excludes the singular point z_0 for the function, $g(z)$.

(C_1, C_2) and two straight lines: (l_1, l_2). The curve C_2 has a small radius, and the two lines l_1 and l_2 are separated by a small distance denoted by ϵ. Everywhere in this region, the function $g(z)$ is analytic as the only singular point, $z = z_0$, is no longer in the region. Then according to Cauchy's theorem, we must have

$$I = \oint_{C'} g(z)dz = \frac{1}{2\pi i} \oint_{C'} \frac{f(z)}{z - z_0} dz = 0. \tag{14.87}$$

This integral can be split into four integrals,

$$\oint_{C'} g(z)dz = \int_{C_1} g(z)dz + \int_{l_2} g(z)dz + \int_{C_2} g(z)dz + \int_{l_1} g(z)dz = 0. \tag{14.88}$$

We are interested in the integral

$$I = \frac{1}{2\pi i} \oint_{C} \frac{f(z)}{z - z_0} dz, \tag{14.89}$$

which is the same as

$$I = \frac{1}{2\pi i} \oint_{C} \frac{f(z)}{z - z_0} dz = \lim_{\epsilon \to 0} \frac{1}{2\pi i} \oint_{C_1} \frac{f(z)}{z - z_0} dz, \tag{14.90}$$

as in the limit $\epsilon \to 0$, integrating along l_1 and l_2 leads to the same function integrated in the opposite direction over the same line, one finds

$$\int_{l_1} g(z)dz = -\int_{l_2} g(z)dz. \tag{14.91}$$

Therefore, equation (14.88) becomes

$$\lim_{\epsilon \to 0} \oint_{C'} g(z)dz = \lim_{\epsilon \to 0}\left[\int_{C_1} g(z)dz + \int_{C_2} g(z)dz\right] = 0,$$

$$\Rightarrow \lim_{\epsilon \to 0} \int_{C_1} \frac{f(z)}{z - z_0} dz = -\lim_{\epsilon \to 0} \int_{C_2} \frac{f(z)}{z - z_0} dz. \tag{14.92}$$

Note that from figure 14.8 the integration over C_1 is counterclockwise whereas the integration over C_2 is clockwise. Now using the equation of a circle on a complex plane, we derived earlier, one can write for point z on the circle C_2

$$z - z_0 = \epsilon e^{i\theta} \Rightarrow z = z_0 + \epsilon e^{i\theta} \Rightarrow dz = i\epsilon e^{i\theta} d\theta, \tag{14.93}$$

One finds for equation (14.92)

$$\lim_{\epsilon \to 0} \oint_{C_1} \frac{f(z)}{z - z_0} dz = -\lim_{\epsilon \to 0} \oint_{C_2} \frac{f(z)}{z - z_0} dz$$

$$= -\lim_{\epsilon \to 0} \int_{2\pi}^{0} \frac{f(z_0 + \epsilon e^{i\theta})}{\epsilon e^{i\theta}} i\epsilon e^{i\theta} d\theta$$

$$= \lim_{\epsilon \to 0} i \int_{0}^{2\pi} f(z_0 + \epsilon e^{i\theta}) d\theta = f(z_0) i \int_{0}^{2\pi} d\theta \tag{14.94}$$

$$\Rightarrow \lim_{\epsilon \to 0} \oint_{C_1} \frac{f(z)}{z - a} dz = 2\pi i f(z_0).$$

In limit $\epsilon \to 0$, we note that the integral over the curve C becomes over C_1, and equation (14.94) becomes

$$\oint_{C} \frac{f(z)}{z - z_0} dz = \lim_{\epsilon \to 0} \oint_{C_1} \frac{f(z)}{z - a} dz = i \int_{0}^{2\pi} f(z_0) d\theta = 2\pi i f(z_0)$$

$$\Rightarrow f(z_0) = \frac{1}{2\pi i} \oint_{C} \frac{f(z)}{z - z_0} dz. \tag{14.95}$$

Equation (14.95) is Cauchy's integral formula.

Example 14.6. Consider the integral

$$I = \int_{z_1}^{z_2} z^2 dz, \tag{14.96}$$

where $z_1 = 1 + i$ and $z_2 = 2 + 4i$.

(a) Evaluate the integral, I, using a straight line joining the points z_1 and z_2 (the green line in figure 14.9).

(b) Evaluate the integral, I, using the two straight lines: the first from the point z_1 to $z_0 = 2 + i$, (the blue line) and the second from the point z_0 to the point z_2 (the pink line in figure 14.9).

(c) Find the integral

$$I = \oint_{C} z^2 dz \tag{14.97}$$

for the right-angled triangular contour, C, shown in figure 14.9 using: (i) Cauchy's theorem and (ii) the results in part (a) and (b).

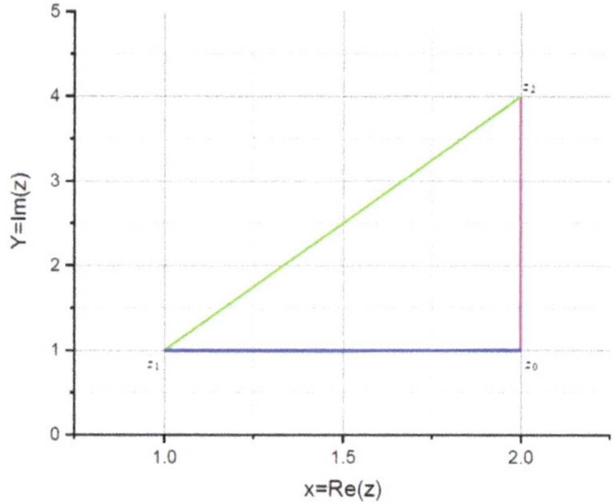

Figure 14.9. Right-angled triangular contour. Note that $z_0 = 2 + i$, $z_1 = 1 + i$, and $z_2 = 2 + 4i$.

Solution:

(a) First we just simply carry out the complex function integral,

$$\int_{z_1}^{z_2} z^2 dz = \frac{z^3}{3} \Bigg|_{z_1 = 1+i}^{z_2 = 2+4i} = \frac{(2 + 4i)^3 - (1 + i)^3}{3} = -\frac{86}{3} - 6i. \tag{14.98}$$

We now integrate using the green line joining the two points $z_1 = 1 + i$ and $z_2 = 2 + 4i$. The equation for the line joining the two points

$$\frac{y - y_1}{x - x_1} = \text{slope} \Rightarrow \frac{y - 1}{x - 1} = 3 \Rightarrow y = 3x - 2. \tag{14.99}$$

Using

$$z = x + iy \Rightarrow dz = dx + idy \tag{14.100}$$

one can write the integral

$$\int_{z_1}^{z_2} z^2 dz = \int_{z_1}^{z_2} (x^2 - y^2 + 2ixy)(dx + idy). \tag{14.101}$$

Using the equation of the line joining the two points in equation (14.99), we have $dy = 3dx$, so that substituting this and $y = 3x - 2$ into equation (14.101), one can write

$$\int_{z_1}^{z_2} z^2 dz = \int_1^2 [x^2 - (3x - 2)^2 + 2ix(3x - 2)](dx + i3dx)$$
$$= \int_1^2 [-8x^2 + 12x - 4 + i(6x^2 - 4x)](dx + i3dx). \tag{14.102}$$

Noting that

$$\int_1^2 [-8x^2 + 12x - 4 + i(6x^2 - 4x)]dx = -\frac{14}{3} + 8i, \tag{14.103}$$

one finds for equation (14.102)

$$\int_{z_1}^{z_2} z^2 dz = -\frac{14}{3} + 8i + i3\left(-\frac{14}{3} + 8i\right) = -\frac{86}{3} - 6i. \tag{14.104}$$

(b) The integral along the first line (the blue line) from the point $z_1 = 1 + i$ to $z_0 = 2 + i$, and the second line (the pink line) from the point z_0 to the point $z_2 = 2 + 4i$ can be expressed as

$$\int_{z_1}^{z_0} z^2 dz + \int_{z_0}^{z_2} z^2 dz = \frac{z^3}{3}\bigg|_{z_1}^{z_0} + \frac{z^3}{3}\bigg|_{z_0}^{z_2} = \frac{z_0^3 - z_1^3 + z_2^3 - z_0^3}{3}$$

$$= \frac{z_2^3 - z_1^3}{3} = \frac{(2 + 4i)^3 - (1 + i)^3}{3} = -\frac{86}{3} - 6i. \tag{14.105}$$

Following the two paths, along the line parallel to the x-axis

$$y = 1 \Rightarrow dy = 0 \Rightarrow dz = dx, \tag{14.106}$$

and then along the line parallel to the y-axis

$$x = 2 \Rightarrow dx = 0 \Rightarrow dz = dy, \tag{14.107}$$

one can also write the integral in terms of the Cartesian coordinates as

$$\int_{1+i}^{2+4i} z^2 dz = \int_1^2 (x^2 - 1 + 2ix)dx + \int_1^4 (2^2 - y^2 + 4iy)idy$$

$$= \left[\frac{x^3}{3} + ix^2 - x\right]_1^2 + i\left[-\frac{y^3}{3} + 2iy^2 + 4y\right]_1^4 = \frac{4}{3} + 3i - 30 - 9i \tag{14.108}$$

$$\Rightarrow \int_{1+i}^{2+4i} z^2 dz = -\frac{86}{3} - 6i,$$

which is the same result as in part (a).

(c) Using Cauchy's theorem

$$\oint_C f(z)dz = 0, \tag{14.109}$$

for the function

$$f(z) = z^2, \tag{14.110}$$

which is analytic everywhere (see equation (14.50)) inside and on the triangle, we can easily find

$$\oint_C z^2 dz = 0. \tag{14.111}$$

Noting that

$$\oint_C z^2 dz = \int_{z_1}^{z_0} z^2 dz + \int_{z_0}^{z_2} z^2 dz + \int_{z_2}^{z_1} z^2 dz$$
$$= \int_{z_1}^{z_0} z^2 dz + \int_{z_0}^{z_2} z^2 dz - \int_{z_1}^{z_2} z^2 dz, \tag{14.112}$$

and using the result in part (a) in equation (14.104) and in part (b) in equation (14.105), we find

$$\oint_C z^2 dz = -\frac{86}{3} - 6i - \left(-\frac{86}{3} - 6i\right) = 0. \tag{14.113}$$

The result in equation (14.113) is the same as the result in equation (14.111) that we found using Cauchy's theorem.

Example 14.7. Evaluate the contour integral

$$I = \oint \frac{e^z}{z(z+1)} dz, \tag{14.114}$$

for the closed path C shown in figure 14.10, which is defined by $|z - 1| = 3$.

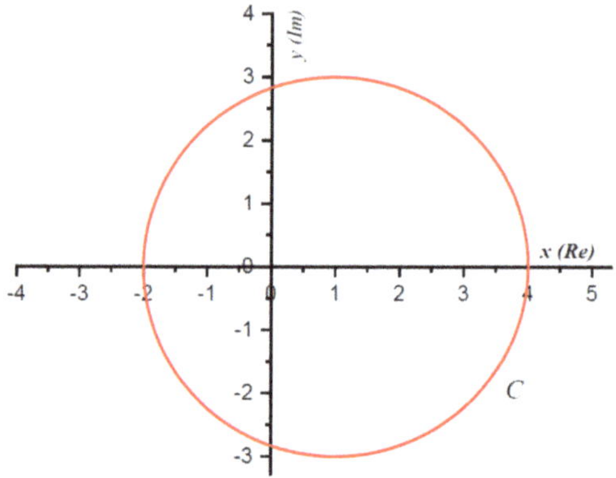

Figure 14.10. A circle with radius $r = 3$ centered at $z = 1$.

Solution: We can easily show

$$\frac{1}{z(z+1)} = \frac{1}{z} - \frac{1}{z+1}, \tag{14.115}$$

so that equation (14.114) can be rewritten as

$$I = \oint_C \frac{e^z}{z(z+1)} dz = \oint_C \frac{e^z}{z-0} dz + \oint_C \frac{e^z}{z-(-1)} dz. \tag{14.116}$$

For the contour defined by the curve $|z - 1| = 3$ shown in figure 14.10, the function

$$f(z) = e^z \tag{14.117}$$

is analytic inside and on the curve bounded by C. Noting that $z_0 = 0$ and $z_0 = -1$ are inside this curve, applying Cauchy's integral formula, one finds

$$\oint_C \frac{e^z}{z - 0} dz = 2\pi i f(0) = 2\pi i e^0 = 2\pi i,$$

$$\oint_C \frac{e^z}{z - (-1)} dz = 2\pi i f(-1) = 2\pi i e^{-1},$$

(14.118)

so that equation (14.116) becomes

$$I = \oint_C \frac{e^z}{z(z + 1)} dz = 2\pi i(1 + e^{-1}).$$

(14.119)

Equation (14.119) is the result of the integral obtained by applying Cauchy's integral formula.

14.6 Laurent's theorem

Consider a region R between two circles C_1 and C_2 centered at the same point, $z = z_0$ shown in figure 14.11. Then a function $f(z)$ that is analytic in the region R can be expanded in *Laurent's series* given by

$$f(z) = a_0 + a_1(z - z_0) + a_2(z - z_0)^2 + \dots$$

$$+ \frac{b_1}{z - z_0} + \frac{b_2}{(z - z_0)^2} + \frac{b_3}{(z - z_0)^3} \dots$$

$$= \sum_{n=0}^{\infty} a_n(z - z_0)^n + \sum_{n=1}^{\infty} \frac{b_n}{(z - z_0)^n},$$

(14.120)

that converges for any value of z within the region R.

Some important notes and terminologies about Laurent's series:

(a) *A regular point:* If $b_n = 0$ for all values of $n = 1, 2, 3\dots$, then $f(z)$ is analytic at $z = z_0$ and z_0 becomes *a regular point* for $f(z)$.

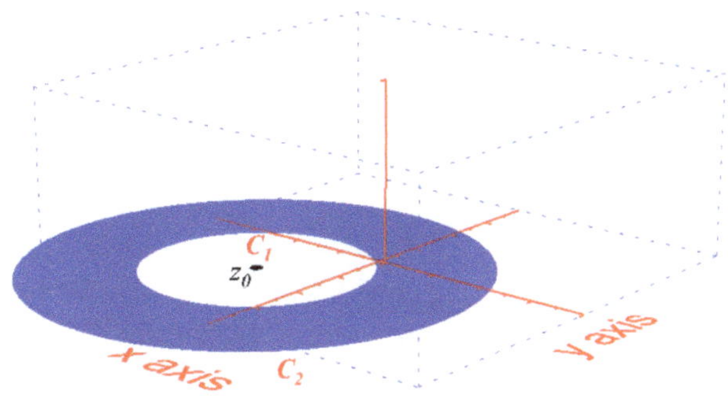

Figure 14.11. A region bounded by two concentric circles, C_1 and C_2.

(b) *The residue:* The coefficient b_1 in *Laurent's series* is called *the residue* for the function $f(z)$.

(c) *A pole of order N:* If the part of the *Laurent's series* with coefficient b_n has terms only up to b_N (i.e., $b_n = 0$ for all $n > N$), then $f(z)$ is said to have *a pole of order N* at the point $z = z_0$.

$$\sum_{n=1}^{\infty} \frac{b_n}{(z-z_0)^n} = \frac{b_1}{(z-z_0)} + \frac{b_2}{(z-z_0)^2} \cdots \frac{b_N}{(z-z_0)^N},$$ (14.121)

where

$$b_{N+1} = b_{N+2} = b_{N+3} \ldots = 0.$$ (14.122)

(d) *A simple pole:* If the part of Laurent's series with coefficient b_n has only the single term b_1 (i.e., $b_n = 0$ for all $n > 1$), then $f(z)$ is said to have *a simple pole* at $z = z_0$.

(e) *Essential singularity:* If the part of Laurent's series with coefficient b_n has $b_n \neq 0$ for all $n \geq 1$, $f(z)$ has an *essential singularity* at $z = z_0$.

14.7 The residue theorem

Let z_k be an *isolated singular point* of $f(z)$ inside a closed curve defined by C, the residue theorem states that

$$\oint_C f(z)dz = 2\pi i R(z_k),$$ (14.123)

where the integral is in a counterclockwise direction and $R(z_k)$ is the residue of the function $f(z)$ at $z = z_k$. In the following, we prove this theorem. To this end, let us consider a small circle of radius, $r = \epsilon_k$, centered about the isolated singular point z_k as shown in figure 14.12. In the figure, the blue region is bounded by the contours C_1, C_2, l_1, and l_2. The function $f(z)$ has no singular point and is analytic everywhere in this region. According to Cauchy's theorem, the integral along these contours must be zero, and one can write

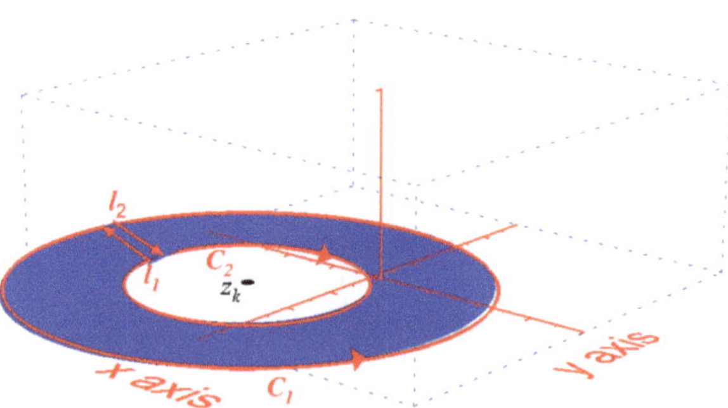

Figure 14.12. A circular region with the isolated singular point, z_k excluded.

$$\int_{C_1} f(z)dz + \int_{l_1} f(z)dz + \int_{l_2} f(z)dz + \int_{C_2} f(z)dz = 0. \qquad (14.124)$$

When the separation between l_1 and l_2 approaches zero, we note that

$$\int_{l_1} f(z)dz = -\int_{l_2} f(z)dz, \quad \int_{C_1} f(z)dz = \oint_{C_1} f(z)dz, \quad \int_{C_2} f(z)dz = \oint_{C_2} f(z)dz,$$

and equation (14.124) becomes

$$\oint_{C_1} f(z)dz = -\oint_{C_2} f(z)dz. \qquad (14.125)$$

Note that from figure 14.12, the integration along C_2 is clockwise (CW) while along C_1 is counterclockwise (CCW). It is customary that, generally, contour integrations are carried out in a CCW direction. Thus changing the CW integration along C_2 to CCW, we find

$$\left[-\oint_{C_2} f(z)dz \right]_{CW} = \oint_{C_2} f(z)dz \Rightarrow \oint_{C_1} f(z)dz = \oint_{C_2} f(z)dz. \qquad (14.126)$$

Since the function $f(z)$ is analytic in the region (blue) shown in figure 14.12, we can expand $f(z)$ in Laurent series about z_k,

$$f(z) = a_0 + a_1(z - z_k) + a_2(z - z_k)^2 + \ldots + \frac{b_1}{z - z_k}$$

$$+ \frac{b_2}{(z - z_k)^2} + \ldots = \sum_{n=0}^{\infty} a_n(z - z_k)^n + \sum_{n=1}^{\infty} \frac{b_n}{(z - z_k)^n}. \qquad (14.127)$$

Using equation (14.127), one can write equation (14.126) as,

$$\oint_{C_1} f(z)dz = \sum_{n=0}^{\infty} a_n \oint_{C_2} (z - z_k)^n d + \oint_{C_2} \frac{b_1}{z - z_k}dz$$

$$+ \oint_{C_2} \frac{b_2}{(z - z_k)^2}dz + \oint_{C_2} \frac{b_3}{(z - z_k)^3}dz \ldots \qquad (14.128)$$

Since the function $(z - z_0)^n$ is analytic everywhere in the region bounded by the curve C_2 for all $n \geq 0$, for equation (14.128), one can find,

$$\oint_{C_2} (z - z_k)^n dz = 0,$$

$$\Rightarrow \oint_{C_1} f(z)dz = \oint_{C_2} \frac{b_1}{z - z_k}dz + \oint_{C_2} \frac{b_2}{(z - z_k)^2}dz \qquad (14.129)$$

$$+ \oint_{C_2} \frac{b_3}{(z - z_k)^3}dz \ldots$$

For a point z that is on the circle with radius ϵ centered about the singular point z_k (i.e., on the contour C_2), we have,

$$z - z_k = \epsilon e^{i\theta} \Rightarrow z = z_k + \epsilon e^{i\theta} \Rightarrow dz = i\epsilon e^{i\theta}d\theta, \tag{14.130}$$

and equation (14.129) can be rewritten as

$$\oint_{C_1} f(z)dz = b_1 \oint_{C_2} \frac{1}{\epsilon e^{i\theta}} i\epsilon e^{i\theta}d\theta$$

$$+ \oint_{C_2} \frac{b_2}{(\epsilon e^{i\theta})^2} i\epsilon e^{i\theta}d\theta + \oint_{C_2} \frac{b_2}{(\epsilon e^{i\theta})^3} i\epsilon e^{i\theta}d\theta \dots$$

$$= 2\pi i b_1 + \sum_{n=2}^{\infty} \frac{ib_n}{\epsilon^{n-1}} \oint_{C_2} e^{-i(n-1)\theta}d\theta \tag{14.131}$$

$$\Rightarrow \oint_C f(z)dz = 2\pi i b_1 + \sum_{n=2}^{\infty} \frac{ib_n}{\epsilon^{n-1}} \int_0^{2\pi} \{\cos\left[(n-1)\theta\right]$$

$$- i \sin\left[(n-1)\theta\right]\} d\theta.$$

Noting that for a positive integer $n \geqslant 2$,

$$\int_0^{2\pi} \cos\left[(n-1)\theta\right]d\theta = -i \int_0^{2\pi} \sin\left[(n-1)\theta\right]d\theta = 0, \tag{14.132}$$

one finds for equation (14.131)

$$\oint_C f(z)dz = 2\pi i b_1 = 2\pi i R(z_k), \tag{14.133}$$

where

$$R(z_k) = b_1, \tag{14.134}$$

is *the residue* of the function $f(z)$ at $z = z_k$. Equation (14.133) is the residue theorem. Although we proved the residue theorem by considering one isolated singular point for the function $f(z)$ in the region bounded by the curve C (the curve C_2 for $\epsilon = 0$), it can be shown that for more than one singular point,

$$\oint f(z)dz = 2\pi i \sum_{k=1}^{N} R(z_k), \tag{14.135}$$

where N is the total number of the singular points inside the region.

14.8 Methods of finding residues

In the last section, we were introduced the residue theorem that we can use to determine contour integrals of complex functions. Such a method of integration requires finding the residue for the complex function. This section will introduce essential methods for finding the residues for a complex function $f(z)$.

Method 1

Expand $f(z)$ about the point z_0 and read the residue value from the series. In chapter 1, we have determined Taylor series expansion

$$f(x) = \sum_{n=0}^{\infty} b_n(x-a)^n, \quad \text{where } b_n = \frac{1}{n!}\frac{d^n f(x)}{dx^n}\bigg|_{x=a}, \tag{14.136}$$

for some essential functions of real variable x. This can be extended to functions of complex variable z.

$$\sin z = \sum_{n=0}^{\infty}\frac{(-1)^n}{(2n-1)!}z^{2n+1} = z - \frac{z^3}{3!} + \frac{z^5}{5!} - \frac{z^7}{7!}\ldots$$

$$\cos z = \sum_{n=0}^{\infty}\frac{(-1)^n}{(2n)!}z^{2n} = 1 - \frac{z^2}{2!} + \frac{z^4}{4!} - \frac{z^6}{6!}\ldots,$$

$$e^z = \sum_{n=0}^{\infty}\frac{1}{n!}z^n = 1 + z + \frac{z^2}{2!} + \frac{z^3}{3!}\ldots, \tag{14.137}$$

$$\ln(1+z) = \sum_{n=1}^{\infty}\frac{(-1)^{n+1}}{n}z^n = z - \frac{z^2}{2} + \frac{z^3}{3} - \frac{z^4}{4}\ldots 1,$$

$$(1+z)^p = \sum_{n=1}^{\infty}\binom{p}{n}z^n = 1 + pz + \frac{p(p-1)}{2!}z^2\ldots$$

where

$$\binom{p}{n} = \frac{p(p-1)(p-2)(p-3)\ldots(p-n+1)}{n!}, \tag{14.138}$$

is the binomial coefficient.

Example 14.8. Find $R(0)$ for the complex function

$$f(z) = \frac{1}{z}\cos(z). \tag{14.139}$$

Solution: The function has an isolated singular point at $z_k = 0$ since the $1/z$ diverges at this point. Thus we need the series expansion for the function

$$g(z) = \cos(z), \tag{14.140}$$

about $z_k = 0$. Recalling that the series expansion for a cosine function about $z_k = 0$ is

$$\cos(z) = 1 - \frac{z^2}{2!} + \frac{z^4}{4!} - \frac{z^6}{6!}\ldots \tag{14.141}$$

Upon dividing this equation by z, we find

$$f(z) = \frac{1}{z}\cos(z) = \frac{1}{z} - \frac{z}{2!} + \frac{z^3}{4!} - \frac{z^5}{6!}\cdots$$
$$= \sum_{n=1}^{\infty} \frac{b_n}{(z - z_k)^n} + \sum_{n=0}^{\infty} a_n(z - z_k)^{2n+1}, \tag{14.142}$$

where

$$z_k = 0, \; a_n = \frac{1}{(2 + 2n)!}, \; b_n = \begin{cases} 1, & \text{for } n = 1, \\ 0, & \text{for } n \neq 1. \end{cases} \tag{14.143}$$

One can then read that

$$R(z_k) = b_1 = 1, \tag{14.144}$$

is the residue for $z_k = 0$.

Example 14.9. Find the residue $R(1)$ for the complex function,

$$f(z) = \frac{e^z}{(z - 1)^2}. \tag{14.145}$$

Solution: The function has an isolated singular point at $z_k = 1$, and we need the series expansion of this function about this point. The more straightforward approach is first to find the series expansion for the function

$$g(z) = e^z, \tag{14.146}$$

about $z_k = 1$, which is given by

$$g(z) = e^z = \sum_{n=0}^{\infty} \left[\frac{1}{n!} \frac{d^n(e^z)}{dz^n} \bigg|_{z=1} \right](z - 1)^n$$
$$\Rightarrow e^z = e\left\{ 1 + \frac{1}{1!}(z - 1) + \frac{1}{2!}(z - 1)^2 + \frac{1}{3!}(z - 1)^3 \right. \tag{14.147}$$
$$\left. + \frac{1}{4!}(z - 1)^4\ldots \right\}.$$

Then dividing this series by $(z - 1)^2$, one can write

$$f(z) = e\frac{1}{(z - 1)^1} + e\frac{1}{(z - 1)^2} + \frac{1}{2!}e(z - 1)^0$$
$$+ \frac{1}{3!}(z - 1) + \frac{1}{4!}(z - 1)^2\ldots \tag{14.148}$$
$$\Rightarrow f(z) = \sum_{n=1}^{\infty} \frac{b_n}{(z - z_k)^n} + \sum_{n=0}^{\infty} a_n(z - z_k)^n,$$

where

$$z_k = 1, \ a_n = \frac{e}{(2 + n!)}, \ b_n = \begin{cases} e, & \text{for } n = 1, 2 \\ 0, & \text{for } n \neq 1, 2. \end{cases} \tag{14.149}$$

There follows that

$$R(z_k) = b_1 = e, \tag{14.150}$$

which is the residue $R(1)$ for the complex function $f(z)$.

Example 14.10. Find $R(\pi)$ for the complex function

$$f(z) = \frac{\sin z}{z - \pi} \tag{14.151}$$

Solution: The function has an isolated singular point at

$$z_k = \pi, \tag{14.152}$$

and we need to expand the function $\sin(z)$ in series about this point,

$$\begin{aligned} \sin(z) &= \sum_{n=0}^{\infty} \left[\frac{1}{n!} \frac{d^n(\sin(z))}{dz^n} \bigg|_{z \, = \, \pi} \right] (z - \pi)^n \\ &= -(z - \pi) + \frac{(z - \pi)^3}{3!} - \frac{(z - \pi)^5}{5!} + \frac{(z - \pi)^7}{7!} \cdots, \end{aligned} \tag{14.153}$$

so that we can write

$$\begin{aligned} f(z) &= \frac{\sin z}{z - \pi} = -1 + \frac{(z - \pi)^2}{3!} - \frac{(z - \pi)^4}{5!} + \frac{(z - \pi)^6}{7!} \cdots \\ &= \sum_{n=1}^{\infty} \frac{b_n}{(z - z_k)^n} + \sum_{n=0}^{\infty} a_n(z - z_k)^{2n}, \end{aligned} \tag{14.154}$$

where

$$z_k = \pi, \ a_n = \frac{(-1)^{2n+1}}{(2n + 1)!}, \ \text{for all } n \geqslant 0, \tag{14.155}$$

$$b_n = 0, \ \text{for all } n \geqslant 1.$$

Since $b_n = 0$ for all $n \geqslant 1$, $R(z_k) = b_1 = 0$ is the residue for $z_k = \pi$.

Method 2
This method is needed when the function $f(z)$ has a *simple pole* (a pole of order 1) at z_k. In such cases, multiply $f(z)$ by $(z - z_k)$ and evaluate the result at $z = z_k$ (or take the limit as z approaches z_k),

$$R(z_k) = \lim_{z \to z_k} [(z - z_k)f(z)]. \tag{14.156}$$

Example 14.11. Find $R(-1)$ for the complex function

$$f(z) = \frac{z}{(z+1)(z+2)}. \tag{14.157}$$

Solution: We note that $f(z)$ has a simple pole at $z_k = -1$. Then using Method 2, equation (14.156) for $z_k = -1$ becomes

$$R(-1) = \lim_{z \to -1}[(z+1)f(z)] = \lim_{z \to -1}\left[\frac{z}{(z+2)}\right] = \frac{-1}{(-1+2)} = -1. \tag{14.158}$$

Example 14.12. Find $R(\pi)$ for the complex function

$$f(z) = \cot(z) = \frac{\cos(z)}{\sin(z)}. \tag{14.159}$$

Solution: This function has a simple pole at $z_k = \pi$. According to Method 2, the residue at $z_k = \pi$, given by equation (14.156), becomes

$$R(\pi) = \lim_{z \to \pi}[(z-\pi)f(z)] \Rightarrow R(\pi) = \lim_{z \to \pi}\left[(z-\pi)\frac{\cos(z)}{\sin(z)}\right]. \tag{14.160}$$

As $z \to \pi$, one finds 0/0. In such cases, we apply L'Hospital rule,

$$R(\pi) = \lim_{z \to \pi}\left[(z-\pi)\frac{\cos(z)}{\sin(z)}\right] = \lim_{z \to \pi}\left[\frac{\dfrac{d}{dz}[(z-\pi)\cos(z)]}{\dfrac{d}{dz}\sin(z)}\right] \tag{14.161}$$

$$\Rightarrow R(\pi) = \lim_{z \to \pi}\left[\frac{\cos(z) - (z-\pi)\sin(z)}{\cos(z)}\right] = \frac{\cos(\pi)}{\cos(\pi)} = 1.$$

We may ask how we knew that $f(z)$ has a simple pole at $z_k = \pi$. In order to find whether a function $f(z)$ has a simple pole at z_k or not, one must find Laurent's series

$$f(z) = \sum_{n=0}^{\infty} a_n(z - z_k)^n + \sum_{n=1}^{\infty} \frac{b_n}{(z - z_k)^n}, \tag{14.162}$$

and show that $b_n \neq 0$ for only $n = 1$. To this end, we make a series expansion for the function

$$f(z) = \cot(z) = \frac{\cos(z)}{\sin(z)}. \tag{14.163}$$

One can show that the series expansion for this function is given by

$$f(z) = \cot(z)$$

$$= \frac{1}{z - \pi} - \frac{z - \pi}{3} - \frac{1}{45}(z - \pi)^3 - \frac{2}{945}(z - \pi)^5 - \frac{(z - \pi)^7}{4725} \dots \tag{14.164}$$

This result indeed shows the function has a simple pole at $z_k = \pi$ since $b_1 = 1 \neq 0$ and $b_n = 0$ for all $n > 1$.

Method 3

This method is essentially the more generalized form of Method 2. In Method 2 we consider when the function $f(z)$ has a simple pole at z_k (order, $n = 1$). Suppose, in general, the function $f(z)$ has a pole of order n at z_k. The residue $R(z_k)$ for the function $f(z)$ is given by

$$R(z_k) = \left[\frac{1}{(m - 1)!} \frac{d^{m-1}}{dz^{m-1}} \{(z - z_k)^m f(z)\} \right]_{z = z_k}, \tag{14.165}$$

where $m \geqslant n$.

Example 14.13. Find $R(3)$ for

$$f(z) = \frac{ze^{zt}}{(z - 3)^2}, \tag{14.166}$$

where t is a parameter (possibly complex).

Solution: We note that $f(z)$ has a pole at $z_k = 3$ of order $n = 2$. Thus we can apply Method 3 in equation (14.165) to find the residue at z_k ., which becomes

$$R(3) = \left[\frac{1}{(2 - 1)!} \frac{d^{2-1}}{dz^{2-1}} \left\{ (z - 3)^2 \frac{ze^{zt}}{(z - 3)^2} \right\} \right]_{z = 3} = \left[\frac{d}{dz}(ze^{zt}) \right]_{z = 3} \tag{14.167}$$

$$\Rightarrow R(3) = [(1 + zt)e^{zt}]_{z = 3} = (1 + 3t)e^{3t},$$

where we chose $m = 2 = n$.

Example 14.14. Find $R(-2)$ for the complex function

$$f(z) = \frac{e^{2z}}{z(z + 2)^3}. \tag{14.168}$$

Solution: The function has a pole at $z_k = -2$ of order $n = 3$. Using Method 3, for $m = n = 3$ and the function $f(z)$, equation (14.165) becomes

$$R(-2) = \frac{1}{(3-1)!}\frac{d^{3-1}}{dz^{3-1}}\left\{(z+2)^3\frac{e^{2z}}{z(z+2)^3}\right\}\Bigg|_{z=-2}$$

$$= \frac{1}{2!}\frac{d^2}{dz^2}\left\{\frac{e^{2z}}{z}\right\}\Bigg|_{z=-2} = \frac{1}{2!}\frac{d}{dz}\left\{\frac{2e^{2z}}{z} - \frac{e^{2z}}{z^2}\right\}\Bigg|_{z=-2}$$

$$= \frac{1}{2!}\left[\frac{4e^{2z}}{z} - \frac{2e^{2z}}{z^2} - \frac{2e^{2z}}{z^2} + \frac{2e^{2z}}{z^3}\right]_{z=-2} \tag{14.169}$$

$$= \frac{2e^{2z}}{z} - \frac{2e^{2z}}{z^2} + \frac{e^{2z}}{z^3}\Bigg|_{z=-2} = \left(-1 - \frac{1}{2} - \frac{1}{8}\right)e^{-4}$$

$$\Rightarrow R(-2) = -\frac{13}{8}e^{-4}.$$

14.9 Applications of the residue theorem

In the last two sections, we have proved the residue theorem and done several examples illustrating the methods for finding the residues for a complex function. The residue theorem states if $\{z_1, z_2, \ldots z_k. \ldots z_n\}$ are singular points of $f(z)$ inside a closed curve defined by C, the contour integral of this function can be determined using the relation

$$\oint_C f(z)dz = 2\pi i \sum_{k=1}^{n} R(z_k), \tag{14.170}$$

where $R(z_k)$ is the residue of the function $f(z)$ at $z = z_k$. We now use this theorem and the various methods introduced to find residues to evaluate several different types of definite integrals. Examples best describe the application of the residue theorem for evaluating such integrals.

Example 14.15. Evaluate the real integral

$$I = \int_0^\infty \frac{dx}{x^4 + 1}. \tag{14.171}$$

Solution: In order to determine such integral using the residue theorem, it is essential to consider the appropriate contour. For such integral, we usually consider a semicircular contour on the x–y plane with the real axis as part of the semicircular contour as shown in figure 14.13.

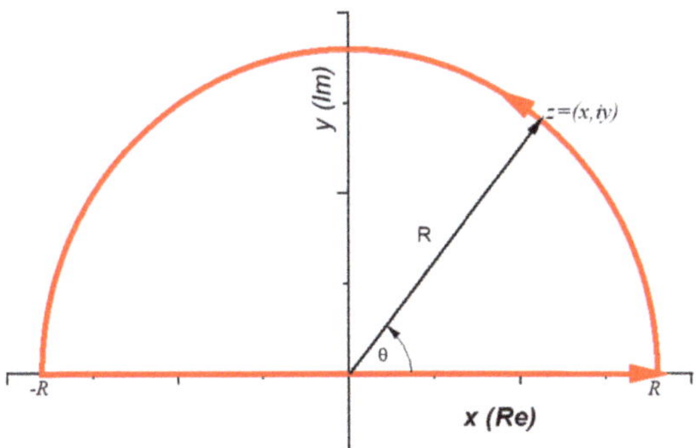

Figure 14.13. A semicircular contour in the upper-half-plane.

This semicircular contour which is on the upper-half complex plane, has a radius of R. Once we chose the appropriate contour, we replace the real variable x with a complex variable z, such that

$$f(z) = \frac{1}{z^4 + 1}, \tag{14.172}$$

and write the complex integral

$$\oint f(z)dz = \oint_C \frac{dz}{z^4 + 1} \tag{14.173}$$

where C is the contour shown in figure 14.13. According to the residue theorem, this integral given by

$$\oint f(z)dz = \oint_C \frac{dz}{z^4 + 1} = 2\pi i \sum_{k=0}^{n} R(z_k), \tag{14.174}$$

where $R(z_k)$ are the residues inside the semicircular contour. On this contour, one can split the integral into two line integrals,

$$\oint \frac{dz}{z^4 + 1} = \int_{-R}^{R} \frac{dx}{x^4 + 1} + \int_{C_1} \frac{dz}{z^4 + 1}, \tag{14.175}$$

where C_1 is the curved part of the contour in figure 14.13. For a complex variable z on C_1, we can express the complex variable z as

$$z = R \exp[i\theta] \Rightarrow dz = iR \exp[i\theta]d\theta, \tag{14.176}$$

where $0 \leqslant \theta \leqslant \pi$. On can then rewrite equation (14.175) as

$$\oint \frac{dz}{z^4 + 1} = \int_{-R}^{R} \frac{dx}{x^4 + 1} + \int_{0}^{\pi} \frac{iR \exp[i\theta]d\theta}{R^4 \exp[4i\theta] + 1}. \tag{14.177}$$

Now let $R \to \infty$, so that for the two integrals on the right side, one finds

$$\lim_{R \to \infty} \int_{-R}^{R} \frac{dx}{x^4 + 1} = \int_{-\infty}^{\infty} \frac{dx}{x^4 + 1}, \quad \lim_{R \to \infty} \int_{0}^{\pi} \frac{iR \exp[i\theta]d\theta}{R^4 \exp[4i\theta] + 1} = 0, \qquad (14.178)$$

and we find

$$\lim_{R \to \infty} \oint \frac{dz}{z^4 + 1} = \int_{-\infty}^{\infty} \frac{dx}{x^4 + 1}. \qquad (14.179)$$

Since $x^4 + 1$ is an even function, we have

$$\int_{0}^{\infty} \frac{dx}{x^4 + 1} = \int_{-\infty}^{0} \frac{dx}{x^4 + 1} \Rightarrow \int_{-\infty}^{\infty} \frac{dx}{x^4 + 1} = 2 \int_{0}^{\infty} \frac{dx}{x^4 + 1} \qquad (14.180)$$

so that from equation (14.179), one can write

$$\int_{0}^{\infty} \frac{dx}{x^4 + 1} = \frac{1}{2} \lim_{R \to \infty} \oint \frac{dz}{z^4 + 1}. \qquad (14.181)$$

The contour integral of the complex function $f(z) = 1/(z^4 + 1)$ on the semicircular contour with infinite radius ($R \to \infty$), according to the residue theorem, is given by equation (14.174). Therefore to find the value for the integral in equation (14.181), we need to find the residues for the function $f(z)$ inside the semicircular contour with infinite radius. However, first, we must find all the poles for the function $f(z)$ inside the contour. Generally, the poles for the function $f(z)$ are the values for z where the $f(z)$ diverges. The function $f(z) = 1/(z^4 + 1)$ diverges when

$$z^4 + 1 = 0 \Rightarrow z = (-1)^{1/4} = [e^{i(\pi + n(2\pi))}]^{1/4}, \qquad (14.182)$$

where $n = 0, 1, 2, 3$ (refer to section 2.3). Therefore, the poles, z_k, are found to be

$$z_0 = e^{i\frac{\pi}{4}} \Rightarrow z_0 = \frac{1}{\sqrt{2}}(1 + i),$$

$$z_1 = e^{i\frac{3\pi}{4}} \Rightarrow z_1 = \frac{1}{\sqrt{2}}(-1 + i),$$

$$z_2 = e^{i\frac{5\pi}{4}} \Rightarrow z_2 = -\frac{1}{\sqrt{2}}(1 + i), \qquad (14.183)$$

$$z_3 = e^{i\frac{7\pi}{4}} \Rightarrow z_3 = \frac{1}{\sqrt{2}}(1 - i).$$

Remember, we are integrating on a semicircular contour in the upper-half of the complex plane and therefore we need only the poles that are inside this contour according to the residue theorem. These poles are shown in figure 14.14. The poles inside the contour on the upper-half complex plane with a phase $0 \leqslant \theta \leqslant \pi$ are

$$z_0 = e^{i\frac{\pi}{4}} = \frac{\sqrt{2}}{2}(1 + i), \quad z_1 = e^{i\frac{3\pi}{4}} = \frac{\sqrt{2}}{2}(-1 + i). \qquad (14.184)$$

We now determine the residue of these poles. We note that the complex function $f(z)$ can then be expressed as

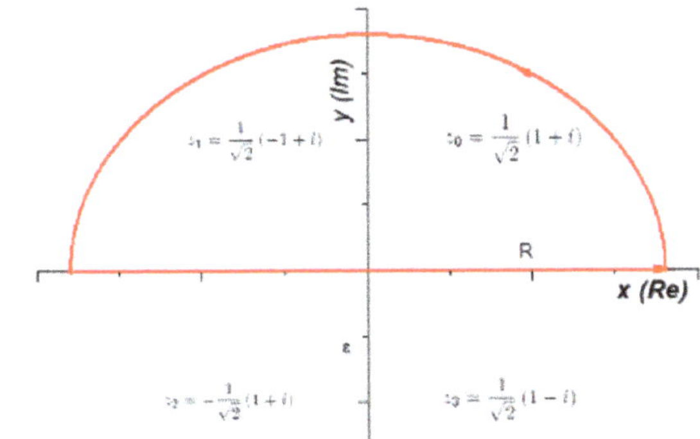

Figure 14.14. The contour of integration and the poles.

$$f(z) = \frac{1}{z^4 + 1} = \frac{1}{(z - z_0)(z - z_1)(z - z_2)(z - z_3)}. \tag{14.185}$$

This shows that the two poles

$$z_0 = e^{i\frac{\pi}{4}} = \frac{1}{\sqrt{2}}(1 + i), \; z_1 = e^{i\frac{3\pi}{4}} = \frac{1}{\sqrt{2}}(-1 + i), \tag{14.186}$$

are simple poles (of order one) and we can use Method 2 to find the residues at these poles

$$R(z_k) = \lim_{z \to z_k}[(z - z_k)f(z)]. \tag{14.187}$$

For z_0, one can write,

$$\begin{aligned}
R(z_0) &= \lim_{z \to z_0}\left[(z - z_0)\frac{1}{(z - z_0)(z - z_1)(z - z_2)(z - z_3)}\right] \\
&= \lim_{z \to z_0}\left[\frac{1}{(z - z_1)(z - z_2)(z - z_3)}\right] \\
\Rightarrow R(z_0) &= \frac{1}{(z_0 - z_1)(z_0 - z_2)(z_0 - z_3)},
\end{aligned} \tag{14.188}$$

and using the values for z_0, z_1, z_2, and z_3 in equation (14.183), we find

$$z_0 - z_1 = \sqrt{2}, \; z_0 - z_2 = \sqrt{2}(1 + i), \; z_0 - z_3 = i\sqrt{2} \tag{14.189}$$

so that upon substituting these values into equation (14.188), we find

$$R(z_0) = \frac{1}{(\sqrt{2})(\sqrt{2}(1 + i))(i\sqrt{2})} = -\frac{1}{2\sqrt{2}(1 - i)}. \tag{14.190}$$

Similarly, the residue at the second pole z_1 is given by

$$R(z_1) = \lim_{z \to z_0}\left[(z - z_1)\frac{1}{(z - z_0)(z - z_1)(z - z_2)(z - z_3)}\right]$$

$$\Rightarrow R(z_1) = \frac{1}{(z_1 - z_0)(z_1 - z_2)(z_1 - z_3)}. \tag{14.191}$$

Using equation (14.183), we have

$$z_1 - z_0 = -\sqrt{2}, \quad z_1 - z_2 = i\sqrt{2}, \quad z_1 - z_3 = \sqrt{2}(-1 + i), \tag{14.192}$$

and substituting these values into equation (14.191) results in

$$R(z_1) = \frac{1}{-\sqrt{2}\,i\sqrt{2}\,\sqrt{2}(-1 + i)} = \frac{1}{2\sqrt{2}(i + 1)}. \tag{14.193}$$

Now using the residue theorem in equation (14.174), and the results in equations (14.191) and (14.193), one can write

$$\lim_{R \to \infty} \oint f(z)\,dz = \lim_{R \to \infty} \oint \frac{dz}{z^4 + 1} = 2\pi i \sum_{k=0}^{1} R(z_k)$$

$$= 2\pi i[R(z_0) + R(z_1)],$$

$$= 2\pi i\left[-\frac{1}{2\sqrt{2}(1 - i)} + \frac{1}{2\sqrt{2}(i + 1)}\right]$$

$$= \pi i\left[\frac{-i - 1 + 1 - i}{2\sqrt{2}}\right] \tag{14.194}$$

$$= \pi i\left[\frac{-2i}{2\sqrt{2}}\right]$$

$$\Rightarrow \lim_{R \to \infty} \oint \frac{dz}{z^4 + 1} = \frac{\pi}{\sqrt{2}}.$$

Substituting equation (14.194) into equation (14.181), we find the result

$$\int_0^\infty \frac{dx}{x^4 + 1} = \frac{1}{2}\lim_{R \to \infty} \oint \frac{dz}{z^4 + 1} \Rightarrow \int_0^\infty \frac{dx}{x^4 + 1} = \frac{\pi}{2\sqrt{2}}, \tag{14.195}$$

which is the result for the integral in equation (14.171).

Example 14.16. Evaluate the real integral

$$I = \int_0^{2\pi} \frac{\cos(3\theta)}{5 - 4\cos(\theta)}\,d\theta. \tag{14.196}$$

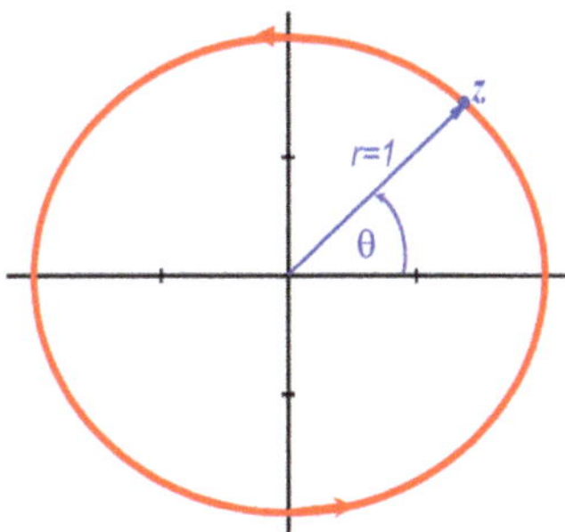

Figure 14.15. A unit circle on a complex plane.

Solution: For such integrals we use a circular contour centered about the origin with radius, $R = 1$ on a complex plane as shown in figure 14.15. Using Euler's formula one can express,

$$\cos(k\theta) = \frac{1}{2}[(e^{i\theta})^k + (e^{-i\theta})^k] = \frac{1}{2}\left[(e^{i\theta})^k + \frac{1}{(e^{i\theta})^k}\right],$$

$$\sin(k\theta) = \frac{1}{2i}[(e^{i\theta})^k - (e^{-i\theta})^k] = \frac{1}{2i}\left[(e^{i\theta})^k - \frac{1}{(e^{i\theta})^k}\right]. \tag{14.197}$$

We note that for a complex number z on a unit circle shown in figure 14.15, we have

$$z = e^{i\theta}, \tag{14.198}$$

so that equation (14.197) becomes

$$\cos(k\theta) = \frac{1}{2}\left[z^k + \frac{1}{z^k}\right] = \frac{1}{2}\left(\frac{(z^k)^2 + 1}{z^k}\right),$$

$$\sin(k\theta) = \frac{1}{2i}\left[z^k - \frac{1}{z}\right] = \frac{1}{2i}\left(\frac{(z^k)^2 - 1}{z^k}\right). \tag{14.199}$$

There follows that for $k = 1$ and $k = 3$,

$$\cos(\theta) = \frac{1}{2}\left(z + \frac{1}{z}\right) = \frac{1}{2}\left(\frac{z^2 + 1}{z}\right),$$

$$\cos(3\theta) = \frac{1}{2}\left(z^3 + \frac{1}{z^3}\right) = \frac{1}{2}\left(\frac{z^6 + 1}{z^3}\right). \tag{14.200}$$

Then one can write for the function $f(\theta)$

$$f(\theta) = \frac{\cos(3\theta)}{5 - 4\cos(\theta)} = \frac{\frac{1}{2}\left(\frac{z^6 + 1}{z^3}\right)}{5 - \frac{2(z^2 + 1)}{z}} = -\frac{z^6 + 1}{2z^2(2z^2 - 5z + 2)}. \tag{14.201}$$

Furthermore, we note that for $z = \exp(i\theta)$,

$$dz = i\exp(i\theta)d\theta \Rightarrow d\theta = -i\exp(-i\theta)dz = -i\frac{dz}{z}. \tag{14.202}$$

Now combining equations (14.201) and (14.202), one can write the integral in equation (14.196) as,

$$I = \int_0^{2\pi} \frac{\cos(3\theta)}{5 - 4\cos(\theta)}d\theta = \oint_C \frac{i(z^6 + 1)dz}{2z^3[2z^2 - 5z + 2]} = \oint_C f(z)dz, \tag{14.203}$$

where C is the unit circle shown in figure 14.15 and

$$f(z) = \frac{i(z^6 + 1)}{2z^3(2z^2 - 5z + 2)}. \tag{14.204}$$

We can then use the residue theorem

$$\oint_C f(z)dz = 2\pi i \sum_{k=1}^{n} R(z_k) \tag{14.205}$$

to find the result for the integral in equation (14.203). To this end, we first find the poles inside for the function $f(z)$. Noting that

$$\begin{aligned} 2z^2 - 5z + 2 &= (2z - 1)(z - 2) \\ \Rightarrow 2z^2 - 5z + 2 &= 2(z - 1/2)(z - 2), \end{aligned} \tag{14.206}$$

one can rewrite $f(z)$ as

$$f(z) = \frac{i(z^6 + 1)}{4(z - 0)^3(z - 1/2)(z - 2)}. \tag{14.207}$$

This expression shows that the function $f(z)$ has poles

$$z_1 = 0, \; z_2 = 1/2, \; z_3 = 2. \tag{14.208}$$

However, only the first two poles ($z_1 = 0$, $z_2 = 1/2$) are inside the unit circle (see figure 14.16). Thus we need only the residues of these two poles, $R(0)$ and $R(1/2)$.

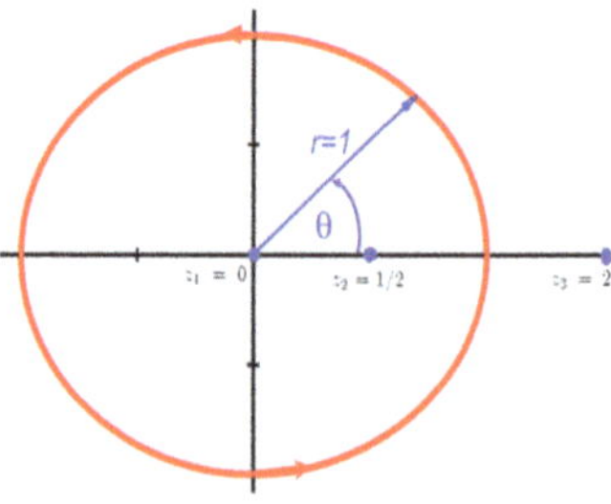

Figure 14.16. The three poles for the function, $f(z)$.

The second pole, $z_2 = 1/2$ is a simple pole since the order is one and the residue can easily be determined to be

$$R(1/2) = \lim_{z \to 1/2} [(z - 1/2)f(z)] = \lim_{z \to 1/2} \left[\frac{i(z^6 + 1)}{4z^3(z - 2)} \right] = \frac{i\left(\frac{1}{64} + 1\right)}{\frac{1}{2}\left(\frac{1}{2} - 2\right)}$$

$$= i\frac{65}{64}\left(-\frac{4}{3}\right) \Rightarrow R(1/2) = -i\frac{65}{48}. \tag{14.209}$$

The first pole $z_1 = 0$, is not a simple pole. It has order $n = 3$. Thus to find the residue for this pole, we use Method 3,

$$R(z_k) = \left[\frac{1}{(m - 1)!} \frac{d^{m-1}}{dz^{m-1}} \{(z - z_k)^m f(z)\} \right]_{z = z_k}, \tag{14.210}$$

where $m \geqslant n$. For $m = 3 = n$ and $z_1 = 0$, one finds

$$R(0) = \frac{1}{(3 - 1)!} \frac{d^{3-1}}{dz^{3-1}} \left\{ (z - 0)^3 \frac{i(z^6 + 1)}{4(z - 0)^3(z - 1/2)(z - 2)} \right\} \Bigg|_{z = 0}$$

$$= \frac{1}{2} \frac{d^2}{dz^2} \left\{ \frac{i(z^6 + 1)}{2(2z^2 - 5z + 2)} \right\} \Bigg|_{z = 0}$$

$$= \frac{1}{2} \left[-\frac{6iz^5(4z - 5)}{(2z^2 - 5z + 2)^2} + \frac{15iz^4}{2z^2 - 5z + 2} \right. \tag{14.211}$$

$$\left. + \frac{i(1 + z^6)(4z - 5)^2}{(2z^2 - 5z + 2)^3} - \frac{2i(1 + z^6)}{(2z^2 - 5z + 2)^2} \right]_{z = 0}$$

$$= \frac{i}{2}\left(\frac{25}{8} - \frac{1}{2} \right)$$

$$\Rightarrow R(0) = \frac{21i}{16}.$$

Therefore, using the results in equations (14.209) and (14.211), for the integral in equation (14.205), one finds

$$\oint f(z)dz = 2\pi i[R(0) + R(1/2)] = 2\pi i\left(-i\frac{65}{48} + \frac{i21}{16} \right) = \frac{\pi}{12}$$

$$\Rightarrow \int_0^{2\pi} \frac{\cos(3\theta)}{5 - 4\cos(\theta)} d\theta = \oint f(z)dz = \frac{\pi}{12}, \tag{14.212}$$

which is the value for the integral in equation (14.196).

14.10 The modified residue theorem

We recall that the residue theorem states that

$$\oint f(z)dz = 2\pi i R(z_k),\qquad (14.213)$$

where z_k are the poles inside the contour of integration. What if there are poles *on* the contour of integration, which we represent by z_k'? This means what if the function $f(z)$ is not analytic at z_k' located on the contour. For example, let us say we are interested to evaluate the real integral

$$I_1 = \int_{-\infty}^{\infty} \frac{dx}{x^2 - 1}.\qquad (14.214)$$

To determine the value for this integral, as we saw in the previous section, we use a semicircular contour shown in figure 14.17 and evaluate the contour integral

$$I_2 = \oint f(z)dz,\qquad (14.215)$$

for the function

$$f(z) = \frac{1}{z^2 - 1},\qquad (14.216)$$

using the residue theorem in the limit, the radius of the contour goes to infinity. However, we know that $f(z)$ has simple poles at $z_1' = -1$ and $z_2' = 1$, which are on the contour. So how can we evaluate I_1 by directly using the residue theorem? We can evaluate such integrals using two different approaches. These approaches either put the poles inside or outside a contour of integration by modifying the semicircular contour. Such approaches lead to modified residue theorem when there are n' poles on the contour of integration and the n poles inside the contour. The modified residue theorem states that,

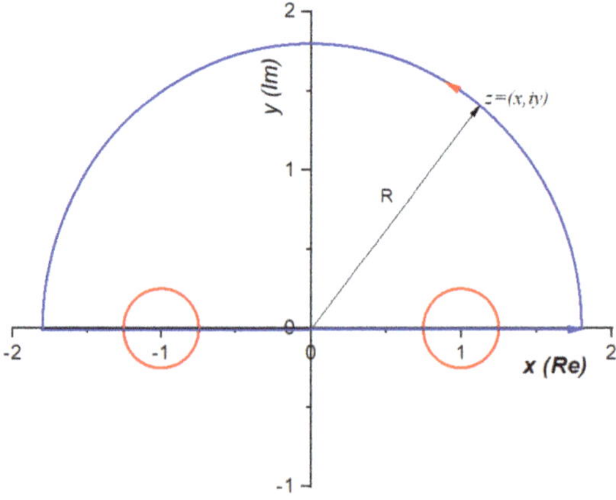

Figure 14.17. A semicircular contour with poles on the contour.

$$\oint f(z)dz = 2\pi i\left[\sum_{k=1}^{n} R(z_k) + \frac{1}{2}\sum_{k=1}^{n} R(z_k')\right], \tag{14.217}$$

where z_k is the pole *inside the contour* and z_k' is *a simple pole on the contour*. In the following, we will prove the modified residue theorem by putting the poles on the curve inside or outside the contour.

The top contour C_1, shown in figure 14.18, keeps the pole inside the contour of integration. In this case, assuming that there are no other poles inside this modified contour, since the pole z_k' now has become a pole inside the contour, we can use the residue theorem and write

$$\oint_{C_1} f(z)dz = 2\pi i R(z_k'). \tag{14.218}$$

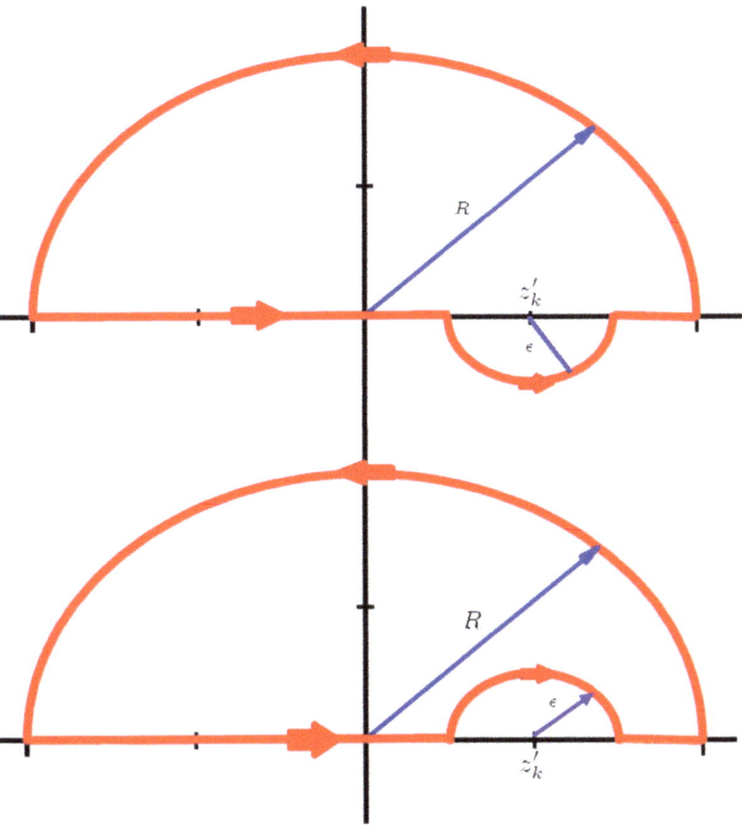

Figure 14.18. A contour of integration that include the pole (top) and that exclude the pole (bottom).

Let us consider another modified contour, C_2 shown at the bottom in figure 14.18. Since now the pole is outside the contour and there is no other pole, according to the residue theorem, we have

$$\oint_{C_2} f(z)dz = 0. \tag{14.219}$$

Now upon adding equations (14.218) and (14.219), we find

$$\oint_{C_1} f(z)dz + \oint_{C_2} f(z)dz = 2\pi i R(z_k'). \tag{14.220}$$

In the limit as $\epsilon \to 0$; we note that

$$\lim_{\epsilon \to 0} \oint_{C_1} f(z)dz = \lim_{\epsilon \to 0} \oint_{C_2} f(z)dz = \oint_C f(z)dz, \tag{14.221}$$

where C is the semicircular contour of integration with the pole on the contour line. Thus one can write

$$2\oint_C f(z)dz = 2\pi i R(z_k') \Rightarrow \oint_C f(z)dz = \pi i R(z_k') = 2\pi i\left[\frac{1}{2}R(z_k')\right]. \tag{14.222}$$

Now including the residues for the poles inside the contour, one can write

$$\oint f(z)dz = 2\pi i\left[\sum_{k=1}^{n} R(z_k) + \frac{1}{2}\sum_{k=1}^{n} R(z_k')\right], \tag{14.223}$$

which is the modified residue theorem.

Example 14.17. Evaluate the integrals

$$I_1 = \int_0^\infty \frac{\sin(x)}{x}dx, \quad I_2 = \int_{-\infty}^\infty \frac{\cos(x)}{x}dx. \tag{14.224}$$

Solution: In order to determine these integrals, taking into account the relation

$$e^{ix} = \cos(x) + i\sin(x) \Rightarrow \frac{e^{ix}}{x} = \frac{\cos(x)}{x} + i\frac{\sin(x)}{x} \tag{14.225}$$

it is worth considering the contour integral

$$\oint_C f(z)dz = \oint_C \frac{e^{iz}}{z}dz, \tag{14.226}$$

where C is a semicircular contour of radius, R centered about the origin on the upper complex plane as shown in figure 14.19. This function $f(z)$ has no poles inside the contour. The only pole is $z_k' = 0$ which is on the contour line. Therefore, applying the modified residue theorem in equation (14.217),

$$\oint \frac{e^{iz}}{z}dz = \pi i R(z_k' = 0). \tag{14.227}$$

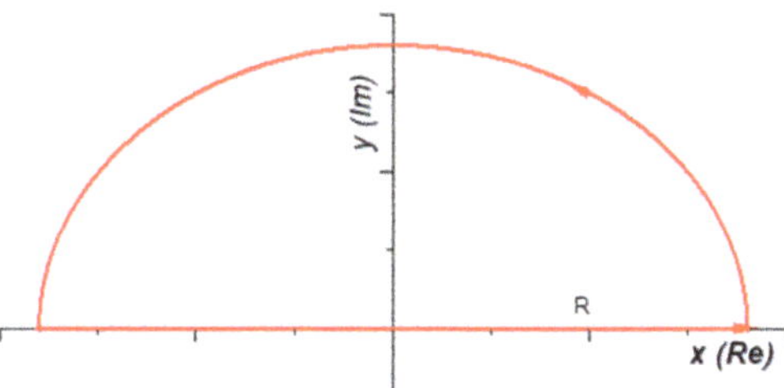

Figure 14.19. A semicircular contour of radius, R centered about the origin on the upper complex plane.

This pole is first order pole, and the residue is given by

$$R(z_k') = \lim_{z \to 0}[(z - z_k')f(z)] = \lim_{z \to 0}\left[z\frac{e^{iz}}{z}\right] = 1. \tag{14.228}$$

Therefore, the integral becomes

$$\oint \frac{e^{iz}}{z}dz = \pi i. \tag{14.229}$$

We note that in the limit $R \to \infty$,

$$\oint \frac{e^{iz}}{z}dz = \int_{-\infty}^{\infty} \frac{e^{ix}}{x}dx + \int_{C'} \frac{e^{iz}}{z}$$

$$= \int_{-\infty}^{\infty} \frac{\cos(x)}{x}dx + i\int_{-\infty}^{\infty} \frac{\sin(x)}{x}dx + \lim_{R \to \infty}\int_{C'} \frac{e^{iz}}{z}dz \tag{14.230}$$

$$= \pi i,$$

where C' is the curved part of the contour in figure 14.19. On this part of the contour z can be expressed as

$$z = R\,e^{i\theta} \Rightarrow dz = iR\,e^{i\theta}d\theta, \tag{14.231}$$

so that

$$\lim_{R \to \infty}\int_{C'} \frac{e^{iz}}{z}dz = \lim_{R \to \infty}\int_0^\pi \frac{e^{iR\,e^{i\theta}}}{R\,e^{i\theta}}iR\,e^{i\theta}d\theta$$

$$= i\lim_{R \to \infty}\int_0^\pi e^{iR\,e^{i\theta}}d\theta$$

$$= i\lim_{R \to \infty}\int_0^\pi e^{iR\cos\theta - R\sin\theta}d\theta \tag{14.232}$$

$$= i\lim_{R \to \infty}\int_0^\pi e^{-R\sin\theta}[\cos(R\cos\theta)$$

$$+ i\sin(R\cos\theta)]d\theta = 0.$$

In view of this result, equation (14.230) becomes

$$\int_{-\infty}^{\infty} \frac{\cos(x)}{x}dx + i\int_{-\infty}^{\infty} \frac{\sin(x)}{x}dx = 0 + i\pi. \tag{14.233}$$

Since

$$f(-x) = \frac{\sin(-x)}{-x} = \frac{-\sin(x)}{-x} = \frac{\sin(x)}{x},$$

(14.234)

we find

$$\begin{aligned}
\int_{-\infty}^{\infty} \frac{\sin(x)}{x}\,dx &= \int_{-\infty}^{0} \frac{\sin(x)}{x}\,dx + \int_{0}^{\infty} \frac{\sin(x)}{x}\,dx \\
&= \int_{\infty}^{0} \frac{\sin(-x)}{-x}\,d(-x) + \int_{0}^{\infty} \frac{\sin(x)}{x}\,dx \\
&= -\int_{\infty}^{0} \frac{\sin(x)}{x}\,d(x) + \int_{0}^{\infty} \frac{\sin(x)}{x}\,dx \\
&= 2\int_{0}^{\infty} \frac{\sin(x)}{x}\,dx \Rightarrow \int_{0}^{\infty} \frac{\sin(x)}{x}\,dx = \frac{\pi}{2}.
\end{aligned}$$

(14.235)

In view of this result, one finds for equation (14.233),

$$\int_{-\infty}^{\infty} \frac{\cos(x)}{x}\,dx + i2\int_{0}^{\infty} \frac{\sin(x)}{x}\,dx = 0 + i\pi$$

$$\Rightarrow \int_{-\infty}^{\infty} \frac{\cos(x)}{x}\,dx = 0, \quad \int_{-\infty}^{\infty} \frac{\sin(x)}{x}\,dx = \frac{\pi}{2},$$

(14.236)

which are the results for the integrals in equation (14.224).

14.11 Mathematica and complex functions

In *Mathematica*, what we can determine is the residue for a complex function. To find the residue for a complex function, one can use the command:

Residue [*expr*, {z, z_0}] finds the residue of *expr* at the point $z = z_0$.

In the following, we determine the residues for the functions we considered in the examples:

- Example 14.8

  Residue$\left[\frac{1}{z}\text{Cos}[z], \{z, 0\}\right]$

  1

- Example 14.9

  Residue$\left[\frac{e^z}{(z-1)^2}, \{z, 1\}\right]$

  e

- Example 14.10

  Residue$\left[\frac{\text{Sin}[z]}{z-\pi}, \{z, \pi\}\right]$

  0

- Example 14.11
  $$\text{Residue}\left[\frac{z}{(z+1)(z+2)}, \{z, -1\}\right]$$
  -1
- Example 14.12
  $$\text{Residue}[\text{Cot}[z], \{z, \pi\}]$$
  1
- Example 14.13
  $$\text{Residue}\left[\frac{ze^{zt}}{(z-3)^2}, \{z, 3\}\right]$$
  $e^{3t}(1 + 3t)$
- Example 14.14
  $$\text{Residue}\left[\frac{e^{2z}}{z(z+2)^3}, \{z, -2\}\right]$$
  $-\dfrac{13}{8e^4}$

For the contour integrals that we determined using the residue theorem, we can verify the results using the command integrate.

Integrate $[f, \{x, x_{\min}, x_{\max}\}]$ gives the definite integral $\int_{x_{\min}}^{x_{\max}} f dx$.

The following are the results for the integrals we determined using the residue theorem:

- Example 14.15
  $$\int_0^\infty \frac{1}{x^4 + 1} dx$$
  $\dfrac{\pi}{2\sqrt{2}}$
- Example 14.16
  $$\int_0^{2\pi} \frac{\text{Cos}[3\theta]}{5 - 4\text{Cos}[\theta]} d\theta$$
  $\dfrac{\pi}{12}$
- Example 14.17
  $$\text{Integrate}\left[\frac{\text{Sin}[x]}{x}, \{x, 0, \infty\}\right]$$
  $\dfrac{\pi}{2}$

14.12 Homework assignment

Problem 1.

(a) Find the real and imaginary parts for the following complex function
$$f(z) = \cos(z^*). \tag{14.237}$$

(b) Find out whether the following functions are analytic or not

    (i)
$$f(x, y) = y + ix, \tag{14.238}$$

(ii)

$$f(z) = \frac{1}{2i + z}.$$ (14.239)

(c) Find the series expansion about $z = 0$ and determine the radius of convergence for the function

$$f(z) = \frac{1}{2i + z}.$$ (14.240)

Problem 2.

(a) Evaluate the contour integral

$$I = \oint_C z^2 dz,$$ (14.241)

for the path shown in figure 14.20.

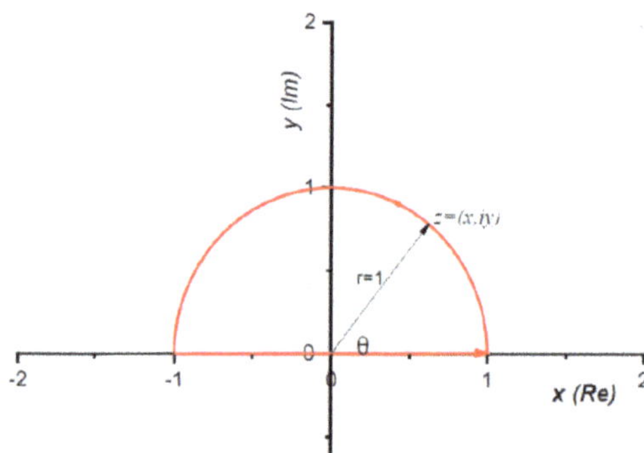

Figure 14.20. A semicircular contour.

(b) Using Cauchy's theorem or integral formula find the integral

$$I_1 = \oint_C \frac{\cosh(z)}{2\ln 2 - z} dz,$$ (14.242)

if C is a circle defined by (i) $|z| = 1$ and (ii) $|z| = 2$.

(c) Show that

$$\frac{d^n}{dz_0^n} f(z_0) = \frac{n!}{2\pi i} \oint_C \frac{f(z)}{(z - z_0)^{n+1}} dz.$$ (14.243)

Problem 3. Find the real and imaginary parts for the functions
(a)

$$f(z) = \frac{1}{z}, \tag{14.244}$$

(b)

$$f(z) = \cos(z). \tag{14.245}$$

Problem 4. Write the power series (about the origin) for the following functions and find the disk of convergence for each series.
(a)

$$f(z) = \frac{1}{2i + z}, \tag{14.246}$$

(b)

$$f(z) = \sqrt{4 + z^2}. \tag{14.247}$$

Problem 5. Evaluate the following integral
(a)

$$I_1 = \oint_C \frac{\sin(2z)dz}{(6z - \pi)^3}, \tag{14.248}$$

where C is a circle of radius $R = 3$ on the complex plane (see figure 14.21).

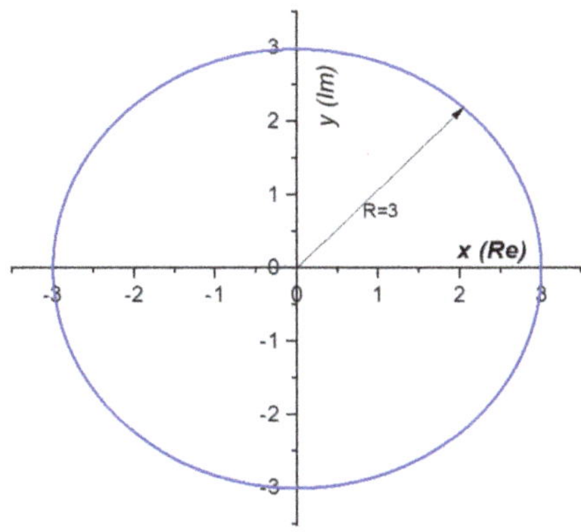

Figure 14.21. A circular contour with radius $R = 3$

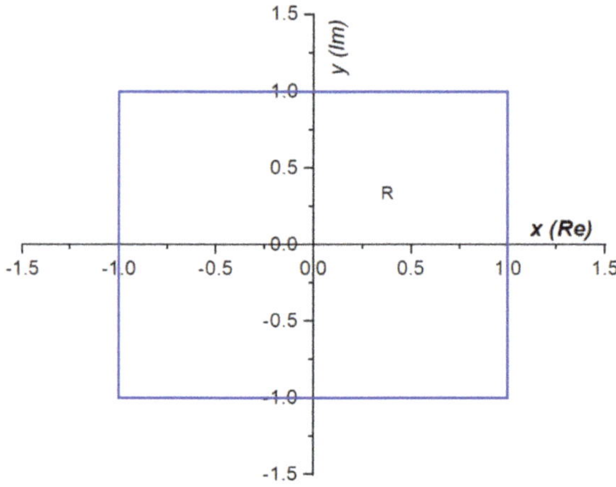

Figure 14.22. A square on the complex plane with vertices (± 1, $\pm i$).

(b)

$$I_2 = \oint_C \frac{e^{3z}dz}{z - \ln 2}, \; I_3 = \oint_C \frac{e^{3z}dz}{(z - \ln 2)^4}, \tag{14.249}$$

where C is a square on the complex plane with vertices $\pm 1 \pm i$ (figure 14.22).

Problem 6. For the following functions, find the first few terms in the Laurent series about $z = 0$ and determine the residue at $z = 0$.
   (a)

$$f(z) = \frac{2 - z}{1 - z^2}, \tag{14.250}$$

(b)

$$f(z) = \frac{z - 1}{z^3(z - 2)}. \tag{14.251}$$

Problem 7. Find Cauchy's integral formula

$$f(z_0) = \frac{1}{2\pi i} \oint_C \frac{f(z)}{z - z_0}dz, \tag{14.252}$$

using the residue theorem.

Problem 8. Find the first few terms in the Laurent series about the given z_0 values and determine the residue of the functions at these values

(a)

$$f(z) = \frac{1 + \cos(z)}{(z - \pi)^2}, \quad z_0 = \pi, \tag{14.253}$$

(b)

$$f(z) = \frac{e^z}{z^2 - 1}, \quad z_0 = 1, \tag{14.254}$$

(c)

$$f(z) = \frac{e^{2z}}{1 + e^z}, \quad z_0 = i\pi. \tag{14.255}$$

Problem 9. Using any of the methods introduced in this chapter, find the residues for the following functions

(a)

$$f(z) = \frac{1 - \cos(2z)}{z^3}, \tag{14.256}$$

(b)

$$f(z) = \frac{z - 2}{z(1 - z)}. \tag{14.257}$$

Problem 10. Evaluate the integrals

(a)

$$I_1 = \int_0^{2\pi} \frac{\sin^2(\theta)d\theta}{5 + 3\cos(\theta)}, \tag{14.258}$$

(b)

$$I_2 = \int_0^{2\pi} \frac{\cos(2\theta)d\theta}{5 + 4\cos(\theta)}. \tag{14.259}$$

Problem 11. Evaluate the integrals
(a)

$$I_1 = \int_{-\infty}^{\infty} \frac{dx}{x^2 + 4x + 5},$$

(b)

$$I_2 = \int_{-\infty}^{\infty} \frac{dx}{(4x^2 + 1)^3}.$$

$$(14.260)$$

$$(14.261)$$

Problem 12. Evaluate the integral

$$I_1 = \int_{-\infty}^{\infty} \frac{dx}{(x^2 + 4)(2 - x)}.$$

$$(14.262)$$

Problem 13. Evaluate the integral

$$I_1 = \int_{0}^{\infty} \frac{\cos(\pi x)dx}{1 - 4x^2}.$$

$$(14.263)$$

IOP Publishing

Studies in Theoretical Physics, Volume 1
Fundamental mathematical methods
Daniel Erenso and Victor Montemayor

Chapter 15

Laplace transform

In this chapter, we begin with the definition of an integral transform for real functions. We then introduce the Laplace transform and find the Laplace transform of some essential functions, including trigonometric, hyperbolic, and exponential functions. One of the applications of Laplace transform combined with inverse transform is to solve differential equations. To this end, we derive the Laplace transform for the nth order homogeneous ordinary differential equation for a function of a single variable. We then see how to solve inhomogeneous ordinary differential equations of a single variable function using this general relation and the results for the Laplace transform for various functions we found. In the last section, we will introduce the basic commands in Mathematica for carrying out Laplace transform and resolve some of the examples we considered using Mathematica.

15.1 Integral transform

In physics, we frequently encounter pairs of functions related by an integral of the form

$$F(p) = \int_a^b K(p, t) f(t) dt. \tag{15.1}$$

The function $F(p)$ is called the (integral) transform of $f(t)$ by the kernel $K(p, t)$. We shall study two types of integral transforms. The first is the Laplace transform, which we introduce in this chapter, and the second is the Fourier transform that we introduce in the next chapter, along with the Fourier series. Integral transform operation is also interpreted as mapping a function $f(t)$ in t-space to a function $F(p)$ in p-space. This interpretation takes on physical significance in the time-frequency relation of the Fourier transform.

 15-1

15.2 The Laplace transform

The Laplace transform $F(s)$ denoted as $\mathcal{L}\,[f(t)]$ is defined by

$$F(s) = \mathcal{L}[f(t)] = \lim_{b\to\infty} \int_0^b e^{-st}f(t)dt = \int_0^\infty e^{-st}f(t)dt. \tag{15.2}$$

The following are the essential properties of the Laplace transform.

i. The Laplace transform for the function $f(t)$ exists even if the integral for this function,

$$I = \int_0^\infty f(t)dt \tag{15.3}$$

does not exist.

ii. For the function $f(t)$ of the Laplace transform to exist, there must be a positive constant M such that

$$\mid e^{-s_0 t}f(t) \mid \leqslant M \tag{15.4}$$

for sufficiently large $t > t_0$ and $s > s_0$. For example, the Laplace transform for the function

$$f(t) = e^{t^2}, \tag{15.5}$$

does not exist since

$$\mid e^{-s_0 t}f(t) \mid = \mid e^{t^2 - s_0 t} \mid, \tag{15.6}$$

diverges for large $t > t_0$.

iii. The Laplace transform fails to exist when the function has a strong singularity as $t \to 0$. For example, the Laplace transform for the function

$$f(t) = t^n \text{ for } n \leqslant -1 \tag{15.7}$$

does not exist as the function diverges when $t \to 0$ for $n \leqslant -1$.

iv. Laplace transform is linear. That means if the Laplace transform for the two functions $f(t)$ and $g(t)$ exist, then we can write that

$$\mathcal{L}[\alpha f(t) + \beta g(t)] = \alpha\mathcal{L}[f(t)] + \beta\mathcal{L}[g(t)], \tag{15.8}$$

where α and β are constants.

Example 15.1. Show that the Laplace transform is linear.

Solution: Let us consider the function

$$h(t) = \alpha f(t) + \beta g(t). \tag{15.9}$$

The Laplace transform for this function is given by

$$H(p) = \mathcal{L}[h(t)] = \int_0^\infty e^{-pt}[\alpha f(t) + \beta g(t)]dt$$

$$= \int_0^\infty e^{-pt}\alpha f(t)dt + \int_0^\infty e^{-pt}\beta g(t)dt = \alpha \int_0^\infty e^{-pt}f(t)dt \quad (15.10)$$

$$+ \beta \int_0^\infty e^{-pt}g(t)dt = \alpha\mathcal{L}[f(t)] + \beta\mathcal{L}[g(t)]$$

$$\Rightarrow H(p) = \mathcal{L}[af(t) + bg(t)] = \alpha\mathcal{L}[f(t)] + \beta\mathcal{L}[g(t)].$$

Note that this is valid when the Laplace transform for the two functions $f(t)$ and $g(t)$ exist.

Example 15.2. Find the Laplace transform of the following functions and specify the conditions (if any specific condition must be satisfied for the transform to exist).
(a)

$$f(t) = 1. \quad (15.11)$$

(b)

$$f(t) = e^{\omega t}. \quad (15.12)$$

(c)

$$f_1(t) = \cosh(\omega t), f_2(t) = \sinh(\omega t). \quad (15.13)$$

(d)

$$f(t) = \cos(\omega t), f(t) = \sin(\omega t). \quad (15.14)$$

Solution:
(a) Recalling that the Laplace transform for a function $f(t)$ is given by

$$F(p) = \mathcal{L}[f(t)] = \int_0^\infty e^{-pt}f(t)dt, \quad (15.15)$$

we find

$$\mathcal{L}[1] = \int_0^\infty e^{-pt}dt = \frac{1}{p}. \quad (15.16)$$

Since the integral is not defined (diverges for negative or zero p), the Laplace transform for this function exist only for $p > 0$.
(b) Using equation (15.15), one can write

$$F(p) = \mathcal{L}[e^{\omega t}] = \int_0^\infty e^{-(p-\omega)t}dt. = \frac{1}{p - \omega}, \quad (15.17)$$

where, for the same reason provided in part (a), the Laplace transform exists when for $p - \omega > 0$, which means $p > \omega$.

(c) Recalling that

$$\cosh(\omega t) = \frac{e^{\omega t} + e^{-\omega t}}{2}, \quad \sinh(\omega t) = \frac{e^{\omega t} - e^{-\omega t}}{2}, \tag{15.18}$$

and taking into account, the Laplace transform is linear, one can see that

$$\mathcal{L}[\cosh(\omega t)] = \mathcal{L}\left[\frac{e^{\omega t} + e^{-\omega t}}{2}\right] = \frac{1}{2}(\mathcal{L}[e^{\omega t}] + \mathcal{L}[e^{-\omega t}]),$$

$$\mathcal{L}[\sinh(\omega t)] = \mathcal{L}\left[\frac{e^{\omega t} - e^{-\omega t}}{2}\right] = \frac{1}{2}(\mathcal{L}[e^{\omega t}] - \mathcal{L}[e^{-\omega t}]). \tag{15.19}$$

Applying the result in equation (15.17), one finds

$$\mathcal{L}[\cosh(\omega t)] = \frac{1}{2}\left(\frac{1}{p - \omega} + \frac{1}{p + \omega}\right) = \frac{p}{p^2 - \omega^2}, \tag{15.20}$$

$$\mathcal{L}[\sinh(\omega t)] = \frac{1}{2}\left(\frac{1}{p - \omega} - \frac{1}{p + \omega}\right) = \frac{\omega}{p^2 - \omega^2}. \tag{15.21}$$

(d) We recall that the trigonometric (sin and cos) and hyperbolic (sinh and cosh) are related by

$$\cos(\omega t) = \frac{e^{i\omega t} + e^{-i\omega t}}{2} = \cosh(i\omega t),$$

$$\sin(\omega t) = \frac{e^{i\omega t} - e^{-i\omega t}}{2i} = -i\sinh(i\omega t), \tag{15.22}$$

and applying the result in equations (15.20) and (15.21), one can easily show that

$$\mathcal{L}[\cos(\omega t)] = \mathcal{L}[\cosh(i\omega t)] = \frac{p}{p^2 - (i\omega)^2} = \frac{p}{p^2 + \omega^2} \tag{15.23}$$

$$\mathcal{L}[\sin(\omega t)] = \mathcal{L}[-i\sinh(i\omega t)] = -i\mathcal{L}[\sinh(i\omega t)] = \frac{-i(i\omega)}{p^2 - (i\omega)^2} = \frac{\omega}{p^2 + \omega^2}. \tag{15.24}$$

Example 15.3. Find the Laplace transform for the unit step function, also called the Heaviside function, defined by

$$U(t - a) = \begin{cases} 0, & \text{for } t < a, \\ 1, & \text{for } t > a. \end{cases} \tag{15.25}$$

Solution: The Laplace transform

$$F(p) = \mathcal{L}[f(t)] = \int_0^\infty e^{-pt} f(t)\, dt \tag{15.26}$$

for the unit step function can be rewritten as

$$\mathcal{L}[U(t-a)] = \int_0^a e^{-pt} U(t-a)\, dt + \int_a^\infty e^{-pt} U(t-a)\, dt$$

$$= \int_a^\infty e^{-pt}\, dt = \left.\frac{e^{-pt}}{-p}\right|_a^\infty \;\Rightarrow\; \mathcal{L}[f(t)] = \frac{e^{-pa}}{p}. \tag{15.27}$$

Example 15.4. Consider a differentiable function $f(t)$. Find the Laplace transform for

(a)

$$\mathcal{L}[f^{(1)}(t)] = \mathcal{L}\left[\frac{df(t)}{dt}\right], \tag{15.28}$$

(b)

$$\mathcal{L}[f^{(2)}(t)] = \mathcal{L}\left[\frac{d^2 f(t)}{dt^2}\right], \tag{15.29}$$

(c)

$$\mathcal{L}[f^{(3)}(t)] = \mathcal{L}\left[\frac{d^3 f(t)}{dt^3}\right]. \tag{15.30}$$

(d) Following the pattern you see in the results in (a), (b), and (c), show that the Laplace transform for the nth derivative of the function $f(t)$ is given by

$$\begin{aligned}
\mathcal{L}[f^{(n)}(t)] &= \mathcal{L}\left[\frac{d^n f(t)}{dt^n}\right] \\
&= p^n \mathcal{L}[f(t)] - p^{n-1} f(0) - p^{n-2} f^{(1)}(0) - p^{n-3} f^{(2)}(0) \\
&\quad - p^{n-4} f^{(3)}(0) - \dots - f^{(n-1)}(0),
\end{aligned} \tag{15.31}$$

where

$$f^{(m)}(0) = \left.\frac{d^m f(t)}{dt^m}\right|_{t=0}, \tag{15.32}$$

for $m = 0, 1, 2. \dots$.

Solution:

(a) The Laplace transform

$$\mathcal{L}[f(t)] = \int_0^\infty e^{-pt} f(t)\,dt, \tag{15.33}$$

for the first derivative of the function $f(t)$

$$f^{(1)}(t) = \frac{df(t)}{dt},$$

becomes

$$\mathcal{L}[f^{(1)}(t)] = \int_0^\infty e^{-pt} \frac{df(t)}{dt}\,dt. \tag{15.34}$$

Applying integration by parts,

$$\int u\,dv = uv - \int v\,du, \tag{15.35}$$

for

$$u = e^{-pt} \Rightarrow du = -pe^{-pt}, \; dv = \frac{df(t)}{dt}\,dt \Rightarrow v = f(t), \tag{15.36}$$

one can rewrite equation (15.34) as

$$\mathcal{L}[f^{(1)}(t)] = e^{-pt}f(t)\,\big|_0^\infty + p \int_0^\infty e^{-pt}f(t)\,dt = -f(0) + p \int_0^\infty e^{-pt}f(t)\,dt$$
$$\Rightarrow \mathcal{L}[f^{(1)}(t)] = p\mathcal{L}[f(t)] - p^0 f(0). \tag{15.37}$$

(b) Similarly for the second derivative

$$\mathcal{L}[f^{(2)}(t)] = \int_0^\infty e^{-pt} \frac{d^2 f(t)}{dt^2}\,dt, \tag{15.38}$$

using integration by parts for

$$u = e^{-pt} \Rightarrow du = -pe^{-pt}, \; dv = \frac{d^2 f(t)}{dt^2}\,dt \Rightarrow v = \frac{df(t)}{dt}, \tag{15.39}$$

on finds

$$\mathcal{L}[f^{(2)}(t)] = e^{-pt} \frac{df(t)}{dt}\,\bigg|_0^\infty + p \int_0^\infty e^{-pt} \frac{df(t)}{dt}\,dt$$
$$= -\frac{df(t)}{dt}\,\bigg|_{t=0} + p \int_0^\infty e^{-pt} \frac{df(t)}{dt}\,dt \tag{15.40}$$
$$\Rightarrow \mathcal{L}[f^{(2)}(t)] = p\mathcal{L}[f^{(1)}(t)] - p^0 f^{(1)}(0).$$

Using the result in equation (15.37) for $\mathcal{L}[f^{(1)}(t)]$, we can rewrite equation (15.40) as

$$\mathcal{L}[f^{(2)}(t)] = p^2\mathcal{L}[f(t)] - pf(0) - p^0 f^{(1)}(0). \tag{15.41}$$

(c) In a similar way for the third derivative, we can show that

$$\mathcal{L}[f^{(3)}(t)] \quad = \quad \int_0^\infty e^{-pt}\frac{d^3f(t)}{dt^3}dt$$

$$= e^{-pt}\frac{d^2f(t)}{dt^2}\Big|_0^\infty + p\int_0^\infty e^{-pt}\frac{d^2f(t)}{dt^2}dt \tag{15.42}$$

$$\Rightarrow \quad \mathcal{L}[f^{(3)}(t)] = p\mathcal{L}[f^{(2)}(t)] - f^{(2)}(0),$$

and using the result in equation (15.41) for $\mathcal{L}[f^{(2)}(t)]$, one finds

$$\mathcal{L}[f^{(3)}(t)] = p^3\mathcal{L}[f(t)] - p^2 f(0) - p^1 f^{(1)}(0) - p^0 f^{(2)}(0). \tag{15.43}$$

(d) In view of the result in equation (15.42), the Laplace transform for the 4th derivative of the function $f(t)$ can be expressed as

$$\mathcal{L}[f^{(4)}(t)] = p\mathcal{L}[f^{(3)}(t)] - f^{(3)}(0), \tag{15.44}$$

and substituting the result in equation (15.43) for $\mathcal{L}[f^{(3)}(t)]$, one finds

$$\mathcal{L}[f^{(4)}(t)] = p^4\mathcal{L}[f(t)] - p^3 f(0) - p^2 f^{(1)}(0) - p^1 f^{(2)}(0) - p^0 f^{(3)}(0). \tag{15.45}$$

For the 5th derivative, similarly, one can show that

$$\begin{aligned}
\mathcal{L}[f^{(5)}(t)] &= p\mathcal{L}[f^{(4)}(t)] - p^0\mathcal{L}[f^{(4)}(0)] \\
&= p^5\mathcal{L}[f(t)] - p^4 f(0) - p^3\mathcal{L}[f^{(1)}(0)] - p^2\mathcal{L}[f^{(2)}(0)] \\
&\quad - p^1\mathcal{L}[f^{(3)}(0)] - p^0\mathcal{L}[f^{(4)}(0)].
\end{aligned} \tag{15.46}$$

Based on the results that we found for the 1st, 2nd, 3rd, 4th, and 5th derivatives, the Laplace transform for the nth derivative of the function $f(t)$, one can write

$$\mathcal{L}[f^{(n)}(t)] = \mathcal{L}\left[\frac{d^n f(t)}{dt^n}\right] = p^n\mathcal{L}[f(t)] - p^{n-1}f(0) - p^{n-2}f^{(1)}(0)$$
$$- p^{n-3}f^{(2)}(0) - p^{n-4}f^{(3)}(0). \ldots - p^1 f^{(n-2)}(0) - p^0 f^{(n-1)}(0), \tag{15.47}$$

where

$$f^{(m)}(0) = \frac{d^m f(t)}{dt^m}\Big|_{t=0}, \tag{15.48}$$

for $m = 0, 1, 2. \ldots$

15.3 Inverse Laplace transform

The inverse of a Laplace transform is represented by

$$\mathcal{L}^{-1}[F(p)] = f(t). \tag{15.49}$$

In order to perform inverse Laplace transform, it is essential to know the Laplace transform for the basic functions that we saw in the examples earlier.

Example 15.5. Evaluate the following inverse Laplace transforms.

  (a)

$$\mathcal{L}^{-1}\left[\frac{5}{p+2}\right]. \tag{15.50}$$

  (b)

$$\mathcal{L}^{-1}\left[\frac{4p-3}{p^2+4}\right]. \tag{15.51}$$

Solution:

  (a) Noting that

$$\mathcal{L}^{-1}\left[\frac{5}{p+2}\right] = 5\mathcal{L}^{-1}\left[\frac{1}{p-(-2)}\right] \tag{15.52}$$

and referring to table 15.1,

$$\mathcal{L}[e^{\omega t}] = \frac{1}{p-\omega}, \tag{15.53}$$

we can write

$$\mathcal{L}^{-1}\left[\frac{5}{p+2}\right] = 5e^{-2t}. \tag{15.54}$$

Table 15.1. Laplace transform for some of the basic functions.

$f(t)$	1	$e^{\omega t}$	$\sin(\omega t)$	$\cos(\omega t)$	$\sinh(\omega t)$	$\cosh(\omega t)$	$U(t-a)$
$\mathcal{L}[f(t)]$	$\dfrac{1}{p}$	$\dfrac{1}{p-\omega}$	$\dfrac{\omega}{p^2+\omega^2}$	$\dfrac{p}{p^2+\omega^2}$	$\dfrac{\omega}{p^2-\omega^2}$	$\dfrac{p}{p^2-\omega^2}$	$\dfrac{e^{-ap}}{p}$

(b) Here also we want to write

$$\mathcal{L}^{-1}\left[\frac{4p-3}{p^2+4}\right] = \mathcal{L}^{-1}\left[\frac{4p-3}{p^2+2^2}\right] = \mathcal{L}^{-1}\left[\frac{4p}{p^2+2^2} - \frac{3}{p^2+2^2}\right]$$

$$= \mathcal{L}^{-1}\left[\frac{4p}{p^2+2^2}\right] - \mathcal{L}^{-1}\left[\frac{3}{p^2+2^2}\right] \tag{15.55}$$

$$\Rightarrow \mathcal{L}^{-1}\left[\frac{4p-3}{p^2+4}\right] = 4\mathcal{L}^{-1}\left[\frac{p}{p^2+2^2}\right] - \frac{3}{2}\mathcal{L}^{-1}\left[\frac{2}{p^2+2^2}\right].$$

From table 15.1, we have

$$\mathcal{L}[\cos(\omega t)] = \frac{p}{p^2+\omega^2}, \quad \mathcal{L}[\sin(\omega t)] = \frac{\omega}{p^2+\omega^2}, \tag{15.56}$$

so that equation (15.55) becomes

$$\mathcal{L}^{-1}\left[\frac{4p-3}{p^2+4}\right] = 4\cos(2t) - \frac{3}{2}\sin(2t). \tag{15.57}$$

Example 15.6. Evaluate the integral

$$f(t) = \int_0^\infty \frac{\sin(tx)}{x}dx, \tag{15.58}$$

applying Laplace and inverse Laplace transform.

Solution: We first apply the Laplace transform to the given integral

$$\mathcal{L}[f(t)] = \int_0^\infty e^{-pt}\left[\int_0^\infty \frac{\sin(tx)}{x}dx\right]dt, \tag{15.59}$$

which can also be expressed by interchanging the order of the integration as

$$\mathcal{L}[f(t)] = \int_0^\infty \frac{1}{x}\left[\int_0^\infty \sin(tx)e^{-pt}dt\right]dx$$

$$= \int_0^\infty \frac{1}{x}\left[\int_0^\infty \left[\frac{e^{itx}-e^{-itx}}{2i}\right]e^{-pt}dt\right]dx$$

$$= \int_0^\infty \frac{1}{x}\left[\int_0^\infty \left[\frac{e^{-(p-ix)t}-e^{-(p+itx)}}{2i}\right]dt\right]dx \tag{15.60}$$

$$= \int_0^\infty \frac{1}{2ix}\left[\frac{e^{-(p-ix)t}}{-(p-ix)} + \frac{e^{-(p+itx)}}{p+ix}\right]_0^\infty dx = \int_0^\infty \frac{1}{2ix}\left[\frac{1}{p-ix} - \frac{1}{p+ix}\right]dx$$

$$\Rightarrow \mathcal{L}[f(t)] = \int_0^\infty \frac{dx}{p^2+x^2} = \frac{\pi}{2p},$$

where we applied the result we determined in using the residue theorem in chapter 14. Noting that

$$f(t) = \mathcal{L}^{-1}[\mathcal{L}[f(t)]], \tag{15.61}$$

and using the result in equation (15.60), one can write

$$f(t) = \int_0^\infty \frac{\sin{(tx)}}{x}dx = \mathcal{L}^{-1}\left[\frac{\pi}{2p}\right] = \frac{\pi}{2}\mathcal{L}^{-1}\left[\frac{1}{p}\right]$$

$$\Rightarrow f(t) = \int_0^\infty \frac{\sin{(tx)}}{x}dx = \frac{\pi}{2}, \text{ for } t > 0. \tag{15.62}$$

Noting that for

$$\sin(-tx) = -\sin{(tx)}, \tag{15.63}$$

one can write

$$f(t) = \int_0^\infty \frac{\sin{(tx)}}{x}dx = -\frac{\pi}{2}, \text{ for } t < 0. \tag{15.64}$$

For $t = 0$, obviously, we find

$$f(t) = \int_0^\infty \frac{\sin{(tx)}}{x}dx = 0. \tag{15.65}$$

Therefore, in view of the results in equations (15.62)–(15.65), one can write

$$\int_0^\infty \frac{\sin{(tx)}}{x}dx = \frac{\pi}{2}[2U(t) - 1] = \begin{cases} \dfrac{\pi}{2}, & t > 0, \\ 0, & t = 0, \\ -\dfrac{\pi}{2}, & t < 0, \end{cases} \tag{15.66}$$

where $U(t)$ is the step function (Heaviside function) in Example 15.3 with the constant 'a' set to zero.

15.4 Applications of Laplace transforms

This section will use Laplace transform and Laplace inverse transforms for solving differential equations. Such application requires the results in table 15.1 and the Laplace transform for nth derivative of the function $f(t)$, which is given by

$$\mathcal{L}[f^{(n)}(t)] = \mathcal{L}\left[\frac{d^n f(t)}{dt^n}\right] = p^n \mathcal{L}[f(t)] - p^{n-1}f(0) - p^{n-2}f^{(1)}(0)$$

$$-p^{n-3}f^{(2)}(0) - p^{n-4}f^{(3)}(0) \ldots - p^1 f^{(n-2)}(0) - p^0 f^{(n-1)}(0), \tag{15.67}$$

Example 15.7. Solve the following differential equation for the initial conditions $y(0) = 0$ and $y^{(1)}(0) = 0$,

$$\frac{d^2y}{dx^2} + y(x) = 1. \tag{15.68}$$

Solution: For the Laplace transform of equation (15.68), one can write

$$\mathcal{L}\left[\frac{d^2y}{dx^2}\right] + \mathcal{L}[y(x)] = \mathcal{L}[1]. \tag{15.69}$$

Using the results in table 15.1 and the relation in equation (15.68) given earlier, one finds

$$\mathcal{L}[1] = \frac{1}{p},$$

$$\mathcal{L}\left[\frac{d^2y}{dx^2}\right] = p^2\mathcal{L}[y(x)] - py(0) - y^{(1)}(0) = p^2\mathcal{L}[y(x)], \tag{15.70}$$

where we used the initial conditions, $y(0) = 0$ and $y^{(1)}(0) = 0$. Upon substituting the results in equation (15.70) into equation (15.69), we find

$$p^2\mathcal{L}[y(x)] + \mathcal{L}[y(x)] = \frac{1}{p} \Rightarrow \mathcal{L}[y(x)] = \frac{1}{p(p^2 + 1)}. \tag{15.71}$$

Now taking the inverse Laplace transform, the solution to the differential equation is found to be

$$y(x) = \mathcal{L}^{-1}\left[\frac{1}{p(p^2 + 1)}\right] = \mathcal{L}^{-1}\left[\frac{1}{p}\right] - \mathcal{L}^{-1}\left[\frac{p}{p^2 + 1}\right] \tag{15.72}$$

$$\Rightarrow y(x) = 1 - \cos(t),$$

where we applied the result from table 15.1,

$$\mathcal{L}^{-1}\left[\frac{1}{p}\right] = 1,$$

$$\mathcal{L}^{-1}\left[\frac{p}{p^2 + \omega^2}\right] = \cos(\omega t) \Rightarrow \mathcal{L}^{-1}\left[\frac{p}{p^2 + 1}\right] = \cos(t). \tag{15.73}$$

Equation (15.72) is solution to the differential equation in equation (15.68). We will verify this result and some of the previous examples in the last section of this chapter using Mathematica.

Example 15.8. For the initial conditions $y(0) = 2$ and $y'(0) = -1$, solve the following differential equation given

$$\frac{d^2y}{dt^2} + 3\frac{dy}{dt} + 2y(t) = 2e^{-t}. \tag{15.74}$$

Solution: Upon taking the Laplace transform of each term in equation (15.74) and imposing the initial conditions $y(0) = 2$ and $y'(0) = -1$, one can write

$$\mathcal{L}\left[\frac{d^2y}{dt^2}\right] = p^2\mathcal{L}[y(t)] - py(0) - y'(0) = p^2\mathcal{L}[y(t)] - 2p + 1,$$

$$\mathcal{L}\left[\frac{dy}{dt}\right] = p\mathcal{L}[y(t)] - y(0) = p\mathcal{L}[y(t)] - 2, \tag{15.75}$$

$$\mathcal{L}[e^{-t}] = \mathcal{L}[e^{-t}] = \frac{1}{p+1}.$$

Substituting the results in equation (15.75) into equation (15.74), we find

$$p^2\mathcal{L}[y(t)] - 2p + 1 + 3(p\mathcal{L}[y(t)] - 2) + 2\mathcal{L}[y(t)] = \frac{2}{p+1}$$

$$\Rightarrow (p^2 + 3p + 2)\mathcal{L}[y(t)] - 2p - 5 = \frac{2}{p+1}$$

$$\Rightarrow (p + 2)(p + 1)\mathcal{L}[y(t)] = 5 + 2p + \frac{2}{p+1} \tag{15.76}$$

$$\Rightarrow \mathcal{L}[y(t)] = \frac{1}{(p+2)(p+1)}\left[5 + 2p + \frac{2}{p+1}\right] = \frac{2p^2 + 7p + 7}{(p+2)(p+1)^2}.$$

Noting that

$$\frac{2p^2 + 7p + 7}{(p+2)(p+1)^2} = \frac{A}{p+2} + \frac{B}{p+1} + \frac{C}{(p+1)^2}$$

$$= \frac{A(p^2 + 2p + 1) + B(p^2 + 3p + 2) + C(p + 2)}{(p+2)(p+1)^2} \tag{15.77}$$

$$= \frac{(A + B)p^2 + (2A + 3B + C)p + A + 2B + 2C}{(p+2)(p+1)^2},$$

one finds the set of equations

$$A + B = 2,\ 2A + 3B + C = 7,\ A + 2B + 2C = 7. \tag{15.78}$$

Upon solving these equations, one finds

$$A = 1,\ B = 1,\ C = 2. \tag{15.79}$$

Now combining equations (15.76)–(15.79), we can write

$$\mathcal{L}[y(t)] = \frac{2p^2 + 7p + 7}{(p + 2)(p + 1)^2} = \frac{1}{p + 2} + \frac{1}{p + 1} + \frac{2}{(p + 1)^2}$$

$$\Rightarrow y(t) = \mathcal{L}^{-1}\left[\frac{1}{p + 2}\right] + \mathcal{L}^{-1}\left[\frac{1}{p + 1}\right] + \mathcal{L}^{-1}\left[\frac{2}{(p + 1)^2}\right]. \tag{15.80}$$

Applying the results in table 15.1, for the inverse Laplace transform for the first two terms in equation (15.80), one finds

$$\mathcal{L}^{-1}\left[\frac{1}{p + 2}\right] = e^{-2t},\ \mathcal{L}^{-1}\left[\frac{1}{p + 1}\right] = e^{-t}. \tag{15.81}$$

For the third term in equation (15.79), we can show that

$$\frac{2}{(p + 1)^2} = -2\frac{d}{d\lambda}\left(\frac{1}{p + \lambda}\right)\Big|_{\lambda=1}$$

$$\Rightarrow \mathcal{L}^{-1}\left[\frac{2}{(p + 1)^2}\right] = \mathcal{L}^{-1}\left[-2\frac{d}{d\lambda}\left(\frac{1}{p + \lambda}\right)\Big|_{\lambda=1}\right]$$

$$= -2\frac{d}{d\lambda}\mathcal{L}^{-1}\left[\frac{1}{p + \lambda}\right]\Big|_{\lambda=1} = -2\frac{de^{-\lambda t}}{d\lambda}\Big|_{\lambda=1} = 2te^{-\lambda t}\,|_{\lambda=1}$$

$$\Rightarrow \mathcal{L}^{-1}\left[\frac{2}{(p + 1)^2}\right] = 2te^{-t}. \tag{15.82}$$

In view of the results in equations (15.81) and (15.82), equation (15.80) becomes

$$y(t) = e^{-2t} + e^{-t} + 2te^{-t} = e^{-2t}(1 + e^t(1 + 2t)). \tag{15.83}$$

Equation (15.83) is the solution to the differential equation in equation (15.74). In the next section we also verify this result using *Mathematica*.

Example 15.9. Consider the circuit shown in figure 15.1. At $t = 0$ the switch S is closed to position A. Find the resulting current, $I(t)$. Use $R = 10\Omega$, $L = 2H$, $\varepsilon(t) = 50 \sin(5t)$.

Solution: Applying Kirchhoff's voltage rule and the given values $R = 10\Omega$, $L = 2H$, and $\varepsilon(t) = 50 \sin(5t)$, we can write

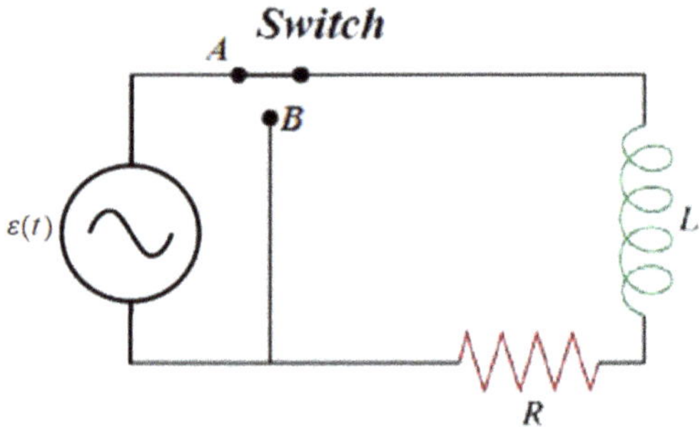

Figure 15.1. AC RL circuit.

$$\varepsilon(t) - IR - L\frac{dI}{dt} = 0 \Rightarrow 50\sin(5t) - 10I - 2\frac{dI}{dt} = 0. \tag{15.84}$$

The Laplace transform can then be written as

$$25\mathcal{L}[\sin(5t)] - 5\mathcal{L}[I] - \mathcal{L}\left[\frac{dI}{dt}\right] = 0$$

$$\Rightarrow \frac{125}{p^2 + 5^2} - 5\mathcal{L}[I] - p\mathcal{L}[I(t)] + I(0) = 0 \tag{15.85}$$

$$\Rightarrow \mathcal{L}[I(t)] = \frac{1}{(p+5)}\left[\frac{125}{p^2+5^2} + 2I(0)\right] = \frac{125}{(p^2+5^2)(p+5)} + \frac{I(0)}{(p+5)},$$

where we used

$$\mathcal{L}[\sin(5t)] = \frac{5}{p^2+5^2}, \quad \mathcal{L}\left[\frac{dI}{dt}\right] = p\mathcal{L}[I(t)] - I(0). \tag{15.86}$$

Noting that

$$\frac{125}{(p^2+5^2)(p+5)} = \frac{A}{p+5} + \frac{BP+C}{p^2+5^2}$$

$$= \frac{Ap^2 + 25A + BP^2 + 5Bp + Cp + 5C}{p^2+5^2}$$

$$= \frac{(A+B)p^2 + (5B+C)p + 25A + 5C}{p^2+5^2} \tag{15.87}$$

$$\Rightarrow A+B = 0, \quad 5B+C = 0, \quad 25A + 5C = 125$$

$$\Rightarrow A = \frac{5}{2}, \quad B = -\frac{5}{2}, \quad C = \frac{25}{2},$$

one can express

$$\frac{125}{(p^2+5^2)(p+5)} = \frac{5}{2}\frac{1}{p+5} - \frac{5}{2}\frac{p}{p^2+5^2} + \frac{25}{2}\frac{1}{p^2+5^2}. \tag{15.88}$$

Substituting equation (15.88) into equation (15.85), one can write

$$\mathcal{L}[I(t)] = \left(\frac{5}{2} + I(0)\right)\frac{1}{p+5} - \frac{5}{2}\frac{p}{p^2+5^2} + \frac{5}{2}\frac{5}{p^2+5^2}$$

$$\Rightarrow I(t) = \left(\frac{5}{2} + I(0)\right)\mathcal{L}^{-1}\left[\frac{1}{p+5}\right] - \frac{5}{2}\mathcal{L}^{-1}\left[\frac{P}{p^2+5^2}\right] \tag{15.89}$$

$$+ \frac{5}{2}\mathcal{L}^{-1}\left[\frac{5}{p^2+5^2}\right],$$

and applying the results in table 15.1, we find

$$I(t) = \left(\frac{5}{2} + I(0)\right)e^{-5t} - \frac{5}{2}\cos(5t) + \frac{5}{2}\sin(5t). \tag{15.90}$$

For the case $I(0) = 0$, equation (15.90) becomes

$$I(t) = \frac{5}{2}e^{-5t} + \frac{5}{2}[\sin(5t) - \cos(5t)]. \tag{15.91}$$

Equation (15.91) is the current in the circuit as function of time. The current $I(t)$ is shown in figure 15.2. Note that for steady state ($t \to \infty$)

$$I(t) = \frac{5}{2}[\sin(5t) - \cos(5t)], \tag{15.92}$$

see figure 15.2.

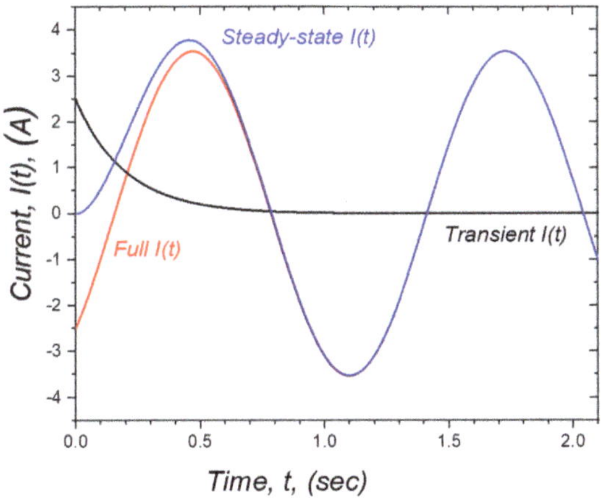

Figure 15.2. The current $I(t)$ as function of time t. Note that blue is for the steady-state.

15.5 Mathematica and Laplace transform

Using Mathematica one can carry out Laplace transform and Laplace inverse transform. The following are the basic commands.

- LaplaceTransform [expr, t, s] gives the Laplace transform of *expr*.
- LaplaceTransform [*expr*, t_1, t_2, ... , s_1, s_2, ...] gives the multidimensional Laplace transform of *expr*.
- InverseLaplaceTransform [*expr*, s, t] gives the inverse Laplace transform of *expr*.
- InverseLaplaceTransform[*expr*, $\{s_1, s_2, ...\}$, $\{t_1, t_2, ...\}$] gives the multidimensional inverse Laplace transform of *expr*.

In the following, we will resolve some of the examples we solved in the previous sections.

- Example 15.2
  LaplaceTransform[1, t, p]
  $$\frac{1}{p}$$
  LaplaceTransform[Exp[ωt], t, p]
  $$\frac{1}{p-\omega}$$
  LaplaceTransform[Sinh[ωt], t, p]
  $$\frac{\omega}{p^2 - \omega^2}$$
  LaplaceTransform[Cosh[ωt], t, s]
  $$\frac{s}{s^2 - \omega^2}$$
  LaplaceTransform[Sin[ωt], t, p]
  $$\frac{\omega}{p^2 + \omega^2}$$
  LaplaceTransform[Cos[ωt], t, p]
  $$\frac{p}{p^2 + \omega^2}$$
  LaplaceTransform[Sin[ωt], t, p]
  $$\frac{\omega}{p^2 + \omega^2}$$

- Example 15.4
  LaplaceTransform [$f'[t]$, t, p]
  $-f[0] + p$LaplaceTransform[$f[t]$, t, p]
  LaplaceTransform
  [$f''[t]$, t, p] $- pf[0] + p^2$ LaplaceTransform[$f[t]$, t, p] $- f'[0]$
  LaplaceTransform
  [$f'''[t]$, t, p] $- p^2 f[0] + p^3$ LaplaceTransform[$f[t]$, t, p] $- pf'[0] - f''[0]$
  LaplaceTransform
  [$f''''[t]$, t, p] $- p^3 f[0] + p^4$ LaplaceTransform[$f[t]$, t, p] $- p^2 f'[0] - pf''[0]$

- Example 15.5

  $\text{InverseLaplaceTransform}\left[\frac{5}{2+p}, p, t\right]$

  $5e^{-2t}$

  $\text{InverseLaplaceTransform}\left[\frac{4p-3}{p^2+4}, p, t\right]$

  $\frac{1}{2}(8\text{Cos}[2t] - 3\text{Sin}[2t])$

- Example 15.6

  $\text{InverseLaplaceTransform}\left[\text{LaplaceTransform}\left[\int_0^\infty \frac{\text{Sin}[tx]}{x}\, dx, t, p\right], p, t\right]$

  $\frac{\pi}{2}$

- Example 15.7

  $\text{LaplaceTransform}[y''[x] + y[x] == 1, x, p]$

  $\left\{\text{LaplaceTransform}[y[x], x, p] + p^2\, \text{LaplaceTransform}[y[x], x, p] - py[0] - y'[0] == \frac{1}{p}\right\}$

  $\text{Solve}\,[\text{LaplaceTransform}[y[x], x, p] + p^2\, \text{LaplaceTransform}[y[x], x, p] - py[0] - y'[0]$

  $\qquad == \frac{1}{p},\ \text{LaplaceTransform}[y[x], x, p]]$

  $\left\{\left\{\text{LaplaceTransform}[y[x], x, p] \rightarrow \frac{1 + p^2 y[0] + py'[0]}{p(1+p^2)}\right\}\right\}$

  $\text{InverseLaplaceTransform}\left[\frac{1 + p^2 y[0] + py'[0]}{p(1+p^2)}, p, t\right]$

  $1 - \text{Cos}[t] + \text{Cos}[t]y[0] + \text{Sin}[t]y'[0]$

  $\text{DSolve}[\{y''[x] + y[x] == 1, y[0] == 0, y'[0] == 0\}, y[x], x]$

  $\{\{y[x] \rightarrow 1 - \text{Cos}[x]\}\}$

- Example 15.8

  $\text{LaplaceTransform}[\{y''[t] + 3y'[t] + 2y[t] == 2\text{Exp}[-t]\}, t, p]$

  $\{2\text{LaplaceTransform}[y[t], t, p] + p^2\, \text{LaplaceTransform}[y[t], t, p] + 3(p$

  $\text{LaplaceTransform}\,[y[t], t, p] - y[0]) - py[0] - y'[0] == \frac{2}{1+p}\}$

  $\text{Solve}[2\text{LaplaceTransform}[y[t], t, p] + p^2\, \text{LaplaceTransform}[y[t], t, p]$

  $\qquad + 3(p\text{LaplaceTransform}[y[t], t, p] - y[0]) - py[0] - y'[0]$

  $\qquad == \frac{2}{1+p},\ \text{LaplaceTransform}[y[t], t, p]]$

  $\left\{\left\{\text{LaplaceTransform}[y[t], t, p] \rightarrow \frac{2 + 3y[0] + 4py[0] + p^2 y[0] + y'[0] + py'[0]}{(1+p)(2+3p+p^2)}\right\}\right\}$

  $\text{InverseLaplaceTransform}\left[\frac{2 + 3y[0] + 4py[0] + p^2 y[0] + y'[0] + py'[0]}{(1+p)(2+3p+p^2)}, p, t\right]$

  $2e^{-t}t + e^{-2t}(2 - y[0] - y'[0]) + e^{-t}(-2 + 2y[0] + y'[0])$

DSolve[{y''[t] + 3y'[t] + 2y[t]== 2Exp[−t], y[0]== 2, y'[0]==−1}, y[t], t]
{{y[t] → e^{-2t}(1 + e^t + 2e^tt)}}

- Example 15.9
  LaplaceTransform[{50Sin[5t] − 10I1[t] − 2I1'[t]==0}, t, p]

$$\left\{ \frac{250}{25 + p^2} - 10\text{LaplaceTransform}[I1[t], t, p] \right.$$

$$\left. - 2(-I1[0] + p\,LaplaceTransform[I1[t], t, p]) == 0 \right\}$$

Solve

$$\left[\frac{250}{25 + p^2} - 10\text{LaplaceTransform}[I1[t], t, p] - 2(-I1[0] + p\,LaplaceTransform[I1[t], t, p]) \right.$$

$$\left. == 0,\ LaplaceTransform[I1[t], t, p] \right]$$

$$\left\{ \left\{ \text{LaplaceTransform}[I1[t], t, p] \to \frac{125 + 25I1[0] + p^2 I1[0]}{(5 + p)(25 + p^2)} \right\} \right\}$$

$$\text{InverseLaplaceTransform}\left[\frac{125 + 25I1[0] + p^2 I1[0]}{(5 + p)(25 + p^2)},\ p,\ t \right]$$

$$\frac{1}{2}e^{-5t}(5 + 2I1[0]) - \frac{5}{2}(\text{Cos}[5t] - \text{Sin}[5t])$$

DSolve[{50Sin[5t] − 10I1[t] − 2I1'[t]==0}, I1[t], t]

$$\left\{ \left\{ I1[t] \to e^{-5t}C[1] + \frac{5}{2}(-\text{Cos}[5t] + \text{Sin}[5t]) \right\} \right\}$$

Note that one can write

$$\left(\frac{5}{2} + I1(0) \right) e^{-5t} = \left(\frac{5}{2} + I1(0) \right) e^{-5t} = c[1]e^{-5t}. \tag{15.93}$$

It is also important to note that in Mathematica since 'I' is used for imaginary, we have used 'I1' for the current.

15.6 Homework assignment

Problem 1. By differentiating both sides of the equation

$$\mathcal{L}[\sin(\omega t)] = \frac{\omega}{p^2 + \omega^2}, \tag{15.94}$$

with respect to ω, show that

$$\mathcal{L}[t \cos(\omega t)] = \frac{p^2 - \omega^2}{(p^2 + \omega^2)^2}. \tag{15.95}$$

Verify your result using Mathematica.

Problem 2. By integrating

$$\mathcal{L}[\cos(\omega t)] = \frac{p}{p^2 + \omega^2}, \tag{15.96}$$

with respect to ω, show that

$$\mathcal{L}\left[\frac{\sin(\omega t)}{t}\right] = \tan^{-1}\left[\frac{\omega}{p}\right]. \tag{15.97}$$

Verify your result using Mathematica.

Problem 3. Show that

$$\mathcal{L}^{-1}\left[\frac{2p - 1}{p^2 - 2p + 10}\right] = 2e^t \cos(3t) + \frac{1}{3}e^t \sin(3t). \tag{15.98}$$

Verify your result using Mathematica.

Problem 4. Consider the differential equation

$$\frac{d^2y}{dt^2} + 4\frac{dy}{dt} + 4y = e^{-2t}, \tag{15.99}$$

with the initial conditions $y(0) = 0$, $y'(0) = 4$.
    (a) Using Laplace transform and inverse transform, show that the solution to
the differential equation is given by

$$y(t) = \left(\frac{t^2}{2} + 4t\right)e^{-2t}. \tag{15.100}$$

    (b) Use Mathematica for Laplace transform and inverse transform and verify
the result in (a).
    (c) Using any of the methods introduced for solving differential equations in
chapter 6, find the solution to the differential equation.
    (d) Verify your results in part (a), (b), and (c) using Mathematica for solving
differential equation.

Problem 5. Consider the set of differential equations

$$\frac{dy(t)}{dt} + 2z(t) = 1, \quad \frac{dz(t)}{dt} - 2y(t) + 2t = 0, \tag{15.101}$$

with initial conditions $y(0) = 0$ and $z(0) = 1$.
    (a) Using Laplace transform and inverse transform, find the solution to this set
of differential equation.
    (b) Use Mathematica for Laplace transform and inverse transform and verify
the result in (a).

(c) Using any of the methods introduced for solving differential equations in chapter six, find the solution to the differential equation. Note that

$$\frac{dy(t)}{dt} + 2z(t) = 1, \Rightarrow z(t) = \frac{1}{2}\left(1 - \frac{dy(t)}{dt}\right)$$

$$\Rightarrow \frac{dz(t)}{dt} - 2y(t) + 2t = 0 \Rightarrow \frac{d^2y(t)}{dt^2} + 4y(t) = 4t.$$

$$(15.102)$$

(d) Verify your results in part (a), (b), and (c) using Mathematica for solving differential equation.

Problem 6. *Damped harmonic oscillator*: Consider a mass m oscillating under the influence of a spring force with, spring constant k and damped by a friction force which is proportional to the velocity ($F_f = -bv$). Assuming that the mass starts from rest at $x(0) = x_0$, $x'(0) = 0$. Find the equation of motion for the mass m and determine the position of the mass as function of time using Laplace transform followed by inverse transform.

IOP Publishing

Studies in Theoretical Physics, Volume 1
Fundamental mathematical methods
Daniel Erenso and Victor Montemayor

Chapter 16

Fourier series and transform

This chapter introduces to Fourier series, Fourier transforms, and Fourier inverse transform. We begin with an introduction to periodic functions and how to find mean and mean-square-values for periodic functions. We then determine the Fourier series for periodic functions and introduce the Dirichlet conditions for a real-valued periodic function to equal the sum of its Fourier series at each point where the function is continuous and discontinuous. In finding the Fourier series for periodic functions, we also consider functions with temporal and spatial arguments. After developing the Fourier transform and inverse transform from the Fourier series, we see the relation to and applications of these transform in Dirac delta function, the wave function for a free particle in quantum mechanics, and the convolution of two or more functions. The last section introduces the basic commands for the Fourier series, Fourier transforms, and inverse transforms in Mathematica by resolving some of the examples we had solved in the previous sections.

16.1 Average and root-mean-sqaure values

Periodic functions: A function $f(x)$ is said to be a periodic function when the function repeats itself after a period T,

$$f(x) = f(x \pm T). \tag{16.1}$$

Average values: The average value for a function $f(x)$ in the interval $a < x < b$ is given by

$$\langle f(x) \rangle = \frac{1}{b-a} \int_a^b f(x)dx. \tag{16.2}$$

For a periodic function with a period T, the average value in a period can then be expressed as

$$\langle f(x) \rangle = \frac{1}{T} \int_{-T/2}^{T/2} f(x)dx. \tag{16.3}$$

Periodic motion: Periodic motion is a motion in which the object continually repeats itself. Such motion can be a back and forth repeated movement about a point or motion in a circular orbit. Since the law of inertia states that an object moves in a straight line unless acted upon by force, some force must act on the object to keep the object under a periodic motion. A periodic motion is characterized by the velocity, period, and amplitude of the motion. The motion of the planets around the Sun, swinging pendulum, and simple harmonic motion of a mass attached to spring are examples of periodic motion.

Example 16.1. Compute $\langle f(x) \rangle$ for $f(x) = \sin(x)$ over the interval $0 < x < \pi$.

Solution: Using the expression

$$\langle f(x) \rangle = \frac{1}{b - a} \int_{a}^{b} f(x)dx, \tag{16.4}$$

we find

$$\langle f(x) \rangle = \frac{1}{\pi} \int_{0}^{\pi} \sin(x)dx = -\frac{1}{\pi} \cos(x)\Big|_{0}^{\pi} \Rightarrow \langle f(x) \rangle = \frac{2}{\pi}. \tag{16.5}$$

Example 16.2. Compute $\langle f(x) \rangle$ for $f(x) = \sin(x)$ over the interval $-\pi < x < \pi$. That is, find $\langle f(x) \rangle$ over one complete period.

Solution: Here we have changed the interval to $-\pi < x < \pi$ and the average should be expressed as

$$\langle f(x) \rangle = \frac{1}{\pi - (-\pi)} \int_{-\pi}^{\pi} \sin(x)dx = -\frac{1}{2\pi} \cos(x)\Big|_{-\pi}^{\pi} = 0. \tag{16.6}$$

Example 16.3. The current in an RLC circuit with a resistor R, an inductor L, a capacitor C, and an oscillating voltage source $\Delta V(t) = \Delta V_m \sin(\omega t)$ (see figure 16.1) is given by $I(t) = I_m \sin(\omega t - \varphi)$, where $\Delta V_m = I_m Z$. The quantity Z is the total impedance of the circuit which is given by

$$Z = \sqrt{R^2 + \left(\omega L - \frac{1}{\omega C}\right)^2}, \tag{16.7}$$

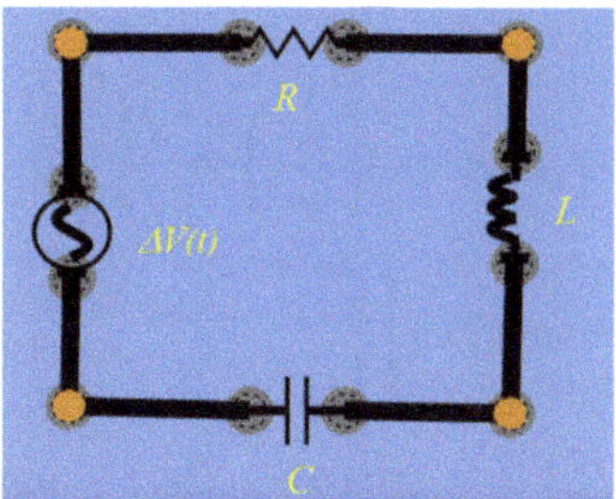

Figure 16.1. An RLC circuit with an ac source voltage.

$\omega = 2\pi f = 2\pi/T$ is the angular frequency for the source voltage. Find an expression for the root-mean-square current I_{rms} over one full period. Note that the root mean square value for a periodic function $f(t)$ with a period T is the square root of the average value for $f^2(t)$ in a period, T

$$f_{rms} = \sqrt{\langle f^2(t)\rangle} = \left[\frac{1}{T}\int_0^T f^2(t)dt\right]^{1/2}. \tag{16.8}$$

Solution: The root mean square current I_{rms} is given by

$$I_{rms} = \sqrt{\langle I^2(t)\rangle} = \left[\frac{1}{T}\int_0^T I^2(t)dt\right]^{1/2}. \tag{16.9}$$

We recall the angular frequency $\omega = 2\pi f = \frac{2\pi}{T}$, and the period, T, can be expressed in terms of ω as

$$T = \frac{2\pi}{\omega} \Rightarrow \frac{1}{T} = \frac{\omega}{2\pi}. \tag{16.10}$$

Then the average $\langle I^2(t)\rangle$ in the interval $0 < t < T$ can be written as

$$\langle I^2(t)\rangle = \frac{1}{T}\int_0^T I^2(t)dt = \frac{\omega}{2\pi}\int_0^T [I_m \sin(\omega t - \varphi)]^2\, dt. \tag{16.11}$$

Introducing the transformation of variable defined by

$$x = \omega t - \varphi \Rightarrow \begin{cases} \dfrac{dx}{\omega} = dt, \\[2mm] t = 0 \Rightarrow x = -\varphi, \\[2mm] t = T \Rightarrow x = \omega T - \varphi = \omega\dfrac{2\pi}{\omega} - \varphi = 2\pi - \varphi, \end{cases} \tag{16.12}$$

equation (16.11) can be rewritten as

$$\langle I^2(t)\rangle = \frac{\omega}{2\pi}\int_{-\varphi}^{2\pi-\varphi} I_m^2 \sin^2(x)\left(\frac{dx}{\omega}\right) = \frac{I_m^2}{2\pi}\int_{-\varphi}^{2\pi-\varphi}\sin^2(x)dx$$

$$= \frac{I_m^2}{2\pi}\left[\frac{x}{2} - \frac{1}{4}\sin(2x)\right]\Bigg|_{-\varphi}^{2\pi-\varphi} \tag{16.13}$$

$$\Rightarrow \langle I^2(t)\rangle = \frac{I_m^2}{2\pi}\left[\pi - \frac{\sin(4\pi - 2\varphi) - \sin(-2\varphi)}{4}\right].$$

Upon substituting

$$\sin(4\pi - 2\varphi) = \sin(4\pi)\cos(2\varphi) - \cos(4\pi)\sin(2\varphi) = -\sin(2\varphi)$$
$$\sin(-\varphi) = -\sin(\varphi), \tag{16.14}$$

one finds from equation (16.13)

$$\langle I^2(t)\rangle = \frac{I_m^2}{2} \Rightarrow I_{rms} = \sqrt{\langle I^2(t)\rangle} = \frac{I_m}{\sqrt{2}}, \tag{16.15}$$

which is then the root mean square current, I_{rms}.

16.2 The Fourier series

Consider a periodic function $f(x)$ with period $T = 2\pi$. The Fourier series for such function can be written as

$$f(x) = \frac{1}{2}a_0 + \sum_{n=1}^{\infty}[a_n \cos(nx) + b_n \sin(nx)], \tag{16.16}$$

where the series expansion coefficients are given by

$$a_n = \frac{1}{\pi}\int_{-\pi}^{\pi} f(x)\cos(nx)dx, \; n = 0, 1, 2, 3. .., $$
$$b_n = \frac{1}{\pi}\int_{-\pi}^{\pi} f(x)\sin(nx)dx, \; n = 1, 2, 3. ... \tag{16.17}$$

Using Euler's formula

$$\cos(nx) = \frac{e^{inx} + e^{-inx}}{2}, \; \sin(nx) = \frac{e^{inx} - e^{-inx}}{2i} \tag{16.18}$$

one can rewrite equation (16.16) as

$$f(x) = \frac{1}{2}a_0 + \sum_{n=1}^{\infty}\left[a_n\left(\frac{e^{inx} + e^{-inx}}{2}\right) + b_n\left(\frac{e^{inx} - e^{-inx}}{2i}\right)\right]$$

$$\Rightarrow f(x) = \frac{1}{2}a_0 + \sum_{n=1}^{\infty}\left(\frac{a_n - ib_n}{2}\right)e^{inx} + \sum_{n=1}^{\infty}\left(\frac{a_n + ib_n}{2}\right)e^{-inx}. \tag{16.19}$$

Replacing n by $-n$ in the second series and noting that $a_0 = a_0 e^{i0(x)}$, one can write

$$f(x) = \frac{1}{2}a_0 e^{i0x} + \sum_{n=1}^{\infty}\left(\frac{a_n - ib_n}{2}\right)e^{inx} + \sum_{n=-1}^{-\infty}\left(\frac{a_n + ib_n}{2}\right)e^{inx} = \sum_{n=-\infty}^{\infty} c_n e^{inx}, \quad (16.20)$$

where

$$c_n = \begin{cases} \dfrac{1}{2}a_0, & \text{for } n = 0, \\[2mm] \dfrac{a_n - ib_n}{2}, & \text{for } n \geqslant 1, \\[2mm] \dfrac{a_n + ib_n}{2}, & \text{for } n \leqslant -1. \end{cases} \quad (16.21)$$

Using the expansion coefficients in equation (16.17), we have

$$\begin{aligned} \frac{a_n - ib_n}{2} &= \frac{1}{2\pi}\int_{-\pi}^{\pi} f(x)[\cos(nx) - i\sin(nx)]dx = \frac{1}{2\pi}\int_{-\pi}^{\pi} f(x)e^{-inx}dx, \\ &\quad \text{for } n = 0, 1, 2, 3. \ldots, \\ \frac{a_n + ib_n}{2} &= \frac{1}{2\pi}\int_{-\pi}^{\pi} f(x)[\cos(nx) + i\sin(nx)]dx = \frac{1}{2\pi}\int_{-\pi}^{\pi} f(x)e^{inx}dx, \\ &\quad \text{for } n = 1, 2, 3. \ldots \end{aligned} \quad (16.22)$$

which leads to a more concise expression for c_n given by

$$c_n = \frac{1}{2\pi}\int_{-\pi}^{\pi} f(x)e^{-inx}dx, \quad (16.23)$$

where $n = \ldots -3, -2, -1, 0, 1, 2, 3. \ldots$

Example 16.4. Find the Fourier series expansion for the function

$$f(x) = \begin{cases} 1, & \text{for } 0 < x < \pi, \\ -1, & \text{for } -\pi < x < 0, \end{cases} \quad (16.24)$$

which is repeated indefinitely with a period of 2π.

Solution: The Fourier series expansion for a periodic function $f(x)$ with a period 2π is given

$$f(x) = \sum_{n=-\infty}^{\infty} c_n e^{inx}, \quad (16.25)$$

where the expansion coefficient expansion c_n is given by equation (16.19)

$$c_n = \frac{1}{2\pi}\int_{-\pi}^{\pi} f(x)e^{-inx}dx. \quad (16.26)$$

For the function in equation (16.24), we find

$$
\begin{aligned}
c_n &= \frac{1}{2\pi}\left[-\int_{-\pi}^{0} e^{-inx}dx + \int_{0}^{\pi} e^{-inx}dx\right] = \frac{1}{2\pi}\left[\int_{0}^{-\pi} e^{-inx}dx + \int_{0}^{\pi} e^{-inx}dx\right] \\
&= \frac{1}{2\pi}\left[-\int_{0}^{\pi} e^{inx}dx + \int_{0}^{\pi} e^{-inx}dx\right] = -\frac{i}{\pi}\int_{0}^{\pi}\left(\frac{e^{inx} - e^{-inx}}{2i}\right)dx \\
&= -\frac{i}{\pi}\int_{0}^{\pi}\sin(nx)dx.
\end{aligned}
\tag{16.27}
$$

$$
c_n = \frac{i}{\pi n}\cos(nx)\Big|_0^{\pi} \Rightarrow c_n = \frac{i}{\pi n}[\cos(n\pi) - 1],
$$

that leads to

$$
c_n = \begin{cases} \dfrac{2i}{\pi n}, & n = 1, 3, 5. \,.. \\[2mm] 0, & n = 0, 2, 4, 6. \,.. \end{cases}
\tag{16.28}
$$

$$
\Rightarrow c_n = \frac{2i}{\pi(2n + 1)}, \quad n = 0, \pm1, \pm2, \pm3. \,..
$$

Substituting the result in equation (16.28) into equation (16.25), one finds (figure 16.2)

$$
f(x) = \sum_{n=-\infty}^{\infty} \frac{2i}{\pi(2n + 1)} e^{i(2n+1)x}.
\tag{16.29}
$$

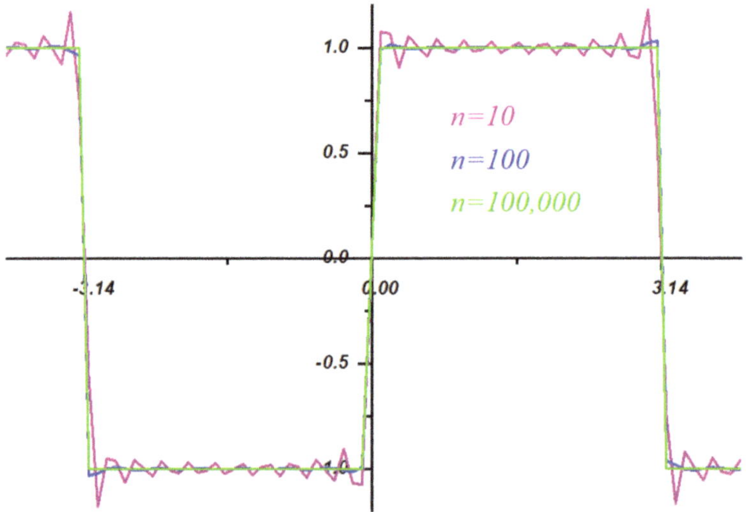

Figure 16.2. The function $f(x)$ when it is approximated by the Fourier series for $n = 10$ (pink), $n = 100$ (blue), $n = 100\,000$ (green).

Noting that

$$\sum_{n=-\infty}^{\infty} \frac{2i}{\pi(2n+1)} e^{i(2n+1)x} = \frac{2i}{\pi}\left[\sum_{n=-\infty}^{-1} \frac{e^{i(2n+1)x}}{2n+1} + \sum_{n=0}^{\infty} \frac{e^{i(2n+1)x}}{2n+1} \right]$$

$$= \frac{2i}{\pi}\left[\sum_{n=1}^{\infty} \frac{e^{i(-2n+1)x}}{-2n+1} + \sum_{n=0}^{\infty} \frac{e^{i(2n+1)x}}{2n+1} \right]$$

$$= \frac{2i}{\pi}\left[-\sum_{n=0}^{\infty} \frac{e^{-i(2n+1)x}}{2n+1} + \sum_{n=0}^{\infty} \frac{e^{i(2n+1)x}}{2n+1} \right] \qquad (16.30)$$

$$= \frac{4}{\pi}\left[\sum_{n=0}^{\infty} \frac{1}{2n+1}\left(\frac{e^{i(2n+1)x} - e^{-i(2n+1)x}}{2i} \right) \right]$$

$$= \frac{4}{\pi} \sum_{n=0}^{\infty} \frac{\sin\left[(2n+1)x\right]}{2n+1},$$

one can write

$$f(x) = \sum_{n=-\infty}^{\infty} \frac{2i}{\pi(2n+1)} e^{i(2n+1)x} = \frac{4}{\pi} \sum_{n=0}^{\infty} \frac{\sin\left[(2n+1)x\right]}{2n+1}. \qquad (16.31)$$

Equation (16.31) is the Fourier series expansion.

16.3 Dirichlet conditions

The Dirichlet conditions are sufficient conditions for a real-valued, periodic function $f(x)$ to equal to the sum of its Fourier series at each point where f is continuous. Moreover, the behavior of the Fourier series at points of discontinuity is determined as well. These conditions are named after Johann Peter Gustav Lejeune Dirichlet. The conditions are

    (a) $f(x)$ must have a finite number of extrema (maxima, minima) in any given interval (period).

    (b) $f(x)$ must have a finite number of discontinuities in any given interval.

    (c) $f(x)$ must be absolutely integrable over a period.

    (d) $f(x)$ must be bounded within that period (that is, it doesn't become infinite).

If all these conditions are met, then the Fourier series will converge to the given function for all values of x where the function is continuous, and will converge to the average value of the function at any value of x where the function has a discontinuity.

Dirichlet's theorem for 1-dimensional Fourier series
Suppose $f(x)$ satisfies Dirichlet conditions. In that case, the Fourier series will converge to the given function for all values of x where the function is continuous

and converge to the average value of the function at any value of x where the function has a discontinuity,

$$\frac{1}{2}a_0 + \sum_{n=1}^{\infty}[a_n \cos(nx) + b_n \sin(nx)] = \sum_{n=-\infty}^{\infty} c_n e^{inx} = \frac{1}{2}(f(x_+) + f(x_-)), \quad (16.32)$$

where the notation

$$f(x_+) = \lim_{x \to x_+} f(x) \quad \text{and} \quad f(x_-) = \lim_{x \to x_-} f(x) \tag{16.33}$$

represents the limiting value for the function $f(x)$ from right and left of x, respectively.

Example 16.5. Verify Dirichlet's theorem for the function in Example 16.4

$$f(x) = \begin{cases} 1, & \text{for } 0 < x < \pi, \\ -1, & \text{for } -\pi < x < 0. \end{cases} \tag{16.34}$$

Solution: The function satisfies all Dirichlet's conditions,

(a) $f(x)$ have a finite number of extrema (maxima, minima) in any given interval $-\pi < x < \pi$,
(b) $f(x)$ has one discontinuity at $x = 0$ in the interval $-\pi < x < \pi$,
(c) $f(x)$ is absolutely integrable over $-\pi < x < \pi$,
(d) $f(x)$ is bounded within $-\pi < x < \pi$.
   and therefore for the Fourier series of the function, we must be able to show that

$$f(x) = \sum_{n=-\infty}^{\infty} c_n e^{inx} = \frac{1}{2}(f(x_+) + f(x_-)). \tag{16.35}$$

Recalling that

$$f(x) = \sum_{n=-\infty}^{\infty} c_n e^{inx} = \frac{4}{\pi} \sum_{n=1}^{\infty} \frac{\sin[(2n+1)x]}{2n+1}, \tag{16.36}$$

for $x = 0$, we have

$$f(0) = \frac{4}{\pi} \sum_{n=1}^{\infty} \frac{\sin[(2n+1)0]}{2n+1} = 0. \tag{16.37}$$

From the given function

$$f(x) = \begin{cases} 1, & \text{for } 0 < x < \pi, \\ -1, & \text{for } -\pi < x < 0, \end{cases} \tag{16.38}$$

for

$$f(0_+) = \lim_{x\to 0_+} f(x) \quad \text{and} \quad f(0_-) = \lim_{x\to 0_-} f(x), \tag{16.39}$$

one finds

$$f(0_+) = 1 \quad \text{and} \quad f(0_-) = -1 \Rightarrow \frac{1}{2}(f(0_+) + f(0_-)) = 0. \tag{16.40}$$

In view of the results in equation (16.36) and (16.40), one can establish the relation

$$f(x) = \sum_{n=-\infty}^{\infty} c_n e^{inx} = \frac{1}{2}(f(x_+) + f(x_-)) = 0, \tag{16.41}$$

for $x = 0$ where the function is discontinuous. This verifies Dirichlet's theorem.

16.4 Fourier series with spatial and temporal arguments

We recall when x is dimensionless, for the periodic function $f(x)$ with period $T = 2\pi$,

$$f(x + 2\pi) = f(x), \tag{16.42}$$

the Fourier series is given by

$$f(x) = \frac{1}{2}a_0 + \sum_{n=1}^{\infty}[a_n \cos(nx) + b_n \sin(nx)], \tag{16.43}$$

where the coefficients are determined using

$$a_n = \frac{1}{\pi} \int_{-\pi}^{\pi} f(x)\cos(nx)dx, \quad \text{for } n = 0, 1, 2, 3. \,..$$
$$b_n = \frac{1}{\pi} \int_{-\pi}^{\pi} f(x)\sin(nx)dx, \quad \text{for } n = 1, 2, 3. \,.. \tag{16.44}$$

More concisely

$$f(x) = \sum_{n=-\infty}^{\infty} c_n e^{inx}, \tag{16.45}$$

where

$$c_n = \frac{1}{2\pi} \int_{-\pi}^{\pi} f(x)e^{-inx}dx \tag{16.46}$$

is the expansion coefficient.

Spatial arguments

For a special periodic function $f(y + \lambda) = f(y)$ that repeats itself after a length λ, introducing the constant $k = 2\pi/\lambda$, one can write for the dimensionless variable

$x = ky$. Replacing $x = ky$ in equations (16.43)–(16.46), the Fourier series for a special function $f(y)$ is given by

$$f(y) = \frac{1}{2}a_0 + \sum_{n=1}^{\infty}[a_n \cos{(nky)} + b_n \sin{(nky)}] = \sum_{n=-\infty}^{\infty} c_n e^{inky}, \qquad (16.47)$$

where the expansion coefficients are given by

$$a_n = \frac{2}{\lambda} \int_{-\lambda/2}^{\lambda/2} f(y) \cos{(nky)}dy, \; n = 0, 1, 2, 3. \, ..,$$

$$b_n = \frac{2}{\lambda} \int_{-\lambda/2}^{\lambda/2} f(y) \sin{(nky)}dy, \; n = 1, 2, 3. \, .., \qquad (16.48)$$

$$c_n = \frac{1}{2\lambda} \int_{-\lambda/2}^{\lambda/2} f(y)e^{-inky}dy.$$

Temporal arguments

For a periodic temporal function $f(t)$ that repeats itself after a period T, $f(t + T) = f(t)$, introducing the constant $\omega = 2\pi/T$ (angular frequency), one can replace the dimensionless variable $x = \omega t$ in equations (16.43)–(16.46) and write the Fourier series for $f(t)$ as

$$f(t) = \frac{1}{2}a_0 + \sum_{n=1}^{\infty}[a_n \cos{(n\omega t)} + b_n \sin{(n\omega t)}] = \sum_{n=-\infty}^{\infty} c_n e^{in\omega t} \qquad (16.49)$$

where

$$a_n = \frac{2}{T} \int_{-T/2}^{T/2} f(t) \cos{(n\omega t)}dt, \; n = 0, 1, 2, 3. \, ..,$$

$$b_n = \frac{2}{T} \int_{-T/2}^{T/2} f(t) \sin{(n\omega t)}dt, \; n = 1, 2, 3. \, .., \qquad (16.50)$$

$$c_n = \frac{1}{2T} \int_{-T/2}^{T/2} f(t)e^{-in\omega t}dt,$$

are the expansion coefficients.

Example 16.6. A voltage signal in the form of a half-rectified sine-wave (see figure 16.3) is fed into a linear circuit whose response to a simple sinusoidal signal is known. In order to determine the response of the circuit to the input signal, we wish to find the Fourier series expansion of that signal. Assume that the input signal is given by

$$\Delta V(t) = \begin{cases} 0, & -\dfrac{T}{2} < t < 0 \\[2ex] \Delta V_m \sin{(\omega t)}, & 0 < t < \dfrac{T}{2} \end{cases} . \qquad (16.51)$$

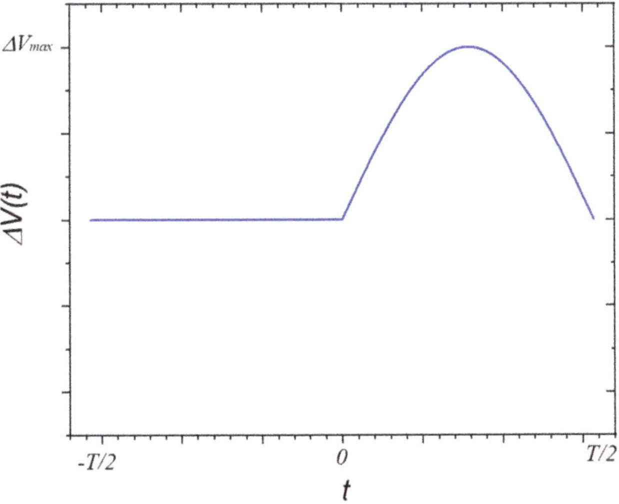

Figure 16.3. A half-rectified sine wave.

Solution: Using the Fourier series expansion coefficients for temporal arguments

$$a_n = \frac{2}{T} \int_{-T/2}^{T/2} f(t) \cos\,(n\omega t)dt,\ n = 0,\,1,\,2,\,3.\,..$$

$$b_n = \frac{2}{T} \int_{-T/2}^{T/2} f(t) \sin\,(n\omega t)dt,\ n = 1,\,2,\,3.\,..$$

(16.52)

for the given function, one can write

$$a_n = \frac{2\Delta V_m}{T} \int_0^{T/2} \cos\,(n\omega t) \sin\,(\omega t)dt,$$

$$b_n = \frac{2\Delta V_m}{T} \int_0^{T/2} \sin\,(n\omega t) \sin\,(\omega t)dt.$$

(16.53)

Applying the trigonometric relations

$$\sin\,(\alpha \pm \beta) = \sin \alpha \cos \beta \pm \cos \alpha \sin \beta,$$

$$\cos\,(\alpha \pm \beta) = \cos \alpha \cos \beta \mp \sin \alpha \sin \beta,$$

(16.54)

one can write

$$\cos\,(n\omega t) \sin\,(\omega t) = \frac{1}{2}\{\sin\,[(n+1)\omega t] - \sin\,[(n-1)\omega t]\},$$

$$\sin\,(n\omega t) \sin\,(\omega t) = \frac{1}{2}\{\cos\,[(n-1)\omega t] - \cos\,[(n+1)\omega t]\},$$

(16.55)

so that

$$a_n = \frac{\Delta V_m}{T}\left\{\int_0^{T/2} \sin\left[(n+1)\omega t\right]dt - \int_0^{T/2} \sin\left[(n-1)\omega t\right]dt\right\},$$

$$b_n = \frac{\Delta V_m}{T}\left\{\int_0^{T/2} \cos\left[(n-1)\omega t\right]dt - \int_0^{T/2} \cos\left[(n+1)\omega t\right]dt\right\}. \tag{16.56}$$

Using the expansion coefficients in equation (16.56), for $n = 0$ and $n = 1$, we find

$$a_0 = \frac{2\Delta V_m}{T}\int_0^{T/2} \sin(\omega t)dt = \frac{2\Delta V_m}{T\omega} = \frac{\Delta V_m}{\pi},$$

$$a_1 = \frac{\Delta V_m}{T}\int_0^{T/2} \sin(2\omega t)dt = \frac{\Delta V_m}{2T\omega} = \frac{\Delta V_m}{\pi}, \tag{16.57}$$

$$b_1 = \frac{\Delta V_m}{2} - \frac{\Delta V_m}{T}\int_0^{T/2} \cos(2\omega t)dt = \frac{\Delta V_m}{2},$$

where we used

$$\omega = \frac{2\pi}{T} \Rightarrow T\omega = 2\pi. \tag{16.58}$$

For $n > 1$, one can also show that a_n

$$
\begin{aligned}
a_n &= \frac{\Delta V_m}{T}\left\{-\frac{\cos\left[(n+1)\omega t\right]}{(n+1)\omega} + \frac{\cos\left[(n-1)\omega t\right]}{(n-1)\omega}\right\}\Bigg|_0^{T/2} \\[2mm]
&= \frac{\Delta V_m}{T\omega}\left\{\frac{1}{n+1} - \frac{\cos\left[\frac{(n+1)\omega T}{2}\right]}{n+1} - \frac{1}{n-1} + \frac{\cos\left[\frac{(n-1)\omega T}{2}\right]}{n-1}\right\} \\[2mm]
&= \frac{\Delta V_m}{2\pi}\left\{-\frac{2}{n^2-1} - \frac{\cos\left[(n+1)\pi\right]}{n+1} + \frac{\cos\left[(n-1)\pi\right]}{n-1}\right\} \\[2mm]
&= \frac{\Delta V_m}{2\pi}\left\{-\frac{2}{n^2-1} + \frac{\cos(n\pi)}{n+1} - \frac{\cos(n\pi)}{n-1}\right\} \\[2mm]
&= -\frac{\Delta V_m}{\pi}\left\{\frac{1}{n^2-1} + \frac{(-1)^n}{n^2-1}\right\} \\[2mm]
&= -\frac{\Delta V_m}{\pi}\frac{1+(-1)^n}{n^2-1} = \begin{cases} 0, & \text{for odd } n > 1 \\[2mm] -\dfrac{\Delta V_m}{\pi}\dfrac{2}{n^2-1}, & \text{for even } n > 0 \end{cases} \\[2mm]
\Rightarrow a_n &= -\frac{\Delta V_m}{\pi}\frac{2}{(2n)^2-1}, \quad n = 1, 2, 3. .
\end{aligned} \tag{16.59}
$$

and b_n

$$b_n = \frac{\Delta V_m}{T}\left\{\int_0^{T/2} \cos\left[(n-1)\omega t\right]dt - \int_0^{T/2} \cos\left[(n+1)\omega t\right]dt\right\}$$

$$= \frac{\Delta V_m}{\omega T}\left\{\frac{\sin\left[(n-1)\omega t\right]}{n-1} - \frac{\sin\left[(n+1)\omega t\right]}{n+1}\right\}\Bigg|_0^{T/2} \tag{16.60}$$

$$\Rightarrow b_n = \frac{\Delta V_m}{2\pi}\left\{\frac{\sin\left[(n-1)\pi\right]}{n-1} - \frac{\sin\left[(n+1)\pi\right]}{n+1}\right\} = 0.$$

Using the results

$$a_0 = \frac{\Delta V_m}{\pi}, \quad a_1 = \frac{\Delta V_m}{\pi}, \quad a_{2n} = -\frac{\Delta V_m}{\pi}\frac{2}{(2n)^2 - 1}, \quad \text{for } n = 1, 2, 3.\,.$$

$$b_1 = \frac{\Delta V_m}{2}, \quad b_n = 0, \quad \text{for } n > 1, \tag{16.61}$$

we can then write

$$f(t) = \frac{\Delta V_m}{\pi} + \frac{\Delta V_m}{\pi}\cos(\omega t) + \frac{\Delta V_m}{2}\sin(\omega t)$$

$$- \frac{2\Delta V_m}{\pi}\sum_{n=1}^{\infty}\frac{\cos(2n\omega t)}{4n^2 - 1}, \tag{16.62}$$

which is the Fourier series for the function $f(t)$.

16.5 The Fourier transform and inverse transform

The Fourier transform
The Fourier transform for the function $f(t)$ that satisfies all Dirichlet's conditions is given by

$$\mathcal{F}[f(t)] = \frac{1}{\sqrt{2\pi}}\int_{-\infty}^{\infty} f(t)e^{-i\omega t}dt = F(\omega). \tag{16.63}$$

The Fourier transform does exist whether the function is a periodic or non-periodic function.

The inverse Fourier transform
We may develop the inverse Fourier transform starting from the Fourier series expansion. Suppose we are considering the interval $[-T, T]$ instead of $[-\pi, \pi]$. The Fourier series expansion for the function $f(t)$ can be expressed as

$$f(t) = \frac{1}{2}a_0 + \sum_{n=1}^{\infty}\left[a_n\cos\left(\frac{\pi n}{T}t\right) + b_n\sin\left(\frac{\pi n}{T}t\right)\right], \tag{16.64}$$

where the *Fourier coefficients* are given by

$$a_n = \frac{1}{T} \int_{-T}^{T} f(t') \cos\left(\frac{\pi n}{T} t'\right) dt', \ \text{for } n = 0, 1, 2, 3. ..$$

$$b_n = \frac{1}{T} \int_{-T}^{T} f(t') \sin\left(\frac{\pi n}{T} t'\right) dt', \ \text{for } n = 1, 2, 3. ...$$

(16.65)

Upon substituting the expansion coefficients in equation (16.65) into equation (16.64), the Fourier series can be written as

$$
\begin{aligned}
f(t) &= \frac{1}{T} \int_{-T}^{T} f(t')dt' + \frac{1}{T} \sum_{n=1}^{\infty} \cos\left(\frac{\pi n}{T} t\right) \int_{-T}^{T} f(t') \cos\left(\frac{\pi n}{T} t'\right) dt' \\
&\quad + \frac{1}{T} \sum_{n=1}^{\infty} \sin\left(\frac{\pi n}{T} t\right) \int_{-T}^{T} f(t') \sin\left(\frac{\pi n}{T} t'\right) dt' \\
&= \frac{1}{T} \int_{-T}^{T} f(t')dt' + \frac{1}{T} \sum_{n=1}^{\infty} \int_{-T}^{T} f(t') \left\{ \cos\left(\frac{\pi n}{T} t\right) \cos\left(\frac{\pi n}{T} t'\right) + \right. \\
&\quad \left. \sin\left(\frac{\pi n}{T} t\right) \sin\left(\frac{\pi n}{T} t'\right) \right\} dt' \\
\Rightarrow f(t) &= \frac{1}{T} \int_{-T}^{T} f(t')dt' + \sum_{n=1}^{\infty} \frac{1}{T} \int_{-T}^{T} f(t') \cos\left[\frac{\pi n}{T}(t - t')\right] dt',
\end{aligned}
$$

(16.66)

where we used the double angle formula for a cosine function. Introducing the variable defined by

$$\omega_n = \frac{\pi n}{T} \Rightarrow \frac{\Delta\omega}{\pi} = \frac{\omega_{n+1} - \omega_n}{\pi} = \frac{1}{T},$$

(16.67)

one can put equation (16.66) in the form

$$
\begin{aligned}
f(t) &= \frac{1}{\pi} \Delta\omega \int_{-T}^{T} f(t')dt' + \frac{1}{\pi} \sum_{n=1}^{\infty} \Delta\omega \int_{-T}^{T} f(t') \cos(\omega_n(t - t'))dt' \\
\Rightarrow f(t) &= \frac{1}{\pi} \sum_{n=0}^{\infty} \Delta\omega \int_{-T}^{T} f(t') \cos(\omega_n(t - t'))dt'.
\end{aligned}
$$

(16.68)

We now let the parameter T approach to infinity such that the interval $[-T, T]$ becomes $(-\infty, \infty)$. In this limit from equation (16.67), we have $\Delta\omega \to d\omega$, $\omega_n \to \omega$ and since the function f satisfies the *Dirichlet conditions*, we can also replace the summation by integration. Thus one can rewrite equation (16.68) as

$$f(t) = \frac{1}{\pi} \int_{0}^{\infty} d\omega \int_{-\infty}^{\infty} f(t') \cos[\omega(t - t')]dt'.$$

(16.69)

Noting that $\cos[\omega(t - t')]$ is an even function and $\sin(\omega(t - t'))$ is an odd function of ω, one can write

$$\frac{1}{2} \int_{-\infty}^{\infty} d\omega \int_{-\infty}^{\infty} f(t') \cos\left[\omega(t - t')\right] dt' = \int_{0}^{\infty} d\omega \int_{-\infty}^{\infty} f(t') \cos\left[\omega(t - t')\right] dt'$$

$$\frac{i}{2} \int_{-\infty}^{\infty} d\omega \int_{-\infty}^{\infty} f(t') \sin\left[\omega(t - t')\right] dt' = 0,$$

(16.70)

so that using the results in equation (16.70), equation (16.69) can be rewritten as

$$f(t) = \frac{1}{\pi} \int_{0}^{\infty} d\omega \int_{-\infty}^{\infty} f(t') \cos\left[\omega(t - t')\right] dt'$$

$$= \frac{1}{2\pi} \int_{-\infty}^{\infty} d\omega \int_{-\infty}^{\infty} f(t')\{\cos\left[\omega(t - t')\right] + i \sin\left[\omega(t - t')\right]\} dt'$$

(16.71)

$$\Rightarrow f(t) = \frac{1}{\sqrt{2\pi}} \int_{-\infty}^{\infty} d\omega \left[\frac{1}{\sqrt{2\pi}} \int_{-\infty}^{\infty} f(t')e^{-i\omega t'} dt'\right] e^{i\omega t}.$$

We recall the Fourier transform for the function $f(t)$ is given by

$$F(\omega) = \mathcal{F}[f(t)] = \frac{1}{\sqrt{2\pi}} \int_{-\infty}^{\infty} f(t)e^{-i\omega t} dt,$$

(16.72)

so that one can establish from equation (16.71) the relation

$$f(t) = \mathcal{F}^{-1}[F(\omega)] = \frac{1}{\sqrt{2\pi}} \int_{-\infty}^{\infty} F(\omega)e^{i\omega t} d\omega.$$

(16.73)

Equation (16.73) is the *inverse Fourier transform*. If $f(t)$ has a discontinuity, the inverse Fourier transform will return the value for $f(t)$ that is at the midpoint of the discontinuity.

Example 16.7. Find and describe the Fourier transform of the function $f(t)$ defined by

$$f(t) = \begin{cases} 1, & |t| < 1 \Rightarrow -1 < t < 1 \\ 0, & |t| > 1 \Rightarrow t > 1 \ \& \ t < -1 \end{cases}$$

(16.74)

which is an even function of t (see figure 16.4). This is the single slit diffraction problem of physical optics. The slit is described by $f(t)$. The amplitude of the diffraction pattern is given by the Fourier transform $F(\omega)$.

Solution: We recall

$$F(\omega) = \mathcal{F}[f(t)] = \frac{1}{\sqrt{2\pi}} \int_{-\infty}^{\infty} f(t)e^{-i\omega t} dt,$$

(16.75)

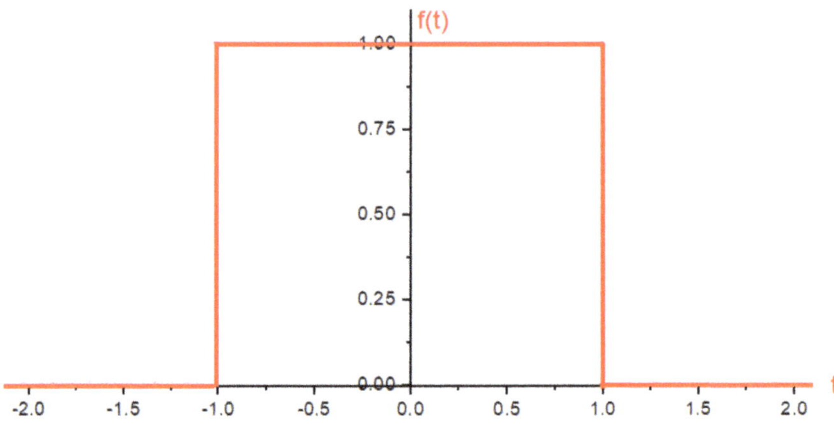

Figure 16.4. The graph for the function $f(t) = |t|$ for $-1 < t < 1$ and $f(t) = |t|$ for $t > 1$ and $t < -1$.

and for the function in equation (16.74), one can write

$$\mathcal{F}[f(t)] = \frac{1}{\sqrt{2\pi}} \int_{-1}^{1} e^{-i\omega t} dt = \frac{1}{\sqrt{2\pi}} \left[-\frac{e^{-i\omega t}}{i\omega} \right]_{-1}^{1} = \frac{2}{\sqrt{2\pi}\,\omega} \frac{e^{i\omega} - e^{-i\omega}}{2i} \tag{16.76}$$

$$\Rightarrow F(\omega) = \mathcal{F}[f(t)] = \sqrt{\frac{2}{\pi}} \frac{\sin(\omega)}{\omega}.$$

The function

$$f(x) = \frac{\sin(x)}{x} \tag{16.77}$$

is known as the *sinc function* commonly expressed as *sinc(x)*. figure 16.5 shows *sinc(x)* versus *x*.

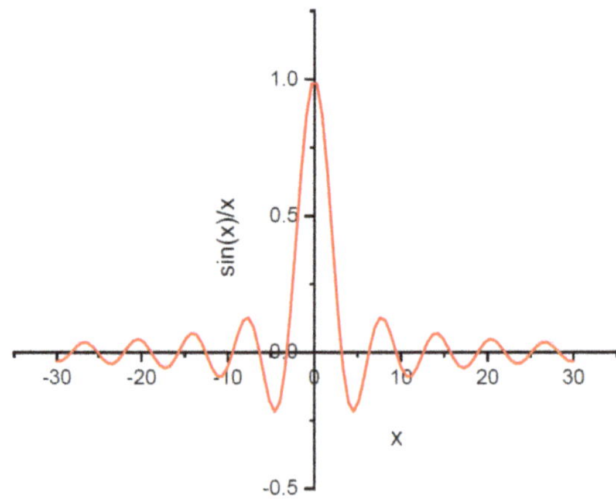

Figure 16.5. The sinc function.

Example 16.8. Find the Fourier transform for the Gaussian function

$$f(x) = e^{-a^2 x^2}, \tag{16.78}$$

where a is a positive real constant. Figure 16.6 shows the graph for the function $f(x)$ for the two different values of a.

Solution: For the function in equation (16.78), the Fourier transform can be written as

$$\mathcal{F}[f(x)] = \frac{1}{\sqrt{2\pi}} \int_{-\infty}^{\infty} e^{-a^2 x^2} e^{-i\omega x} dx = \frac{1}{\sqrt{2\pi}} \int_{-\infty}^{\infty} e^{-(a^2 x^2 + i\omega x)} dx \tag{16.79}$$

and using

$$a^2 x^2 + i\omega x = a^2 \left[\left(x + \frac{i\omega}{2a^2} \right)^2 + \left(\frac{\omega}{2a^2} \right)^2 \right] = a^2 \left(x + \frac{i\omega}{2a^2} \right)^2 + \frac{\omega^2}{4a^2} \tag{16.80}$$

equation (16.79) can be put in the form

$$\mathcal{F}[f(x)] = \frac{e^{-\frac{\omega^2}{4a^2}}}{\sqrt{2\pi}} \int_{-\infty}^{\infty} e^{-a^2 \left(x + \frac{i\omega}{2a^2} \right)^2} dx. \tag{16.81}$$

We can determine the value for the integral in equation (16.81) applying the methods for contour integration we introduced in chapter 14. To this end, we consider the contour integral

$$I = \oint_C e^{-a^2 z^2} dz \tag{16.82}$$

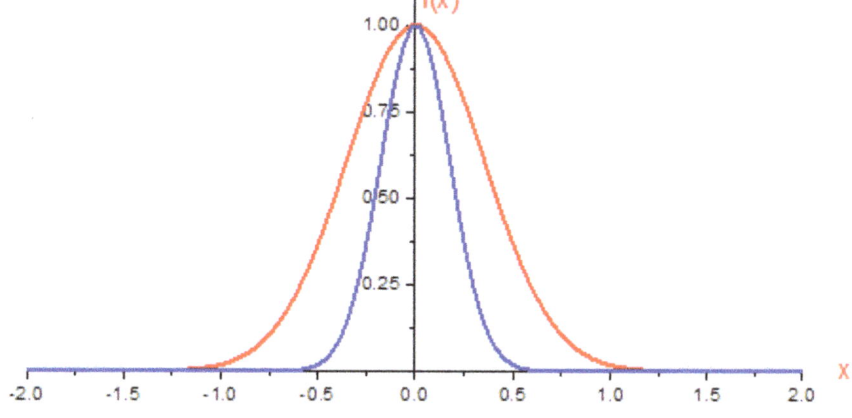

Figure 16.6. The graph for the function $f(x) = \exp(-a^2 x^2)$ for $a = 2$ (red) and $a = 4$ (blue).

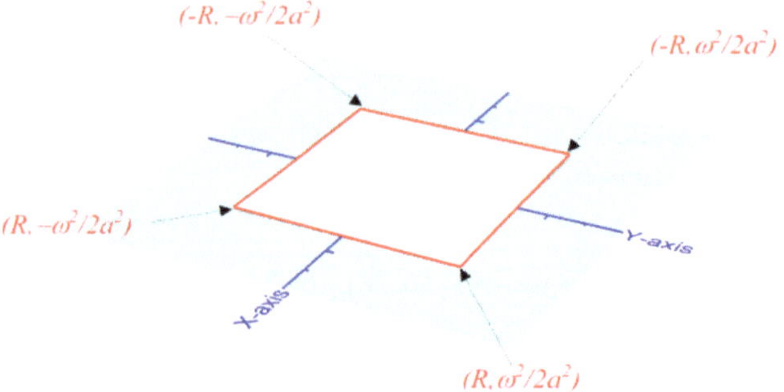

Figure 16.7. A rectangular contour on a complex plane.

for the rectangular closed loop C on the complex plane with vertices $(-R, \omega/2a^2)$, $(-R, -\omega/2a^2)$, $(R, -\omega/2a^2)$, and $(R, \omega/2a^2)$ shown in figure 16.7. For this contour one can write the complex integral in equation (16.82) as

$$\oint e^{-a^2z^2}dz = \int_{-R}^{R} e^{-a^2x^2}dx + i\int_{0}^{\frac{\omega}{2a^2}} e^{-a^2(R+iy)^2}dy$$

$$+ \int_{R}^{-R} e^{-a^2\left(x+\frac{i\omega}{2a^2}\right)^2}dx + i\int_{-\frac{\omega}{2a^2}}^{0} e^{-a^2(-R+iy)^2}dy$$

$$= \int_{-R}^{R} e^{-a^2x^2}dx + i\int_{0}^{\frac{\omega}{2a^2}} e^{-a^2(R+iy)^2}dy$$

$$+ \int_{R}^{-R} e^{-a^2\left(x+\frac{i\omega}{2a^2}\right)^2}dx + i\int_{-\frac{\omega}{2a^2}}^{0} e^{-a^2(R-iy)^2}dy. \tag{16.83}$$

Noting that in the limit $R \to \infty$,

$$\int_{0}^{\frac{\omega}{2a^2}} e^{-a^2(R+iy)^2}dy = \int_{-\frac{\omega}{2a^2}}^{0} e^{-a^2(R-iy)^2}dy = 0, \tag{16.84}$$

and there is no poles inside the curve one finds, from equation (16.83),

$$\oint_C e^{-a^2z^2}dz = 0 \Rightarrow \int_{-\infty}^{\infty} e^{-a^2x^2}dx + \int_{\infty}^{-\infty} e^{-a^2\left(x+\frac{i\omega}{2a^2}\right)^2}dx = 0$$

$$\Rightarrow \int_{-\infty}^{\infty} e^{-a^2x^2}dx - \int_{-\infty}^{\infty} e^{-a^2\left(x+\frac{i\omega}{2a^2}\right)^2}dx = 0 \tag{16.85}$$

$$\Rightarrow \int_{-\infty}^{\infty} e^{-a^2\left(x+\frac{i\omega}{2a^2}\right)^2}dx = \int_{-\infty}^{\infty} e^{-a^2x^2}dx = \frac{\sqrt{\pi}}{a}.$$

Therefore, in view of the result in tequation (16.85), equation (16.81) becomes

$$F(\omega) = \mathcal{F}[f(x)] = \frac{e^{-\frac{\omega^2}{4a^2}}}{a\sqrt{2}}. \tag{16.86}$$

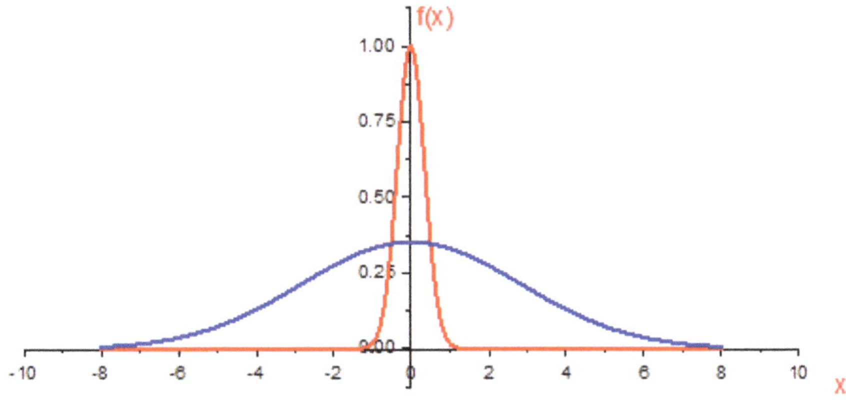

Figure 16.8. The graphs for the function $f(x) = \exp(-a^2x^2)$ for $a = 2$ (red) and its Fourier transform in equation (16.86) (blue).

Equation (16.86) is the Fourier integral transform for the function in equation (16.78). Figure 16.8 displays the function in equation (16.78) and its Fourier transform in equation (16.86).

16.6 The Dirac delta function and the Fourier inverse transform

We recall the inverse Fourier transform $f(t) = \mathcal{F}^{-1}[F(\omega)]$ for the function $F(\omega) = \mathcal{F}[f(t)]$ which is given by

$$f(t) = \frac{1}{\sqrt{2\pi}} \int_{-\infty}^{\infty} F(\omega)e^{i\omega t}d\omega. \tag{16.87}$$

Upon substituting

$$F(\omega) = \mathcal{F}[f(t)] = \frac{1}{\sqrt{2\pi}} \int_{-\infty}^{\infty} f(t')e^{-i\omega t'}dt', \tag{16.88}$$

into equation (16.87), one can write

$$f(t) = \int_{-\infty}^{\infty} f(t')\left[\frac{1}{2\pi} \int_{-\infty}^{\infty} e^{-i\omega(t-t')}d\omega\right]dt'. \tag{16.89}$$

Using the property of the Dirac delta function from chapter 11

$$f(t) = \int_{-\infty}^{\infty} f(t')\sigma(t - t')dt', \tag{16.90}$$

and comparing this expression with equation (16.89), one can establish the relation

$$\sigma(t - t') = \frac{1}{2\pi} \int_{-\infty}^{\infty} e^{-i\omega(t-t')}d\omega. \tag{16.91}$$

The Dirac delta function has different forms. One of these forms can be determined by carrying out the integration in equation (16.91),

$$\sigma(t - t') = \frac{1}{2\pi}\lim_{a\to\infty}\int_{-a}^{a} e^{-i\omega(t-t')}d\omega = \frac{1}{2\pi}\lim_{a\to\infty}\left[\frac{e^{ia(t-t')} - e^{-ia(t-t')}}{i(t - t')}\right]$$

$$\Rightarrow \sigma(t - t') = \lim_{a\to\infty}\frac{\sin\left[a(t - t')\right]}{\pi(t - t')}. \tag{16.92}$$

Using equation (16.90), for the case where the function $f(t') = 1$, we find one of the properties of the Dirac delta function from chapter 11,

$$\int_{-\infty}^{\infty} \sigma(t - t')dt' = 1. \tag{16.93}$$

Let us examine the Dirac delta function in equation (16.92) for different values of a at $t' = 0$. The results are shown in figure 16.9. From the graphs in figure 16.9, as $a \to \infty$, we can establish the relation

$$\sigma(t - t') = \lim_{a\to\infty}\frac{\sin\left[a(t - t')\right]}{\pi(t - t')} = \begin{cases} \infty, & t = t', \\ 0, & t \neq t'. \end{cases} \tag{16.94}$$

In view of the result in equation (16.94) and the graphs displayed in figure 16.9

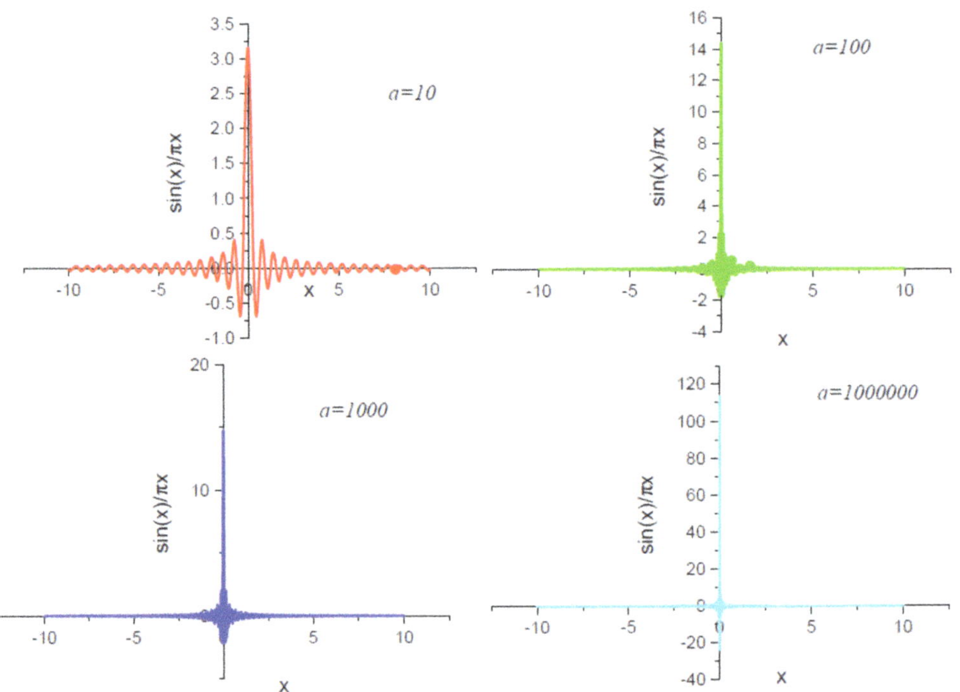

Figure 16.9. The graphs for the function $f(t) = \sin\left[a(t - t')\right]/\pi(t - t')$ at $t' = 0$ for different values of a.

$$\int_{t_1}^{t_2} f(t')\sigma(t - t')dt' = \begin{cases} f(t), & t_1 < t < t_2, \\ 0, & \text{otherwise.} \end{cases} \tag{16.95}$$

The result in equation (16.95) shows that the limit of integration does not necessarily be from $-\infty$ to ∞ for the Dirac delta function.

16.7 Applications of the Fourier transform

In chapter 13 we were introduced to the partial differential equations that we often see in different branches of physics. One of these equation is the time-dependent Schrödinger equation,

$$-\frac{\hbar^2}{2m}\nabla^2\Psi(\vec{r}, t) + U(\vec{r})\Psi(\vec{r}, t) = i\hbar\frac{\partial\Psi(\vec{r}, t)}{\partial t}. \tag{16.96}$$

Using separation of variable, we have shown that for a free particle with mass m traveling in the x-axis, where $U(\vec{r}) = 0$, the time-independent Schrödinger equation is given by

$$-\frac{\hbar^2}{2m}\frac{d^2\psi(x)}{dx^2} = E\psi(x) \Rightarrow \frac{d^2\psi(x)}{dx^2} = -\frac{2mE}{\hbar^2}\psi(x), \tag{16.97}$$

where E is the energy of the particle. Using De Broglie wavelength of the particle, we can relate the momentum of this particle p with the magnitude of the wave vector $k = 2\pi/\lambda$ by

$$\lambda = \frac{h}{p} \Rightarrow p = \frac{h}{\lambda} = \frac{2\pi}{\lambda}\frac{h}{2\pi} \Rightarrow p = k\hbar. \tag{16.98}$$

Then the total energy for a free particle in terms of the magnitude of the wave vector can be written as

$$E = \frac{p^2}{2m} = \frac{k^2\hbar^2}{2m}. \tag{16.99}$$

Using equation (16.99), one can rewrite equation (16.97) as

$$\frac{d^2\psi(x)}{dx^2} = -k^2\psi(x) \Rightarrow \frac{d^2\psi(x)}{dx^2} + k^2\psi(x) = 0 \tag{16.100}$$

the solution of which is given by

$$\psi(x) = A(k)e^{ikx}. \tag{16.101}$$

Since for a free particle k takes continuous value, the general solution to equation (16.100) can be expressed as

$$\psi(x) = \int_{-\infty}^{\infty} A(k)e^{ikx}dk. \tag{16.102}$$

We do not know anything about $A(k)$. What we do know, from a pure mathematical point of view, it is a constant resulting from integrating a second-order differential equation. From a physical point of view, it is determined by quantum mechanical conditions that need to be met for a free particle. In quantum mechanics everything is probabilistic which is determined by the wave function. The probability of finding the particle anywhere $(-\infty < x < +\infty)$ must be one and therefore

$$\int_{-\infty}^{\infty} |\psi(x)|^2\, dx = \int_{-\infty}^{\infty} \psi^*(x)\psi(x)dx = 1. \tag{16.103}$$

Using equation (16.102), we can rewrite equation (16.103) as

$$\int_{-\infty}^{\infty}\left[\int_{-\infty}^{\infty} A^*(k')e^{-ik'x}dk'\right]\left[\int_{-\infty}^{\infty} A(k)e^{ikx}dk\right]dx = 1$$

$$\Rightarrow \int_{-\infty}^{\infty} A(k)\int_{-\infty}^{\infty} A^*(k')\left(\frac{1}{2\pi}\int_{-\infty}^{\infty} e^{-ix(k-k')}dx\right)dk'\,dk = \frac{1}{2\pi}. \tag{16.104}$$

Applying the Dirac delta function

$$\sigma(t - t') = \frac{1}{2\pi}\int_{-\infty}^{\infty} e^{-i\omega(t-t')}d\omega, \tag{16.105}$$

we can put equation (16.104) in the form

$$\int_{-\infty}^{\infty} A(k)\left[\int_{-\infty}^{\infty} A^*(k')\sigma(k - k')dk'\right]dk = \frac{1}{2\pi}. \tag{16.106}$$

In view of the relation

$$f(t) = \int_{-\infty}^{\infty} f(t')\sigma(t - t')dt' \tag{16.107}$$

equation (16.106) becomes

$$\int_{-\infty}^{\infty} A(k)A^*(k)dk = \frac{1}{2\pi} \Rightarrow \int_{-\infty}^{\infty} |A(k)|^2\, dk = \frac{1}{2\pi}. \tag{16.108}$$

Noting that we can expressing $A(k)$ as

$$A(k) = \frac{1}{\sqrt{2\pi}}\phi(k), \tag{16.109}$$

where $\phi(k)$ is known as the momentum eigenfunction, which you will see in quantum mechanics class. For now, it is a well-defined function of k. Using this function, equation (16.108) can then be rewritten as

$$\int_{-\infty}^{\infty} |\phi(k)|^2\, dk = \int_{-\infty}^{\infty} \phi^*(k)\phi(k)dk = 1. \tag{16.110}$$

Using equation (16.109), one can express the wave function in equation (16.102) as

$$\psi(x) = \frac{1}{\sqrt{2\pi}} \int_{-\infty}^{\infty} \phi(k)e^{ikx}dk. \tag{16.111}$$

Comparing equation (16.111) with equation (16.73) we can see that wave function for a free particle is the inverse Fourier transform of the momentum eigenfunction $\phi(k)$. Therefore, one can then write for the momentum eigenfunction

$$\phi(k) = \mathcal{F}[\psi(x)] = \frac{1}{\sqrt{2\pi}} \int_{-\infty}^{\infty} \psi(x)e^{-ikx}dx, \tag{16.112}$$

which is the Fourier transform of the wave function.

Example 16.9. For a particle described by a Gaussian function

$$\psi(x) = Ae^{-ax^2} = \left(\frac{2a}{\pi}\right)^{1/4}e^{-ax^2} \tag{16.113}$$

    (a) Show that the momentum eigenfunction is given by

$$\phi(k) = \frac{A}{\sqrt{2a}}e^{-k^2/4a} \tag{16.114}$$

    (b) Show that the expectation values for the magnitude of the wave vector k and k^2 are given by

$$\langle k \rangle = 0, \ \langle k^2 \rangle = \alpha. \tag{16.115}$$

    (c) Find the expectation values $\langle x \rangle$ and $\langle x^2 \rangle$.
    (d) Show that the Heisenberg Uncertainty Principle

$$\Delta x \Delta p_x \geq \frac{\hbar}{2} \tag{16.116}$$

    is satisfied.

In quantum mechanics expectation values for physical observables like the position, momentum, or wave vector are determined using the wave function or the momentum eigenfunction. If the observable (which is an operator) is represented in position space by $f(x)$, the expectation value for this physical observable is given

$$\langle f(x) \rangle = \int_{-\infty}^{\infty} \psi^*(x)f(x)\psi(x)dx. \tag{16.117}$$

On the other hand if it is represented in momentum space by $g(k)$, the expectation value can be determined using the momentum eigenfunction

$$\langle g(k) \rangle = \int_{-\infty}^{\infty} \phi^*(k)g(k)\phi(k)dk. \tag{16.118}$$

Since in quantum mechanics what we can measure is the most probable value for the physical observable, there is an inherent uncertainty. The uncertainties in quantum measurement for a physical observable, for example, for the position of the particle x and its momentum p are given by

$$\Delta x = \sqrt{\langle x^2 \rangle - \langle x \rangle^2}, \ \Delta p = \sqrt{\langle p^2 \rangle - \langle p \rangle^2}, \tag{16.119}$$

respectively.

Solution:

(a) We recall the momentum eigenfunction $\phi(k)$ is given by

$$\phi(k) = \frac{1}{\sqrt{2\pi}} \int_{-\infty}^{\infty} \psi(x)e^{-ikx}dx, \tag{16.120}$$

so that upon substituting the wave function

$$\psi(x) = Ae^{-ax^2} = \left(\frac{2a}{\pi}\right)^{1/4} e^{-ax^2}, \tag{16.121}$$

we find

$$\phi(k) = \frac{A}{\sqrt{2\pi}} \int_{-\infty}^{\infty} e^{-ax^2}e^{-ikx}dx = \frac{A}{\sqrt{2\pi}} \int_{-\infty}^{\infty} e^{-a\left(x^2 + \frac{ik}{a}x\right)}dx. \tag{16.122}$$

Noting that

$$x^2 + \frac{ik}{a}x = \left(x + \frac{ik}{2a}\right)^2 + \frac{k^2}{4a^2} \tag{16.123}$$

we can write

$$\phi(k) = \frac{A}{\sqrt{2\pi}}e^{-\frac{k^2}{4a}} \int_{-\infty}^{\infty} e^{-a\left(x + \frac{ik}{2a}\right)^2}dx. \tag{16.124}$$

Using the result in Example 16.7, we can easily show that

$$\phi(k) = \frac{A}{\sqrt{2a}}e^{-\frac{k^2}{4a}}. \tag{16.125}$$

(b) In the momentum space, one can express the expectation values

$$\langle k \rangle = \int_{-\infty}^{\infty} \phi^*(k)k\phi(k)dk, \ \langle k^2 \rangle = \int_{-\infty}^{\infty} \phi^*(k)k^2\phi(k)dk, \tag{16.126}$$

and using the momentum eigenfunction we found in part (a), one finds

$$\langle k \rangle = \left(\frac{A}{\sqrt{2\alpha}} \right)^2 \int_{-\infty}^{\infty} k e^{-\frac{k^2}{2\alpha}} dk = 0, \tag{16.127}$$

and

$$\langle k^2 \rangle = \left(\frac{A}{\sqrt{2\alpha}} \right)^2 \int_{-\infty}^{\infty} k^2 e^{-\frac{k^2}{2\alpha}} dk. \tag{16.128}$$

Using the transformation of variable, one can show that

$$x = \frac{k}{\sqrt{2\alpha}} \Rightarrow dk = \sqrt{2\alpha}\, dx \Rightarrow \langle k^2 \rangle = \left(\frac{A}{\sqrt{2\alpha}} \right)^2 \int_{-\infty}^{\infty} 2\alpha x^2 e^{-x^2} \sqrt{2\alpha}\, dx$$

$$= \left(\frac{A}{\sqrt{2\alpha}} \right)^2 (2\alpha)^{3/2} \int_{-\infty}^{\infty} x^2 e^{-x^2} dx = \left(\frac{A}{\sqrt{2\alpha}} \right)^2 (2\alpha)^{3/2} \frac{\sqrt{\pi}}{2} \tag{16.129}$$

$$\Rightarrow \langle k^2 \rangle = \frac{\sqrt{2\alpha}}{\sqrt{\pi}} (\sqrt{2\alpha}) \frac{\sqrt{\pi}}{2} = \alpha,$$

where we used

$$\int_{-\infty}^{\infty} y^2 e^{-y^2} dy = \frac{\sqrt{\pi}}{2}, \quad A = \left(\frac{2\alpha}{\pi} \right)^{1/4}. \tag{16.130}$$

(c) It is easier to find these expectation values in position space rather than in momentum space,

$$\langle x \rangle = \int_{-\infty}^{\infty} \psi^*(x) x \psi(x) dx, \quad \langle x^2 \rangle = \int_{-\infty}^{\infty} \psi^*(x) x^2 \psi(x) dx. \tag{16.131}$$

Using the wave function

$$\psi(x) = A e^{-\alpha x^2} = \left(\frac{2\alpha}{\pi} \right)^{1/4} e^{-\alpha x^2} \tag{16.132}$$

we can write the expectation values as

$$\langle x \rangle = A^2 \int_{-\infty}^{\infty} x e^{-2\alpha x^2} dx = 0, \tag{16.133}$$

and

$$\langle x^2 \rangle = A^2 \int_{-\infty}^{\infty} x^2 e^{-2\alpha x^2} dx. \tag{16.134}$$

Introducing the transformation of variables, we can show that

$$y = \sqrt{2\alpha}\,x \Rightarrow dx = \frac{dy}{\sqrt{2\alpha}} \Rightarrow \langle x^2 \rangle = \frac{A^2}{\left(\sqrt{2\alpha}\right)^{3/2}} \int_{-\infty}^{\infty} y^2 e^{-y^2} dy$$

$$\Rightarrow \langle x^2 \rangle = \frac{1}{\left(\sqrt{2\alpha}\right)^{3/2}} \frac{\sqrt{2\alpha}}{\sqrt{\pi}} \frac{\sqrt{\pi}}{2} = \frac{1}{4\alpha},$$

(16.135)

where here also we used the relations in equation (16.130).

(d) The uncertainty in position and momentum

$$\Delta x = \sqrt{\langle x^2 \rangle - \langle x \rangle^2}, \quad \Delta p_x = \hbar \Delta k = \hbar \sqrt{\langle k^2 \rangle - \langle k \rangle^2},$$

(16.136)

using the results in part (b) and (c), becomes

$$\Delta x = \frac{1}{2\sqrt{\alpha}}, \quad \Delta p_x = \hbar \sqrt{\alpha}.$$

(16.137)

Then the uncertainty principle

$$\Delta x \Delta p_x = \frac{1}{2\sqrt{\alpha}} \hbar \sqrt{\alpha} = \frac{\hbar}{2} \geq \frac{\hbar}{2}$$

(16.138)

has been satisfied. In fact the result indicates that the Gaussian wave function represent the minimum uncertainty state.

16.8 Fourier transform and convolution

This section introduces the convolution of two or more functions and its relation to the inverse Fourier transform of these function.

Convolution

Let us consider two functions $f_1(x)$ and $f_2(x)$ for which the corresponding Fourier transforms $F(\omega)$ and $G(\omega)$ exist. The convolution of these two functions (denoted by $f*g$) over the interval $(-\infty, \infty)$, which is commutative, is given by

$$f_1(x)*f_2(x) = \frac{1}{\sqrt{2\pi}} \int_{-\infty}^{\infty} f_1(y) f_2(x - y) dy$$

$$= \frac{1}{\sqrt{2\pi}} \int_{-\infty}^{\infty} f_2(y) f_1(x - y) dy = f_2(x)*f_1(x).$$

(16.139)

This form of integrals appears in probability theory in the determination of the probability density of two random, independent variables. One example of convolution is the solution to Poisson's equation, which gives the electric potential $V(\vec{r})$ at a point in free space described by the position vector $\vec{r}$ due to a localized charge with charge density $\rho(\vec{r}')$,

$$V(\vec{r}) = \frac{1}{4\pi\epsilon_0} \int \rho(\vec{r}') \frac{1}{|\vec{r} - \vec{r}'|} d^3r'.$$

(16.140)

Equation (16.140) can be interpreted as convolution of the charge density function $\rho(\vec{r}')$ (as $g(y)$) and a weighting function $(4\pi\epsilon_0 \,|\, \vec{r} - \vec{r}'|)^{-1}$ (as $f(x - y)$). The operation of convolution sometimes referred as the Faltung, the German term for 'folding'.

The convolution theorem and inverse Fourier transform
Let us consider the convolution of the function $f_1, f_2, f_3. ..f_n$ with the corresponding Fourier transform $F_1(\omega), F_3(\omega), F_3(\omega), ...F_n(\omega)$. The convolution theorem states that the convolution of these functions is equal to the Fourier inverse transform to the product of the Fourier transform for each functions,

$$f_1 * f_2 * f_3 . ..f_n = \mathcal{F}^{-1}[F_1(\omega)F_3(\omega)F_3(\omega). .F_n(\omega)]. \tag{16.141}$$

Next we prove the convolution theorem using the convolution of two function. To this end, we recall

$$f_1 * f_2 = \frac{1}{\sqrt{2\pi}} \int_{-\infty}^{\infty} f_2(y)f_1(x - y)dy. \tag{16.142}$$

Let the Fourier transform of the function $f_1(x - y)$ is $F_1(\omega)$ so that one can write for the inverse Fourier transform

$$f_1(x - y) = \mathcal{F}^{-1}[F_1(\omega)] = \frac{1}{\sqrt{2\pi}} \int_{-\infty}^{\infty} F_1(\omega)e^{i\omega(x-y)}d\omega. \tag{16.143}$$

Using equation (16.143), one can rewrite the convolution integral in equation (16.142) as

$$\begin{aligned}
f_1 * f_2 &= \frac{1}{\sqrt{2\pi}} \int_{-\infty}^{\infty} f_2(y)\frac{1}{\sqrt{2\pi}}\left(\int_{-\infty}^{\infty} F_1(\omega)e^{i\omega(x-y)}d\omega \right)dy \\
&= \frac{1}{\sqrt{2\pi}} \int_{-\infty}^{\infty} F_1(\omega)\frac{1}{\sqrt{2\pi}}\left(\int_{-\infty}^{\infty} f_2(y)e^{-i\omega y}dy e^{i\omega x} \right)d\omega \\
&= \frac{1}{\sqrt{2\pi}}\left[\int_{-\infty}^{\infty} F_1(\omega)F_2(\omega)e^{i\omega x}d\omega \right] \Rightarrow f_1 * f_2 = \mathcal{F}^{-1}[F_1(\omega)F_2(\omega)],
\end{aligned} \tag{16.144}$$

where we used the relation

$$F_2(\omega) = \int_{-\infty}^{\infty} f_2(y)e^{-i\omega y}dy. \tag{16.145}$$

The result in equation (16.144) proves the convolution theorem.

Example 16.10. Discuss, compute, and plot the convolution of the functions $f_1(x)$ and $f_2(x)$ given by

$$f_1(x) = \begin{cases} 1, & -10 < x < 10, \\ 0, & \text{otherwise,} \end{cases}, f_2(x) = \begin{cases} \dfrac{1}{10}x, & 0 < x < 10, \\ 0, & \text{otherwise.} \end{cases} \tag{16.146}$$

Figure 16.10 displays the graphs for the functions described by equation (16.146).

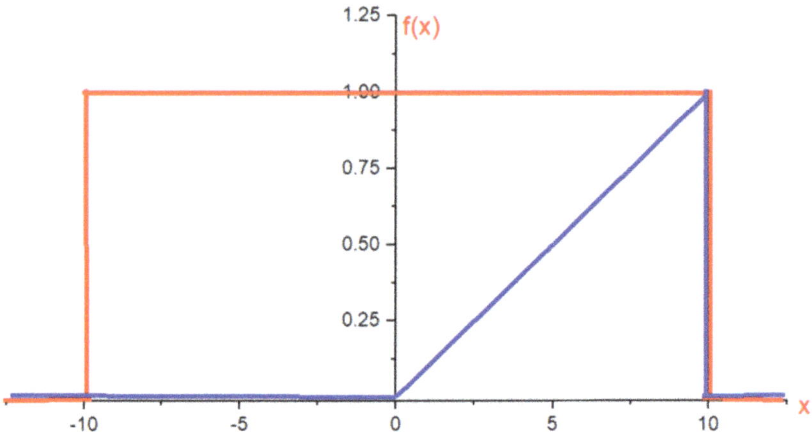

Figure 16.10. The graph for the functions $f_1(x)$ and $f_2(x)$.

Solution: For the two functions described by the equation (16.146), one can write

$$f_2(x - y) = \begin{cases} \dfrac{1}{10}(x - y), & 0 < x - y < 10 \\ 0, & \text{otherwise} \end{cases}$$

$$= \begin{cases} \dfrac{1}{10}(x - y) & x - 10 < y < x \\ 0 & \text{otherwise} \end{cases},$$

(16.147)

so that the convolution of the two functions becomes

$$f_1 * f_2 = \frac{1}{10\sqrt{2\pi}} \int_{x-10}^{x} f_1(y)(x - y)\,dy. \tag{16.148}$$

We need to be careful in evaluating this integral because of the function $f(y)$.

Case 1: $x \leqslant -10$
For this case the range of y would be $(-\infty, -20)$. In this interval we have $f_1(y) = 0$, and we find

$$f_1 * f_2 = 0. \tag{16.149}$$

Case 2: $-10 < x < 0$
The range of y would be $(-20, 0)$. In this interval we have $f_1(y) = 0$ for $y < -10$ and $f_1(y) = 1$ for $-10 < y < 0$, and we find

$$f_1*f_2 = \frac{1}{10\sqrt{2\pi}} \int_{-10}^{x} (x - y)dy = \frac{1}{10\sqrt{2\pi}}\left(xy - \frac{y^2}{2}\right)\Bigg|_{-10}^{x}$$

$$\Rightarrow f_1*f_2 = \frac{1}{\sqrt{2\pi}}\left(\frac{x^2}{20} + x + 5\right). \tag{16.150}$$

Case 3: $0 < x < 10$

The range of y would be $(-10, 10)$. In this interval, we have $f_1(y) = 1$ and convolution becomes

$$f_1*f_2 = \frac{1}{10\sqrt{2\pi}} \int_{x-10}^{x} (x - y)dy = \frac{1}{10\sqrt{2\pi}}\left(xy - \frac{y^2}{2}\right)\Bigg|_{x-10}^{x}$$

$$= \frac{1}{10\sqrt{2\pi}}\left(\frac{x^2}{2} - \left(\frac{x^2 - 100}{2}\right)\right) \Rightarrow f_1*f_2 = \frac{5}{\sqrt{2\pi}}. \tag{16.151}$$

Case 4: $10 < x < 20$

For this case the range of y would be $(0, 20)$. In this interval, we have $f_1(y) = 0$ for $10 < y < 20$ and $f_1(y) = 1$ for $0 < y < 10$, and we find

$$f_1*f_2 = \frac{1}{10\sqrt{2\pi}} \int_{x-10}^{10} (x - y)dy = \frac{1}{10\sqrt{2\pi}}\left(xy - \frac{y^2}{2}\right)\Bigg|_{x-10}^{10}$$

$$\Rightarrow f_1*f_2 = \frac{1}{\sqrt{2\pi}}\left(x - \frac{x^2}{20}\right). \tag{16.152}$$

The graph of the convoluted functions is shown in figure 16.11.

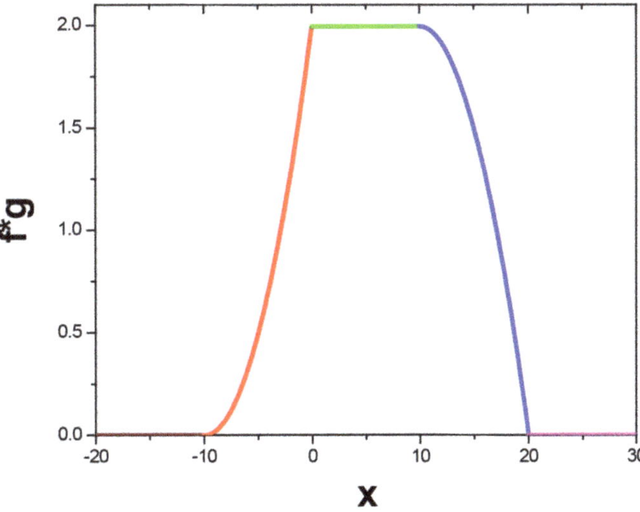

Figure 16.11. The graph for the convolution of the functions $f_1(x)$ and $f_2(x)$.

Important properties of convolution
(a) $f(x)*g(x) = g(x)*f(x)$.
(b) Convolution tends to smooth out $f(x)$.
(c) Convolution tends to spread out $f(x)$.

16.9 Mathematica, Fourier series, transform, and inverse transform

- FourierSeries [*expr,t,n*] gives the *n*th-order Fourier series expansion of *expr* in *t*.
- FourierSeries[*expr*, $\{t_1, t_2, ...\}$, $\{n_1, n_2, ...\}$] gives the multidimensional Fourier series.
- NFourierTransform[*expr*, *t*, ω] gives a numerical approximation to the Fourier transform of *expr* evaluated at the numerical value ω, where *expr* is a function of *t*.
- NInverseFourierTransform[*expr*,ω,*t*] gives a numerical approximation to the inverse Fourier transform of *expr* evaluated at the numerical value *t*, where *expr* is a function of ω.
- NFourierSinTransform[*expr,t,ω*] gives a numerical approximation to the Fourier sine transform of *expr* evaluated at the numerical value ω, where *expr* is a function of *n*.
- NInverseFourierSinTransform[*expr,ω,t*] gives a numerical approximation to the inverse Fourier sine transform of *expr* evaluated at the numerical value *t*, where *expr* is a function of ω.
- NFourierCosTransform[*expr,t,ω*] gives a numerical approximation to the Fourier cosine transform of *expr* evaluated at the numerical value ω, where *expr* is a function of *n*.
- NInverseFourierCosTransform[*expr,ω,t*] gives a numerical approximation to the inverse Fourier cosine transform of *expr* evaluated at the numerical value *t*, where *expr* is a function of ω.

Bibliography

[1] Boas M L 2006 *Mathematical Methods in the Physical Sciences* 3rd edn (New York: Wiley)
[2] Arfken G B and Weber H-J 2005 *Mathematical Methods for Physicists* 6th edn (Boston, MA: Elsevier)
[3] Arfken G B 1985 *Mathematical Methods for Physicists* 3rd edn (Orlando, FL: Academic Press)